DATE DUE

DRAFTING AND DESIGN FOR
ARCHITECTURE

DRAFTING AND DESIGN FOR
ARCHITECTURE

Eighth Edition

DANA J. HEPLER

Principal, Hepler Associates,
Architects and Land Planners, New
York City and Massapequa, New
York, and Adjunct Professor of
Architecture, New York Institute of
Technology, Westbury, New York

PAUL ROSS WALLACH

Architecture Instructor, Cañada College,
Redwood City, California, and Technical
Writing Consultant, Burlingame,
California

DONALD E. HEPLER

President, Technical Writing and
Design Service Inc., Somers,
Connecticut

THOMSON

DELMAR LEARNING

Africa • Australia • Canada • Denmark • Japan • Mexico • New Zealand •
Philippines • Puerto Rico • Singapore • Spain • United Kingdom • United States

THOMSON

DELMAR LEARNING

DRAFTING AND DESIGN FOR ARCHITECTURE
by Dana J. Hepler, Paul Ross Wallach, and Donald E. Hepler

Vice President, Technology and Trades SBU:
Alar Elken

Editorial Director:
Sandy Clark

Senior Acquisitions Editor:
James Devoe

Senior Developmental Editor:
John Fisher

Marketing Director:
Dave Garza

Channel Manager:
Bill Lawrenson

Production Director:
Mary Ellen Black

Production Manager:
Andrew Crouth

Production Editor:
Stacy Masucci

Art/Design Specialist:
Mary Beth Vought

Technology Project Manager:
Kevin Smith

Technology Project Specialist:
Linda Verde

Editorial Assistant:
Tom Best

Library of Congress Cataloging-in-Publication Data:
Hepler, Donald E.
   Architecture drafting and design / Donald E. Hepler, Paul Ross Wallach, Dana J. Hepler.—8th ed.
      p. cm.
   Includes index.
   Includes index.
   ISBN 1-4018-7995-0
   1. Architectural drawing. 2. Architectural design. I. Wallach, Paul Ross. II. Hepler, Dana J. III. Title.

NA2700.H4 2005
780'.28'4—dc22   2004063731

NOTICE TO THE READER

CONTENTS

PART 1

INTRODUCTION TO ARCHITECTURE 1

PART 2

ARCHITECTURAL DRAFTING FUNDAMENTALS 27

PART 3

BASIC AREA DESIGN 79

PART 4
BASIC ARCHITECTURAL DRAWINGS   193

PART 5

PRESENTATION METHODS   345

PART 6

FOUNDATIONS AND CONSTRUCTION SYSTEMS   383

APPENDIXES

PREFACE

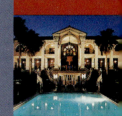

Drafting and Design for Architecture is a comprehensive textbook designed for use in a first course in architectural or construction drafting. Its purpose is to help students learn the fundamental skills and concepts necessary for architectural planning, designing, and drawing.

ORGANIZATION AND CONTENT

Material in this eighth edition has been organized into nine major parts consisting of thirty-seven chapters plus appendixes.

Parts

- **Part One**, "Introduction to Architecture," provides background information on the history and development of major architectural styles, with excellent examples shown of both past and present designs. It also covers the basic principles and elements of architectural design.

- **Part Two**, "Architectural Drafting Fundamentals," provides basic information on the use of scales, drafting instruments, and CAD systems, and explains the various architectural drafting conventions used in creating working drawings. Information in this part is needed to apply the information covered in subsequent specific drafting and design chapters. CAD information has been rewritten, updated, and expanded for this edition.

- **Part Three**, "Basic Area Design," covers the environmental and functional design factors needed to plan specific areas of a structure. This includes the design considerations necessary for effective solar orientation, efficient energy use, and ergonomic and ecological planning. Major considerations include the function, location, decor, size, and shape of the various areas.

- **Part Four**, "Basic Architectural Drawings," presents the design process and drafting methods used to combine areas into composite, functional, and effective architectural plans. Procedures for designing and drawing floor plans, elevations, sectional, detail, cabinetry, and site development drawings are explained. Guidelines for designing structures for persons with physical impairments are also included in this part. The information on site development and design factors and procedures has been completely revised and updated. New information on cabinetry drawings has been added.

- **Part Five**, "Presentation Methods," shows the different methods used to present architectural designs to nontechnical personnel such as marketing staffs, financial supporters, and prospective buyers. Step-by-step instructions for preparing one- and two-dimensional drawings, renderings, and three-dimensional models are provided.

■ **Part Six**, "Foundations and Construction Systems," begins with an overview of the basic scientific and modular principles on which construction systems are based. Each major construction system is then explained as students are introduced to the specialized drawings needed to complete detailed descriptions of the structural design. Types of drawings included are those used to describe foundations and fireplaces and wood-frame, masonry, concrete, steel, and reinforced-concrete systems. This part includes a chapter that covers disaster prevention design features needed to reduce structural failure due to earthquakes, tornadoes, floods, and hurricanes.

■ **Part Seven**, "Framing Systems," explains and shows in detail how to design and draw the framing systems for the major construction components of a building: floors, walls, and roofs.

■ **Part Eight**, "Electrical and Mechanical Design Drawings," includes the principles and procedures for preparing working drawings to describe the electrical, comfort control (HVAC), and plumbing systems of a structure. Passive and active solar heating and cooling systems are also explained.

■ **Part Nine**, "Drawing Management and Support Services," describes how architectural plans are checked and combined into sets and how drawings are interrelated to other drawings, details, and documents such as schedules, specifications, cost estimates, financial plans, codes, and contracts. A complete set of working drawings is presented.

The appendixes include material that is applicable and frequently used in the study and practice of architectural drafting, design, and construction but does not completely or sequentially fall exclusively into one of the thirty-seven basic content chapters. Coverage here includes architectural and construction career information, types of mathematical calculations, standard abbreviations used on architectural drawings and by professional organizations, architectural synonyms employed in different related fields and or geographical areas, and a glossary of architectural and construction terms.

■ **Appendix A** describes the many careers that require knowledge of some aspect of architectural drafting, design, and construction. This information includes educational requirements, essential skills and knowledge needed, job descriptions, and activities. Careers covered here include not only basic architectural drafting and design areas, but also related construction careers for which a knowledge of architectural drafting and design is important or helpful.

■ **Appendix B** contains a summary of the mathematical calculations most frequently needed and used in architectural drafting and design and in related construction fields. This includes calculations in basic arithmetic and geometry and also the mathematical operations necessary to calculate structural forces and materials as applied to light construction.

■ **Appendix C** lists abbreviations that are used on all types of architectural drawings and documents to conserve space while providing consistency in interpretation and communication. It also lists the acronyms used by professional organizations that serve the architecture and construction fields.

■ **Appendix D** lists synonyms that are used by drafting and design professionals. Because the use of some terms varies between different fields and geographical areas, an understanding of the various use of terms is essential.

■ **Appendix E** is a glossary of architectural and construction terms that allows students to cross-reference terms found in various parts of this text and which may be found on all types of architectural drawings and documents.

Chapter Organization

The sequence of chapters is generally organized in the order usually practiced in the architectural design and development process. However, this order can be rearranged to accommodate different course goals and priorities.

Chapter concepts are organized such that the student progresses from the familiar to the unfamiliar and from the simple to the complex. Where possible, related drawings are shown together or cross-referenced to provide a broader understanding of the relationship between drawings and documents. This is done on a single illustration, within the same chapter, and among cross-reference drawings in different chapters. To further reinforce the relationship of drawings, numerous symbol charts are provided. These show the plan and elevation symbol, abbreviation, and a pictorial drawing of each architectural feature.

Each chapter is introduced with listings of the major objectives and important terms defined and explained in the chapter. Each chapter ends with a set of exercises.

Because communication in the field of architectural drafting and design depends largely on understanding the vocabulary of architecture, new terms, abbreviations, and symbols are defined or explained where they first appear. This learning is reinforced throughout the remainder of the text.

OBJECTIVES

considerable time and detailed application of the principles of architectural drafting and design. Exercises that require original design work are marked with a house symbol. Completion of these exercises by the student will result in the creation of a complete set of related architectural plans and documents.

Drafting and Design for Architecture is illustrated with more than *1,500* drawings, photographs, and charts. Every illustration has been specifically selected and/or prepared to reinforce and amplify the principles and procedures described in the text.

Special Features

Math and science concepts are integral to the study of architecture. Math formulas and calculations are presented as applied throughout the text. The sections in which they appear are marked with the symbol shown at the right.

Math Connection

Science concepts are explained as appropriate. Those sections are marked with this symbol:

Science Connection

By relating architectural drafting and design practices to specific math and science principles, the student is introduced to the importance of these concepts in the preparation of architectural drawings and documents.

"Using CAD" boxed articles occur in some chapters to illustrate how this valuable software program can assist in drafting and design.

Exercises

The exercises that appear at the end of each chapter are organized to provide the maximum amount of reinforcement of the concepts covered. Exercises are flexible, ranging from the very simplest, which can be completed in a few minutes, to the more complex, which require

EXERCISES

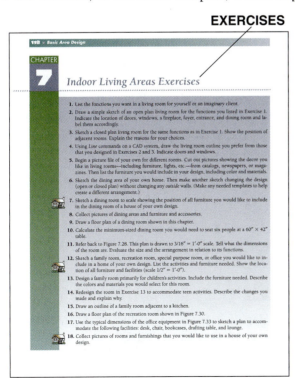

USING CAD

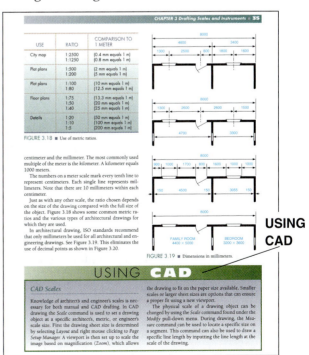

ANCILLARY MATERIALS

Ancillary materials related to this edition include:

The Student Workbook

The contents of the Student Workbook are related to the text by chapter and include 250 drawing exercises to be used as tests, assignments, or teaching aids. Two Auto-CAD tutorials to be used by students to complete two basic plans are also provided. (ISBN 1401879977)

Solutions Manual

A solutions manual is available with an introduction to presenting the text, teaching strategies and answers, information on using the Workbook, an architectural drawings test, and an architectural drawing set that relates to the test. (ISBN 1301879985)

The Instructor's E.Resource CD

This educational resource creates a truly electronic classroom. It is a CD-ROM containing tools and instructional resources that enrich the classroom and make the instructor's preparation time shorter. The elements of the e.resource link directly to the text to provide a unified instructional system. With the e.resource you can spend your time teaching, not preparing to teach. (ISBN 1401879969)

Features contained in the e.resource include the following:

- **Course Outlines.** The skills and material to be learned are outlined by text chapters for a nine, eighteen, and 36-week course. Recommended electives for vocational and professional career preparation are also provided.

- **Program Development.** This section includes instructional suggestions, information on classroom organization, and developing student skills. Work-school relationships, methods of conducting a course, and out-of-class activities are also covered.

- **Handout Masters.** The handout masters are designed to be printed, photocopied, and distributed to students. They extend and enrich the text content, sometimes with a drawing, sometimes with written information. Each handout is keyed to a specific text chapter. In the "Teaching Strategies" section, you will find tips for using the handouts in your classes.

- **PowerPoint Presentations.** These slides provide the basis for a lecture outline to present concepts and material. Key points and concepts can be graphically highlighted for student retention.

- **Optical Image Library.** This database of key images (all in full color) taken from the text can be used in lecture presentations, tests and quizzes, and Power-Point presentations. Additional Image Masters and Color Image Masters, which tie directly to lesson plans provided in the "Teaching Strategies" section, are also provided.

- **Exam View Test Bank.** Questions of varying levels of difficulty are provided in true/false, multiple choice, fill-in-the-blank, and short answer formats so you can assess student comprehension. This versatile tool enables the instructor to manipulate the data to create original tests.

- **Feng Shui.** Because of the increased popularity of this approach to architectural design, an introduction is provided which includes a collection of feng shui design features.

- **Metric Applications in Architecture.** An introduction to the metric system as applied to architectural drafting plus metric drafting conventions is presented here. A set of metric working drawings and related questions and answers is also provided.

- **Professional References.** These references include professional organizations related to all phases of architecture, construction, and engineering. A list of accredited schools of architecture in the U.S. is provided.

Videos

Two video sets, containing four 20-minute tapes each, are available. The videos correspond to the topics addressed in the text:

- Set #1, ISBN 0766830942
- Set #2, ISBN 0766830954

Video sets are also available on the Interactive Video CD-ROM.

- Set #1, ISBN 0766831167
- Set #2, ISBN 0766831175

ABOUT THE AUTHORS

Dana J. Hepler received his bachelor's degree from Ohio State University and master's degree in architecture from New York Institute of Technology. He has been associated with several of the largest architectural firms in the world as designer, director of planning, and construction manager. He has received national and international awards for his designs. Presently he is principal of Hepler Associates, Architects and Land Planners, New York City and Massapequa, New York, and adjunct professor of architecture at the New York Institute of Technology, Westbury, New York.

Paul Ross Wallach received his undergraduate education at the University of California at Santa Barbara and did his graduate work at California State University, Los Angeles. He has acquired extensive experience in the drafting, designing, and construction phases of architecture and has taught architecture and engineering drawing for many years in Europe and California at the secondary and postsecondary levels. He currently teaches architecture at Cañada College in Redwood City, California, and also does technical writing and consulting.

Donald E. Hepler completed his undergraduate work at California State College, California, and his graduate work at the University of Pittsburgh, Pennsylvania. He has been an architectural designer and drafter for several architectural firms, has served as an officer with the United States Army Corps of Engineers, and has taught architecture, design, and drafting at both the secondary and college levels. He is the former publisher of McGraw-Hill's technical education program and is currently devoting full time to technical authorship and consulting.

ACKNOWLEDGMENTS

The publisher and authors gratefully acknowledge the cooperation and assistance received from many individuals and companies during the development of *Drafting and Design for Architecture*. Thanks are due to the drafting teachers who reviewed the work and helped us develop excellent educational materials into a more comprehensive, up-to-date, and effective architectural drafting and design program: Susan Campbell, Glendale Community College, Glendale, Arizona; Robert J. Duering, Central Community College, Grand Island, Nebraska; Elizabeth H. Dull, High Point University, High Point, North Carolina; Donald W. Hain, Orleans/Niagara BOCES, Sanborn, New York; Catherine L. Kendall, Pellissippi State Technical Community College, Knoxville, Tennessee; Robert Potts, Schuykill Institute of Business and Technology, Pottsville, Pennsylvania; Phillip A. Reed, Old Dominion University, Norfolk, Virginia; and Joe Swantek, Schuykill Institute of Business and Technology, Pottsville, Pennsylvania. Credits for contributing architects, illustrators, and photographers are shown directly under each illustration. A special thanks is given to contributors of multiple illustrations. These include John B. Schols, David Karram, Alfred Karram, Home Planners Inc., John Henry, Jenkins and Shue, The Western Pennsylvania Conservancy, Lindal Cedar Homes, Scholz Designs, Marc Michaels, Interior Design, and Trus-Joist MacMillan.

PART 1

Introduction to Architecture

1 CHAPTER

Architectural History and Styles

OBJECTIVES

In this chapter you will learn to:

■ recognize historical architectural styles and identify several distinct characteristics of each style.

■ relate how the development of materials and construction methods influenced architectural styles.

TERMS

arch
bearing walls
buttress
dome
Early American style
English style

French style
Gothic arch
Italian style
keystone
Mediterranean style

Mid-Atlantic style
New England Colonial
post-and-lintel construction
ranch style
vault

INTRODUCTION

The study of architectural history is more vast and complex than can be covered in one textbook—much less in one chapter. However, an overview of architectural forms is an excellent way to begin. Architecture is dynamic. As societies change and develop, so does architecture. This chapter gives a background for evaluating and studying the broad range of architectural styles and forms.

 THE DEVELOPMENT OF ARCHITECTURAL FORMS

When humans were nomadic, shelter consisted of natural caves or portable tents made of animal skins. As people began to settle in fixed locations, there was a greater need to draw or plan the construction of dwellings. They began to construct permanent tents and adobe huts and to modify caves or shelters with existing natural materials. This all began at least 10,000 years ago and by 7800 B.C. the first known town of Jericho had developed on the north end of the Dead Sea. By 5500 B.C. towns of domed huts were built as shown in Figure 1.1.

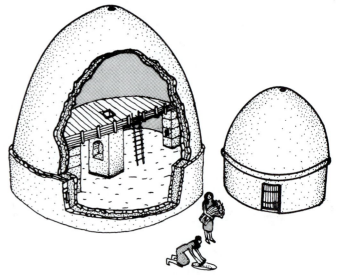

FIGURE 1.1 ■ Early domed huts.

The addition of more permanent dwellings near fertile areas gave rise to villages. Village life created a need for still more planning, such as for public areas. The art and science of architecture began with the planning and construction of the first dwellings and public areas. The field of architectural drafting began when people first drew the outline of a shelter or a village in the sand or dirt. They planned how to build structures with existing materials. The following dates show the earliest developments in architecture and construction:

- ■ *9000 to 8000* B.C. Last use of caves as habitats.
- ■ *8000 to 7000* B.C. Building of open-air huts and houses.
- ■ *7000 to 6500* B.C. First permanent villages and fortifications.
- ■ *6500 to 5000* B.C. Building of rulers' castles and fortifications.
- ■ *5000 to 4000* B.C. Beginning of religious architecture.
- ■ *4000 to 3000* B.C. Art forms and writings added to buildings. Beehive huts built with grass and reeds.
- ■ *3000 to 2000* B.C. Building of the great pyramids of Egypt.

As centuries passed and civilizations developed, human needs expanded. Lifestyles and cultures began to develop and change. More complete, accurate, and detailed drawings became necessary, and basic principles of architecture began to be developed. The first known architect was Imhopted, who designed the step pyramid in Egypt in 2780 B.C. This pyramid was an unoccupied solid structure, but most early occupied buildings were bearing wall types. **Bearing walls** are solid walls that provide support for each other and for the roof. Figure 1.2 shows an early Egyptian application of bearing wall construction. This is the temple built at Ereda in 3500 B.C. Note

FIGURE 1.2 ■ Bearing wall construction in 3500 B.C.

the comparison of this design with many contemporary structures. One of the first major problems in architectural design and construction was how to provide door and window openings in these supporting walls without sacrificing the needed support.

Post and Lintel

One solution to the problem was simply to place a horizontal beam, called a *lintel*, across two vertical posts. This early type of construction became known as **post-and-lintel construction**. This method was used by the Egyptians and later by the ancient Greeks. See Figure 1.3.

Because most ancient people used stone as the primary building material, architectural designs were limited.

FIGURE 1.3 ■ The Parthenon (447 B.C.) is an early post-and-beam structure. *North Wind Picture Archives*

Because of the great weight of the stone, stone post-and-lintel construction could not support wide openings. Therefore, many posts (or columns) were placed close together to provide the needed support. The Greeks and, later, the Romans expanded their architectural designs by creating several styles of these columns. See Figure 1.4.

Oriental architects also made effective use of the post and lintel. They were able to construct buildings with greater space between the posts under the lintel because they used lighter materials, such as wood. With these lighter materials, they developed a style of architecture that was very open

and graceful. Classical oriental homes are built of wood, paper, plaster, and stone. Roofs are covered with tile, thatch, or wood shingles. Sliding wall panels (*shoji*), which are used for doors, and ornamental gardens, are also features of oriental architecture as shown in Figure 1.5.

The Arch

The Romans, who used stone, began a new trend in the design of wall openings when they developed the **arch**. Arch construction overcame several limitations of the

FIGURE 1.4 ■ Greek and Roman column orders of architecture.

FIGURE 1.5 ■ Oriental architecture. *Yo Shin So*

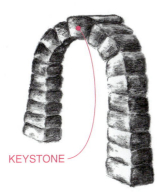

KEYSTONE

FIGURE 1.6 ■ A keystone supports both sides of an arch.

FIGURE 1.7 ■ A barrel vault is a series of connected arches.

post and lintel. Arches were easier to erect because they were constructed from many smaller, lighter blocks of stone. Each stone is supported by leaning on the keystone in the center. The **keystone** is a wedge-shaped stone that locks the other stones in place. See Figure 1.6. This construction has the advantage of spanning greater areas instead of being limited by the size of the one stone used for the lintel. The Romans combined arches and columns extensively in their architecture.

The Vault

The success of the arch led to the development of the vault. A **vault** (Figure 1.7) can be viewed as a series of arches that forms a continuous arched covering. The term may also refer to an arched underground passage-

way or a space, such as a room, that is covered by arches. When two barrel vaults intersect as shown in Figure 1.8, a cross vault is created.

The Dome

A **dome** is a further refinement of the arch. A dome is made of many arches arranged so that their bases form a circle and the tops meet in the center. See Figure 1.9. The Romans viewed the dome as a symbol of power. Throughout the world, domes have often been used in religious and governmental structures, as in the U.S. Capitol building (Figure 1.10).

FIGURE 1.8 ■ A cross vault is the intersection of two barrel vaults.

FIGURE 1.10 ■ The U.S. Capitol building is a contemporary domed structure. *David R. Frazier Photolibrary, Inc./Mark Burnett*

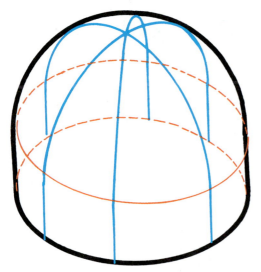

FIGURE 1.9 ■ Arches spaced in a circle form a dome.

The Gothic Arch

A variation of the arch was a defining characteristic of the Gothic style of architecture that spread throughout Europe during the Middle Ages. The pointed arch was called the **Gothic arch** and became a very popular feature in cathedrals. The emphasis on vertical lines created a sense of height and aspiration. However, the pointed arch posed the same problem as did other arches: spreading at the bottom because of the weight above.

To add support to an arch or bearing wall, a protruding structure called a **buttress**, or pilaster, was added at the base, as shown in Figure 1.11. As the style evolved, buttresses were connected to higher areas of the walls and came to be called flying buttresses. A flying buttress helps support the sides of a wall without adding additional weight. Thinner walls and more windows could be used. Large structures such as the high-arched cathedral in Figure 1.12 incorporate flying buttresses in their design.

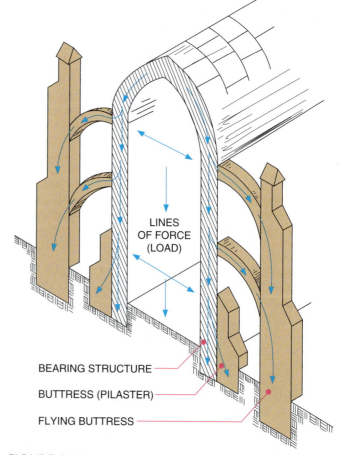

LINES OF FORCE (LOAD)

BEARING STRUCTURE

BUTTRESS (PILASTER)

FLYING BUTTRESS

FIGURE 1.11 ■ Buttresses (pilasters) support bearing walls. Arrows represent lines of force.

FIGURE 1.12 ■ Flying buttresses support the thin walls of Notre Dame Cathedral. *David R. Frazier Photolibrary, Inc.*

DEVELOPMENT OF ARCHITECTURAL STYLES

The Gothic (pointed) arch is one of many features that were developed over time. The development of one architectural solution and a resulting style in one culture often causes changes in the architecture of another culture. Transitions occur from one time period to another as well as from one part of the country and world to another.

Few structures, past or present, are pure examples of one specific style. In fact, to identify an architectural style as one that originated in only one country or in only one time period is difficult, if not inaccurate. Nonetheless, architectural styles *are* categorized and labeled by their most common and outstanding features. The label *Early American style,* for example, is something of a misnomer, because all styles that found their way to America during colonial times can be labeled Early American. The overlapping of characteristics of architectural design is also typical among European styles. Nonetheless, labels are applied and used as a frame of reference with which to study and compare architectural styles.

The term *Colonial* can be applied to most early American homes. The progression of design developed in the following order:

1. *Early Colonial* is the first true Colonial style of homes in America.
2. *Classical Colonial* is the adoption of the ancient and classical architectural forms.
3. *Traditional Colonial* is the adoption of the older style of Colonial architecture with modernizing features.
4. *Contemporary Colonial* is the adoption of clean contemporary lines with much glass and an open (informal) floor plan.

INFLUENCES ON EARLY AMERICAN ARCHITECTURE

To understand the development of American architecture, an overview of the following European styles provides an important background. The English, French, Spanish, and Italians have provided the most significant influences on Early American architecture.

FIGURE 1.13 ■ Elizabethan style. *Masonite Corp.*

English Architecture

English style architecture includes several variations of some common architectural features. For example, some features that many English structures share are high-pitched roofs, massive chimneys, half-timber siding, small windows, and exterior stone walls. Within this frame of reference, variations in architectural style range from the very simple to the very lavish. Wood may replace the stone on the exterior walls. The most commonly adapted English styles include the Elizabethan and the Tudor.

An example of Elizabethan style with its characteristic half-timber walls is shown in Figure 1.13. Figure 1.14A shows a Tudor-style residence with its multiple gables, small leaded windows, and large free-standing chimneys.

French Architecture

Because the **French styles** were brought to this country much later than the English styles, their impact on Colonial residential architecture was far less pronounced. However, some French styles, such as Regency, Mansard, Provincial, and Chateau were accepted and used in many areas. French Mansard architecture can be identified by the Mansard roof. See Figure 1.14B. This roof design was

FIGURE 1.14A ■ Tudor style. *Scholz Design*

FIGURE 1.14B ■ Contemporary application of a Mansard roof. *Home Planners, Inc.*

FIGURE 1.15 ■ French Provincial style. *Country Club of the South*

developed by the French architect Francois Mansard. On a French Provincial house the roof is high pitched with steep slopes and rounded dormer windows projecting from the sides. The features of a French Provincial design are shown in Figure 1.15.

Spanish and Italian Architecture

Spanish and Italian architecture share several similarities—arches, low-pitched roofs of ceramic tile, and stucco exterior walls, as shown in Figure 1.16. A distinguishing feature of a Spanish home is an open courtyard patio. Two-story Spanish homes contain open balconies often with grillwork trim.

Although **Italian style** architecture is very similar to Spanish architecture, a few features are particular to Italian styles. Columns and arches are generally part of an entrance,

FIGURE 1.16 ■ Southwest style derived from Spanish Architecture. *Lee Caudill, Coldwell Banker*

FIGURE 1.17 ■ Contemporary adaptation of Italian architecture. *John Henry, Architect*

as in Figure 1.17, and windows or balconies open onto a loggia. A loggia is an open passage covered by a roof.

Classical moldings around first floor windows also help to distinguish the Italian style from the Spanish. Despite these distinguishing features, both of these styles are generally classified as **Mediterranean style** or Southern European style architecture.

EARLY AMERICAN STYLES

The early colonists came to this country from many different cultures and were familiar with many different styles of architecture. **Early American style** architecture refers to all styles that developed in various regions of the colonies.

The European styles that primarily dominated Early American residential architecture were brought from England and France. French styles, however, were brought to this country in the eighteenth century, much later than English styles. The English styles had greater impact on colonial residential architecture.

New England Colonial

The colonists who settled the New England coastal areas brought the strong influence of English architectural styles with them. Because they lacked time and equipment and depended on the locally available building materials, the colonists had to greatly simplify the English styles. This adapted style came to be known as **New England Colonial** architecture.

One of the most popular of the New England Colonial styles was called the Cape Cod. See Figure 1.18. This one-and-one-half-story gabled-roof house has a central front entrance, a large central chimney, exterior walls of clapboard or beveled siding, and may include dormers. The floor plan is generally symmetrical. Cold New England winters also influenced the development of varied design features for added warmth, such as window shutters, small window areas, and enclosed breezeways.

Mid-Atlantic Colonial

The availability of brick, a seasonal climate, and the influence of the architecture of Thomas Jefferson led to the development of the **Mid-Atlantic style** of architecture. In colonial days, Mid-Atlantic style buildings, located from Virginia to New Jersey, were formal, massive, and ornate. This style is also known as *classical revival* because it was influenced by early Greek and Roman architecture. It also included adaptations of many urban English designs, such as the symmetrical, hip-roofed Mid-Atlantic style shown in Figure 1.19. The *Georgian* style is a simplified version of this design that includes many elements of the New England Colonial style.

FIGURE 1.18 ■ Cape Cod style. *Home Planners, Inc.*

FIGURE 1.19 ■ Mid-Atlantic Colonial style. *Scholz Design*

Dutch Colonial

Many colonists from the Netherlands and Germany settled in New York and Pennsylvania. A gambrel roof was a typical part of their Colonial style of architecture. This style was originally described as "Deutsch." However, the German term was given the English form by the colonists who called the masonry farmhouse style "Dutch" Colonial.

Southern Colonial

When the early settlers migrated south, warmer climates and outdoor living activities led them to develop the Southern Colonial style of architecture. The house became the center of plantation living. Southern Colonial homes were usually larger than most English houses. A second story was added, often with a veranda or porch. See Figure 1.20.

FIGURE 1.20 ■ Southern Colonial style. *Chris McDevitt, Remax*

LATER AMERICAN STYLES

After the colonial period, other architectural styles continued to evolve. Styles were influenced by climate, availability of land, and industrial developments—as well as by other architectural styles.

Victorian Era

During the early reign of Queen Victoria (1839 to 1901), architecture reflected the past with Greek and Gothic revival and Renaissance styles. Later more ornate designs such as Queen Anne became popular. The new machinery developed during the Industrial Revolution in the late eighteenth century led to the addition of intricate house decorations (gingerbread). Ornate finials, lintels, parapets, and balconies were added to existing designs, as shown in Figure 1.21.

Ranch Style

As settlers moved west of the colonies, they adapted architectural styles to meet their needs. The availability of land eliminated the need for second floors. Because the needed space was spread horizontally rather than vertically, a rambling plan called **ranch style** resulted.

The Spanish and Mexican influence in the Southwest led to the popularization of the *southwestern ranch,* with a U-shaped plan and a patio in the center. One-story Spanish/Mexican homes were the forerunners of the present ranch-style homes that were developed in the Southwest. See Figure 1.22A. Figure 1.22B shows a contemporary adaptation of the rambling ranch style.

Influences on Contemporary Styles

Advances in architecture throughout history have depended on using available building materials. In American colonial times, builders had only wood, stone, and some ceramic materials, such as glass, with which to work. Early American architecture reflects the reliance on these materials, and they continue to be used in contemporary buildings.

FIGURE 1.21 ■ This design contains styles characteristic of the Queen Anne–Victorian era. *Home Planners, Inc.*

FIGURE 1.22A ■ Southwestern ranch style. *Home Planners, Inc.*

FIGURE 1.22B ■ Contemporary ranch style. *Home Planners, Inc.*

With today's improved technological developments, lighter and safer buildings can be designed in forms, sizes, and shapes never before possible. See Figure 1.23. With so many choices, designers can create many new combinations of styles and materials. These must be carefully combined using the basic principles and elements of design presented in Chapter 2.

Historical styles continue to influence contemporary architecture. Advancements in technology, however, have freed designers from many design restrictions of the past. Present-day designers must often decide how many contemporary features can or should be incorporated into the design of a particular architectural style. The house design shown in Figure 1.24 contains many historic architectural elements yet incorporates numerous contemporary features such as large glass areas.

Contemporary Materials

Advances in architecture throughout history have depended on using locally available building materials. As recently as American colonial times, builders had only wood, stone, and ceramic materials with which to work. Early American architecture reflects the use of these materials. These materials and bearing wall construction continue to be used in contemporary buildings; however, great changes are being seen in buildings with the development of steel, aluminum, structural glass, reinforced and prestressed concrete, wood laminates, plastics, and other new synthetics.

Today's buildings are stronger, safer, and more maintenance free than they were just a few years ago. New developments in central heating and cooling and energy efficiency have enabled buildings to meet the heavy demands of new and more complex appliances and electronic

FIGURE 1.23 ■ Contemporary style incorporating new materials and shapes. *Barry Sugerman, AIA*

FIGURE 1.24 ■ Traditional design incorporating contemporary components. *Isleworth—Tom Price, Architect; Phil Eschbach, Photographer*

devices. Due to the development of new building materials and the elimination of dangerous materials such as asbestos and lead, contemporary buildings provide a more healthy environment for the occupants. New materials and construction methods also make roofs and floors stronger, safer, and quieter than ever before under high-wind or earthquake conditions. Even devices for sensing problems such as detectors for smoke, radon, and break-ins have become much more efficient and effective.

Contemporary Construction Methods

The development of new materials is usually not possible without the development of new construction methods.

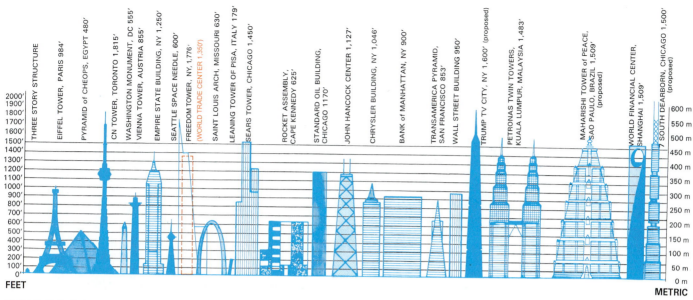

FIGURE 1.25 ■ Comparison of the world's largest structures.

For example, large glass panels could not have been used in the eighteenth century even if they had been available, because no large-span lintel-support system had been developed. Only when both new materials and new methods exist can the design be completely flexible.

Present-day structures are usually a combination of old and new. In a modern building, examples of the old post-and-lintel method can be used together with skeleton-frame, curtain-wall, or cantilevered construction.

Today, large premanufactured components can be used. Also, contemporary structures tend to have fewer structural restrictions. Lines may be simpler, bolder, and less cluttered. Other contemporary buildings can be constructed with more diversified structural shapes. No longer are architectural shapes simply squares or cubes. Shapes such as triangles, octagons, pentagons, circles, and spheres are now used extensively. Without these advancements, the major structures of today, like those shown in Figure 1.25, could not be built—and the planned buildings of tomorrow (Figure 1.26) extend the boundaries even further.

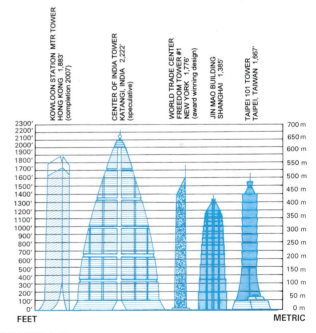

FIGURE 1.26 ■ Proposed buildings of the future.

NEW CHALLENGES IN ARCHITECTURE

Today's architecture stems from a rich historical background. Architectural styles continually change and develop. The future of architecture will certainly continue to be influenced by the development of new materials and new construction methods, as well as by the way people live in society.

With technological advancements, designers have greater freedom of choice to create diverse and exciting architecture. Greater freedom, however, and more choices also mean greater challenges. One of the primary challenges of architectural design is to blend art and technology. The role of the architect is to create a relationship between art and technology that enables all types of buildings to be both technically appropriate and aesthetically acceptable.

CHAPTER

1

Architectural History and Styles Exercises

1. Explain what is meant by the term *architectural styles*.

2. Describe post-and-lintel construction. Give examples of it in three different cultures. Compare it to the development of the arch.

3. Find examples of several different styles of architecture in your area. Photograph or sketch the structures and list their distinguishing characteristics.

4. Which European countries and styles had the greatest impact on Early American architecture? List two or three characteristics of each architectural style.

5. Describe a contemporary structure—imagined or real. Tell why you consider it to be a contemporary style.

 6. Considering the styles in this chapter, which do you prefer for your home? Explain a few advantages and disadvantages of building this style.

7. Do research to identify and prepare a report on an architectural style not mentioned in this text. Share what you learn with your class.

8. Identify houses in your town or city that are examples of various architectural styles.

9. Compare the tallest structure in your area to the structures in Figure 1.25.

10. List the major construction styles of ancient and modern structures.

CHAPTER 2

Fundamentals of Design

OBJECTIVES

In this chapter you will learn to:

- relate design concepts to architecture.
- understand why form follows function.
- identify six elements of design.
- apply design principles to a work of architecture.

TERMS

aesthetic value
eclectic design
elements of design
functionalism

hue
intensity
organic design
principles of design

proportion
transition
unity
value

INTRODUCTION

The reason one building is considered attractive and another unattractive is related to how well the fundamental principles and elements of design are applied. These fundamentals are used in developing architectural designs that are both attractive and functional. This applies to all types and styles of architecture although the applications may vary depending on individuality and creativity.

ARCHITECTURE AND DESIGN

Design activities are either formal or informal. Informal design occurs when a product is made by the designer without the use of a plan. Formal design involves the complete preparation of a set of working drawings. The working drawings are then used in constructing the product. Architectural design is nearly always of the formal type.

Ideas in the creative state are recorded by sketching basic images. These sketches are then revised until the ideas are crystallized and given final form. First sketches rarely produce a finished design. Usually, many revisions are necessary.

A basic idea, regardless of how creative and imaginative, is useless unless the design can be constructed successfully. Designing involves the transfer of basic sketches into architectural working drawings. Every use-

ful building must perform a specific function. Every part of the structure should also be designed to perform a specific function. Today's buildings are designed to be functional and aesthetic and for both purposes the elements and principles of design are applied.

Louis Sullivan, an important American architect in the late 1800s and early 1900s, wrote, "Our architecture reflects us as truly as a mirror." Architecture reflects the people, society, and culture of a given time. For example, modern architecture reflects our freedom and our technological advances.

Form Follows Function

"Form follows function" is a design concept conceived by Louis Sullivan but largely identified with Frank Lloyd Wright, probably the most famous of all twentieth century architects. This concept has now been accepted by most designers.

"Form follows function" means that any architectural form (shape, object) should have an intended practical purpose and should perform a function. This concept distinguishes architecture from other art forms, such as sculpture. A sculpture's primary purpose is its **aesthetic value**. Its value is in the *appreciation* of its form, its beauty, or its uniqueness. **Functionalism** is the quality of being useful, of serving a purpose other than adding beauty or aesthetic value. Functionalism in architecture led to the development of the organic concept. In the **organic design**

FIGURE 2.1A ■ Exterior lines designed for consistency with interior line shapes. *John Henry, Architect*

FIGURE 2.1B ■ Interior lines designed for consistency with the exterior. *John Henry, Architect*

approach all materials, functions, forms, and surroundings are completely coordinated and in harmony with nature.

Interior Design

The overall architectural style of a structure is an important consideration in developing the design of the interior. This means that the design elements and principles are matched to achieve a finished product that is consistent in style both inside and outside. Notice how the repeated use of the pointed arch in Figures 2.1A and 2.1B creates a design that efficiently relates the interior to the exterior.

Eclectic design can be very interesting and attractive although it may appear inconsistent if architectural periods or themes are mixed in one area. Whenever this mixing is poorly done, the results are often more eccentric than a pleasing combination of periods. When the elements and principles of design are followed, however, the results can be very effective. Eclectic design is not eccentric design.

Designers need to consider that styles and individual tastes change. An effective, creative designer recognizes the difference between *trends* (general developments) and *fads* (temporary popular fashions).

Creativity in Architectural Design

Creativity in architecture involves the ability to imagine forms before they exist. Creative imagination often involves arranging familiar objects and patterns in new ways. Creativity and imagination are both needed to bring many isolated and unrelated factors together into arrangements of cohesive unity and beauty. Every part of a building should be designed in relation to its function. Architects and interior designers apply the elements and principles of design to a building's function to make it aesthetically pleasing as well.

ELEMENTS OF DESIGN

Like a mixture composed of many ingredients, a design is composed of many elements. The basic **elements of design** are the tools of design. These include: *line, form, space, color, light* (value), *texture,* and *materials.*

Line

Lines enclose space and provide the outline or contour of forms. Straight lines are either horizontal, vertical, or diagonal. Curved lines have an infinite number of variations. The element of *line* can produce a sense of movement or produce a greater sense of length or height.

Horizontal lines emphasize width as the eye moves horizontally. These lines suggest relaxation and calm. Vertical lines create the impression of height because they lead the eye upward. These lines create a feeling of strength and alertness. Diagonal lines create a feeling of restlessness or transition. Vertical and horizontal lines tend to dominate architectural designs, giving a sense of stability. Curved lines indicate soft, graceful, and flowing movements. The curved and repeated arch lines of the ceiling in Figure 2.2 blend well with the floor-level horizontal lines to create a dramatic effect. As in any art form, it is often the combination of straight and curved lines in patterns that create the most pleasing design.

Form

Lines joined together can produce a *form* and create the *shape* of an area. More than two straight lines joined together can produce triangles, rectangles, squares, and other geometric shapes. Closed curved lines can form circles, ovals, and ellipses, as well as free-form closed curves. The relationships of these forms or shapes is an important factor in design.

Circles and ovals convey a feeling of completeness. Squares and rectangles produce a feeling of mathematical

FIGURE 2.3 ■ Square and rectangular shapes create the effect of precision. *Shakertown Corp.*

precision. The masonry vertical rectangles combined with the large wooden horizontal rectangles create an extremely strong and precise impression of the building shown in Figure 2.3. Whether the form of an object is closed, open, solid, or hollow, the form of the structure should always be determined by its function.

Space

Space surrounds form and is contained within it. A design can create a feeling of space. Architectural design includes the art of defining space and space relationships. Space is as important a consideration as the actual objects and materials.

Color

Choices of *color* have a strong influence on the final appearance of any design. In architecture, color can strengthen or diminish interest. It can also distinguish one part from another. Color may be an integral part of an architectural material such as natural wood. Manufactured products, such as synthetic wall panels, may have color added to create a desired effect. To create effective designs, designers need to understand the nature and relationships of colors.

The Color Spectrum

Colors in the spectrum are divided into primary, secondary, and tertiary colors. *Primary* colors are red, yellow, and blue. These cannot be made from any other color. The primary colors and combinations of colors are illustrated on the color wheel in Figure 2.4.

A *secondary* color can be made from equal mixtures of two primaries. Green is a combination of yellow and

FIGURE 2.2 ■ Dramatic use of repeated ceiling arch design. *Marc-Michaels Interior Design, Inc.; Sargent, Photographer*

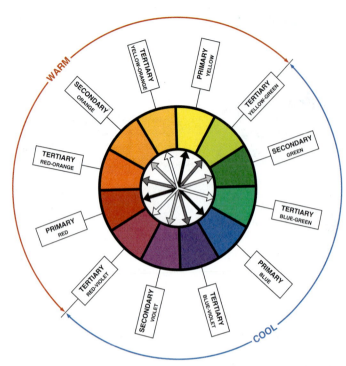

FIGURE 2.4 ■ Color wheel based on the triadic color system of the three primary colors: red, yellow, and blue.

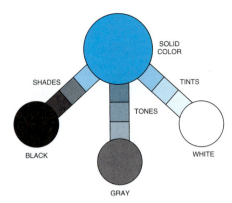

FIGURE 2.5 ■ Color combined with black or white creates tints, tones, or shades.

blue. Violet is a combination of blue and red. Orange is a combination of red and yellow.

A *tertiary* color is the combination of a primary color and a neighboring secondary color. The tertiary colors are red-orange, yellow-orange, yellow-green, blue-green, blue-violet, and red-violet.

A *neutral* does not show color in the ordinary sense of the word. The neutrals are white, gray, and black. The three primary colors, if mixed in equal strengths, will produce black. When colors cancel each other out in this manner, they are neutralized.

Color Quality

For greater accuracy in describing a color's exact appearance, colorists distinguish three qualities: hue, value, and intensity.

The **hue** of a color is its basic consistent identity. A color hue may be identified as being yellow, yellow-green, blue, blue-green, and so forth. Even when a color is made lighter or darker, the hue remains the same.

The **value** of a color refers to the lightness or darkness of a hue. See Figure 2.5. A great many degrees of value can be obtained. Varying the value of colors can dramatically change the mood of a room.

A *tint* is lighter (or higher) in value than the normal value of a color. It is produced by adding white to a color. A lighter tint of a hue will make a room look larger.

A *shade* is darker (or lower) in value than the normal value of the color. A shade is produced by adding black to the normal color. A dark shade will often make a room look smaller.

A *tone* is usually produced by adding gray to the normal color. Each color on a color wheel can have a value that is equivalent to another color if they have the same amount of gray in them.

The **intensity** (strength) of a color is its degree of purity (or brightness), that is, its freedom from neutralizing factors. This quality is also referred to in color terminology as *chroma*. A color entirely free of neutral elements is called a saturated color. The intensity of a color can be changed, without changing the color's value, by mixing that color with a gray of the same value.

Color *harmonies* are groups of colors that relate to each other in a predictable manner. The basic color harmonies are *complementary, monochromatic,* and *triadic.* Complementary colors are opposite each other on the color wheel. Monochromatic colors are side by side. Triadic colors are three colors that are an equal distance apart on the color wheel.

Uses of Color

The use of color has a very strong effect on the atmosphere of a building in several ways. The perceived level of formality, temperature, and mood are all influenced by the color design. The combination of soft pastel yellows and oranges creates a relaxed and quiet atmosphere in the room shown in Figure 2.6. Colors such as red, yellow, and orange create a feeling of warmth, informality, cheer, and exuberance. Colors such as blue and green create a feeling of quiet, formality, restfulness, and coolness.

Color is also used to change the apparent visual dimensions of a building. It is used to make rooms appear higher or longer, lower or shorter. Warm bold colors, such as red, create the illusion of advancement.

FIGURE 2.6 ■ Pastels and soft materials and lighting create a quiet atmosphere. *Marc-Michaels Interior Design, Inc.; Sargent, Photographer*

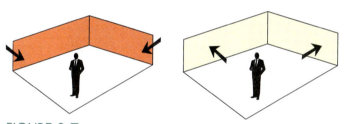

FIGURE 2.7 ■ Bold colors advance—pale colors recede.

FIGURE 2.8 ■ Computer study of the different effects of light on a structure. *Integraph Corp.*

Cool and pale colors (including pastels) tend to recede. See Figure 2.7.

Light and Shadow

Light reflects from the surfaces of forms. Shadows appear in areas that light cannot reach. Light and shadow both give a sense of depth to any structure. The effective designer plans the relationship of light and dark areas accordingly.

To achieve a dramatic effect, the designer must consider which surfaces reflect (instead of absorb) light and which surfaces refract (bend) light as it passes through materials. The designer must also remember that with continued exposure to light, visual sensitivity decreases. People become adapted to degrees of darkness or lightness after extended exposure. Thus, a designer should plan for a variety of levels of light in a room or building.

Light and shadow patterns at different times of the day or night must also be considered. Computer programs al-

low designers to study the relationship of design elements to the environment with high levels of accuracy and flexibility. For example, the three images of the structure shown in Figure 2.8 simulate the effects of sunlight and artificial light on an architectural design.

FIGURE 2.9 ■ Masonry textures dominate this design. *Kentuck Knob, Western Pennsylvania Conservancy; Robert Ruscuak, Photographer*

Texture and Materials

Materials are the raw substances with which designers create. Materials possess their own unique properties, such as color, form, dimension, degree of hardness, and texture. *Texture* is a significant factor in the selection of appropriate materials. *Texture* refers to the surface finish of an object—its roughness, smoothness, coarseness, or fineness. Surfaces of materials such as concrete, stone, and brick are rough and dull and suggest strength and informality. Smoother surfaces, such as those of glass, aluminum, and plastics, create a feeling of luxury and formality. The designer must be careful not to include too many textures of a similar nature, such as stone and brick. When positioned close together, they tend to compete with each other. Textures, such as wood and stone, are more pleasing when contrasted with other surfaces. The rough texture of stone is the dominant feature of the view shown in Figure 2.9.

Rough surfaces reduce the apparent height of a ceiling or distance of a wall and may appear darker. Smooth surfaces increase the apparent height of a ceiling or wall and reflect more light, thus making colors appear brighter.

PRINCIPLES OF DESIGN

The basic **principles of design** are the guidelines for how to combine the elements of design. For buildings to be aesthetically pleasing, as well as functional, the basic principles of design should be applied. These are *balance, rhythm, repetition, emphasis, subordination, proportion, unity, variety, opposition,* and *transition.*

Balance

Equilibrium (feeling of stability) in design is known as balance. Buildings are *informally balanced* if they are asymmetrical and *formally balanced* if they are symmetrical. The balance scale in Figure 2.10 illustrates this difference. The exterior of the building shown in Figure 2.11 is formally balanced, and the exterior of the building in Figure 2.12 is

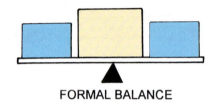

FORMAL BALANCE

INFORMAL BALANCE

FIGURE 2.10 ■ Two types of balance.

FIGURE 2.11 ■ Formally balanced design. *John Henry, Architect*

informally balanced. Figure 2.13 shows an interior room that is formally balanced.

Whether a design is formal or informal, balance requires a harmonious relationship in the distribution of space, form, line, color, light, and materials.

FIGURE 2.12 ■ Informally balanced design. *Randall E. Staff, Architect; Bryason Homes Inc., Builder; Hoffman Illustrations, Rendering*

FIGURE 2.13 ■ Formally balanced living room. *Marc-Michaels Interior Design, Inc.; Sargent, Photographer*

Rhythm and Repetition

When lines, planes, or surface treatments are repeated in a regular sequence, the order or arrangement creates a sense of rhythm. *Rhythm* creates motion and carries the viewer's eyes to various parts of the space. This may be accomplished by the repetition of lines, colors, and patterns. *Repetition* is designed into the structure of the building shown in Figure 2.14 by repeating structural window and roof line shapes.

Emphasis and Subordination

The principle of *emphasis,* or giving something importance, means drawing a viewer's attention to an area or subject. In architectural design, some emphasis or *focal point* (center of attention) should be designed into each exterior and interior space. Directing attention to a point of emphasis, the focal point, can be accomplished by arrangement of features, contrast of colors, line direction, variations in light, space relationships, and changes in materials or texture. See Figure 2.15.

Subordination occurs when emphasis is achieved through design. Other features become subordinate. They have less emphasis or importance.

Proportion

Proportion means the relationship of one part to another, or ratio. The early Greeks found that the proportions of a rectangle in the ratio of 2 to 3, 3 to 5, 5 to 8, and 6 to 10 were more pleasing than other ratios. For example, a room or a rug with dimensions of 9′ × 15′ or 10′ × 16′ will have the proportions 3 to 5 and 5 to 8.

FIGURE 2.14 ■ Repeated structural components create rhythm. *Western Wood Products Association*

FIGURE 2.15 ■ Screen position creates a focal point for this media-oriented room. *Audio Tec Design*

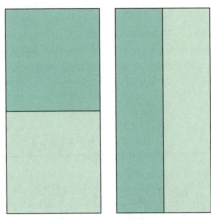

FIGURE 2.17 ■ The two rectangles are the same although proportionally they appear different.

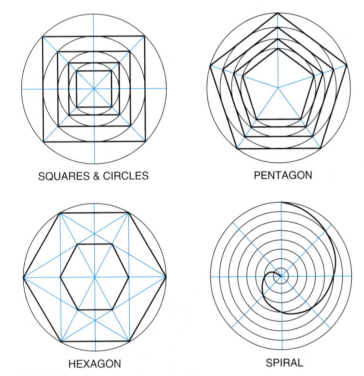

SQUARES & CIRCLES   PENTAGON

HEXAGON   SPIRAL

FIGURE 2.16 ■ Proportional systems used in two-dimensional design.

Figure 2.16 shows several classical systems used to create desirable proportions in a design. In the sixteenth century scientists observed that many natural features related to each other at a ratio of 1 to 1.618, which they called the "golden rule" or Phi. Closely related proportions have been used by artists and architects for centuries to create attractive shapes.

The proportion (ratio) of interior space, furniture, and accessories should be harmonious. Large bulky components in small rooms should be avoided, just as small components in large rooms should not be used. Areas can appear completely different depending on how the proportional division of space within the area is allocated. The total volume of space within the two rectangles shown in Figure 2.17 is the same; however, the proportions appear different due to the division of space.

Unity

Unity is the expression of the sense of wholeness in the design. Every structure should appear complete. No parts should appear as appendages or afterthoughts. Designers achieve unity through the use of consistent line and color, even though the building is composed of many different parts. Unity, or harmony, as the name implies, is the joining together of the basic elements of good design to form one harmonious, unified whole. Unity can be achieved by using any of the elements of design consistently throughout the entire design.

Unity is often achieved through the use of basic geometric shapes. For centuries, the architectural development of livable structures has been restricted by the use of the square and the cube. Other shapes, such as the triangle, octagon, parabola, pyramid, pentagon, circle, and sphere are now used extensively. This is possible due to the development of materials that are stronger and lighter and have a variety of uses, such as those used in the geodesic dome shown in Figure 2.18.

FIGURE 2.18 ■ A geodesic dome structure possesses mass unity. *James Eismont, Photographer*

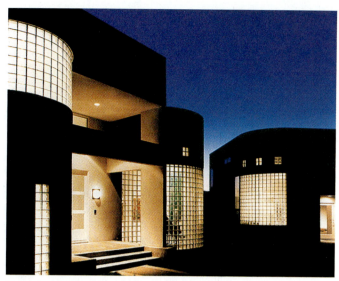

FIGURE 2.19 ■ Opposition created with light. *Pittsburgh Corning Corp.*

Variety and Opposition

Too much unity, too much rhythm, or too much repetition can ruin a sense of *variety* or contrast. Likewise, too little of any of the elements of design will also result in a lack of variety. Without variety, any area can become dull and tiresome to the eye of the observer.

Variety can often be achieved with *opposition*. Opposition involves contrasting elements such as short and long, thick and thin, straight and curved, light and dark. Changes in color are also a means to achieve variety. The building shown in Figure 2.19 illustrates opposition created by lighting effect. The building lines in Figure 2.20 provide line opposition.

Transition

The change from one color to another or from a curved to a straight line, if done while maintaining the unity of the design, is known as **transition.** Transition may be the change from a curved molding on the floor to the flat wall or a change from one floor covering to another in adjoining rooms.

Transition in architecture is also considered in relation to changes in the surroundings. For example, where extreme climate changes occur, the transition between sea-

FIGURE 2.20 ■ Opposition created with building lines. *Diane Kingston, Photographer*

sons needs to be considered so that a design functions well under all conditions. For example, Frank Lloyd Wright's design for the "Fallingwater" residence is just as dramatic in winter (Figure 2.21) as in summer (Figure 2.22).

FIGURE 2.21 ■ "Fallingwater" in winter. *Western Pennsylvania Conservancy; Thomas A. Heinz, Photographer*

FIGURE 2.22 ■ "Fallingwater" in summer. *Western Pennsylvania Conservancy; Christopher Little, Photographer*

CHAPTER

2

Fundamentals of Design Exercises

1. Choose a figure from Chapter 1 or 2 and discuss how the concept of "form follows function" relates to it.

2. Choose an architectural design from Chapter 1 or 2. Describe the six elements of design in the figure you selected.

3. How would you make a small room look larger and a large room look smaller? Explain the reasons for your decision.

4. Describe a building in your neighborhood or in a magazine in terms of design. Tell what elements and principles of design are its most outstanding features. (You might include a photograph or magazine illustration.)

 5. List each element of design and describe your preference for applying each to a residence of your own design.

6. Describe the difference between a *shade*, a *tone*, and a *tint*.

7. Describe primary, secondary, and tertiary colors.

8. Describe complementary, monochromatic, and triadic colors.

9. List the basic elements of design.

10. List the basic principles of design.

PART 2

Architectural Drafting Fundamentals

3 CHAPTER

Drafting Scales and Instruments

OBJECTIVES

In this chapter you will learn to:

- measure and prepare drawings with different scales.
- draw with drafting instruments.
- select and use appropriate types of paper and other drafting supplies.
- use time-saving devices.

TERMS

architect's scale
civil engineer's scale
compass
dividers
drafting machine
flexible curve
full scale

metric scale
overlays
parallel slide
protractor
stampat
technical pens

templates
trace
triangles
T square
underlays
vellum

INTRODUCTION

A thorough understanding of the various architect's scales and drafting instruments is necessary to prepare architectural drawings. This is vital whether drawings are prepared using normal drafting instruments or a computer-aided design (CAD) drafting system. This knowledge enables the drafter to create drawings that are accurate and can be read by other construction professionals. Manual instruments are covered in this chapter and computer-aided drafting is introduced in Chapter 5.

 SCALES

In architectural drawing the term *scale* may refer to the proportional (ratio) size of a drawing or to an architectural measuring instrument.

When making a scaled (proportional) drawing, one measurement is used to represent another. Scaled drawings allow objects of all sizes to be proportionally re-

duced or enlarged to show the correct relationship of all parts. Different ratios are required depending on the size of the object and the drawing format size.

Without the use of *reduced* scales no object larger than a sheet of paper could be accurately drawn. For example, if the earth were drawn to the same scale used on a typical architectural drawing, 1/4″ = 1′-0″ (or 1″ = 4′-0″, a ratio of 1 to 48), the drawing sheet would need to be 165 miles wide! See Figure 3.1.

Conversely, small objects, even those that may be invisible or barely visible to the human eye, can be proportionally enlarged to show details. In architectural drawing, a small item such as a hinging mechanism may need to be drawn at an *enlarged* scale.

In the preparation of scaled drawings, instruments called scales are used. The three main types of scales are the architect's scale, civil engineer's scale, and metric scale.

Architect's Scale

The ability to use **architect's scales** accurately is required not only in preparing drawings but also in checking

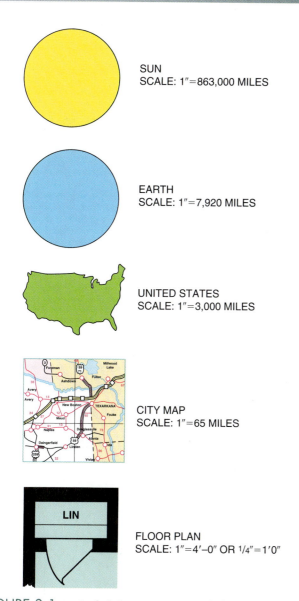

FIGURE 3.1 ■ Scaled drawings are needed to show the size and shape of large objects.

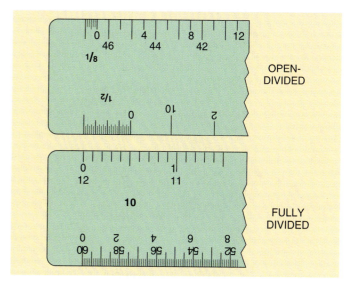

FIGURE 3.2 ■ Open-divided and fully divided scales.

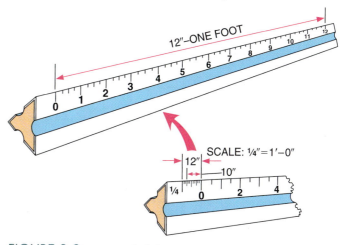

FIGURE 3.3 ■ In scaled drawings, one measurement represents another.

existing architectural plans and details. The architect's scale is also needed in a variety of related architectural jobs such as bidding, estimating, and model building.

Whether to reduce a structure's size so that it can be drawn to fit on paper or to enlarge a small detail for clarity and accurate dimensions, a drafter needs to use the appropriate scale divisions.

Architect's scales are either open divided or fully divided. In fully divided scales, each main unit on the scale is fully subdivided into smaller units along the full length scale. On open-divided scales, only the main units of the scale are graduated (marked off) all along the scale. There is a fully subdivided unit at the start of each scale. See Figure 3.2.

The main function of an architect's scale is to enable the architect, designer, or drafter to plan accurately and make drawings in proportion to the actual size of the structure. For example, when a drawing is prepared to a reduced scale of 1/4″ = 1′-0″, a line that is drawn 1/4″ long is thought of by the drafter as 1′-0″, not as 1/4″. See Figure 3.3.

Using a Scale

Architect's scales may be either bevel or triangular style. See Figure 3.4. Note that the triangular style has three sides and six edges. It accommodates 11 different scales (marked units of measure). One edge is a full-size scale of 12″, divided into 16 parts per inch. The other five edges contain open-divided scales paired to include 3/16 with 3/32, 1/4 with 1/8, 3/4 with 3/8, 1 with 1/2, and 3 with 1 1/2. Locating two scales on each edge maximizes the use of space. One scale reads from left to right. The opposite scale, which is twice as large, reads from right to left. For example, the 1/4″ scale and the 1/8″ scale are placed

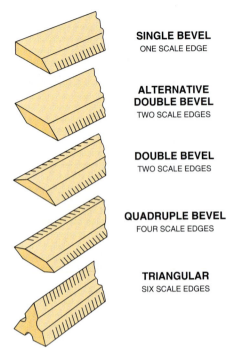

FIGURE 3.4 ■ Architectural scale shapes.

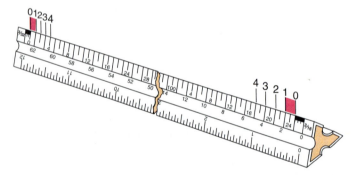

FIGURE 3.5 ■ Scales that read from right to left are twice the size as those that read from left to right.

on the same edge but are read from opposite directions. Be sure you are reading the scale numbers in the correct direction when using an open-divided scale. Otherwise, your measurement could be wrong. See Figure 3.5.

The architect's scale can be used to make the divisions of the scale equal 1″ or 1′-0″. For example, 1/2″ can equal 1″ or 1′-0″ or any unit of measurement such as yards or miles. See Figure 3.6.

Because buildings are large compared to the size of a person or appliance, most major architectural drawings use a scale that relates the parts of an inch to 1′-0″. Architectural details, such as cabinet construction and joints, often use the parts of an inch to represent 1″. On open-divided scales, the divided section at the end of the scale is not a part of the numerical scale. This divided section is an additional length to show smaller subdivisions of the larger unit. When measuring with the scale, start at zero,

If ½″ = 1″, then ¼″ = ½″.

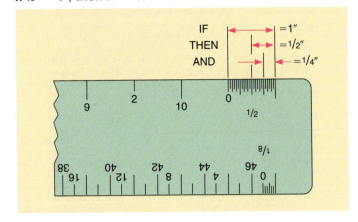

If ½″ = 1′-0″, then ¼″ = 6″.

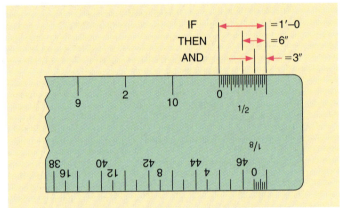

FIGURE 3.6 ■ On a half-inch scale, 1/2″ may equal 1″ or 1′-0″.

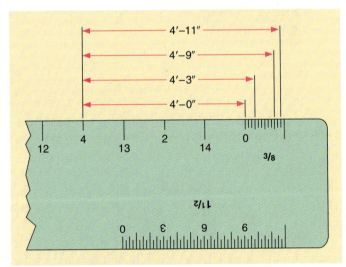

FIGURE 3.7 ■ Subdivisions are used for inches or inch fractions.

not at the end of the fully divided section. First measure the number of larger units (for example, feet) and then measure the additional smaller units (inches) in the subdivided area. Look at Figure 3.7. The distance of 4′-11″ is scales, they may equal a fractional part of an inch. Figure 3.8 shows a further use of the architect's scale.

To further understand the architect's scale, compare one specific distance shown on different scales. Compare the actual length of the 5′-6″ dimension on the four different scales shown in Figure 3.9. Figure 3.10 shows the reduction on six different scales.

The architect's scale is only as accurate as its user. In using the scale, always lay out the overall dimensions of the drawing first. If the width and length are correct, only minor errors in subdimensions may occur. Moreover, if overall dimensions are correct, it's easier to check subdimensions. If one is inaccurate, another will be also.

Remember, an architect's scale is a measuring device, not a drawing instrument. Never use a scale as a straightedge for drawing. The fine increment lines on a scale will be worn down or removed if misused, making accurate measuring difficult.

Selecting a Scale (Proportion)

If the structure to be drawn is extremely large, a small scale must be used. Small structures can be drawn to a larger scale, because they will not take up as much space on the drawing sheet. Most plans that show major parts of residences (floor plans, elevations, and foundation plans) are drawn to 1/4″ scale. Construction details pertaining to these drawings are often drawn to 1/2″, 3/4″, or 1″ = 1′-0″.

Remember that as the scale changes, not only does the length of each line increase or decrease, but the width of each wall also increases or decreases, as shown in Figure 3.11. A wall drawn to the scale of 1/16″ = 1′-0″ is small

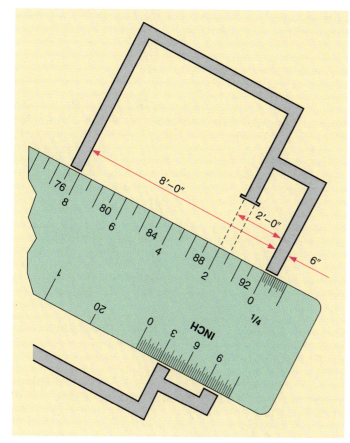

FIGURE 3.8 ▪ Scale used to measure feet and inches.

established by measuring from the division line 4 to 0 for feet. Then, measure on the subdivided area 11″ past 0. On this scale, each line in the subdivided part equals 1″. On smaller scales, these lines may equal only 2″. On larger

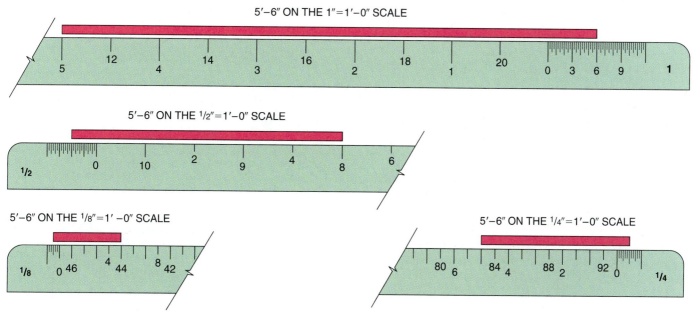

FIGURE 3.9 ▪ The distance 5′-6″ measured on different scales.

1″ = 1′-0″ will **decrease** drawing size by 12 times								
1½″ = 1′-0″	″	″	″	″	″	″	8	″
3″ = 1′-0″	″	″	″	″	″	″	4	″
½″ = 1′-0″	″	″	″	″	″	″	24	″
¼″ = 1′-0″	″	″	″	″	″	″	48	″
⅛″ = 1′-0″	″	″	″	″	″	″	96	″

FIGURE 3.10 ■ Resulting reduction of the different scales.

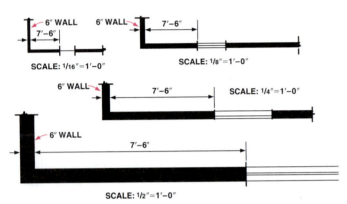

FIGURE 3.11 ■ Floor plan corner wall drawn at different scales.

and little detail can be shown. The 1/2″ = 1′-0″ wall would probably cover too large an area on the drawing if the building were very large. Therefore, the 1/4″ and 1/8″ scales are used most often for drawing floor plans and elevations.

The **full scale** (1) is used to draw objects full size 1″ = 1″ or 1′ = 1′ or to a scale of 1″ = 1′-0″. An open-divided

full scale is 12″, and each inch divided into 16 units throughout.

Civil Engineer's Scale

Although the architect's scale is used for most architectural drawings, the **civil engineer's scale** is often used for plans that show the size and features of the land surrounding a building (plot plans, site plans, landscape plans). A civil engineer's scale divides the inch into decimal parts. These parts are 10, 20, 30, 40, 50, and 60 parts per inch. See Figure 3.12. Each one of these units can represent any distance, such as an inch, a foot, a yard, or a mile, depending on the final drawing size.

The civil engineer's scale can also be used to draw floor plans. The scale 1/4″ = 1′-0″ (a 1:48 ratio) is the same ratio as 1″ = 4′-0″ (also a 1:48 ratio). A civil engineer's scale does not use feet and inches. A civil engineer's scale includes feet and the decimal parts of a foot. Thus 2′-6″ reads 2.5′ and 7′-3″ reads 7.25′ on an engineer's scale. See Figure 3.13.

A civil engineer's scale of 1″ = 10′ is normally used for small sites. If a land site is very large, a scale of 1″ = 20′ or 1″ = 30′ may be needed to allow the plan to fit the sheet size. See Figure 3.14.

Selecting a Scale (Instrument)

Different scales are designed for a broad range of applications. Before a drawing is started, determine the actual size of the area to be covered. Then select the scale—whether an architect's or civil engineer's scale—that will provide the greatest detail and yet fit completely on the drawing sheet. See Figure 3.15.

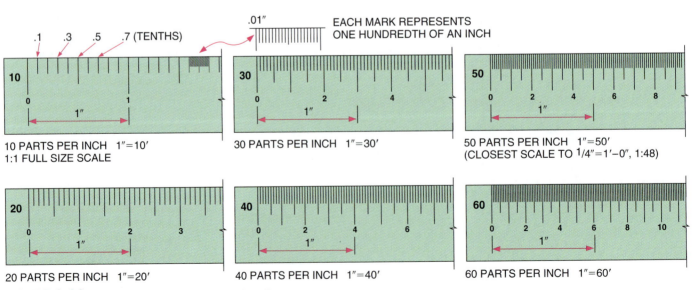

FIGURE 3.12 ■ Civil engineer's (decimals) scales.

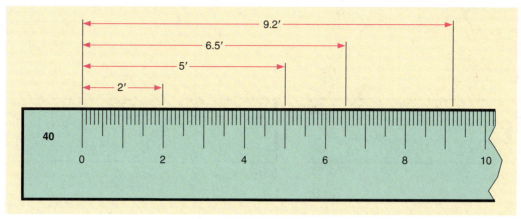

FIGURE 3.13 ■ The number 40 on an engineeer's scale represents the same ratio as 1/4″ = 1′-0″ on an architect's scale.

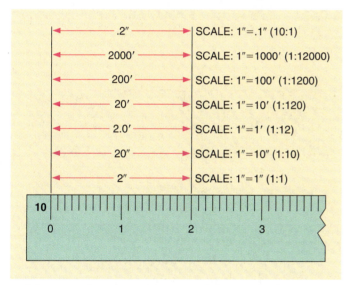

FIGURE 3.14 ■ Distances represented by 2″ on different engineer's scales.

Metric Scales

Math Connection

Metric scales such as those shown in Figure 3.16 are used in the same manner as the architect's scale to prepare reduced-size drawings. Metric scales, however, use ratios in increments of 10 rather than the fractional ratios of 12 used in architect's scales. The metric system of measure is a decimal system. Units are related by tens.

Most measurements used on architectural drawings are linear distances. The basic unit of measure in the metric system for distance is the *meter* (m). Prefixes are used to change the base (meter) to larger or smaller amounts by *units of 10*. Prefixes that represent subdivisions of less than one meter are deci-, centi-, and milli-. A decimeter equals one-tenth (0.1) of a meter. A centimeter equals one one-hundredth (0.01) of a meter. A millimeter equals one one-thousandth (0.001) of a meter. See Figure 3.17. The most commonly used subdivisions of a meter are the

DRAWING TYPE	U.S. CUSTOMARY ARCHITECT'S SCALES (FEET/INCHES)	ISO METRIC SCALES (MILLIMETERS)	CIVIL ENGINEER'S SCALES (FEET/DECIMAL)
Site plans	1/8″ = 1′–0″ THRU 1/32″ = 1′–0″	1:100 THRU 1:500	1″ = 10′ THRU 1″ = 200′–0″
Floor plans	1/4″ = 1′–0″ or 1/8″ = 1′–0″	1:50 or 1:00	1″ = 10′ THRU 1″ = 30′–0″
Foundation plans	1/4″ = 1′–0″ or 1/8″ = 1′–0″	1:50	1″ = 10′ THRU 1″ = 30′–0″
Exterior elevations	1/4″ = 1′–0″ or 1/8″ = 1′–0″	1:50 or 1:00	1″ = 4′–0″ or 2′–0″
Interior elevations	1/2″ = 1′–0″	1:20	1″ = 4′–0″ or 2′–0″
Construction details	1 1/2″ = 1′–0″ THRU 3/4″ = 1′–0″	1:5 THRU 1:10	1″ = 1′–0″ or 2′–0″
Cabinet details	1/2″ = 1′–0″	1:20	1″ = 2′–0″

FIGURE 3.15 ■ Range of scales used on architectural drawings.

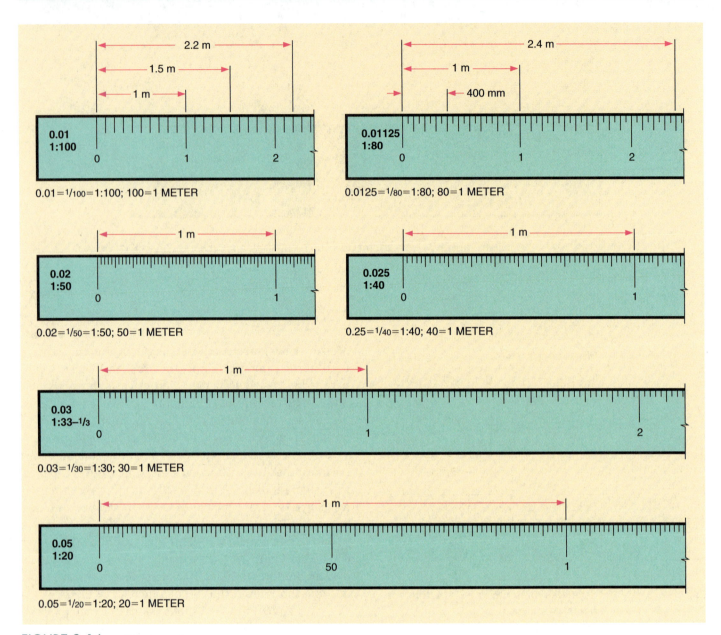

0.01=1/100=1:100; 100=1 METER

0.0125=1/80=1:80; 80=1 METER

0.02=1/50=1:50; 50=1 METER

0.25=1/40=1:40; 40=1 METER

0.03=1/30=1:30; 30=1 METER

0.05=1/20=1:20; 20=1 METER

FIGURE 3.16 ■ Metric scales.

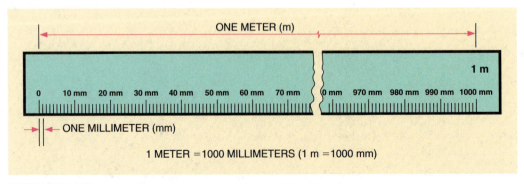

FIGURE 3.17 ■ A millimeter is one one-thousandth of a meter.

USE	RATIO	COMPARISON TO 1 METER
City map	1:2500 1:1250	(0.4 mm equals 1 m) (0.8 mm equals 1 m)
Plat plans	1:500 1:200	(2 mm equals 1 m) (5 mm equals 1 m)
Plot plans	1:100 1:80	(10 mm equals 1 m) (12.5 mm equals 1 m)
Floor plans	1:75 1:50 1:40	(13.3 mm equals 1 m) (20 mm equals 1 m) (25 mm equals 1 m)
Details	1:20 1:10 1:5	(50 mm equals 1 m) (100 mm equals 1 m) (200 mm equals 1 m)

FIGURE 3.18 ■ Use of metric ratios.

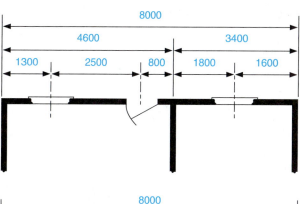

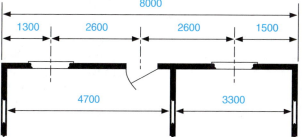

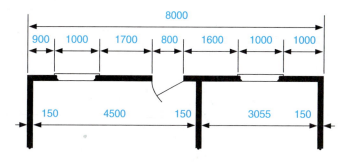

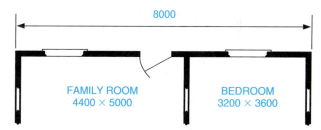

FIGURE 3.19 ■ Dimensions in millimeters.

centimeter and the millimeter. The most commonly used multiple of the meter is the *kilo*meter. A kilometer equals 1000 meters.

The numbers on a meter scale mark every tenth line to represent centimeters. Each single line represents millimeters. Note that there are 10 millimeters within each centimeter.

Just as with any other scale, the ratio chosen depends on the size of the drawing compared with the full size of the object. Figure 3.18 shows some common metric ratios and the various types of architectural drawings for which they are used.

In architectural drawing, ISO standards recommend that only millimeters be used for all architectural and engineering drawings. See Figure 3.19. This eliminates the use of decimal points as shown in Figure 3.20.

USING CAD

CAD Scales

Knowledge of architect's and engineer's scales is necessary for both manual and CAD drafting. In CAD drawing the *Scale* command is used to set a drawing object at a specific architect's, metric, or engineer's scale size. First the drawing sheet size is determined by selecting *Layout* and right mouse clicking to *Page Setup Manager.* A viewport is then set up to scale the image based on magnification (*Zoom*), which allows the drawing to fit on the paper size available. Smaller scales or larger sheet sizes are options that can ensure a proper fit using a new viewport.

The physical scale of a drawing object can be changed by using the *Scale* command found under the *Modify* pull-down menu. During drawing, the *Measure* command can be used to locate a specific size on a segment. This command can also be used to draw a specific line length by inputting the line length at the scale of the drawing.

Some drawings prepared with fractional dimensions need to be converted to metric dimensions. Figure 3.21 shows the conversion of inches to millimeters. To convert inch dimensions to millimeter dimensions, use the following formula:

Formula: in. × 25.4 = mm
(inches × 25.4 = millimeters)

Example: 6'-6" = 78"
78" × 25.4 = 1981.2 mm

Prepare all drawings in a set using either metric ratios or the customary fractional system. *Do not mix metric and customary units.* If approximate conversion from one system to the other is necessary, refer to the appendix. When very accurate conversion from customary to metric units is necessary, consult a handbook or use the *Metric Practice Guide* from ASME.

GUIDES FOR STRAIGHT LINES

T Square

The **T square** serves several purposes. It is used primarily as a guide for drawing horizontal lines. It also serves as a base for a triangle that is used to draw vertical and inclined lines. The T square is particularly useful for drawing extremely long lines that deviate from the horizontal plane. Common T-square lengths for architectural drafting are 18", 24", 36", and 42".

T squares must be held tightly against the edge of the drawing board, and triangles must be held firmly against the T square to ensure accurate horizontal and vertical lines. See Figure 3.22A. Because only one end of the T

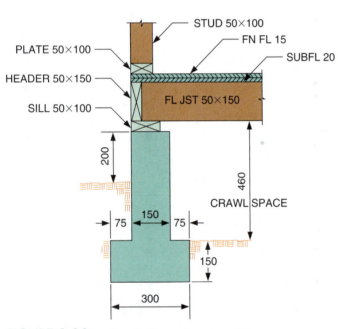

FIGURE 3.20 ■ Detail dimensions in millimeters.

ALL DIMENSIONS FOR ARCHITECTURAL DRAWINGS ARE IN MILLIMETERS (mm)		
CONVERSIONS		
Convert from:	To:	Multiply by:
Feet	Millimeters	304.8 (305)
Millimeters	Feet	25.4 (25)

FIGURE 3.21 ■ Metric conversion table – feet and millimeters.

A. Right-handed method.

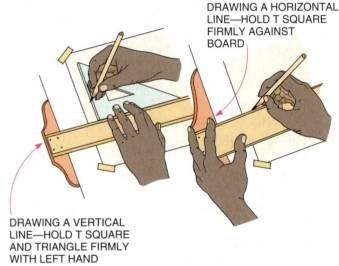

DRAWING A HORIZONTAL LINE—HOLD T SQUARE FIRMLY AGAINST BOARD

DRAWING A VERTICAL LINE—HOLD T SQUARE AND TRIANGLE FIRMLY WITH LEFT HAND

B. Left-handed method.

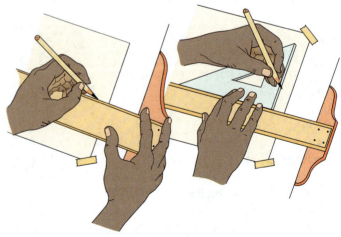

FIGURE 3.22 ■ Methods of T-square drawing.

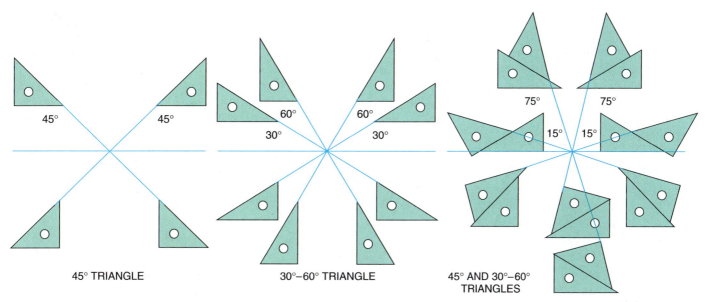

FIGURE 3.23 ■ Angles possible using 45°, 60°, and 30° triangle combinations.

square is held against the drawing board, some sag may occur when long T squares are not held securely. Figure 3.22B shows the difference between right-handed and left-handed use of a T square. Horizontal lines are drawn from left to right by right-handed people and right to left by left-handed people. Vertical lines are made by pulling the pencil or pen upward.

Parallel Slide

The **parallel slide** (or parallel rule) performs the same function as the T square. Extremely long lines are common in many architectural drawings such as floor plans and elevations. Because most of these lines should be drawn continuously, the parallel slide is used extensively by architectural drafters.

A parallel slide is anchored at both sides of a drawing board. This attachment eliminates the possibility of sag at one end, which is a common objection to the use of the T square. Another advantage of using the parallel slide is that the drawing board can be tilted to a very steep angle without causing the slide to slip to the bottom of the board. If the parallel slide is adjusted correctly, it will stay in the exact position in which it is placed.

Triangles

Triangles are used to draw vertical and diagonal or inclined lines with either a T square or other horizontal guide. A variety of combinations produces numerous angles, as shown in Figure 3.23.

The 8-inch 45° triangle and the 10-inch 30°–60° triangle are preferred for architectural work. Adjustable trian-

gles are used to draw angles that cannot be laid out by combining the 45° and 30°–60° triangles.

Drafting Machine

A **drafting machine** is a mechanical tool that can serve as an architect's scale, triangle, protractor, T square, or parallel slide all in one. A drafting machine consists of a "head" to which two graduated scales are attached perpendicular to each other. The scales (arms) of the drafting machine are usually made of aluminum or plastic. The horizontal scale performs the function of a T square or parallel slide in drawing horizontal lines. The vertical scale performs the function of a triangle in drawing vertical lines.

USING CAD

CAD Line Tasks

Preparing architectural drawings requires the use of the *Line* command more than any other command. Click on the *Draw* pull-down menu, then select *Line*. Lines can have a wide variety of characteristics applied to them including colors, weights, style, or type. The creation of the line is done by picking a starting point, pointing in a specific direction for the line to follow, and typing in a distance. This is known as a distance and direction style of input. A line can also be created by picking points to begin and end the line segment. The type size and color is determined when the *Line* command is used.

Large drawings can be made using track drafting machines in which a protractor head is mounted on a movable track. These machines are smoother and faster than elbow-type machines. However, the operation of the head is identical. The protractor head of a track machine is mounted on a vertical track that is attached to a horizontal track.

Protractor

A **protractor** provides a graduated scale applicable to all 360° of a circle. To draw an angle by compass degree, align the center of the protractor on the vertex of the angle and mark the degree located on the circumference of the protractor. Then draw a line between the center and the degree mark.

INSTRUMENTS FOR CURVED LINES

Compasses

A **compass** is used in architectural work to draw circles, arcs, radii, and parts of many symbols. Small circles are drawn with a *bow compass.*

Large circles on architectural drawings, such as those used to show the radius of driveways, walks, patios, and stage outlines, are drawn with a large *beam compass.* Very small circles on architectural drawings are drawn with either a *drop-bow compass* or a *circle template.*

Dividers

Dividing an area into an equal number of parts is a common task performed by architectural drafters. In addition to the architect's scale, **dividers** are used for this purpose.

To divide an area equally by the trial-and-error method, first adjust the dividers until they appear to represent the desired division of the area. Then place one point at the end of the area and "step off" the distance with the dividers. If the divisions turn out to be too short, increase the opening on the dividers. If the divisions are too long, decrease the setting. Repeat the process until the line is equally divided.

Dividers are also used frequently to transfer dimensions and to enlarge or reduce the size of a drawing.

Irregular Curve Instruments

Many architectural drawings contain irregular lines. A **flexible curve** is used to repeat irregular curves that have no true radius or series of radii and cannot be drawn with a compass. Curved lines that are not part of an arc can also be drawn with a French (irregular) curve.

DRAFTING PENCILS AND PENS

Drafting pencils and pens are used with other drafting instruments to produce accurate, readable, and consistent architectural lines and symbols.

Drafting Pencils

Pencils used for drafting are either wood encased or mechanical, as shown in Figure 3.24. The width and density of the line produced depends on the degree of hardness and the point of the pencil's lead. Although referred to as

USING CAD

CAD Arcs, Circles, and Curves

Arcs, circles, and curves are used frequently on architectural drawings for windows, doors, pool shapes, electrical connections, and so forth. Arcs can be created by choosing the *Draw* pull-down menu, selecting *Arc,* and then choosing the specific style of arc to create. These styles include the options for angle, length of chord, and radius. Locating the start point, center point, and end point will also yield an arc.

The specific type of arc style that is selected will determine how the arc is created. Arcs require three points to create and are created in a counterclockwise rotation. Irregular curves are drawn using the *Spline* command and locating several points on a curve with a cursor. A spline is a smooth curve that passes through a series of connected points. Increasing the number of points will improve the accuracy and smoothness of the curve. Ellipses are drawn by locating the ends of the major and minor axis points. The *Ellipse* command is then used to select the center of the ellipse, point in the major axis direction, and type the major axis radius, then the minor radius.

"lead," the core of a drawing pencil is composed mainly of graphite. Hard pencils (3H, 4H) are often used to begin architectural layout work. Medium pencils (2H, H, F) are used for most of the lines in a completed drawing. Soft pencils (HB, B, 2B) are used for lettering and thick cutting-plane lines, as well as for shading in pictorial drawings. Figure 3.25 shows the range of lead hardness and the resulting line weights.

Drafting pencils can be sharpened to several types of points depending on the type of line desired. Regardless of the type of pencil point used, care must be taken to produce an even point. When uneven points are used, such as a chisel point, uneven lines will result. See Figure 3.26. Pencil lines used on architectural drawings vary in width, but should not vary in density. Thin lines should be just as black and dense as thick lines.

Technical Pens

Ink pens used for drafting are called **technical pens.** Their points range in thickness from .13 mm to 2.00 mm. See Figure 3.27. One reason for using pens rather than pencils is to create very dense and consistent lines. Working with pens manually tends to slow down drawing speed, and ink lines are difficult to erase.

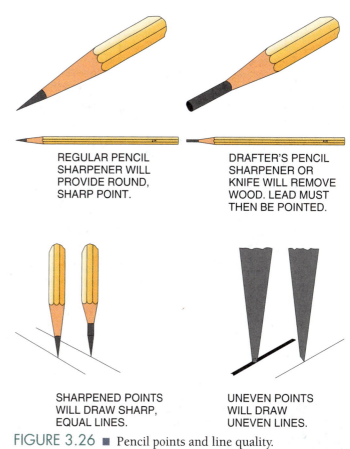

REGULAR PENCIL SHARPENER WILL PROVIDE ROUND, SHARP POINT.

DRAFTER'S PENCIL SHARPENER OR KNIFE WILL REMOVE WOOD. LEAD MUST THEN BE POINTED.

SHARPENED POINTS WILL DRAW SHARP, EQUAL LINES.

UNEVEN POINTS WILL DRAW UNEVEN LINES.

FIGURE 3.26 ■ Pencil points and line quality.

WOOD-BONDED CASE

2 H

GRADE MARK

WOODEN PENCILS

STANDARD SIZE LEAD

STANDARD LEAD HOLDER

METAL OR PLASTIC CASE

THIN LEAD (REQUIRES NO SHARPENING)

THIN LEAD HOLDER

MECHANICAL PENCILS

FIGURE 3.24 ■ Drafting pencil types.

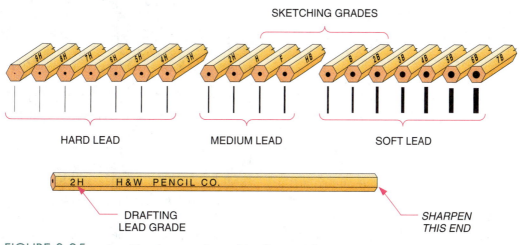

SKETCHING GRADES

HARD LEAD

MEDIUM LEAD

SOFT LEAD

2H   H&W PENCIL CO.

DRAFTING LEAD GRADE

SHARPEN THIS END

FIGURE 3.25 ■ Lead hardness and matching line weights.

WIDTH	SIZE
.13	6 × 0
.18	4 × 0
.25	3 × 0
.30	00
.35	0
.50	1
.70	2½
1.00	3½
1.40	6
2.00	7

FIGURE 3.27 ■ Technical pen widths in millimeters.

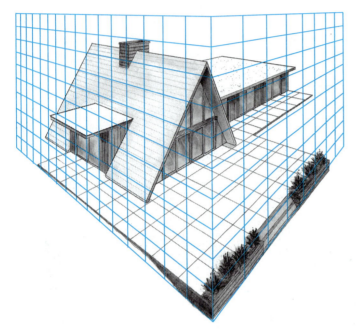

FIGURE 3.28 ■ Use of preprinted perspective grid.

PAPERS AND DRAWING SURFACES

Most architectural drawings are prepared on **vellum** or a good-quality tracing paper. Preliminary design work and progressive sketches are usually done on extremely thin tracing paper ("bum wad," "flimsy," "trash"). These preliminary drawings are eventually discarded.

A wide variety of drafting papers is available. Some vellum papers have nonreproducible grid lines that do not show when the original drawing is duplicated. Grid papers are also printed with nonreproducible angles and lines for perspective drawings (pictorial drawings). See Figure 3.28.

The size of the drawing surface is determined at the beginning of a project. The drawing format selected should be larger than the largest drawing in the set. Figure 3.29 shows the standard sizes of paper or vellum used for architectural drawings.

CUSTOMARY (INCHES)	METRIC (mm)
8″ × 10″	
8″ × 11″	
*8.5″ × 11″ (A size)	210 × 297 mm (A4)
*9″ × 12″	
11″ × 14″	297 × 420 mm (A3)
*11″ × 17″ (B size)	
*12″ × 18″	
14″ × 17″	
15″ × 20″	
*17″ × 22″ (C size)	420 × 594 mm (A2)
*18″ × 24″	
19″ × 24″	
21″ × 27″	
*22″ × 34″ (D size)	594 × 841 mm (A1)
*24″ × 36″	
*34″ × 44″ (E size)	841 × 1189 mm (A0)
*36″ × 48″	

* Most commonly used.

FIGURE 3.29 ■ Standard drawing paper sizes.

The type of paper, as well as the drawing instrument, greatly affects the line quality. Different pencil grades of hardness or softness are needed for different papers. Weather conditions also affect line quality. During periods of high humidity, harder pencils must be used. Drawing paper with a hard surface helps to produce distinct, clean lines, especially when using technical pens. Soft surfaces absorb too much ink and result in feathered lines.

CORRECTION EQUIPMENT

Mistakes and corrections are part of every drawing process. Designers employ a variety of erasers and ways of keeping drawings and sketches clean. *Basic erasers* are used for gen-

USING CAD

Making Changes on CAD Drawings

Lines, blocks, or layers can be deleted by using the *Undo* command. This command can be used to erase the last entry or previous entries by clicking the *Undo* command's backward arrow until the desired entries are eliminated. This command can also be used to eliminate an entire group or block. To reverse deletions and restore the entries as first drawn, the *Redo* command—or the forward arrow—can be used. After deletions are made, drawing can proceed as before.

eral purposes. *Gum erasers* are used for light lines. *Electric erasers* are very fast and do not damage the surface of the drawing paper. A very light touch is used to eradicate lines. *Kneaded erasers* pick up loose graphite by dabbing.

To keep drawings and sketches clean, *dry cleaner bags* are used to remove smudges. *Powder* sprinkled on the drawing reduces smudging and keeps instruments clean. It also enables drafting instruments to move freely.

Erasing shields are thin pieces of metal or plastic with a variety of small, different-shaped openings. The appropriate opening is positioned over a line to be erased. The shield covers the surrounding area. Lines can be erased without disturbing nearby lines that are to remain on the drawing.

A *drafting brush* is used periodically to remove eraser and graphite particles and to keep them from being re-distributed on the drawing. Do not blow on a drawing or use your hand to remove debris.

TIME-SAVING AIDS AND DEVICES FOR DRAFTING

Construction often begins immediately upon completion of the working drawings. Under these conditions, speed in the preparation of drawings is of utmost importance. To work quickly, many time-saving devices are employed by architectural drafters. These devices eliminate unnec-essary time on the drawing board without sacrificing the quality of the drawing.

Architectural Templates

Templates are usually made of sheet plastic. Openings in the template are shaped to represent various symbols and fixtures. A symbol or fixture is traced on the drawing by following the outline with a pencil or pen. This proce-dure eliminates the repetitive task of measuring and lay-ing out the symbol each time it is to be used on the drawing. Note that the template scale must always be the same as the scale of the drawing.

A wide assortment of templates is available. Many are used to draw only one type of symbol. Some are designed specifically for furniture, doors, windows, or landscape features. Others provide electrical or plumbing symbols. Some serve as lettering, circle, or ellipse guides.

Overlays

An **overlay** is any sheet that is placed over an original drawing. The information placed on an overlay becomes a visual part of the original drawing. Some overlays re-main separate sheets and some are permanently affixed to the base drawing.

Stampat

A solid overlay cannot be used on a drawing unless the drawing is to be photocopied only. A solid overlay on vel-lum will appear as a solid blue surface in some reproduc-tion processes.

To add an overlay to a vellum or any transparent drawing media, a **stampat** must first be prepared. A stampat is a transparent sheet onto which a drawing has been photocopied on one surface. The other surface is adhesive. The stampat sheet is adhered to the original drawing surface before the drawing is reproduced. The stampat drawing will appear on the reproduced print looking as if it had always been part of the original drawing. This method is widely used to attach typical details on a drawing so that they do not have to be re-peatedly redrawn.

When using a CAD system, typical details are stored in a symbol library and then added to a drawing or later re-vised to suit a specific need.

Sheet Overlays

Most separate *sheet overlays* (**trace**) are made by drawing on transparent or translucent material such as acetate or drafting film or translucent material such as vellum. Overlays are used in the design process to add to or change features of the original drawing, without marking the original drawing.

Overlays are also used to add features to a drawing that would normally complicate the original drawing. Lines that are hidden and many other details can be made clear by drawing this information on an overlay. See Figure 3.30. Some sheet overlays are also used to aid in the pre-sentation of an architectural design concept.

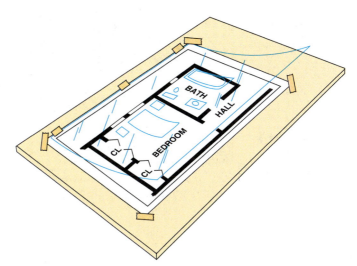

FIGURE 3.30 ■ Use of transparent or translucent trace overlay.

Pressure-Sensitive Overlays

Pressure-sensitive overlays adhere directly to the surface of the drawing. Details that are often repeated on other drawings or projects are frequently reproduced on pressure-sensitive stampat "appliqué paper."

The transparent appliqué is then attached to any drawing without repeated redrawing.

Tapes

Many types of *printed pressure-sensitive tapes* can be substituted for drawn lines and symbols on architectural drawings. These are used to produce lines and symbols that otherwise would be difficult and time consuming to construct. When using a CAD system, these symbols are located in the line symbol library.

Drafting tape is used to attach drawings to a drawing board. In addition, strips of drafting tape can also help ensure the equal length of lines when ruling many close lines. Strips of tape are placed on the drawing to mask the areas not being lined. The lines are then drawn on the paper and extended onto the tape. When the tape is removed, the ends of the lines are even and sharp.

Stamps

For architectural features that are often repeated, stamps are effective time-savers. Stamps can be used with any color ink. Stamps are used most often for symbols that do not require precise positioning on the drawing, such as landscape features, people, and cars, as shown in Figure 3.31. Stamps

FIGURE 3.31 ■ Typical press-on or stamp patterns used on architectural drawings.

may also be used for furniture outlines and labels. Stamps that use nonreproducible blue ink are used to stamp outlines on drawings, which are then traced over, altered, or rendered.

Underlays

Underlays are drawings or parts of drawings that are placed under the original drawing and traced onto the original. Architects often use them as master drawings. To be effective, a master drawing must be prepared to the correct scale and aligned carefully each time it is used.

Many symbols and features of buildings are drawn more than once. Many drafters prepare a series of underlays of the features repeated most often on their drawings. These are traced exactly or altered on the original drawing. When using a CAD system, these symbols or images can be called up from the graphics library and positioned anywhere on a drawing at any size or angle. Underlays are commonly prepared for doors, windows, fireplaces, trees, walls, and stairs.

Guidelines for lettering are frequently prepared on underlays. When placed under the drawing, the drafter can trace the lines instead of measuring each one.

Squared Paper

Graph or squared paper often serves as underlay guidelines for architectural drawings. These grid sheets are printed in gradations of 4, 8, 16, and 32 squares per inch. Squared paper is also available in decimal-divided increments of 10, 20, and 30 or more squares per inch.

Grid underlays are optional when drawing on a CAD system. Grids can be added and removed at any time by assigning a layer to a grid that is different than the drawing layers.

Burnishing Plates

Burnishing plates are embossed sheets with raised areas that represent an outline of a symbol or texture. The plates are placed under a drawing. Then a soft pencil is rubbed over the raised portions of the plate onto the surface of the drawing. The use of burnishing plates allows the drafter to quickly create consistent texture lines throughout a series of drawings.

CHAPTER

3

Drafting Scales and Instruments
Exercises

1. Draw the following four lines using a scale of 1/4″ = 1′0″:
   a. 5′-0″
   b. 7′-6″
   c. 9′-10″
   d. 11′-3″

2. Measure the distances you drew in Exercise 1, using the 1/8″ = 1″ scale.

3. Using a 2D CAD system, learn the commands for producing different scaled drawings.

4. Measure a book, desk, car, and room using a metric scale. Record your results. Compare your measurements with customary measurements.

5. Convert the following dimensions to millimeters: 5′-6″, 6′-8″, 10′-4″, 11′-7″, 15′-3″.

6. List the scales (proportional measures) you will use in drawing plans of a residence you are designing. Explain why you chose a particular scale in relation to the size of the paper you are using.

7. Draw the walls shown in Figure 3.11. Use a scale of 1/4″ = 1′-0″.

8. Practice drawing lines using all pencil grades on vellum and paper. Compare the results.

9. Select the drawing sheet size you will use to draw the residence you are designing. Explain how you determined the selected size.

Architectural Drawing Conventions

OBJECTIVES

In this chapter you will learn to:

- differentiate between the types and purposes of architectural drawings.
- produce the line conventions used on architectural drawings.
- develop good lettering techniques.
- sketch lines, patterns, and a floor plan.

TERMS

coding system
construction documents
detail drawings
elevations

layering
line conventions
models
plans

renderings
sections
title block
working drawings

INTRODUCTION

Builders follow a set of working drawings in order to make a designer's idea a reality. In order to clearly communicate information about the project to be built, standards and conventions in the preparation of these drawings have been established. Drafters and designers must understand and apply these conventions to make the plans consistently readable and understandable.

Other design and construction professionals also need to understand the language of architectural drawing to effectively use sets of plans in their work.

ARCHITECTURAL DRAWINGS

The design of a structure is interpreted through the use of several types of architectural drawings, including floor plans, elevations, details, and pictorial drawings. Drawings may vary from simple to complex. The number of drawings needed to construct a building depends on the complexity of the structure and on the degree to which the designer needs or wants to control the methods and details of construction. A *minimum* set of plans provides the builder with great latitude in selection of materials and processes. A *maximum* set of plans will ensure, to the greatest degree possible, agreement between the wishes of the designer and the final constructed building. See Figure 4.1.

Even though a building may be relatively simple, as many drawings as necessary should be prepared. Any detailed working drawing that is omitted forces the builder into the role of the designer. For some buildings and some builders, this may be acceptable. For others, this is highly unacceptable. The more plans, details, and specifications developed accurately for a structure, the closer the finished building will be to what was conceived by the designer.

Types of Drawings

Architectural drawings are often called "the plans." However, specific architectural drawings show certain views

DRAWINGS AND DOCUMENTS	SIZE OF SET OF PLANS		
	Min.	Aver.	Max.
Floor plans	X	X	X
Front elevation	X	X	X
Rear elevation		X	X
Right elevation	X	X	X
Left elevation		X	X
Auxiliary elevations			X
Interior elevations		X	X
Exterior pictorial renderings		X	X
Interior renderings			X
Plot plan (site)	X	X	X
Landscape plan			X
Survey plan	X	X	X
Full section	X	X	X
Detail sections		X	X
Floor-framing plans			X
Exterior-wall framing plans			X
Interior-wall framing plans			X
Stud layouts			X
Roof-framing plan			X
Electrical plan		X	X
Air-conditioning plan			X
Plumbing diagram			X
Schedules			X
Specifications			X
Cost analysis			X
Scale model			X
Reflected ceiling plan			X
Code compliance document		X	X

FIGURE 4.1 ■ Types of drawings in a plan set.

CAD Layering

Layering is a method of organizing a drawing in which specific line types, colors, or entities are drawn separately so they can be removed from or added to a drawing. These include dimensions, construction lines, details, symbols, grids, furnishings, equipment, materials, and fixtures. Preparing many layers provides ease and flexibility in managing and manipulating drawings and details later in the design and drawing process.

Different line weights, types, and colors can be assigned to each layer. Each color can also be assigned a line weight or type. This results in a colored line shown on the monitor, but the assigned line will print or plot. Each layer can be printed separately or in any combination.

Layers within layers are also possible. For example, a floor plan may contain many layers representing a series of floor plan levels—second, third, and so on—that is used to ensure the alignment of features. First activate the *Layer (LA)* command from the *Format* pull-down menu. Then you can create a new layer or edit an existing layer using the *Layer Properties Manager*. The changeable items include the line weights, line types, color, plotting visibility, and screen visibility.

of a structure and only some of those details are called *plans*. The following list includes brief descriptions of the various types of architectural drawings:

- **Plans** (or plan view) are views from the top down, a "bird's-eye" view. An example is the floor plan shown in Figure 4.2.

- **Elevations** are flat two-dimensional views of all vertical walls. See Figure 4.3.

- **Sections** show a view of one "slice" of a planned structure. It's as if an imaginary line were cut vertically at a particular place, showing the parts of the structure and components along the plane of that cut. See Figure 4.4.

- **Detail drawings** are prepared at a larger scale than other types of drawings. They are drawn to reveal precise information about construction

FIGURE 4.2 ■ Floor plan.

methods and materials. Details may be prepared in plan view, in pictorial form, or as sections. See Figure 4.5.

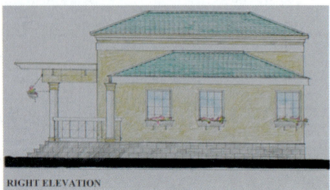

FIGURE 4.3 ■ Elevation drawings.

■ **Renderings** are usually one-, two-, or three-point perspective drawings. Often called *pictorials*, these show how the finished product is expected to look.

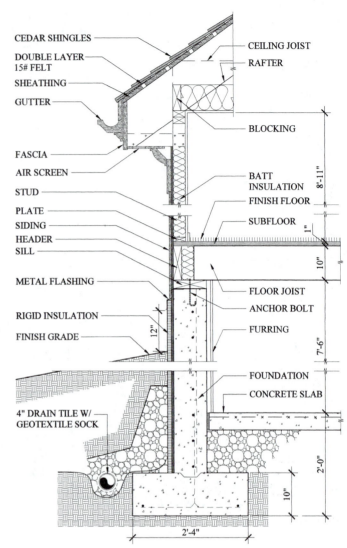

FIGURE 4.4 ■ Sectional drawing.

Renderings are made of both the building and its site. See Figures 4.6 and 4.7.

■ **Models** are constructed as three-dimensional reduced-scale replicas of a structure or structures as shown in Figure 4.8 and detailed in Chapter 21. CAD-generated 3D drawings, as shown in Figure 4.9, are also classified as models although prepared on a two-dimensional surface.

Drawings and Documents

Information about an architectural design and its construction is provided basically in three ways: general-purpose drawings, working drawings, and construction documents.

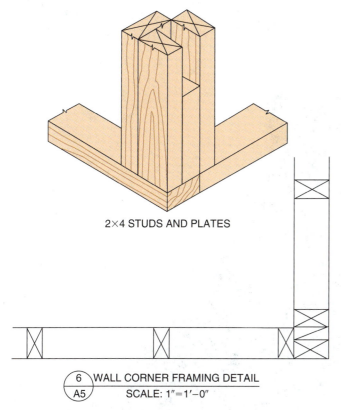

2×4 STUDS AND PLATES

⑥ WALL CORNER FRAMING DETAIL
A5 SCALE: 1″=1′–0″

FIGURE 4.5 ■ Detail drawing.

FIGURE 4.6 ■ Exterior rendering. *Home Planners, Inc.*

Architectural drawings used for sales promotion or preliminary planning purposes are known as general-purpose drawings. Drawings of this type usually consist of only approximate room sizes and dimensions on a single-line floor plan. See Figure 4.10. Pictorial drawings of exterior and/or front elevation views are also used for these purposes.

Drawings used during the building process are known as **working drawings**. Working drawings should contain all the information needed to completely construct a building: the dimensions, materials, and drawings of the building's shape. Complete floor plans and elevations are required. A full set of working drawings also includes specialized drawings, such as framing, electrical, plumbing, and landscape plans.

Even with a set of drawings, the building information is still incomplete. So much information is necessary for building a structure that not all of it can be put on a set of working drawings. For this reason **construction documents** are prepared that contain hundreds of facts and figures, plus legal and financial information related to the building process. Documents such as building specifications and schedules eliminate guesswork and specify exactly which processes, materials, and building components are to be used.

Reading Architectural Drawings

A small number of working drawings and documents may be sufficient and easy to use for one residence. However, for a very large project, several sets of complicated plans may be needed that require many different views, as well as details and documents containing very specific information.

Coding System

To make a large number of drawings manageable and easy to use, a **coding system** is often necessary. The coding system identifies every specific drawing and detail. It is also a method of keeping similar drawings together and organized in a working drawing set.

Most architects follow the American Institute of Architects' (AIA) alphanumeric coding system. In the AIA's coding system, drawings are identified by letters and numbers for ease of referencing. Figure 4.11 illustrates the breakdown of these letters into codes, which identify the group and drawing number. For example, in the code shown in Figure 4.11, "A" indicates that the drawing belongs to a set of architectural working drawings. The number "2" identifies the group to which the drawing belongs within the set, and the "1" after the period shows that the drawing is the first one in the group.

The group number always remains the same, no matter how many drawings are within it. More drawings may be added within groups without interrupting the alphanumerical order in the set.

FIGURE 4.7 ■ Rendering of an estate master plan. *M. K. Morrison Associates and Hepler Associates PC*

USING CAD

Coding CAD Drawings

Use of the AIA alphanumeric coding system on CAD drawings enables these categories to be used to create layers assigned to each discipline, drawing, or group. Numbering systems for mechanical drawings are coordinated primarily through the individual companies in accordance with standards set by American National Standards Institute standards.

Title Blocks

Similar to other kinds of written information, architectural drawings are identified by titles. In any drawing system **title blocks** identify drawings in a consistent and convenient format. Figure 4.12 lists the information that appears on most title blocks, and Figure 4.13 shows a typical completed title block. Title blocks are usually located on the bottom and/or right side of each drawing sheet.

Because the number of revisions varies, revision entries are made in sequence from bottom to top. Revisions

FIGURE 4.8 ■ Presentation models.

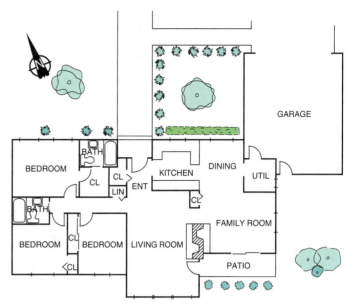

FIGURE 4.10 ■ Single-line floor plan.

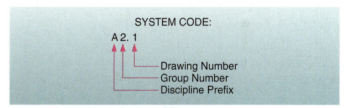

FIGURE 4.11 ■ Coding system used to identify discipline, drawing number, and group.

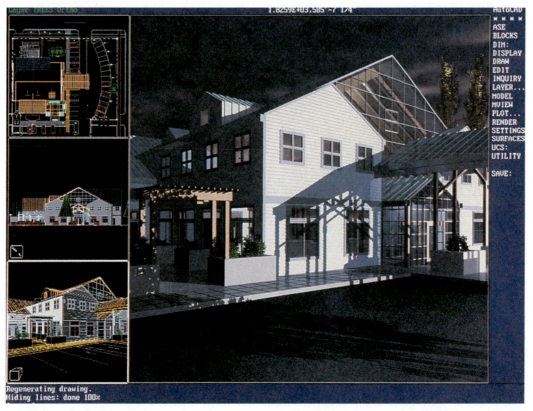

FIGURE 4.9 ■ CAD-created and -rendered model. *Robert McNeel & Associates*

- Project title and number
- Drawing sheet title
- Name and address of client
- Name and address of architect or firm
- Name and address of contractor (if known)
- Initials of designer
- Initials of drafter
- Initials of checker
- Revision block including
    - Title
    - Number
    - Preparer
- Professional seal space
- Scale
- Date
- Sheet number (using AIA code & showing number of sheets in the set)
- Key plan (if needed to identify location)

FIGURE 4.12 ■ Title block information.

are shown with a number inside a triangle on each drawing change.

Cross-Referencing

Drawing all views, sections, or details of features on one floor plan or one elevation is usually impossible. For example, floor plans do not show height details and dimensions. Elevation drawings do not show all horizontal dimensions. It is therefore often necessary to provide cross-references in order to guide the reader from one drawing to another. Numbered symbols are generally used for this purpose.

As shown in Figure 4.14, a circle with a directional arrow is drawn on a plan view. The arrow points in the direction of the area to be referenced elsewhere and shows the drawing sheet number and the detail number, which helps the user locate the referenced elevation or detail.

Layering is a method of aligning related plan drawings to ensure accuracy and eliminate much duplication of effort. First, a base drawing is prepared. In architectural work the base drawing is usually a floor plan. Then related drawings are prepared directly over the base and aligned. Layering is sometimes called *overlay drafting* or *pin drafting*.

Aligning specialized plans in a set of drawings, as shown in Figure 4.15, ensures that all structural features such as walls and columns align. It also eliminates the potential problems of overlapping mechanical, electrical, piping, and other facilities on the same drawing base. When this aligning is done on a CAD system, the layering task is used. Most complete CAD systems allow for the use of 255 layers.

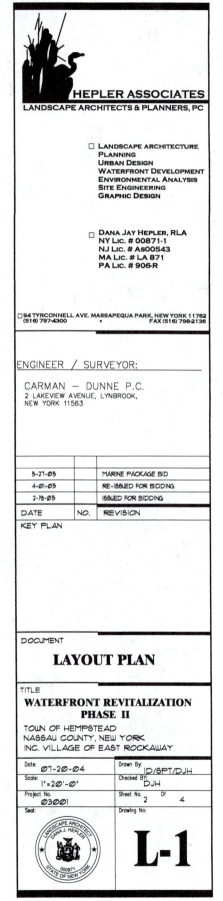

FIGURE 4.13 ■ Typical title block.

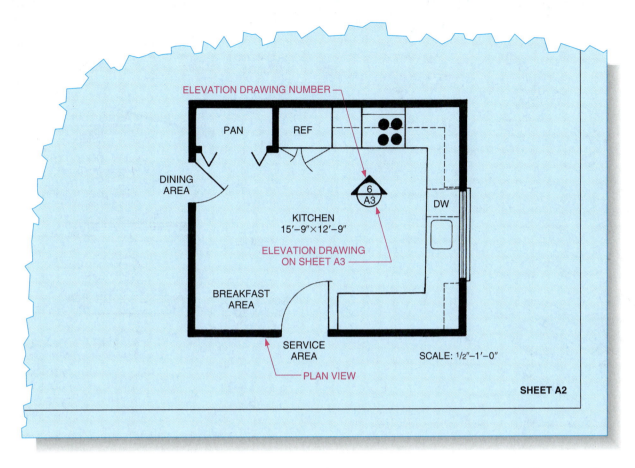

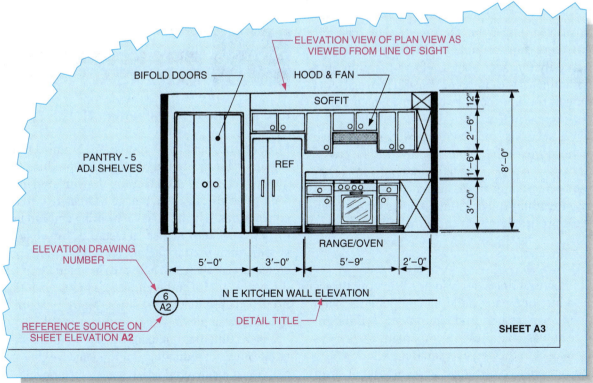

FIGURE 4.14 ■ Use of cross-referencing symbols.

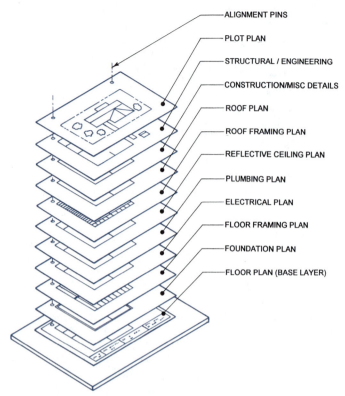

FIGURE 4.15 ■ Layering of plan drawings in a set.

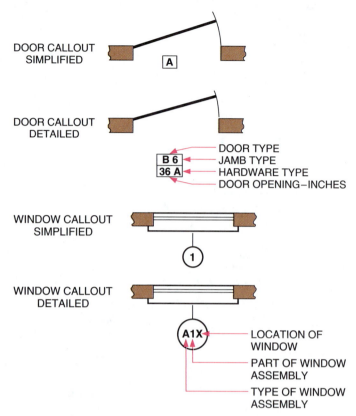

FIGURE 4.16 ■ Use of callout symbols on doors and windows.

Layering also simplifies the interpretation of drawings by subcontractors. For example, a plumbing contractor can be given only the floor plan level and the plumbing level. This eliminates the clutter of HVAC, electrical, etc., plans, although the contractor may be given a complete set of plans for cross referencing.

Pages with sections and detail drawings that have been referenced *from* other plans also need to be cross-referenced *back* to the original drawing. In other words, cross-referencing needs to work two ways, so that a person reading the drawings knows where each drawing belongs in relation to the entire structure.

Callouts

A set of architectural drawings also needs to contain information to identify many building components, such as doors, windows, rooms, and equipment. A different geometric form designates each component. For example, a door may be indicated as a square, a window as a small circle, a room as a rectangle, and equipment as an octagon. These shapes become labels known as *callouts*. See Figure 4.16. To show visually separate components, numbers or letters are usually shown inside the geometric shape.

ARCHITECTURAL CONVENTIONS

Architectural Line Conventions

Most drafting and CAD equipment is aimed toward helping an architect, drafter, or designer produce the highest quality line work. Architectural drawings are mainly communicated through a language of lines referred to as **line conventions.** The lines have meaning and can be read like the letters of an alphabet. In fact, the term *alphabet of lines* is sometimes used to denote line conventions used on architectural drawings. These are shown in Figure 4.17 with the pencil grades and technical or plotter pen thicknesses needed to produce these lines. Many types of lines are found on a single drawing. See Figure 4.18.

Just as different line patterns are used to represent certain features of a drawing, various *line weights* are used to emphasize or de-emphasize areas of a drawing. Architectural line weights are standardized to provide for consistent interpretation of architectural drawings. All architectural line weights must be very dark (opaque) so they will make clear reproduction copies. The only lines that should re-

NAME OF LINES	LINE SYMBOLS	LINE WIDTH	PENCIL		PEN SIZES
1. Object lines		Thick	H,F	2	0.50 mm
2. Hidden lines		Medium	2H,H	0	0.35 mm
3. Center lines		Thin	2H,3H,4H	0	0.35 mm
4. Long break lines		Thin	2H,3H,4H	0	0.35 mm
5. Short break lines		Thick	H,F	2	0.50 mm
6. Phantom lines		Thin	2H,3H,4H	0	0.35 mm
7. Stitch lines		Thin	2H,3H,4H	0	0.35 mm
8. Border lines		Very thick	F,HB	3	0.80 mm
9. Extension lines		Thin	2H,3H,4H	00	0.25 mm
10. Dimension lines					
11. Leader lines		Thin	2H,3H,4H	00	0.25 mm
12. Cutting plane lines		Very thick	F,HB	3	0.80 mm
13. Section lines		Thin	2H,3H,4H	00	0.25 mm
14. Layout lines		Very thin light	4H		
15. Guidelines					
16. Lettering	ARCHITECTURAL	Thick	H,F	1	0.40 mm

FIGURE 4.17 ■ Architectural line conventions.

main very light are layout and guide lines so they will *not* be seen on the reproductions. Following is a list of types of lines in the alphabet of lines:

1. *Object lines,* also called *visible lines,* are used to show the main outline of the building, including exterior walls, interior partitions, porches, patios, driveways, and walls. These lines should be drawn wide to stand out on the drawing.

2. *Dimension lines* are thin unbroken lines on which building dimensions are placed.

3. *Extension lines* extend from the object lines to the dimension lines. They are drawn thin to eliminate confusion with the object outlines.

4. *Hidden lines* are used to show areas that are not visible on the surface, but that exist behind the plane of projection. Hidden lines are also used in floor plans to show objects above the floor section, such as wall cabinets, arches, and beams. Hidden lines are drawn thin.

5. *Center lines* denote the centers of symmetrical objects such as exterior doors and windows. These lines are usually necessary for dimensioning purposes. Centerlines are drawn thin.

6. *Cutting-plane lines* are very wide lines used to denote an area to be sectioned. In this case, the only part of the cutting-plane line drawn is the extreme ends of the line. This is because the cutting-plane line would interfere with other lines on the drawing.

7. *Break lines* are used when an area cannot or should not be drawn entirely. A ruled line with freehand breaks is used for long, straight breaks. The long break line is thin. A wavy, uneven freehand line is used for smaller, irregular breaks. The short break line is wide.

8. *Phantom lines* are used to indicate alternate positions of moving parts, adjacent positions of related parts, and repeated detail. The phantom line is thin.

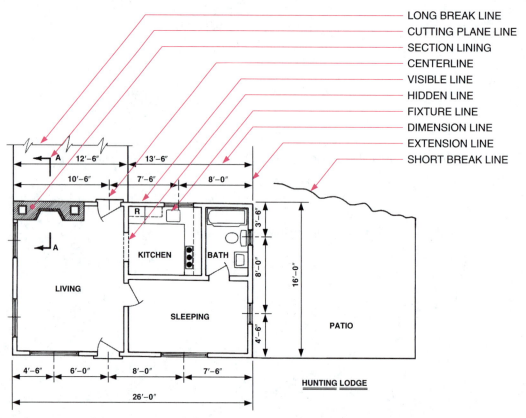

LONG BREAK LINE
CUTTING PLANE LINE
SECTION LINING
CENTERLINE
VISIBLE LINE
HIDDEN LINE
FIXTURE LINE
DIMENSION LINE
EXTENSION LINE
SHORT BREAK LINE

FIGURE 4.18 ■ Application of line conventions.

9. *Fixture lines* outline the shape of kitchen, laundry, and bathroom fixtures, or built-in furniture. These lines are thin to eliminate confusion with object lines.

10. *Leaders* are used to connect a note or dimension to a symbol or to part of the building. They are drawn thin and sometimes are curved to eliminate confusion with other lines.

11. *Section lines* are used to indicate the cut surface in sectional drawings. A different symbol pattern is used for each building material. The section lining patterns are drawn thin.

12. *Border lines* are the heaviest lines used on a drawing and are often preprinted with the title block. Border lines define the active area of a drawing sheet.

13. *Guidelines* are drawn to provide a horizontal guide for lettering to keep letters and numbers aligned. These are very light lines so they do not reproduce on a finished blueprint.

14. *Construction lines* are very light preliminary layout lines that do not become part of the finished drawing when reproduced. These lines are the lightest on any drawing.

Architectural Lettering

Styles

Without lettering, a plan does not communicate a complete description of the materials, type, size, and location of the various components. All labels, notes, dimensions, and descriptions must be legible on architectural drawings if they are to be an effective means of graphic communication.

USING CAD

CAD Line Weights and Types

CAD software programs allow the user to change the individual line weight (the thickness of the line in either inches or millimeters) and the style of the line type (center, hidden, dashed phantom, and many others). In addition, the user can choose from more than 16 million color choices. The changes should be coordinated through the *Layer* command tools for improved manageability of drawings.

Straight

ABCDEFGHIJKLMNOPQRSTUVWXYZ   1234567890

Inclined

ABCDEFGHIJKLMNOPQRSTUVWXYZ   *1234567890*

FIGURE 4.19 ■ The American National Standard Alphabet.

Architectural designs are often personalized. Likewise, lettering styles may reflect the individuality of various architects and drafters. Architectural drafters often develop their own style of lettering to work quickly, yet maintain accurate and attractive drawings. Nevertheless, personalized styles are all based on the American National Standard Alphabet shown in Figure 4.19.

No personalized style should be used that is difficult to read or easily misinterpreted. Errors of this type can be very costly, especially if numbers used for dimensioning are misread.

Developing Lettering Skills

Practice is necessary to develop the skills needed to letter effectively. Although architectural lettering styles may be very different, all professional drafters follow certain basic techniques for lettering. Although finished drawings may be prepared with a CAD program, readable and understandable labels and dimensions are nevertheless essential for design and field sketches.

1. Always use guidelines when lettering. See Figure 4.20.
2. Choose one style of lettering, and practice the formation of the letters of that style until you master it.
3. Make letters bold and distinctive. Avoid a delicate, fine touch.
4. Make each line quickly from the beginning to the end of the stroke. Do not try to develop speed at first. Make each stroke quickly, but take your time between letters and between strokes until you have mastered each letter.
5. Practice with larger letters (about 1/4″, or 6 mm), and gradually reduce the size until you can letter effectively at 1/8″ (3 mm).
6. Aim for uniform and even spacing of areas between letters by practicing words and writing sentences, not alphabets.
7. Practice lettering whenever possible—as you take notes, address envelopes, or write your name.
8. Use only the CAPITAL alphabet. Lowercase letters are rarely used in architectural work.

USE GUIDELINES FOR GREATER

ACCURACY IN LETTERING.

LETTERING WITHOUT GUIDELINES

LOOKS LIKE THIS.

FIGURE 4.20 ■ Use of lettering guidelines.

Proportions for fractions

4′−3¹/₂″ 6′−6³/₄″ 8′−9¹/₄″

Alternate fraction style used to conserve space

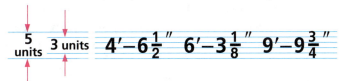

FIGURE 4.21 ■ Lettering fractions.

9. If your lettering has a tendency to slant in one direction or the other, practice making a series of vertical and horizontal guidelines.
10. If slant lettering *is* desired, practice slanting the horizontal strokes at approximately 68°.
11. Letter the drawing last to avoid smudges and overlapping with other areas of the drawing. This procedure will enable you to space out your lettering and to avoid lettering through important drawing details.
12. Use a medium-soft pencil, preferably an HB or F. A medium-soft lead pencil will glide and is more easily controlled than a hard lead pencil.
13. Numerals used in architectural drawing should be adapted to the same style as the letters. Fractions also should be made consistent with the style. Fractions are 1 2/3 times the height of the whole number. The numerator and the denominator of a fraction are each 2/3 of the height of the whole number as shown in Figure 4.21. Notice also that in the expanded style, the fraction is slashed to conserve vertical space. The fraction takes the same amount of space as the whole number.
14. The size of the lettering should be related to the importance of the labeling. See Figure 4.22.
15. Specialized lettering templates can also be used.

CAD and Typeset Lettering

Various kinds of typeset lettering are available. Pressure-sensitive letters are applied one at a time. These letters are used primarily for major labels on architectural drawings.

$\frac{1}{4}''$ **TITLES–LABELS**

$\frac{3}{16}''$ **TITLES–LABELS**

$\frac{1}{8}''$ GENERAL LETTERING

$\frac{1}{16}''$ NOTES IN SMALL AREAS

FIGURE 4.22 ■ Letter height related to label importance.

PREFERRED   ACCEPTABLE   NOT ACCEPTABLE

FIGURE 4.23 ■ Intersection line standards.

In CAD, lettering is produced using the keyboard and the CAD program's "text" features. Menu items include the option of selecting type font (style), slope angle, line weight (pen size), width, height, and spacing of characters.

ARCHITECTURAL DRAWING TECHNIQUES

Drawing (as a verb) is an overall term for the creation of all types of graphic forms. *Drafting* is drawing with the use of mechanical devices.

Using the Appropriate Pencil

Floor plans and elevation drawings are prepared primarily for the builder. These drawings must be accurately scaled and dimensioned. The accuracy, effectiveness, and appearance of a finished drawing depend largely on

USING CAD

CAD Lettering

Two types of lettering are available in most CAD programs: multiline text and single-line text. Multiline text looks and acts like text produced by a word processor. The popup interface allows all aspects of the text to be modified or enhanced including underlining, bolding, italicizing, and font style, and letter height changes. The single-line text tool is the older style text with fewer graphics but the functionality is still available. Both text options are found under the *Draw* pull-down menu under *Text*.

the selection of the correct pencil and the point of that pencil.

The degrees of hardness of drawing pencils range from 9H, extremely hard, to 7B, extremely soft. Pencils in the hard range are used for layout work. Basic architectural drawings are usually drawn with pencils in a medium range. If the pencil is too soft, it will produce a line that smudges. Mechanical pencils with leads of 0.3, 0.5, 0.7, and 0.9 mm are very thin and do not need to be sharpened because the lead width matches the correct line width.

Care must be taken when drawing corner intersections. Overlapping corner lines may intersect another material part or dimension and create confusion. When corner lines do not meet, no corner exists for interpretation or measurement. See Figure 4.23 for the preferred method.

Rendering and Sketching Techniques

In addition to the precise technical line work on floor plans and elevations, other line techniques are used for rendering and sketching. Some drawings are *rendered* to provide the prospective customer with a better idea of the final appearance of the building. These drawings show no dimensions but may include items such as plantings, floor surfaces, shade and shadows, and material textures.

Some of the line techniques used for renderings are simply variations—in the distance between lines, the width of lines, or the blending of lines. Dots, gray tones, or solid black areas are other ways of showing materials, texture, contrast between areas, or light and shadow patterns. See Figure 4.24. Pictorial renderings are covered in Chapter 20.

Sketching is a means of communicating that is used constantly by designers. In fact, most designers begin with a sketch. Sketches, or rough drafts drawn freehand, are used to record dimensions and the placement of existing objects and features prior to beginning a final drawing. Many times alternatives to a design problem are

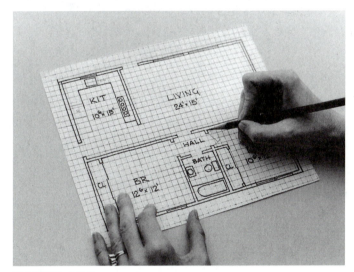

FIGURE 4.24 ■ Shading methods.

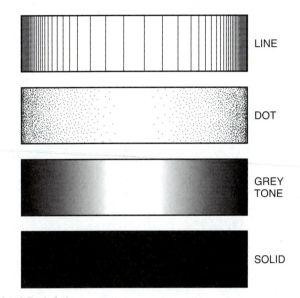

FIGURE 4.25 ■ Graph paper sketching.

shown with sketches. Sketches also help record ideas on the job site and help the designer remember unique features about a structure or site. Then the actual design activity can continue in a different location.

When sketching, use a soft pencil. Hold the pencil comfortably. Draw with the pencil; do not push it. Position the paper so your hand can move freely. Sketch in short, rapid strokes. Long, continuous lines tend to arc when drawn freehand. Sketching on graph paper helps increase speed and accuracy, as shown in Figure 4.25.

Drafting Media

Architectural drawings are prepared on paper, vellum, or polyester film depending on whether pencil or ink is used. CAD drawings require inking vellum or polyester film for plotting. Drawing sheets are available in two American Standard size series, as shown in Figure 4.26. Because architectural drawings are prepared on such a large reduction scale, large sheets, usually D or E sizes, are required. For example a 120'-0" building drawn on a 1/4" = 1'-0" scale requires 30" of actual drawing space.

LETTER SIZES	STANDARD SIZES	
A	9x12	8.5x11
B	12x18	11x17
C	18x24	17x22
D	24x36	22x34
E	36x48	34x44
F	28x40	

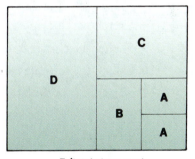

E (total sheet size)

FIGURE 4.26 ■ Drafting media sizes.

CHAPTER

4

Architectural Drawing Conventions Exercises

1. Describe six types of architectural drawings in terms of the type of information that is communicated in each type. List the ones you would use for your own set of drawings.

2. How are drawings used during the planning and construction of a building?

3. Explain the purpose of a coding system and cross-referencing.

4. Using a CAD system, draw a reference symbol and a callout. Label them.

5. Find and obtain a sample title block that is used by a local design or architectural office. Design your own version of a title block.

6. Select a lettering style from a CAD text menu and create all the data you would use in a title block.

7. Practice drawing each of the lines shown in Figure 4.17.

8. Use three different grades of pencil to draw five of the lines in Figure 4.17. Compare your results.

9. Copy the rules for lettering using any lettering style you choose.

10. Select a lettering style to be used on your own set of plans. Complete three practice sheets. Critique and improve each.

11. Draw the line and shade forms shown in Figure 4.24.

12. Practice drawing the architectural line conventions shown in Figure 4.17 freehand using a soft lead pencil.

13. On quarter-inch grid paper design and sketch a small floor plan.

14. Sketch the floor plan symbols shown in Figure 4.18.

15. Sketch the corner framing detail in Figure 4.5.

16. Sketch the four elevations in Figure 4.3.

17. Sketch the sectional detail drawing in Figure 4.4.

CHAPTER 5

Introduction to Computer-Aided Drafting and Design

OBJECTIVES

In this chapter you will learn:

- the different kinds of CAD hardware and software and their functions.

- the basic CAD drawing commands.

- how a CAD system is used to create architectural drawings.

TERMS

Cartesian coordinate system
central processing unit (CPU)
commands
compact disk (CD)
computer-aided drafting (CAD)

entity
hardware
modem
networking
rendering

software
solid model
surface model
symbol library
wireframe drawing

INTRODUCTION

Computer-aided drafting (CAD) and design is a process through which architectural and engineering drawings and documents are prepared on a computer. Basically, a CAD system is a combination of computer software (programming) and hardware (equipment) that allows designers and drafters to create drawings and store them electronically. All phases of the design and drawing process, from preliminary concept drawings through the completion of final working drawings and documentation, can be completed using a CAD system.

This chapter introduces the basic principles and practices used in computer-aided drafting for architecture. More specific reference information is provided throughout the text where applications are needed. Nevertheless, this coverage is not intended to replace a user's manual, which should be the student's guide of choice to the step-by-step sequences and procedures necessary for successful CAD operations.

CAD CHARACTERISTICS

A drawing prepared using a CAD system should appear identical to a good-quality drawing prepared manually.

However, other factors must be considered. A computer-aided drafting system is an electronic drafting tool that uses the speed and accuracy of a computer to produce clear, accurate, and consistent drawings. This is done by performing repetitious tasks such as symbol insertion, line drawing, automatic dimensioning, and lettering. By automating these tasks, drawing productivity is greatly increased. This results in the preparation of more drawings with fewer errors, leaving more time available for the creative process.

The ease and speed with which drawings can be electronically stored and retrieved result in fast, clear, and accurate changes and revisions. This is accomplished through the use of editing functions that allow drawings to be quickly deleted, redrawn, rotated, mirrored, stretched, or otherwise manipulated electronically. The speed with which two-dimensional drawings can be used to create three-dimensional drawings is one of the most convenient and time-saving capabilities of CAD. The greatest feature of CAD is the capacity to quickly, accurately, and consistently alter, combine, and add to drawings.

In addition to producing drawings, CAD systems can electronically produce related construction documents such as schedules, specifications, budgets, structural calculations, and graphics analyses. These can all be transmitted worldwide via the Internet.

CAD systems and humans are compatible. Each excels in what the other cannot do well. Humans can think, create, visualize, design, reason, and make decisions—CAD systems cannot. Humans are slow, inaccurate, inconsistent, and error prone compared to CAD systems, which are extremely fast, accurate, and consistent. However, CAD systems are only as accurate as the information supplied. Consequently, to use a CAD system effectively a thorough understanding of the principles and practices of architectural drafting and design is essential. This includes a working knowledge of the design process, drafting standards and procedures, projection methods, drawing types, documentation, and construction systems.

Always remember that CAD systems cannot design and draw any more than keyboards can write stories. People design and draw *using* drawing tools such as CAD.

CAD FUNCTIONING

CAD systems function through the interaction among CAD hardware, software, and the operator. Information entered into a computer is knows as *input*. A CAD drafter communicates with the computer through the use of input devices, usually a keyboard, mouse, and cursor. When a drafter requests a specific function such as a line, arc, or numeral, the computer memory switches to that function. The operator then indicates specifically—through a keyboard, grid system, mouse, or stylus pick—where the line, arc, or numeral is to be placed on the monitor. Once the function is completed on the monitor's screen, the next task is performed in the same manner. Each task is performed in sequence until the drawing is complete.

CAD COMPONENTS

CAD Hardware

The **hardware** for a CAD system includes the **central processing unit (CPU)**, input devices such as a keyboard and a mouse, storage devices, and output devices such as monitors and printers. Figure 5.1 shows the typical hardware components of a CAD stand-alone workstation. Figure 5.2 illustrates the relationship and interaction among the CPU, storage, and input and output functions.

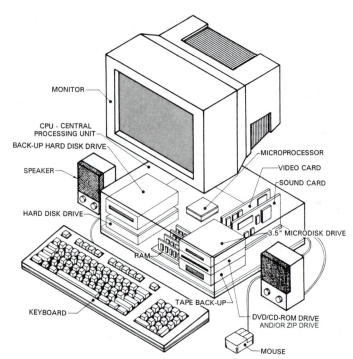

FIGURE 5.1 ▪ Typical hardware components of a CAD system.

CPU

The central processing unit is the "brain" or engine in all CAD systems. The main component of the CPU is a microprocessor chip. This chip controls the speed and power of the system, both of which are important for adequately operating the software.

The Monitor

A monitor is a hardware device resembling a television screen. It allows the operator to see the results of commands given to the computer. CAD systems should be equipped with the largest monitors possible to avoid eyestrain while working with complex drawings. A 17-inch monitor is the smallest monitor recommended for long-term CAD work. Many CAD workstations have larger monitors.

The resolution of a monitor depends on the number of pixels (dots of light). More pixels are needed to produce a sharper line image. A 19-inch or larger screen should have 1600×1200 or more pixels for CAD work. Color monitors are needed to effectively use many CAD functions. The colors selected to show on a monitor can be assigned specific layers. This allows drawings to be plotted with various line weights and types assigned to each color. Most monitors consist of cathode-ray tubes (CRTs). These are inexpensive but occupy a large desk surface. Liquid-crystal display (LCD) monitors are more

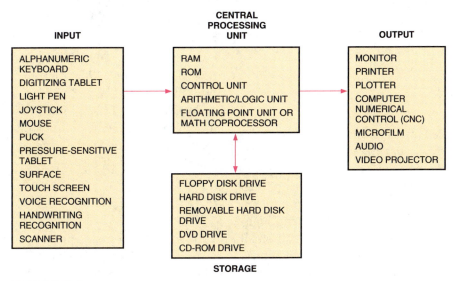

FIGURE 5.2 ■ Relationship among CAD system components.

expensive than CRTs but occupy a smaller area because of their flat panel screens.

Memory

Large amounts of graphics, numerical data, and text can be stored in a computer's memory. Computers contain two types of memory: ROM (read-only memory) and RAM (random-access memory). ROM contains the fixed data that the computer uses while it is operating. It contains instructions that keep the "operating system," which coordinates instructions between the software and hardware, operating smoothly. RAM is the computer's temporary memory. The amount of RAM determines the amount of software data the CPU can process at one time.

Data Storage

Storage capacity is not the same as memory. Memory enables a computer to function, whereas storage devices are used to electronically store software programs and CAD drawing files. *Flash drives* are portable storage devices with memory chips on which data are stored. The most common storage devices are CD drives and hard disk drives. These are standard equipment on almost all computer models.

Hard disk drives are generally the main storage device on a computer. They store the main operating system, the software, and any electronic files, such as drawing files, that the operator creates and saves. CAD programs generally take up a large amount of hard disk space, and architectural CAD files can take up a large percentage also. Therefore, most CAD systems include a minimum of an 18-GB (gigabyte) disk drive. Hard disk capacities of up to 250 gigabytes are common on large CAD systems.

Compact disks (CDs) are used for added storage and/or data backup for hard disk drives. The use of laser disk technology provides greater speed and increased capacity. DVD technology is similar to that of CDs but through compression, can store 8 to 16 times more data than a CD depending on whether single-layer or dual-layer DVDs are used.

Safety-conscious companies keep a complete, current backup of the files on their hard disks. Then, if a hard disk becomes corrupted or fails, the company can retrieve its important files from the backup. To solve the need for comprehensive backups, many companies use an additional hard drive or CD (CDR, CDRW, DVD), which can hold from 600 MB to 4.7 GB of memory.

Input Devices

Data or information (*input*) can be entered into the computer using various input devices. The alphanumeric *keyboard* is a standard input device that resembles a typewriter. In addition to using the keyboard to type words and numerals on drawings or documents, the keyboard can also be used to enter drawing commands. Most computer systems also include a "pointing device" called a *mouse* that can be used to move the arrow (called the *cursor*) around the screen. The mouse is also used to control the crosshair pointer's location on the monitor.

A newer development in computer technology provides another means of entering data—by *voice*. Voice recognition devices enable a computer operator to enter commands by speaking into a microphone connected to the computer. Voice recognition also provides an alternative for people with physical challenges who cannot easily manipulate a mouse or keyboard.

Output Devices

Drawings are reproduced on paper or vellum through the use of printers or plotters. Drawing sheet size is the determining factor in selecting the output device. *Plotters* produce high-quality drawings using ink-jets, ballpoints, or pencils of various colors.

Laser printers offer a fast method of producing high-quality graphics and typeset reproductions. Their small format, however, is a drawback for most architectural work that requires large sheets. Most laser printers can print an area no larger than 11 × 14 inches. Large laser printers are available but extremely expensive.

Ink-jet printers offer good graphic quality at low cost. These printers spray tiny jets of black and/or colored ink to produce a good-quality drawing. Desktop ink-jet printers are also restricted to small sizes and cannot handle the large D or E size sheets needed to reproduce most architectural drawings. This restricts their architectural use to small-scale drawings, details, and check prints. Large ink-jet plotters use both sheet-fed and roll media, which allows the duplication of large sizes.

Modems

A **modem** is a telecommunications device that allows computer operators to send and receive information over standard telephone or optic cable lines. See Figure 5.3. In this sense, modems are both input devices and output devices. Modems are often installed in CAD systems, particularly in architectural offices that may have more than one location. They allow large CAD drawings to be sent directly from one branch to another or from the designer to a major client for approval. This reduces the amount of time involved in creating a hard copy and transporting it to its destination. The **networking** of computers can use either wired or unwired wireless technology.

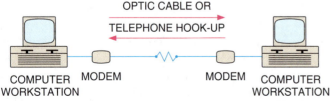

OPTIC CABLE OR
TELEPHONE HOOK-UP

COMPUTER   MODEM   MODEM   COMPUTER
WORKSTATION   WORKSTATION

FIGURE 5.3 ■ Modems provide long-distance links between workstations.

Multimedia Computers

Multimedia is a term used to indicate the combining of more than one medium. Computers can now incorporate text, photographs, audio, graphics, animation, and full-motion video.

To expand a conventional CAD system into a multimedia system, several components must be added. These include a DVD read/write drive and also a digital camera and digital video drives.

CAD Software

CAD **software** contains programs that instruct the hardware to perform specific tasks as directed by the drafter. Therefore, drafters must give directions to the software that will produce a functional design on a technically correct drawing. This not only requires proficiency in CAD operation but also a working knowledge of the principles and practices of architectural and construction design and drafting.

CAD software is judged by the number of functions (tasks) the program can perform, ease of operation, accuracy, line resolution, and flexibility.

USING CAD SYSTEMS

Commands provide the link between drafters and software. Commands are used to activate specific CAD functions, such as drawing lines or circles, in precise locations. Commands can be entered using a keyboard or pointing device such as a mouse. A mouse controls a cursor or crosshair pointer on a monitor. Most CAD software contains on-screen menus, which are used with a cursor to select commands, as shown in Figure 5.4.

Different software programs use different terms to describe the same function or command. To be consistent, AUTOCAD terms are used throughout this text. Three general categories of commands contained in a CAD system are drawing commands, editing commands, and utility commands.

Drawing Commands

Drawing commands are the commands used to create geometry. Individual geometric forms, such as lines, circles, and polygons, are known as **entities** or *objects*. Most CAD programs have hundreds of commands. The commands presented here are the necessary ones that are used most often. Many seldom used commands are

FIGURE 5.4 ■ Typical on-screen CAD menu.

not essential, but increase the speed and convenience of operation.

Many procedures can be used to complete a given task. The most basic are included in the following descriptions of drawing commands:

■ The *Line* command is probably the most used CAD command. Using this command, the drafter can draw lines by connecting points with a cursor (controlled by a mouse) or by entering the coordinates at the keyboard. Two general categories of lines are available in most CAD systems: *single lines* and *polylines*. A wide variety of line types, widths, and styles can be selected from the *Object-Properties* toolbar. Figure 5.5 shows several types of lines.

■ The *Circle* command creates circles of any size using various methods. For example, drafters can specify a circle using the center and a radius or diameter, or by specifying two or three points on the edge of the circle. See Figure 5.6.

■ The *Arc* command creates arcs of any radius and any length. To create an arc, drafters can specify the center and angle or radius, or they can specify the start, middle, and end points of the arc. Most CAD software provides several other options in addition to these basic ones. See Figure 5.7.

■ The *Text* command is used for adding notes and dimensions to a drawing by positioning the cursor on

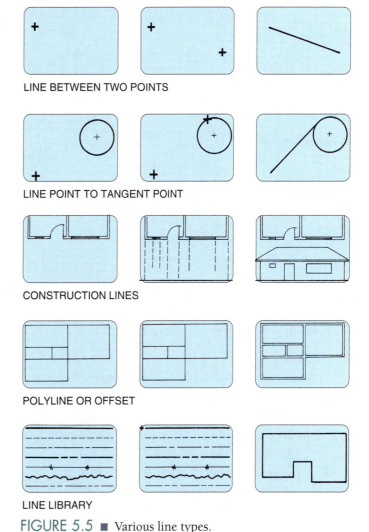

LINE BETWEEN TWO POINTS

LINE POINT TO TANGENT POINT

CONSTRUCTION LINES

POLYLINE OR OFFSET

LINE LIBRARY

FIGURE 5.5 ■ Various line types.

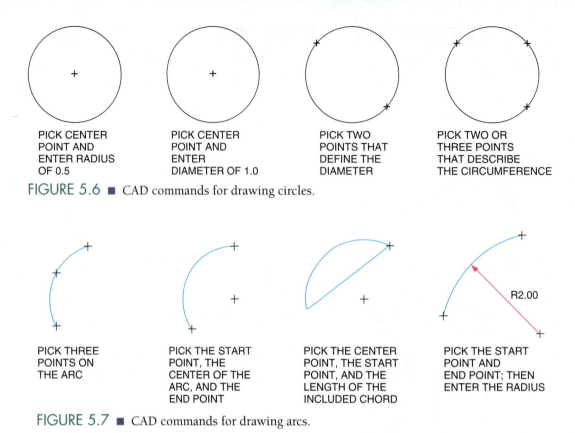

PICK CENTER
POINT AND
ENTER RADIUS
OF 0.5

PICK CENTER
POINT AND
ENTER
DIAMETER OF 1.0

PICK TWO
POINTS THAT
DEFINE THE
DIAMETER

PICK TWO OR
THREE POINTS
THAT DESCRIBE
THE CIRCUMFERENCE

FIGURE 5.6 ■ CAD commands for drawing circles.

R2.00

PICK THREE
POINTS ON
THE ARC

PICK THE START
POINT, THE
CENTER OF THE
ARC, AND THE
END POINT

PICK THE CENTER
POINT, THE START
POINT, AND THE
LENGTH OF THE
INCLUDED CHORD

PICK THE START
POINT AND
END POINT; THEN
ENTER THE RADIUS

FIGURE 5.7 ■ CAD commands for drawing arcs.

FIGURE 5.8 ■ Examples of text insertion.

BOXED EAVE DETAIL

the selected location and inputting letters and/or numerals with the keyboard, as shown in Figure 5.8.

■ *The Spline* command is used to produce curves by locating a series of cursor points and connecting them with lines as shown in Figure 5.9.

■ The *Ellipse* command is used to complete an ellipse inside a rectangle by inputting the major and minor axis of the ellipse as shown in Figure 5.10.

■ Rectangles are drawn with the *Rectangle* command by locating two opposite corners (Figure 5.11), by typing the width and length, or by dragging the cursor to opposite corners.

■ Polygons are completed with the *Polygon* command by locating the center and radius and inputting the number of sides (Figure 5.12) or by specifying an inscribed or circumscribed polygon.

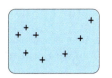 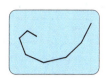 

FIGURE 5.9 ■ Method of drawing curves.

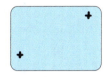 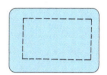 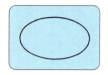

FIGURE 5.10 ■ Method of drawing an ellipse.

  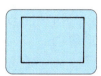

FIGURE 5.11 ■ One method of drawing a rectangle.

  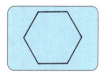

FIGURE 5.12 ■ One method of drawing a polygon.

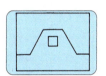 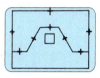 

FIGURE 5.13 ■ Method of identifying hatched areas.

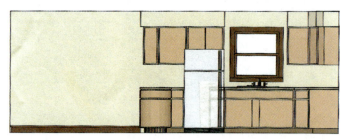

FIGURE 5.14 ■ Hatched interior elevation drawing.

EDUCATIONAL TESTING SERVICE
CAMPUS EXPANSION
LAWRENCE TOWNSHIP   NEW JERSEY

SITE PLAN

FIGURE 5.15 ■ Hatched site plan. *The Hiller Group*

- The *Hatch* command is used to fill in areas of a drawing with surface patterns such as crosshatching or solid colors. Hatching is done by using a pointing device on all border lines of the area to be hatched, as shown in Figure 5.13. Figure 5.14 shows an interior elevation with hatched shading, and Figure 5.15 is an example of site plan hatching. Figure 5.16 shows a typical variety of the color patterns available for hatched areas.

- Dimensioning commands control the dimension style (architectural or engineering), positioning (vertical, angular, or horizontal), and mode (customary, metric, decimal or fractional). Dimensions are entered by placing the cursor on the end points of the distance to be dimensioned as shown in Figure 5.17.

FIGURE 5.16 ■ Color hatching library options.

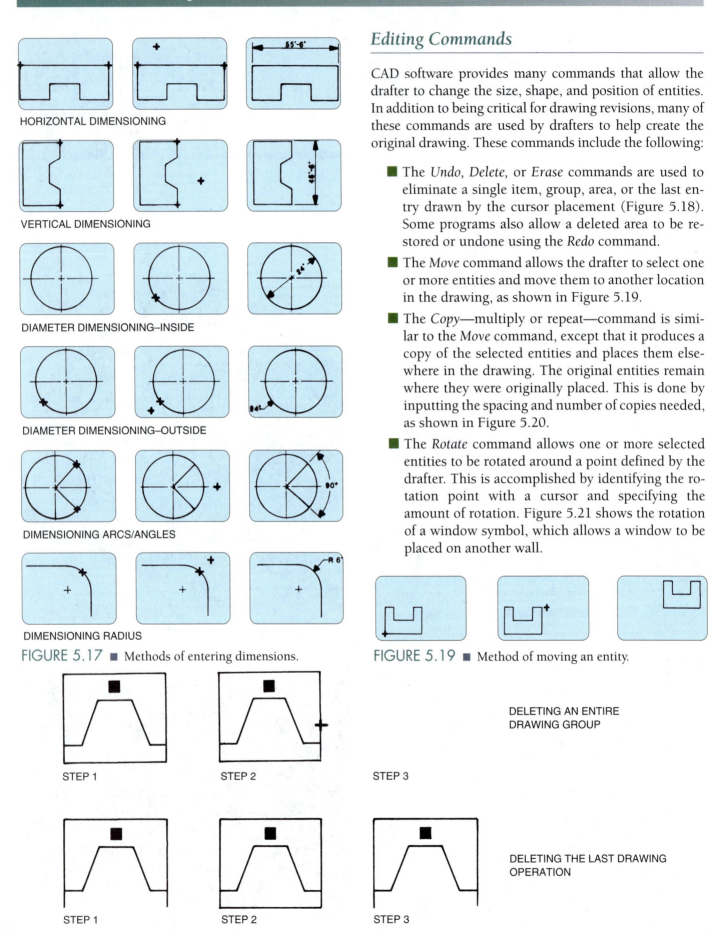

HORIZONTAL DIMENSIONING

VERTICAL DIMENSIONING

DIAMETER DIMENSIONING–INSIDE

DIAMETER DIMENSIONING–OUTSIDE

DIMENSIONING ARCS/ANGLES

DIMENSIONING RADIUS

FIGURE 5.17 ■ Methods of entering dimensions.

STEP 1   STEP 2   STEP 3

DELETING AN ENTIRE DRAWING GROUP

STEP 1   STEP 2   STEP 3

DELETING THE LAST DRAWING OPERATION

FIGURE 5.18 ■ Methods of deleting a drawing, group, or area.

Editing Commands

CAD software provides many commands that allow the drafter to change the size, shape, and position of entities. In addition to being critical for drawing revisions, many of these commands are used by drafters to help create the original drawing. These commands include the following:

■ The *Undo, Delete,* or *Erase* commands are used to eliminate a single item, group, area, or the last entry drawn by the cursor placement (Figure 5.18). Some programs also allow a deleted area to be restored or undone using the *Redo* command.

■ The *Move* command allows the drafter to select one or more entities and move them to another location in the drawing, as shown in Figure 5.19.

■ The *Copy*—multiply or repeat—command is similar to the *Move* command, except that it produces a copy of the selected entities and places them elsewhere in the drawing. The original entities remain where they were originally placed. This is done by inputting the spacing and number of copies needed, as shown in Figure 5.20.

■ The *Rotate* command allows one or more selected entities to be rotated around a point defined by the drafter. This is accomplished by identifying the rotation point with a cursor and specifying the amount of rotation. Figure 5.21 shows the rotation of a window symbol, which allows a window to be placed on another wall.

FIGURE 5.19 ■ Method of moving an entity.

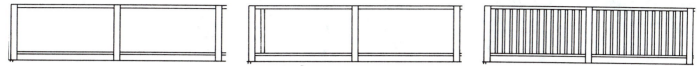

FIGURE 5.20 ■ Method of multiplying entities.

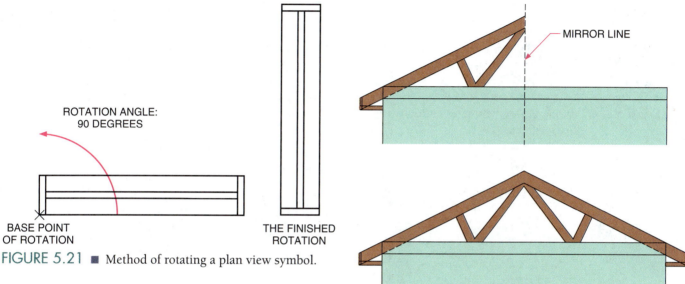

ROTATION ANGLE:
90 DEGREES

BASE POINT
OF ROTATION

THE FINISHED
ROTATION

FIGURE 5.21 ■ Method of rotating a plan view symbol.

MIRROR LINE

FIGURE 5.23 ■ Roof truss mirroring.

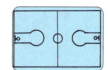

FIGURE 5.22 ■ Plan view mirroring.

THE SWIMMING
POOL OUTLINE

THE COMPLETED
POOL AFTER
OFFSETTING THE
OUTLINE 6 INCHES
TO THE INSIDE

FIGURE 5.24 ■ Use of *Offset* command to create parallel curved lines.

■ The *Mirror* command produces an exact mirror image of selected entities. A mirror image can be drawn and reflected left-right, up-down, or on any X-Y axis. Figure 5.22 shows a plan application, and Figure 5.23 shows a roof truss that has been mirrored.

■ The *Offset* command produces an exact parallel copy of a line at a specified distance from a given line at all points on the line. Figure 5.24 shows a curved offset line that is used to create the ledge thickness of a swimming pool.

■ The *Align* command enables objects to be aligned in a straight row by cursor identification (Figure 5.25).

■ The *Array Polar* command allows objects to be repeated and evenly distributed in a circular or rectangular pattern by entering the center location and the number of entries as shown in Figure 5.26.

■ The *Stretch* command is used to elongate a feature. The cursor is used to select the portions to be lengthened (Figure 5.27).

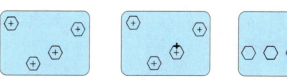

FIGURE 5.25 ■ Method of aligning objects.

■ The *Join* command creates a corner where two lines are aligned to intersect by placing the cursor on both lines (Figure 5.28).

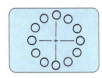
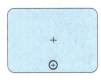

FIGURE 5.26 ■ Method of repeating objects in a circular pattern with the *Array Polar* command.

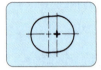
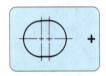
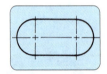

FIGURE 5.27 ■ Method of lengthening an entity with the *Stretch* command.

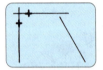
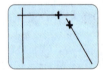
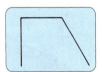

FIGURE 5.28 ■ Method of completing corners with the *Join* command.

- The *Lengthen* command can either shorten or lengthen a segment of a line by placing the cursor on the part of the line to be changed. Figure 5.29 shows the use of the *Delete Segment, Shorten Segment,* and *Extend Segment* functions.
- The *Gap* or *Trim* command is used to interrupt a solid line. Multiple gaps at regular intervals are produced by inputting the gap size, spacing, and number (Figure 5.30).
- The *Break* command interrupts a portion of a drawing by identifying the ends of the area to be interrupted, as shown in Figure 5.31.

Utility Commands

Some of the commands available in a CAD program do not manipulate the drawing directly. Instead, they assist the drafter in various ways. Examples of these utility commands are listed next:

- The *Zoom* command magnifies or reduces the size of a drawing on the screen. See Figures 5.32A and 5.32B. It does not affect the actual dimensions of the drawing, however. The drafter can *zoom in* to view details that are too small to see when the en-

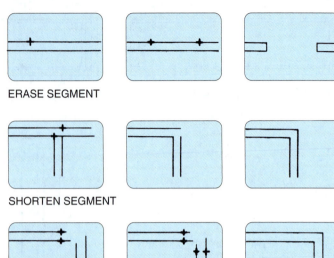

ERASE SEGMENT

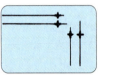

SHORTEN SEGMENT

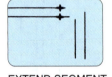

EXTEND SEGMENT

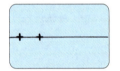

DELETE SEGMENT

FIGURE 5.29 ■ Methods of using the *Segment* functions of the *Lengthen* command.

FIGURE 5.30 ■ Method of creating gaps in a line.

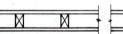
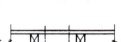

FIGURE 5.31 ■ Method of creating a break in a drawing.

tire drawing shows on the screen, or *zoom out* to see more of the drawing at one time.

- The *Pan* command moves the drawing horizontally and vertically on the screen without changing the current zoom percentage. Figures 5.33A and 5.33B show examples of how the *Pan* command can be used on an elevation and floor plan. This moves the viewport, which is the rectangular window through which a drawing is viewed. The viewport is controlled by using the *Zoom* and *Pan* commands together.

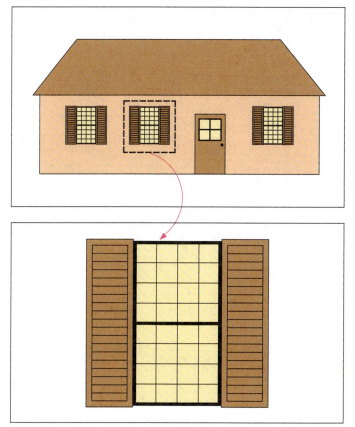

FIGURE 5.32A ■ *Zoom* command used to magnify (zoom in) a window detail of a drawing.

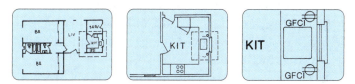

FIGURE 5.32B ■ Steps in zooming in.

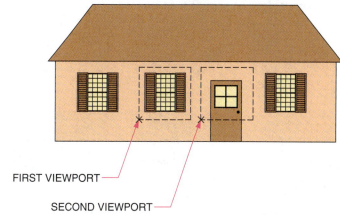

FIRST VIEWPORT

SECOND VIEWPORT

FIGURE 5.33A ■ Use of *Pan* command to move a drawing segment.

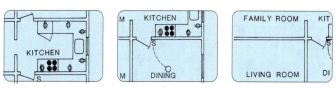

FIGURE 5.33B ■ Panning a floor plan.

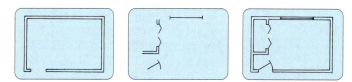

FIGURE 5.34A ■ Application of the *Layering* command to floor plan segments.

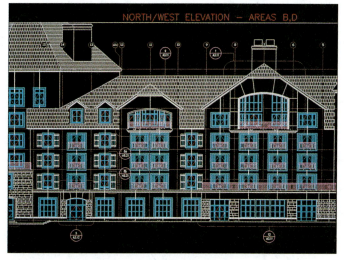

FIGURE 5.34B ■ Color layers used to separate building components and dimensions.

■ The *Plot-Print* command allows the drawing to be plotted with a pen plotter or printer. Entire drawings, blocks, or layers can be plotted together or separately.

■ The *Layer* command allows different segments of a drawing to be drawn separately then combined into a single drawing (Figure 5.34A). An individual layer can be used for separate parts of a drawing such as dimensions, grids, furnishings, or labels. Multiple building levels and specialized plans can also be layered using the same base drawing, as shown in Figure 5.34B.

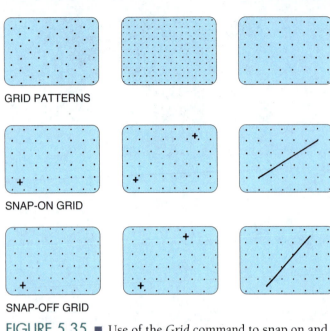

GRID PATTERNS

SNAP-ON GRID

SNAP-OFF GRID

FIGURE 5.35 ■ Use of the *Grid* command to snap on and snap off grids.

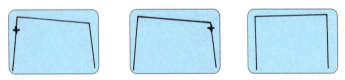

FIGURE 5.36 ■ Use of the *Ortho* command to align (orthogonalize) on the X and Y axis at 90° angles.

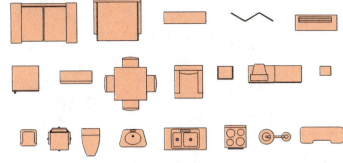

FIGURE 5.37A ■ Typical CAD symbol library.

  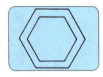

FIGURE 5.37B ■ Identifying and positioning library items on a plan.

ORIGINAL SCALE   SCALE DOWN   SCALE UP

FIGURE 5.38 ■ Use of the *Scale* command to increase and decrease the original drawing size.

- The *Grid* command creates Cartesian coordinate X and Y dots to be superimposed on a drawing. The *Snap-On* function moves the cursor to the nearest grid intersection or dot. The *Snap-Off* function allows the cursor to be moved to any location regardless of grids, as shown in Figure 5.35.

- The *Ortho* command (orthogonalize) adjusts all lines to a vertical or horizontal position at 90° angles, as shown in Figure 5.36.

- The *Block* command allows access to symbols or detail blocks electronically stored in many categories as plan, section, and elevation views. Some libraries are embedded in the original CAD software (Figure 5.37A); others can be created and stored by the drafter. The size and position of library items can be changed to match the scale and view of the drawing. Figure 5.37B shows how library items are positioned.

- The *Scale* command is used to establish the scale of each drawing or increase or decrease the scale of an existing drawing (Figure 5.38). Sheet size should be determined when the scale is established.

- The *Drag* command allows a line or entity, such as a symbol, to be moved (dragged) to a different location by the cursor.

- The *Save* command allows drawings to be filed on the hard disk or CD.

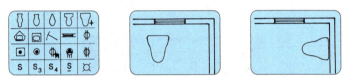

TYPES OF CAD DRAWINGS

The type of CAD drawing a drafter creates depends on the purpose of the drawing. The two major types of drawings are two-dimensional (2D) drawings and three-dimensional (3D) drawings.

Two-Dimensional Drawings

Traditional architectural working drawings are usually two dimensional. Only two dimensions (width and length, width and height, or length and height) are shown on one

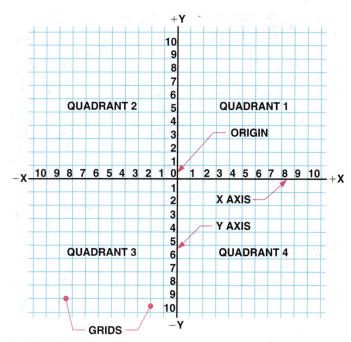

FIGURE 5.39 ■ Parts of a Cartesian coordinate system.

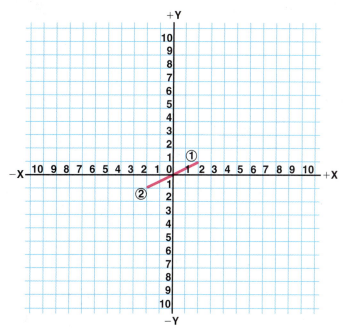

FIGURE 5.40 ■ Locating points on a Cartesian coordinate grid.

drawing. Two-dimensional drawings created on CAD systems include floor plans, elevation drawings, detail drawings, and site drawings.

Cartesian Coordinate Systems

Two-dimensional CAD drawings are based on a **Cartesian coordinate system**, which is based on an *X axis* and *Y axis*. These two axes are at right angles to each other. The point at which they meet is called the *origin*. See Figure 5.39. Note also that each axis has a positive (+) side and a negative (−) side.

In Cartesian geometry, the axes divide a drawing into four imaginary quadrants. To locate a point in any of these quadrants, you need only specify two numbers. The two numbers are called a *coordinate pair.* Point 1 in Figure 5.40 is located in quadrant 1 two units to the right of the origin on the X axis and one unit above the origin on the Y axis. Point 2 is located in quadrant 3, two units to the left of the origin on the X axis and one unit below the origin on the Y axis. Note that the X coordinate is always the first number in a coordinate pair, and the Y coordinate is always the second number. Polygrams, as shown in Figure 5.41, are constructed by repeating this process in the sequence shown in the figure's data table. The data table shows the coordinates of the four lines on the drawing.

Unlike polygons, the development of curves requires the use of more closely spaced coordinate points. With that exception, the drawing process is the same as in the

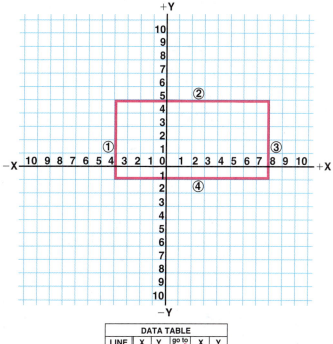

DATA TABLE					
LINE	X	Y	go to	X	Y
1	−4	−1	→	−4	+5
2	−4	+5	→	+8	+5
3	+8	+5	→	+8	−1
4	+8	−1	→	−4	−1

FIGURE 5.41 ■ Using the Cartesian coordinate system to plot a rectangle.

Spline command. Figure 5.42 shows a curve and a related data table containing coordinates for the development of a 20-point curve. The use of more points produces a

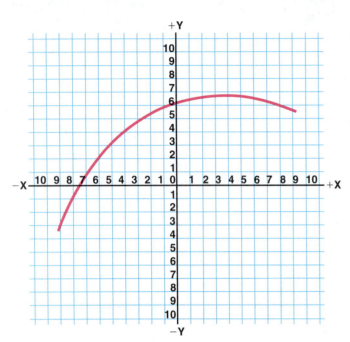

DATA TABLE				
SEGM	X	Y	go to →	X	Y
1	−9	−3	→	−8.7	−2
2	−8.7	−2	→	−8.2	−1
3	−8.2	−1	→	−7.7	0
4	−7.7	0	→	−7	1
5	−7	1	→	−6	2.1
6	−6	2.1	→	−5	3.1
7	−5	3.1	→	−4	4
8	−4	4	→	−3	4.8
9	−3	4.8	→	−2	5.5
10	−2	5.5	→	−1	5.8
11	−1	5.8	→	0	6.2
12	0	6.2	→	1	6.4
13	1	6.4	→	2	6.6
14	2	6.6	→	3	6.8
15	3	6.8	→	4	6.9
16	4	6.9	→	5	6.8
17	5	6.8	→	6	6.6
18	6	6.6	→	7	6.4
19	7	6.4	→	8	6
20	8	6	→	9	5.5

FIGURE 5.42 ■ Using the Cartesian coordinate system to plot a curve.

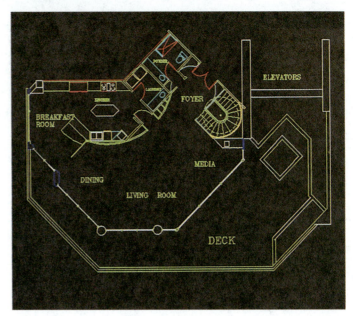

FIGURE 5.43 ■ Concept floor plan showing color layers for different components. *Alfred Karram Design*

smoother curve. The use of frequent points produces a flatter, segmented curve.

CAD Floor Plans

CAD floor plans range from fully dimensioned working drawings, as shown in Figure 14.31 in Chapter 14, to multicolored, layered concept plans (Figure 5.43) or surface textured and rotated plans, as shown in Figure 14.2a. Site, HVAC, plumbing, floor framing, and electrical plans are specialized floor plans and are prepared as outlined in Chapters 14 through 18. These plans are drawn using the *Line* command for walls; the *Block* command for symbols such as doors, windows, and fixtures; the *Dimension* command to describe sizes; and the *Text* command for notes and dimension numerals. Colored lines as seen on a monitor may be converted to different line weights and types when plotted.

CAD Elevation Drawings

Elevation drawings are two dimensional, and they represent length and height compared to floor plans, which show the width and length of a structure. CAD elevations are created from a floor plan in the manner described in Chapter 16.

First, the horizontal ends of doors, windows, chimneys, offsets, and building ends are projected from the floor plan using the *Construction Line* command on a separate layer. This is done in a distinct color and on a separate layer. Next, different color horizontal construction lines are drawn, representing the heights of these features plus the heights of the ground line, floor lines, and ridge lines. Symbols representing doors and windows are brought from the **symbol library** and placed in the construction line intersections. Object lines are then added and the construction lines removed using the *Erase* command. Then siding patterns, as shown in Figure 5.44, can be added using the *Hatch* command. Last, dimensions and notes are added using the *Text* and *Dimension* commands.

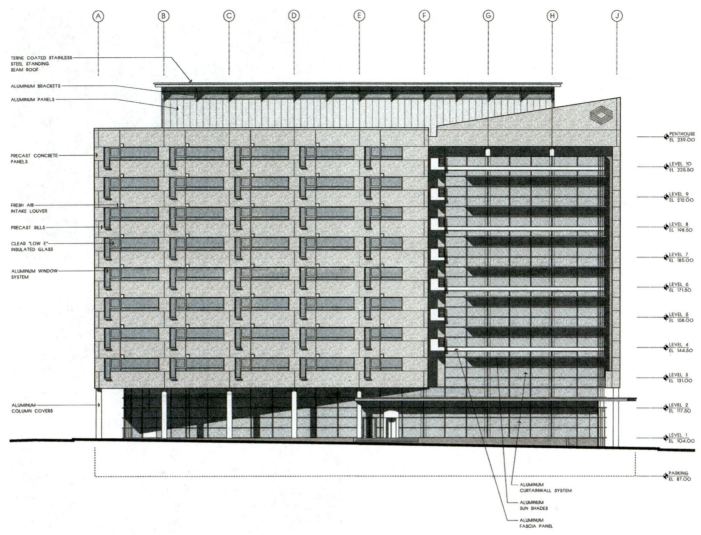

FIGURE 5.44 ■ Elevation drawing with hatch layers added for siding, shading, and shadows. *The Hiller Group*

CAD Detail Drawings

Multicolored layers are used extensively in detail drawings to show different materials, notes, and dimensions. The *Hatch* command is often used to show material sections as shown in Figure 5.45. See Figure 4.4 for the working drawing of this section.

Three-Dimensional Drawings

Most current architectural software programs can create three-dimensional drawings directly from two-dimensional drawings. This is done by adding a third axis (Z) to the Cartesian coordinate system, as shown in Figure 5.46. The Z coordinate allows the addition of depth to the drawing. Once a floor plan is completed on a CAD system, a 3D drawing can be created of the exterior or interior by using 3D commands. This also applies to creating 3D plumbing, HVAC, electrical, and site profiles from plan views. Figure 5.47 shows two pictorial views of the highway intersection plan on the right.

Three different types of three-dimensional drawings can be produced on a CAD system. These include wireframe drawings, surface models, and solid models.

Wireframe Drawings

Wireframe drawings are basically see-through stick drawings in which some or all hidden areas are exposed. They show an object's width, length, and height (X, Y, and Z) dimensions. These can be difficult to understand unless some or all of the hidden lines are "removed" (temporarily not displayed).

The wireframe drawing shown in Figure 5.48 shows the width, length, and height (X, Y, and Z) dimensions

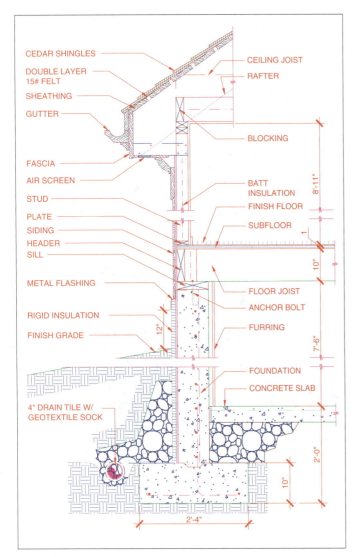

CEDAR SHINGLES
DOUBLE LAYER 15# FELT
SHEATHING
GUTTER
FASCIA
AIR SCREEN
STUD
PLATE
SIDING
HEADER
SILL
METAL FLASHING
RIGID INSULATION
FINISH GRADE
4" DRAIN TILE W/ GEOTEXTILE SOCK

CEILING JOIST
RAFTER
BLOCKING
BATT INSULATION
FINISH FLOOR
SUBFLOOR
FLOOR JOIST
ANCHOR BOLT
FURRING
FOUNDATION
CONCRETE SLAB

8'-11"
10"
7'-6"
12"
2'-0"
10"
2'-4"

FIGURE 5.45 ■ Detail drawing with hatched areas and object lines, dimensions, and notes in different color layers.

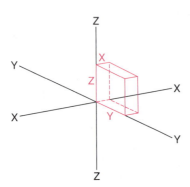

FIGURE 5.46 ■ X, Y, and Z axes are needed to produce three-dimensional drawings.

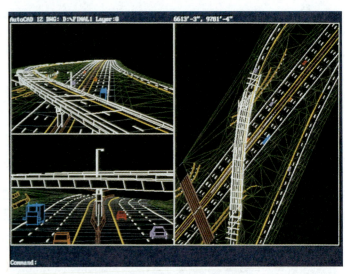

FIGURE 5.47 ■ Pictorial drawings created from a plan view.

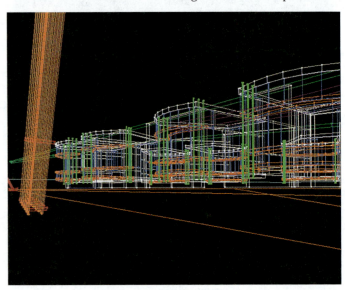

FIGURE 5.48 ■ Wireframe drawing with all lines shown. *David Karram*

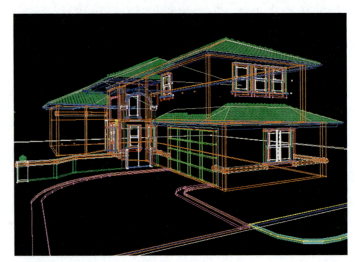

FIGURE 5.49 ■ Wireframe drawing with hidden lines removed. *David Karram*

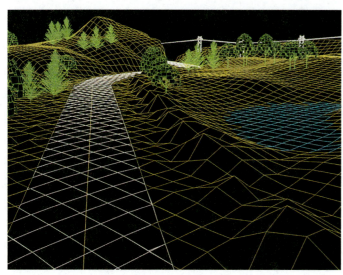

FIGURE 5.50 ■ Wireframe drawing showing site contours. *Autodesk, Inc.*

with all hidden lines exposed. Figure 5.49 shows a wireframe drawing with hidden areas removed. In addition to building drawings, wireframe perspectives are also used extensively to show site plan contours, as in Figure 5.50.

Surface Models

Surface models are drawings that consist of solid plane surfaces instead of the connected lines used in wireframe drawings.

Surface models are often rotated to reveal all surfaces in 3D, as shown in Figure 5.51. The buildings represented by the surface model shown in Figure 5.52 are the same as the wireframe building in Figure 5.48. The house shown as a surface model in Figure 5.53 is the same house as shown in the wireframe model in Figure 5.49.

Solid Models

3D **solid models** are the same as surface models except the interior of the structure can be shown as in Figure 5.54. Although a solid model may look similar to a surface model or a wireframe with hidden lines removed, it is actually quite different. A solid model has mass properties, which other forms of 3D drawings do not have. For example, you can assign a material, such as copper, aluminum, or steel, to a solid model. Then the CAD software can measure the mass, density, and other properties of the model. Architects often use solid models to perform structural analyses on buildings and other structures.

In addition to their mass properties, solid models can be rendered so that they look almost like photographs, as

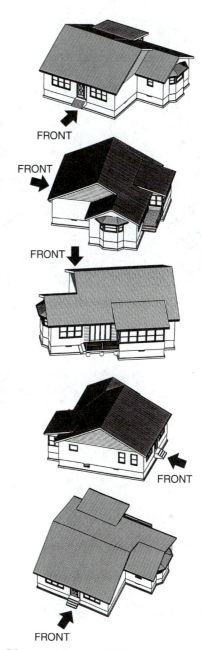

FIGURE 5.51 ■ Surface model drawing showing rotation options.

shown in Figure 5.55. In CAD, **rendering** is a process of adding shading and lights to a drawing so that it looks more realistic.

Virtual Reality Systems

Virtual reality is a natural extension of three-dimensional drawing technology. A *virtual reality* system creates a computer-generated "world" or "reality" in which the user seems to be immersed in the computer images.

A virtual reality (VR) program is created by completing a 3D model drawing of a building, then rendering

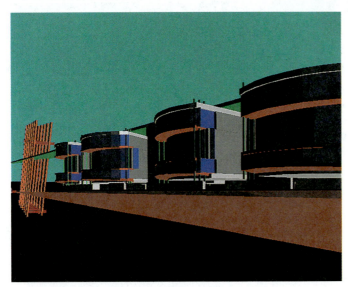

FIGURE 5.52 ■ Surface model drawing of the wireframe building in Figure 5.48. *David Karram*

FIGURE 5.53 ■ Surface model of the wireframe house in Figure 5.49. *David Karram*

walls and adding details using specialized software. The user, wearing a helmet with a viewing screen and a data glove, can move freely (as if in the building) to inspect walls, heights, and the location of features such as doors and windows. VR is rarely used due to its great expense and complexity.

Some architectural applications allow the architect to make real-time changes to the building. For example, if a client decides a window is too high, the architect can "reach out" within the virtual environment and move it down. The change is reflected in the CAD drawing file as well as in the virtual environment.

 ## OTHER COMPUTER APPLICATIONS

In addition to preparing architectural drawings, CAD and other computer systems are used to produce many documents related to architecture. These include the processing of information to produce schedules, specifications, budgets, contracts, and construction management forms. See Chapters 35 through 37 for the application of these systems.

More sophisticated programs are used to produce engineering-related data and simulations such as structural analyses, environmental effects, and energy requirements. Figure 5.56 shows a light and shadow computer simulation of an identical view in full sunlight, at dusk, and at night.

Manufacturers of building materials and components often provide designers with compact disks which contain drawings of their product. Some details can be stored and inserted into any drawing and altered for consistency with the conventions of the base drawings. Some imported files cannot be altered and must be used as received. Using manufacturers websites in this manner shortens and improves the design process. In addition to manufacturers, many municipalities store their code information including zoning, planning and land-use regulations on a website. Master municipal contracts and bidding documents can be downloaded by approved licensed professionals.

Computer spreadsheets are used extensively to prepare building materials and components schedules. Information such as cost, square footage, volume, model numbers, colors and materials are inputted. The totals for each item and/or location can then be calculated. By storing the costs of all building materials, components, and labor costs, schedules can be used to calculate the total cost of each project. Drawings can also be coded to provide a database which can automatically generate project estimates. Computer programs are also used to store and retrieve specification data. Programs can be developed by designers or master specifications, which can be inserted and altered, and are available from the Construction Specifications Institute (CSI).

FIGURE 5.54 ■ Solid model drawing showing hatched interior.

FIGURE 5.55 ■ Solid model rendering showing a variety of CAD-rendered textures. *Mardian Development Company*

A. Full sunlight.

B. Moonlight or dusk.

C. Artificial light on a dark night.

FIGURE 5.56 ■ Light and shadow patterns in (A) full sunlight, (B) at dusk, and (C) at night.

CHAPTER

5

Introduction to Computer-Aided Drafting and Design Exercises

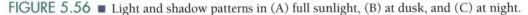

1. List the hardware components in a typical CAD system.

2. Explain why the basic principles and practices in architectural drafting must be learned in addition to CAD operations.

3. List the major commands used for drawing and editing.

4. Describe three utility commands.

5. Describe the difference between memory and storage.

6. Define the following terms: software, hardware, input, output, computer, flashdrive, hard disk, menu, cursor, plotter, enter, command, mode, monitor, hard copy, ram, mouse, library, block, Cartesian coordinates.

PART 3

Basic Area Design

Environmental Design Factors

OBJECTIVES

In this section you will learn:

- to orient a house on a lot to take best advantage of solar energy and features of the lot.

- to design structures ergonomically.
- ways to prevent pollution (ecology).

TERMS

active solar design
earth-sheltered homes

ecological planning
ergonomics

orientation
passive solar systems

INTRODUCTION

A wide range of factors must be considered to develop a fully functional architectural design—from a building's geographical area to the dimensions of an average adult. The designer must carefully study such environmental factors as the climate, the land, and energy sources. A building's design is also influenced by the needs of persons who will occupy the building. Considering these factors, the designer should also attempt to protect and improve the environment.

 ORIENTATION

A building must be positioned to maximize desirable features and minimize the negative aspects of the environment. This is accomplished through effective **orientation**. A building's orientation is its relationship to its environment.

Energy Orientation

Local resources and climatic conditions have always affected uses of energy—heating, cooling, and lighting. For example, early Native Americans built adobe houses un-

der overhanging cliffs. The cliffs provided protection from the hot sun during summer and the cool wind during winter. See Figure 6.1. The adobe material in these houses absorbed heat during the day and released it at night to warm the area. People soon found they could use other natural resources for heat. By burning renewable fuel such as wood, twisted grass, or blubber, people became less dependent on architectural designs and materials for protection from harsh environmental conditions. Later, when inexpensive fossil fuels (e.g., coal) appeared to be an endless energy source, people began to rely almost exclusively on those. Designers controlled inside environments artificially.

Today we know that the supply of fossil fuels is finite. There is a need to apply energy-efficient principles in building design to control indoor environments.

Energy must be obtained from sources and methods besides the burning of fossil fuels. Solar, wind, hydroelectric, and nuclear energy are available for use today. The combustion of natural materials such as reclaimed waste, wood, and other organic materials can also be used. Except for energy from the sun, all of these sources require special equipment to effectively heat and cool buildings. Mechanical or electrical devices can be added to control the sun's energy. Such systems are called **active solar design** systems. However, carefully designed buildings can use the power of the sun without mechanical devices and still provide much environ-

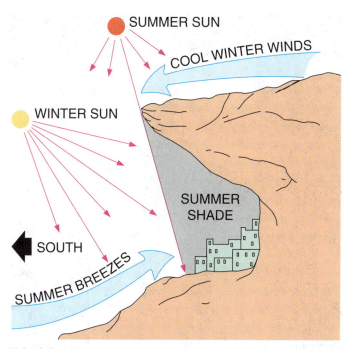

FIGURE 6.1 ■ Early methods of solar heating and cooling.

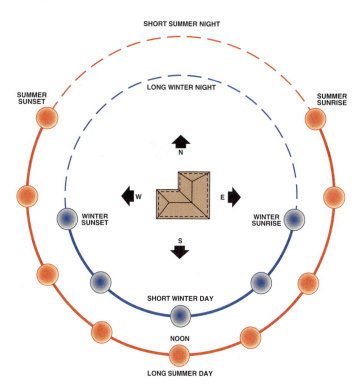

FIGURE 6.3 ■ Basis of seasonal solar planning.

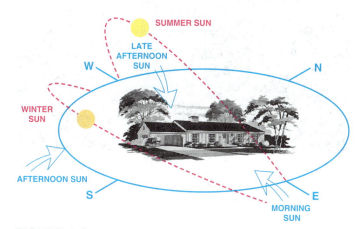

FIGURE 6.2 ■ Summer and winter sun paths.

mental control. These systems are called **passive solar systems.** They use only the design features and orientation of a building to gain and control the sun's energy. (Detailed discussions of both types of solar systems are presented in Chapter 32.)

Solar Orientation

Science Connection

The first step in designing the orientation of a building is to consider its relationship to the sun. Because the earth's axis is tilted, the angle of sunlight changes from summer to winter as the earth revolves around the sun. See Figure 6.2. In the northern hemisphere,

the south and west sides of a structure are warmer than the east and north sides. The south side of a building is the warmest side, because of the nearly constant exposure to the sun during its periods of intense radiation. Ideally, a building should be oriented and windows positioned to absorb this southern-exposure heat in winter and to repel the excess heat in summer. See Figure 6.3.

A structure needs to be located and oriented to ensure that the areas requiring the most solar exposure will be correctly positioned in reference to the sun. Figure 6.4 shows a model that can be used to check sun and shade patterns produced at different hours or during different seasons. Keep in mind that effective solar orientation should not only provide the greatest heating or cooling effect, but should also be planned to provide the greatest amount of natural sunlight where needed. For example, designers should consider the placement of windows and skylights as a source of both heat and light. See Figure 6.5.

Walls, floors, and furniture can absorb and store heat from the sun. This heat is naturally released when the temperature is cooler or at night. Open interiors and high ceilings encourage ventilation and cooler temperatures. Low ceilings and closed floor plans tend to increase temperatures. Chapter 32 contains more detailed

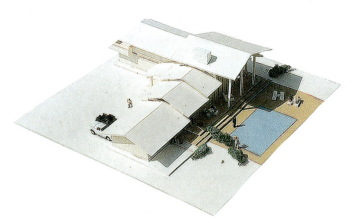

FIGURE 6.4 ■ Model used to study shade patterns.

FIGURE 6.5 ■ Extensive use of skylights. *Klumb residence—Preston Bolton Architects; Balthazar Karab, Photographer*

information on features that take advantage of passive solar principles in the design of structures.

Room and Outdoor Area Locations

Rooms should be located to absorb the heat of the sun or to be baffled (shielded) from the heat of the sun. Consider not only the function of the room, but the seasons and the time of day the room will be used most. The location of each room should also make maximum use of the light from the sun. See Figure 6.6.

Generally, sunshine should be available in the kitchen and dining areas during the early morning and should reach the living room by afternoon. To accomplish this, kitchen and dining areas should be placed on the south or east side of the house. Living room areas are placed on the south or west side to receive the late-day rays of the sun, when the room is most likely to be used. The north side is the most appropriate side for placing sleeping areas. It provides the greatest darkness in the morning and evening and is also the coolest side. Northern light is also consistent and diffused and has little glare.

The same principles apply to planning outdoor living areas. Those areas that require sun in the morning should be located on the east side. Those requiring the sun in the evening should be placed on the west side. For both indoor and outdoor areas, remember that in the northern hemisphere, the north sides of buildings receive no direct sunlight.

Math Connection

Overhang and Baffle Protection

Roof overhangs and baffles (shields) should be designed to allow the maximum amount of sunlight and heat to penetrate the inside of a building in winter. Conversely, the maximum amount of sun and heat should be shielded from entering the interior during a summer midday.

Because the angle of the sun differs in summer and in winter, roof overhangs can be designed with a length, height, and angle that will shade windows in summer and allow the sun to enter during the winter. See Figure 6.7. The edge of the overhang also needs to be related to the height of the window. As shown in Figure 6.8, more summer sun rays will reach the window area if the overhang is smaller. Larger overhangs result in fewer sun rays reaching the window area. Figure 6.9 shows several effective overhang baffling systems. Baffling should be accomplished without blocking out natural light.

These considerations apply primarily to the *south* side. The morning sun (east) and the late afternoon sun (west)

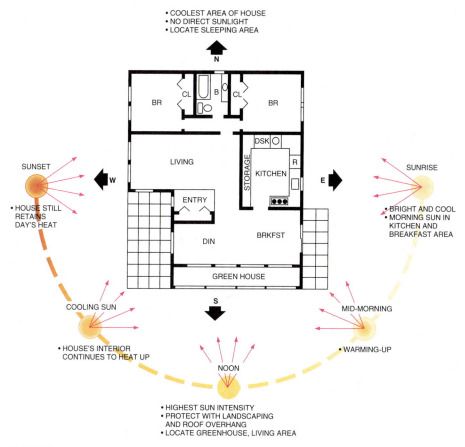

FIGURE 6.6 ■ Guidelines for room orientation.

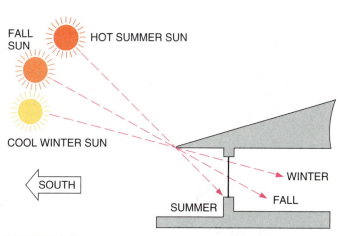

FIGURE 6.7 ■ Overhang designed to control seasonal sun exposure.

will not be as affected by the sun, and, as you know, the north side receives no direct sunlight.

Figure 6.10A shows how fixed louvers can be used to block sunlight. Figure 6.10B shows a trellis system that allows the slats to be moved 90° to block summer sun or admit winter rays.

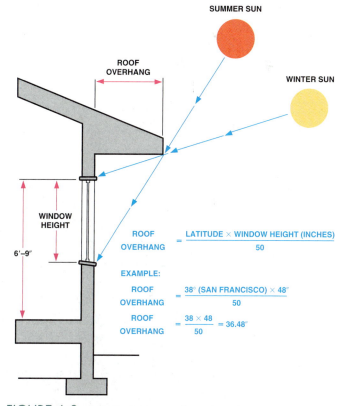

$$\text{ROOF OVERHANG} = \frac{\text{LATITUDE} \times \text{WINDOW HEIGHT (INCHES)}}{50}$$

EXAMPLE:

$$\text{ROOF OVERHANG} = \frac{38° \text{ (SAN FRANCISCO)} \times 48''}{50}$$

$$\text{ROOF OVERHANG} = \frac{38 \times 48}{50} = 36.48''$$

FIGURE 6.8 ■ Calculating roof overhang.

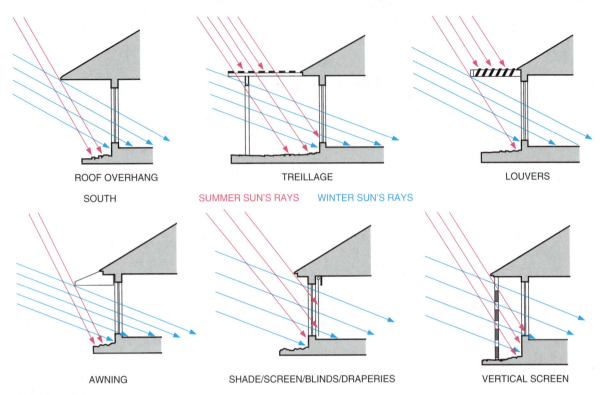

ROOF OVERHANG
SOUTH

TREILLAGE

LOUVERS

SUMMER SUN'S RAYS   WINTER SUN'S RAYS

AWNING

SHADE/SCREEN/BLINDS/DRAPERIES

VERTICAL SCREEN

FIGURE 6.9 ■ Types of sun baffling.

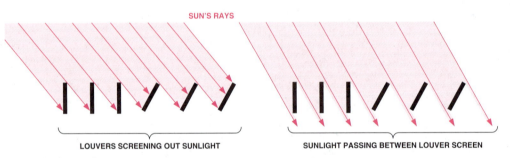

SUN'S RAYS

LOUVERS SCREENING OUT SUNLIGHT

SUNLIGHT PASSING BETWEEN LOUVER SCREEN

FIGURE 6.10A ■ Use of louvers to screen sunlight.

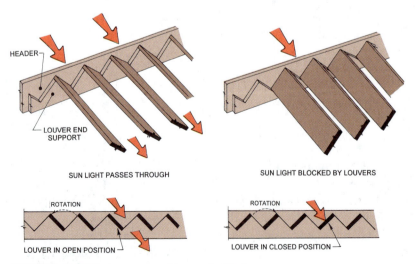

HEADER

LOUVER END SUPPORT

SUN LIGHT PASSES THROUGH

SUN LIGHT BLOCKED BY LOUVERS

ROTATION

LOUVER IN OPEN POSITION

ROTATION

LOUVER IN CLOSED POSITION

FIGURE 6.10B ■ Convertible sun baffle louvers.

Land and a Structure

A particular plan may be compatible with one site and yet appear totally out of place in another location. The success of a design depends on how well the structure is integrated with its surroundings.

Characteristics of a Site

Every structure should be designed as an integral part of the site, regardless of the shape or size of the terrain (land surface). Buildings should not appear as appendages ("add-ons") to the land but as a functional part of the landscape. For the indoor and outdoor areas to function effectively as parts of the same plan, the building and the site must be designed together.

Before orienting and designing a structure, consider the specific physical characteristics of the site, such as hills, valleys, fences, other buildings, and trees. Physical features such as these may affect wind patterns and the amount and direction of available sunlight in different seasons.

Large bodies of water may also affect air temperature and air movements. Also, surrounding pavement areas and buildings can raise or lower temperatures, because concrete and asphalt collect and store the sun's heat.

Consider the view options in orienting the house. Orientation of specific areas toward the best view, or away from an objectionable view, usually means careful planning of the position of the various areas of the building. The house shown in Figure 6.11 is oriented to take advantage of sunset vistas. The plan of this design is shown in Chapter 18.

Lot Areas

Building sites are sold and registered as *lots*. The size and shape of a lot affects the flexibility of choice in locating structures. For planning purposes, lots are divided into three areas according to their function: public, private, and service areas. See Figure 6.12. The placement of all structures on a lot determines the relative size and relationship of the three areas. Remember, the features of the lot should be an integral part of the total design. The site design is as important as the basic floor plan design of a structure.

Earth-Sheltered Homes

Characteristics of the land (site) and soil conditions are of utmost importance in the design of **earth-sheltered homes**. These types of homes are designed to be partially covered with earth. See Figure 6.13.

The thought of living partly underground may at first seem oppressive. However, with careful planning and proper orientation, adequate light can be achieved.

FIGURE 6.11 ■ Orientation designed to capture sunset vistas.

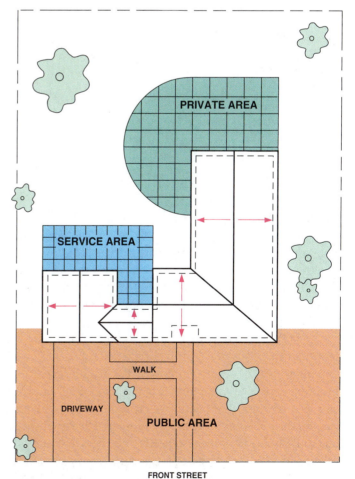

FIGURE 6.12 ■ Residential lot areas.

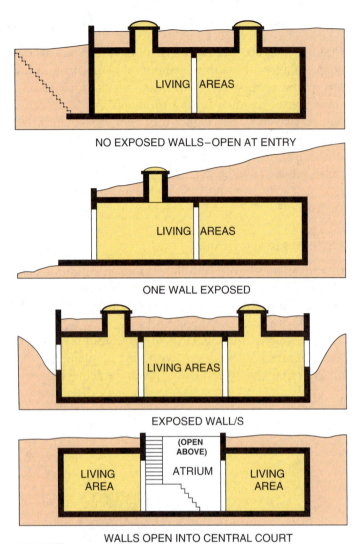

NO EXPOSED WALLS–OPEN AT ENTRY

ONE WALL EXPOSED

EXPOSED WALL/S

WALLS OPEN INTO CENTRAL COURT

FIGURE 6.13 ■ Types of earth-sheltered homes.

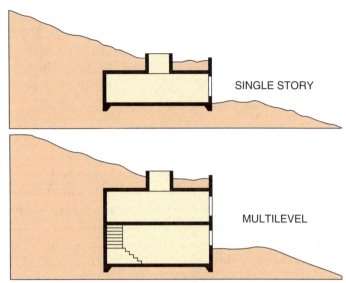

SINGLE STORY

MULTILEVEL

FIGURE 6.14 ■ Single-level and multilevel earth-sheltered homes on sloping lots.

Earth-sheltered homes have several advantages. Regardless of how high or low the outside temperature is, the soil just a short distance below the surface remains at a comfortable and constant temperature. Heating and cooling units can be smaller and are used less often than those in conventional structures. The underground location also avoids the problems of wind resistance and winter storm winds.

Construction costs for earth-sheltered homes can be less than those for conventional types of construction, if experienced builders are employed. The major concern in building earth-sheltered homes is waterproofing, but this can be accomplished with waterproof paint, sealants, membrane blankets, and proper drainage. The structure of an earth-sheltered home also needs to be heavier than that of conventional buildings, to support the heavy soil loads.

Earth-sheltered structures generally require little maintenance. Groundwater accounts for most maintenance problems. However, with careful study of drainage patterns during the planning phase and effective waterproofing during construction, such problems can be avoided.

Soil is an important consideration for designers of earth-sheltered buildings. Soils used to cover roofs and walls must be carefully selected to avoid frost heave, excessive swelling, and runoff tendencies. Soil that supports vegetation (organic soil) must be used for surface areas.

The best site location for an earth-sheltered home is on a gentle downward slope. See Figure 6.14. The exposed walls should make maximum use of glass to capture as much light and heat as possible. This requires a southern exposure. Windowed areas should also be oriented away from prevailing winds but should face the best possible view.

Rooms requiring the most light, such as the living room, should be located in windowed areas. Seldom-used rooms or rooms not requiring windows can be located in ground-locked areas as shown in Figure 6.15. Skylights can be used in otherwise dark areas if they are effectively sealed and drained.

Although sloping lots are best, earth-sheltered designs can be placed on relatively level lots. Mounds of earth called *berms* can be created to provide protection.

Vegetation

Vegetation of all types greatly aids in heat, light, wind, humidity, and noise control. Trees, shrubs, and ground-cover foliage, when effectively used, can also baffle undesirable views and enhance attractive scenes.

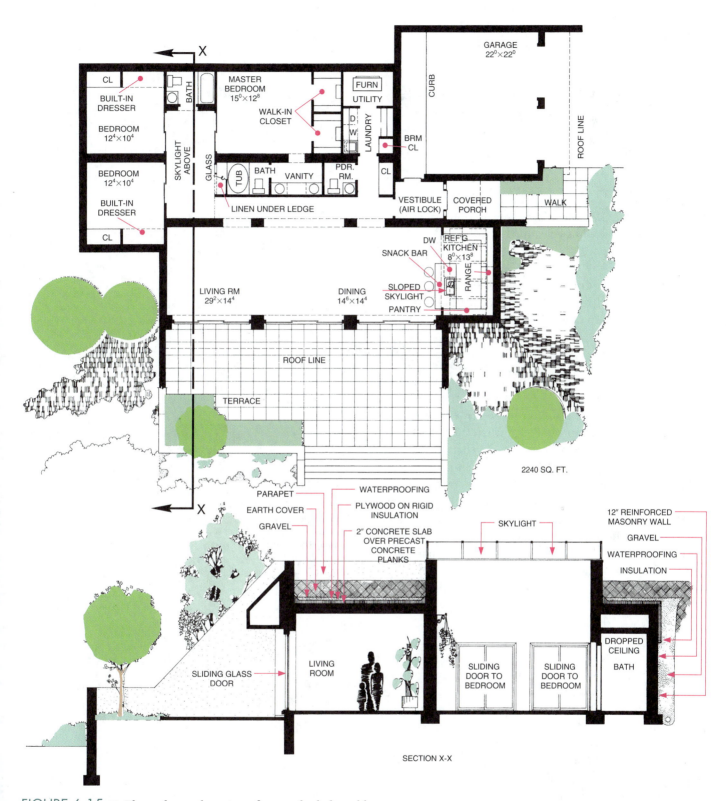

FIGURE 6.15 ■ Floor plan and section of an earth-sheltered home.

Deciduous trees maximize summer cooling and winter heating. They provide shade in the summer and then lose their leaves in winter, which allows the sun's warmth to penetrate the building. See Figure 6.16. Dense, coniferous (evergreen) trees and shrubs are most effective for blocking or redirecting north or northwest storm winds to help protect a building during all seasons.

SUMMER   WINTER

FIGURE 6.16 ■ Deciduous trees help maximize summer cooling and winter solar heating.

Vegetation is an important design consideration. However, it should not be used as a substitute for appropriate orientation design to control a building's environment.

Wind Control

Science Connection

One of the functions of effective orientation is wind control. Although the sun can provide natural energy to a structure, wind can easily diminish the sun's effect. Outdoor living can also be seriously curtailed by excessive wind.

Existing indoor heat can be lost very rapidly when cold air is forced into buildings through minute crevices, usually around doors and windows. Heat also escapes by *windchill* loss through walls, windows, roofs, and foundations. The windchill effect is the loss of internal stored heat. Building orientation, vegetation, and the features of the land can be used to control or minimize the effects of prevailing winds.

Wind Patterns

Air movements at a site should be studied during different parts of the day and during different seasons. Once wind patterns are known, buildings should be oriented to take full advantage of (or offer full protection from) the cooling effect of wind directions.

Some wind patterns are relatively common. Desirable summer breezes usually flow from one direction and winter winds from the opposite direction. Also, cool air will always move to replace rising warm air. For this reason, air above a body of water usually moves toward land during the day and from land toward water at night. See Figure 6.17. The wind pattern on southern sloping sites is generally a movement up the slope during the day and down the slope at night. See Figure

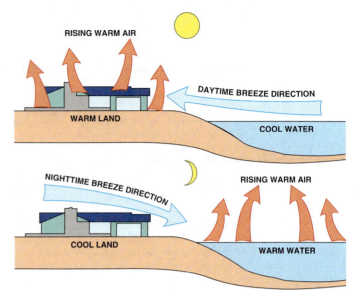

FIGURE 6.17 ■ Effect of large bodies of water on air movement.

Uphill during the day.

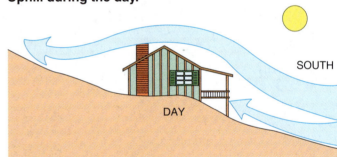

Downhill during the night.

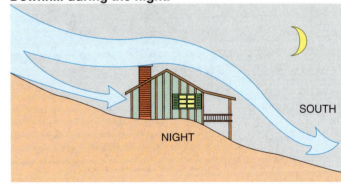

FIGURE 6.18 ■ Air movement on a southern sloping site.

6.18. Strong prevailing winds can change these movements, however.

Protective Measures

Gentle breezes are usually desirable, but harsh winds are not. Protection from wind can be provided by locating buildings in sheltered valleys or opposite the windward side of hills. Existing wooded areas can have a baffling effect on wind. See Figure 6.19. If no wooded areas exist, or if vegetation is young or not available, construction baffles such as fences or walls may be necessary. Figure 6.20 illustrates several methods of baffling and the effect on wind patterns.

A building can be oriented at an angle to present a narrow side or angle to the wind and avoid direct, right-angle wind impact. This effect is shown in Figure 6.21. Low roof angles can also help deflect wind over a structure.

Wind Effects

In planning urban buildings or isolated clusters of buildings, care must be taken to avoid the creation of turbulent wind eddies. This effect is caused by high-velocity winds striking the upper floors of high-rise buildings and being forced downward and back against lower buildings, creating turbulence on the surface. See Figure 6.22. This effect could also be caused on a smaller scale by winds trapped in courtyards and patios.

In urban situations, the *venturi effect* also complicates wind control. The venturi (wind-tunnel) effect is created as large amounts of moving air are forced into narrow openings. The reduced area through which the wind must pass creates a partial vacuum, and the air picks up

FIGURE 6.19 ■ Wooded areas help baffle wind.

Detached

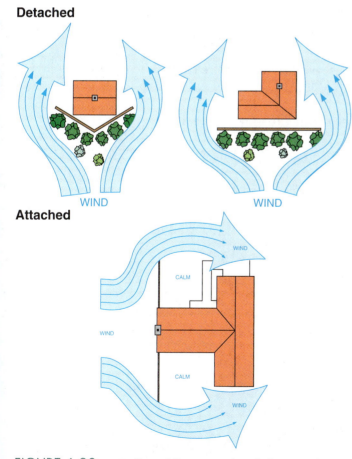

Attached

FIGURE 6.20 ■ Walls and fences used to deflect winds.

NARROW SIDE OF HOUSE EXPOSED TO WIND
—LIMIT WINDOWS ON WINDWARD SIDE

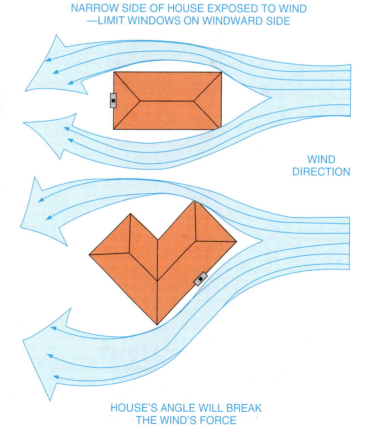

WIND
DIRECTION

HOUSE'S ANGLE WILL BREAK
THE WIND'S FORCE

FIGURE 6.21 ■ Use of building position to reduce wind effect.

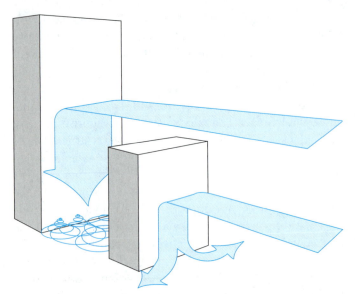

FIGURE 6.22 ■ Structure size and position affect wind patterns.

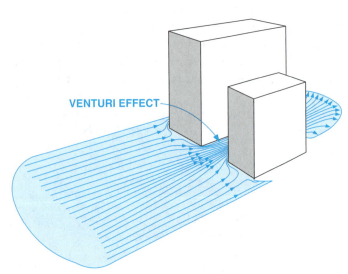

FIGURE 6.23 ■ Venturi effect created by building positions.

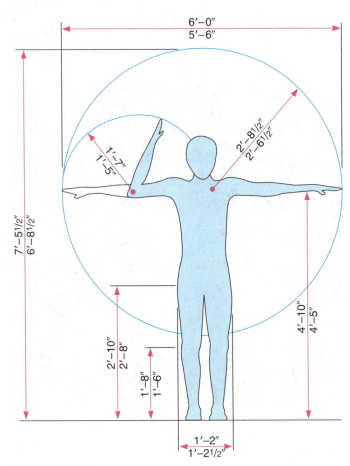

FIGURE 6.24 ■ Average human dimensions.

speed as it is pulled through the opening. See Figure 6.23. The venturi effect can be partially controlled by avoiding the alignment of streets or buildings with the direction of prevailing winds.

ERGONOMIC PLANNING

Buildings are for people. Therefore, buildings must be ergonomically (biotechnically) planned. **Ergonomics** is a science that deals with designing and arranging things that people use. In architectural design, this means the

design must match the size, shape, reach, and mobility of all residents. See Figure 6.24.

Human Dimensions

Human dimensions are especially critical in planning the size and position of cabinets, shelves, and work counters. Traffic areas, door openings, and windows are all based on human dimensions. When buildings (such as schools) are designed primarily for children, obviously the scale must be adjusted.

Ergonomic planning also applies to persons with physical disabilities, such as those with hearing, speech, visual, or mobility impairments. For example, door openings need to be wider and countertops lower for convenient use by persons in wheelchairs. (Detailed design requirements for persons with disabilities are provided in Chapter 13.)

Safety Factors

Safety factors in design cover a vast array of concerns for any person who occupies a building, whether it be a home or a public or commercial structure. The designer must ensure that no design feature creates a health or safety risk, such as hidden steps or low headroom clearances. Air for an environmentally safe building should be electronically filtered to provide for the elimination of harmful pollutants. Safety precautions include specifying appropriate mechanical equipment, such as gas furnaces, electrical wiring, devices, and machinery. Hazardous materials (such as asbestos) and accident-causing materials, such as extra smooth floors, thin glass, or unstable ceiling coverings, must be avoided.

Design also involves the safe arrangement of outdoor traffic areas. Adequate vehicular turning angles, fire lanes, and exit signs are a few examples. Of course, safety in design also implies that a building will be structurally sound and will adequately support all anticipated weight.

The increasing use of technology is creating another concern. Before the design process is begun, an assessment must be made to determine what special technology needs are anticipated. Automated buildings (called "smart buildings") now contain such built-in electronic features and accommodations as TV cables, high- and low-voltage circuits, and computer equipment. Magnetic or radio-wave interferences may need to be blocked to enable sensitive electronic equipment to operate effectively. Distance between persons and electromagnetic forces should be considered for safety.

 ## ECOLOGY

In the past one hundred years, the number of people on the earth has more than tripled. This population increase, combined with the shift from an agrarian (farming) society to an industrial one, has led to the creation of environmental problems previously unknown. The ever-increasing material needs of our technological economy have created enormous pollution problems that must be solved if humanity is to survive. Designers must plan in ways that eliminate or reduce pollutants.

A prime requirement in the creation of every design is to preserve our supply of clean air, pure water, and fertile land. The contemporary architect or designer must be sure that structures do not interfere with natural ecological balances. Good **ecological planning** means protecting or improving the environment without sacrificing the qualities of good design. It requires knowledge of the problems of pollution and possible solutions.

Land Integrity

Land is polluted by the discharge of solid and liquid wastes on land surfaces. Pollutants originate from industrial, agricultural, and residential waste. When they exist in excessive quantities, pollutants create health hazards, contribute to soil erosion, cause unpleasant odors, and overwork sewage-treatment plants. Land is also degraded by excessive removal of vegetation and by destroying the natural contours of a site.

Achieving solutions to these problems starts during the design process and continues through all construction phases. Design and construction professionals can prevent, correct, or avoid these problems in many ways, such as these:

- Specify natural untreated building materials wherever possible.
- Reserve open, green spaces in site development projects as described in Chapter 18.
- Minimize the removal of trees and major vegetation.
- Recycle topsoil removed during the building process.
- Minimize recontouring of natural land slopes, which results in tree root destruction.
- Use engineered landfills for disposing of excess land material.
- Preserve natural wildlife habitats wherever possible.
- Minimize the use of insecticides and pesticides and/or use natural biodegradable products.
- Erect barriers to prevent soil erosion during the land contouring process.
- Integrate buildings into the site with minimum contour changes as shown in Figure 6.25.

Sanitary landfilling is a practice that involves compacting solid waste material into layers (about 10' thick), and covering it with clean soil. Thorough landfill projects also include the planting of ground cover—trees and shrubs—on the filled area. Such practice restores the land to its original condition (or better), both ecologically and aesthetically.

Air Quality

The two types of air pollutants are those that originate outside a building or site and those that originate inside a building or site. Outside pollutants include industrial and vehicular emissions and naturally occurring irritants

FIGURE 6.25 ■ Residence integrated into a wooded site.

such as vegetation pollen, wind-borne dust, and mold spores and mildews from ponds, lakes, and streams.

Pollutants found inside a structure include toxic gases such as carbon monoxide and nitrous dioxide derived from unvented gas heaters, leaking chimneys and furnaces, chimney downdrafts, and gas stoves. Another gas, radon, may come from the earth beneath buildings and from unfiltered well water.

Another source of dangerous interior pollutants includes the airborne particles that evaporate from some plywoods, particle boards, foam insulation, textiles, and glues. These sources include formaldehyde, asbestos, wood preservatives, and paints. Biological pollutants—mostly mold and dust mites—can also be created in wet or moist areas and breathing them can lead to serious respiratory aliments.

Building occupants often create their own pollutants by adding tobacco smoke, aerosol spray, and cleaning agent vapors to the interior environment.

Designers and construction managers can reduce or eliminate many air pollution problems through the following actions:

- Specify products that do not include excessive amounts of vapor-producing formaldehyde.
- Thoroughly insulate integral garages from the remainder of the building.
- Specify non-lead-based paints.
- In renovation work, ensure that insulation does not contain asbestos before removal.
- Provide maximum ventilation, air conditioning, and dehumidifiers in areas where moisture and mold may accumulate such as laundries, indoor pools, or spas.

- Specify and indicate the location of all smoke, carbon monoxide, and radon alarms on plan and elevation drawings.
- If radon is present in the area, provide an internal exhaust from beneath the foundation to above the roof.
- If possible, include a mud room and half bath at a service entrance to eliminate the transfer of outside debris and pollutants to the inside.
- Design or specify fireplaces with maximum draw and air supply to eliminate downdrafts.
- Orient the structure to minimize the exposure to any excessive outside pollutants.
- Provide maximum air filtration to collect outside pollutants, especially airborne pollens.
- Design maximum exhaust to areas that may contain secondhand smoke.
- Design adequate hood exhausts over cooktops.
- Adhere to U.S. Environmental Protection Agency standards for emission and material use.
- Vent all furnaces to the outside and ensure sufficient air supply.
- Provide proper drainage and thoroughly seal foundation walls.
- Employ as many passive solar features as possible.
- Add active solar components as appropriate.

Specific details relating to implementing these design suggestions are covered in Chapters 13, 18, 27, 32, and 33.

Water Quality

The quality of water entering and surrounding a building is often compromised by many factors. Water pollution is often caused by the dumping or leakage of sewage, industrial chemicals, and agricultural wastes into oceans, lakes, rivers, and streams. These wastes include pathogens (such as disease-causing bacteria), unstable organic solids, mineral compounds, plant nutrients, and insecticides. Water pollution results in the destruction of marine life and presents very serious potential health hazards to animal and human life.

The excessive use of insecticides and pesticides as well as existing forms of natural bacteria and heavy metals in surrounding aquifers also contribute to water pollution. Even safe water can become hazardous if surface water accumulation results in flooding or provides a stagnant pool ideal for the growth of mosquito larva, harmful bacteria, and mold.

Designers can help maintain good water quality through the following actions:

- Design the site topography by directing water flow away from all structures by general contouring or, if necessary, through the use of swales, trenches, or berms.
- Design gutters and downspouts to exhaust water away from structures.
- Specify drainage tile around the perimeter of foundation walls.
- For large areas install area drains to interrupt massive water runoffs.
- Install the largest possible septic tank and field based on the lot size.
- Design gray-water receptacles to diminish the amount of dishwasher and bath water exhaust into septic or municipal sewer systems.
- Locate the septic system as far away as possible from bodies of water.
- In designing the plumbing system, include water filter devices such as charcoal and ultraviolet filters and water softeners.
- Include a water purification system specifically for drinking water.
- In locating structures on a site, avoid wetlands that offer a habitat for wildlife.
- Avoid removal of vegetation from very steep slopes to prevent excessive runoff.

Visual Attraction

Many air, water, and land pollutants, such as unsanitary garbage dumps and smog-producing agents, are not only unhealthy but also visually undesirable. Other sources of visual pollution include junkyards, exposed utility lines, public litter, barren land, and large billboards. They are aesthetically objectionable and should be avoided in the architectural design process. Designers can ensure maximum visual attraction for structures and sites through adhering to the following:

- Follow the basic elements and principles of design as outlined in Chapter 2.
- Design building elevations that are consistent with the architectural character of the community as illustrated in Chapters 1 and 15.

- Develop the site plan according to the guidelines shown in Chapter 18.
- Use as much of the natural terrain and vegetation as possible.
- Specify underground lines for power, phone, and cable delivery.
- Design trash storage facilities to eliminate visible exterior trash.
- Use foliage to block undesirable sights.
- Include a total landscape and planting plan with the original plan set.

Noise Abatement

Sound requires a path. Sound energy travels in all directions from a source. It weakens through distance or by the interruption of its path with a physical screen.

Sound levels are measured in *decibels*. Exposure to excessive decibel levels, over 80, or to even moderate levels, over 60, for long periods creates stress and can result in neurosis, irritability, and hearing loss. Unwanted sound is called noise! Excessive noise can also create hazardous environments by eliminating people's ability to identify and discriminate between sounds, especially those that warn of danger.

Some of the methods designers can use to keep noise at acceptable levels include the following:

- Orient floor plans to locate quiet areas away from outside noise sources.
- Use foliage to provide sound barriers and buffers.
- Add insulation and thickness to interior walls that house noise sources.
- Maximize the use of sound-absorbing materials such as carpets, drapes, upholstery, and tapestries.
- Specify double or triple glazed windows.

Electronic Hazards

In addition to the danger of electrical shock from faulty electrical products or connections, continuous exposure to excessive amounts of electromagnetic force is considered unsafe. Living in proximity to high-power voltage lines, faulty microwave appliances, and concentrations of television cables and wiring may pose a threat to occupants' health.

Homes should be located safe distances from voltage lines. Designing safe distances or shields between people and radiation fields is the best architectural solution to these potential problems.

Preventive Measures

Many pollution problems can be prevented through effective architectural design. For example, effective landscape planning that preserves existing vegetation can help maintain acceptable oxygen-nitrogen cycles, provide wildlife habitats, and preserve and enhance the natural beauty of the site. Effective landscaping can also provide summer shade, help retain groundwater, and reduce noise. Architectural design professionals can also help solve or prevent many pollution problems by becoming involved with local planning boards to help initiate realistic, environmentally related ordinances.

CHAPTER
6
Environmental Design Factors Exercises

1. Explain why certain side(s) of a house receive the most light and heat. Compare winter and summer changes.

2. Draw a rectangle to represent a floor plan of a house. Label each side N, E, S, W. Then indicate where you would orient and place each room.

3. Draw the outline of an earth-sheltered home.

4. Sketch a 75′ × 110′ property. Label one side as a hill with a 10° slope upward. Sketch a house on the property in the most desirable location.

5. Tell how you would use solar planning, wind control, and vegetation for a residence of your own design.

6. Find magazine and newspaper articles or advertisements about products for buildings that are designed to control pollution.

7. Identify buildings that emit pollutants into the air, water, or land. List ways of correcting these conditions.

8. List the ecological factors to be considered in planning a residence of your own design.

9. Make a bar graph showing the decibel levels of deafening and very loud noises, pleasant sounds, and quiet.

CHAPTER 7

Indoor Living Areas

OBJECTIVES

In this chapter you will learn:

- to identify the functions of indoor living areas.
- to design the location, decor, size, and shape of indoor living areas.
- how a room's orientation, walls, floors, windows, ceilings, lighting, and furniture can contribute to room function and appearance.
- to design indoor living areas and work them into a convenient floor plan.

TERMS

closed plan	home office	recreation room
decor	home theater	serving walls
egress	living area	studio
entertainment room	media room	study
family room	open plan	templates (furniture)
great room/gathering room	partition	

INTRODUCTION

Your first impression of a home is probably the image you retain of the *living area.* In fact, this is the only indoor area of the home that most guests observe. The **living area** is where the family entertains, relaxes, dines, listens to music, watches television, enjoys hobbies, and participates in other recreational activities.

LIVING AREA PLANS

In most two-story dwellings, the living area is normally located on the first floor and is adjacent to the foyer or entrance. However, in split-level homes or one-story homes with basements, part of the living area may be located on the lower level.

The total living area is divided into rooms or smaller areas that serve specific purposes. These subdivisions may include the living room, dining room, family room, great room, recreation room, and special-purpose rooms such as a den, office, or studio. In some homes, particularly smaller ones, rooms may serve two or more func-

tions. For example, the living room and dining room are often combined. In other homes, the entire living area may be one room.

The subdivisions of a living area are called rooms even though they are not always separated by a **partition** or a wall. The subdivision areas perform the function of a room, whether there is a complete separation, a partial separation, or no separation.

When partitions do not totally divide the rooms of an area, the arrangement is called an **open plan**. When rooms are completely separated by partitions and doors, the plan is known as a **closed plan**.

Open Plan

In an open plan living area, the living room, dining room, and entrance may be part of one open area. Instead of walls, separation of the living room from other rooms is accomplished in different ways. Figure 7.1 shows an example of the open plan living area in Frank Lloyd Wright's "Fallingwater."

Placement of area rugs or furniture will not completely separate areas of a room, but will create a functional and visual separation. If occasional privacy is desired in an

FIGURE 7.1 ■ Open plan living area. (*Fallingwater, Mill Run, Pennsylvania, Western Pennsylvania Conservancy—Frank Lloyd Wright, Architect; Christopher Little, Photographer*)

FIGURE 7.2 ■ Glass walls create an open plan effect. (*Alfred Karram Design*)

FIGURE 7.3 ■ A two-sided fireplace used to separate rooms. (*James Eismont, Photographer*)

open plan, folding doors can be installed to close off part of the area.

An open plan effect can be designed with glass walls to separate functions. See Figure 7.2. The glass walls separate the living room from the outodoor patio and yet maintain the open view and allow light to come into the room.

In the open plan shown in Figure 7.3, a two-sided fireplace and two-story chimney is used to separate the living room and dining room. Figure 7.4 shows an open floor plan with wide openings and levels used to separate the living room, dining room, family room, and foyer.

Closed Plan

In a closed plan, rooms are completely separated from the other rooms by means of walls. See Figure 7.5. Access is through doors, arches, or relatively small openings in partitions. Closed plans are found most frequently in traditional or period-type homes.

Combined Plans

Some large contemporary living area plans include both closed and open rooms. In combined plans there is a closed, formal living room and an open plan living area that functions as an all-purpose *great room*. (Great rooms are discussed later in this chapter.) The living area shown in Figure 7.6 has no separations between the living, dining, and kitchen areas although the kitchen is not directly visible from the living room.

LIVING ROOMS

The living room is the formal center of the living area in most homes. Thus its functions, location, decor, size, and shape are extremely important and affect the design and appearance of the entire residence.

Function

A designer begins a living area design by first determining the functions of rooms required in terms of the residents' needs. A living room is the key room in the design because it can serve many purposes. For example, it can be an entertainment center, a library, a social room, and perhaps a dining center. Its particular functions depend on the living habits of the residents. A living room needs to be designed in relation to its functions and activities.

Sometimes a living room can be designed by the process of elimination. For example, if a separate recreation room is planned, then planning for a TV in the living room might not be necessary. If a separate den or study is provided for reading and for storing large numbers of books, then this function might also be eliminated from the living room. On the other hand, if many living area activities are to be combined in one room, then a great room may be designed. Regardless of the exact, specific activities anticipated, the living room should be planned as an integral part of the home for family and guests.

Location

Ideally, the living room should be centrally located and adjacent to the outside entrance. See Figure 7.7. In smaller residences, the entrance may open directly into the living room. Whenever possible, however, this arrangement should be avoided. The living room should not be the only "traffic lane" to the sleeping and service areas of the house. Guests could be disturbed.

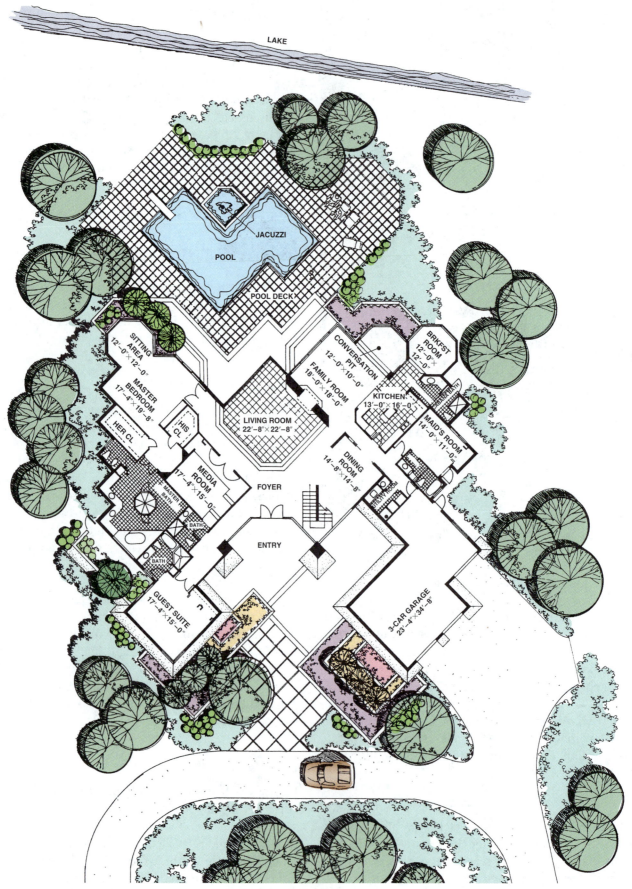

FIGURE 7.4 ■ Use of openings and levels to separate room functions. (*Alfred Karram Design*)

FIGURE 7.5 ■ A closed plan living room. (*John Berenson Interior Design Inc.*)

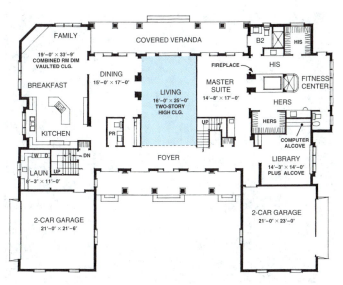

FIGURE 7.7 ■ Centrally located living room. (*Scholz Homes Inc.*)

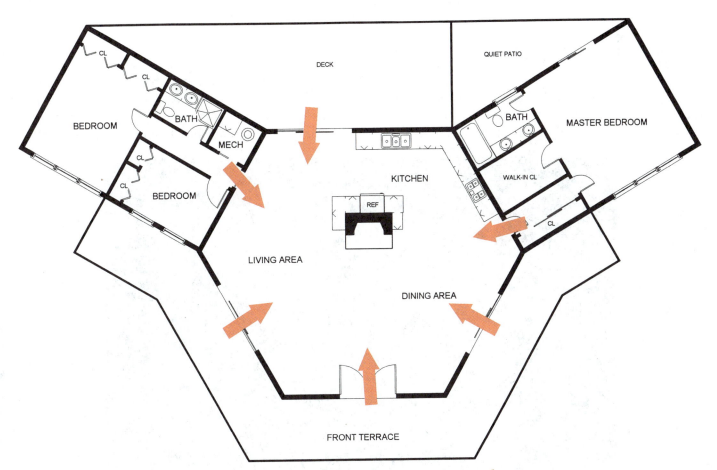

FIGURE 7.6 ■ Completely open great room. (*Home Planners Inc.*)

Orientation

Careful consideration should be given to the placement of the living room in relation to its surroundings, including other rooms. The living room should be oriented to take full advantage of the position of the sun and the most attractive view. Because the living room is used primarily in the afternoon and evening, it should be located to receive the afternoon sun in the southwest sky. See Figure 7.8.

FIGURE 7.8 ■ A sun and view-oriented living area. (*Eagle Windows*)

Decor

The general **decor** (pattern of decoration) of the living room depends primarily on the tastes, habits, and personalities of the residents. If their tastes are contemporary, then the wall, ceiling, window, and floor treatments should be consistent with the clean, smooth lines often found in con-

temporary architecture and contemporary furniture. If the residents prefer colonial or another style of architecture, then this style should be the theme of the decor.

For example, the decor of the living room in Figure 7.9 is consistently traditional; the living area in Figure 7.10 is distinctly contemporary with hard reflective surfaces. Both are totally different examples of the effective use of the principles and elements of good design presented in Chapter 2.

Appropriate color choices, effective lighting techniques, and the tasteful selection of materials for walls, floor covering, and ceiling can make a room appear inviting. The selection and placement of well-designed furniture can contribute to the appearance of comfort in a living room. The use of mirrors, floor-to-ceiling drapes, and arrangements of furniture can create a spacious effect in a relatively small room. Decorating a room, like selecting clothing, should minimize faults and emphasize good points.

Walls

The appearance of walls depends on more than wall coverings. The design and placement of doors, windows, fireplaces, chimneys, or built-in furniture along the walls of the living room will influence the entire room's appearance

FIGURE 7.9 ■ An example of excellence in traditional design and decor. (*Marc Michaels Design*)

FIGURE 7.10 ■ An example of excellence in contemporary design and decor. (*Alfred Karram Design*)

and should be designed as integral parts of the room. A designer considers these features in conjunction with the kind of wall-covering materials selected. Many materials are available, including plaster, gypsum, wallboard, wood paneling, brick, stone, tile, plastics, paper, and glass.

Windows

Just as the placement of a window in a living room wall should become an integral part of a wall, the view from the window should become part of the living room decor. Notice how the window shapes in Figure 7.11 conform to the ceiling line and also provide an uninterrupted transition to the outside through the sliding glass doors. When planning windows, also consider the various seasonal changes in landscape features.

Although windows themselves can be decorative items, the primary function of a window is to admit light and to provide a pleasant view of the landscape. Under some conditions, however, only the admission of light is desirable. If the view from the window is unpleasant or is restricted by other buildings, translucent glass can be used to allow the natural light to enter without showing the unwanted view.

Fireplaces

The primary function of a fireplace is to provide heat, but it can also be used as a room partition or as a major decorative feature. A fireplace and the chimney masonry can cover an entire wall and can become the focal point of the living area, as shown in Figure 7.12. A massive free-standing fireplace, however, also can function as a partition between rooms in an open floor plan. Like all elements of a room's decor, a fireplace should correlate with the architectural style of the room.

Floors

The living room floor should reinforce and blend with the color scheme, textures, and overall style of the living room. Exposed hardwood flooring, room-size carpeting, wall-to-wall carpeting, throw rugs, and sometimes polished flagstone and tile are appropriate for living-room use.

Ceilings

Most conventional ceilings are flat surfaces covered with plaster or gypsum board. However, new building materials,

FIGURE 7.11 ■ Effective relationship between ceiling lines and window shapes. (*Lindal Cedar Homes Inc.*)

FIGURE 7.12 ■ Fireplace and chimney used as a major focal point in living area design. (*Lindal Cedar Homes Inc.*)

such as laminated beams and arches, and new construction methods have resulted in greater varieties and improvements in ceiling design. Higher ceilings allow for better air circulation and warm air exhaust. They also create a feeling of spaciousness that a low ceiling over the same amount of floor space does not. Crown molding also adds a decorative design element to a ceiling.

Lighting

Appropriate lighting is essential to a room's atmosphere and comfort. Living room lighting generally comprises three types of lighting arrangements: general lighting, local lighting, and decorative lighting.

General lighting refers to illuminating the entire room and is often accomplished through the use of ceiling fixtures, wall spotlights, and cove lighting. *Local lighting* is light for a specific purpose, such as reading, drawing, or sewing. Local lighting may be provided by table lamps, wall lamps, pole lamps, or floor lamps. *Decorative lighting* is used to improve the appearance of a room, create

a mood, or to enhance a particularly attractive feature in the room.

Furniture

Furniture for the living room should reflect the motif (theme) and architectural style of the home. Whether freestanding or built-in, furniture should maintain lines consistent with the entire wall treatment and blend functionally into the total decor of the room.

A living room designed primarily for conversation is often called a *formal* living room (formerly called a *parlor*). This type of living room is usually closed and small and would, of course, require furniture different from an *informal* living room designed for television viewing, dining, and other activities. Houses that have formal living rooms usually also have family rooms with informal furniture.

Size and Shape

Ideally, when designing a living room, the type and amount of living room furniture should be determined *before* the size of the living room is established. One of the most difficult aspects of planning the size and shape of a living room, or any other room, is to provide sufficient wall space for the effective placement of furniture. Continuous wall space is needed for the placement of many kinds of furniture, especially musical equipment, bookcases, chairs, and sofas. The placement of fireplaces, doors, or openings to other rooms should be planned to conserve as much wall space as possible for furniture placement.

Figure 7.13 shows typical ranges of size for living, room furniture. In addition to standard sizes, furniture may be custom designed (specially made) for specific locations and shapes. See Figure 7.14.

To ensure that the room will accommodate the necessary furniture, **templates**, as shown in Figure 7.15, and a room plan are made to the same scale. The templates represent the width and length of each piece of furniture. The room plan represents the size and shape of the room or area. By seeing the amount of space occupied by the furniture, a designer can then more effectively evaluate the design of the entire room. Rectangular rooms are generally easier to plan and to place furniture in than are square rooms.

Living rooms vary greatly in size. A room 12′ × 18′ would be considered a small or minimum-sized living room. A living room of average size is approximately 16′ × 20′, and a large living room would be 20′ × 26′ or more.

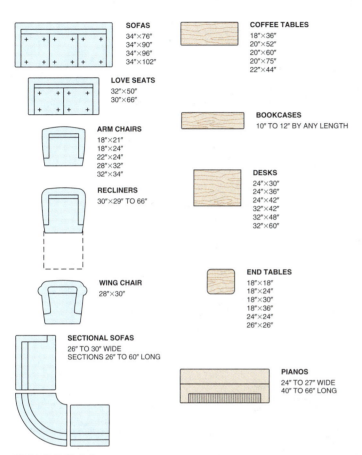

SOFAS
34″×76″
34″×90″
34″×96″
34″×102″

LOVE SEATS
32″×50″
30″×66″

ARM CHAIRS
18″×21″
18″×24″
22″×24″
28″×32″
32″×34″

RECLINERS
30″×29″ TO 66″

WING CHAIR
28″×30″

SECTIONAL SOFAS
26″ TO 30″ WIDE
SECTIONS 26″ TO 60″ LONG

COFFEE TABLES
18″×36″
20″×52″
20″×60″
20″×75″
22″×44″

BOOKCASES
10″ TO 12″ BY ANY LENGTH

DESKS
24″×30″
24″×36″
24″×42″
32″×42″
32″×48″
32″×60″

END TABLES
18″×18″
18″×24″
18″×30″
18″×36″
24″×24″
26″×26″

PIANOS
24″ TO 27″ WIDE
40″ TO 66″ LONG

FIGURE 7.13 ■ Typical sizes of living room furniture.

FIGURE 7.14 ■ Furniture designed to match wall and window contours. (*Eagle Windows*)

 DINING ROOMS

The design of dining facilities for a residence depends greatly on the dining habits of the family. A separate dining room may be large and formal. An informal dining area may consist of a dining alcove in a living area or even a breakfast nook in the kitchen. Large homes may contain several dining facilities in different areas.

Function and Location

The function of a dining area is to provide a place for the family and guests to eat breakfast, lunch, or dinner, whether in casual or formal situations. When possible, a separate dining area capable of seating from 6 to 12 persons for dinner should be provided in addition to breakfast and lunch facilities.

Dining facilities may be located in many different areas, depending on the residents' needs and preferences. Regardless of the exact position of the dining area, it should be adjacent to the kitchen. The ideal dining location is one that requires few steps from the kitchen to the dining table.

In a closed plan, a separate dining room is usually located between the living room and kitchen, as shown in Figure 7.16. In an open plan, several different dining locations are possible. Open-area dining facilities can be in the kitchen (Figure 7.17) or the living room. However, the preparation of food and other kitchen activities should not be in direct view from the dining area. Although directly visible, notice how level and distance

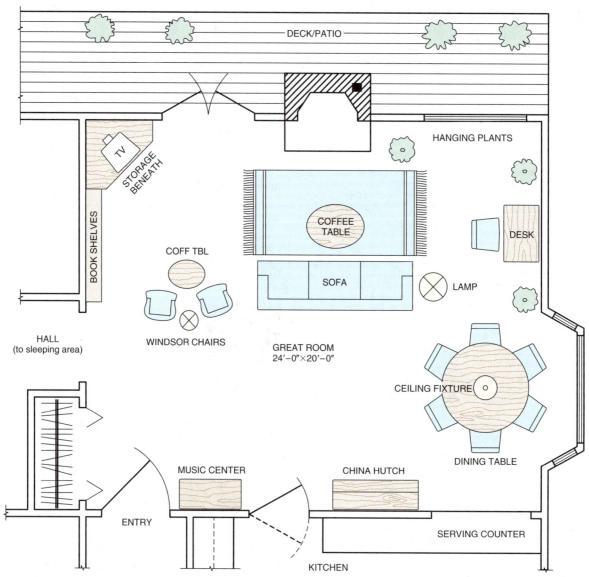

FIGURE 7.15 ■ Furniture templates used to plan room size and shape.

are used to separate the dining area from the living area as shown in Figure 7.18.

To baffle (separate or hide) an area, many design options are possible. For example, folding or sliding doors can separate and hide the kitchen from the dining area. See Figure 7.19. A two-sided fireplace offers another option for separating the dining room from other parts of the living area.

When dining facilities are not located in the living room, the dining area should be located adjacent to it. Family and guests normally enter the dining room from the living room and use both rooms jointly.

A partial separation of the dining room and the living room can be accomplished by different floor levels or by dividing the rooms with common half walls. **Serving walls** also provide a functional separation between the dining room and kitchen. Fireplaces, entertainment enclosures, and storage cabinets are also effective in partially separating the dining and living areas as shown in Figure 7.20.

Some families enjoy dining outdoors on a porch or a patio. If so, the porch, deck, or patio should be near the kitchen and directly accessible to it. Locating the patio or dining porch directly outside the dining room or kitchen provides maximum use of the facilities. This location minimizes the distance from the kitchen and possible inconveniences of outside dining facilities.

FIGURE 7.16 ■ Effective location of a dining room in a closed plan. (*Home Planners Inc.*)

FIGURE 7.17 ■ Dining area located in a kitchen. (*Marvin Windows*)

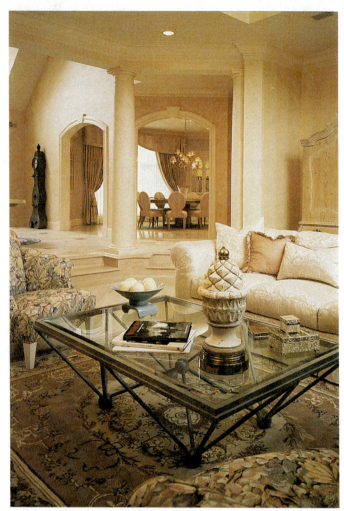

FIGURE 7.18 ■ Dining area separated by level and distance. (*Daphine Weiss Inc., Interior Designer; Maxwell MacKenzie, Photographer*)

FIGURE 7.19 ■ Sliding doors used as optional separators between the dining room and the kitchen.

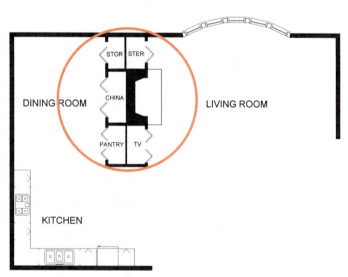

FIGURE 7.20 ■ Use of fireplace and cabinetry to separate areas. (*Home Planners Inc.*)

Decor

The decor of the dining room should blend with the remainder of the house. The floor, walls, and ceiling treatment of the dining area should work well with the decor of the living area. If a dining porch or a dining patio is used, its decor should also be considered part of the dining room decor.

To create a partially closed dining area, the decor of any kind of partial divider wall should be considered as well as its purpose. A divider may be a planter wall; glass wall; half wall of brick, stone, or wood panels; fireplace; or grillwork.

Controlled lighting is another means to greatly enhance the decor of the dining room. General illumination that can be subdued or intensified can provide the appropriate atmosphere for any occasion. This type of light-

ing is controlled by a *rheostat,* commonly known as a dimmer switch. In addition to general illumination, local lighting should be provided directly over the dining table. See Figure 7.21.

Size and Shape

The size and shape of the dining area are determined by the size of the family, the size and amount of furniture, and the clearances and traffic areas needed between pieces of furniture. The dining area should be planned for the largest group that will dine in it. There is little advantage in having a dining room table that expands if the

FIGURE 7.21 ■ The effective use of local and general lighting in a dining area. (*Alfred Karram Design*)

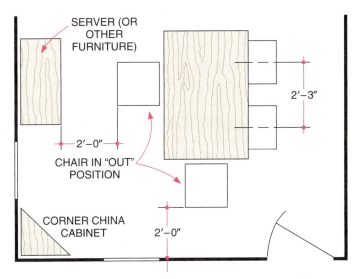

FIGURE 7.22 ■ Dining area space allowances.

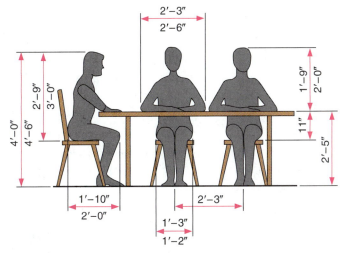

FIGURE 7.23 ■ Space requirements for dining room seating.

room is not large enough to accommodate the expansion. One advantage of the open plan is that the dining facilities can be expanded into the living area.

Dining room furniture may include an expandable table, side chairs, armchairs, buffet, server or serving cart, china closet, and serving bar. In most situations, a rectangular dining room will accommodate the furniture better than a square dining room.

Regardless of the furniture arrangement, a *minimum* space of 2′ (610 mm) should be allowed between a chair and the wall or other furniture when the chair is pulled

out. This allowance will permit serving traffic behind chairs, and will allow persons to approach or leave the table without difficulty.

Another space consideration is the distance *between* people when seated at the table. This spacing is accomplished by allowing 27″ (686 mm) from the centerline of one chair to the centerline of another. See Figure 7.22. Figure 7.23 illustrates the average space required for adults when seated for dining. The shapes and sizes of typical dining room furniture are shown in Figure 7.24.

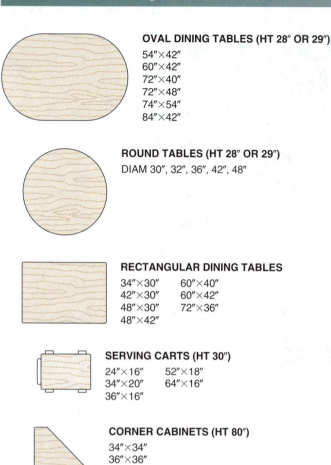

OVAL DINING TABLES (HT 28″ OR 29″)

54″×42″
60″×42″
72″×40″
72″×48″
74″×54″
84″×42″

ROUND TABLES (HT 28″ OR 29″)

DIAM 30″, 32″, 36″, 42″, 48″

RECTANGULAR DINING TABLES

34″×30″   60″×40″
42″×30″   60″×42″
48″×30″   72″×36″
48″×42″

SERVING CARTS (HT 30″)

24″×16″   52″×18″
34″×20″   64″×16″
36″×16″

CORNER CABINETS (HT 80″)

34″×34″
36″×36″
38″×38″

BUFFETS (HT 31″)

36″×16″   52″×18″
48″×16″   60″×20″

CHINA CABINETS (HTS 60″ TO 72″)

48″×16″   62″×16″
50″×20″   36″×18″

DINING ROOM CHAIRS (SEAT HT 16″)
(BACK HTS 29″ TO 36″)

17″×19″
20″×17″
22″×19″
24″×21″

FIGURE 7.24 ■ Typical sizes of dining room furniture.

A dining room that would accommodate the minimum amount of furniture—a table, four chairs, and a buffet—would be approximately 10′ × 12′ (3048 mm × 3658 mm). A minimum-sized dining room that would accommodate a dining table, six or eight chairs, a buffet, a china closet, and a server would be approximately 12′ × 15′ (3658 mm × 4572 mm). A large din-

FIGURE 7.25 ■ Dining room furniture templates.

ing room would be 14′ × 18′ (4267 mm × 5486 mm) or larger. Figure 7.25 shows a dining room layout using furniture templates.

FAMILY ROOMS

Because of the trend toward more informal living, the majority of homes today are designed to include a family room.

Function

The purpose of the **family room**, as the name implies, is to provide facilities for family-centered activities. In extremely large residences, special-purpose rooms may be provided for specific types of activities, such as playing music, sewing, or painting. Typically however, additional rooms are not part of the plan, and facilities and equipment must be provided for a wide variety of activities. For this reason, the family room is also known as the *activities* or *multiactivities* room.

Location

Activities in the family room often result in the accumulation of hobby materials and clutter. Thus, the family

FIGURE 7.26 ■ Kitchen extension into a family room. (*Whirlpool Corp.*)

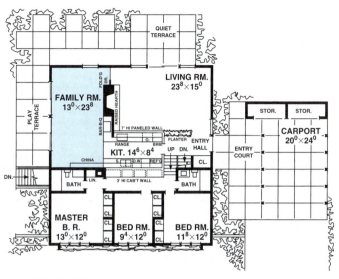

FIGURE 7.27 ■ Family room accessible from the living room and kitchen.

room should be easily accessible, but not visible, from the rest of the living area.

Commonly, the family room is located adjacent to the kitchen. This location revives and expands the idea of the old country kitchen as the room in which most family activities were centered. See Figure 7.26.

When the family room is located adjacent to the living room or dining room, it becomes an extension of those rooms for social affairs. In this location, the family room is often separated from the other rooms by folding doors, screens, or sliding doors. See Figure 7.27.

Another popular location for the family room is between the service area and the living area. This location is especially appropriate if a home workshop is assigned to the family room.

Decor

Decoration of the family room should provide a vibrant atmosphere. Ease of maintenance should be one of the chief considerations. Furniture materials such as plastic, leather, and wood are easy to care for and provide great flexibility in color and style. Family room furniture should be informal and suited to all members of the family.

The floor should be resilient—able to keep its original shape or condition despite hard use. Linoleum or tile made of asphalt, rubber, or vinyl will best resist the abuse normally given a family room floor. If rugs are used, they should be the kind that will stand up under rough treatment. They should also be washable.

For walls, soft, easily damaged materials such as wallpaper and gypsum board should be avoided. Materials such as tile and paneling are most functional. Chalkboards, bulletin boards, cupboards, and toy-storage cabinets should be used when appropriate. Work areas that fold into the wall when not in use conserve space. Such areas perform a dual function if the wall cover can also be used as a chalkboard or a bulletin board.

Because a variety of hobby and game materials will be used in the family room, sufficient storage space must be provided. This includes the use of built-in facilities such as cabinets, closets, and drawer storage.

Acoustical ceilings are recommended. These help keep the noise of the various activities from spreading to other parts of the house. This feature is especially important if the family room is located on a lower level.

Size and Shape

The size and shape of the family room depend on the equipment needed for the planned activities. The room may vary from a minimum-sized room of approximately 150 sq. ft. to a very spacious family room of 300 sq. ft. or more. Most family rooms require a size that ranges between these two extremes. See Figure 7.28.

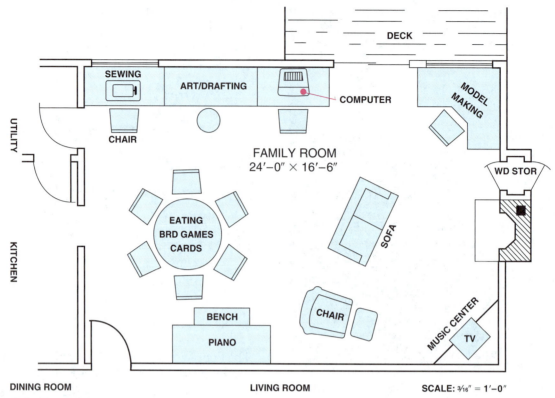

FIGURE 7.28 ■ Multiactivity family room.

GREAT ROOMS

Rooms that combine the functions of a living room, dining room, family room, and sometimes the kitchen are called the **great room** or sometimes the **gathering room**. The great room shown in Figure 7.29 includes space for these rooms plus library and entertainment room facilities.

SPECIAL PURPOSE ROOMS

In addition to the traditional living room, dining room, family room, and great room, the living area may also include rooms devoted to specific activities. These rooms are designed for recreation, work, or entertainment, as discussed next.

Recreation Rooms

The **recreation room** may also be called a game room or playroom. As the name implies, it is a room designed specifically for active play, exercise, and recreation.

Recreation rooms may also include facilities for crafts and hobbies that do not require large power tools or equipment. Exercise equipment may also be included if a separate exercise room is not planned.

Function

The function of the recreation room often overlaps that of the family room. Overlapping occurs when a multipurpose room is designed to provide for recreational activities such as table tennis and billiards, but also includes facilities for quieter activities such as knitting, model building, and other hobbies.

The design of the recreation room depends on the number and arrangement of the facilities needed for the various pursuits. Activities for which many recreation rooms are designed include billiards (Figure 7.30), chess, checkers, table tennis, darts, watching television, eating, and dancing. See Figure 7.31.

Location

The recreation room should be located away from the quiet areas of the house. Most often, it is located in the basement or on the ground level.

A basement location uses space that might otherwise be wasted. Also, basement recreation rooms can often

FIGURE 7.29 ▪ Great room including kitchen exposure. (*Home Planners Inc.*)

FIGURE 7.30 ▪ Recreation room featuring billiards and home theater facilities. (*Isleworth Real Estate—Tom Price, Architect; Laurence Taylor, Photographer*)

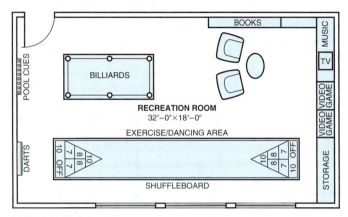

FIGURE 7.31 ■ Multiactivity recreation room.

FIGURE 7.32 ■ Home office with built-in storage facilities. (*Mica Products & Wood*)

provide more space for activities that require large areas and equipment, such as table tennis, billiards, and shuffleboard. A good ground-level location would allow activities to be expanded onto a patio or terrace.

Decor

Designers take more liberties in decorating the recreation room than with any other room. They do so primarily because of the active, informal atmosphere that characterizes the recreation room. This atmosphere lends itself readily to unconventional furniture, fixtures, and color schemes.

Bright, warm colors can reflect a party mood. Furnishings and accessories can be used to accent a variety of central themes. Regardless of the theme, recreation room furniture should be comfortable and easy to maintain. The same decorating guidelines that apply to the family room also apply to recreation room walls, floors, and ceilings.

Size and Shape

The size and shape of a recreation room may depend on whether the room occupies basement space or an area on the main level. If basement space is used, the only restrictions on the size are the other facilities there, such as the laundry, utility, or workshop areas.

HOME OFFICES

The most common work-related room in a residence is the **home office** (Figure 7.32). This can be a professional office or an office used for personal business. A home office may function as a **study** or **studio** depending on use. Large residences may include both; smaller homes may combine a home office with a family room, great room, or guest bedroom.

Major considerations in planning a home office include furniture size, type, and placement, and lighting, storage, and electronic facilities. Figure 7.33 shows the outline of typical office furniture and equipment. The number and size of each item depends on the type of activity planned. It is also important to consider the location of the office in relation to other rooms including the entry, **egress**, and traffic flow. If the home office is to be used by clients or guests, an outside or foyer access should be planned, as shown in Figure 7.34.

Electronic facilities for a home office range widely depending on the office function. For example, computer configurations are determined by the requirement of size, speed, drives, speakers, keyboards, mouse, modem, printer, scanner, and wire management needs. Wiring amperage must be adequate to serve all office functions simultaneously. Electrical surge protection is also included because office machines, especially computers and printers, have sensitive circuits that can be damaged by power fluctuations. (See Chapter 31.) Although the office of the future may be wireless, today wires and cables should be channeled in accessible walls or under floors or baseboard units to avoid the dangers of exposed wiring. Furniture may include plug-ins for phones, fax machines, modem, and satellite, cable TV, Internet, and computer networks.

Although the maximum amount of natural lighting is desirable, general artificial, well-diffused lighting should

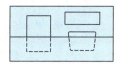 **WORKSTATION CABINETS**
HT VARIES
DEPTH 26″–30″
LENGTH VARIES

 DRAFTING TABLES
HT 36″ - (SLANT 15°)
30″×40″ TO 48″×84″

 COMPUTER STATION
TYP. 18″×30″

 DRAFTING STOOLS
TYP. 16″×16″

 PRINTER STANDS
TYP. 24″×24″

 TYPEWRITERS
TYP. 18″×15″

 COMPUTER CARTS
TYP. 32″×24″

 TELEPHONE ANSWERING MACHINES
TYP. 8″×6″

 FAX MACHINES
TYP. 14″×11″

 DESKS
HT 30″
DEPTH 28″ TO 30″
LENGTH VARIES

 TELEVISIONS
AVE. 20″×20″

 DESK SWIVEL CHAIRS
20″×18″

 FILE CABINETS
HT VARIES
DEPTH 15″ TO 18″
LENGTH 18″ TO 27″

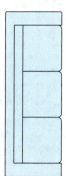 **SOFA/BEDS**
TYP. 36″×80″

 STORAGE CABINETS
HT 42″ TO 78″
DEPTH 18″ TO 24″
LENGTH 36″

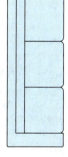

BOOKCASES
HT VARIES
DEPTH 12″
LENGTH VARIES

 SCANNER
HT 3″ TO 5″
DEPTH 10″ TO 15″
LENGTH 15″ TO 20″

 **DESK TOP
PHOTOCOPIER**
HT 8″ TO 12″
DEPTH 13″ TO 15″
LENGTH 16″ TO 21″

 PHOTO PRINTER
HT 6″
7″ × 8″

 LIGHT TABLE
HT 36″
30″×40″ TO 48″×84″

 FLAT DRAWING FILE
HT VARIES
DEPTH 18″ TO 48″
LENGTH 36″ TO 72″

 COFFEE MAKER
HT 11″
10″×6″

 TELEPHONE
9″×6″

FIGURE 7.33 ■ Typical sizes of office furniture and equipment.

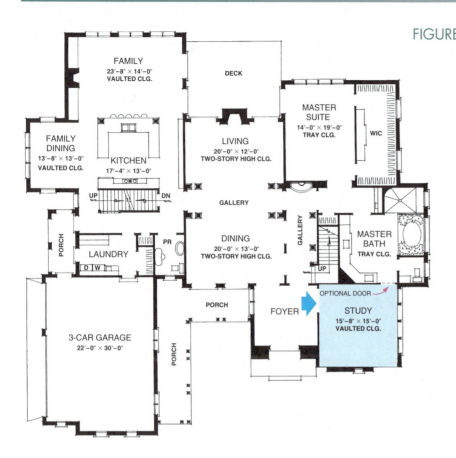

FIGURE 7.34 ■ Home office (study) located for convenient outside access. (*Scholz Design Inc.*)

be designed for nighttime use. Glare-proof local task lighting is also needed at each task station.

ENTERTAINMENT ROOMS

Entertainment rooms are also known as **media rooms** or **home theaters** and include big-screen TVs, VCRs, DVDs, stereo systems, and sometimes a piano or organ. Small residences may include these facilities in a family room or great room; large residences may have a separate home theater or music room. The music center shown in Figure 7.35 is located in a separate area of a living room. The entertainment unit shown in Figure 7.36 is located on a great room wall with electronic components exposed. The entertainment center in Figure 7.37 is designed as a wall unit in a living room with all components concealed in a bookshelf cabinet design.

Entertainment or media rooms should be located with direct access to the living area and away from bedrooms or other private areas. Preplanning of all wiring, lighting, acoustics, soundproofing, seating, and built-in cabinetry is extremely important in designing an entertainment room or center.

FIGURE 7.35 ■ Music center located in a living area. (*Marc Michaels, Designer; Dan Forer, Photographer*)

FIGURE 7.36 ■ Entertainment center with exposed components. (*Mastered Concepts & Design*)

FIREPLACES

Fireplaces can be designed as an aesthetic and functional part of any living or sleeping area.

Because masonry fireplaces and chimneys are exceptionally heavy, they cannot be supported by the normal building footings. Special provisions must be made. The design of the fireplace influences the type of foundation that is needed. In designing fireplaces, the style, type, support, framing, size, materials, components, ratio of the opening and firebox, and the height of the chimney must all be considered. Fireplaces are classified by fuel type, type of opening, construction, and architectural style (contemporary, Spanish, colonial, etc.).

Fuel Types

Fireplaces burn either wood, natural gas, or synthetic materials.

Wood

Oxygen is the vital ingredient needed for effective wood burning and proper functioning of the fireplace. Because warm air rises, air in the room is drawn into the fireplace, supplying the fire with needed oxygen. Because cold air continually replaces rising warm air, much of the heat produced by many fireplaces goes up the chimney. To reduce this heat loss and redirect some of the heat back inside, warm-air outlets that balance the inlet of cold air are effective. Use of outlets of this type allows heat to reenter the room, while smoke, debris, and toxic fumes are directed outside through the chimney.

FIGURE 7.37 ■ Entertainment center with concealed components. (*Casework Specialities Inc.*)

Natural Gas

A gas fireplace offers maximum operating convenience and warm-air circulation. Concealed circulating fans keep air moving around the firebox and expel heated air into the room. Gas fireplaces have automatic pilot lights that are quickly turned off or on. Units can also be thermostatically controlled. "Logs" used in gas fireplaces resemble wood in appearance but are made of noncombustible ceramic or masonry materials.

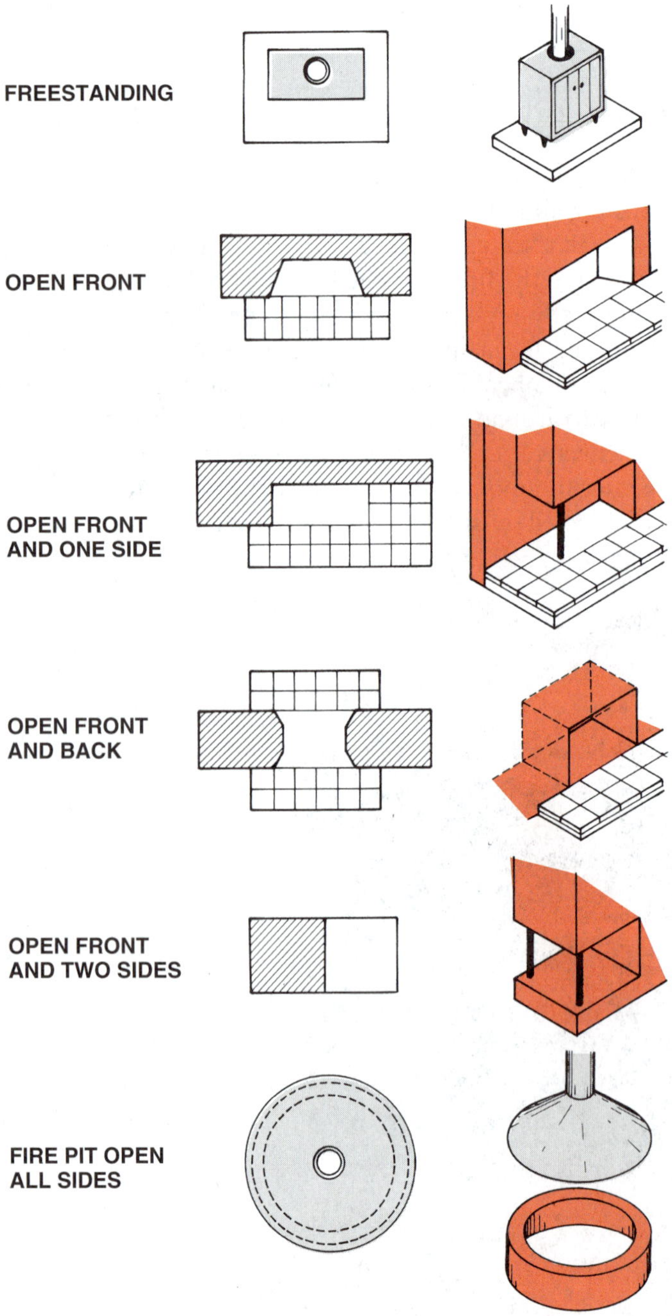

Requiring neither heavy masonry and foundations nor front hearth, prefabricated gas fireplaces can rest on any type of flooring with no limitations on enclosure size or trim. Gas fireplaces require no direct vertical flue. Therefore, the area above the firebox need not be a chimney flue.

Synthetic Materials

Some fireplaces are designed to burn synthetic materials, such as gelled alcohol. These fireplaces require no venting to the outside, no hearth, and no special flooring.

Types of Fireplace Openings

Fireplaces are divided into six basic types of openings as shown in Figure 7.38. A flush opening is also called a single face and is shown in Figures 7.11 and 7.12. A two-sided or corner fireplace is used in L-shaped areas

FREESTANDING

OPEN FRONT

OPEN FRONT AND ONE SIDE

OPEN FRONT AND BACK

OPEN FRONT AND TWO SIDES

FIRE PIT OPEN ALL SIDES

FIGURE 7.38 ■ Types of fireplace openings.

FIGURE 7.39 ■ Two-sided corner fireplace. (*Heatilator, Inc.*)

FIGURE 7.40 ■ Three-sided, peninsula fireplace. (*Majestic Fireplaces*)

FIGURE 7.41 ■ See-through gas fireplace. (*Heatilator, Inc.*)

FIGURE 7.42 ■ Open-pit fireplace. (*John Henry, Architect*)

as shown in Figure 7.39. A three-sided or peninsula fireplace is used to partially separate two rooms as shown in Figure 7.40. The see-through fireplace allows the fireplace to be viewed from opposite sides. The see-through fireplace in Figure 7.41 is a gas fireplace without a vertically aligned chimney. Four-sided or open-pit fireplaces as sketched in Figure 7.42 require large hoods to capture smoke and fumes.

Freestanding metal or ceramic fireplaces are available in a variety of shapes as shown in Figure 7.43. They are relatively light wood-burning stoves and therefore need no concrete foundation for support. A stovepipe leading into the chimney provides the exhaust flue. Because metal units reflect more heat than masonry, metal fireplaces are much more heat efficient, especially if centrally located. For safety, fire-resistant materials such as concrete, brick, stone, or tile must be used beneath and around these fireplaces.

Wood-burning stoves are freestanding fireplaces that provide heat through the conduction of heated metal, which radiates warm air convection currents in an enclosed space.

See Chapter 23 for fireplace construction details.

FIGURE 7.43 ■ Freestanding fireplaces and stoves. (*Majestic Fireplaces, Art Mac Dillos, Gary Skillestad*)

CHAPTER

7

Indoor Living Areas Exercises

1. List the functions you want in a living room for yourself or an imaginary client.

2. Draw a simple sketch of an open plan living room for the functions you listed in Exercise 1. Indicate the location of doors, windows, a fireplace, foyer, entrance, and dining room and label them accordingly.

3. Sketch a closed plan living room for the same functions as in Exercise 1. Show the position of adjacent rooms. Explain the reasons for your choices.

4. Using *Line* commands on a CAD system, draw the living room outline you prefer from those that you designed in Exercises 2 and 3. Indicate doors and windows.

5. Begin a picture file of your own for different rooms. Cut out pictures showing the decor you like in living rooms—including furniture, lights, etc.—from catalogs, newspapers, or magazines. Then list the furniture you would include in your design, including color and materials.

6. Sketch the dining area of your own home. Then make another sketch changing the design (open or closed plan) without changing any *outside* walls. (Make any needed templates to help create a different arrangement.)

7. Sketch a dining room to scale showing the position of all furniture you would like to include in the dining room of a house of your own design.

8. Collect pictures of dining areas and furniture and accessories.

9. Draw a floor plan of a dining room shown in this chapter.

10. Calculate the minimum-sized dining room you would need to seat six people at a 60″ × 42″ table.

11. Refer back to Figure 7.28. This plan is drawn to 3/16″ = 1′-0″ scale. Tell what the dimensions of the room are. Evaluate the size and the arrangement in relation to its functions.

12. Sketch a family room, recreation room, special purpose room, or office you would like to include in a home of your own design. List the activities and furniture needed. Show the location of all furniture and facilities (scale 1/2″ = 1′-0″).

13. Design a family room primarily for children's activities. Include the furniture needed. Describe the colors and materials you would select for this room.

14. Redesign the room in Exercise 13 to accommodate teen activities. Describe the changes you made and explain why.

15. Draw an outline of a family room adjacent to a kitchen.

16. Draw a floor plan of the recreation room shown in Figure 7.30.

17. Use the typical dimensions of the office equipment in Figure 7.33 to sketch a plan to accommodate the following facilities: desk, chair, bookcases, drafting table, and lounge.

18. Collect pictures of rooms and furnishings that you would like to use in a house of your own design.

Outdoor Living Areas

OBJECTIVES

In this chapter you will learn to:

- design and sketch a porch, patio, and lanai.
- design and sketch a swimming pool.
- calculate the area and volume of swimming pools.

TERMS

balcony	marquee	stoop
court	patio	swim-out
deck	porch	veranda
lanai		

INTRODUCTION

A home's living areas may be extended to the outdoors. Porches, patios, decks, pools, and other features provide space for dining, entertaining, playing, exercising, or relaxing. When planning an outdoor living area, consider the area's function, location, decor, size, and shape. A well-designed area will look and function like a natural extension of the interior.

 PORCHES

A **porch** is a covered platform leading into an entrance of a building. Porches are commonly enclosed by glass, screen, or posts and railings. Balconies and decks are actually elevated porches. (A **deck**, however, usually refers to an open, elevated platform.) Similar to a porch, a **stoop** is a projection from a building. However, a stoop does not provide sufficient space for activities. It provides only shelter and an access to or landing surface for the entrance of the building.

Porches are often confused with patios. Although a patio may also be adjacent to a house and seemingly attached to it, a patio is directly on the ground, even if it has a finished surface. The main difference between a porch and a patio is that a porch is attached *structurally* to a house.

Function and Types

The classic front porch and back porch that characterized most homes built in this country during the 1920s and 1930s were built merely as places in which to sit. Little effort was made to use the porch for any other activities. However, Southern Colonial homes were designed with **verandas**, large porches, extending around several sides of the home. See Figure 8.1. Outdoor life often centered on the veranda. The multiple decks shown in Figure 8.2 serve a variety of purposes including dining, relaxing, and regulating traffic flow. These decks are connected at different levels to the living–dining area, entrance, bedrooms, and a wading pool.

A **balcony** is a porch suspended from an upper level of a structure. There is usually no access from the outside. Balconies often extend a living area. See Figure 8.3. Others serve as a private extension of a bedroom. The balcony protects the lower level from the sun and precipitation.

Hillside lots lend themselves to vertical plans and allow maximum flexibility for such outdoor living areas. See Figure 8.4. Spanish- and Italian-style architecture is typically characterized by numerous balconies that integrate indoor and outdoor living areas. New developments in building materials have increased the recent popularity of balconies in many styles of architecture.

FIGURE 8.1 ■ Example of a home with a two-level veranda. (*Dixie Pacific*)

FIGURE 8.2 ■ Multilevel decks used for different functions. (*Western Pennsylvania Conservancy—Thomas A. Heinz, Photographer*)

Location

The location of the porch depends on its purpose or function. The family's preferences for use of the porch should also be considered when designing its orientation. For example, if daytime use is anticipated for the porch and direct sunlight is desirable, then a southern exposure should be planned. If little sun is wanted during the day,

a northern exposure is preferable. If morning sun is desirable, an eastern exposure is best, and for the afternoon sun and sunset, a western exposure.

A continuous porch is often designed to function with the living area and/or with the sleeping area. The porch shown in Figure 8.5 continues on three sides of the house to include all three areas: living, sleeping, and service areas.

FIGURE 8.3 ■ Multiple balconies extend from each living area. (*Eagle's Nest—Robert Boward, Architect; E. Joyce Reesh—Arvida*)

FIGURE 8.4 ■ Balconies serving as living and sleeping areas. (*David Garris*)

Decor

The porch should be designed as an integral and functional part of the total structure. A blending of roof styles and major lines of the porch roof and house roof is especially important.

A porch can be made consistent with the rest of the house by extending the lines of the roof to provide sufficient roof overhang, or projection, over the porch area. See Figure 8.6. A similar consistency should characterize the vertical columns or support members of the porch and the railings.

FIGURE 8.5 ■ Continuous porch with patio below. (*Lindal Cedar Homes Inc.*)

Porch railings can provide adequate ventilation and also offer semiprivacy and safety. Various materials and styles can be used, depending on the degree of privacy or sun and wind protection needed. Railings on elevated porches, such as balconies, should be higher than 3' (914 mm) for general safety, as well as to discourage the use of the top rail as a place to sit. By code, most *balusters* (vertical posts) must be spaced closely enough (usually 4") to prevent a child's head from going through. See Figure 8.7.

Porch furniture should withstand any kind of weather. The covering material should be waterproof, stain resistant, and washable. Nonetheless, protection from wind and rain should be planned.

Size and Shape

Porches range in size from the very large veranda to rather modest-sized stoops. A porch approximately 6' × 8' (1829 mm × 2438 mm) is considered minimum sized. An 8' × 12' (2438 mm × 3658 mm) porch is about average. Porches larger than 12' × 18' (3658 mm × 5486 mm) are considered large. The shape of the porch depends greatly on how the porch can be integrated into the overall design of the house.

PATIOS

The word **patio** is Spanish for courtyard, an open space enclosed wholly or partly by buildings. Courtyard living was an important part of Spanish culture. Therefore,

FIGURE 8.6 ■ Continuous deck with overhanging sun baffles. (*L'Ermitage at Grey Oaks, Naples, Florida*)

FIGURE 8.7 ■ Porch with railing and supported overhang. (*Lindal Cedar Homes Inc.*)

courtyard design was an important component of early Spanish architecture.

Function and Types

The patio may perform all the functions outdoors that the living room, dining room, recreation room, kitchen, and family room perform indoors. The patio may be referred to by other names, such as *loggia, breezeway, court,* and *terrace.*

Patios are divided into three main types according to function: living patios (including dining), play patios, and quiet patios. See Figure 8.8. Regardless of the type of patio, it should be secluded from the street and from neighboring residences if possible.

Where a patio is enclosed on three sides, as also shown in Figure 8.8, it is called a **court.** Courts (or courtyards) are a characteristic of Spanish architecture. When all four sides are enclosed, the patio actually becomes an interior atrium as described in Chapter 7. The patio shown in Figure 8.9 combines the facilities of a quiet patio and a dining patio.

Location

The type of patio affects its location in relation to other rooms in the home. For instance, living patios should be located close to the living room or dining room. When dining is anticipated on the patio, access should be provided from the kitchen or dining room.

A children's play patio, or play terrace, for physical activities is not necessarily associated with the living area. Sometimes a play patio is located next to the service area so that it can double as a service terrace. Children's play patios should be located to allow for easy adult observation.

A quiet patio can become an extension of the bedroom for relaxation or sleeping. A quiet terrace should be secluded from the normal traffic of the home.

Often the design of the house will allow these separately functioning patios to be combined into one large, continuous patio. See Figure 8.10. Similar to a continuous porch, this type of patio may be accessible from the

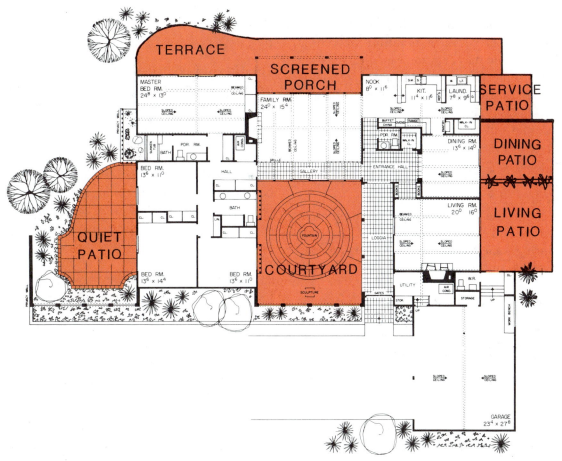

FIGURE 8.8 ■ Floor plan with different types of patios. (*Home Planners Inc.*)

FIGURE 8.9 ■ Dining patio combined with lanais. (*Wood residence—Taylor and Taylor, Architects*)

FIGURE 8.10 ■ Continuous patio designed for multiple functions. (*John B. Scholz, Architect*)

playroom, living room, bedrooms, and/or kitchen. Other designs divide large patios and porches into different areas by different levels.

Patios can be placed at the end of a building, between corners, or along the exterior form of the structure. They may also be placed in the center, such as in the center of a U-shaped house. A courtyard patio offers complete privacy from all sides.

A patio can be located completely apart from the house. A shady, wooded area, a beautiful view, or a unique feature of terrain may determine an ideal location for a patio. The designer should take full advantage of the most pleasing view and should restrict the view of undesirable sights. If the patio is located a short distance from the house, it should still be easily accessible to the house with paths or walkways.

When the patio is placed on the north side of the house, the house itself can be used to shade the patio. If sunlight is desired, the patio should be located on the south side of the house. By locating patios on different sides of a building, sun exposure and sun protection can be available during some part of the day. Protection from the sun may also be controlled with planned landscaping and fences.

Patios located adjacent to a large body of water, as shown in Figure 8.11, must be designed to withstand storm surges and sea spray. They may also require boat docking facilities to be included in the design.

Decor

The materials used in the decor of the patio should be consistent with the lines and materials used in the home. Patios should not appear to be designed as an after-

FIGURE 8.11 ■ Patio adjacent to a seawall. (*Jack and Barbara Irvin residence—Caldwell Banker–Coastco; Houston/Tyner, Architects*)

thought. They should appear and function as an integral part of the total design.

Patio Surface

The patio surface, or deck, should be constructed from materials that are permanent and maintenance free. Flagstone, redwood, concrete, engineered wood, reconstituted wood chips, plastic, and brick are among the best materials for use on patio decks. Wood creates a warm appearance. On some wood decks, slats may be spaced to provide drainage.

Brick-surface patios are very popular because bricks can be placed in a variety of arrangements. The area between the bricks may be filled with concrete, gravel, sand, or grass. A concrete patio is effective when a smooth, unbroken surface is desired. Concrete works well for patios where bouncing-ball games are played or where a poolside cover (roof) is desired. However, where patio surfaces also function as pool decks, a non-heat-absorbing material is preferable.

Patio Cover

The manner in which a patio is covered, or not covered, is closely related not only to the decor but also to the sunlight. Patios need not be covered if the house naturally shades the patio. Because a patio is designed to provide outdoor living, too much cover can defeat the purpose of the patio.

Coverings can be graded, or tilted, to allow light to enter the patio when the sun is high and block the sun's rays when the sun is lower. Straight or slanted louvers can be placed to admit the high sun and block the low sun or vice versa. See Chapter 7.

Plastic, fiberglass, and other translucent materials used to cover patios admit sunlight and yet provide protection from the direct rays of the sun and from rain. When such translucent coverings are used, it is often desirable to cover only part of the patio. This arrangement provides sun for part of the patio and shade for other parts and also allows rising heat to escape. Balconies can also be used effectively to provide shade and control light on a patio.

Patio Walls and Baffles

Patios are designed for outdoor living, but outdoor living does not mean living in public. Some privacy is usually desirable. Natural landforms can sometimes provide privacy. Walls can often be used effectively to baffle, or shield, the patio from a street view, from wind, and from the long shadows created by low rays of the sun. Baffling devices include solid or slatted fences, concrete blocks, post-and-rail, brick, or stone

FIGURE 8.12 ■ Protected patio. (*Julius Shulman*)

FIGURE 8.13 ■ Patio designed for night use. (*Western Wood Products Association*)

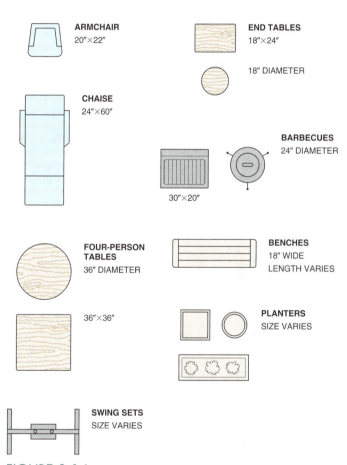

FIGURE 8.14 ■ Typical sizes of common patio furniture.

walls, and hedges or other landscaping. Figure 8.12 shows a patio with maximum protection both from the sides and overhead.

A solid baffle wall is often undesirable because it restricts the view, eliminates the circulation of air, and makes the patio appear smaller. When possible, natural vegetation or a sloped baffle is preferred.

In mild climates, a patio may be enclosed with solid walls to make the patio function as another room. In such an enclosed patio, some opening should nonetheless be provided to allow light and air to enter. Grillwork openings are an effective and aesthetically pleasing solution to this problem. Where wind and blowing sand or dust may be a problem, glass windscreens can be used to protect the patio.

Lighting

The patio should be designed so that it can be used both day and night. This means using general and local electrical lighting as well as natural lighting. If the windows and doors between the house and patio are designed correctly, light from inside the house can be utilized on the patio at night. See Figure 8.13. The combination of internal and external lighting design can extend the number of hours the patio can be used.

Size and Shape

As with other rooms, the function influences the size of a patio. Patios vary greatly in size. An oriental garden terrace, for example, can be small because it often has no furniture and is designed primarily to provide a baffle and a beautiful view. A patio used for recreation needs to be large enough for equipment and furnishings, such as picnic tables and benches, lounge chairs, serving carts, game apparatus, and barbecue pits. See Figure 8.14. Adequate space for the storage of games, apparatus, and fixtures also needs to be considered.

Patios tend to vary more in length than in width. Some patios may extend along the entire length of the house. A patio 12′ × 12′ (3658 mm × 3658 mm) is considered a minimum-sized patio. Patios with dimensions of 20′ × 30′ (6096 mm × 9144 mm) or more are considered large.

LANAIS

Lanai is the Hawaiian word for porch, but it also refers to a covered exterior passageway. Large lanais often double as patios.

Function

Lanais actually function as exterior hallways. They provide shelter for the exterior passageways of a building.

Lanais that are parallel to exterior walls are usually created by extending the roof overhang to cover a traffic area where people walk. A typical lanai eliminates the need for more costly interior halls. Lanais are used extensively in warm climates. Figure 8.15 shows a lanai plan, and Figure 8.16 shows a photograph of the finished area. Note that the lanais shown in Figures 8.17 and 8.9 double as lounge areas.

Location

In residence planning, a lanai can be used most effectively to connect opposite areas of a home. Lanais are commonly located between the garage and the kitchen, the patio and the kitchen or living area, and the living area and service area. U-shaped buildings are especially suitable for lanais because it is natural to connect the ends of the U.

When lanais are carefully located, they can also function as sheltered access from inside areas to outside facilities such as patios, pools, outdoor cooking areas, or courtyards. A covered or partially covered patio is also considered a lanai when it doubles as a major access path from one area of a structure to another.

A lanai can also be semi-enclosed and provide not only traffic access but privacy, as well as sun and wind shielding. When a lanai connects the building with the street it is called a **marquee**.

Decor

The lanai should be a consistent, integral part of the structure's design. The lanai cover may be an extension of the roof overhang or supported by columns. Columns also provide a visual boundary without blocking the view. If glass is placed between the columns, the lanai becomes an interior hallway rather than an exterior one.

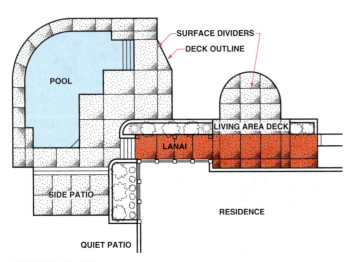

FIGURE 8.15 ■ Lanais connecting living area to pool deck.

FIGURE 8.16 ■ Photograph of the lanais shown in Figure 8.15.

FIGURE 8.17 ■ Lanais designed for multiple use. (*John B. Scholz, Architect*)

This feature is sometimes the only difference between a lanai and an interior hall.

It is often desirable to design and locate the lanai to provide access from one end of an extremely long building to the other end. The lines of this kind of lanai strengthen and reinforce the basic horizontal and vertical lines of the building.

If a lanai is to be utilized extensively at night, effective lighting must be provided. Light from within the house can be used when drapes are open, but additional lighting fixtures are used for the times when drapes are closed.

Size and Shape

Lanais may extend the full length of a building and may be designed for maximum traffic loads. They may be as small as the area under a roof overhang. However, a lanai at least 4′ (1219 mm) wide is desirable. The length and type of cover is limited only by the location of areas to be covered.

SWIMMING POOLS

Swimming pools are becoming more popular as an integral part of residential design. Designed for exercise and relaxation, pools can also enhance the design of a house.

Pools add much to the initial cost of a residence. Pools also require expensive and continual maintance. The addition of a pool and/or spa to a site design plan is there-fore more common in warmer climates where year-round use is possible.

Function

The ideal pool should provide for all functions: exercise, relaxation, and enhancement of the site decor. Although pools are primarily used during daytime hours, a lighted pool, such as the very dramatic one in Figure 8.18, expands the living area by making the pool area inviting and usable at night. Different colors of lights can be used to create specific effects, as shown in Figure 8.19.

Location and Orientation

Several factors affect the location of residential pools: the relationship with the house, sun exposure, and privacy. The pool should be located as close to the living area as

FIGURE 8.18 ■ Lighting enables pool area to be used at night. (*Greg Wilson, Photographer*)

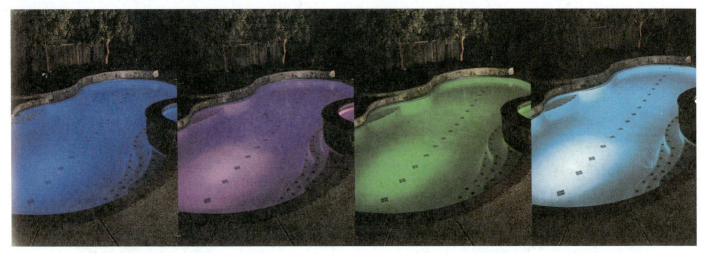

FIGURE 8.19 ■ Color lighting options used to create different effects. (*Fiberstars*)

possible, allowing pool deck and patio space between the house and the pool.

Most building codes require controlled access to the pool. This means you must enter the pool area from the house or through a fence gate.

The orientation of a pool should be considered in relation to the sun. A pool should be positioned to allow the option of full sun exposure or partial shade and an adjacent shade-escape area. Shade for the pool deck or connecting patio may come from the north side of a house. On wooded sites, the orientation can be designed so trees can supply the needed shade. See Figure 8.20. However, the trees should not block the sun from the pool during most of the day.

Enclosure is a major consideration in the design of pools for privacy and safety. The pool in Figure 8.20 is built into an area surrounded by the natural vegetation of the dense woods, providing privacy as well as shade.

Screened walls and overhead enclosures have the advantage of blocking bugs and debris, but they also reduce the amount of direct solar heat on the water. Enclosures should be planned during the floor plan and elevation design phases (Chapters 13 and 15) to ensure consistency with the lines of the house.

A pool located near a large natural body of water must be designed to avoid interfering with the natural water table. Pools in some areas, as shown in Figure 8.21, must be seawall secured to prevent collapse and to keep out wildlife, such as alligators.

Pool Construction

Materials

The frames of pools are constructed with concrete, wood, or steel. Pool surfaces that you see beneath the water may be covered with any of a variety of materials: vinyl sheets, marble composite, pebble aggregate, or paint.

Pool decks are constructed using non-heat-absorbing concrete mixtures, acrylic composites, or wood slats on elevated surfaces. The use of spray-on deck surfaces that are heat resistant, slip resistant, waterproof, and "mildew proof" are the most popular. Flagstone and pure concrete are not recommended for pool deck surfaces because these materials retain heat and become slippery when wet.

Pool Shapes

With the development of dry-mix concrete, pool walls of almost any shape, including free-form, can be created.

FIGURE 8.20 ■ Pool located in wooded area, which provides shade and privacy.

FIGURE 8.21 ■ Pool located on a lake front.

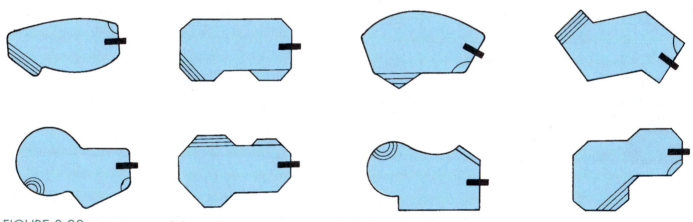

FIGURE 8.22 ■ Various pool shapes showing access steps, diving boards, and swimouts.

See Figure 8.22. More than one shape can fit a pool site, depending on the contour of the site. Patio shapes around or near the pool also need to be considered.

Pools can also be designed and built to create the illusion of a waterfall as viewed from the top. This is done through the use of a recirculating spillway at one end, as shown in Figure 8.23.

In addition to the surface shape of a pool, the depth must be considered. It must first be determined if the pool is to be all shallow (3′ to 4′ deep), all deep (6′ to 10′), or a combination. Combinations of 3′ to 4′ on one end dropping to depths of 6′ to 10′ on the opposite end are most popular. If a diving board is to be included, then the deep end must include a *diving well* at least 8′ deep

FIGURE 8.23 ■ Pool designed with recirculating waterfall. (*Endless Pools*)

extending out 10′ horizontally from the end of the board as shown in Figure 8.24.

Calculating Pool Sizes

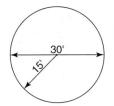

Math Connection

Residential pools range in size from about 200 sq. ft. to 800 sq. ft. or more. For example, a small 12′ × 18′ pool is 216 sq. ft. and a large 20′ × 40′ pool is 800 sq. ft. To calculate the area of a rectangular pool, use the following formula:

$$W \times L = A$$
(width × length = area)
Example: 14′ × 28′ = 392 sq. ft.

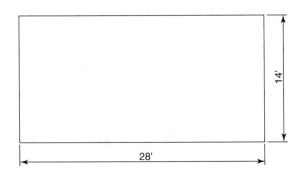

To calculate the area of a round pool, use the formula for determining the area of a circle:

$$\pi r^2 = A$$
(pi × radius squared = area)
Example: 3.14 × (15′ × 15′) = 706.5 sq. ft.

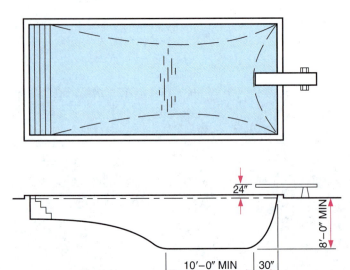

FIGURE 8.24 ■ Pool dimensions needed for diving board use.

To calculate the area of a pool, such as the one in Figure 8.25, that has a combination of circular and rectangular areas, divide the entire pool area into smaller round and rectangular segments. First calculate the area of each part, round to even square feet, and add them together. For example, to find the area of the pool shown in Figure 8.25:

(Rectangle) Area A	16	×	30	= 480
(Rectangle) Area B	6	×	10	= 60
(Half-circle) Area C	3.14	×	64 × 1/2 =	100
TOTAL SQ. FT.				640

For estimating cost and water capacity, cubic area is used to define the size of a pool. For example, an 18′ × 38′ pool that is 6′ deep contains 18′ × 38′ × 6′, or 4,104 cu. ft., of space. The same size pool 8′ deep contains 5,472 cu. ft. To find the volume of a container, use the following formula:

$$V = W \times L \times D$$
volume (cu. ft.) = width ×
length × depth
Example: 14′ × 28′ × 8′ = 3,136 cu. ft.

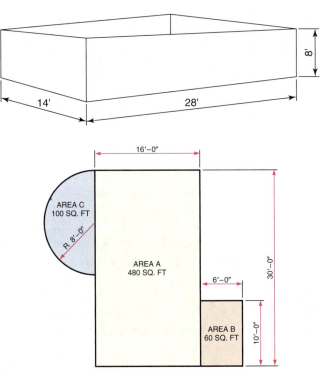

FIGURE 8.25 ■ Calculation of square footage for an irregularly shaped pool.

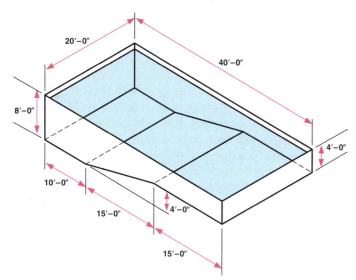

FIGURE 8.26 ■ Measurements needed to calculate pool volume.

FIGURE 8.27 ■ A house can provide part of a pool safety barrier. (*Signature Properties*)

Because pools usually slope from shallow to deep, most cubic foot calculations involve dividing the pool into segments that have the same depth, or using average depth, as shown in Figure 8.26.

To determine the volume of a pool with a combination of cylindrical and cubic shapes, divide the pool area into separate cylindrical and cubic segments according to identifiable shapes. Then calculate the volume of each and add them together. For example, to find the volume of the pool shown in Figure 8.25, use the appropriate formula for each segment area. Remember, you can multiply in any sequence. Assume the bottom of each area is flat. The depth of area A is 8′, B is 3′, and C is 6′.

(Rectangle)	Area A	8′ × 16′ × 30′	=	3,840
(Rectangle)	Area B	3′ × 6′ × 10′	=	180
(Half-circle)	Area C	6′ × 3.14 × 32′	=	603
	TOTAL CU. FT.			4,623

Safety Devices

More than 600 people drown in residential pools each year. Visibility from house to pool is important, but certain pool design features can help reduce this number dramatically. These include minimum 4′ fencing, self-latching gates, latched house doors and windows, strong pool covers, alarms, ladders, steps and/or swim-outs. A **swim-out** is an elevated platform below the water level that allows the swimmer to get out of the pool without using a ladder. In combination with fencing, a house can function as part of the barrier between the pool and site intruders as shown in Figure 8.27. If children live in the

house that has the pool, other safety measures must be taken.

Other safety devices and equipment beyond the basic design features include clip-on child alarms, rope and float line, filter basket cover, posted emergency information, outside telephone, and portable infant fences.

Pool Equipment

Pools are simply cavities in the ground filled with water. To make a pool function properly, water must be circulated, filtered, purified, and sometimes heated. All of these functions require operating equipment.

Pool water is circulated through a series of filters and purifiers to keep the water sufficiently pure and clean for swimming. A water pump pulls water from the pool through a series of pipes connected to a skimmer device and drain. The pump moves the water through the filter, purifier, and sometimes a heater. After the water passes through these devices the pure, clean, and heated water returns to the pool through pipes. These pipes are connected to *outlets* in the pool walls *under* the waterline. The number of outlets spaced throughout the pool determines the amount and balance of water circulation. Small pools may need only one outlet, while larger pools may need four or more.

Timing devices are recommended to control the amount of time the pump operates each day. Normally the pump is set to operate during the daylight hours because the pool equipment produces some noise. For this reason, it is better to locate the equipment away from lounging areas. The functioning of the plumbing system required to circulate, filter, and purify pool water is covered in Chapter 33.

Additional (luxury) features to consider when designing a pool include a diving board, whirlpool spas, screened enclosures, and decorative fountains. Diving boards require additional foundation thickness.

Spas can be designed into the total pool layout. See Figure 8.28. They can be included in the same pool circulation system. However, if extremely hot water is anticipated (hot tub), a bypass system needs to be used to avoid overheating the pool water.

FIGURE 8.28 ■ Spa integrated with pool design. (*IMHOPTEP—Alfred Karram*)

CHAPTER 8

Outdoor Living Areas Exercises

1. From catalogs, newspapers, and magazines, cut out pictures of porch furniture that you particularly like (that is, that you would choose for your own porch).

2. Plan a porch and/or patio for a house of your own design. Sketch the basic outline and the facilities.

3. Draw the outline of the patio shown in Figure 8.8.

4. Draw the outline of a lanai you would plan for a U-shaped home of your own design.

5. Draw the outline of the lanais you designed in Exercise 4.

6. Sketch a floor plan of your own home. Add a lanai to connect two of the areas, such as the sleeping and living areas.

7. Explain the purposes of a lanai and describe two different plans where lanais would function well.

8. Name the required operating equipment needed for residential pools.

9. List the factors to consider in designing a pool.

10. Design a pool deck and patio area for a home you are designing. Locate the position of all operating equipment.

11. Draw the outline of a free-form pool shape with deck and patio areas.

CHAPTER 9

Traffic Areas and Patterns

OBJECTIVES

In this chapter you will learn:

- to determine the effectiveness of a traffic pattern in a house.
- to plan hallways that function efficiently.
- guidelines for designing stairs.
- to calculate the correct space needed for stairways and stairwells.

- the kinds and functions of entrances.
- guidelines for entrance design.
- to design a foyer and entry.

TERMS

apron
dividers
foyer

landings
riser
stairs

traffic areas
tread

INTRODUCTION

The **traffic areas** of any building provide passage from one room or area to another and within a room or area. Planning the traffic areas of a residence is not extremely complex because relatively few people are involved. Nevertheless, efficient allocation of space is important. The main traffic areas of a residence include the halls, entrances or foyers, stairs, and areas of rooms that are part of a traffic pattern.

TRAFFIC PATTERNS

Traffic patterns of a residence should be carefully considered when designing room layout. A minimum amount of space should be devoted to traffic areas. Extremely long halls and corridors should be avoided. These are difficult to light and provide no living space. Traffic patterns that require passage through one room to get to another should also be avoided, especially in the sleeping area.

The traffic pattern shown in the plan in Figure 9.1 is efficient and functional. It contains a minimum amount of hall space without creating a boxed-in appearance. It also provides access between areas and from the entrance without passing through other areas. Compare this pattern with that of the poorly designed plan in Figure 9.2.

One method of determining the effectiveness of the traffic pattern of a house is to imagine yourself moving through the house by placing your pencil on the floor plan and tracing your route through the house as you perform a whole day's activities. Do the same for other members of the household. You will be able to see graphically where the heaviest traffic occurs and whether the traffic areas have been planned effectively.

ENTRANCES

Entrances are divided into several different types: the site entrance, the main building entrance, the service entrance, and special-purpose entrances. House entrances usually have an outside waiting area (porch, marquee, lanai), a separation (door), and an inside waiting area (**foyer**, entrance hall). See Figure 9.3.

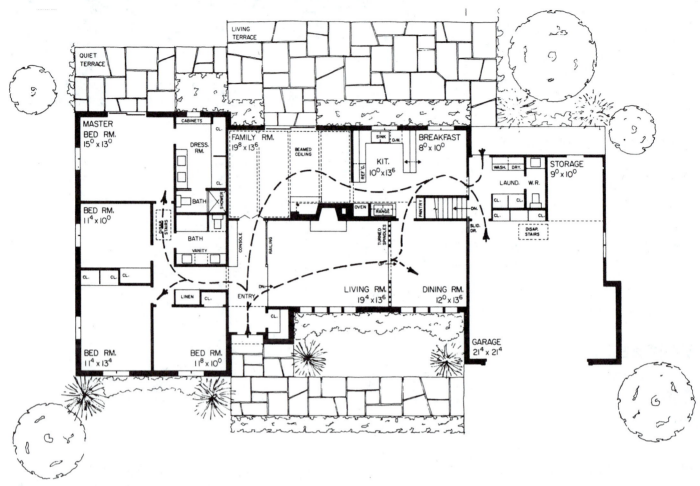

FIGURE 9.1 ■ Effective traffic pattern. (*Scholz Homes Inc.*)

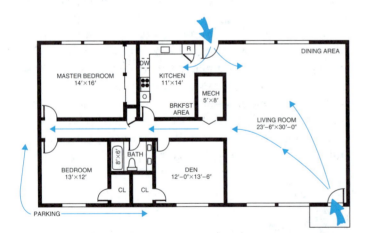

FIGURE 9.2 ■ Inefficient traffic pattern.

Function and Types

Entrances provide for and control the flow of traffic into and out of a building. Different types of entrances have different functions depending on the design of the structure.

Site Entrance

To design a site entrance, attention must first be given to the space from the street or road to the house. A site entrance includes the driveway, walkway, and adjacent parking or turnaround space for vehicles.

Driveways connect the street or road to a walkway and to a garage or carport and should be easily identified from the street. Driveways may be designed to lead directly to a garage, as in Figure 9.4. Some driveways are designed to intersect with the front entrance and connect with a detached, side, or rear garage, as in Figure 9.5. Driveways may be straight, curved, or circular. Figure 9.6 shows a circular driveway. Circular or semicircular driveways allow a car to return to the street without driving in reverse or turning around.

A turning and parking **apron** (area leading to garage) provides a means to exit a driveway without backing up onto the street. See Figure 9.7. To avoid double backing, the turning radii shown must be strictly followed.

Some driveway entrances need to be gated for security reasons. For example, if a pool is not separately

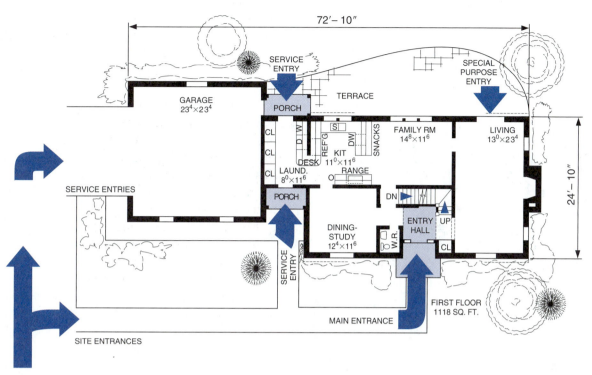

FIGURE 9.3 ■ Types of entrances. (*Home Planners Inc.*)

FIGURE 9.4 ■ Straight driveway used as a residential entrance.

FIGURE 9.5 ■ Curved driveway leading to detached garage. (*Alfred Karram Design*)

fenced, then a perimeter fence along the property borders is required to have a drive entrance gate. See Figure 9.8.

Walkways leading to a front entrance may either connect the house entrance directly with the street or sidewalk, or lead to the driveway, or both. See Figure 9.9.

Main House Entrance

The main entrance provides access to the house. It is the entrance through which guests are welcomed and from which all major traffic patterns radiate. The main entrance should be readily identifiable. It should provide shelter for anyone awaiting entrance.

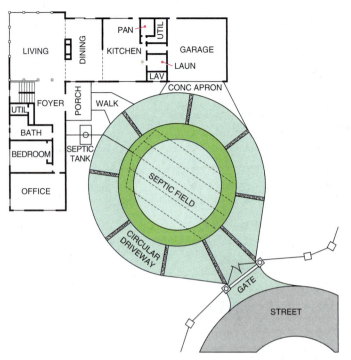

FIGURE 9.6 ■ Circular drive connected to entry and garage.

Some provision should be made in the main entrance wall to see callers from the inside before admitting them. Side panels, lights (panes) in the door or windows that face the side of the entrance are ways to view someone outside.

The main entrance should be planned to create a desirable first impression. A direct view of other areas of the house from the foyer should be baffled but not sealed off. The link between the main entrance and other interior areas is the foyer. Where space allows, a foyer should be designed to allow traffic to flow separately to the living, service, and sleeping areas. Figure 9.10, shows several foyer plans designed for this purpose.

The entrance foyer should include a closet for the storage of outdoor clothing. This foyer closet should be large enough for both family and guests to use.

A foyer arrangement must allow for the swing of the entrance door or doors. If the foyer is too shallow, passage will be blocked when the door is open, and only one person can enter at a time.

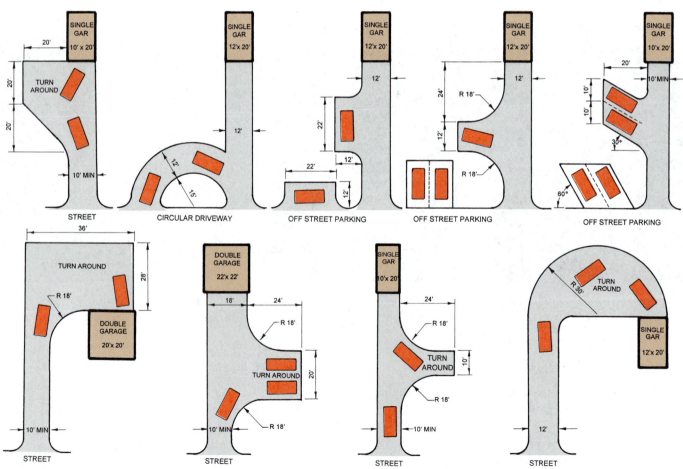

FIGURE 9.7 ■ Driveway dimensions for turning and parking.

FIGURE 9.8 ■ Gated drive. (*Glenn Wright Construction; J. Brian Acker, Photographer*)

FIGURE 9.9 ■ Overhang covered walkway connecting driveway and entrance.

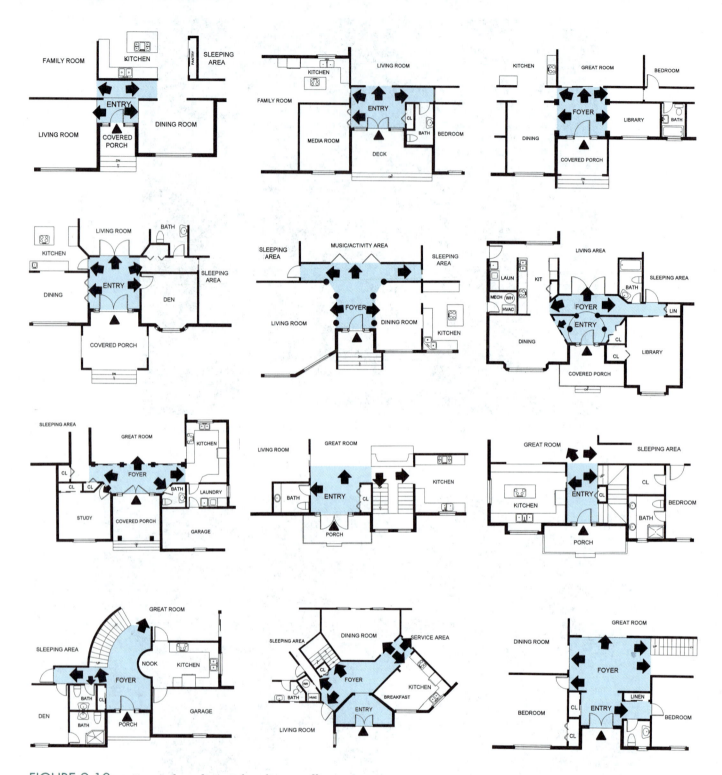

FIGURE 9.10 ■ Foyer plans designed to direct traffic.

Use this checklist for the design of a main entrance:

- Adequate space to handle traffic flow
- Access to all three areas of a home
- A guest closet
- Bathroom access for guests

- Consistent decor
- Outside weather protection
- Effective lighting day and night
- Avoid traffic through the living room center

Service Entrance

The service entrance is to be used for any entry or exit that would be inappropriate and inconvenient at the main entrance. A person should be able to pass through the service entrance and enter parts of the service area, such as the garage, mud room, laundry, or workshop. Supplies can also be delivered to the service areas without going through other parts of the house.

Special-Purpose Entrances

Special-purpose entrances and exits do not provide for outside traffic. Instead they are intended for movement from the inside living area of the house to the outside living areas. A sliding door from the living area to the patio is a special-purpose entrance. It is not an entrance through which street or sidewalk traffic would have access. Figure 9.11 shows the location of service entrances and special-purpose entrances. Figure 9.12 shows a porch lounge adjacent to a bedroom special entrance.

Location

The main entrance should be centrally located to provide easy access to each area. It should be conveniently accessible from driveways, sidewalks, or street.

The service entrance should be located close to the driveway and garage, and near the kitchen or food-storage areas.

Special-purpose entrances and exits are often located between the bedroom and the quiet patio, between the living room and the living patio, and between the dining room or kitchen and the dining patio.

Decor

To create a desirable first impression, a main entrance should be easily identifiable and yet be an integral part of the architectural style.

The total design of the entrance should be consistent with the overall design of the house. That means the design of the door, the side panel, and the deck and cover should be directly related to the lines of the house.

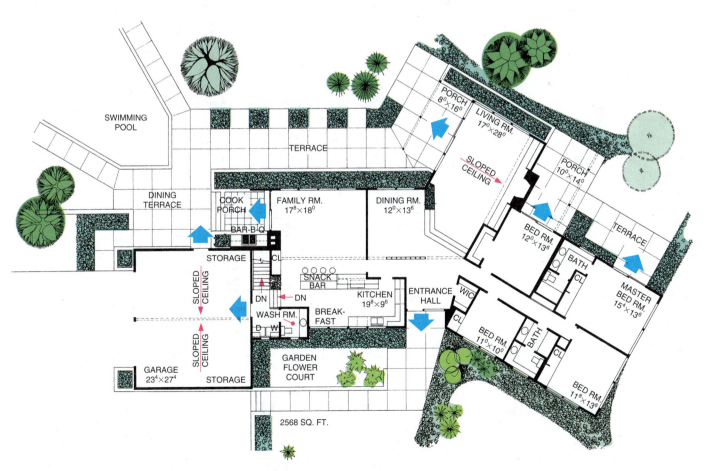

FIGURE 9.11 ■ Special-purpose and service entrances. (*Home Planners Inc.*)

FIGURE 9.12 ■ Porch adjacent to bedroom entrance. (*California Redwood Association*)

FIGURE 9.13 ■ Open plan foyer. (*John Henry, Architect*)

Open and Closed Planning

Open planning is desirable for entrances. This means the view from the main entrance to the living area should be baffled without creating a boxed-in appearance. The foyer should not appear to be a dead end. The extensive use of glass, effective lighting, and carefully placed baffle walls can create an open and inviting impression. See Figure 9.13.

Open planning between the entrance foyer and the living areas can also be accomplished by the use of louvered walls or planter walls. These provide a relief or change in the line of sight but not a complete separation. Lowering or elevating the foyer or entrance can also produce the desired effect of separation without enclosing the area. Foyers are not bounded in open planning. In formal or closed plans, the foyer is either partially or fully closed off.

Surface Materials

The outside portion of the entrance should be weather-resistant wood, stone, brick, or concrete. The foyer deck should be easily maintained and be resistant to mud, water, and dirt brought in from the outside. Asphalt, vinyl or rubber tile, stone, flagstone, marble, and terrazzo are most frequently used for the foyer deck. The use of a different material in the foyer area helps to define the area when no other separation exists.

Paneling, masonry, murals, and glass are used extensively for entrance foyer walls. The walls of the exterior portion of the entrance should be consistent with the other materials used on the exterior of the house.

Lighting

An entrance must be designed to function day and night. Natural lighting should be planned for lighting entrance areas during daylight hours. General lighting, spot lighting, and all-night lighting are effective after dark.

Lighting can be used to accent distinguishing features or to illuminate the pattern of a wall. This type of lighting actually provides more light by reflection and helps to identify and accentuate the entrance at night. See Figure 9.14.

Size and Shape

The size and shape of the areas inside and outside the entrance depend on the budget and the type of plan. The outside covered portion of the entrance should be large

FIGURE 9.14 ■ Illuminated entrance and facade. (*John Henry, Architect*)

enough to shelter several people and at the same time provide the amount of space needed to open a storm door. Outside shelter areas are the same range in size and shape as porches and patios. (Refer back to Chapter 8.)

The inside of the entrance foyer should be sufficiently large to allow several people to enter at the same time, remove their coats, and put their things in the closet. A 6′ × 6′ (1829 mm × 1829 mm) foyer is considered minimum for this function. A foyer 8′ × 10′ (2438 mm × 3048 mm) is average, but a more desirable size is 8′ × 15′ (2438 mm × 4572 mm).

HALLS

Halls are the highways and streets inside the home. They provide a controlled path that connects the various areas of the house. Halls should be planned to eliminate or minimize the passage of traffic through rooms. Long, dark, tunnel-like halls should be avoided. Halls should be well lighted, light in color and texture, and planned with the decor of the whole house in mind.

Where hall space is limited, adequate lighting is essential, as shown in Figure 9.15, where the hall also doubles as a gallery and library. Even in very large hallways, as seen in Figure 9.16, an abundance of well-defused light is necessary.

FIGURE 9.15 ■ Hallway doubling as a library and gallery.

FIGURE 9.16 ■ Large dramatically lighted hallway and foyer. (*Marc Michaels, Design; Sargent Architectural, Photography*)

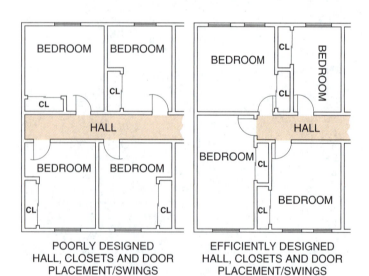

POORLY DESIGNED
HALL, CLOSETS AND DOOR
PLACEMENT/SWINGS

EFFICIENTLY DESIGNED
HALL, CLOSETS AND DOOR
PLACEMENT/SWINGS

FIGURE 9.17 ■ Methods of minimizing hall length.

Minimum hall widths are determined by building codes. Halls must also be wide enough for furniture movement and for wheelchair access.

One method of channeling hall traffic without the use of solid walls is with the use of **dividers**. Planters, half walls, louvered walls, and even furniture can be used as dividers. The plans in Figure 9.17 illustrate some of the basic principles of efficient hall design.

STAIRS

Stairs are inclined hallways that provide access from one level to another. Stairs may lead without a change of direction, or they may turn 90° or 180° by means of **landings**. There are several types of stairs. See Figure 9.18. Figure 9.19 shows an example of a classical curved stair system with an intermediate landing and ornamental iron railing.

Materials and Lighting

With the use of newer, stronger building materials and new techniques, stairs can now be supported by many different devices. Stairs no longer need to be enclosed in areas that restrict light and ventilation.

Stairwells (areas for stairs) should be lighted at all times when in use. Natural light is the most energy efficient. Thus windows should be utilized to provide natural light for stairs wherever possible. Three-way switches should be provided at the top and bottom of the stairwell to control the stair lighting. (For details, see Chapter 31.)

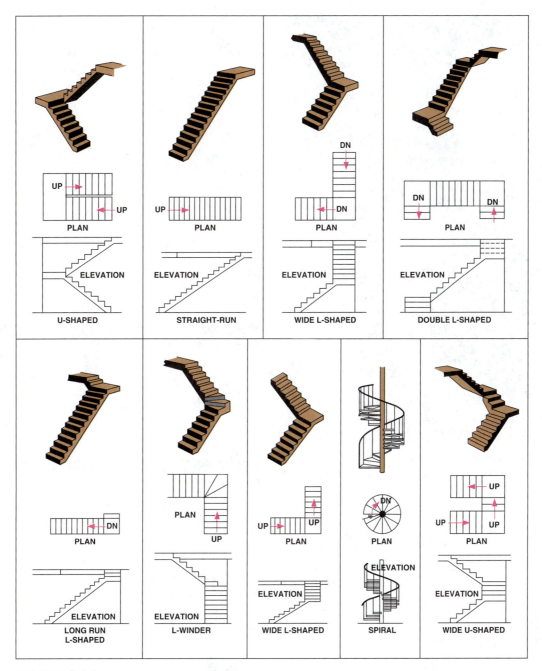

FIGURE 9.18 ■ Common types of stair systems.

Size and Shape

There are many variables to consider in designing stairs. The **tread** is the horizontal part of the stair, the "step," or the part on which you walk. Treads must be made of or covered with nonslip surfaces. The average depth, or distance from front to back, of the tread is 10″ (254 mm). The **riser** is the vertical part of the stair. The average riser height is 7 1/4″ (184 mm). See Figure 9.20.

The overall width of the stairs is the distance between the stair railings. A minimum of 3′ (914 mm) should be allowed for the total stair width. However, a width of 3′-6″ (1067 mm) or even 4′ (1219 mm) is preferred to accommodate the movement of furniture. See Figure 9.21.

Headroom is the vertical distance between the top of each tread and the top of the stairwell ceiling. A minimum headroom distance of 6′-6″ (1981 mm) should be allowed. However, distances of 7′ (2134 mm) are more desirable.

The tread width, the riser width, the width of the stairwell opening, and the headroom all help to determine the total length of the stairwell. Landing dimensions are

FIGURE 9.19 ■ Classical curved stair system. (*Marc Michaels, Design; Sargent Architectural, Photography*)

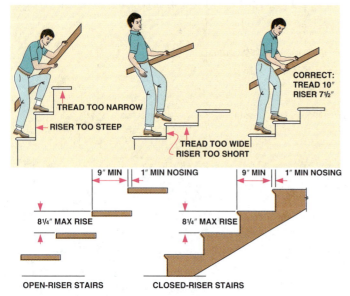

TREAD TOO NARROW

RISER TOO STEEP

TREAD TOO WIDE
RISER TOO SHORT

CORRECT:
TREAD 10"
RISER 7½"

9" MIN | 1" MIN NOSING | 9" MIN | 1" MIN NOSING

8¼" MAX RISE | 8¼" MAX RISE

OPEN-RISER STAIRS | CLOSED-RISER STAIRS

FIGURE 9.20 ■ The importance of correct tread and riser dimensions.

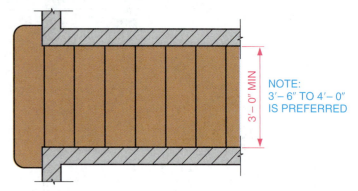

3'–0" MIN

NOTE:
3'–6" TO 4'–0"
IS PREFERRED

FIGURE 9.21 ■ Minimum stair width.

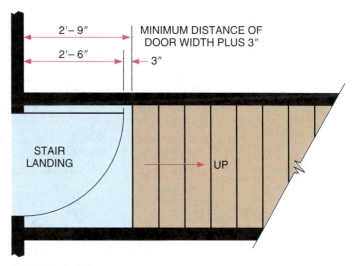

2'–9"

2'–6"

MINIMUM DISTANCE OF
DOOR WIDTH PLUS 3"

3"

STAIR
LANDING

UP

FIGURE 9.22 ■ Minimum landing dimensions.

generally determined by the size of the stairs and the space for the stairwell. More clearance must be allowed where a door opens onto a landing. See Figure 9.22. A landing should be planned for stair systems that have more than 16 risers. It should be located at the center between levels to eliminate long runs.

ELEVATORS

The use of elevators in homes and light construction buildings is increasing significantly. A minimum of 3' × 4' of space on a floor plan is required. Elevators are designed into a floor plan primarily for use by people with physical disabilities. Smaller units (dumbwaiters) can be included for the vertical movement of household goods, food, fireplace wood, etc.

CHAPTER

9

Traffic Areas and Patterns Exercises

1. Sketch the floor plan of a home of your design. Plan the most efficient traffic pattern by tracing the route of your daily routine.

2. Draw the plan view of one of the stair systems shown in this chapter.

3. Draw or sketch a plan view of a stair system in your home or school.

4. Name the types of stairs that turn 90°, 180°, 360°, and 0°.

5. List the types of entrances and tell the function of each type.

6. Redesign an entrance shown in this chapter. Add space that is consistent with the main lines of the house.

7. Redesign and enlarge the foyer for the living area shown in Figure 9.3. Label the materials you select for the outside deck, overhang, access walk, foyer floor, and walls.

 8. Add a foyer to the plan of a house you are designing.

9. Draw a plan for the foyer you redesigned in Exercise 7.

10. Redesign the floor plan shown in Figure 9.23 to include foyer with access to all areas.

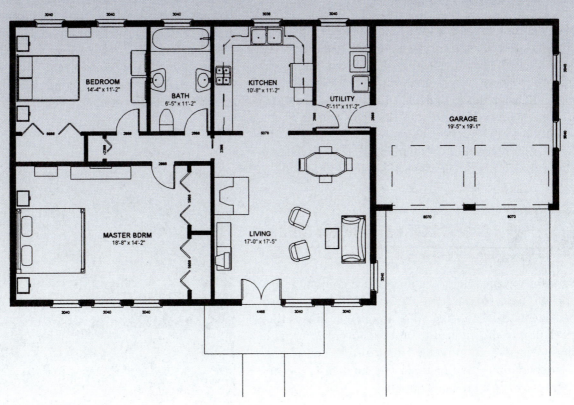

FIGURE 9.23 ■ Floor plan to be redesigned.

Kitchens

OBJECTIVES

In this chapter you will learn to:

- apply guidelines to efficient kitchen design.
- determine the best shape, size, and location for the kitchen.
- plan a work triangle for a kitchen.
- design an aesthetically consistent decor for a kitchen.
- sketch small and large kitchens of the basic kitchen shapes.

TERMS

corridor kitchen
family kitchen
island kitchen

L-shaped kitchen
one-wall kitchen
peninsula kitchen

U-shaped kitchen
work triangle

INTRODUCTION

A well-planned kitchen is one that functions efficiently, and yet is attractive and easy to maintain. To design an efficient kitchen, the designer must consider the room's function, location, decor, size, and shape. However, because a kitchen requires so much equipment, the design of a kitchen entails additional considerations and decisions. In addition to the kitchen design factors covered in this chapter, the integration of plumbing, electrical, heating, and ventilation requirements must also be considered.

KITCHEN DESIGN CONSIDERATIONS

Kitchen design involves planning space configurations, work surfaces, storage requirements, and the number, type, size, and location of all components. In addition, the relation of the kitchen to other rooms or areas, light sources, vistas, and traffic flow must also be considered.

Understanding the functions of a kitchen is the first step in planning a kitchen's design.

Functions

Food preparation is, of course, the primary function of the kitchen. However, the kitchen may also be used as a dining area. The proper placement of appliances is important in a well-planned kitchen. Locating appliances in an efficient pattern eliminates wasted motion. An efficient kitchen has three basic areas or centers: the storage center, the cooking center, and the cleanup center. A fourth area, preparation, is combined into one or more of the others, usually storage. See Figure 10.1.

FIGURE 10.1 ■ Kitchen functional areas: cooking, preparation, cleanup, storage. (*Zeyko*)

Storage and Preparation Center

The refrigerator is the major appliance in the storage and preparation center. The refrigerator may be freestanding, built-in, or even suspended from a wall. Cabinets for the storage of utensils and food ingredients, as well as a countertop work area for preparing food, are also included at this center.

Cooking Center

The major appliances in the cooking center are the range (cooktop), microwave oven, and oven. The range and oven may be combined into one appliance or be separated into two appliances, with the burners installed in the countertop (cooktop) as one appliance and an oven built into a cabinet. The cooking center should have countertop work space, as well as storage space for minor appliances and cooking utensils. An adequate supply of electrical or gas outlets for using appliances is necessary.

Cleanup Center

At the cleanup center, the sink is the major appliance. Sinks are available in one-, two-, or three-bowl models with a variety of cabinet arrangements, countertops, and drainboard areas. The cleanup center may also include a waste disposal unit, an automatic dishwasher, a waste compactor, and cabinets for storing cleaning supplies.

The Work Triangle

If you draw a line connecting the three centers of the kitchen, a triangle is formed. See Figure 10.2. This is called the **work triangle**. The perimeter of an efficient kitchen work triangle should be no more than 22′ (6706 mm) or less than 12′ (3658 mm). Although the size of the work triangle is an indication of kitchen efficiency, the triangle is primarily useful as a starting point in kitchen design.

The arrangements of the three areas of the work triangle may vary greatly. However, efficient arrangements can be designed in each of the seven basic types of kitchens.

Types of Kitchens

U-shaped Kitchen

The **U-shaped kitchen** is very efficient and popular. The sink is located at the bottom of the U, and the range and the refrigerator are at the opposite ends. In this arrangement, traffic passing through the kitchen is completely separated from the work triangle. The open space in the U between the sides should be 4′ (1219 mm) or 5′ (1524 mm). This arrangement produces a very efficient small kitchen. See Figure 10.3. Figure 10.4 shows various U-shaped kitchen designs and the planned work triangles.

When designing U-shaped kitchens, special attention must be given to door hinges and drawer positions. Design cabinet doors and drawers to open without interfering with each other, especially at cabinet corners.

Peninsula Kitchen

The **peninsula kitchen** is similar to the U-shaped kitchen, but one end of the U is not adjacent to a wall. It projects into the room like a piece of land (peninsula)

FIGURE 10.2 ■ The kitchen work triangle. (*Norcraft Companies, Inc.*)

FIGURE 10.3 ■ U-shaped kitchen. (*Frigidare Corp.*)

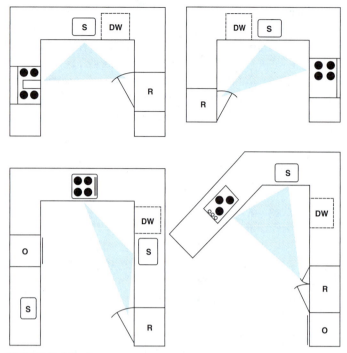

FIGURE 10.4 ■ U-shaped kitchen arrangements.

FIGURE 10.5 ■ Peninsula kitchen.

into a body of water. This peninsula is often used for the cooking center. However, it may serve several other functions as well. The peninsula is often used for an eating area as well as for food preparation. See Figure 10.5. It may join the kitchen to the dining room or family room. Figure 10.6 shows various arrangements of peninsula kitchens and the resulting work triangles.

Most peninsula kitchens contain large countertops for work space. Peninsulas may contain only lower or base cabinets, but some may include upper cabinets suspended from ceilings.

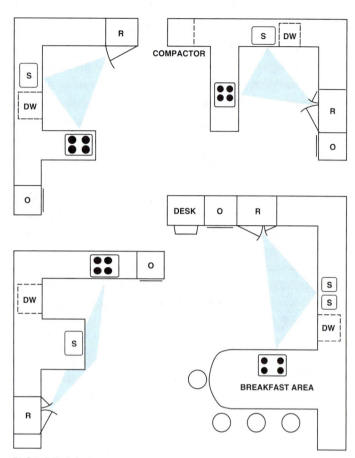

FIGURE 10.6 ■ Peninsula kitchen arrangements.

L-shaped Kitchen

The **L-shaped kitchen** has continuous counters, appliances, and equipment located on two adjoining, perpendicular walls. Two work centers are usually located on one wall and the third center is on the other wall. See Figure 10.7. The work triangle is not in the traffic pattern. If the walls of an L-shaped kitchen are too long, the compact efficiency of the kitchen is destroyed.

An L-shaped kitchen requires less space than the U-shaped kitchen. The remaining open space often created by an L-shaped arrangement can serve as an eating area adjacent to a family room, without taking space from the work areas. If the center area is used for eating, a minimum of 36″ (914 mm) must be allowed as an aisle between cabinets and chairs.

Corridor Kitchen

Two-wall **corridor kitchens** are very efficient arrangements for long, narrow rooms. See Figure 10.8. They are very popular for small apartments, but are used extensively anywhere space is limited. A corridor kitchen pro-

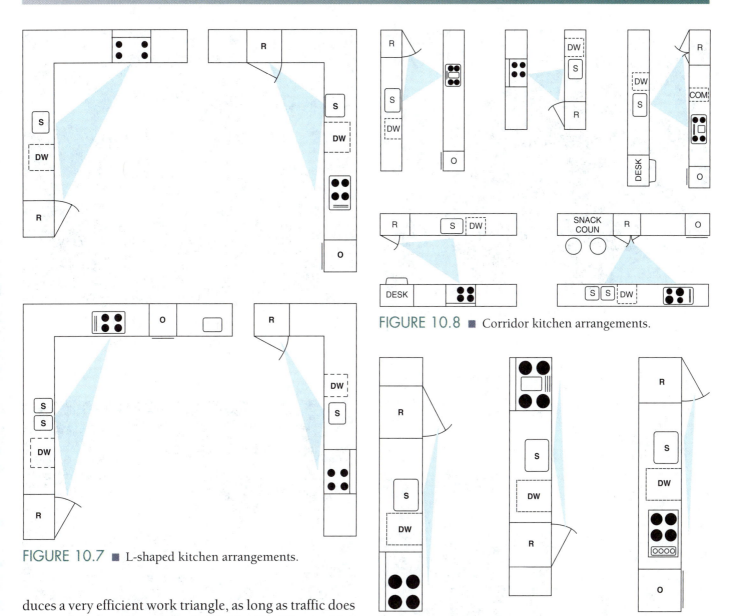

FIGURE 10.7 ■ L-shaped kitchen arrangements.

FIGURE 10.8 ■ Corridor kitchen arrangements.

FIGURE 10.9 ■ One-wall kitchen arrangements.

duces a very efficient work triangle, as long as traffic does not need to pass through the work triangle. The corridor space between cabinets (not walls) should be no smaller than 4′ (1219 mm). One of the best work arrangements locates the refrigerator and sink on one wall and the range on the opposite wall.

One-wall Kitchen

A **one-wall kitchen** is an excellent plan for small apartments, cabins, or houses in which little space is available. The work centers are located along one line rather than in a triangular shape, but this design still produces an efficient arrangement. See Figure 10.9.

When planning a one-wall kitchen, the designer must be careful to avoid creating walls that are too long. Adequate storage facilities need to be well planned also, because space is often limited in a one-wall kitchen.

Island Kitchen

The **island kitchen**, another geographically named arrangement, has a separate, freestanding structure that is usually located in the central part of the kitchen. An island in the kitchen is accessible on all sides. It usually has a rangetop or sink, or both. Other facilities are sometimes located in the island, such as a mixing center, work table, serving counter, extra sink, and/or snack center.

The kitchen shown in Figure 10.10 contains an island sink, work counter, and a lower level eating area. The island in Figure 10.11 includes a range, "prep" sink, and eating area. The kitchen shown in Figure 10.12A includes two islands. One island contains a sink and large work counter; the other contains extra burners and a

FIGURE 10.10 ■ Island including sink, work counter, and eating area. (*Allmilmo Corp.*)

FIGURE 10.12A ■ Kitchen with two islands. (*McGuire Real Estate; Russell MacMasters, Photographer*)

FIGURE 10.11 ■ Island with range, sink, and eating area. (*New England Cabinet Co.*)

FIGURE 10.12B ■ Ninety-degree view of kitchen in Figure 10.12A. (*McGuire Real Estate; Russell MacMasters, Photographer*)

large work counter that doubles as a serving counter. You can visualize this kitchen better by observing the 90° view in Figure 10.12B.

Figure 10.13 shows examples of other island facilities. The island design is especially convenient when two or more persons work in the kitchen at the same time.

When an island contains a range or grill, allow at least 16″ (406 mm) on the sides for utensil space. Also consider the use of a downdraft exhaust system that pulls vapors down and out rather than up to eliminate the need for overhead hooded vents. Allow at least 42″ (1067 mm) on all sides of an island. If used for eating, also add the depth of the chair or stool.

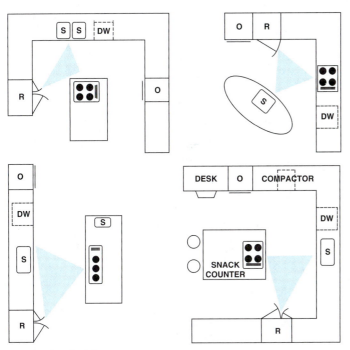

FIGURE 10.13 ■ Island kitchen arrangements.

FIGURE 10.14 ■ Family kitchen. (*Merrillat*)

Family Kitchen

The **family kitchen** is an open kitchen using any kitchen shape. The function of an open kitchen, however, is to provide a meeting place for the entire family—in addition to the usual kitchen services. A family kitchen often appears to have two parts in one room. The three food preparation work centers comprise one section. The dining area and family room facilities comprise another section. See Figure 10.14. Figure 10.15 shows several typical arrangements for family kitchens. Open plan family room kitchens can also be adapted to serve as part of an open plan great room design. In this type of design, a complete visual opening between the living and dining facilities is planned. Kitchens of this type are designed in many configurations, as shown in Figure 10.16.

Regardless of its shape, the kitchen is the core of the service area and should be located near the service entrance as well as near the waste disposal area. The kitchen must be adjacent to eating areas, both indoors and outdoors. The children's play area should also be visible or easily accessible from the kitchen.

Family kitchens must be rather large to accommodate the necessary facilities. An average size for a family kitchen is 225 sq. ft. (20 sq. m). Eating areas can be designed with either tables and chairs or with chairs and/or

stools at a counter. When counters are used for eating, allow at least 12″ (305 mm) for knee space between the end of the counter and the face of the base cabinet.

Decor

Kitchens cost more per square foot than any other room. Most of this cost relates to the selection of appliances, cabinetry, and fixtures. By selecting the least expensive models of appliances, hardware, and cabinetry, the same kitchen design can often be built for one-fourth the cost of a kitchen that contains the most expensive features.

Even though most kitchen appliances are produced in contemporary designs, some clients and designers prefer to decorate kitchens with a traditional style as a motif or theme. The cabinets, floors, walls, and accessory furniture would then be selected according to that chosen theme. Designing a totally harmonious kitchen is made easier by the wide variety of appliance sizes, colors, and styles. Compare the bold and simple lines of the contemporary styled kitchen shown in Figure 10.17, page 154, with the elements of country style found in the kitchen in Figure 10.18, page 154. All aspects of each design are consistent with the overall decor.

Regardless of the style, the kitchen walls, floors, countertops, and cabinets should require a minimum amount of maintenance. Materials that are relatively maintenance free include stainless steel, stain-resistant plastic, ceramic tile, washable wall coverings, washable paint, vinyl, molded and laminated plastic countertops, doors, drawers, and cabinet bases.

ISLAND FAMILY KITCHEN

ONE-WALL FAMILY KITCHEN

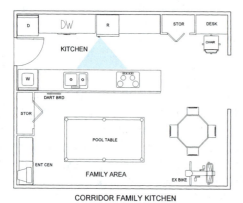

CORRIDOR FAMILY KITCHEN

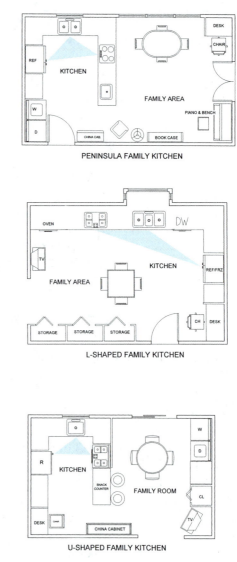

PENINSULA FAMILY KITCHEN

L-SHAPED FAMILY KITCHEN

U-SHAPED FAMILY KITCHEN

FIGURE 10.15 ■ Family kitchen arrangements.

Options in kitchen design have broadened because of new synthetic and composite materials and new construction methods for cabinets and countertops.

Size and Shape

Average human dimensions as described in Chapter 6 are a key factor in selecting the size and shape of kitchen components. Reaching distances are most important. Figure 10.19 shows typical reaching considerations from a standing position. Many component heights must be altered to accommodate wheelchair occupants as shown in Figure 10.20. Fortunately, components are manufactured in a wide range of sizes to accommodate almost any design configuration. Figure 10.21 includes standard horizontal (width) dimensions used in kitchen design. Figure 10.22 shows the dimensions of standard wall and base cabinets. These are the most commonly used sizes. Figure 10.23, page 156, shows the total range of cabinet sizes, and Figure 10.24, page 157, lists the range of sizes of major kitchen appliances.

Kitchen Drawings

Drawing kitchen plans is covered in Chapter 14, Drawing Floor Plans. Drawing kitchen wall elevations is covered in Chapter 16, Drawing Elevations.

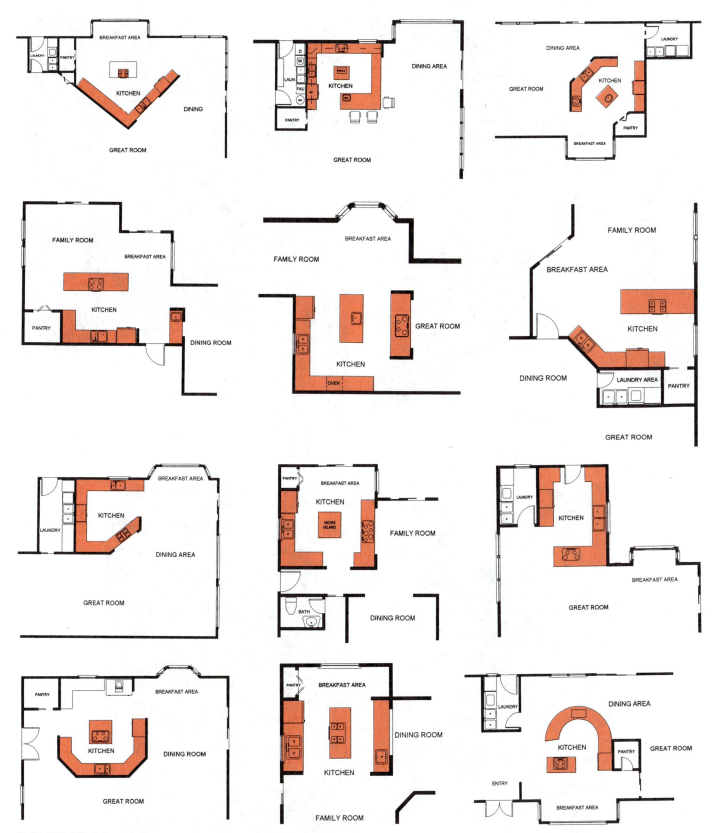

FIGURE 10.16 ■ Open plans for a great room kitchen.

FIGURE 10.17 ■ Contemporary styled kitchen. (*Mannington Resilient Floors*)

FIGURE 10.18 ■ Colonial styled kitchen. (*Armstrong World Industries*)

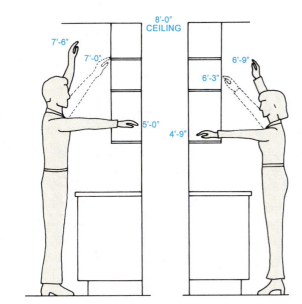

FIGURE 10.19 ■ Typical kitchen reaching heights.

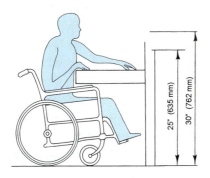

Counter top heights and thigh clearance.

FIGURE 10.20 ■ Wheelchair reaching heights for kitchens.

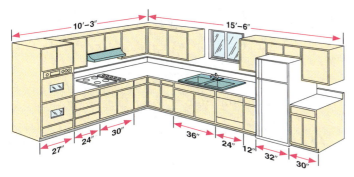

FIGURE 10.21 ■ Standard kitchen horizontal dimensions. (*William Wagoner*)

KITCHEN PLANNING GUIDELINES

Remember the following guidelines for designing efficient kitchens.

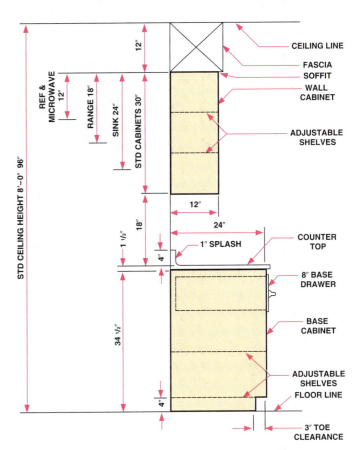

FIGURE 10.22 ■ Standard wall and base cabinet dimensions.

Function

1. The food preparation centers include the storage and mixing center, the cooking center, and the cleanup center.
2. Each work area includes all necessary appliances and facilities.
3. Adequate storage facilities need to be provided throughout the kitchen.

Location

4. Traffic lanes are clear of the work triangle.
5. The kitchen is located adjacent to the dining area.
6. The kitchen should be located near the children's play area.

Size and Space

7. The work triangle measures no more than 22′ (6706 mm) or less than 12′ (3658 mm).
8. Lapboard heights are 26″ (660 mm).
9. Working heights for counters are 36″ (914 mm).
10. Working heights for tables are 30″ (762 mm).

BASE CABINET SIZES (IN INCHES)		
HEIGHT	DEPTH	WIDTH
34 1/2	24	9
		12
		15
		18
		21
		24
		27
		30
		33
		36
		39
		42
		46
		48

WALL CABINET SIZES (IN INCHES)		
HEIGHT	DEPTH	WIDTH
12	12	36 39 42
15	12	30 33 36 39 42
18	12	24 27 30 33 36 39 42
24	12	15 18 21 24 27 30 33 36 39 42
30	12	12 15 18 21 24 27 30 33 36 39 42

FIGURE 10.23 ■ Range of cabinet sizes.

11. Adequate counter space is provided for meal preparation.
12. Allow at least 12″ (305 mm) for knee space if counters are used as eating areas.
13. If space allows, include a pantry to store food staple quantities.
14. Allow at least 4′ (1219 mm) for aisle space between cabinets or appliances.
15. Allow at least 15″ (381 mm) on each side of an island cooktop for utensil storage.
16. Keep shelves within a reachable height (maximum height 84″, or 2134 mm).
17. Counter space is provided next to each appliance.

Utilities

18. An adequate number of electrical outlets are provided for each work center.
19. Shadowless and glareless light is provided and is concentrated on each work center.
20. Plumbing lines are planned for sink(s), icemaker, and any purified water system.
21. Provide adequate ventilation through overhead hoods, downdraft cooking-fume exhausts, circulating ceiling fans, and/or adequate heating/air-conditioning systems.

Appliances

22. The oven and range are separated from the refrigerator by at least one cabinet.
23. Cabinet and appliance locations are planned according to manufacturer's recommendations. Figures 10.23 and 10.24 show standard dimensions.
24. Allow for door swing on all appliances.
25. The direction of door openings on appliances and cabinets should allow easy access from the work triangle. See Figure 10.25.
26. The refrigerator door should open toward the food preparation work space.
27. If a microwave oven is included, add a heat-resistant countertop between the cooktop and microwave.
28. Position the dishwasher next to the sink for easy loading.
29. The base cabinets, wall cabinets, and appliances create a consistent standard unit without gaps or awkward depressions or extensions.

Design considerations for persons with physical impairments are provided in Chapter 13, Designing Floor Plans.

REFRIGERATORS
HT 56" TO 66"
DEPTH 24"
WIDTH 24" TO 32"

REF/FREEZERS
HT 56" TO 66"
DEPTH 24" 'TO 28"
WIDTH 30" TO 42"

SINGLE SINKS
DEPTH 20" TO 22"
WIDTH 24" TO 30"

DOUBLE SINKS
DEPTH 20" TO 22"
WIDTH 32" TO 42"

TRIPLE SINKS
DEPTH 20" TO 22"
WIDTH 42" TO 55"

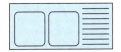

DBL SINKS/DRAIN BRDS
DEPTH 20" TO 22"
WIDTH 50" TO 60"

TYPICAL CORNER SINKS
20"
14"
14"
20"

COMPACTORS
TYP. 12" × 21"

SINK DISPOSAL UNITS
TYP. HT 14", 8" DIAM

FREESTANDING RANGES
HT 36"
DEPTH 24" TO 27"
WIDTH 20" TO 40"

DROP-IN RANGES
TYP. 22" × 30"

COOKTOPS/GRILLS
TYP. 21" × 18"

4-BURNER COOKTOPS
GAS/ELEC
TYP. 21" × 26"

4-BURNER COOKTOPS/GRILLS
TYP. 21" × 30"

6-BURNER COOKTOPS
TYP. 21" × 36"

SINGLE OVENS
HT 30" (DBL OVENS HT 50" TO 70")
DEPTH 24"
WIDTH 27" TO 30"

MICROWAVE OVENS
HT 18" TO 20"
DEPTH 14" TO 20"
WIDTH 20" TO 30"

RANGE HOODS
HT VARIES
DEPTH 17" TO 24"
WIDTH 30" TO 72"

WASHERS/DRYERS
DEPTH 24" TO 27"
WIDTH 24" TO 29"

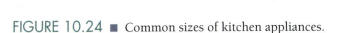

FIGURE 10.24 ■ Common sizes of kitchen appliances.

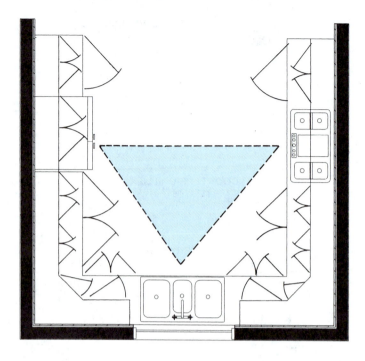

FIGURE 10.25 ■ Cabinet doors should open away from work locations.

10

Kitchens Exercises

1. List the seven types of kitchen shapes and give at least one advantage and one disadvantage of each.

2. Sketch a floor plan of one of the U-shaped kitchens shown in Fig. 10.4. Show the position of the dining area in relation to this kitchen, using the scale 1/2″ = 1′-0″.

3. Sketch a family kitchen using any of the seven kitchen types in your design.

4. Sketch a floor plan of the kitchen in your own home. Prepare a revised sketch to show how you would propose to redesign this kitchen. Try to make the work triangle more efficient.

 5. Sketch a floor plan of a kitchen you would include in a house of your own design, using the scale 1/2″ = 1′−0″.

6. Draw one plan of any of the kitchen shapes shown in Figure 10.16.

7. Calculate the space needed and plan a kitchen with a work triangle in a 14′ × 16′ peninsula kitchen and an L-shaped kitchen, 8′ × 14′.

8. Collect pictures of kitchens shown in magazines. Identify the kitchen type and find the work triangle in each. List good points and bad points of each kitchen design.

9. List the cabinets and appliances shown in Figure 10.26. Estimate the size of each.

FIGURE 10.26 ▪ List all cabinets and appliances in this kitchen. Estimate the size of each under three centers: storage and preparation, cooking, and cleanup. *(DAL-Tile)*

11 CHAPTER
General Service Areas

OBJECTIVES

In this chapter you will learn to:

- determine what kinds of equipment are included in a utility room.
- evaluate the best location for a utility room.
- sketch a garage and a carport.
- design storage facilities for a garage.
- calculate the area needed for garages and driveways.
- design and sketch an efficient and safe workshop area.
- design and sketch storage facilities.

TERMS

carport
detached garage
dropleaf workbench
hand tools

integral garage
peninsula workbench
power tools
utility room

ventilated shelving
walk-in closet
wall closet
wardrobe closet

INTRODUCTION

General service areas may include utility rooms, garages and carports, workshops, pet facilities, and storage areas. Because a great number of different activities are related to these areas, they should be designed for great efficiency. Service areas should include facilities for the maintenance and servicing of the other areas of the home.

UTILITY ROOMS

The **utility room** may include facilities for washing, drying, ironing, sewing, and storing household cleaning equipment. It may contain heating and air-conditioning equipment and/or even pantry shelves for storing groceries. Other names for this room are *service room* or *all-purpose room*.

Function

The major function of most utility rooms is to serve as a laundry area. They may also accommodate water heating, water purification, heating, and air-conditioning equip-

ment. Figure 11.1 shows utility rooms designed to function as a laundry, storage, or mechanical equipment room.

Laundry

To make laundry work as easy as possible, the appliances and working spaces in a laundry area should be located in the order in which they will be used. Such an arrangement will save time and effort during the steps in the process of laundering: receiving and preparing, washing, drying, ironing, storage, and, often, sewing.

The first step in laundering—receiving and preparing the items—requires hampers or bins, as well as counters on which to collect and sort the articles. Storage facilities should be located nearby for laundry products such as detergents, bleaches, and stain removers.

The next step is the actual washing. It takes place in the area containing the washing machine (washer) and laundry tubs, trays, or sinks.

The equipment needed for the third step includes a dryer, indoor drying lines, and space to store clothespins. Dryers require either a 220-volt (220V) outlet or an access to gas.

For the last step of the process, the required equipment consists of a counter for folding, an iron and ironing board, and a rack on which to hang finished ironing.

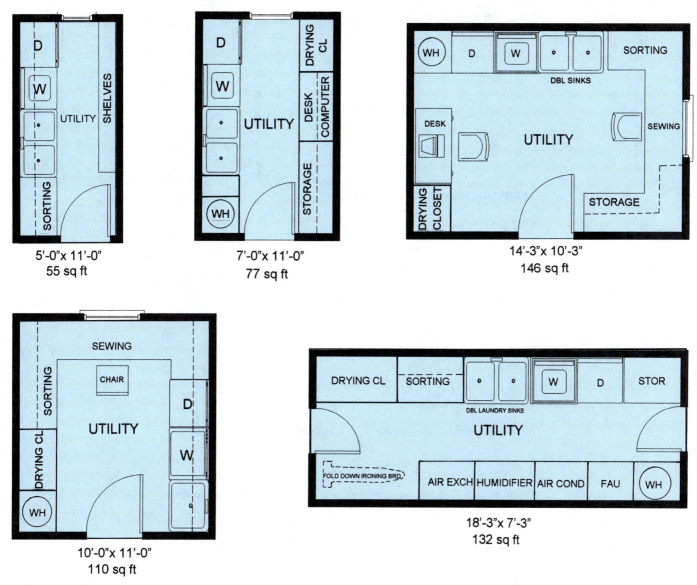

FIGURE 11.1 ■ Utility rooms designed for multiple functions.

Facilities for sewing and mending are often included. A sewing machine may be portable or it may fold into a counter or wall.

Heating and Air Conditioning

If the utility room is used for heating and air conditioning, additional space must be planned for the furnace, heating and air-conditioning ducts, water heater, and any related equipment such as humidifiers or air purifiers.

Location

A separate utility room is desirable because all laundry functions and maintenance equipment can be centered in one place. Space is not always available for a separate utility room, however. Laundry facilities may need to be located in some other area. Plans for the location of laundry facilities in the garage, closet, service, or sleeping areas are shown in Figure 11.2.

Placing the laundry appliances in or near the kitchen puts them in a central location and near a service entrance. Plumbing facilities are nearby, and some kitchen counters may be used for folding. However, these advantages may be offset by noise from the machines and odors from detergents, bleaches, and softeners near the kitchen. It is also desirable to keep laundering away from the preparation of food.

Other locations to consider for laundry appliances are less used areas such as the service porch, basement,

FIGURE 11.2 ■ Laundry room locations.

garage, or carport. A closet or a family room may also work well for laundry facilities. Because clothing is changed in the sleeping area, laundry facilities in that area may be convenient.

Style and Decor

Style and decor in a utility room depend on the function of the appliances, which are themselves an important fac-

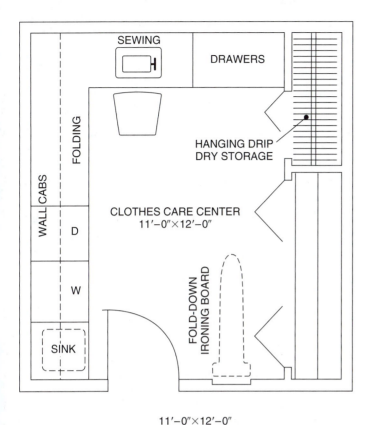

FIGURE 11.3 ▪ Laundry including clothing maintenance functions.

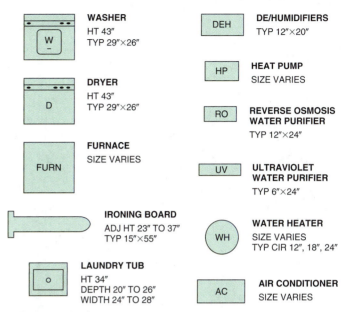

FIGURE 11.4 ▪ Utility room equipment sizes.

tor in the appearance of the room. Simplicity, straight lines, and continuous counter spaces produce an orderly effect and permit work to progress easily. Such features also make the room easy to clean.

An important part of the decor is the color of the paint used for walls and cabinet finishes. Colors should harmonize with the colors used on the appliances. All finishes should be washable. The walls may be lined with sound-absorbing material.

The lighting in a utility room should be 48″ (1219 mm) above any equipment used for washing, ironing, and sewing. However, lighting fixtures above work areas and laundry sinks can be farther from the worktop area. Lighting should be located to avoid shadows on work surfaces caused by the body of the person using the area.

Size and Shape

When space is available, all phases of clothing maintenance can be located in the laundry area. See Figure 11.3. When space is limited, portable work center units can be used. For storage, the same sizes of cabinets used in kitchens and bathrooms can also work well in laundry areas.

Depending on what equipment is included, the shapes and sizes of utility rooms differ. See Figure 11.4. The average floor space required for appliances, counters, and storage areas is 100 sq. ft. (9.3 m²). However, this size may also vary according to the budget or needs of the household.

GARAGES AND CARPORTS

Areas for parking and storing vehicles often make up a large percentage of the space available on a property. Therefore, the maximum utilization of space is important to consider when designing garages, carports, and driveways.

Function and Location

A garage is an enclosed structure designed primarily to shelter an automobile. It may be used for many secondary purposes—as a workshop, as a laundry room, or for storage space. A garage may be connected with the house, as an **integral garage** (Figure 11.5), or it may be a separate building as a **detached garage**. In any case, there should be easy access from the garage to the service area of the house. Figure 11.6 shows several possible garage locations in relation to the house.

Many building codes require an elevation change (step-up) of 16″ (.406m) to 18″ (.457m) from an integral

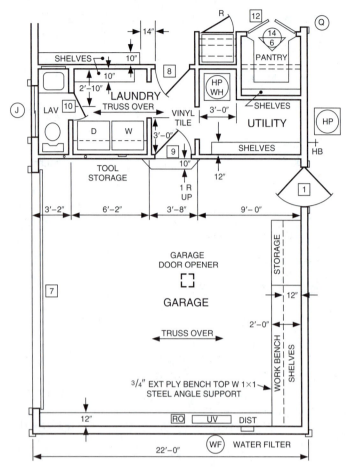

FIGURE 11.5 ■ Integral garage and adjacent utility room.

A covered walkway or breezeway from the garage or carport to the house should be provided if the garage is detached. Often a patio or porch is planned for this area to also integrate the detached garage with the house.

A **carport** looks like a garage with one or more of the exterior walls removed. It may be completely separate from the house, or it may be built against the existing walls of the house or garage. Carports are most acceptable in mild climates where complete protection from cold weather is not needed. They offer protection primarily from sun and moisture.

Both the garage and the carport have distinct advantages. The garage is more secure and provides more shelter. However, carports lend themselves to open planning techniques and are less expensive to build than garages.

Decor

The lines of the garage or carport should be consistent with the major building lines and the architectural style of the house. See Figure 11.7. The garage or carport must never appear to be an afterthought.

Floor

The garage floor must be solid and easily maintained. A concrete slab 4″ (102 mm) thick and reinforced with welded wire mesh provides the best deck surface for a garage or carport. A vapor barrier of waterproof materials should be provided under the slab. The garage floor must have adequate drainage either to the outside or through drains located inside the garage. See Figure 11.8.

garage floor to a residence floor. This is to prevent gasoline fumes or leakage from entering a house. Fireproof doors and fireshield sheathing or drywall are also required in common house–garage walls and ceilings.

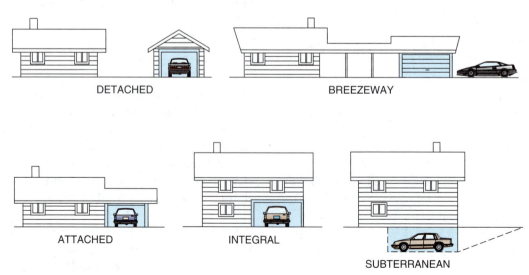

FIGURE 11.6 ■ Garage locations.

FIGURE 11.7 ■ Garage style matching home style. (*Slattery & Root, Architects; E. Abraben, Photographer*)

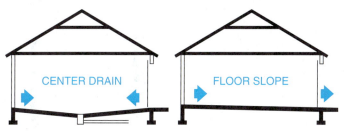

FIGURE 11.8 ■ Garage floor drainage options.

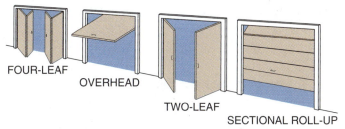

FIGURE 11.9 ■ Common garage door types.

Doors

The design of the garage door greatly affects the appearance of the house. Several types of garage doors are available: two-leaf swinging, overhead, four-leaf swinging, and sectional roll-up. See Figure 11.9. Electronic devices are available for opening the door of the garage from the car.

For overhead doors, ceiling clearance must be planned to avoid interference between the opened door and light fixtures or other projections. Sectional roll-up doors require a 16″ (406 mm) clearance between the top of the door and the ceiling. Solid overhead doors require more clearance depending on the height of the door. Horizontal ceiling space must be at least 6″ (152 mm) above the

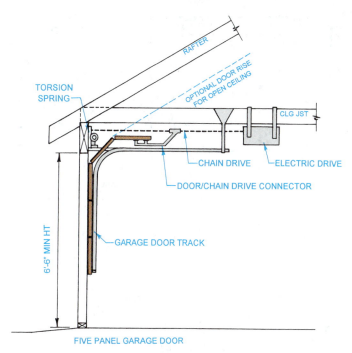

FIGURE 11.10 ■ Space requirements for overhead garage door openers.

height of the door. Figure 11.10 shows typical clearances required for overhead doors.

Materials for garage doors include steel, aluminum, fiberglass, or wood—usually redwood or cedar. Metal doors are insulated or hollow. Solid overhead or roll-up doors are either manually operated or electronically controlled by a radio transmitter that activates a motor and lights. Garage doors of all types are available in a variety of patterns, styles, and sizes. They can be purchased with or without windows.

Storage Design

Most garages are also used for storage space. Storage facilities and even living space can be created in otherwise wasted space.

Cabinets should be elevated above the floor several inches to avoid moisture and to facilitate cleaning the garage floor. Garden-tool cabinets can be designed to open from the outside of the garage.

Size

The size of the garage depends on the number of vehicles to be parked, plus space for any workshop facilities and storage. For example, storage space for bicycles, lawnmowers, and other lawn and landscape maintenance equipment must be considered as well as space for vehicles. Typical garage sizes are shown in Figure 11.11. To

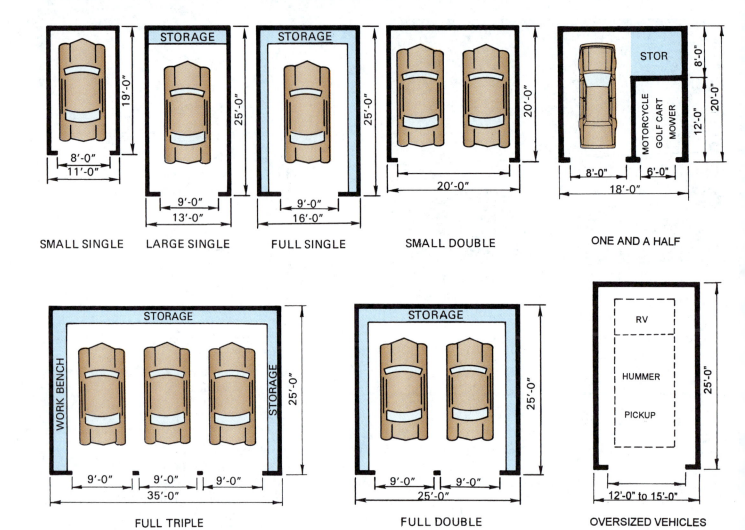

SMALL SINGLE   LARGE SINGLE   FULL SINGLE   SMALL DOUBLE   ONE AND A HALF

FULL TRIPLE   FULL DOUBLE   OVERSIZED VEHICLES

FIGURE 11.11 ■ Typical garage sizes.

allow for side mirror clearance, garage doors designed to house SUVs, pickup trucks, or large sedans must be at least 9′ (2.743m) wide and 8′ (2.438m) high. Height and widths up to 18′ (5.486m) are available for RV storage.

Although some variations exist among manufacturers, standard garage door heights are 6′−6″, 6′−8″, 6′−9″, 7′−0″, 7′−6″, and 8′−0″. Standard garage door widths are 7′, 8′, 9′, 10′, 12′, 14′, 15′, 16′, 17′, 18′, and 20′.

MUD ROOMS

A mud room in the service area allows access from the outside without passing through other rooms. The mud room often connects the garage with the utility room and should provide space for changing and storing outer garments. Figure 11.12 shows a typical mud room configuration.

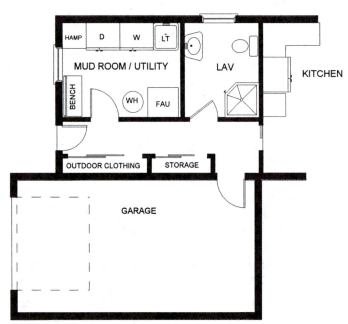

FIGURE 11.12 ■ Typical mud room configuration.

DRIVEWAYS

The main functions of a driveway are to provide access to all entrances and to the garage and to provide temporary parking space. See Figure 11.13. However, a driveway can serve other purposes, too. A wide apron at the door of the garage can become a useful area, whether for car washing or for children's games. It can also enable cars to turn around without backing out onto a main street.

The driveway should be of brick, stone, asphalt, or concrete construction. Concrete should be reinforced with welded-wire fabric to maintain rigidity and prevent cracking. Masonry pavers (Figure 11.14) are often used over a concrete base or compacted base material. The driveway should be designed at least several feet wider than the track of a car, which is approximately 5′–0″ (1524 mm). Slightly wider driveways of approximately 7′ to 9′ (2134 to 2743 mm) width are desirable for access and pedestrian traffic. To comfortably accommodate wheelchairs, a width of 10′ (3048 mm) may be needed. Sufficient space in the driveway should be provided for parking guests' cars. (Review Chapter 9, Traffic Areas and Patterns, for more details concerning driveway configurations and sizes.)

WORKSHOPS

The workshop is an area planned for working with equipment, tools, and materials.

FIGURE 11.13 ■ Parking and turning aprons in the service area.

FIGURE 11.14 ■ Driveway surface of masonry pavers. (*Cornerstone Developers, Inc.*)

FIGURE 11.15 ■ Home workshop for a variety of activities. (*Lisanti, Inc.*)

Function and Location

A home workshop is designed for activities ranging from hobbies to home maintenance. See Figure 11.15. As part of the service area, a workshop may be located in the garage, in the basement, in a separate room, or even in an adjacent building.

Workbench space, power tools, hand tools, and the storage areas should be systematically planned to allow for the maximum amount of work space. Tools and equipment for working with large materials should be placed where the material can be handled easily. Any flammable finishing material, such as turpentine or oil-based paint, should be stored in metal cabinets.

Workbench

A workbench, usually with a vise for holding materials in place, is a major component of a home workshop area. The average workbench is 36″ (914 mm) high.

Workbenches are available to suit different needs. A movable workbench is appropriate for large projects. A **peninsula workbench** has three working surfaces with storage compartments on each of the three sides. A **dropleaf workbench** is excellent for work areas where a minimum amount of space is available. The side portions, or "drop leaves," can be extended for increased work space or folded down for storage in a small space.

Hand Tools

Certain **hand tools** are necessary for any type of hobby or home maintenance work. These basic tools include a claw hammer, carpenter's square, files, hand drills, screwdrivers, planes, pliers, chisels, scales, wrenches, saws, a brace and bit, mallets, and clamps.

Hand tools may be safely stored in cabinets that keep them dust free or hung on appropriate hooks on *perforated hardboard*. Tools too small to be hung should be kept in drawers.

Power Tools

Electrically operated tools are called **power tools**. Those commonly used in home workshops include electric

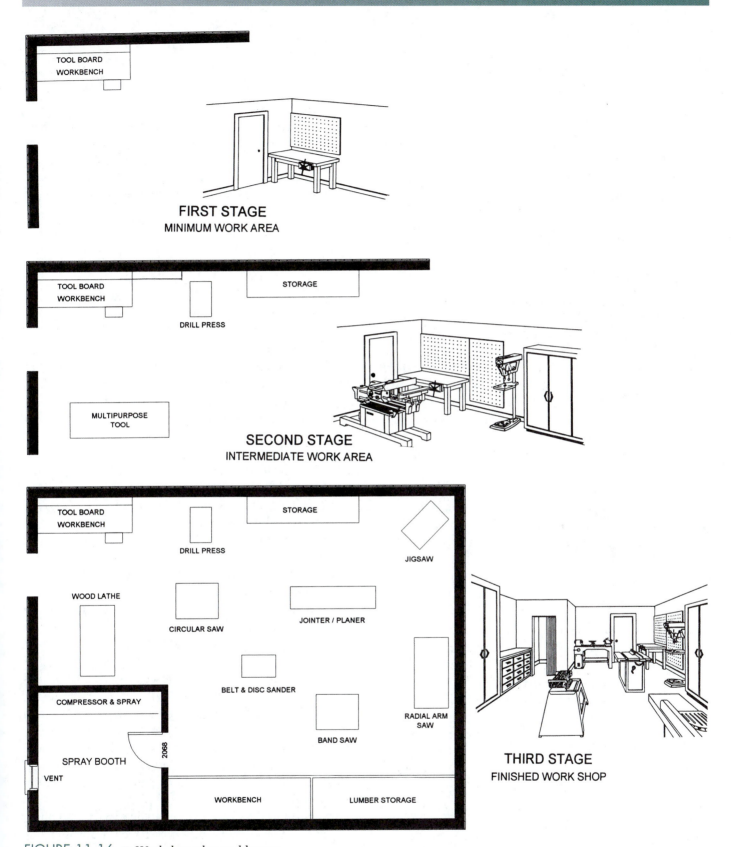

FIGURE 11.16 ■ Workshop plan and layout.

drills, saber saws, routers, band saws, circular saws, radial-arm saws, jointers, belt sanders, lathes, and drill presses.

To conserve motors, separate-drive motors can be used to drive more than one piece of power equipment. Separate 110V and 220V electrical circuits for lights as well as power tools should be planned for the home workshop area. Figure 11.16 shows work space clearances and arrangements necessary for safe and efficient machine operation.

Multipurpose machines that can perform a variety of operations are convenient in a home workshop. Less equipment is necessary, and less space is required.

Decor

The work area should be as maintenance free as possible. Glossy paint, paneling, or tiles over drywall can retard the accumulation of shop dust on the walls. Exhaust fans help eliminate much of the dust and the gases produced in the shop. The shop floor should be concrete or linoleum. For safety, abrasive strips on floors around machines will eliminate the possibility of slipping.

Noise is a concern. Do not locate noisy equipment near the sleeping areas. Interior walls and ceilings should be soundproofed by offsetting studs and adding adequate insulation to produce a sound barrier. See Figure 11.17.

Light and color are very important factors in designing the work area. Pastel colors, which reduce eyestrain, should be used for the general color scheme of the shop. Extremely light colors that produce glare and extremely dark colors that reduce effective illumination should be avoided. Choose colors that not only create a pleasant atmosphere in the shop but also help to provide the most efficient and safe working conditions.

Workshops should be well lighted. General lighting should be provided in the shop to a high intensity level on machines and worktable tops. (Refer to Chapter 31 for electrical design considerations.)

Size

The size of the work area depends on the size and number of tools, equipment, the workbench, and the storage facilities provided or anticipated. Plan the size of the work area for maximum expansion. At first, only a workbench and a few tools may occupy the area. If space is planned for the maximum amount of facilities, new equipment, when added, will fit appropriately into the basic plan. The designer must also anticipate the types and amounts of materials that will require storage space in the future and design the space accordingly.

STORAGE AREAS

Areas should be provided for general storage as well as for specific storage within each room.

Function and Types

Storage facilities—whether closets, cabinets, furniture, or room dividers—should be designed for convenient retrieval of stored articles. Those articles that are used daily or weekly should be stored in or near the room where they are needed. Those used only seasonally should be placed in more permanent storage areas. Areas that would otherwise be considered wasted space can become general storage areas. Parts of the basement, attic, or garage often fall into this category.

Closets

There are three basic types of closets: wardrobe, walk-in, and wall. A **wardrobe closet** is a shallow clothes closet built into the wall. The minimum depth for this closet is 24″ (610 mm).

Guest closets, normally located in the foyer area, are wardrobe closets designed to hold outdoor apparel, as shown in Figure 11.18. Wardrobe closets and walk-in closets, which also include space for stacked clothing, shoes, and drawer items, are usually located in bedrooms.

Depths of more than 24″ (762 mm) will make reaching the back of the closet difficult. Swinging or sliding doors should expose all parts of the closet that need to be within reach. A disadvantage of the wardrobe closet is the amount of wall space required for the doors.

A **walk-in closet,** as the name implies, is a closet large enough to enter and turn around. The area needed for this type of closet is equal to the amount of space needed to hang clothes plus space for a walkway. (See Chapter 12 for more details about these types of closets.) Although some space is wasted because of the walkway, the closet area takes up less wall space in the room. Only one closet door is needed.

A **wall closet** is a shallow closet in the wall for cupboards, shelves, and drawers. Wall closets are normally 18″ (457 mm) deep. This size provides access to all stored items without using an excessive amount of floor area.

Protruding closets that create an offset in a room should be avoided. By filling the entire wall of a room with closet space, a square or rectangular room can be designed without the use of offsets.

Doors on closets should be sufficiently wide to allow easy accessibility. Swing-out doors have the advantage of providing extra storage space on the back of the door. However, space must be allowed for the door swing. For this reason, sliding doors are often preferred. All closets, except very shallow linen closets, should be provided with lighting.

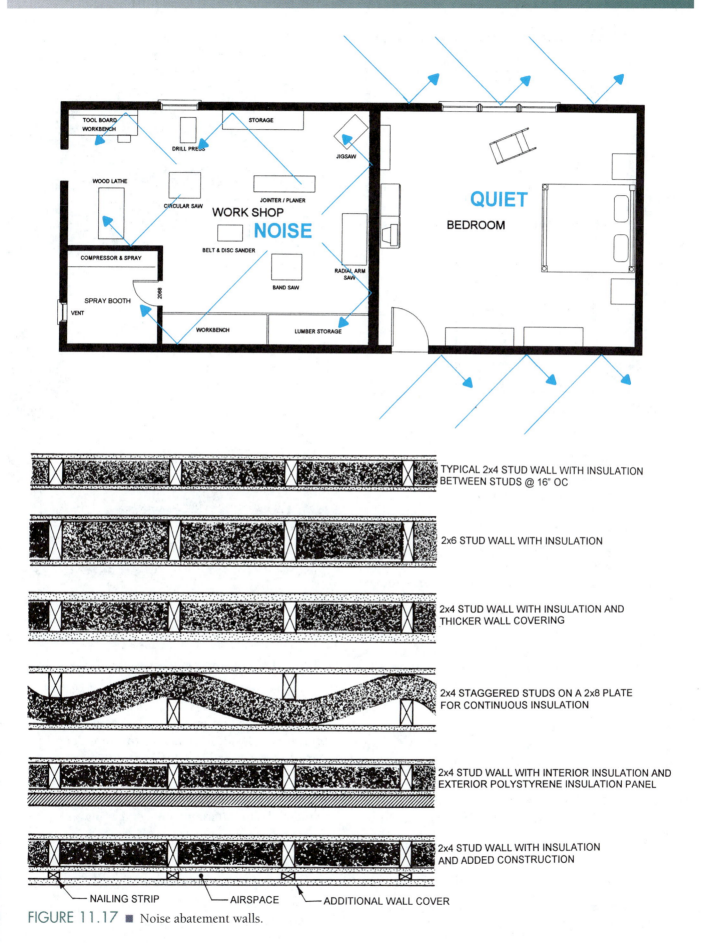

FIGURE 11.17 ■ Noise abatement walls.

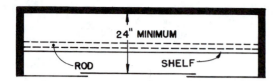

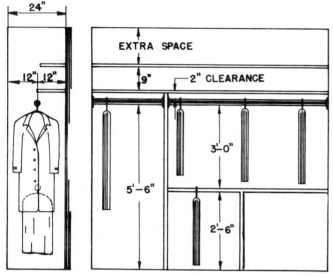

FIGURE 11.18 ■ Wardrobe closet dimensions.

FIGURE 11.19 ■ Typical storage configuration. (*California Closets*)

Furniture and Built-In Features

Chests and dressers are freestanding pieces of furniture used for storage, generally in the bedroom. They are available in a variety of sizes and usually contain shelves and/or drawers.

Window seats are hollow, chest-like structures that are built in below windows for persons to sit on. The hinged tops are often padded and can be raised to allow storage of items inside.

A room divider often doubles as a storage area. Room dividers often extend from the floor to the ceiling but may also be only several feet high. Many room dividers include shelves and drawers on both sides.

Shelves

Shelves for storage areas are available in a variety of sizes and materials including solid lumber, plywood, and hollow-core plywood. **Ventilated shelving** is made by welding steel rods together at 1/2″ or 1″ intervals. The rods are then coated with vinyl. Ventilated shelving is available in 9″, 12″, or 16″ depths and up to 12′ lengths.

Location

Different types and configurations of storage facilities are located throughout a dwelling. Figure 11.19 shows

FIGURE 11.20 ■ Entertainment center storage. (*Mirage Home Theatre*)

a typical storage configuration designed for many different areas. Besides furniture, the most appropriate types of storage facilities for each room in the house are as follows:

Living room: room divider, wall cabinets, bookcases, window seats, entertainment center (Figure 11.20).

Dining area: room divider, closet.

Family room: built-in wall storage, window seats.

Recreation room: built-in wall storage.

Porches: storage under porch stairs.

Patios: sides of barbecue, storage shed.

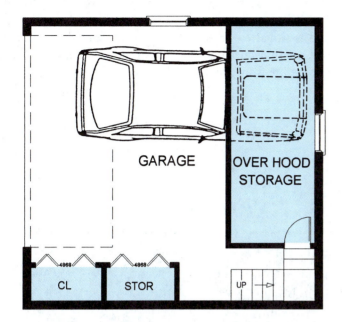

FIGURE 11.21 ■ Garage storage options.

Outside: storage areas built into the side of the house.

Halls: wall closets, ends of blind halls, bookshelves.

Entrance: room divider, closet.

Den: wall closet, bookcases.

Kitchen: wall and floor cabinets, room divider, wall closets.

Utility room: floor and wall cabinets.

Garage: cabinets built above the hood of a car, wall closets along sides, added construction on the outside of the garage. Figure 11.21 shows an over-hood storage plan, and Figure 11.22 shows a plan for the maximum use of an upper garage level for living quarters.

Workshop: tool board, closets, cabinets.

Bedroom: closets, storage under, at foot, and at head of bed; cabinets; shelves.

Bathroom: cabinets, room dividers.

SPECIALIZED AREAS

Other facilities that must be considered in designing the service area include exercise equipment, pet support, and trash and garbage disposal. Exercise equipment may occupy a space with weights, racks, and floor mats or a

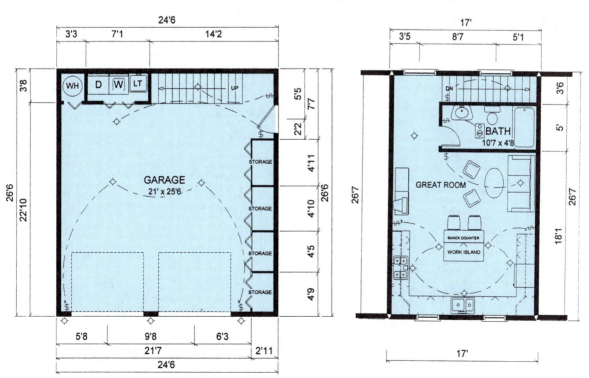

FIGURE 11.22 ■ Living quarters designed into a garage upper level.

separate room equipped with cardio machines, massage table, TV, water fountain, and shower. Pet areas should include equipment for food and water dispensing, sleeping, and, for cats, a litter box. In developing architectural plans, an often forgotten area is the temporary storage and removal of trash and garbage. A specific area complete with secured containers should be planned that is not visible from any of the living or sleeping areas.

11

General Service Areas Exercises

1. Design a utility room including a complete laundry facility within an area of 12′ × 12′ (144 sq. ft.). Show the location in relation to other areas of a house.

2. Design a utility room for the house you are planning.

 3. Design a full double garage and driveway for the house of your design. Include storage, laundry facilities, and a workbench. Identify the type of door you would use.

4. Draw a plan of a garage and/or a workshop.

5. Design a work area for the house you are planning.

6. Add storage facilities to the house of your design.

7. Draw a walk-in closet plan.

CHAPTER 12
Sleeping Areas

OBJECTIVES

In this chapter you will learn to:

- plan and draw bedrooms for a sleeping area.
- plan and draw baths appropriate to the size and arrangement of the floor plan.
- design an efficient bath.

TERMS

bidet	half-bath	walk-in closet
central bath	lavatory	wall storage cabinet
compartment plan	master bath	wardrobe closet
fixtures	master bedroom	water closet

INTRODUCTION

Approximately one-third of our time is spent sleeping. Therefore, the sleeping area should be planned to provide facilities for maximum comfort and relaxation. The sleeping area should be located in a quiet part of the house and include bedrooms and baths.

BEDROOMS

Houses are often categorized by the number of bedrooms. For example, a house may be described as a three-bedroom home or a four-bedroom home. A single person or couple with no children may require only a one-bedroom home. Three-bedroom homes are most common, since they accommodate most families. See Figure 12.1.

Function

The primary function of a bedroom is to provide facilities for sleeping. However, some bedrooms may also provide facilities for writing, reading, watching TV, listening to music, or relaxing. As with other rooms, the size and shape of each bedroom depends on the occupants, activities, and furniture designated for that room.

For babies and very young children, a bedroom may serve as a nursery, with a crib and related furniture and equipment. For older children, a bedroom may be a double room with twin beds and other furniture such as desks and entertainment equipment. A **master bedroom** for adults not only has a large bed or beds and other furniture, it usually also has an adjacent bath and, perhaps, a separate dressing room. See Figure 12.2.

Location

For sleeping comfort and privacy, bedrooms should be grouped in a quiet part of the house, as far from the living area as possible. If a further separation is wanted between the master bedroom and children's or guest bedrooms, locating the master bedroom on a different level is recommended for multilevel dwellings. Regardless of the location, all bedrooms must have access to a hall from which a bath is also accessible.

If space permits, some area, preferably a separate room, can be planned for solitary use. This can be for

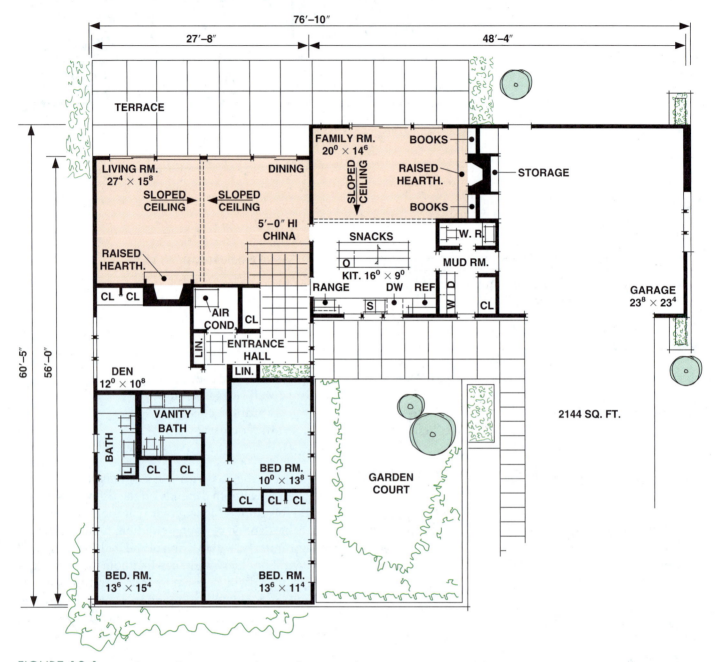

FIGURE 12.1 ■ Bedroom sleeping areas. (*Home Planners Inc.*)

reading, meditation, or quiet study. The sleeping area is ideal for this type of room or area since it is located away from the active areas of the house.

Guidelines for Noise Control

Because noise contributes to fatigue, location is particularly important for a restful bedroom area. To eliminate as much noise as possible, planned locations and well-selected materials can help accomplish this. See Figure 12.3. The following guidelines are valuable for designing bedrooms that are quiet and restful:

1. The bedroom should be in the quiet part of the house, away from major street noises.

2. Air is a good insulator. Therefore, closets can be located to provide sound buffers. Clothing and other items stored in closets can also help muffle sound.

3. Carpeting or porous wall and ceiling panels absorb noises. Rooms above bedrooms should be carpeted.

4. Floor-to-ceiling draperies help to reduce noise.

FIGURE 12.2 ■ Master bedroom suite.

5. Acoustical tile in the ceiling is effective in reducing noise.

6. Trees and shrubbery outside the bedroom help absorb sounds.

7. The use of double-glazed insulating glass for windows and sliding doors helps to reduce outside noise.

8. The windows of an air-conditioned room should be kept closed during hot weather. This eliminates noise and aids in keeping the bedroom free from dust and pollen.

9. In extreme cases when complete soundproofing is desired, the wall structure and materials may be designed to provide continuous sound insulation.

10. Placing rubber pads under appliances such as refrigerators, dishwashers, washers, and dryers often eliminates vibration and noise throughout the house.

Wall Space

The bedroom entrance door, closet doors, and windows should be grouped to conserve wall space whenever possible. By minimizing the distance between doors and windows, the amount of usable wall space is expanded. Long stretches of wall space are best for efficient furniture placement.

Doors

Several types of doors may be used in bedrooms. Pocket, bypass (sliding), and bifold doors are used for closets. Swinging and pocket doors can be used for the entrance door. A swinging door should always swing into the bedroom against an adjacent wall, and not outward into the hall. See Figure 12.4. The door connecting the bedroom with an outside deck, balcony, or patio should be French (glazed), swinging, or sliding glass to maximize light and ventilation. Bedroom closet doors can be specified as

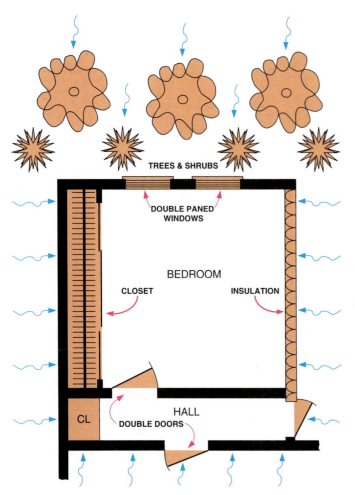

FIGURE 12.3 ■ Methods of minimizing noise.

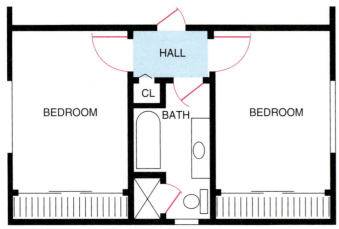

FIGURE 12.4 ■ Bedroom doors should swing into the bedroom and toward a wall.

mirrored, which provides a dual function—as a door and as a full-length mirror—as shown in Figure 12.5.

Standard door heights range from 6'-8" to 8'-0". Standard door widths for bedroom doors range from 2'-6" to

FIGURE 12.5 ■ Mirrored closet door. *(Gar Schmitt & Associates, Inc.)*

3'-0". A width of at least 2'-8" is needed to allow passage of some furniture, and a minimum width of 2'-9" is needed for wheelchair entry.

Windows and Ventilation

Bedroom windows should be placed to provide air circulation, light, and solar heat from the south. Similar to door placement, designers must also consider the efficient use of wall space for window placement. One method of conserving wall space for bedroom furniture is to use high windows. High, narrow-strip windows, called ribbon windows, provide space for furniture to be placed underneath. They also ensure some privacy for the bedroom. Building and fire codes require an escape-size window in each room if no outside door exists.

Proper ventilation is necessary in bedrooms and is conducive to sound rest and sleep. When air conditioning is available, the windows and doors may remain closed. Central air conditioning and humidity control provide constant levels of temperature and humidity and are efficient methods of providing ventilation and air circulation.

Without air conditioning, windows and doors must provide the ventilation. Bedrooms should have cross

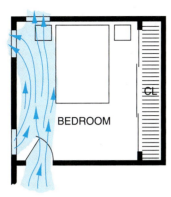

FIGURE 12.6 ■ Cross ventilation to avoid draft over bed.

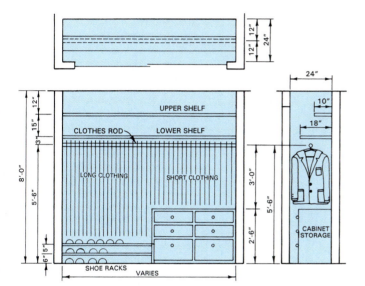

FIGURE 12.7 ■ Wardrobe closet.

FIGURE 12.8 ■ Wall storage closet (*Closet Design Group Inc.*)

ventilation. However, the draft must not pass over the bed. See Figure 12.6. High ribbon windows provide cross ventilation without causing a draft on the bed. Jalousie windows are also effective, because they direct air upward.

Storage Space

Storage space placed in or adjacent to bedrooms is needed primarily for clothing and personal accessories. Furniture, closets, cabinets, and dressing rooms serve as storage facilities. These should be located within easy reach and should be easy to maintain.

Furniture storage space in the bedroom area is usually found in dressers, chests, vanities, and dressing tables. However, most storage space should be provided in the closets. Closet types were introduced in Chapter 11. Figure 12.7 shows a bedroom **wardrobe closet** or-

ganized to store hanging and stacked clothing, shoes, and drawer items. Average dimensions are also included. Figure 12.8 is an example of a bedroom **wall storage cabinet,** which can be used to replace or reduce the amount of bedroom furniture. These closets should be located to avoid the creation of awkward wall offsets. See Figure 12.9.

The minimum amount of space required for bedroom **walk-in closets** includes the storage space plus the amount of space to walk and turn around. Note the dimensions in Figure 12.10. Figure 12.11 shows several walk-in closet configurations. Figure 12.12 shows a portion of a walk-in closet devoted to shelf, cabinet, and drawer storage. Compare the wardrobe and walk-in closet space in Figure 12.13.

Dressing Areas

A dressing area is usually located adjacent to the master bedroom. It may be a separate room, an alcove, or part of the bedroom separated by a divider that also provides

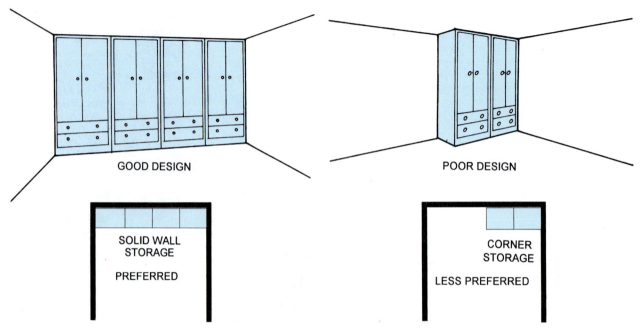

FIGURE 12.9 ■ Closets should not create room offsets.

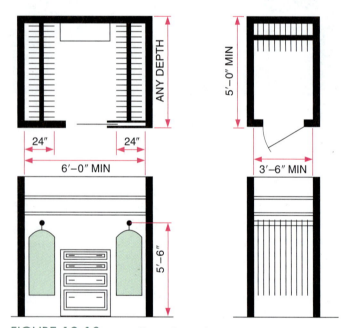

FIGURE 12.10 ■ Walk-in closet dimensions.

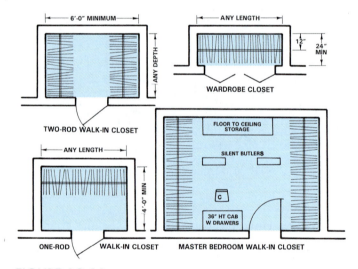

FIGURE 12.11 ■ Walk-in closet and wardrobe configurations.

storage space. Figures 12.14A and 12.14B, show a dressing room, bath, balcony, and bedroom, which comprises a master bedroom suite.

Children's Rooms

Children's bedrooms and nurseries must be planned to be comfortable, quiet, and sufficiently flexible to allow change as the child grows and matures. See Figure 12.15. For example, storage shelves and rods in closets should be adjustable so that they may be raised as the child becomes taller. Light switches should be placed low for small children and have a delay switch that allows the light to stay on for some time after the switch has been turned off.

Adequate facilities for study and hobby activities should be provided, such as a desk and worktable. Storage space for books, models, and athletic equipment is also desirable. Chalkboards and bulletin boards on the walls help make the child's room usable.

Grouping or dividing children's rooms into convertible double rooms is one method of designing for future change as children grow. (See Figure 12.16.) Eventually the divided room can become a single large guest bedroom or study.

The use of pull-down Murphy beds is effective for youth or guest bedrooms. This design, as shown in Figure 12.17, provides space for other activities when the bed is in the upright wall position. The use of attics (Figure 12.18) as a location for youth or guest bedrooms also makes maximum use of otherwise less used space.

FIGURE 12.12 ■ Walk-in closet drawer and cabinet storage. *(Closet Design Group Inc.)*

FIGURE 12.13 ■ Wardrobe and walk-in closet comparison.

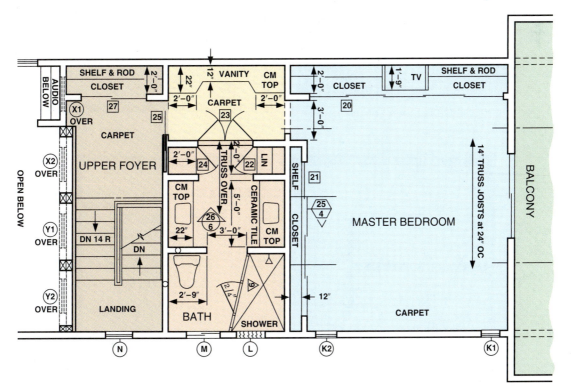

FIGURE 12.14A ■ Master suite with dressing areas and adjoining bath.

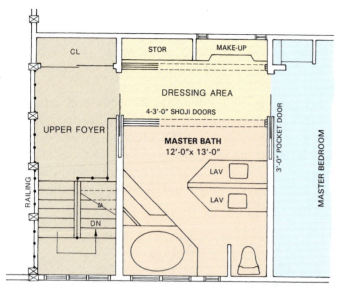

FIGURE 12.14B ■ Alternative design for plan shown in Figure 12.14A.

FIGURE 12.15 ■ Child-oriented bedroom. (*Tony Stone Images/David Rigg*)

Decor

In general, bedrooms should be decorated in quiet, restful tones. Matching or contrasting bedspreads, draperies, and carpets help accent the color scheme. Uncluttered

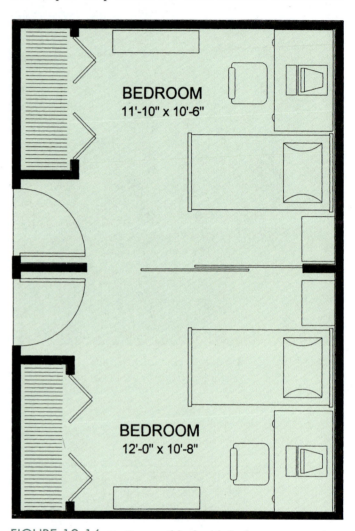

FIGURE 12.16 ■ Convertible double bedroom.

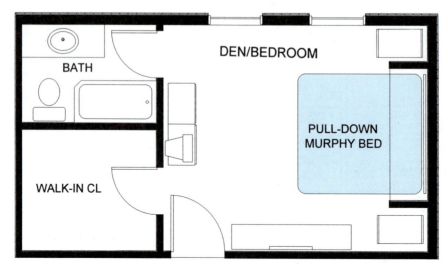

FIGURE 12.17 ■ Use of pull-down bed to maximize the use of space.

FIGURE 12.18 ■ Use of attic as youth or guest bedroom.
(*Western Wood Products Association*)

FIGURE 12.19 ■ Effective use of fabrics, colors, lighting,
and architectural details in a bedroom.
(*Fran Murry Interiors*)

furniture with simple lines also helps to develop a restful atmosphere in the bedroom. The bedroom shown in Figure 12.19 illustrates the effective blending of fabrics, colors, lighting, and architectural details to create a warm and relaxing environment.

Size and Shape

The type, size, and style of furniture to be included in the bedroom should be chosen before the size of the bedroom is established. In a preliminary design, the sizes and amount of furniture determine the size of the room and not the reverse. Because the bed or beds require the most space, the room size must provide adequate space for and around the bed. A minimum-sized bedroom should accommodate at least a single bed, bedside table, and dresser. In contrast, a larger, master bedroom suite may include a separate dressing area, vanity area, master bath, TV, VCR, radio, king or queen bed, dressers, armoire, chaise, chairs, walk-in closet, and built-in storage cabinets. Bedroom furniture and components may also be built in. Note how the built-in components in the bedroom in Figure 12.20 blend with the architectural style.

The size of the furniture and the space between the furniture needs to be considered to determine the di-

Bedrooms and all livable rooms need to be a minimum of 100 sq. ft. An average size bedroom is between 100 sq. ft. and 200 sq. ft. Bedrooms over 200 sq. ft. are considered large.

BATHS

Baths (or bathrooms) must be planned to be functional, attractive, and easily maintained. They can vary widely in size depending on the space available and additional areas included.

Function

Designing a bath involves the appropriate placement of fixtures, cabinets, accessories, and plumbing lines. Adequate ventilation, heating, and lighting also need to be planned. In addition to the normal functions, baths may also provide facilities for dressing, exercising, or laundering. Some may include a sauna and whirlpool bath. See Figure 12.22. Where space exists, more elaborate

FIGURE 12.20 ■ Built-in components that blend with architectural style. (*La Strada Properties*)

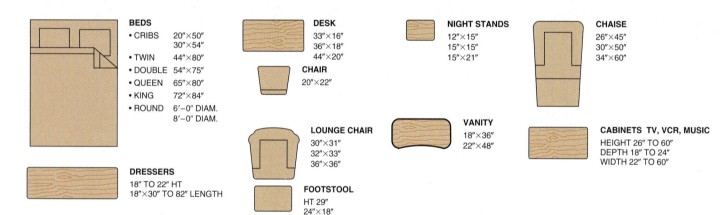

BEDS
- CRIBS 20″×50″
  30″×54″
- TWIN 44″×80″
- DOUBLE 54″×75″
- QUEEN 65″×80″
- KING 72″×84″
- ROUND 6′-0″ DIAM.
  8′-0″ DIAM.

DRESSERS
18″ TO 22″ HT
18″×30″ TO 82″ LENGTH

DESK
33″×16″
36″×18″
44″×20″

CHAIR
20″×22″

LOUNGE CHAIR
30″×31″
32″×33″
36″×36″

FOOTSTOOL
HT 29″
24″×18″

NIGHT STANDS
12″×15″
15″×15″
15″×21″

VANITY
18″×36″
22″×48″

CHAISE
26″×45″
30″×50″
34″×60″

CABINETS TV, VCR, MUSIC
HEIGHT 26″ TO 60″
DEPTH 18″ TO 24″
WIDTH 22″ TO 60″

FIGURE 12.21 ■ Typical bedroom furniture sizes.

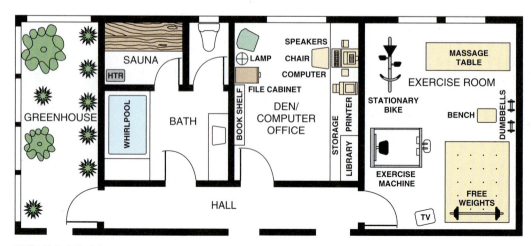

FIGURE 12.22 ■ Bath with sauna and whirlpool.

FIGURE 12.23 ▪ Large lounge-type bath in a colonial style. *(Eljer)*

baths can be designed to function as relaxing lounges, as shown in Figure 12.23.

Fixtures

The basic bath **fixtures** (items connected to plumbing lines) included in most baths are a **lavatory** (sink), a **water closet** (toilet fixture), and a bathtub or shower. The convenience of the bath depends on the arrangement of these fixtures.

Lavatories, or sinks, are available in a wide variety of colors, materials, sizes, and shapes. They are manufactured with porcelain-covered steel or cast iron, stainless steel, brass, copper, or acrylic or other composite materials such as cultured marble. Sinks are either set into an opening, molded with the countertop into one piece, or made into a freestanding fixture on a pedestal. Sink areas should be well lighted and free of traffic, and include a mirror on the wall over the sink. Side mirrors provide additional angles for viewing. A comfortable sink height for most people is between 34″ and 36″. For wheelchair access, a width of 36″ is usually required.

Water closets are available in either one-piece or two-piece models. One-piece models are either mounted on the floor or on a wall. Water closets need a minimum of 18″ distance from the center to a side wall or to other fixtures. For wheelchair access, a clear opening ranging from 2′-6″ to 4′-0″ is required in front and on the sides of a water closet. If space allows, the water closet should not be visible when the door to the bath is open. Figure 12.24 shows common clearances recommended for bath fix-

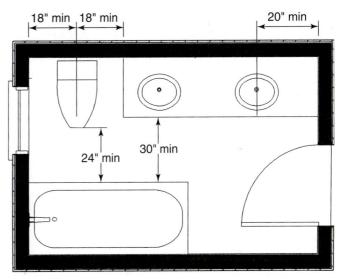

FIGURE 12.24 ▪ Recommended minimum bath fixture clearances.

tures. For wheelchair access, the two walls adjacent to the water closet must be a minimum of 6′-2″.

The variety of tubs and showers available in squares, rectangles, or irregular shapes allows a great amount of flexibility in fixture arrangements. Many small or average-size baths include a showerhead on a tub wall. This tub-shower combination then requires an enclosure or a shower curtain rod. Figure 12.25A shows a picture of the bath plan in Figure 12.25B. This plan includes a large oval whirlpool tub with shower and matching curved shower curtain enclosure. Grab bars and other fixtures installed for people with impairments must be secured to studs.

FIGURE 12.25A ■ Tub-shower-whirlpool combination.

FIGURE 12.26 ■ Tub located for optimum outside exposure. (*American Standard*)

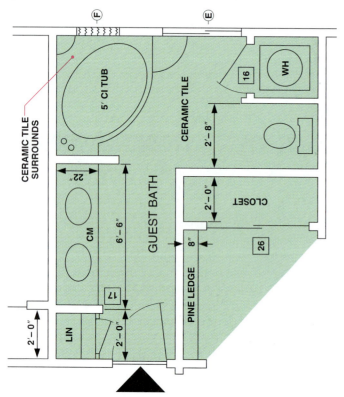

FIGURE 12.25B ■ Plan view of Figure 12.25A.

Tubs without a showerhead can be located in open areas. See Figure 12.26. Large tubs can include whirlpool outlets, jets, and controls for hydrotherapy. Special tubs with side openings that seal when closed are available for persons with physical impairments. Grab bars installed beside the tub are simple, but effective safety features, particularly for older people and those with physical impairments.

A separate shower stall is often preferred or substituted for a tub. One-piece prefabricated units are available made of materials such as fiberglass, acrylic, coated steel, or aluminum. Many shower stalls are constructed using ceramic tile, marble, or synthetic materials. Glass shower walls, used for panels and doors, are made from shatterproof glass mounted between metal frames as shown in Figure 12.27. Where space is available, showers can be designed without a door. See Figure 12.28.

Some showerheads can now equalize hot and cold water pressure. For example, if cold water pressure is reduced during shower use, the hot water pressure is also automatically reduced to avoid scalding.

FIGURE 12.27 ■ Shower door enclosure. (*Century Shower Door Co.*)

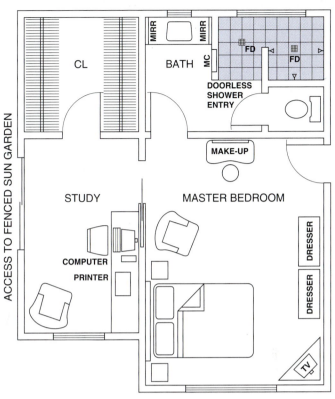

FIGURE 12.28 ■ Shower design without doors.

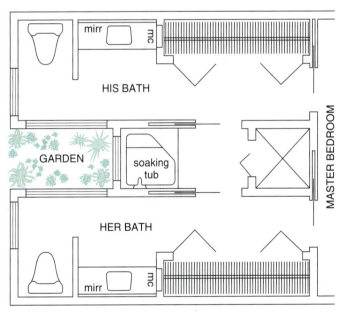

FIGURE 12.29 ■ Two-compartment bath.

Accessories

In addition to the three basic fixtures, many accessories help improve a bath's functions. Accessories range from various furnishings and plumbing devices to lights and heating devices. *Bath furnishings* include such items as a medicine cabinet, extra mirrors, a magnifying mirror, extra counter space, a dressing table, shelves for linen storage, and a clothes hamper. A bath designed for, or used in part by, children should include a low or tilt-down mirror, benches for reaching the lavatory, low towel racks, and shelves for bath toys.

Plumbing accessories such as a whirlpool bath or **bidet** might be installed. Figure 12.23 shows a bidet installed beside the water closet. Foot-pedal controls for water and single-control faucets are other types of accessories.

Layout

There are two basic types of bath layouts: the compartment plan and the open plan. In the **compartment plan**, partitions (sliding doors, glass dividers, louvers, or even plants) are used to divide the bath into compartments: the water closet area, the lavatory area, and the bathing area. See Figure 12.29. In the open plan, all bath fixtures are completely or partially visible.

FIGURE 12.30 ■ Fixtures located to minimize plumbing lines.

Ventilation

Baths should have either natural ventilation from a window or forced ventilation from an exhaust fan. Care should be taken to place windows in a position where they will not cause a draft on the tub or interfere with privacy. A bath can be designed without windows. However, substitute sources of light and ventilation are then needed. This can be achieved by installing a light-ventilating fan combination that is controlled by a single switch.

Lighting

Lighting should be relatively shadowless in the area used for grooming. Shadowless general lighting can be achieved with fluorescent tubes installed on the ceiling and covered with glass or plastic panels. Skylights can also help provide general illumination.

Heating

Heating in the bath is especially important. In addition to the conventional heating outlets, an electric heater or heat lamp is often used to provide instant heat. The source of heat should be placed under the window to eliminate drafts. All gas and oil heaters should be properly ventilated. Heat lamps should be controlled by a timer to avoid overheating.

Location

Two factors in locating baths are positioning of plumbing lines and accessibility from other areas of the house.

Plumbing Lines

The plumbing lines that carry water to and from the fixtures should be concealed and minimized as much as possible. When two baths are located side by side, placing the fixtures back to back on opposite sides of the plumbing wall reduces the length of plumbing lines. See Figure 12.30. In multiple-story dwellings, efficient use of plumbing lines can be accomplished if the baths are placed one directly above another. When a bath is placed on a second floor, a plumbing wall must be provided through the first floor for the soil and water pipes.

Accessibility

Ideally, a private bath should be located adjacent to each bedroom, as shown in Figure 12.31. Often, this is not possible, and a **central bath** is designed to meet the needs of the entire family. See Figure 12.32. The central or general bath should be in the sleeping area, accessible from all the bedrooms. A bath for general use plus a bath adjacent to the master bedroom is a desirable compromise. A **master bath** is accessible only from the master bedroom.

Bathing or showering facilities are usually not needed in the living or service areas. **Half-baths** containing only a lavatory and water closet are designed for these areas, unless a full bath is conveniently located nearby.

Decor

Today's baths need not be strictly functional and sterile in decor. They can be planned and furnished in a variety of

FIGURE 12.31 ■ Sleeping areas with baths designed for each bedroom.

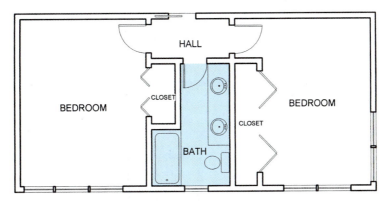

FIGURE 12.32 ■ Central bath serving two bedrooms.

styles. Baths should be decorated and designed to provide the maximum amount of light and color.

Materials

Materials used in the bathroom should be water resistant, easily maintained, and easily sanitized. Tiles, linoleum, marble, plastic laminate, and glass are excellent materials for bath use. If wallpaper or wood paneling is used, it should be waterproof. If plaster or drywall construction is exposed, a gloss or semigloss paint should be used on the surface.

Fixtures are now available in a variety of colors, so that they can be coordinated or even matched with accessories. Matching countertops and cabinets are also available.

New materials and components enable the designer to plan baths with modular units that range from one-piece molded showers and tubs to entire bath modules. In these units, plumbing and electrical wiring are connected after the unit is installed.

Size and Shape

Bath sizes and shapes are influenced by the size and spacing of basic fixtures and accessories. The type of plan—whether compartmentalized or open—and the relationship to other rooms in the house also influence size and shape. Additional space may be required to accommodate wheelchairs. Figure 12.33 shows a variety of bath shapes and arrangements.

The minimum size bath that can include all three basic fixtures is 5′ × 8′. Sizes range from this minimum to luxury compartmentalized baths. Figure 12.34 shows the standard sizes of bath cabinets. Although cabinets may be custom made to any size, choosing standard cabinet dimensions saves considerable time and expense. Figure 12.35 shows the various standard sizes of bath fixtures.

HALF BATH 25 SQ FT +

HALF BATH 35 SQ FT +

HALF BATH 30 SQ FT +

SMALL FULL BATH 50 SQ FT +

FULL BATH 100 SQ FT +

PARTITIONED BATH 80 SQ FT +

BACK TO BACK BATHS 120 SQ FT +

LUXURY BATH 100 SQ FT +

LARGE PARTITIONED BATH 100 SQ FT +

LARGE LUXURY BATH 220 SQ FT +

LARGE BATH WITH PEDESTAL BATH 140 SQ FT +

LARGE BATH 150 SQ FT +

LARGE LUXURY BATH 160 SQ FT +

HERS & HIS BATH WITH GARDEN 200 SQ FT +

LARGE LUXURY BATH 350 SQ FT +

FIGURE 12.33 ■ *A variety of bath shapes and configurations.*

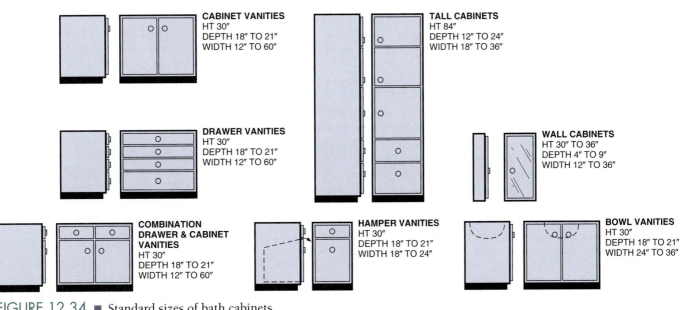

CABINET VANITIES
HT 30″
DEPTH 18″ TO 21″
WIDTH 12″ TO 60″

TALL CABINETS
HT 84″
DEPTH 12″ TO 24″
WIDTH 18″ TO 36″

DRAWER VANITIES
HT 30″
DEPTH 18″ TO 21″
WIDTH 12″ TO 60″

WALL CABINETS
HT 30″ TO 36″
DEPTH 4″ TO 9″
WIDTH 12″ TO 36″

COMBINATION DRAWER & CABINET VANITIES
HT 30″
DEPTH 18″ TO 21″
WIDTH 12″ TO 60″

HAMPER VANITIES
HT 30″
DEPTH 18″ TO 21″
WIDTH 18″ TO 24″

BOWL VANITIES
HT 30″
DEPTH 18″ TO 21″
WIDTH 24″ TO 36″

FIGURE 12.34 ■ Standard sizes of bath cabinets.

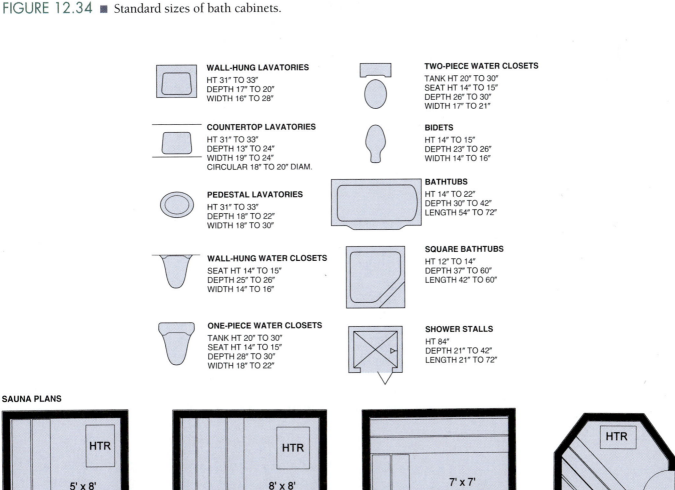

WALL-HUNG LAVATORIES
HT 31″ TO 33″
DEPTH 17″ TO 20″
WIDTH 16″ TO 28″

TWO-PIECE WATER CLOSETS
TANK HT 20″ TO 30″
SEAT HT 14″ TO 15″
DEPTH 26″ TO 30″
WIDTH 17″ TO 21″

COUNTERTOP LAVATORIES
HT 31″ TO 33″
DEPTH 13″ TO 24″
WIDTH 19″ TO 24″
CIRCULAR 18″ TO 20″ DIAM.

BIDETS
HT 14″ TO 15″
DEPTH 23″ TO 26″
WIDTH 14″ TO 16″

PEDESTAL LAVATORIES
HT 31″ TO 33″
DEPTH 18″ TO 22″
WIDTH 18″ TO 30″

BATHTUBS
HT 14″ TO 22″
DEPTH 30″ TO 42″
LENGTH 54″ TO 72″

WALL-HUNG WATER CLOSETS
SEAT HT 14″ TO 15″
DEPTH 25″ TO 26″
WIDTH 14″ TO 16″

SQUARE BATHTUBS
HT 12″ TO 14″
DEPTH 37″ TO 60″
LENGTH 42″ TO 60″

ONE-PIECE WATER CLOSETS
TANK HT 20″ TO 30″
SEAT HT 14″ TO 15″
DEPTH 28″ TO 30″
WIDTH 18″ TO 22″

SHOWER STALLS
HT 84″
DEPTH 21″ TO 42″
LENGTH 21″ TO 72″

SAUNA PLANS

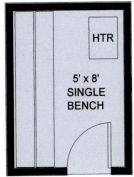

5' x 8'
SINGLE
BENCH

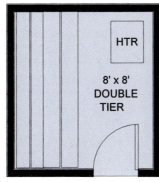

8' x 8'
DOUBLE
TIER

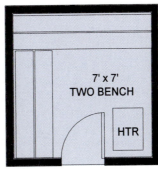

7' x 7'
TWO BENCH

HTR

5' x 5' OCTAGON
DOUBLE TIER

FIGURE 12.35 ■ Standard sizes of bath fixtures.

CHAPTER

12

Sleeping Areas Exercises

1. Design a bedroom, 100 sq. ft. in size, for a 6-year-old child.

2. Design a bedroom, 150 sq. ft. in size, for a teenager.

3. Design a master bedroom with an adjoining bath that is 200 sq. ft. in size.

4. Draw a bedroom shown in this chapter.

 5. Plan the bedroom areas for the home you are designing.

6. Using dimensions provided in this chapter, calculate the minimum size of a bedroom that could accommodate a king size bed, built-in TV, a dresser, and a lounge chair.

7. Collect pictures of bedrooms you like. Identify good and bad design features.

8. Draw a plan with a master bath and a central bath.

9. Draw a plan of a bath you think is poorly designed. Then draw a plan for remodeling the bath to make it more functional.

10. Draw the plans for the bath areas in the home of your own design.

11. Design and draw a bath that is 12′ × 8′.

12. Calculate the dimensions needed for a bath you design with: one cabinet vanity, two drawer-vanities, one hamper vanity, one wall cabinet, a countertop lavatory, a bathtub, and a one-piece water closet. (Use the standard sizes given in Figures 12.34 and 12.35 for reference.)

PART  4

Basic Architectural Drawings

13 CHAPTER

Designing Floor Plans

OBJECTIVES

In this chapter you will learn to:

- gather information from a client that is needed to design an architectural project.
- analyze a building site.
- use the design process to prepare for drawing accurate and functional floor plans.
- create floor plan sketches.
- design floor plans to accommodate the needs of persons with physical impairments.

TERMS

base map
conceptual design
easements
feng shui
floor plans

idealized drawings
room template
setbacks
single-line drawing

site analysis
site-related drawing
situation statement
user analysis

INTRODUCTION

The most commonly used architectural drawings are floor plans. Many examples are shown in Part 3. This chapter outlines procedures used in developing floor plan designs, beginning with a presentation of the steps in the design process. Guidelines for developing architectural designs to accommodate special needs are also included.

FLOOR PLAN DEVELOPMENT

Final **floor plans** contain more specific information about an architectural design than any other type of drawing. They include descriptions of locations, sizes, materials, and components contained in the design. Floor plans serve as a point of reference for other drawings in a set. For this reason the first phase of the architectural design process leads to the development of *basic floor plans*.

THE DESIGN PROCESS

The architectural design process involves many personal, social, economic, and technical variables to create detailed working drawings. To effectively apply the principles and elements of design to an architectural project, established design sequences and procedures must be followed. This process is a logical sequence of thought and activities that begin with an inventory and analysis of the project. This process continues through the design of basic site and floor plans and proceeds through the completion of all working drawings. See Figure 13.1.

Defining the Project

The first step in the design process is to define the project. An agreement established between a client and a designer involves the purpose, theme, scope of the project, budget, and schedules. This agreement is then translated into a **situation statement** as shown in Figure 13.2. This statement

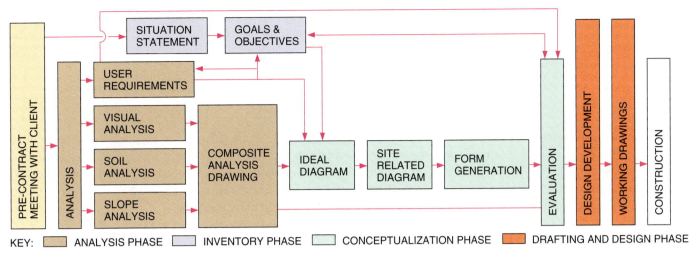

FIGURE 13.1 ■ The architectural design process.

Situation

Mr. and Mrs. John Smith have acquired a five acre parcel of wooded land. They want the site developed as a residence and also for home office use. The Smith family of four have strong feelings for environmental preservation and are very fitness oriented. They want a high degree of privacy and need the capacity to entertain weekend guests. The total cost cannot exceed $60,000 more than the sale price of their existing home. Completion of construction (closing) must be no later than one year from the signing date of the design contract.

FIGURE 13.2 ■ Design situation statement.

Needs	Wants
• Living area open plan	1. Jogging track
• Contemporary design	2. Shop area near garage
• Home office or study which can be shared	3. Badminton court
• Living area fireplace	4. Minimum lawn area
• Courtyard patio	5. Bridge connecting deck and dock
• Pool with large living deck	6. Whirlpool
• Boat dock	7. Basketball court
• Drive apron for parking	8. Minimize tree removal
• Private MBR deck	9. Maximize solar use
• Privacy from road	10. Separate home office for Mr. & Mrs. Smith
• Three bedrooms including master suite	11. MBR fireplace
• Maximize view of mountains	12. Greenhouse
• Dining facilities for 10 guests maximum	13. Billiards room
	14. Cabana near pool

FIGURE 13.3 ■ Design wants and needs list.

identifies and records the client's major requirements and any special design requirements and problems. Any subsequent design drawings relate to this situation statement.

Needs and Wants

The success of any design depends on how well the finished product meets the needs of the residents. During the entire process, the needs of the residents—and their "wants" as well—must be kept clearly in mind. This includes physical and lifestyle considerations.

A *need* is an absolute requirement and must be implemented in the design. A *want* is a desirable feature, but not absolutely required. Wants can be compromised because of budget constraints, space, or code restrictions. An effective design must meet all of the client's needs and as many wants as possible.

The designer and the client prepare needs and wants lists. See Figure 13.3. If the "wants" are listed in priority order, items on the list can be cut beginning at the bottom of the list and proceeding to the top, if necessary. In this way, what is wanted most is more likely to remain in the final design.

Goals and Objectives

Once the client and designer agree on a situation statement and create a list of the client's needs and wants, major goals and specific objectives can be developed. This is an important step. Constant reference is made to these goals and objectives during all phases of the design process. They provide the focus of the project for the designer and become a basis for evaluating all aspects of the design. See Figure 13.4. Once this is completed the analysis phase can begin.

Analyzing the Project

After the project is defined, an organized and sequenced analysis must be made of user wants and needs, site features, soil conditions, the slope of the land, and the views. These separate analyses lead to the development of a comprehensive analysis.

Major goals

Design a contemporary residence for the existing site with good visual profiles, aesthetic appeal and emphasis on functional, non-destructive use of all site features, including maximum use of solar energy. Plan working, living, and recreation areas to conform to space and priority needs. Position all facilities so that all are not visible from one vantage point.

Objectives

1. Provide stimulating, casual, and open atmosphere.
2. Locate private and public areas to avoid user conflicts.
3. Position all facilities for minimum environmental impact and minimum maintenance.
4. Orient structures for maximum solar use.
5. Building areas to blend with existing site landform.
6. Relate interior living areas to exterior space.
7. Residence not to be completely visible from access road.
8. Plan circulation patterns for both vehicular and pedestrian traffic with parking area.
9. Plan facilities for badminton, swimming, basketball, jogging, and whirlpool.
10. Provide courtyard for seasonal use.
11. Interior and exterior dining facilities for maximum 10 guests.
12. Provide Mr. Smith with an office for evening and weekend use.
13. Provide Mrs. Smith with an accessible office area to meet clients daily.
14. Use natural contemporary lines and materials consistent with site.
15. Design a focal point fireplace for living area.
16. Design gradual realization for vehicular approaching traffic.
17. Keep total cost within limits established by clients.
18. Boat dock to have access from deck area.
19. Provide three bedrooms, including master suite.

FIGURE 13.4 ■ Design goals and objectives.

User Analysis

In a **user analysis**, each goal is further refined into descriptions of space elements, usage, size, and the relationships between areas. With a user analysis, a designer can break down each design element into manageable parts. To make evaluation, verification, and discussion easier, a chart is usually prepared. See Figure 13.5.

The user analysis has great influence on the development of a design. No area should be omitted. If the user analysis is inadequate or contains erroneous information, the final design will not reflect the major goals and objectives of the project.

The user analysis is particularly important when designing a project that will be used by a person or persons

Space elements	Primary users	Min. size	Notes and relationships
Living room	8–16 adults	16' X 22'	Pool view–fireplace Access to foyer & dr.
Dining room	6–12 adults	14' X 20'	Access to kit & lr.
Study #1	Mr. Smith	14' X 20'	Private–quiet.
Study #2	Mrs. Smith	12' X 14'	Private–client Accessible–joint office w/Mr. S ?
Entry	Family–guests	8' X 10'	Visible from drive–baffle from street.
Parking	2 family cars 6 guest cars	9' X 10' stalls	Access to main entry.
Decks or terrace	Family–guests	6' X 20'	Overlook pool–next to living area.
Courtyard	Family–guests	200 sq. ft.	For casual entertainment next to kit.
Kitchen	Family	12' X 16'	Access to deck, lr, & laundry.
Garage	Family	20' X 24'	Access to kit–convert to shop.
Service pickup	Service pers.	40 sq. ft.	Screen from living areas.
Master bedroom	2 adults	16' X 24'	Morning sun–king bed–access to pool–suite w/ bath–quiet area.
Bedrooms	2 children	16' X 18'	Plan for teen growth–away from living & master br.
Baths	Children & guests	2' 8' X 10'	Access from children s rooms & guests.
Guest bedroom	Guests	12' X 16'	Bath access–or convertible study ?
Site considerations	All	Entire site	Solve sitting water problem. Use rock formations & add foliage for visual appeal.
Solar considerations	All	Bldgs. & site	Use passive techniques–care in orientation of facilities.
Recreation facilities	All	Courts, pool	Orient w/sun & screen from residence.

FIGURE 13.5 ■ Building user analysis.

with impairments. Design considerations and guidelines for developing designs to accommodate special needs are presented at the end of this chapter.

Site Analysis

An architectural project should be developed to take advantage of a site's positive features and minimize its negative features. Completing a **site analysis** not only helps the designer make proper design decisions, but also helps ensure appropriate land use.

Three types of site analyses are used to develop a final site analysis drawing: soil analysis, slope analysis, and visual analysis. Each of these three distinct factors affects the potential use of the different areas of the site. There are five phases in the development of a site analysis.

■ *Phase 1—Development of a base map:* A **base map** shows all fixed factors related to the site that must be accommodated in the site plan. It includes topographical features; the outline and location of property lines, adjacent streets, existing structures, walkways, paths,

terraces, and utility lines; easements; setback limits; and the north compass direction. **Easements** are rights-of-way across the land, such as for utility lines. **Setbacks** are minimum distances structures must be located from property lines as set by the local government. (See Chapter 18 for setback details.)

Base maps are usually prepared to a scale of 1″ = 10′, 1″ = 20′, 1″ = 30′ on an engineers scale; or 1/8″ = 1′-0″ up to 1/32″ = 1′-0″ on an architect's scale. Scale selection depends on the site size and drawing format. Many copies of this map will be used during the design process for analysis and development. See Figure 13.6.

■ *Phase 2—Soil analysis:* Soil is composed of rocks, organic materials, water, and gases. Variations in the percentage of these ingredients determine the physical characteristics of the soil and its capacity to support the weight of structures. In general, coarse-grained soils, because of their drainage and bearing capacity, are preferred for buildings but not for plants. Conversely, fine-grained soils with high or-

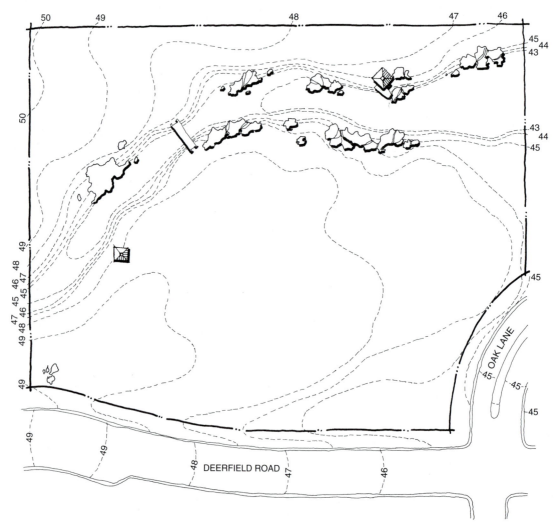

FIGURE 13.6 ■ Property base map.

ganic content are preferred for plants but not for buildings. The U.S. Department of Agriculture (the USDA) classifies four types of soil according to their quality for building a structure:

Type 1. *Excellent:* Coarse-grained soils—no clays, no organic matter.

Type 2. *Good to fair:* Fine, sandy soils (minimum organic and clay content).

Type 3. *Poor:* Fine-grained silts and clays (moderate organic content).

Type 4. *Unsuitable:* Organic soils (high clay and peat content).

To prepare a soil analysis drawing, follow these steps:

1. Obtain a soil classification for the site from a county soil survey or from private borings.

2. Draw areas on the base map representing the different soil types, as shown in Figure 13.7.

3. Note the bearing capacity and depth to bedrock for each soil category. This information is given in kilo-

pounds (kips, or K) in the USDA survey book (1 K = 1,000 lb. for 1 sq. ft. of soil area).

4. Provide a legend showing the categories of soil types and describe the soil characteristics of each type. On the drawing, note where soil conditions can or cannot be used for building.

5. Color-code each soil capacity type in the legend to match the drawing.

■ *Phase 3—Slope analysis:* The slope of a particular site greatly affects the type of building that can or should be designed for it. The slope percentage also determines what locations are acceptable, preferred, difficult, or impossible for building. See Figure 13.8. The cost of building may be greatly affected by excessive slope angles.

To complete a slope analysis drawing, refer to Figure 13.9 and complete the following steps:

1. To the base map, add *contour lines* (lines connecting points that have the same elevation) derived from a

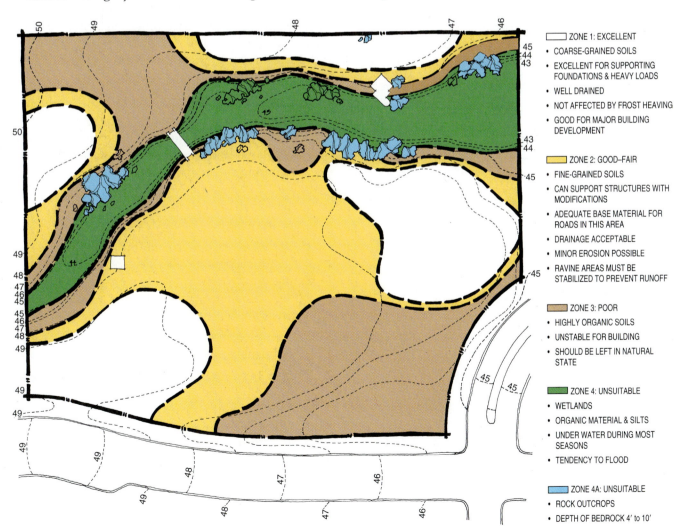

FIGURE 13.7 ■ Soil analysis drawing.

0–2% FLAT

2–3%
SLIGHT
SLOPE

3–7%
MODERATE
SLOPE

7–10%
MEDIUM
SLOPE

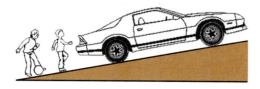

10–15%
STEEP
SLOPE

15–30%
VERY
STEEP SLOPE

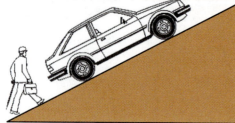

30–50%
EXTREMELY
STEEP SLOPE

FIGURE 13.8 ■ Slope percentages.

U.S. Geological Survey (USGS) map of the area. If the site is very hilly, a surveyor may add more closely spaced contours to provide a more detailed description of the slope of the site. Existing contour lines are dashed, since the finished contour grade lines will later be drawn solid. (See Chapter 18 for more information on contour lines.)

2. Identify the classification of slopes on the drawing:

■ 0% to 5%—excellent

■ 5% to 10%—good to fair

■ 10% to 25%—poor

■ over 25%—unsuitable

3. Identify each slope category using colors or tones to show the degree of development potential of each section. Generally, light colors are used for areas suitable for development and dark colors are used for less suitable areas.

4. Provide a color-keyed legend of slope categories and note both potential and constraints for development for each slope category. Note assets and/or limitations for each slope category. Note erosion or drainage problems, if any, for each category.

■ *Phase 4—Visual analysis:* Analyze the aesthetic and environmental potential of a site visually. Because visual observations and aesthetic qualities are often subjective and elusive, an organized method of recording and analyzing is important. Refer to Figure 13.10 and follow these steps to prepare a visual analysis drawing that can be used to provide input for future design phases:

1. On the base map, locate the direction of the best views from each important viewer position. Label the nature of each view, and rate it as good, fair, or poor. Make recommendations for the treatment of each view such as "enhance" or "screen."

2. Identify existing structures on the base map and describe their condition as good, fair, poor, unsound, or hazardous. Note suggestions to enhance, remove, or rehabilitate.

3. Draw the outline and location of all existing and significant plant material, such as large shrubs and trees. Label the type, and indicate the condition of each as good, fair, or poor. Also locate, draw, and indicate large stands of ground vegetation to be saved.

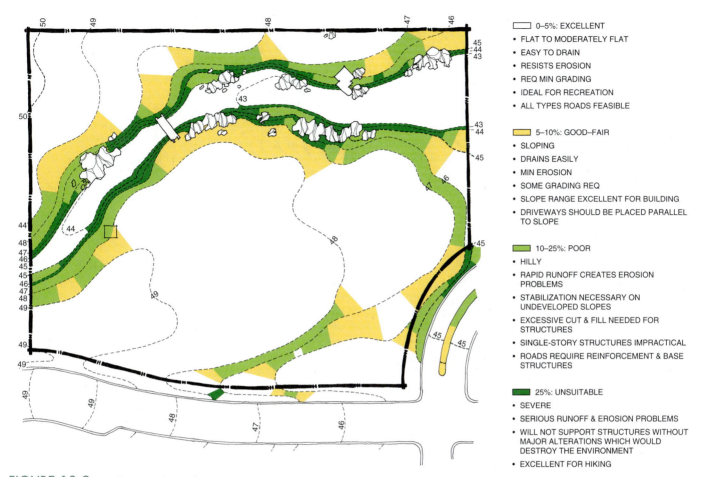

0–5%: EXCELLENT
- FLAT TO MODERATELY FLAT
- EASY TO DRAIN
- RESISTS EROSION
- REQ MIN GRADING
- IDEAL FOR RECREATION
- ALL TYPES ROADS FEASIBLE

5–10%: GOOD–FAIR
- SLOPING
- DRAINS EASILY
- MIN EROSION
- SOME GRADING REQ
- SLOPE RANGE EXCELLENT FOR BUILDING
- DRIVEWAYS SHOULD BE PLACED PARALLEL TO SLOPE

10–25%: POOR
- HILLY
- RAPID RUNOFF CREATES EROSION PROBLEMS
- STABILIZATION NECESSARY ON UNDEVELOPED SLOPES
- EXCESSIVE CUT & FILL NEEDED FOR STRUCTURES
- SINGLE-STORY STRUCTURES IMPRACTICAL
- ROADS REQUIRE REINFORCEMENT & BASE STRUCTURES

25%: UNSUITABLE
- SEVERE
- SERIOUS RUNOFF & EROSION PROBLEMS
- WILL NOT SUPPORT STRUCTURES WITHOUT MAJOR ALTERATIONS WHICH WOULD DESTROY THE ENVIRONMENT
- EXCELLENT FOR HIKING

FIGURE 13.9 ■ Slope analysis drawing.

4. Identify any wildlife population and habitat areas to be saved. Indicate animal food and water sources.

5. With directional arrows, show the direction of prevailing winter winds. Also show the direction of prevailing summer breezes.

6. Find and label the source of any desirable fragrances and/or undesirable odors. For the latter, indicate possible solutions, such as minimizing with aromatic vegetation, screening, or removal of the source.

7. Locate and label exposed open space, semi-enclosed public space, and private space.

■ *Phase 5—Composite analysis:* Once the soil, slope, and visual analysis drawings are completed, this information is combined into a single composite analysis drawing. A composite analysis drawing is prepared to determine the best location zones (ar-

eas) for the placement of structures on the site. Location zones are judged and numbered for development potential:

1. Excellent

2. Good or fair

3. Poor

4. No development potential

To prepare a composite analysis drawing refer to Figure 13.11 and follow these steps:

1. Place the soil analysis drawing directly over the slope analysis. Align the property lines with the base map and tape the base map to the drawing board.

2. Attach tracing paper over the slope and soil drawings and trace a line around each distinct area.

3. Determine which development zone each outlined area represents. One example would be if a 0 to

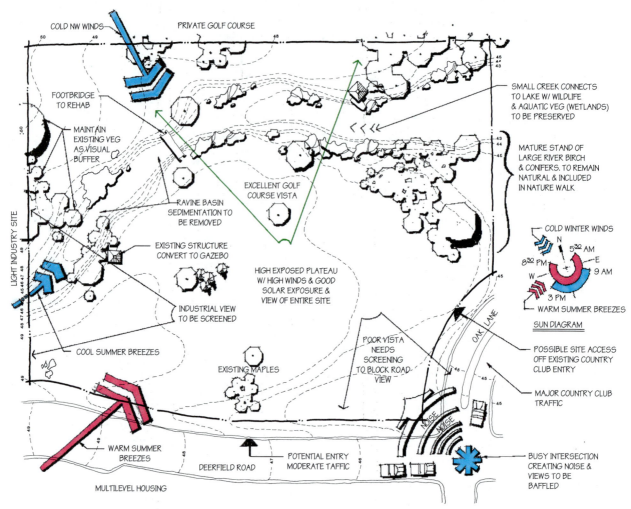

FIGURE 13.10 ■ Visual analysis notes.

5 percent slope area overlaps with a coarse-grained soil area, the zone is labeled "1" (excellent potential). Another example would be a poor, clay soil area overlaps with a 20 percent slope area, the zone is labeled "3" (poor). Label each zone on an overlay drawing.

4. Place the overlay drawing over the visual analysis drawing and repeat the same outlining of areas covered in step 3 to complete the composite analysis drawing as shown in Figure 13.11. Apply judgment concerning priorities when there is an overlapping area conflict.

Developing a Conceptual Design

A **conceptual design** represents the best response to the information on the site analysis and in the user analysis chart. Two types of sketches are created to develop a conceptual design: idealized and site-related.

Idealized Drawings

Idealized drawings or diagrams are a series of study sketches (usually drawn on inexpensive tracing paper, nicknamed "trash" or "bum wad"). These designate ideal spatial relationships of the user elements from the user analysis. Ideal diagrams are freehand, bubble-like sketches that show how the separate user elements fit together. The bubbles are not used to show the *sizes* of the areas, only their *spatial* relationships. Designers usually complete a number of studies or sketches. They do as many sketches as necessary until they achieve one study sketch that provides the best possible spatial relationship between the different elements. See Figure 13.12.

Site-Related Drawings

A **site-related drawing**, such as that shown in Figure 13.13, is one that matches the idealized drawing to the site and introduces size requirements. The main effort at

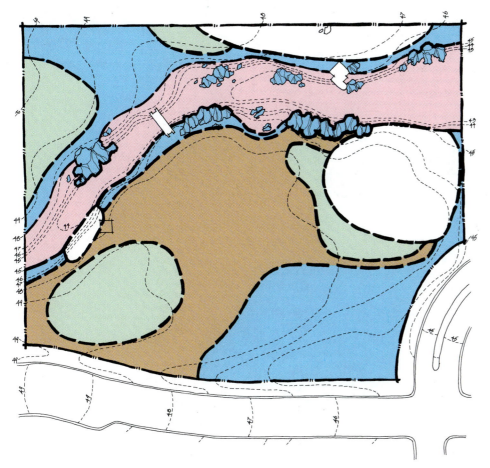

ZONE 1: EXCELLENT
- PRIME DEVELOPMENT LOCATION
- COMPACTABLE SOILS
- EXCELLENT DRAINAGE
- SLOPE 0–5%
- MINIMUM EROSION PROBLEMS
- ROAD & SERVICES EXCELLENT
- SOLAR ORIENTATION OF STRUCTURES POSSIBLE

ZONE 2: GOOD–FAIR
- GENTLE SLOPES 0–10%
- COMPACTABLE SOIL
- GOOD DRAINAGE
- EROSION PROBLEMS MINIMAL EXCEPT NEAR RAVINE
- PRIME AREA FOR SITING STRUCTURES

ZONE 3: GOOD TO FAIR WITH MODIFICATIONS
- VARIABLE SLOPE 0 TO 10%
- SOIL SUITABLE FOR STRUCTURES WITH MODIFICATIONS
- HIGH RUNOFF RATES NEAR BANKS
- EROSION RISKS AT BANKS
- VEG REMOVAL AT CREEK BANKS COULD DAMAGE ECOTONE
- EXCELLENT FOR NATURAL USE FOR PATHS

ZONE 4: RESTRICTED DEVELOPMENT
- ORGANIC SOIL
- POOR DRAINAGE
- GENERALLY UNDER WATER
- FOG POCKET DANGERS
- LEAVE UNDERDEVELOPED AND AS NATURAL AS POSSIBLE

ZONE 4A: RESTRICTED DEVELOPMENT
- ESTABLISHED WOODED STANDS
- WILDLIFE HABITAT & FEEDING GROUNDS AT WATERS EDGE
- PROVIDES BUFFER FROM ROAD

FIGURE 13.11 ■ Composite analysis of site.

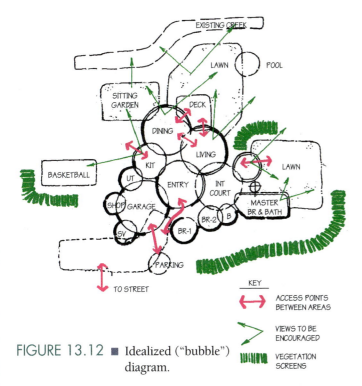

FIGURE 13.12 ■ Idealized ("bubble") diagram.

KEY
↔ ACCESS POINTS BETWEEN AREAS
↗ VIEWS TO BE ENCOURAGED
〰 VEGETATION SCREENS

this stage of design is concentrated on "fitting" all the various elements of the user analysis onto the site, while maintaining the most ideal spatial relationships.

The scale (size) of each element is first introduced at this phase of the design process. The approximate position, size, and shape of each room, area, or feature are sketched on tracing paper placed over the composite site analysis drawing. Now the floor plan design and position begin to take physical form in relation to the land and the surroundings. Several site-related studies are usually completed which integrate the design with the site. At this stage, a designer needs to focus on the specific characteristics of the site—its constraints and opportunities.

From all the site-related diagrams and sketches, one is chosen that becomes the floor plan conceptual design. The designer now begins to generate the form of the design. Drawings are refined into a loose graphic drawing for evaluation. See Figure 13.14. This drawing may not be exactly to scale, but becomes a preliminary floor plan.

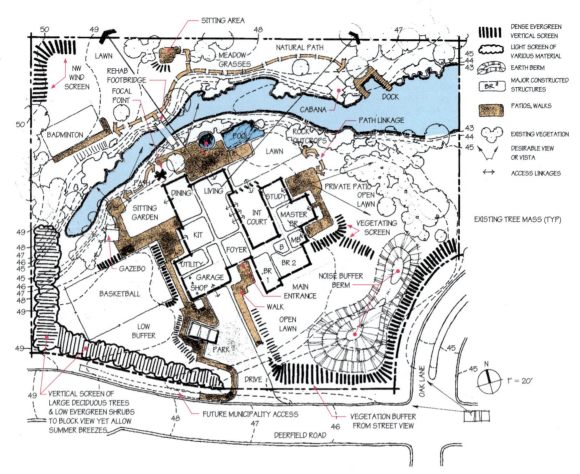

FIGURE 13.13 ■ Site-related floor plan sketch.

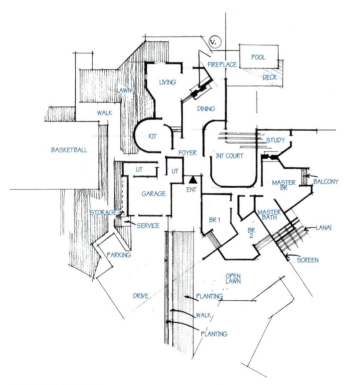

FIGURE 13.14 ■ Floor plan conceptual study drawing.

Evaluating the Design

Evaluation is always needed to determine the degree of excellence of a design. Self-evaluation of a design is critical and necessary. This requires checking the quality to see if it measures up to the predetermined goals and objectives and to the user analysis requirements.

The conceptual design must be evaluated and necessary revisions made before beginning the final design development phase. To redesign some elements at the conceptual design stage is easier, cheaper, and more time efficient than later in the process.

Many details and sizes will not be determined at this point. However, the position of the structures and the relationship between the design elements should not change significantly after this evaluation is completed.

To evaluate a design, the contents of conceptual design must be compared with *each* specific goal and objective in the user analysis. If a goal has not been accomplished, then that part of the design must be altered to achieve the desired result. A well-developed design will contain very few discrepancies between the goals of the user analysis and the conceptual design.

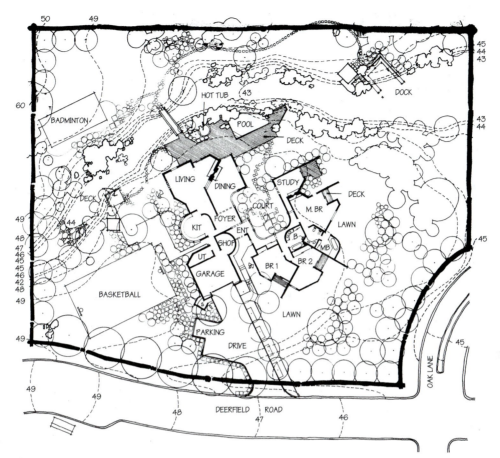

FIGURE 13.15 ■ Scaled floor plan and site drawing.

Design Development

After the necessary changes have been made in the conceptual design as a result of evaluation and client feedback, the final design development phase begins. During this phase, details are added to the site-related diagram in progressive sketches. Sketches are redone until the outlines of the design parts fit together without overlapping and without awkward offsets.

Once the design is "smoothed out," a scaled, **single-line drawing** is prepared, as shown in Figure 13.15. This drawing includes both floor plan and site features. After this drawing is completed, the designer can concentrate on refining interior building floor plans. See Figure 13.16.

FUNCTIONAL SPACE PLANNING

Once a conceptual plan is developed, the exact space needed for each area needs to be finalized. Experienced designers can determine the relationships of areas and

record design ideas through the use of progressive sketches. Students and inexperienced designers should proceed through a process of determining final space requirements through the use of templates. The work done with templates, as presented in this chapter, can be performed manually or on a CAD system.

Floor Plan Sketches

A floor plan design sketch must satisfy all original goals and objectives. Once this is accomplished, many more sketch versions of the floor plan may be necessary. Think of the first sketch as only the beginning. Many sketches are usually necessary before a designer achieves an acceptable floor plan. Through successive sketches costly and poor design features can be eliminated. By planning to use standard building materials and furnishings, many sizes are established. The exact positions and sizes of doors, windows, closets, and halls should be determined at this point. You may wish to consider both open and closed types of floor plans. Refer back to Chapter 7.

FIGURE 13.16 ■ Revised and refined floor plan.

Further refinement of the design is done by resketching until a satisfactory design is reached. Except for very minor changes, making a series of sketches is always better than erasing and changing the original sketch. Many designers use tracing paper to trace the acceptable parts of the design and then add design improvements on the new sheet. This procedure provides the designer with a record of the total design process. Early sketches sometimes contain ideas and solutions to problems that might develop later in the final design process.

A final scaled sketch should be prepared on grid paper to provide a more accurate and detailed sketch. It should also include the locations of shrubbery, trees, patios, walks, driveways, pools, and gardens. Once a final sketch is complete, three-dimensional conceptual CAD models may also be developed to aid in interpreting the conceptual design.

After plans at this stage are approved by the client, contractor, and zoning and building departments, the preparation of working drawings can begin. Working drawings and documents are prepared to further refine the basic design concepts into very exact plan sets that can be used for bidding, budgeting, and construction purposes. You will learn more about these processes later in this text.

Variations in Developing Floor Plans

The design process and sequence of preparing floor plans have been considered from the "inside-outside" point of view. The needs of the inside areas determine the size and shape of the outside. However, some design situations require a plan to be developed within a given, predetermined outside area. Apartment, modular units, mobile, and manufactured home design fall

FIGURE 13.17 ■ Manufactured home designs must fit into a predefined space.

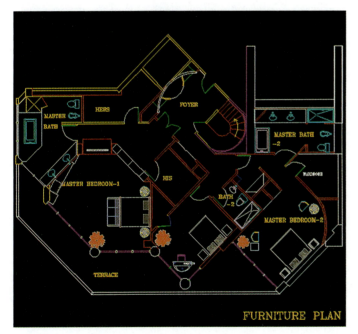

FIGURE 13.18 ■ Apartment floor plan fitted into a pre-defined space. *(Alfred Karram Design)*

FIGURE 13.19 ■ Three-stage expansion floor plans. *(Home Planners, Inc.)*

into this category. Figure 13.17 shows how "outside-in" applies to mobile home design. Figure 13.18 shows a floor plan developed to fit within a predetermined apartment space.

Because of limitations of time or money, it may be desirable to construct a house over a period of time. A house can be built in several steps. The basic part of the house can be constructed first. Then additional rooms (usually bedrooms) can be added in future years as the need develops.

When future expansion of the plan is anticipated, the complete floor plan should be drawn before the initial construction begins, even though the entire plan set may not be complete at that time. If only part of the building is planned and built and a later addition is made, the addition will invariably look "tacked on." This appearance can be avoided by designing the original floor plan for expansion. See Figure 13.19.

Planning Space for Rooms and Areas

Furniture and Equipment

Selecting the style, size, and amount of furniture needed for a room is the first step in determining a room's space requirements. Furniture should be chosen according to the needs of the residents—whether that means including a piano for someone interested in music or a large amount of bookcase space for an avid reader. The artist, drafter, or engineer may require drawing furniture and/or computer hardware in the den or study. These individual pieces of furniture affect each room's specific size and shape. Figure 13.20 shows the common sizes of residential furniture.

ITEM	LENGTH, IN (mm)	WIDTH, IN (mm)	HEIGHT, IN (mm)
COUCH	72(1829)	30(762)	30(762)
	84(2134)	30(762)	30(762)
	96(2438)	30(762)	30(762)
LOUNGE	28(711)	32(813)	29(737)
	34(864)	36(914)	37(940)
COFFEE TABLE	36(914)	20(508)	17(432)
	48(1219)	20(508)	17(432)
	54(1372)	20(508)	17(432)
DESK	50(1270)	21(533)	29(737)
	60(1524)	30(762)	29(737)
	72(1829)	36(914)	29(737)
STEREO CONSOLE	36(914)	16(406)	26(660)
	48(1219)	17(432)	26(660)
	62(1575)	17(432)	26(660)
END TABLE	22(559)	28(711)	21(533)
	26(660)	20(508)	21(533)
	28(711)	28(711)	20(508)
TV CONSOLE	38(965)	17(432)	29(737)
	40(1016)	18(457)	30(762)
	48(1219)	19(483)	30(762)
SHELF MODULES	18(457)	10(254)	60(1524)
	24(610)	10(254)	60(1524)
	36(914)	10(254)	60(1524)
	48(1219)	10(254)	60(1524)
DINING TABLE	48(1219)	30(762)	29(737)
	60(1524)	39(914)	29(737)
	72(1829)	42(1067)	28(711)
BUFFET	36(914)	16(406)	31(787)
	48(1219)	16(406)	31(787)
	52(1321)	18(457)	31(787)
DINING CHAIRS	20(508)	17(432)	36(914)
	22(559)	19(483)	29(737)
	24(610)	21(533)	21(787)

ITEM	DIAMETER, IN (mm)	HEIGHT, IN (mm)
DINING TABLE (ROUND)	36(914)	28(711)
	42(1067)	28(711)
	48(1219)	28(711)

FIGURE 13.20 ■ Common furniture sizes.

USING CAD

Designing with CAD Symbols

Drawing symbols are entities or groups that are used repeatedly and are stored in libraries as blocks. Typical items that are blocks include windows, doors, map symbols, and fixtures. A block is a group of entities that is stored as a single object typically as a separate drawing file. Drawings stored only within the current drawing that cannot be used outside the current drawing are created by the *Block* command. Drawings can also be stored as a separate drawing file that can be used in any drawing; this is accomplished by using the *Wblock* command.

All drawing files (.dwg) can be used like a *Wblock*. Blocks (generic term used for both *Block* and *Wblock* creations) are placed back into drawings by using the *Insert* command. Choose the file to be inserted, enter the scale of the drawing (typically 1″ = full size) and then the insertion point on the drawing. The insertion point on the inserted object was created as the *Block* insertion point in the *Block/Wblock* command. Blocks can also be used to replicate whole drawings like an apartment building. Autocad programs include libraries of both architectural and engineering drawing symbols.

After furniture dimensions are established, furniture templates can be made and arranged in functional patterns. As you learned in previous chapters, furniture templates are thin pieces of paper, cardboard, plastic, or metal that are used to determine exactly how much floor space each piece of furniture will occupy. See Figure 13.21.

Templates are always selected to the scale that will be used in the final drawing of the floor plan. The scale most frequently used on floor plans is 1/4″ = 1′-0″. Scales of 3/16″ = 1′-0″ and 1/8″ = 1′-0″ are sometimes used for larger buildings.

Wall-hung furniture, or any projection from furniture, even though it does not touch the floor, should be included as a template. This is necessary because the floor space under this furniture is not usable for any other purpose.

FIGURE 13.21 ■ Templates representing furniture width and length.

Figure 13.22 shows the use of furniture templates on a floor plan. Furniture templates are placed in the arrangement that will best fit the living pattern anticipated for the room. Space must be allowed for the free flow of traffic, as well as for opening and closing doors, drawers, and windows.

Room Sizes and Shapes

After suitable furniture arrangements have been established, room dimensions can be determined by drawing an outline around the scaled furniture templates. Then a **room template** can be made by cutting around the outline of the room.

Because the cost of a home is largely determined by the size and number of rooms, room sizes must also be adjusted to conform to an acceptable price range. Figure 13.23 shows area sizes for each room in small, medium, and large dwellings. These areas represent only average sizes. Even where no financial restrictions exist, room sizes should be limited by the functional requirements of the room. Just as a room can be too small, it can also be too large to function well for its intended purpose.

Visualizing the exact amount of real space that will be occupied by furniture or that should be allowed for traffic through a room is sometimes difficult. One device used to give a point of reference is a template of a human

figure. See Figure 13.24. This template will help you see how a person would move throughout the room. With a human figure template, you can check the appropriateness of furniture size, number, placement, and the adequacy of traffic allowances.

Combining Areas into a Floor Plan

Students and inexperienced designers often prefer to create floor plans through the use of templates rather than use the idealized (bubble) diagram method shown in Figure 13.12. Figure 13.25, page 212, shows the sequence of creating or evaluating space, starting with furniture needs through the development of a scaled sketch. As areas are combined, adjustments are made to allow space for such features as fireplaces, traffic flow, and storage space. Unlimited design variations may be possible within one defined area. The next step is to sketch these areas and rooms into a floor plan in the same manner as "fitting" progressive sketches into an idealized diagram.

DEVELOPING PLANS TO ACCOMMODATE SPECIAL NEEDS

Special design provisions must often be made to enable people with physical impairments that restrict mobility, hearing, or vision to use areas and facilities within and around buildings. Design requirements for full accessibility in public buildings are found in building codes. Information for special residential designs can be obtained from federal, state, and local governmental agencies and some private organizations and companies. Information can also be accessed on-line and from libraries. Following these guidelines also aids in designing residences that promote "aging in place." This concept allows older citizens to live in their homes for many added years.

Following are two lists of some of the many design guidelines for planning buildings that can be fully, safely, and conveniently used by persons with impairments. One list applies to public buildings and the other to residences. However, the same general principles and practices often apply to both. Always follow federal regulations and check with the state and local agencies in the area in which the structure will be built for regulations that apply to the development of plans for any specific project. Also, talk with people who have impairments. Find out what kinds of problems they encounter. Then find ways

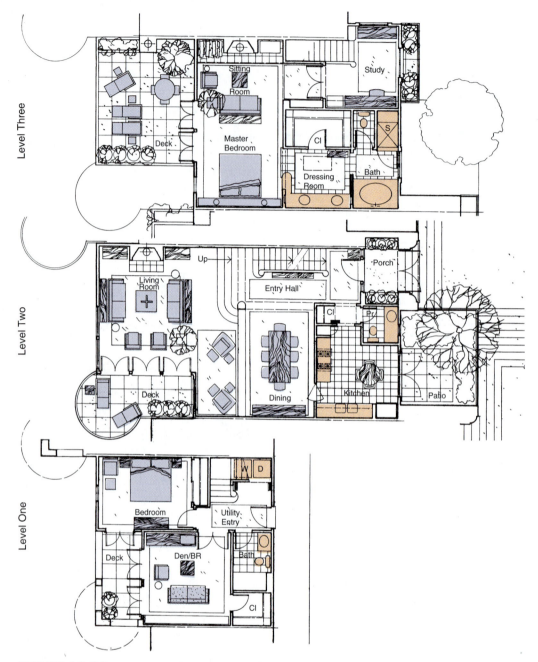

Level Three

Level Two

Level One

FIGURE 13.22 ■ Furniture templates used to check adequate room sizes.

to eliminate the problems by utilizing safe and acceptable design alternatives.

Public Buildings

All building codes contain design requirements for public buildings that prohibit the use of structural barriers or infringements on the comfort and safety of persons with impairments.

Outdoor Considerations

1. A passenger loading zone at least 4′ × 20′ must be provided near an accessible entrance.

2. The minimum width of ramps is 3′-0″, with a maximum slope of 1:12 and a maximum rise of 2′-6″. Handrails must be provided for ramps longer than 6′-0″, and at least 6′-0″ of level area must be provided at the top of each ramp. Nonskid surfaces must be used on all ramps.

3. Parking facilities must be provided in the parking area nearest the building and marked with the international access symbol. This area must be out of the main traffic flow and connected to the building by a ramp if the level changes. Parking spaces for persons with impairments must be at least 8′-0″ wide and have an access area of at least 4′-0″ or 5′-0″ wide depending on local code. See Figure 13.26.

4. For people with visual impairments, walkways must be at least 4′-0″ wide, with no less than 6′-8″ headroom clearances. Walks must be level and ramps must be used when it is necessary to change levels. Walks must be free of obstructions and be surfaced with nonskid material. At least 32″ must be allowed for a cane sweep width, as shown in Figure 13.27.

Entrances and Doors

5. At least one entrance must be accessible to wheelchair traffic and provide access to the entire building.

6. Doors must open at least 90° and be 2′-8″ (min.) to 3′-0″ wide, with threshold heights no more than 3/4″ on exterior doors and 1/2″ on interior doors. Walls or objects can be no closer than 4′-0″ from the door hinge. See Figure 13.28.

7. Doors should be provided with handles that can be opened with a closed fist.

8. Thresholds should not be higher than 1/2″ on swinging doors and 3/4″ on sliding doors.

9. Door kickplates should cover the bottom 10″ of a door accessed by wheelchair users.

Floors and Pathways

10. Texture changes using raised strips, grooves, rough, or cushioned surfaces should be used to warn people with visual impairments of an impending danger area, including ramp approaches.

ROOM	SMALL	MEDIUM	LARGE
Formal Living	160	200	400 +
Dining	120	200	300 +
Kitchen	100	160	250 +
Utility-Laundry	40	60	120 +
Master Bedroom	150	250	320 +
Bath	40	80	120 +
Den/office	80	120	200 +
Family-great room	190	250	470 +
Foyer	60	100	160 +
Porch	50	100	200 +
Closet	10	20	30 +
Walk-in Closet	30	50	100 +
Halls	3′ wide	3′-6″	4′-0″ +
2-Car Garage	440	540	680 +
3-Car Garage	640	800	940 +
Bedrooms	120	170	250 +

FIGURE 13.23 ■ Common room sizes (in square feet).

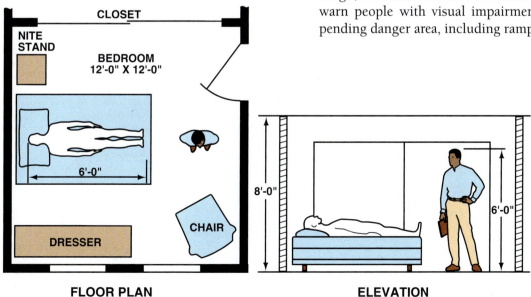

FLOOR PLAN   ELEVATION

FIGURE 13.24 ■ Human templates used to check space needs.

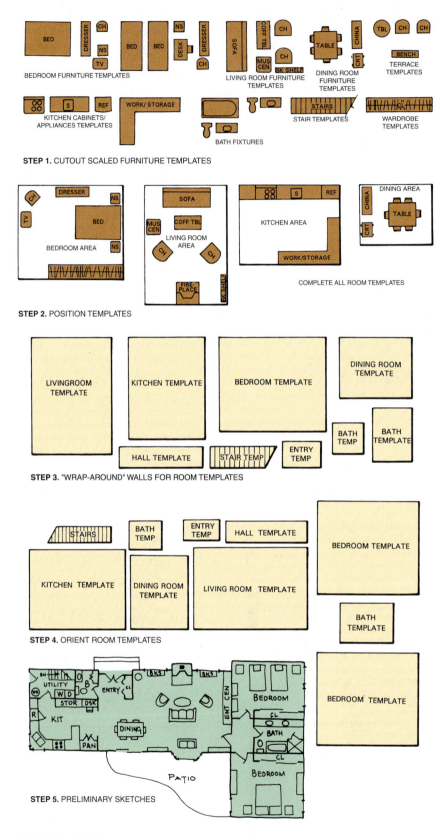

FIGURE 13.25 ■ Sequence of designing with floor plan templates.

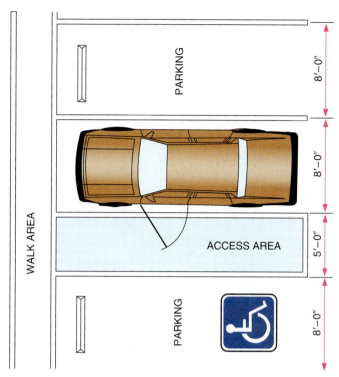

FIGURE 13.26 ■ Standard handicapped parking and access space.

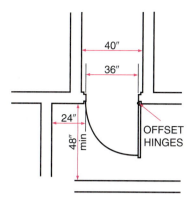

FIGURE 13.28 ■ Door and wall clearance for people with impairments.

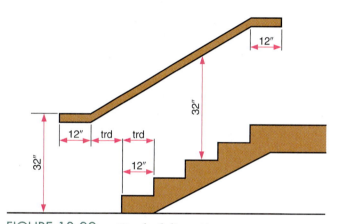

FIGURE 13.29 ■ Handrail dimensions.

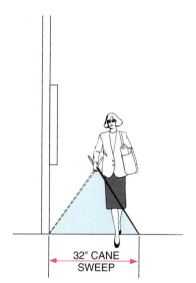

FIGURE 13.27 ■ Cane sweep distance.

11. Floors should have nonslip surfaces even when wet, or be covered with carpeting with a pile thickness of no more than 1/2″. Floors should have contrasting color borders to warn people with limited vision of a solid object ahead.

Indoor Traffic Areas

12. Hall widths must be at least 3′-0″, with 5′-0″ provided in all turning areas. Halls must provide access to all areas of the building without the need to pass through other rooms.

13. There should be a minimum of three treads in a series of stairs. Treads and risers should be uniform and treads should have a minimum width of 11″ with round nosings. A landing should be planned for stair systems that contain more than 16 risers. Handrails must be at least 32″ above the floor and tread height. See Figure 13.29.

14. At least one ramp (or elevator) must be provided as an alternative to stairs.

15. Wall projections, if located between 27″ and 80″ from the floor, cannot extend more than 4″ from a wall. Objects mounted below 27″ from the floor

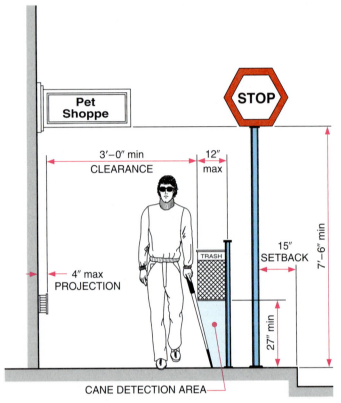

FIGURE 13.30 ■ Projection limits of objects from walls.

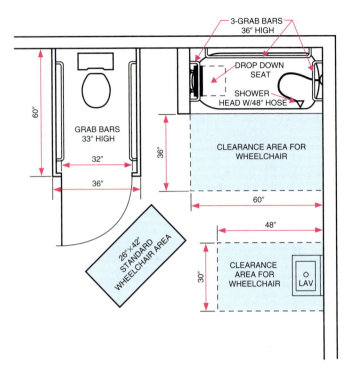

FIGURE 13.31 ■ Lavatory clearances for people with impairments.

may project any amount. Freestanding objects between 27″ and 80″ may only project 12″ from their support, as shown in Figure 13.30.

Signs, Alarms, Lighting

16. Public phone amplifiers, eye-level warning lights to augment audio alarms, and high-frequency alarms should be provided for people with hearing impairments.

17. Braille signs, level-change warning surfaces, and restrictions on wall protrusions over 4″ must be provided for people with visual impairments. Elevator buttons must also include braille.

18. Protruding signs must be at least 7′-6″ from the floor.

19. Emergency warning alarms should be both visible and audible.

20. Lighting should be free of glare or deep shadows.

Lavatory Facilities

21. Lavatory facilities must include a 32″ wheelchair turning radius. Water closet seat tops must be 1′-6″ from the floor. A 27″ minimum knee-room

height must be provided under sinks and drinking fountains. See Figure 13.31. Grab bars must be provided near water closets, sinks, and bath areas.

22. Sinks should be mounted no closer to the floor than 30″ and should extend a minimum of 17″ from the wall to provide adequate knee space, as shown in Figure 13.32. A minimum floor space area of 2′-6″ × 4′-0″ must be provided around sinks for wheelchair access. See Figure 13.33.

23. The controls for all fixtures and appliances should be within reach from a wheelchair.

24. Door handles and controls for fixtures should be the lever type, which allows people with little strength to operate them.

Residences

Designers must carefully follow building codes when designing a residence for someone with an impairment. Beyond that, working closely with the client is important. Except in extreme cases, people with impairments have abilities as well. By identifying the specific needs and capabilities of the client, "overdesigning" can be avoided and a safe, convenient, and comfortable design can be achieved.

Design features that affect people with mobility limitations usually relate to providing for wheelchair use. Because

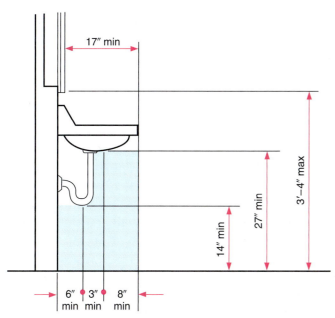

FIGURE 13.32 ■ Undersink knee-space requirements for wheelchair use.

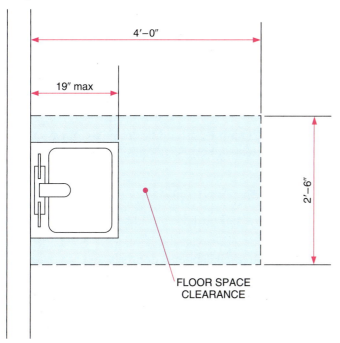

FIGURE 13.33 ■ Floor space clearances needed around sinks for wheelchair use.

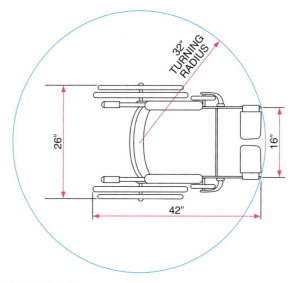

FIGURE 13.34 ■ Wheelchair turning radius.

Some general guidelines for designing residences are provided in the following list. Unless stated otherwise, design guidelines apply to wheelchair use.

Outdoor Entrances/Exits

1. To accommodate wheelchairs, the pathway or ramp to the entrance should be 36″ to 48″ wide and have a nonslip surface such as outdoor carpeting or sand paint.

2. If a ramp is used, a minimum landing platform of 5′ × 5′ should be located in front of the door. Plan the shape of the platform to accommodate the door swing and still allow easy access. For long ramps, more landings may be needed. A covering for protection from the weather is recommended. If room is unavailable for a ramp, consider planning for the use of a mechanical lift.

3. The vertical rise of a ramp should be 1:12. Handrails should be 32″ to 36″ high and extend 1′-0″ past the end of the ramp.

4. The height of the doorknob from the floor should be 36″ or less. Threshold height should be 1/2″ or less.

Indoor Traffic Areas and Floors

5. Levers on doors can be operated more easily than doorknobs.

6. Doorways need to be 32″ to 36″ wide. If a turn is required for a wheelchair to pass through a doorway, be sure the doorway is wide enough or provide extra turn space in front of the door.

wheelchairs require the greatest amount of space compared to other disability apparatus, plans that accommodate wheelchairs will easily function for other design conditions. Figure 13.34 shows the dimensions and turning radius of a *standard* wheelchair. Wheelchairs vary in size, however. When designing a floor plan for use by someone in a wheelchair, use the dimensions of the largest model.

7. For ease of use by people in wheelchairs and people with limited vision, doorways should not have raised thresholds.

8. Hallways must be 36″ to 48″ wide.

9. Hardwood floors or tiled surfaces are best. If carpeting is preferred, use carpet that has short, dense pile.

Living Areas

10. Rooms should have 5′-0″ or more of clear area for turning a wheelchair.

11. Furniture planning and placement should allow adequate area for wheelchairs to move through the room. People with limited vision should have clear passageway through rooms and not be required to walk around articles of furniture.

12. Height of tables and work areas, such as desks, should be approximately 30″.

Kitchens

13. Allow adequate work area space for turning a wheelchair or for using crutches or a walker, usually 16 sq. ft. to 25 sq. ft. Important factors are the shape of the kitchen and the arrangement of appliances that create the work triangle.

14. Appliance cooking controls should be placed in front of the burners. Ovens should have side-hinged doors. All controls must be operable with a closed fist.

15. Braille control panels for people who are blind or have extremely limited vision are available from some appliance manufacturers.

16. A clear 28″, 31″, or 36″ floor space should be provided under selected base cabinets or next to appliances.

17. Countertops should be 30″ to 33″ from the floor, with 27″ to 29″ for knee clearance. Cabinet pulls should be recessed.

18. Dishwashers, washers, and dryers should be front loading and have front controls.

19. Side-by-side refrigerator/freezers are most convenient.

20. Sinks need to be 34″ or less from the floor.

Baths

21. Water closet seats should be 1′-6″ from the floor.

22. Lavatories should not be higher than 34″ from the floor and should have adjoining counter space. They need to be open underneath. Exposed pipes should be insulated. A single faucet with a lever control is preferred.

23. The bottom of the mirror should be 40″ from the floor. (Consider the eye level of a person in a wheelchair.)

24. The top of the medicine cabinet should not be more than 50″ from the floor.

25. Space should be allowed near the bathtub to allow for transfer from a wheelchair. Reinforced grab bars should be installed.

26. A shower should be at least 5′ × 4′ in size, and the floor should have a nonslip surface. Reinforced grab bars should be installed. There should be no lip on the floor surface entrance to the shower.

Bedrooms and Storage Areas

27. Bedrooms should be designed to allow wheelchair maneuverability and access to the bed, as shown in Figure 13.35. The top surface of bedroom furniture should be a maximum of 34″ above the floor.

28. Sliding doors are preferred for closets. Hang rods should be 4′-6″ or less from the floor.

29. Storage facilities should be designed for easy reach from a wheelchair. See Figure 13.36.

Electrical Considerations

30. Switches and outlets should be 40″ from the floor.

31. Lights or other visual cues (aids) should be connected to the telephone, doorbell, and other devices as needed for people with hearing impairments.

32. Locate switches on the latch side of doors.

33. Plan a switch to control at least one light in each room.

34. Outlets should average one for every six feet of wall space.

35. Kitchen appliance outlets should average one for every four feet of wall space.

36. GFCI outlets should be installed near any water source.

37. All outside outlets must be weatherproof outlets.

38. Specify alarms to detect smoke, gas, sound, and movement.

39. Provide a Braille marked master distribution panel.

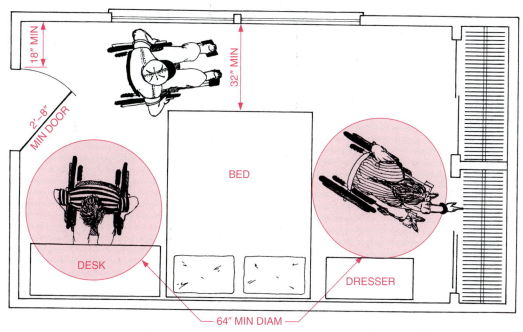

FIGURE 13.35 ■ Bedroom wheelchair clearances.

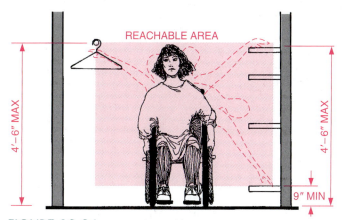

FIGURE 13.36 ■ Reaching distances from wheelchairs.

FENG SHUI

The design process presented in this chapter is based on analyzing personal wants and needs, site characteristics, and the development of spatial relationships in a floor plan. Other design practices such as feng shui are also used in developing floor plans. **Feng shui** is an ancient form of design based on the concepts of energy flow, balance, and harmony in the relationship of natural elements. This involves the systematic orientation and arrangement of many design components such as furniture placement, land forms, waterways, materials, color, and floor plan layout.

CHAPTER 13

Designing Floor Plans Exercises

1. List the design steps necessary to design a residence through the development of a conceptual design.
2. Prepare a situation statement and set goals and objectives for a house of your own design.

3. Explain how a composite analysis is prepared and used to create a plan of a design.

4. Prepare room templates and use them to make a functional arrangement for the living area, service area, and sleeping area of a house.

5. Arrange templates for a sleeping area, service area, and living area of your own design in a total plan.

6. Make room templates of each room in your own home. Rearrange these templates according to a remodeling plan, and make a sketch.

7. Make a list of furniture you would need for a home you will design. The list should include the number of pieces and size (width and length) of each piece of furniture.

8. Make a furniture template 1/4″ = 1′-0″ for each piece of furniture you will include in a home of your design.

9. Develop a floor plan for a family of four, including two small children and someone who uses a wheelchair.

10. Gather pictures of floor plans from real estate magazines and home-planning catalogs. Evaluate them in terms of their space planning arrangements.

11. Choose a floor plan in Chapter 14 and redesign it in two ways. One plan should accommodate a person in a wheelchair and the other one a person who is visually impaired.

12. Change the first floor master suite in Fig. 13.37 to a recreation room. Design and sketch a second floor master suite.

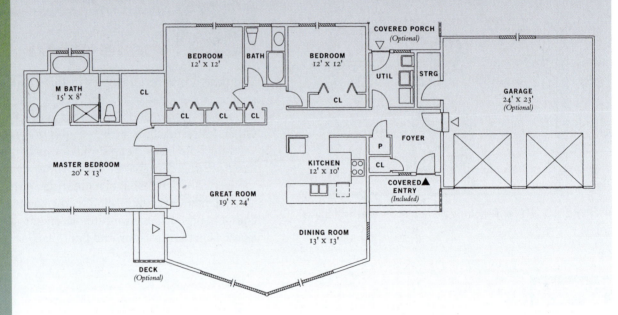

FIGURE 13.37 ■ H Floor plan to be revised and second floor added. (*Lindal Cedar Homes*)

CHAPTER 14

Drawing Floor Plans

OBJECTIVES

In this chapter, you will learn to:

- use information on a scaled floor plan to draw a complete floor plan.

- name and explain the types of floor plans.

- use graphic symbols to communicate information on a floor plan.

- draw a floor plan according to a sequence of steps.

- draw dimensions that convey precise, accurate information for builders.

TERMS

abbreviated floor plans
break line
dimension lines
floor plan sketches
multiple-level floor plans

mullions
muntins
object lines
overall dimensions
reflected ceiling plans

reversed plans
schedules (door/window)
subdimensions
symbols
working drawings

INTRODUCTION

A complete floor plan is a scaled drawing of the outline and partitions of a building as seen if the building were cut (sectioned) horizontally about 4′ (1219 mm) above the floor line. See Figure 14.1. There are many types of floor plans, ranging from very simple sketches to completely dimensioned and detailed floor plan working drawings.

TYPES OF FLOOR PLANS

Floor plans are classified by the amount and type of information each conveys. **Floor plan sketches** (Figure 13.25) are used in the design process or on a construction site and contain minimal details. Single-line floor plans as shown in Figure 13.14 are used for the same purpose but are more accurate in scale. **Abbreviated floor plans**, which include minimal dimensions or labels, are used primarily for marketing and display purposes. See

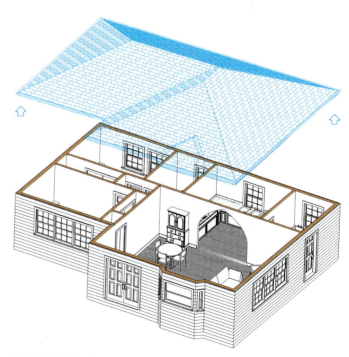

FIGURE 14.1 ■ A floor plan is a section through a building.

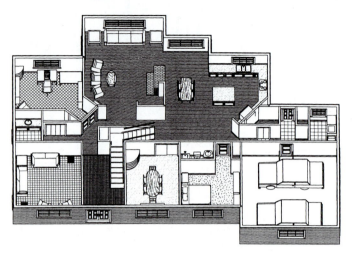

FIGURE 14.2A ■ A presentation floor plan with furniture and floor materials shown.

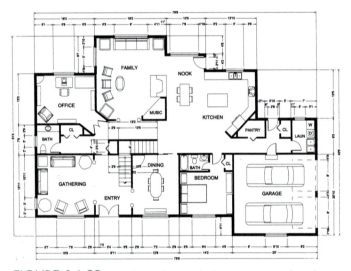

FIGURE 14.2B ■ Floor plan with dimensions and without textures.

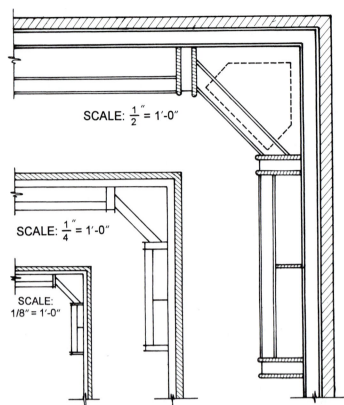

FIGURE 14.3 ■ Floor plan details at different scales.

furniture, appliances, cabinets, connecting walks, patios, lanais, or decks. Wall and surface construction materials are also shown.

The prime function of floor plans is to communicate information to building contractors. Complete working-drawing floor plans prevent misunderstandings between designers and builders. Contractors should be able to correctly interpret working drawings without consultation.

Specialized floor plans are developed from basic working-drawing floor plans. For construction and installation of electrical, plumbing, and HVAC (heating, ventilating, air-conditioning) systems, separate specialized plans are drawn with specific symbols added. On very small projects, these symbols may all be included on the basic plan. However, this often makes the drawing too crowded and difficult to read. For most projects, electrical, HVAC, and plumbing plans are separate drawings. These are covered in detail in Chapters 31, 32, and 33, respectively.

The amount of detail used in drawing floor plans depends on the scale of the plan. Figure 14.3 shows the difference between the amount of detail possible on two plans at different scales. If more information is needed than the scale of the plan allows, a removed detail drawing should be prepared.

Figure 14.27 later in this chapter. Abbreviated floor plans are often drawn with furniture and surface treatments to add realism. The floor plan in Figure 14.2A is identical to the plan in Figure 14.2B, but the plan in Figure 14.2A is a pictorial that includes furniture and surface textures, whereas Figure 14.2B has no textures added and is dimensioned.

Drawings that contain all information needed to construct a structure are called **working drawings**. Completely dimensioned and accurately scaled floor plans are working drawings necessary for construction. Basic information included shows the size and position of all exterior walls, interior partitions, fireplaces, doors, windows, stairs, built-in

FLOOR PLAN SYMBOLS

On drawings, drafters use **symbols** to identify construction materials such as fixtures, doors, windows, stairs, and partitions. The use of symbols saves time and space. Imagine trying to repeat a description every time that a material or component is used!

Common symbols used on floor plans include symbols for walls, doors, windows, appliances, fixtures, sanitation facilities, and building materials. See Figure 14.4. Floor plan symbols for plumbing, heating, air-conditioning, and electrical components are covered in later chapters. Architectural symbols are standardized. However, some variations of symbols are used in different parts of the country.

Wall Symbols

Different types of wall construction are represented by different floor plan wall symbols. On simple plans, walls are represented by single lines. However, on working-drawing floor plans, the actual scaled width of each wall is drawn. Figures 14.5 and 14.6 show methods of representing different types and variations of wall construction on floor plans.

Door Symbols

Floor plan door symbols show the top view of a door and the width of each doorway. Door symbols usually show each door open 30° to 90° and connected to an arc that represents the door swing. Figure 14.7 illustrates a door symbol and methods of representing wall openings (or doorways) that do not include a door. Notice that the outline of the doorsill is added to all exterior door symbols on floor plans.

Door Types

Interior doors, those located within interior partitions, and exterior doors, those that lead outdoors, are generally flush, paneled, or louvered. Interior flush doors have a hollow core covered with a thin wood, plastic, or metal veneer. The core of exterior flush doors is solid to

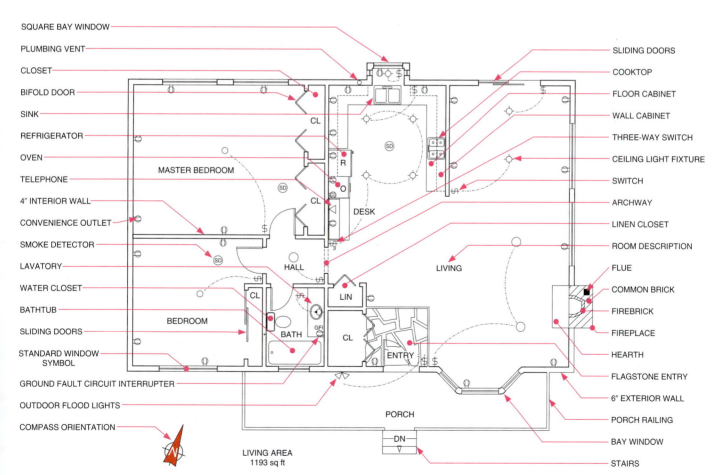

FIGURE 14.4 ■ Floor plan symbols in place.

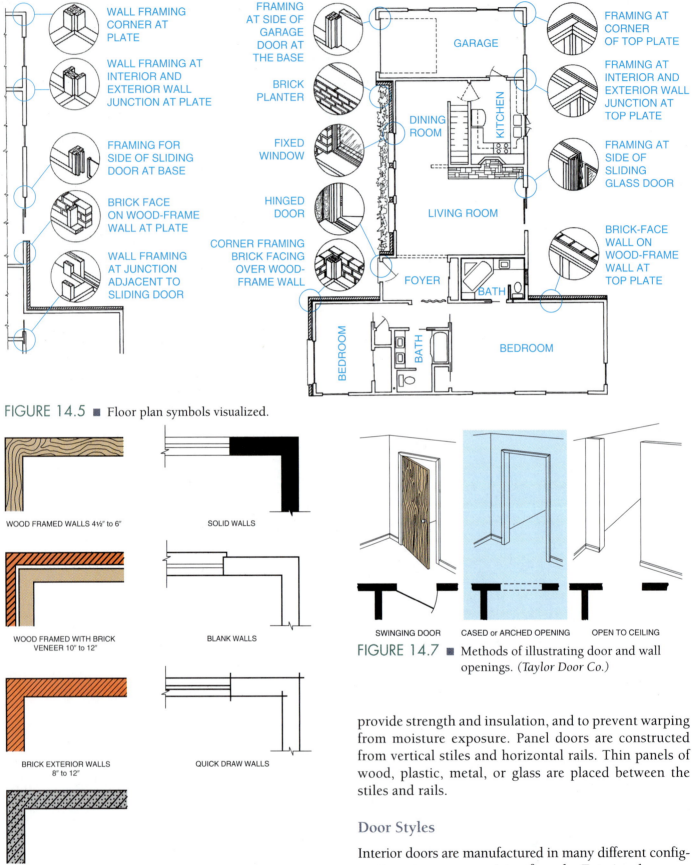

FIGURE 14.5 ■ Floor plan symbols visualized.

WALL FRAMING CORNER AT PLATE

WALL FRAMING AT INTERIOR AND EXTERIOR WALL JUNCTION AT PLATE

FRAMING FOR SIDE OF SLIDING DOOR AT BASE

BRICK FACE ON WOOD-FRAME WALL AT PLATE

WALL FRAMING AT JUNCTION ADJACENT TO SLIDING DOOR

FRAMING AT SIDE OF GARAGE DOOR AT THE BASE

BRICK PLANTER

FIXED WINDOW

HINGED DOOR

CORNER FRAMING BRICK FACING OVER WOOD-FRAME WALL

GARAGE

DINING ROOM

KITCHEN

LIVING ROOM

FOYER

BATH

BEDROOM

BATH

BEDROOM

FRAMING AT CORNER OF TOP PLATE

FRAMING AT INTERIOR AND EXTERIOR WALL JUNCTION AT TOP PLATE

FRAMING AT SIDE OF SLIDING GLASS DOOR

BRICK-FACE WALL ON WOOD-FRAME WALL AT TOP PLATE

WOOD FRAMED WALLS 4½" to 6"

WOOD FRAMED WITH BRICK VENEER 10" to 12"

BRICK EXTERIOR WALLS 8" to 12"

CONCRETE BLOCK WALLS 4" to 12"

SOLID WALLS

BLANK WALLS

QUICK DRAW WALLS

FIGURE 14.6 ■ Methods of drawing different wall types.

SWINGING DOOR   CASED or ARCHED OPENING   OPEN TO CEILING

FIGURE 14.7 ■ Methods of illustrating door and wall openings. *(Taylor Door Co.)*

provide strength and insulation, and to prevent warping from moisture exposure. Panel doors are constructed from vertical stiles and horizontal rails. Thin panels of wood, plastic, metal, or glass are placed between the stiles and rails.

Door Styles

Interior doors are manufactured in many different configurations to serve a variety of needs. Exterior doors are generally single- or double-swing doors for entrances and bypass sliding glass doors for patio or deck traffic. See

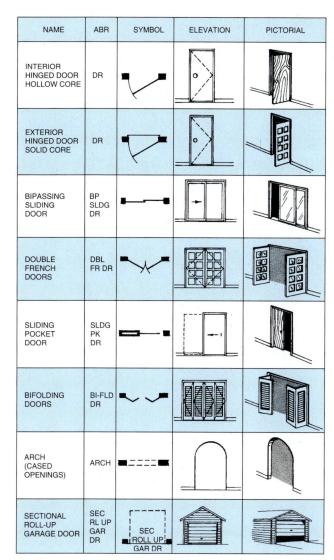

FIGURE 14.8 ■ Common door symbols.

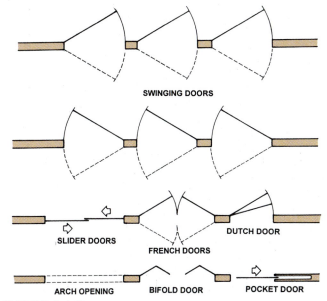

FIGURE 14.9 ■ Swinging, sliding, and folding door configurations.

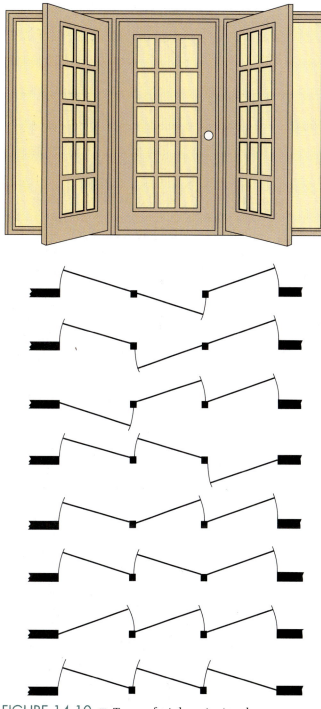

FIGURE 14.10 ■ Types of triple swinging door configurations.

Figure 14.8. Figure 14.9 shows swinging, sliding, and folding door patterns. Many added configurations are available for three unit components, as illustrated in Figure 14.10.

Door styles are also indicated on door schedules and on elevation drawings that are cross-referenced with floor plans. **Schedules** are detailed lists that contain information such as size and type. Often on a floor plan, a number in a square near the door symbol identifies the door style. See Figure 14.11.

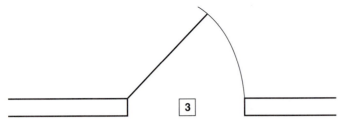

FIGURE 14.11 ■ Door schedule key on a floor plan.

Door Sizes

Different door types and styles are available in many size ranges for width, height, and thickness. See Figure 14.12. If a door schedule is not used, door width and height dimensions are often shown directly on the door symbol. When this is done, the foot and inch dimensions are abbreviated, as shown in Figure 14.13. Exact door framing information should always be determined from the manufacturer's data.

Most building codes require that exterior doors be solid and at least 3'-0" wide. Interior doors must be at least 2'-6" wide but can be hollow core. Bathroom doors must be at least 2'-2" but wider widths are recommended. The minimum width of wheelchair access doors is 2'-8", with 3'-0" being preferable.

Window Symbols

Floor plan window symbols show the outline of the sash, glass position, and any mullions and muntins. A **mullion** is a vertical member separating multiple windows. **Muntins** are vertical and horizontal framing strips that separate window panes (lights). Mullions and muntins are often confused. See Figure 14.14. Windows are often distinguished by the manner in which they open. For example, on casement windows, the direction of swing is indicated much like it is in a door symbol. On awning windows, the outline of the open window position is shown with dashed lines. On small-scale drawings, often only the sash outline or glass position is shown. Figure 14.15 shows the plan and elevation symbols for common window types.

Only the width of windows is needed to draw window symbols on floor plans. Height dimensions are shown either on elevation drawings or stated in a window schedule that contains size, style, type, and manufacturer's information. Figure 14.16 shows common window sizes. Exact sizes for rough framing must always be secured from manufacturers' data.

On a floor plan, a letter in a circle is provided as a key or cross-reference to the window schedule.

TYPE	WIDTH	HEIGHT
Exterior Swing	2/6 2/8 3/0 3/6 4/0	6/8 7/0 8/0
Sliding Glass: Single Unit Double 3 Panel 4 Panel	3/6 3/0 3/4 4/0 4/4 5/0 6/0 8/0 7/0 8/0 9/0 12/0 10/0 12/0 16/0	6/8 6/10 7/0 8/0 6/8 6/10 7/0 8/0 6/8 6/10 7/0 8/0 6/8 6/10 7/0 8/0
Sidelights Transom	1/0 1/2 1/4 1/6 2/6 2/8 3/0 3/6 4/0 6/0 8/0 9/0 10/0 12/0	6/8 7/0 8/0 1/0 1/2 1/6
Double French	3/0 4/0 5/0 6/0	6/8 7/0 8/0
Garage Roll-up	8/0 9/0 10/0 12/0 14/0 15/0 16/0 17/0 18/0 20/0	6/6 6/8 6/9 7/0 7/6 8/0
Interior Swing	2/4 2/6 2/8 3/0 3/6 4/0	6/8 7/0 8/0
Closet Slide Panel	1/0 1/2 1/4 1/6 1/8 1/10 2/0 2/4 2/6 2/8 3/0 3/6 4/0	6/8 7/0
Bifold	2/0 2/4 2/6 2/8 3/0	6/8 7/0
Pocket	2/6 2/8 2/10 3/0	6/8
Folding	4/0 6/0 8/0 12/0 16/0 20/0 24/0	6/8 7/0

FIGURE 14.12 ■ Common light construction door sizes.

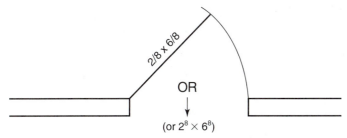

FIGURE 14.13 ■ Methods of labeling door sizes on floor plans.

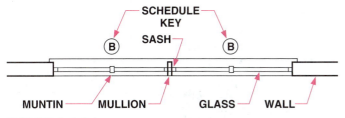

FIGURE 14.14 ■ Elements of a floor plan window symbol.

USING CAD

Drawing Floor Plans on CAD

Floor plan walls are drawn using the *Line* or *Multiline* command. The *Multiline* command task can only be used to draw a linear wall style. Creating a wall using the *Multiline* command allows the designer to lay out the specifics (wall thickness) of the wall at one time. There is little flexibility in creating floor plans using the Multiline feature.

The preferred method is to use the *Line* command to draw the exterior or interior wall and then use the *Offset* command, after setting the wall thickness, to copy the line at a preset distance to complete the two-line wall system.

The *Join* command is also used to connect two perpendicular lines to create a corner by placing the cursor on both lines. After walls are drawn, symbols for doors, windows, and fixtures are added from symbol libraries or blocks. Dimensions are then added using the *Dim* command, and the *Text* command is used to add notes.

Appliance and Fixture Symbols

Figure 14.17 shows the plan and elevation symbols used for common appliances and fixtures. Overall

NAME	ABR	SYMBOL	ELEVATION	PICTORIAL
DOUBLE CASEMENT WINDOW	DBL CSMT WDW			
45° BAY WINDOW DOUBLE-HUNG	BAY WDW DHW			
BOW CASEMENT BAY WINDOW	BOW CSMT WDW			
DOUBLE-HUNG WINDOW	DHW			
HORIZONTAL SLIDING WINDOW	SLD WDW			
SWINGING CASEMENT WINDOW	CSMT WDW			
HOPPER WINDOW	HOP WDW			
DOUBLE DOUBLE-HUNG WINDOW	DBL DHW			

FIGURE 14.15 ■ Common window symbols.

width and length dimensions (as shown previously in Chapters 10, 11, and 12) can be helpful for drawing floor plan symbols.

When preparing detail drawings, manufacturers' specifications must always be used. These show the exact dimensions of each unit plus the cutout dimensions for needed clearance. Appliance and fixture details are listed on schedules, as well as on floor plans. Schedules provide more information for purchasing, related cabinet design, and installation.

TYPE	WIDTH	HEIGHT
Double-Hung	1/10 2/0 2/4 2/6 2/8 2/10 3/0 3/4 3/6 3/8 3/10 4/0 4/6	2/6 3/0 3/2 3/6 4/0 4/6 4/10 5/0 5/6 5/10 6/0
Horizontal Slider	3/8 4/8 5/8 6/0 6/6	2/10 3/6 4/2 4/10 5/6 6/2
Casement	1/6 2/0 3/0 3/4 4/0 6/0 8/0 10/0 12/0	1/0 2/0 3/0 3/4 3/6 4/0 4/6 5/0 5/4 5/6 6/0
Fixed	1/0 1/6 2/0 2/6 3/0 4/0 4/6 5/0 5/10 6/0	4/6 4/10 5/6 6/6 7/0 7/6 8/0
Hopper	2/0 2/8 3/6 4/0	1/4 1/6 1/8 2/0 4/0 6/0
Jalousie	1/8 2/0 2/6 3/0	3/0 4/0 5/0 6/0
Bay-Bow	4/0 6/0 8/0 10/0 12/0	3/0 3/4 4/0 5/0 6/0
Awning	2/0 2/6 2/8 3/0 3/4 4/0 5/4 6/0 6/8 8/0 10/0 12/0	1/6 2/0 3/0 3/6 4/0 5/0 6/0
Half-Elliptical	5/0 6/0 8/0	1/6 1/0 1/10
Half-Round Top	2/0 2/4 2/6 2/10 3/0 3/6 4/0 4/8 5/0 5/4 6/0 R	—
Quarter-Round	1/6 2/0 2/6 3/0 R	—

FIGURE 14.16 ■ Common light construction window sizes.

Bathroom Symbols

Figure 14.18 shows symbols for common bath fixtures. Symbols for freestanding units are usually drawn using a fixture template. Fixtures that align with cabinets must be carefully positioned. The type, style, and size specifications for each fixture must be taken from manufacturers' data to ensure proper fitting. This information is included in the appliance schedules and/or specifications.

Furniture Symbols

Complete working-drawing floor plans do not usually include furniture symbols because they interfere with construction notes and dimensions. Furniture symbols are used mostly by interior designers on abbreviated floor plans to represent the width and length of each furniture piece. See Figure 14.19. These symbols are either drawn with drafting instruments, furniture templates, or obtained from a computer software library.

USING CAD

CAD Floor Plan Symbols Blocks

After wall and other line work is completed, symbols such as doors, windows, fixtures, and fireplaces can be inserted. These symbols are added by selecting each symbol from a symbol block using the *Block* command and locating the position of each on the drawing with the cursor. A block can be part of the CAD software or created by drawing and filing it in a group block. Figure 14.20 shows a typical CAD floor plan symbol block.

Some blocks contain different levels of detail. Compare the CAD simplified fireplace symbol with the detailed fireplace plan, both of which are shown in Figure 14.21. Symbol libraries can be accessed by clicking on the *Tools* menu and clicking on *AutoCAD Design Center*.

NAME	ABR	SYMBOL	ELEVATION	PICTORIAL
SINK	S			
FLOOR CABINETS	FL CAB			
WALL CABINETS	W CAB			
RANGE	R			
REFRIGERATOR	REF			
DISHWASHER	DW			
OVEN BUILT-IN	O			

NAME	ABR	SYMBOL	ELEVATION	PICTORIAL
WASHER	W			
DRYER	D			
LAUNDRY TUB	LT			
WATER HEATER	WH			
COOK TOP RANGE	CK TP			
RANGE WITH OVEN COVER	R			
FOLD-UP IRONING BOARD	I BRD			

FIGURE 14.17 ■ Common appliance and fixture symbols.

NAME	ABR	SYMBOL	ELEVATION	PICTORIAL
LAVATORY FREESTANDING	LAV FR STN			
LAVATORY WALL HUNG	LAV WL HNG			
LAVATORY COUNTERTOP	LAV CNT TP			
LAVATORY CORNER	LAV COR			
WATER CLOSET ONE PIECE	WC 1 PC			
WATER CLOSET TWO PIECE	WC 2 PC			
WATER CLOSET WALL HUNG	WC WL HNG			

NAME	ABR	SYMBOL	ELEVATION	PICTORIAL
BATH TUB RECESSED	BT REC			
BATH TUB CORNER	BT COR			
BATH TUB ANGLE	BT ANG			
SHOWER HEAD	SH HD			
SHOWER SQUARE	SH SQ			
SHOWER CORNER	SH COR			

FIGURE 14.18 ■ Common bath fixture symbols.

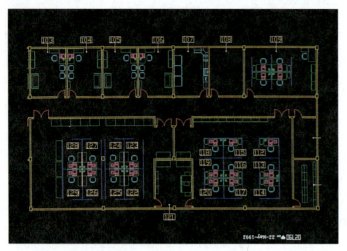

FIGURE 14.19 ■ Use of CAD library furniture symbols.

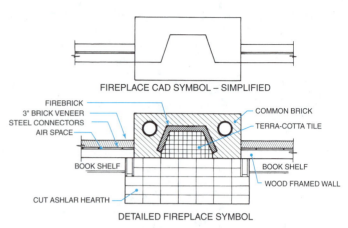

FIREPLACE CAD SYMBOL – SIMPLIFIED

FIREBRICK
3" BRICK VENEER
STEEL CONNECTORS
AIR SPACE

COMMON BRICK
TERRA-COTTA TILE

BOOK SHELF

BOOK SHELF

WOOD FRAMED WALL

CUT ASHLAR HEARTH

DETAILED FIREPLACE SYMBOL

FIGURE 14.21 ■ Standard fireplace symbol compared to a detailed symbol.

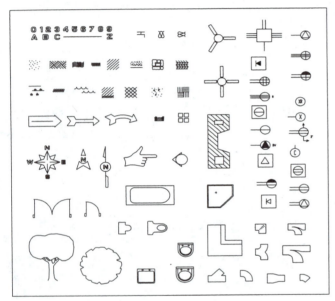

File Name: 4000-ART.DWG

ARCHITECTURAL SYMBOL LIBRARY
DOORS AND WINDOWS (PLAN VIEW)

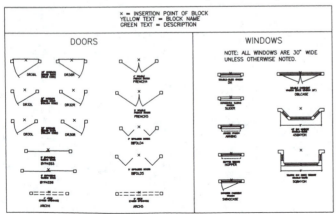

File Name: DRWNLIB.DWG

FIGURE 14.20 ■ Typical CAD library symbols.

STEPS IN DRAWING FLOOR PLANS

For maximum speed, accuracy, and clarity, the following steps should be observed in laying out and drawing floor plans. See Figure 14.22.

1. Block in the overall dimensions of the house, and add the thickness of the outside walls with a hard pencil (4H).

2. Lay out the position of interior partitions with a 4H pencil.

3. Locate the position of doors and windows by centerline and by their widths (4H). Double check all dimensions. When establishing window dimensions, be sure the window square footage is at least 10 percent of the room's square footage.

4. Darken the **object lines** (visible lines), such as the main exterior walls and interior partitions, with an F pencil.

5. Add door and window symbols with a 2H pencil. Draw the door to swing open toward a perpendicular wall to provide the most convenient access. See Figure 14.23.

6. Add symbols for stairwells, if applicable.

7. Erase extraneous layout lines if they are too heavy. If they are extremely light, they can remain.

8. Draw the outlines of kitchen and bath fixtures.

9. Add the symbols and sections for any masonry work, such as fireplaces and planters. The outline of the chimney, firebox, and hearth with material

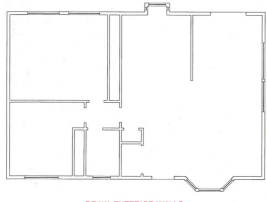

DRAW: EXTERIOR WALLS
INTERIOR WALLS
WINDOWS
DOORS

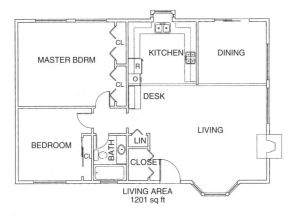

DRAW: FLOOR CABINETS   DOOR SWINGS
WALL CABINETS   FIREPLACE
APPLIANCES   ADD ROOM NAMES
FIXTURES

LIVING AREA
1201 sq ft

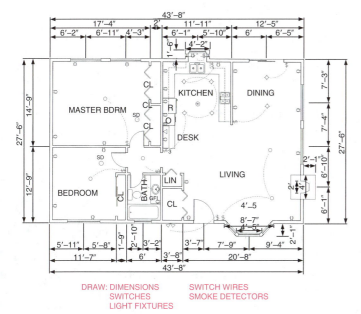

DRAW: DIMENSIONS   SWITCH WIRES
SWITCHES   SMOKE DETECTORS
LIGHT FIXTURES

FIGURE 14.22 ■ Sequence of drawing floor plans.

CORRECT ACCESS   DIFFICULT ACCESS

FIGURE 14.23 ■ Doors should swing open toward a perpendicular wall.

USING CAD

CAD Floor Plan Grids

Square or rectangular floor plans can be drawn using a two-dimensional grid system that aligns the plan on X and Y Cartesian coordinate axis lines. Grid lines or dots that represent grid intersections can be superimposed on the drawing using different color lines or line weights. Assigning grid lines to a different layer helps separate the grid lines from other elements of the plan. Grid line spacing must be in increments that are compatible with the scale of the drawing. Grid size and spacing can be set to any size.

Grids are frequently used in preparing modular drawings. The most commonly used grid spacing is 12″. The use of grids allows the design process to proceed without the use of scales. Lines can also be drawn on absolute X and Y coordinates when exact X and Y coordinates are known. Relative coordinates can be used by referencing the last used coordinates as a base.

There are two types of CAD grid systems; *Snap Grid* and *Display Grid.* The *Snap On* command allows the crosshairs to move from one grid point to the nearest grid intersection. The *Snap-Off* command allows a point to remain exactly where placed by the cursor without regard to the grid lines. The *Display Grid* command allows the grids to remain on the monitor while other drawing operations are performed. The displayed grid can be turned on or off repeatedly during the drawing process. The use of the *Ortho* command will align all drawing lines either vertically or horizontally (90°) without the use of a grid. *Polar* increments can also be set to draw/snap angular lines to any selected degrees of a circle.

USING CAD

Floor Plan Layering

All elements of a floor plan can and should be layered so that layers can be turned on, off, moved, or become a separate drawing at any time. The entire floor plan can also be a layer that contains sublayers such as doors, windows, dimensions, fixtures, furniture, and even movable objects such as people and automobiles. Using the *Layer* command allows designers to group entities together in layers to make identification and editing easier, control what will be sent to the plotter, and control the coloring and line types used on entities. Floor plans that align vertically with other plans can also be layered (first and second floors of a house) by specific colors related to what is being drawn. This usually involves outside walls, stairwells, and chimney locations. Preparing layered plans eliminates errors of alignment between plan drawings, such as placement of load-bearing walls and supports from the foundation to the first floor to the second floor through to the roof. Specialized floor plans such as plumbing, electrical, and HVAC plans are also layered to ensure agreement with the basic plan by preventing lines and fixtures from being located in the same space.

---

symbols are shown on floor plans. Fireplace construction is shown on detail drawings.

10. Dimension the drawing as instructed later in this chapter.

Size and Scale in Floor Plans

Floor plans for large commercial or industrial buildings may be drawn at a scale of 1/8″ = 1′-0″ or less. Most residential floor plans are drawn to a scale of 1/4″ = 1′-0″.

Figure 14.24A shows a complete floor plan. This plan was prepared at a scale of 1/4″ = 1′-0″. The size was reduced to fit the book page. A portion of this drawing is shown at the original 1/4″ = 1′-0″ scale in Figure 14.24B. Figure 14.24C is a photograph with the plan area indicated by a red arrow.

MULTIPLE-LEVEL FLOOR PLANS

Drawing Separate Plans

Bilevel, two-story, one-and-one-half-story, and split-level homes require a separate floor plan for each level. The separate plans of **multiple-level floor plans** are prepared on tracing paper and drawn at the same scale as the first floor plan. The tracing paper needs to be placed directly over the first-floor plan to ensure alignment of exterior walls, partitions, and vertical features. Once the major outline has been traced, the first-floor plan is removed.

Alignment of features, such as stairwell openings, outside walls, plumbing walls, vents, and chimneys, is critical in preparing second-floor plans. Where no second floor exists over part of a first floor, the outline of the first-level roof is shown. See Figure 14.25.

Figure 14.26 shows a typical second-floor plan of a one-and-one-half-story house. This drawing, in addition to revealing the second-floor plan, shows the outline of the roof as a single line, and the outline of the building as dotted lines under the roof.

Figure 14.27 shows a first-, second-, and third-floor plan of a three-level house. Visualize the position of each level by referring to the pictorial rendering of this house. Note that the three different levels of this plan do not stack evenly on top of each other. Upper floors are smaller in area. Compare the alignment of the first-level floor plan in Figure 14.24A with the upper level floor plan shown in Figure 12.14A. In drawing multiple floor plans with a CAD computer program, the layering command places each plan on a separate layer, which can be viewed separately or combined with other layers.

Calculating Dimensions for Stair Systems

Math Connection

Floor plan stair symbols show the width and depth of each tread beginning at the plan level. A **break line** (see Figure 4.17) is used to eliminate the need to draw every stair to the next level, either up or down. (See Chapter 9 for information on different types of stairs.)

When drawing multiple-level floor plans, stair systems must align on all levels. Also the number and width of treads (or steps) must be calculated and shown on all

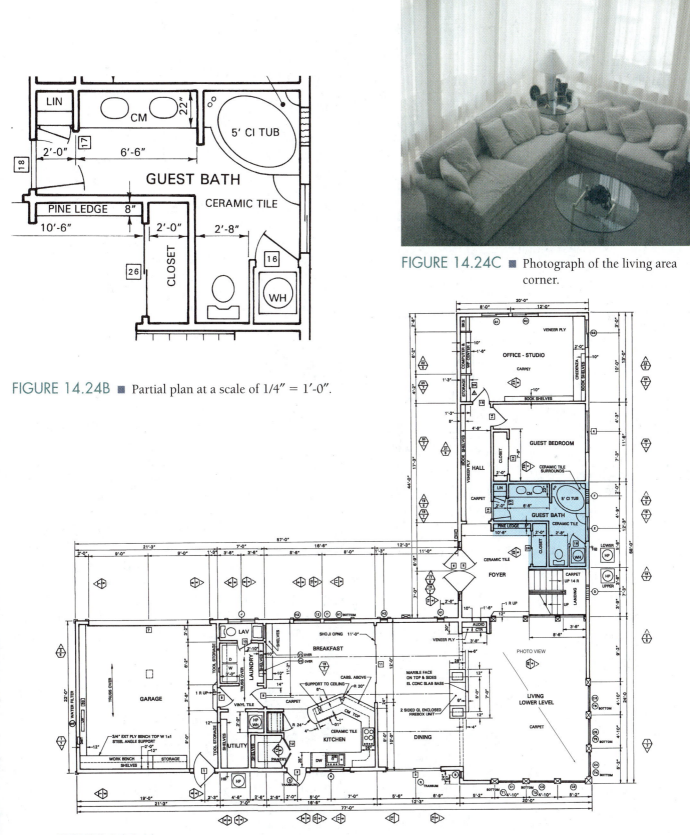

FIGURE 14.24C ■ Photograph of the living area corner.

FIGURE 14.24B ■ Partial plan at a scale of 1/4″ = 1′-0″.

FIGURE 14.24A ■ Reduced scale floor plan.

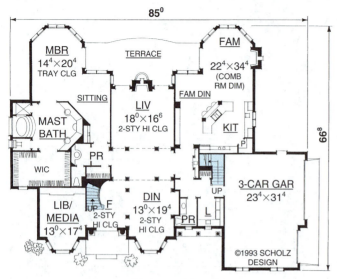

FIRST FLOOR PLAN

SECOND FLOOR PLAN

FIGURE 14.25 ■ Alignment of first- and second-floor plans. (*Scholz Homes Inc.*)

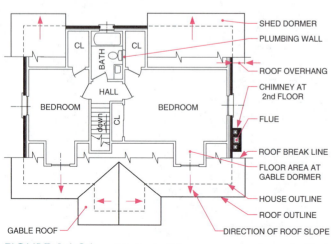

FIGURE 14.26 ■ Method of drawing one-and-one-half-story floor plan.

levels. To calculate stair tread width use the following formula:

$$\frac{\text{stair run (inches)}}{\text{number of treads}} = \text{tread width (inches)}$$

Example:
$^{144"}/_{15} = 9.6"$

In addition to indicating the stair system, plans of second floors or higher levels must also show the outline of the stairwell opening. To determine the size of the stairwell opening, the complete stair system should be designed before the final floor plan is prepared.

Figure 14.28 shows methods of determining the exact dimensions of a stair system.

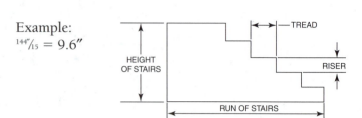

REVERSED PLANS

Floor plans offered as options in the development stage are often reversed to provide more plan choices. Reversing plans alters the appearance of a house and relocates rooms to avoid or take advantage of environmental and orientation factors and street locations. **Reversed plans** are accomplished by turning a plan over to provide a mirror image, as shown in Figure 14.29. The plan can be traced in this position or a print created by feeding the drawing into a print machine with the front side facing down. On a finished plan, this should be done before lettering to avoid printing reversed letters. On a CAD system, a mirror command is used to produce a reversed plan. The two versions are often labeled left-hand or right-hand plans.

REFLECTED CEILING PLANS

Complex ceiling designs and multiple-lighting fixtures or levels often require the preparation of a **reflected ceiling plan.** See Figure 14.30. These plans are drawn using a floor plan as a base. Floor plan walls and partitions are traced and symbols of ceiling features are drawn as the ceiling would be viewed if the floor were a mirror.

FIGURE 14.27 ■ Alignment of a three-level plan. (*Winter Park Design*)

USING CAD

Reversing Plans on CAD

If you place a mirror on its edge on a drawing you will see the reverse image of the drawing in the mirror. CAD drawings can be reversed on the X or Y axis by clicking the *Modify* menu, then the *Mirror* command. Plans are often mirrored to create variety and options in a development, to adjust a plan to better fit a site, or to reduce the amount of drafting time. To ensure that labels are not reversed the *Set Variable* command and the *Mirror Text* setting should be used. This will return the text portion of the drawing to its original readable position. Otherwise the mirror command must be done before labeling or dimensioning.

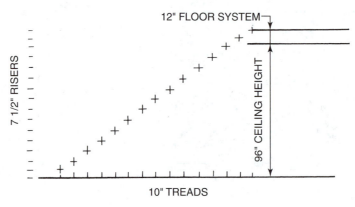

FIGURE 14.28A ■ Tread and riser layout.

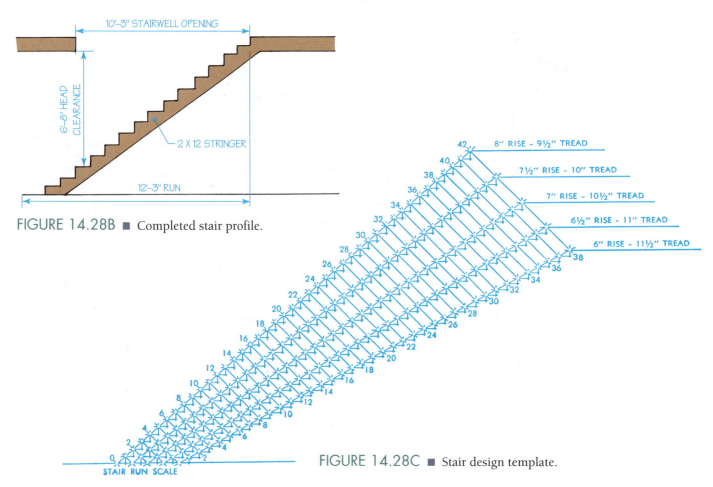

FIGURE 14.28B ■ Completed stair profile.

FIGURE 14.28C ■ Stair design template.

USING CAD

Reflecting Plans on CAD

Reflected plans such as a reflected ceiling plan are prepared by drawing ceiling features on a layer aligned with the floor plan. The layer is then printed separately to produce a separate plan.

FLOOR PLAN DIMENSIONING

A completely dimensioned drawing is necessary to complete any building exactly as designed. Dimensions on the floor plan show the builder the width and length of the building. They show the location of doors, windows, stairs, fireplaces, planters, and so forth. Just as symbols and notes show exactly what materials are to be used in

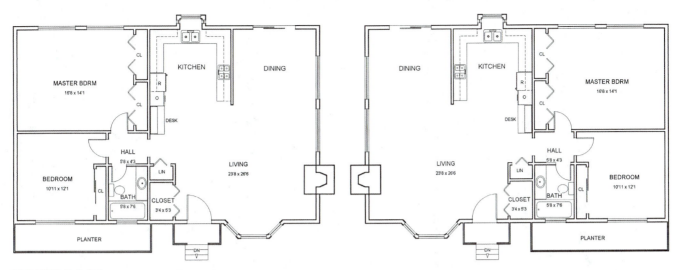

FIGURE 14.29 ■ Reversed floor plan.

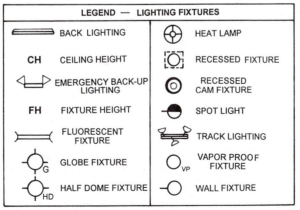

LEGEND — LIGHTING FIXTURES	
BACK LIGHTING	HEAT LAMP
CH CEILING HEIGHT	RECESSED FIXTURE
EMERGENCY BACK-UP LIGHTING	RECESSED CAM FIXTURE
FH FIXTURE HEIGHT	SPOT LIGHT
FLUORESCENT FIXTURE	TRACK LIGHTING
GLOBE FIXTURE G	VAPOR PROOF FIXTURE VP
HALF DOME FIXTURE HD	WALL FIXTURE

SEE LIGHTING SCHEDULE FOR MANUFACTURE'S DESCRIPTION AND ORDERING

FIGURE 14.30 ■ Reflected ceiling plan.

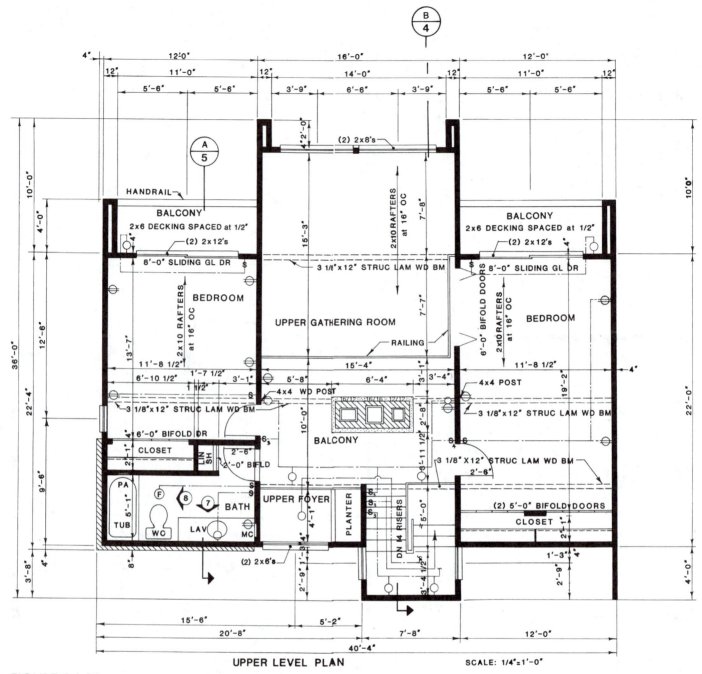

UPPER LEVEL PLAN

SCALE: 1/4"=1'-0"

FIGURE 14.31 ■ Fully dimensioned floor plan.

the building, dimensions show the sizes of materials and exactly where they are located.

Because a large building must be drawn on a relatively small sheet, a small scale such as 1/4" = 1'0" or 1/8" = 1'0" must be used. The use of such a small scale means that many dimensions must be crowded into a very small area. Therefore, only major dimensions such as the overall width and length of the building and of separate rooms, closets, halls, and wall thicknesses are usually

shown on the floor plan. Figure 14.24A shows a fully dimensioned first-floor plan and Figure 14.31 shows a second-level floor plan with many details and dimensions. Dimensions too small to show directly on the floor plan are described either by a note on the floor plan or by separate, enlarged details.

Enlarged details, or detail drawings, are sometimes merely enlargements of a portion of the floor plan. They may also be section drawings cross-referenced on the floor plan. Separate drawings are usually necessary to communicate adequately dimensions for fireplaces, planters, built-in cabinets, door and window details, stairframing details, or any unusual construction methods.

Selection Dimension

A floor plan must be completely dimensioned to ensure that the house will be constructed precisely as designed. Complete dimensions convey the exact wishes of the architect and owner to the builder. If adequate dimensions are not provided, the builder is placed in the position of a designer. A good builder is not expected to be a good designer. Supplying complete dimensions will eliminate the need for the builder to guess or interpret the size and position of the various features of this plan. All needed information about each room, closet, partition, door, and window is given.

A floor plan with only limited dimensions shows just the overall building dimensions and the width and length of each room. It is sufficient to summarize the relative sizes of the building and its rooms for the prospective owner, but insufficient for building purposes.

Math Connection

GUIDELINES FOR DIMENSIONING

Many construction mistakes result from errors made in architectural drawings. Most errors in architectural drawings are the result of mistakes in dimensioning. Dimensioning errors are therefore costly in time, efficiency, and money. Familiarization with the following guidelines for dimensioning floor plans will eliminate much confusion and error.

These guidelines are illustrated by the numbered arrows in Figure 14.32.

1. Architectural **dimension lines** are unbroken lines with dimensions placed above the line. Dimension lines should be located no closer than 1/4" from object lines. Up to 1" is preferred. Dimension lines should be spaced at least 1/4" between rows.

2. Foot and/or inch marks are normally used on architectural dimensions. Sometimes these marks are omitted and a dash or slash is used. For example 8-4 or 8/4 means 8'4". If metric measures are used, the dimensions are always in millimeters.

Therefore size unit notations are not needed. However, a note should be placed on the plan stating that all dimensions are in millimeters.

3. Dimensions over 1' are expressed in feet and inches. Detail drawings often contain only inch dimensions regardless of size.

4. Dimensions less than 1' are shown in inches.

5. A slash is often used with fractional dimensions to conserve vertical space.

6. Vertical dimensions should be placed to read from the right side of the drawing. Horizontal dimensions read from the bottom of the drawing.

7. **Overall dimensions** for the length and width of a building are placed outside other dimensions.

8. Line and arrowhead weights for architectural dimensioning are thin and dark as shown in Chapter 19. Arrowhead styles are optional. See Figure 14.33.

9. Room sizes may be shown by stating width and length on abbreviated plans.

10. When the area to be dimensioned is small, numerals may be placed outside the extension lines.

11. Framed interior walls are dimensioned to the center of partitions. Figure 14.34 illustrates the methods of dimensioning wood-framed walls, openings, and partitions.

12. Window and door sizes may be shown directly on the door or window symbol, or may be indexed to a door or window schedule with a reference callout.

13. Solid concrete walls are dimensioned from wall to wall, exclusive of wall coverings. See Figure 14.35.

14. Curved leaders are sometimes used to eliminate confusion with other dimension lines.

15. When areas are too small for arrowheads, dots may be used to indicate dimension limits.

16. The dimensions of brick or stone veneer must be added to the framing dimension. See Figure 14.36, page 240.

17. When the space is small, arrowheads may be placed outside the extension lines.

18. A dot on the end of a leader refers to the entire area noted.

19. Dimensions that cannot be seen on the floor plan or those too small to place on the drawn object are noted with leaders for easier reading.

20. In dimensioning stairs, the number of risers is placed on a line with an arrow indicating the direction down (DN) or up (UP).

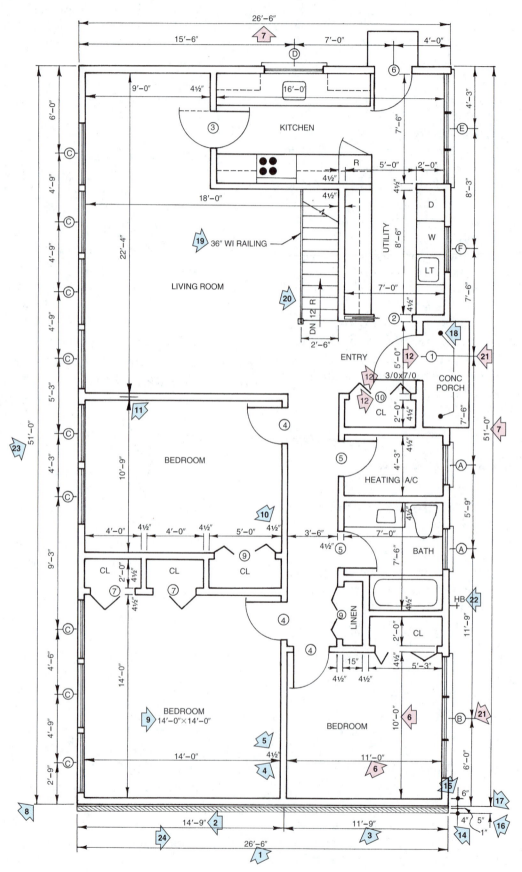

FIGURE 14.32 ■ Illustrated guidelines for dimensioning.

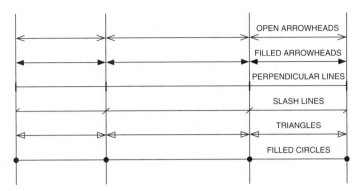

FIGURE 14.33 ■ Dimensional arrowhead styles.

21. Windows, doors, pilasters, beams, construction members, and areaways are dimensioned to their centerlines. (Areaways are the sunken areas in front of basement doors and windows that allow light and air to reach the basement or crawl space.)

22. Use notes or abbreviations when symbols do not show clearly what is intended.

23. Architectural dimensions always refer to the actual size of the building regardless of the scale of the drawing. The building in Figure 14.32 is 51′-0″ in length.

24. **Subdimensions** must add up to overall dimensions. For example: 14′-9″ + 11′-9″ = 26′-6″. Most rows of dimensions include both feet and inches and may include fractional inches. In adding rows of mixed numbers such as these, add the inches separately, convert the inch total to feet and inches, and then re-add the foot total as follows:

■ Total number of inches ÷ 12 = feet and inch fractions

■ LCD = lowest common denominator in a series into which all denominators can be divided.

Example:	To add:	Follow these steps:
	1′-7⅞″	Step 1: 1′-7¹⁴/₁₆″
	2′-8¼″	2′-8⁴/₁₆″
	6′-10⁹/₁₆″	6′-10⁹/₁₆″
	11′-2¹¹/₁₆″	9′-25²⁷/₁₆″
		Step 2: 9′-26¹¹/₁₆″
		Step 3: 11′-2¹¹/₁₆″

The following guidelines are not illustrated in Figure 14.32.

25. When framing dimensions alone are desirable, rooms are dimensioned by distances to the outside face of the studs in the partitions. See Figure 14.37.

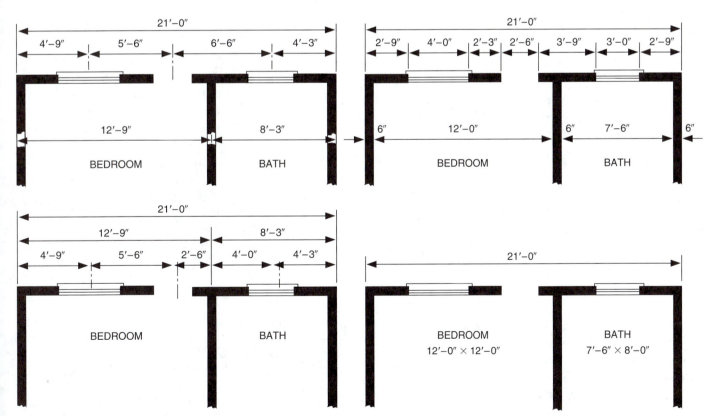

FIGURE 14.34 ■ Methods of dimensioning frame construction.

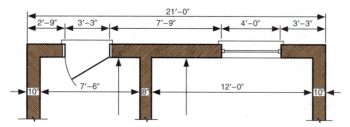

FIGURE 14.35 ■ Methods of dimensioning concrete walls.

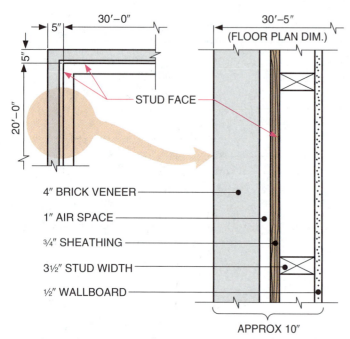

4" BRICK VENEER
1" AIR SPACE
¾" SHEATHING
3½" STUD WIDTH
½" WALLBOARD

STUD FACE

30'-0"
5"
20'-0"

30'-5"
(FLOOR PLAN DIM.)

APPROX 10"

FIGURE 14.36 ■ Method of dimensioning a brick veneer wall.

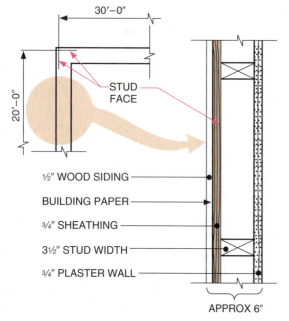

30'-0"
20'-0"

STUD FACE

½" WOOD SIDING
BUILDING PAPER
¾" SHEATHING
3½" STUD WIDTH
¾" PLASTER WALL

APPROX 6"

WOOD FRAME WALL

FIGURE 14.37 ■ Method of dimensioning to the stud face on a wood-framed wall.

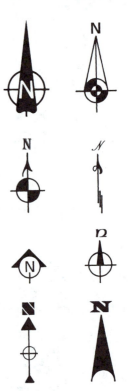

FIGURE 14.38 ■ Styles of north arrows used on plan views.

26. Because building materials vary somewhat in size, first establish the thickness of each component of the wall and partition, such as plaster, brick, or tile thicknesses. Add these thicknesses together to establish the total wall thickness. Common thicknesses of wall and partition materials are shown in Figures 14.36 and 14.37.

27. The scale of each drawing can be noted in the title block. If separate drawings on a single drawing sheet are drawn at different scales, each drawing must be labeled with the appropriate scale.

28. An arrow showing the direction of north is placed on each floor plan unless the plan is prepared without any site identification. Various types of north arrows are shown in Figure 14.38.

METRIC DIMENSIONING

If a building is designed using metric sizes, all dimensions are shown in millimeters (mm), as shown in Figure 14.39. Refer to Chapter 3 to review metric scales.

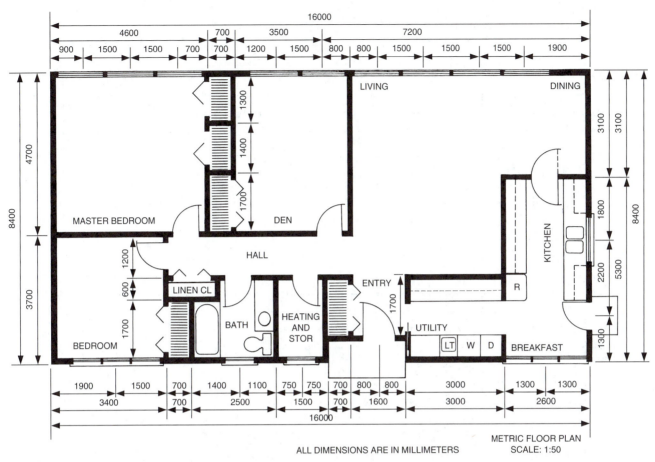

ALL DIMENSIONS ARE IN MILLIMETERS

METRIC FLOOR PLAN
SCALE: 1:50

FIGURE 14.39 ■ Metric floor plan dimensioning.

USING CAD

CAD Dimensioning

Floor plans are dimensioned by using a combination of tools that are found under the *Dimension* pull-down menu. The dimension style needs to be set up using standards from the AIA. Other setups are used for mechanical, structural, and electrical styles of drawings. The settings found in the dimensional style area are extensive—everything from arrowhead styles and text types down to the space from the end of the dimensional line to the text can be set.

Linear dimensions are the most common for all types of drawings. These are applied by selecting a line and selecting the dimension placement or by picking two extension lines and then the dimensional placement. The value of the dimension is automatically placed so accuracy when creating the drawing is important. Circular objects are dimensioned using the radius or diameter tools; angles are dimensioned using the *Angular Dimension* command.

MODULAR DIMENSIONS

An attempt should always be made to establish major dimensions to conform to 16", 24", and 48" incre-ments. However, if a building is designed to totally conform to modular grids, all dimensioning *must* conform to these standards as covered in Chapter 22. Figure 14.40 illustrates how a floor plan is fitted onto a modular grid.

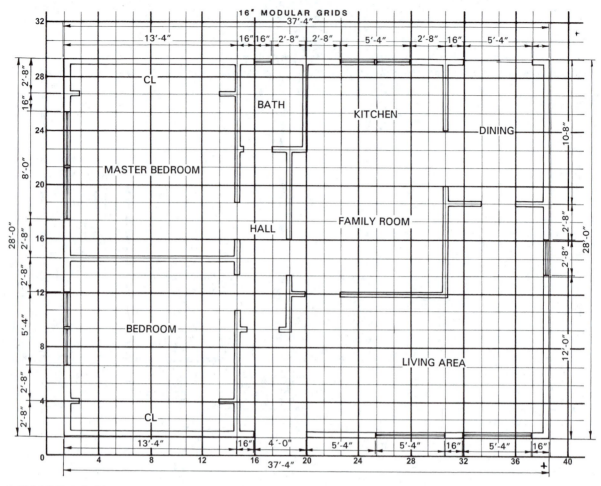

FIGURE 14.40 ■ Floor plan fitted to a 16″ modular grid.

CHAPTER 14

Drawing Floor Plans Exercises

1. Draw a complete floor plan, using a sketch of your own design as a guide, and using the scale 1/4″ = 1'0″.

2. Dimension an original scaled floor plan that you have completed for a previous assignment.

3. Draw and dimension the floor plan of your own home.

4. Design and draw a studio apartment under 800 square feet.

5. Design and draw a cabin plan under 800 square feet.

6. Design and draw a three-bedroom, two-bath home under 1,500 square feet.

7. Design and draw a two-story, four-bedroom home under 2,500 square feet.

CHAPTER 15

Designing Elevations

OBJECTIVES

In this chapter you will learn to:

- apply the principles and elements of design (Chapter 2) to creating elevation drawings.

- recognize different roof styles as options for roof design.

- select and design window styles in relation to elements of design and window functions.

- locate doors on an elevation design, considering style, size, and types of doors.

TERMS

eave line
elevation drawings
fenestration
gable

ground line
overhang
pitch
ridge line

rise
run
slope (roof)

INTRODUCTION

Elevation drawings, or elevations, show the vertical surfaces of a structure. Exterior elevation drawings show the entire front, sides, and rear of a structure. Interior elevations, discussed later in Chapter 16, show interior walls. In this chapter, you will learn how to apply the elements of design to creating the exterior form of a building, including the selection of roof, window, and door styles.

Designing the elevations of a structure is only one part of the total design process. However, the elevation design reflects the part of the building that people see. The entire structure may be judged by the elevations.

various factors. Roof style, overhang, grade-line position, and the relationship of windows, doors, and chimneys to the building line must be considered. Choosing a desirable elevation design is not an automatic process that follows the floor plan design, but a creative process that requires knowledge and imagination.

The designer should keep in mind that only horizontal distances can be established on the floor plan. However, on an elevation, vertical heights, such as heights of windows, doors, and roofs, must also be shown. As these vertical heights are established, the appearance of the outside and the way that the heights affect the internal functions of the building must be considered.

FLOOR PLAN RELATIONSHIP

Because a structure is designed from the inside out, the design of the floor plan normally precedes the design of the elevation. The complete design process requires a continual relationship between the elevation and the floor plan.

Flexibility is possible in the design of elevations, even in those designed from the same floor plan. Once the location of doors, windows, and chimneys has been established on the floor plan, the development of an attractive and functional elevation for the structure depends on

ELEMENTS OF DESIGN AND ELEVATIONS

Creating floor plans is a process of allocating interior space to meet functional needs. Designing elevations involves combining the elements of design to create functional and attractive building exteriors.

The principles and elements of design were defined in Chapter 2. In this chapter, the elements of design (line, form, space, color, light, materials, and texture) are applied to the creation of elevations. The total appearance of an elevation depends on the relationship among its component parts, such as surfaces, roofs, windows,

FIGURE 15.1 ■ Horizontal emphasis in design. (*"Falling-water," Western Pennsylvania Conservancy—Christofer Little, Photographer*)

FIGURE 15.2 ■ Vertical emphasis in design. (*Lindal Cedar Homes*)

doors, and chimneys. The balance of these parts, the emphasis placed on various components of the elevation, the texture of the surfaces, the light, the color, and the shadow patterns all greatly affect the general appearance of an elevation.

Lines

The lines of an elevation tend to create either a horizontal or vertical emphasis. The major horizontal lines of an elevation are the **ground line**, **eave line**, and **ridge line**. If these lines are accented, the emphasis will be placed on the horizontal as shown in Figure 15.1. The deck lines of this building add to the horizontal emphasis. In contrast the dominance of the vertical window mullions and the extreme high pitch of the roof give the building in Figure 15.2 a vertical emphasis. In general, a low building will usually appear longer and even lower if the design consists mostly of horizontal lines. The reverse is true for tall buildings with vertical lines.

Building lines should be consistent. The lines of an elevation should appear to flow together as one integrated line pattern. Continuing a line through an elevation for a long distance is usually better than breaking the line and starting it again. Rhythm can be developed with lines, and lines can be repeated in various patterns.

When additions are made to an existing design, care must be taken to ensure that the lines of the addition are consistent with the established lines of the structure. The lines of the component parts of an elevation should relate to each other, and the overall shape should reflect the basic shape of the building.

Because many factors affect the appearance of an elevation, the relationship of these factors is important. For

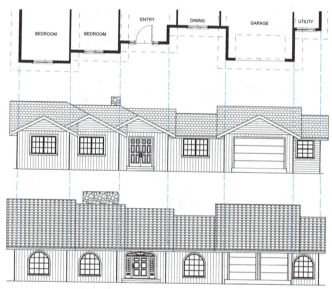

FIGURE 15.3 ■ Many factors affect elevation appearance.

example, observe how the elements of design are combined in Figure 15.3 to produce two different elevations, although both are projected from the same floor plan.

Form and Space

Lines combine to produce form and create the geometric shape of an elevation. Elevation shapes should be balanced. The term *balance* refers to the symmetry of the elevation. Formal balance is used extensively in Colonial and period styles of architecture. Informal balance is more widely used in contemporary residential architecture and in ranch and split-level styles. (Refer back to Chapter 2.)

Vertical or horizontal emphasis, or accent, can be achieved by several different devices. An area may be accented by mass, color, or material.

In addition to the elements of design, the basic architectural style of a building needs to be considered when designing elevations. A building's style is closely identified by the elevation design.

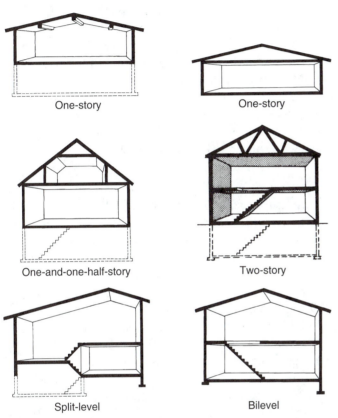

FIGURE 15.4 ■ Basic types of residential structures. *(NLMA)*

One-story

One-story

One-and-one-half-story

Two-story

Split-level

Bilevel

The type of building structure must also be compatible with the architectural style of the elevation. However, within basic styles of architecture, there is considerable flexibility in the type of structure. Figure 15.4 shows the basic types of residential structures.

Elevations should appear as one integral and functional facade, rather than as a surface in which holes have been cut for windows and doors and other structural components. Doors, windows, and chimney lines should be part of a continuous pattern of the elevation and should not appear to exist alone. See Figure 15.5.

Light and Color

An elevation that is composed of all light areas or all dark areas tends to be uninteresting and neutral. Some balancing of light, shade, and color is desirable in most elevations. Shadow patterns can be created by depressing specific areas, using overhangs, texturing, and varying colors. Door and window trim, columns, battens (strips covering joints), and overhangs can be used to create most shadows.

Materials and Texture

An elevation may contain various types of materials, such as glass, wood, masonry, and ceramics. These must be carefully and tastefully balanced for the design to be effective. An elevation composed of too many similar materials is ineffective and neutral. Likewise, an elevation that uses too many different materials is equally objectionable. In choosing materials for elevations, designers should not mix horizontal and vertical siding or different types of masonry. If brick is the

FIGURE 15.5 ■ Door and window lines related to building angles. *(Karram Designs)*

primary masonry, brick should be used throughout. It should not be mixed with stone.

ELEVATION DESIGN SEQUENCE

The first step in elevation design is to choose an architectural style. (Refer to Chapter 1.) Then sketch the outline of an exterior wall showing the roof shape and the position of doors, windows, and other key features such as chimneys or dormers.

Next, create a series of progressive sketches to develop an elevation design. Experiment with different roof styles, door and window designs, siding materials for the exterior walls, overhangs, chimney shapes, roof materials, and trim variations. Sketches can also show various architectural styles derived from the same floor plan. The sketches shown in Figure 15.6 represent part of the elevation design process, which resulted in the design shown later in Figure 15.16.

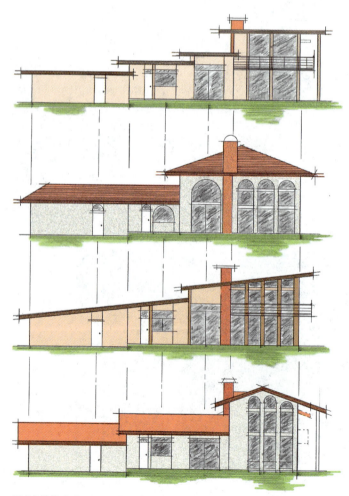

FIGURE 15.6 ■ Elevation design sketches.

Roof Types

To design elevations, a designer needs to know roof styles and which style best matches the building's overall style. There are many styles of roofs. The gable, hip, flat, and shed styles are the most popular. See Figure 15.7. Other features that affect the appearance of the roof must also be considered. These include the size and shape of dormers, skylights, vents, chimneys, and cupolas. In addition to style, the overhang size and the roof pitch (angle) must be determined during the design process.

Roof framing plans are subsequently developed from the basic roof design that is developed. A roof plan shows the outline of the top view of a roof with solid ridge, valley, and chimney lines. Dashed lines represent the outline of the floor plan under the roof. Small arrows show the slope direction. (Detailed information on roof framing drawings is presented in Chapter 30.)

Gable Roofs

A **gable** is the triangular end of a building. Roofs that fit over this area are gabled roofs. Gable roofs are the most common roof style because of their adaptability to a wide variety of architectural styles, from Colonial to contemporary. They also drain and ventilate easily. Figure 15.8 illustrates a gable roof in a Colonial structure, and Figure 15.9 shows a contemporary gable application.

Variations of gable roofs include A-frames, winged, and pleated gables. A-frame roofs extend to the floor line, creating continuous ceilings and walls inside. Winged gable roofs are created by extending the ridge overhang further than the overhang at the corners. See Figure 15.9. Pleated (folded plate) roofs consist of a series of aligned and connected small gable roofs.

Hip Roofs

Hip roofs provide eave-line protection around the entire perimeter of a building. The hip roof overhangs shade windows that would not be shaded at a gabled end. For this reason, hip roofs are very popular in warm climates. See Figure 15.10. Another variation of the hip roof is the Dutch (Deutsch) hip. A Dutch hip is created by extending the ridge outward to make a partial gable end at the top of the hip.

Flat Roofs

When a low building silhouette is desirable, flat roofs are ideal. Flat roofs have a slight slope (1/8″ to 1/2″ per foot) for drainage, unless water is used as an insulator. They can also function as decks on multilevel structures. Flat roofs do not have the structural advantage normally gained by rafters leaning on a ridge board. Heavier rafters

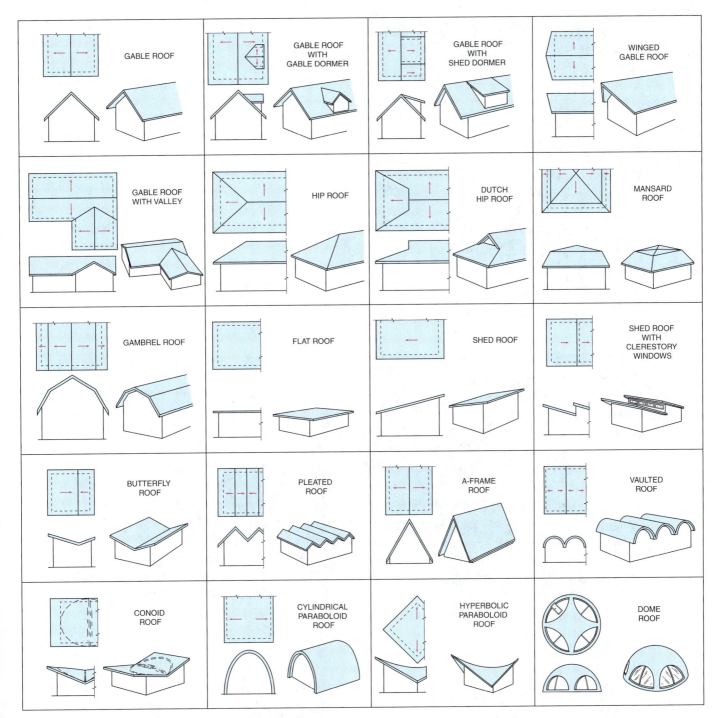

FIGURE 15.7 ■ Residential roof types.

(ceiling joists) are needed. Because of snow-load problems in cold climates, flat roofs are used more often popular in warm climates. See Figure 15.1.

Shed Roofs

A shed roof is a flat roof that is slanted. If the down slope faces south, shed roofs are ideal for solar panels. When clerestory windows are added between offsetting shed roofs, light can be provided for the center of a building. Many industrial buildings use multiple shed roofs and clerestory windows in a sawtooth pattern to maximize center light.

Butterfly Roofs

Two shed roofs that slope to the center create a butterfly roof. This roof style allows for higher outside walls, which can provide more light access.

FIGURE 15.8 ■ Colonial-style gable roof.

Gambrel Roofs

Gambrel roofs are double-pitched roofs. They are also known as barn roofs. Gambrel roofs are the distinguishing feature of Dutch Colonial houses. They are used to create more headroom in one-and-one-half-story homes.

Mansard Roofs

Mansard roofs are double-pitched hip roofs with the outside constructed at a very steep pitch. This type of roof is used on French Provincial homes.

Vaulted Roofs

Vaulted roofs are curved panel roofs. They are composed of a series of manufactured curved panels that are erected side by side between two bearing walls. This arrangement allows for larger open areas since the curved construction is structurally stable.

FIGURE 15.9 ■ Contemporary winged gable roof.

FIGURE 15.10 ■ Elevation drawing with hip and gable roofs. (*John Henry, Architect; Jones Clayton, Construction*)

FIGURE 15.11 ■ Contemporary style residence with Bermuda roofs. *(Edna Carin, Remax, Danville, CA)*

Bermuda Roofs

Bermuda roofs originated on the island of Bermuda in the Caribbean Sea. There the large fascia areas characteristic of the design are used to collect rainwater. This design effect is often used as a feature in other areas of the world. See Figure 15.11.

Dome and Dome-Shaped Roofs

As you learned in Chapter 1, domes have been used in architecture for centuries. Dome roofs, like A-frames, provide both roof and walls in one structurally sound unit. Because there is no need for internal support walls or columns, completely open floor space and flexible room sizes are possible.

The geodesic dome, developed by R. Buckminster Fuller, can be inexpensively mass produced at relatively low cost. Actually, geodesic domes are not true "domes." They are series of triangles that are combined to form hexagons and pentagons.

Several restrictions must be kept in mind when designing dome structures. The use of domes for residential construction restricts the design to a predetermined area. Working with walls that are not plumb (vertical) creates problems in fitting cabinets, fixtures, appliances, and furniture effectively into the design.

New Technology and Roof Styles

The development of new building materials and methods in molded plywood, plastics, and reinforced concrete has led to the development of many shapes of roofs. The conoid, cylindrical parabolic, and hyperbolic parabolic roofs are among the most recent designs.

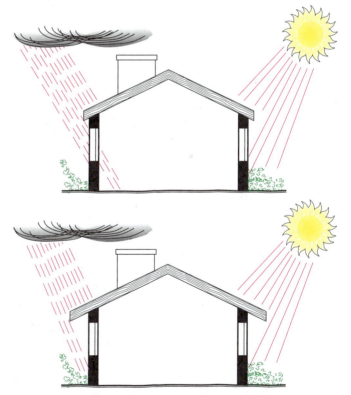

FIGURE 15.12 ■ Effects of long and short overhangs.

Roof Overhang

The **overhang** is the portion of the roof that projects past the outside walls. Sufficient roof overhang should be provided to afford protection from the sun, rain, and snow. See Figure 15.12. The length and angle of the overhang greatly affect a roof's appearance and ability to provide protection. Figure 15.13 shows that when the pitch

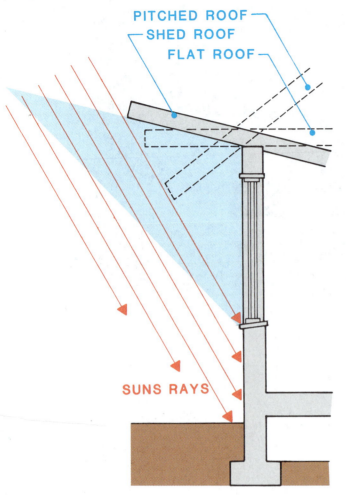

FIGURE 15.13 ■ Overhang effect at different angles.

FIGURE 15.14 ■ Overhang openings admit additional light.

is low, a larger overhang is needed to provide protection. However, if the overhang of a high-pitch roof is extended to equal the protection of the low-pitch overhang, it may block the view from the building. To provide protection and, at the same time, allow sufficient light to enter the windows, slatted overhangs may be used. The roof design shown in Figure 15.14 includes openings that allow light to penetrate while providing extensive roof overhang protection for the structure.

The fascia edge of an overhang does not always need to parallel the sides of a house. See Figure 15.15. The amount of overhang is also determined by architectural style. Large gable end overhangs, such as the 9′ overhang shown in Figure 15.16, must be supported by columns. Some gable overhangs are enclosed to form a soffit as shown earlier in Figure 15.8.

Roof Pitch

In designing roofs, the pitch or angle of the roof must be determined. The **slope** is the relation of the horizon-

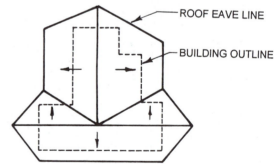

FIGURE 15.15 ■ Roof outline not aligned with building perimeter.

tal distance (run) to the vertical distance (rise). The **run** is the horizontal distance between the ridge and the outside wall. The **rise** is the vertical distance between the top of the wall and the ridge. The run is always shown in units of 12. Therefore a slope of 6/12 means the roof rises 6″ for every 12″ of run. The **pitch** is the

FIGURE 15.16 ■ Large overhang requiring column support.

ratio of the rise over the span. Pitch is covered in more detail in Chapters 16 and 30.

Dormers

The design of dormers greatly affects the silhouette of elevations. Dormer design is influenced by the floor plan outline and by the need for added light, space, and ventilation. Figure 15.17 shows the most common types of residential dormers. Dormer construction is covered in Chapter 30.

Chimneys

The outline of any structure is influenced by chimney size and shape. Large chimneys create the feeling of power and mass. In Figure 15.18 note how chimney size and shape, combined with roof pitch, overhang, and grade-line differences, can create totally different elevations using the same basic floor plan. Chimney details are covered in Chapter 23.

Skylights

Roof windows are commonly called skylights. Skylights are either domed, or flat as shown in Figure 15.19. Skylights require a cathedral ceiling unless access walls are built to connect the roof opening with a flat ceiling. Light shafts can be used to transmit light from the roof through an attic or crawl space to a flat ceiling. Figure 15.20 shows several types of light shafts.

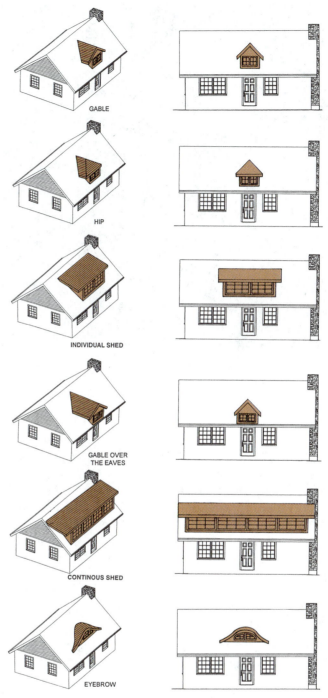

FIGURE 15.17 ■ Common dormer types.

FIGURE 15.18 ■ Different effects of chimney profile, grade line, and roof pitch on an elevation's appearance.

FIGURE 15.19 ■ Skylights provide added light where needed. (*Potlatch*)

Decks, Porches, and Balconies

Decks, porches, and balconies are primarily designed along with the floor plan. If they are covered or extend beyond the outside walls of a structure, they totally change the silhouette of a building. The pictorial elevation view of the site in Figure 15.21 reveals the scope of the total design not apparent in the Figure 15.1 view.

Window Styles

Windows are designed and located to provide light, ventilation, a view, and—in some climates—heat. To accomplish these goals, the size, location, and shape of each window must be planned according to the following guidelines:

1. Relate window lines to the elevation shape, as shown in Figure 15.5, to avoid a tacked-on look.

2. Plan window height to allow for furniture and built-in components that are placed near windows. See Figure 15.22.

3. Plan window sizes to match available standard sizes.

4. Decide which windows need to open for ventilation and which should be fixed.

5. Be sure each window functions from the inside as required.

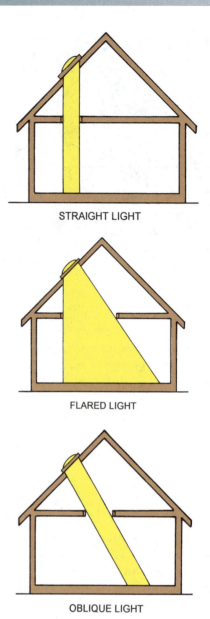

STRAIGHT LIGHT

FLARED LIGHT

OBLIQUE LIGHT

FIGURE 15.20 ■ Types of light shafts. (*Anderson Windows*)

6. Position windows to access the best views. Avoid window placement that exposes undesirable views. Avoid mullions and muntins if they restrict views. See Figure 15.23.

7. In warm climates, minimize the amount of window space on the south and maximize north-facing windows. Do the reverse in cold climates.

8. Keep the window style consistent with the architectural style of the house.

9. Where possible, align the tops of all windows and doors in each elevation.

10. If the building has more than one level, vertically align the sides of windows where possible.

FIGURE 15.21 ■ Elevation view of "Fallingwater." (*Western Pennsylvania Conservancy—Thomas A. Heinz, Photographer*)

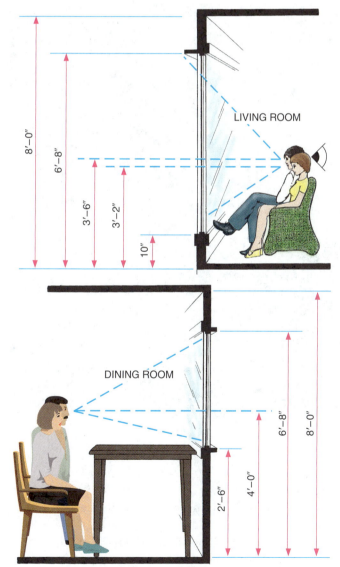

FIGURE 15.22 ■ Window height effect on view. (*Small Homes Council*)

11. Don't allow small areas between windows and other major features. Balance the wall spaces between windows, doors, and chimneys according to the principles of balance and proportion described in Chapter 2.

12. If windows are to provide the entire light source during daylight hours, 20 percent of the room's floor area should be windowed. Ten percent is considered minimum.

13. Windows that provide ventilation should be located to capture prevailing breezes and provide the best air circulation.

14. If possible, locate windows on more than one wall in each room, to provide for the best distribution of light and ventilation.

15. **Fenestration** is the arrangement of windows or openings in a wall. Arrange fenestration patterns to conform to the elements of design. See Figure 15.24.

Window Types

Windows slide, swing, pivot, or remain fixed. Choosing the right window type for each need requires a knowledge of the function and operation of each type. See Figure 15.25. Figure 15.26 shows the elevation symbols for the most common window types.

Door Styles

The style, size, and location of doors do not have as great an effect on elevation design as windows. This is because of the limited options among door sizes and for door locations. Usually an elevation will either

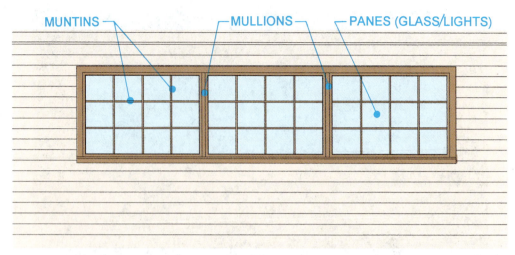

FIGURE 15.23 ■ Mullions, muntins, and panes illustrated.

FIGURE 15.24 ■ Integrated door and window wall design. (*Integrity Windows and Doors*)

have only one door or will contain no doors at all. Nevertheless, the principles of placement relative to other elevation features are the same for doors as for windows.

Door types fall into three main categories: exterior, interior, and garage doors. Exterior doors provide security and visual privacy. Interior doors provide privacy and sound control between rooms. Figure 15.27 shows an elevation view of the most common types of interior and exterior doors.

Detailed information about doors is contained in a door schedule and cross-referenced to floor plans and/or elevations. Door framing information is presented in Chapter 29.

CATHEDRAL

CLERESTORY RIBBON

CORNER

PICTURE

WALL

IN-SWING CASEMENT

AWNING

SLIDERS

DOUBLE HUNG

JALOUSIE

FIXED ARCH

RIBBON

BAY

FRENCH

OUT-SWING WOOD FRAME

FIXED DORMER

DOUBLE-DOUBLE HUNG

BOW

FIGURE 15.25 ■ Interior view of common window types.

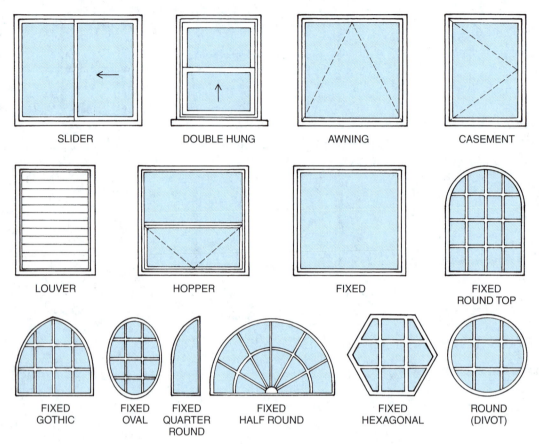

SLIDER   DOUBLE HUNG   AWNING   CASEMENT

LOUVER   HOPPER   FIXED   FIXED ROUND TOP

FIXED GOTHIC   FIXED OVAL   FIXED QUARTER ROUND   FIXED HALF ROUND   FIXED HEXAGONAL   ROUND (DIVOT)

FIGURE 15.26 ■ Elevation symbols for the common window types.

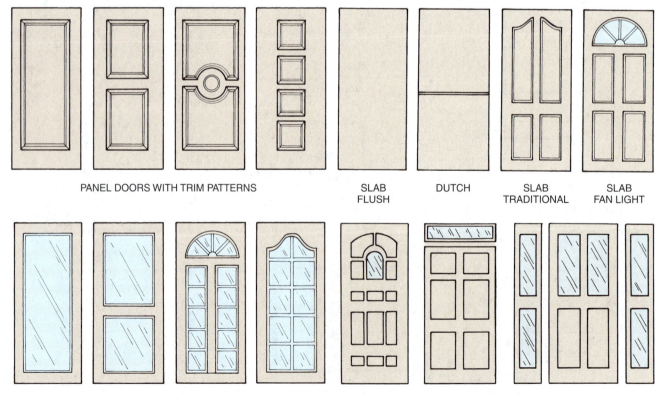

PANEL DOORS WITH TRIM PATTERNS   SLAB FLUSH   DUTCH   SLAB TRADITIONAL   SLAB FAN LIGHT

GLASS PANEL DOORS   ENTRY DOORS

FIGURE 15.27 ■ Common exterior door styles.

CHAPTER

15

Designing Elevations Exercises

1. Sketch an elevation of your own design. Trace the elevation four times drawing in a flat roof, gable roof, shed roof, and butterfly roof. Choose the one you like best and the one that is most functional for your design. Explain why you made that choice.

2. Sketch the front elevation of your home or a home you like. Change the roof style, but keep it consistent with the major lines of the elevation. Move or change the doors and windows to improve the design.

3. Redesign the elevation from Exercise 2, moving the doors and windows and changing the materials. Be sure the door and window lines relate to the major lines of the building.

4. Collect pictures of roofs, windows, and doors that you particularly like. Try to identify house styles for which they are best suited.

Drawing Elevations

OBJECTIVES

In this chapter you will learn:

- to follow steps to project elevations from a floor plan and complete an elevation drawing.
- to draw accurately scaled and dimensioned elevations.
- to mathematically establish the pitch of a roof.
- symbols used on elevations.
- shading and rendering techniques to use on elevations.

TERMS

auxiliary elevation
datum line
exterior elevation drawings
finished dimensions

foreshortened
framing dimensions
interior elevation drawings
orthographic projection

presentation drawings
profile drawings
slope diagram
span

INTRODUCTION

The main features of the outside of a building are shown on elevation drawings. **Exterior elevation drawings** are two-dimensional orthographic representations of the exterior of a structure. These drawings are prepared to show the design, materials, dimensions, and final appearance of the structure's exterior components. In a building, these components include doors, windows, the surfaces of the sides, and the roof. Interior as well as exterior elevation drawings are projected from floor plans. Dimensions are used to show sizes, and elevation symbols are used to indicate various features on the drawings.

ELEVATION PROJECTION

In **orthographic** (multiview) **projection**, related views of an object are shown as if they were on a two-dimensional, flat plane. To visualize and understand orthographic projection, imagine a building surrounded by a transparent box, as shown in Figure 16.1. If you draw the outline of the structure on the transparent planes that make up the box, you may create several orthographic views. For example, the front view is on the front plane, the side view on the side plane, and the top view on the top (horizontal) plane.

If the planes of the top, bottom, and sides were hinged and swung out away from the box, as shown in Figure 16.2, six views of the house would be created. Note how each view is positioned on an orthographic drawing. Study the position of each view as it relates to the front view. The right side is to the right of the front view. The left side is to the left, the top (roof) view is on the top, and the bottom view is on the bottom. The rear view is placed to the left of the left-side view, since, if this view were hinged around to the back, it would fall into this position.

Notice that the length of the front view, top (roof) view, and bottom view are exactly the same as the length of the rear view. Notice also that the heights and alignments of the front view, right side, left side, and rear view are the same. Memorize the position of these views and remember that the lengths of the front, bottom, and top views are *always* the same. Similarly, the heights of the rear, left, front, and right side are *always* the same.

As with all orthographic drawings, surfaces that recede at an angle other than 90° from the projection plane are **foreshortened**. That is, the distance between the two ends are not true size. Figure 16.3 shows how the vertical and horizontal distances from roof ridge to eave are not true on the top and front views, but are true on the side view.

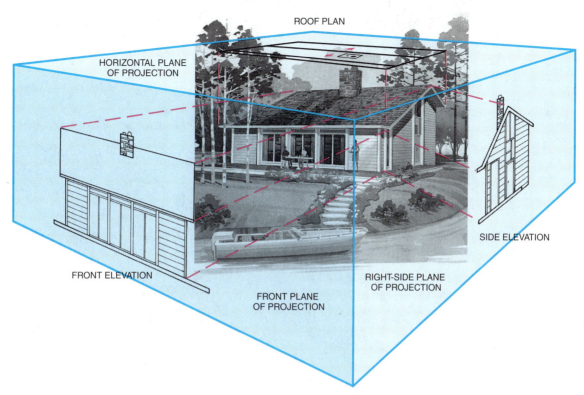

FIGURE 16.1 ▪ Visualizing elevation planes through a projection box. (*Home Planners, Inc.*)

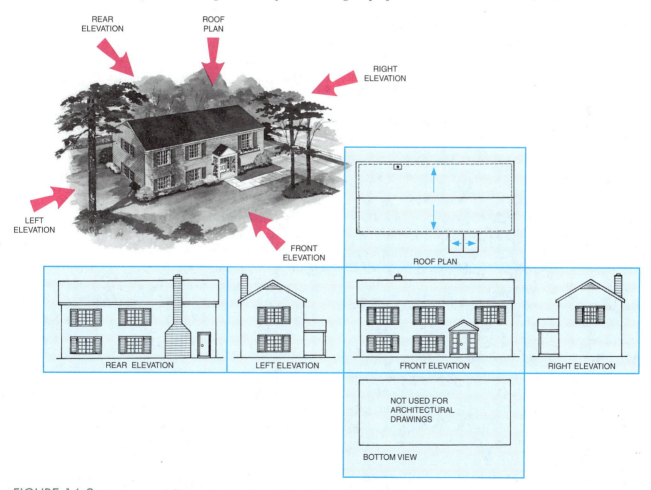

FIGURE 16.2 ▪ Open and flat projection box sides. (*Home Planners, Inc.*)

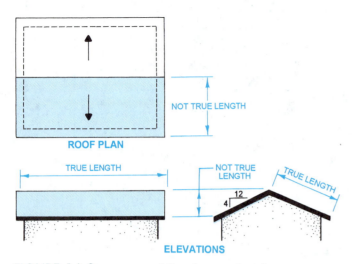

FIGURE 16.3 ■ True and foreshortened surface views.

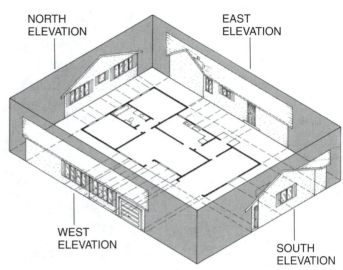

FIGURE 16.4 ■ Floor plan projected to elevation planes.

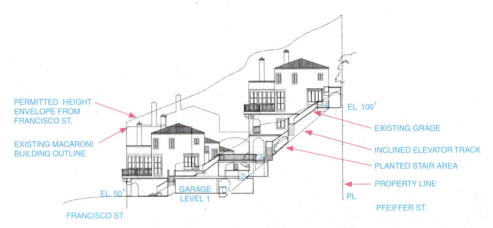

FIGURE 16.5 ■ Site profile. *(Bracken, Arrigoni, & Ross, Inc.)*

All six views are rarely used to depict architectural structures. Instead, only four elevations (sides) are usually shown. The top roof view is used to create floor plans. The roof plan is developed from the top view. The bottom view of a floor is not developed. Instead the foundation underneath the structure is described by foundation plan and construction sections.

Figure 16.4 shows how elevations are projected from the floor plan. The positions of the chimney, doors, windows, overhang, and building corners are projected directly from the floor plan to the elevation plane.

Elevation drawings are used to show the design of the finished appearance of a structure. Elevations are drawn to an exact scale, usually the same as the floor plan. Elevations accurately represent all height dimensions that are not shown on floor plans. The style of windows, doors, and siding are also indicated on elevation drawings. The vertical position of all horizontal planes, such as ground lines, floor lines, ceiling lines, deck or patio lines, and roof lines, are only revealed on elevation drawings. Lines below the ground line such as foundation and footing lines are drawn by dashed lines.

Only through the use of elevation drawings can the vertical relationship of buildings be visualized. For example, on the site profile in Figure 16.5, heights are shown by elevation notations. However, little height detail is apparent without elevation drawings. Elevation drawings of a site are known as **profile drawings**. These drawings show a section cut through the terrain.

PROJECTING ELEVATIONS

Think of an elevation as a drawing placed on a flat, vertical plane. Figure 16.6 shows how a vertical plane is related to and projected from a floor plan.

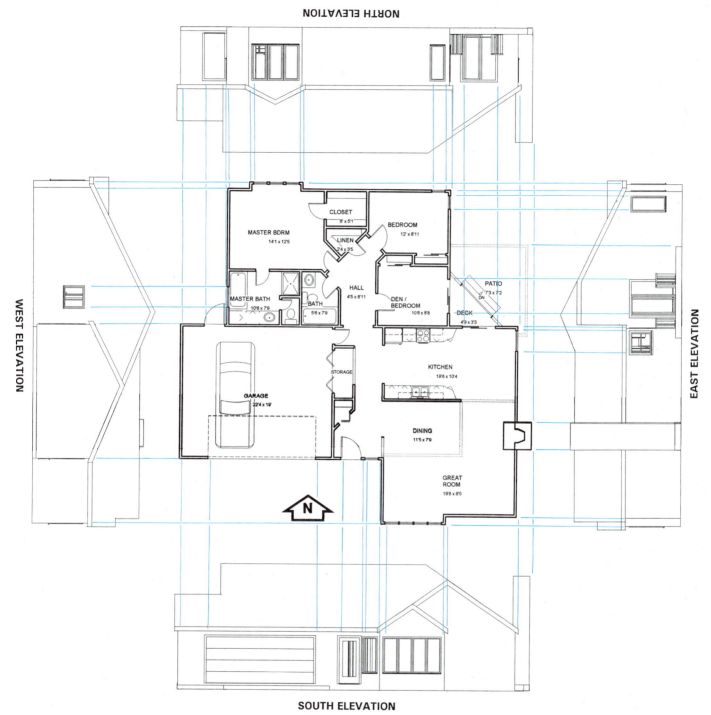

FIGURE 16.6 ■ Projection of floor plan features to elevation drawings.

Orientation

Four elevations are normally projected by extending lines outward from each wall of the floor plan. When these elevations are classified according to their location, they are called the front, rear (or back), right, and left elevation. When these elevations are projected on the same drawing sheet, the rear elevation appears to be upside down and the right and left elevations appear to rest on their sides. See Figure 16.7. Because of the large size of most combined floor plan and elevation drawings, and because of the need to show elevations as normally seen, the elevation drawing is rotated so each elevation can be drawn with the ground line on the bottom. See Figure 16.8.

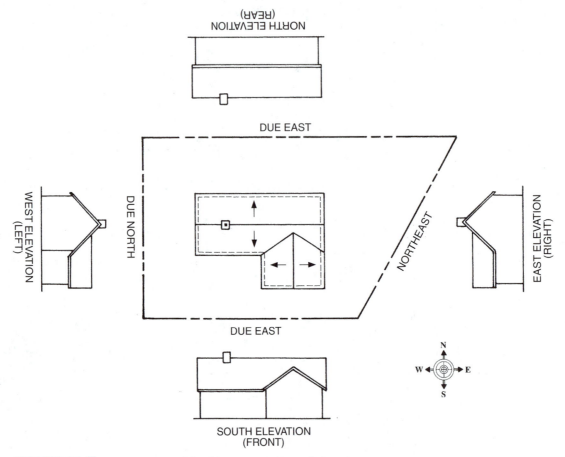

NORTH ELEVATION
(REAR)

DUE EAST

WEST ELEVATION
(LEFT)

DUE NORTH

NORTHEAST

EAST ELEVATION
(RIGHT)

DUE EAST

SOUTH ELEVATION
(FRONT)

FIGURE 16.7 ■ Compass and building orientation of elevations.

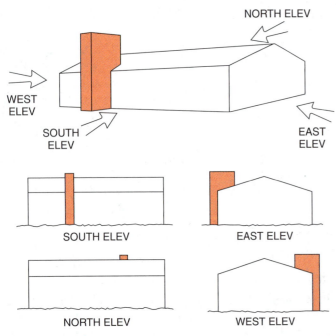

FIGURE 16.8 ■ Ground-line positioning of elevations.

The north, east, south, and west compass points are often used to describe and label elevation drawings. This method is preferred because it reduces the chance of elevation callout error. When this method is used, the north arrow on the floor plan or site plan is the key. For example, in Figure 16.7, the rear elevation is facing north. Therefore, the rear elevation is also the north elevation. The front elevation is the south elevation, the left elevation is the west elevation, and the right elevation is the east elevation.

When elevations do not align exactly with the four major compass points, a split compass reading may be used. Figure 16.9 shows a split labeled Southwest Elevation. This illustration also shows the relationship of the elevation drawing to the construction process and to the finished building.

Auxiliary Elevations

A floor plan may have walls at angles that deviate from the normal 90°. Then some lines and surfaces on the elevations may appear shortened because of the receding angles. An **auxiliary elevation** view may then be necessary to clarify the true size of the elevation. To project an auxiliary elevation, follow the same projection procedures as for other elevation drawings. When an auxiliary elevation is drawn, it is prepared in addition to—and

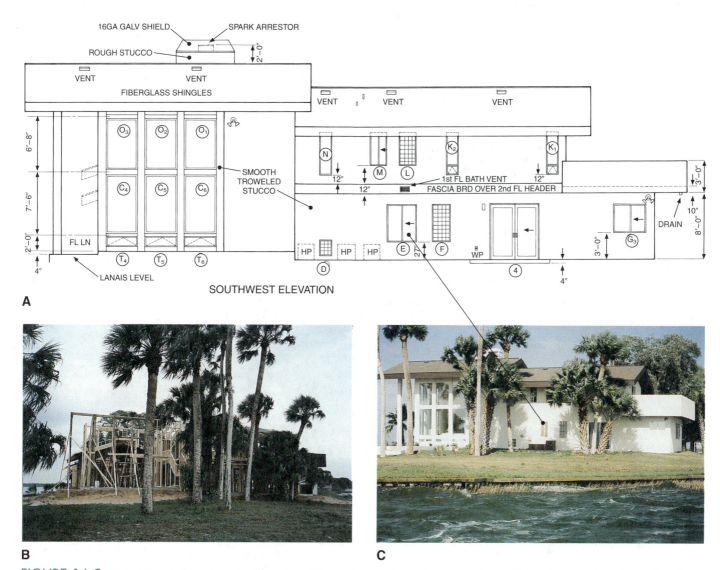

FIGURE 16.9 ■ (A) Southwest elevation drawing. (B) Southwest view of construction. (C) Southwest view of completed structure. (B) (*Diane Kingston, Photographer*)

does not replace—other standard elevation drawings. Where inclined surfaces intersect more than one right-angled elevation, as in Figure 16.10, several auxiliary elevations may be needed to show the true length of all surfaces. Auxiliary elevations are also used to describe vertical design features of separate structures which cannot be shown on plan views. Figure 16.11 shows a separate elevation drawing that details the gate pictured in Figure 18.37d.

Elevations and Construction

Framing elevations show the position, type, and size of members needed for constructing the framework of a structure. When these are not prepared, builders rely solely on exterior and interior elevations for the height of framing members. This means that precise dimensions on the elevation drawings are crucial for accurate construction.

Steps in Projecting Elevations

The major lines of an elevation drawing are derived by projecting vertical lines from the floor plan to the elevation drawing plane and measuring the position of horizontal lines from the ground line. To develop an elevation drawing that exactly reflects the features of a floor plan, refer to Figure 16.12 and follow these steps:

STEP 1 Using the floor plan, *project the vertical lines* that represent the main lines of the building. These lines show the overall length or width of the building. They also show the width of doors, windows, and the major parts or offsets of the building.

When projecting an elevation on a CAD system, use the *Grid* function to project the major lines from the floor plan to the elevation plane.

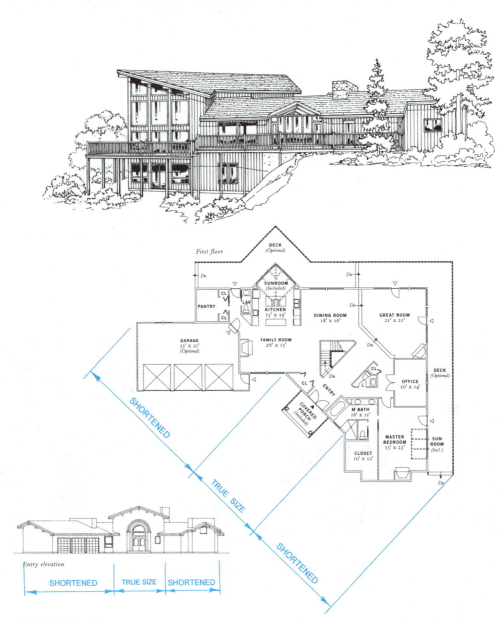

FIGURE 16.10 ■ Auxiliary view with true and foreshortened distances. (*Lindal Cedar Homes*)

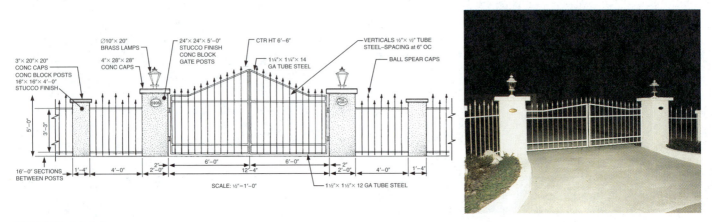

FIGURE 16.11 ■ (A) Dimensioned elevation of gate and fence. (B) Completed gate from Figure 18.37D.

Step 1

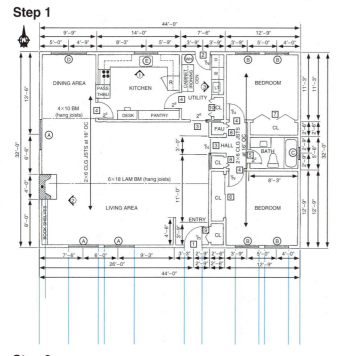

Step 2

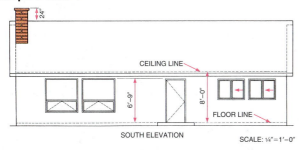

Steps 3 and 4

CEILING LINE

6'-9"

8'-0"

FLOOR LINE

SOUTH ELEVATION

SCALE: ¼"=1'-0"

Steps 5 and 6

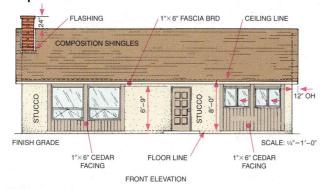

24"
FLASHING
1"×6" FASCIA BRD
CEILING LINE
COMPOSITION SHINGLES
STUCCO
6'-9"
8'-0"
STUCCO
12" OH
FINISH GRADE
SCALE: ¼"=1'-0"
1"×6" CEDAR FACING
FLOOR LINE
1"×6" CEDAR FACING
FRONT ELEVATION

FIGURE 16.12 ■ Steps in projecting and drawing elevations.

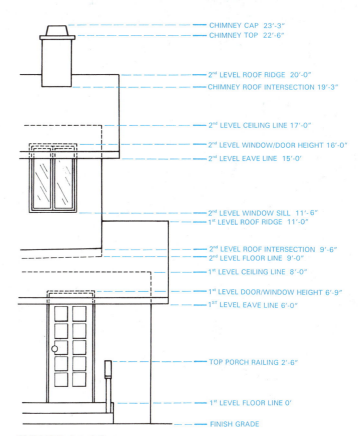

CHIMNEY CAP 23'-3"
CHIMNEY TOP 22'-6"

2ⁿᵈ LEVEL ROOF RIDGE 20'-0"
CHIMNEY ROOF INTERSECTION 19'-3"

2ⁿᵈ LEVEL CEILING LINE 17'-0"

2ⁿᵈ LEVEL WINDOW/DOOR HEIGHT 16'-0"
2ⁿᵈ LEVEL EAVE LINE 15'-0'

2ⁿᵈ LEVEL WINDOW SILL 11'-6"
1ˢᵗ LEVEL ROOF RIDGE 11'-0"

2ⁿᵈ LEVEL ROOF INTERSECTION 9'-6"
2ⁿᵈ LEVEL FLOOR LINE 9'-0"
1ˢᵗ LEVEL CEILING LINE 8'-0"

1ˢᵗ LEVEL DOOR/WINDOW HEIGHT 6'-9"

1ˢᵗ LEVEL EAVE LINE 6'-0"

TOP PORCH RAILING 2'-6"

1ˢᵗ LEVEL FLOOR LINE 0'

FINISH GRADE

FIGURE 16.13 ■ Key elevation height lines.

During the drawing process, floor plans can be rotated 90° to position each elevation with the ground line on the bottom during the drawing process. High-end architectural software can create elevations from floor plans automatically if height dimensions are input.

STEP 2 *Measure and project horizontal lines* that represent the height of the ground line, footing, doors, tops and bottoms of windows, chimney, siding, breaks, planters, and other key features. To eliminate the repetition of measuring each of these lines for each elevation, a sheet showing the scaled lines is often prepared. See Figure 16.13.

STEP 3 *Complete the basic elevation.* First, develop the roof elevation projection. To determine the height of the eave and ridge line, the roof slope (angle) must be established. On a high-slope roof, there is a greater distance between the ridge line and the eave line than on a low-slope roof. See Figure 16.14. Pitch is the angle of the roof described in terms of the ratio of the rise over the span (rise/span). Span is the horizontal distance covered by a roof. Rise is the vertical distance. The run is always expressed in units of 12. The span is the run

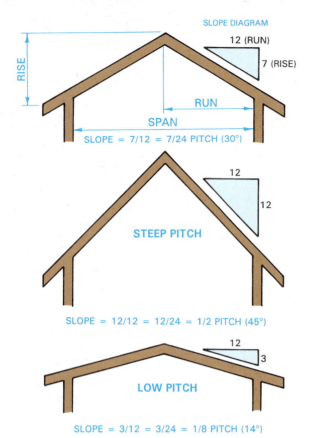

FIGURE 16.14 ■ Roof pitch description.

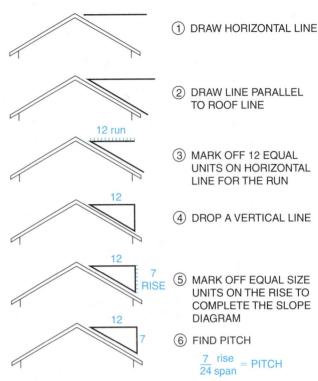

FIGURE 16.15 ■ Steps in drawing a roof slope diagram.

doubled. Therefore the span is always expressed in units of 24.

After the pitch is established, a **slope diagram** must be drawn on the elevation, as shown in Figure 16.15. The slope diagram is developed on the working drawings by the drafter. The carpenter must work with the pitch fraction (ratio) to determine the angle of the rafters from a pitch angle table, so the ends of the rafters can be correctly cut. Double the run to find the span. The **span** is the distance between the supports of the roof. It is a constant of 24. Place the rise over the span (24) and reduce if necessary. This fraction is used by the carpenter to determine the rafter angle in degrees.

A roof elevation can be projected from a roof plan. See Figure 16.16. Note that the end of every eave, and every valley and ridge intersection is projected at a right angle to the plan view outline. Figure 16.17 shows a comparison of X, Y, and Z roof dimensions between elevation views of common roof types.

STEP 4 *Establish the intersection of all vertical and horizontal lines,* including the eave and ridge line.

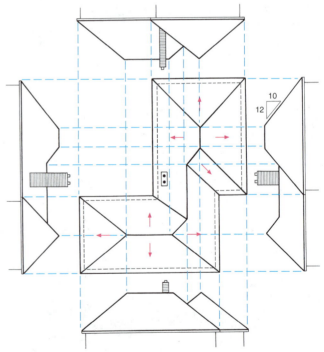

FIGURE 16.16 ■ Relation of roof plan and roof elevations.

These represent the outline of all features to be shown on the elevation. After they are established, darken the lines to identify the position of each.

PROJECTION AND RELATIONSHIP FOR ELEVATIONS AND ROOF PLANS

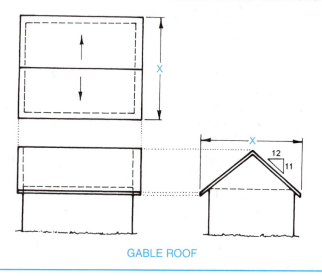

GABLE ROOF

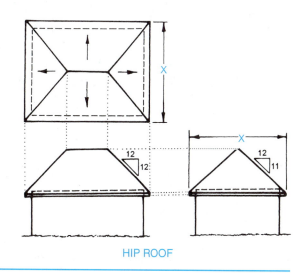

HIP ROOF

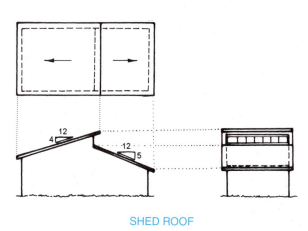

SHED ROOF

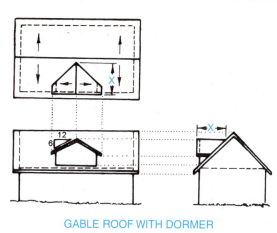

GABLE ROOF WITH DORMER

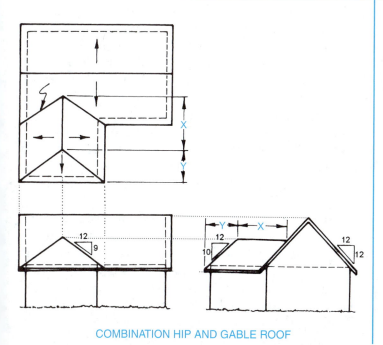

COMBINATION HIP AND GABLE ROOF

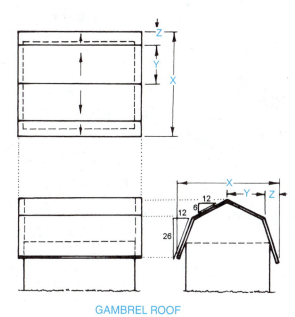

GAMBREL ROOF

FIGURE 16.17 ■ Related distances on roof style elevations.

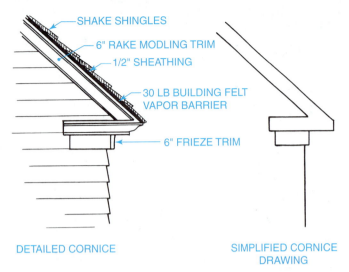

DETAILED CORNICE

SHAKE SHINGLES
6" RAKE MODLING TRIM
1/2" SHEATHING
30 LB BUILDING FELT VAPOR BARRIER
6" FRIEZE TRIM

SIMPLIFIED CORNICE DRAWING

FIGURE 16.18 ■ Detailed and simplified cornice drawing.

STEP 5 *Add details and symbols,* such as indicating door and window trim, mullions, muntins, siding, and roofing materials. The amount of detail included on an elevation depends on the scale and on whether separate details are to be prepared. Figure 16.18 shows two extremes in drawing cornice molding.

STEP 6 *Add final dimensions, labels, and notes.*

Many different elevation styles can be projected from one floor plan. The roof style, pitch, overhang, grade level, windows, chimney, and doors can all be manipulated to create different effects.

USING CAD

Exterior Elevation Projection on CAD

Projection (*Construction*) lines are used to project an elevation drawing from a floor plan by creating a projection line layer using the *Draw* and then the *Construction Line* commands. First the horizontal lines of the elevation must be drawn with construction lines. These represent the location of all elevation heights above grade. Then key intersections from the floor plan are located with the cursor and projected to each related horizontal line. These represent the sides of window, door, chimney, and other offsets on the plan. A heavier or different color line task is then used to connect the major lines of the elevation over the construction line layer. Elevation features that should align can be automatically aligned using the *Align* command.

ELEVATION SYMBOLS

Symbols are needed to clarify and simplify elevation drawings. They help to describe the basic features of an elevation. Some symbols identify door and window styles and positions. Standardized patterns of dots, lines, and shapes show the types of building materials used on exterior walls. These kinds of symbols on an elevation make the drawing appear more realistic.

Material Symbols

Most standard architectural symbols resemble the material they represent. However, in many cases the symbol does not show the exact appearance of the material. For example, the symbol for brick does not include all the lines shown in a pictorial drawing. Representing brick on an elevation drawing exactly as it appears is a long, laborious, and unnecessary process. Many elevation symbols appear as if the material were viewed from a distance.

When using a CAD system, elevation symbols such as doors and windows can be stored in symbol libraries. For example, the *Material Symbols* function can be used to add siding material symbols on elevation surfaces. See Figure 16.19.

Window Symbols

The position and style of windows greatly affect the appearance of elevations. Windows are, therefore, drawn on the elevation with as much detail as the scale of the drawing permits. Parts of windows that should be shown on all elevation drawings include the sill, sash, mullions, and muntins, if any. See Figures 16.20 and 14.15.

In addition to showing the parts of a window, it is also necessary to show the direction of the hinge for casement and awning windows. Dotted lines are used on elevation drawings, as shown in Figure 16.20. The point of the dashed line shows the part of the window to which the hinge is attached.

Many different styles of windows are available. Refer back to Chapter 15. These illustrations show the normal amount of detail used in drawing windows on elevations. An alternative method of showing window styles on elevation drawings is used to include one window detail for each style on the plan drawn to a larger scale. See Figure 16.21, page 271. When the elevation drawing is prepared, the size and outlined position of the window are shown with a letter or number to refer to a detail

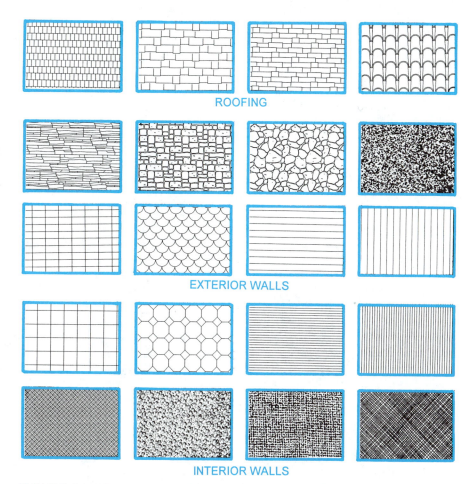

ROOFING

EXTERIOR WALLS

INTERIOR WALLS

FIGURE 16.19 ■ Common elevation material symbols.

drawing. This detail drawing is also indexed to a window schedule that contains complete purchasing, framing, and installation data for each window. Many window types contain optional muntins. Some muntins snap onto the surface of the glass and do not actually separate panes of glass. Figure 16.22A shows common muntin grid designs. Figure 16.22B shows the common types of muntin grid patterns. (See Chapter 35 for examples of schedules.)

Door Symbols

Doors are shown on elevation drawings by methods similar to those used for illustrating window styles and positions. They are either drawn completely, if the scale permits, or shown in abbreviated form. Sometimes door codes are indexed to a door schedule. See Figure 16.23. The complete drawing of the door, whether shown on the elevation drawing sheet or as a detail drawing on a separate sheet, should show the division of panels and lights, sill, jamb, and head-trim details.

USING CAD

CAD Elevation Symbol Libraries

Elevation symbols are of two types: individual and surface symbols. Individual symbols such as doors and windows are stored in *Blocks* and inserted (*Insert* command) onto the elevation drawing in the same manner as floor plan symbols are applied. Surface symbols such as siding and roofing materials are applied using the *Hatch* command. Hatch boundaries must totally enclose a space with no gaps. The hatch pattern is then selected from a hatch library. Hatch blocks can be drawn and stored by the user or selected from the system's library. Many component and material manufacturers supply software containing blocks of their products that can be stored and used in combination with other blocked symbols.

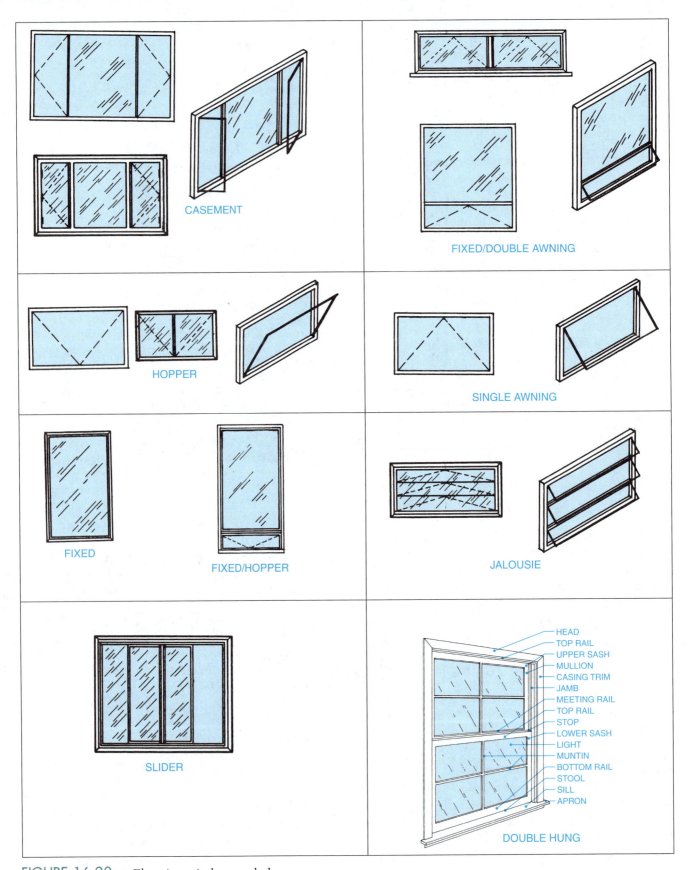

FIGURE 16.20 ■ Elevation window symbols.

COMPLETED WINDOW DETAIL –
ONE DRAWING FOR EACH
TYPE OF WINDOW USED
ON THE STRUCTURE

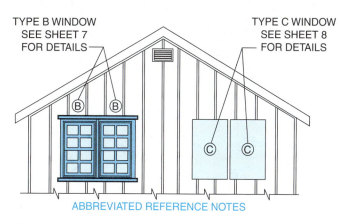

TYPE B WINDOW
SEE SHEET 7
FOR DETAILS

TYPE C WINDOW
SEE SHEET 8
FOR DETAILS

ABBREVIATED REFERENCE NOTES

FIGURE 16.21 ■ A single window detail may be prepared to show a window style.

FIGURE 16.22A ■ Common muntin grid designs.

Many exterior door styles are available. Refer back to Figure 15.27. The total relationship of the door and trim to the entire elevation cannot be seen unless the door trim is also shown. Exterior doors are normally wider than interior doors. Exterior doors must provide access for larger amounts of traffic and be large enough to permit the movement of furniture. They must also be thick enough to provide adequate safety, insulation, and sound barriers. Refer back to Chapter 14 for common door sizes.

INTERIOR ELEVATIONS

Just as exterior elevations illustrate the outside walls, **interior elevation drawings** are necessary to show the design of interior walls (vertical planes). Because of the need to show cabinet height and counter arrangement detail, interior wall elevations are most often prepared for

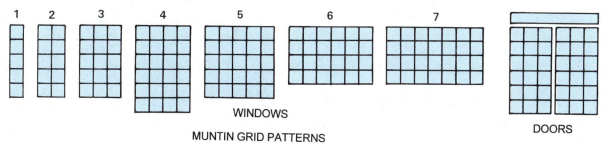

1 2 3 4 5 6 7

WINDOWS

DOORS

MUNTIN GRID PATTERNS

FIGURE 16.22B ■ Common muntin grid patterns.

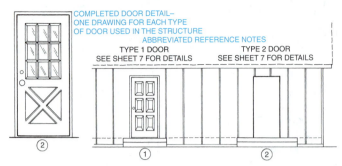

FIGURE 16.23 ■ Use of door codes to identify door styles.

STEP 1 Draw the floor plan outline and project lines outward from each wall.

STEP 2 Connect projected lines to represent the floor line and ceiling line of each elevation.

STEP 3 Add details, cabinets, and fixtures to the floor plan and project lines to represent the height of each feature on each elevation.

kitchen and bathroom walls. See Figure 16.24. An interior wall elevation shows the appearance of the wall as viewed from the center of the room.

A coding system is used to identify the walls on the floor plans for which interior elevations have been prepared. The code symbol tells the direction of the view, the elevation detail number, and the page or sheet number. See Figure 16.25. If only a few interior elevations are prepared, then the title of the room and the compass direction of the wall are the only identification needed.

Where one plane is located behind another, the forward plane is often split to reveal parts of both planes, as shown in Figure 16.26. It is preferable, however, to show a view of both planes if the elevation is to be used for construction purposes.

Steps in Drawing an Interior Elevation

The following steps in drawing an interior elevation are outlined in Figure 16.27.

Projecting the interior elevation in this manner is appropriate for accurate drawing, but results in an elevation drawn on its side or upside down. Therefore, interior elevation drawings, like exterior elevations, are not left in the position as they were originally projected from the floor plans. Interior elevations are repositioned so that each floor line appears on the bottom as a room would normally be viewed.

Once the features of the wall are projected to the elevation from the floor plan, dimensions, instructional notes, and additional features can be added to the drawing. See Figure 16.28.

Interior elevations provide a great amount of detail: the height of all cabinets, shelving, ledges, railings, wall lamps, fixtures, valances, mirrors, chair rails, electrical outlets, switches, landings, and stair profiles. Elevation drawings also include wall surface treatment labeling. Using a common floor line and ceiling line for several elevations eliminates much layout work. In some situations, an interior elevation can span several levels. Check the floor plan in Figure 14.24A for the sources of elevations shown in Figures 16.29.

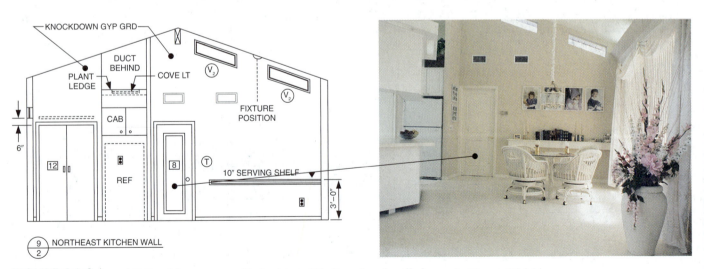

FIGURE 16.24 ■ (A) Breakfast room wall elevation. (B) Completed wall shown in Figure 16.24A.

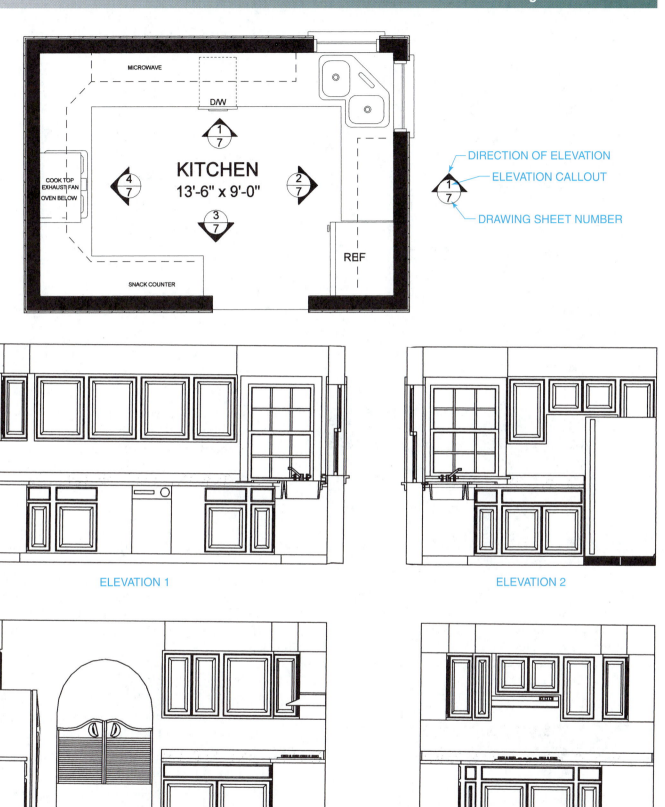

FIGURE 16.25 ■ Numerical codes used to identify wall elevations.

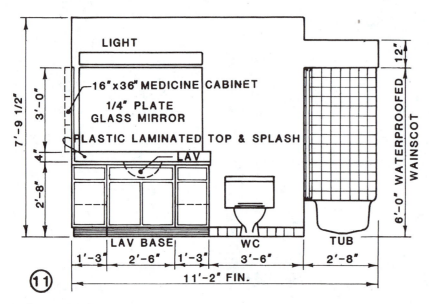
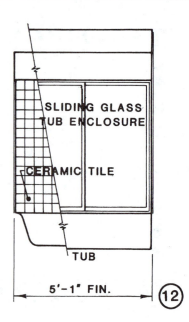

FIGURE 16.26 ■ Interior elevation splitting two planes. (*Home Planners, Inc.*)

USING CAD

Interior Elevation Projections on CAD

Interior wall elevation drawings are prepared on CAD by treating a specific floor plan wall the same as an outside wall and by following the same drawing sequence. This involves assigning a separate layer to the floor plan and each elevation. Here the height of all features—doors, windows, ledges, railings, cabinets, counters, and so on—must be established and drawn before floor plan key intersections are projected.

ELEVATION DIMENSIONING

The vertical (height) dimensions are as important on elevation drawings as horizontal (width and length) dimensions are on floor plans. Many dimensions on elevation drawings show the vertical distance from a datum line. The **datum line** is a reference that remains constant. Sea level is commonly used as the datum or basic reference for many drawings. However, any given line can be conveniently used as a base or datum line for vertical reference.

Dimensions on elevation drawings show height above the ground line. They also show the vertical distance from the floor line to the ceiling and roof ridge and eave lines, and to the tops of chimneys, doors, and windows. Distances below the ground line are shown by dotted lines.

Standards and Guidelines for Dimensioning

Elevation dimensions must conform to basic standards to ensure consistency of interpretation. The numbered arrows on the elevation drawing in Figure 16.30 show the applications of the following guidelines for elevation dimensioning:

1. Vertical elevation dimensions should be read from the right side of the drawing.
2. Levels to be dimensioned should be labeled with a note, abbreviation, or term.
3. Room heights are shown by dimensioning from the floor line to the ceiling line.
4. The depth of footings is dimensioned from the ground line.
5. Heights of windows and doors are dimensioned from the floor line to the top of the windows or doors.
6. Elevation dimensions show only vertical distances (height). Horizontal distances (length and width) are shown on floor plans.

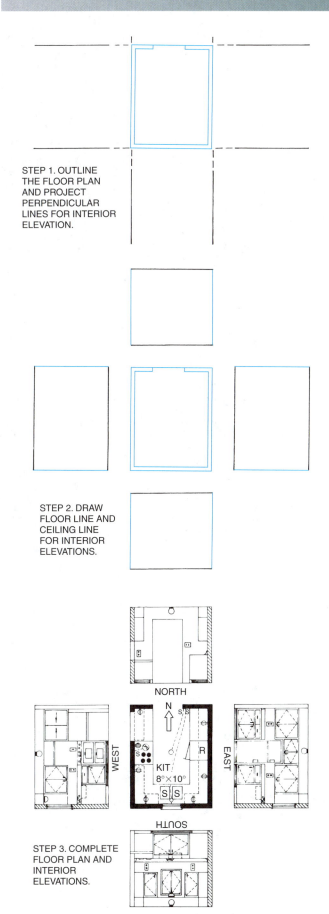

STEP 1. OUTLINE THE FLOOR PLAN AND PROJECT PERPENDICULAR LINES FOR INTERIOR ELEVATION.

STEP 2. DRAW FLOOR LINE AND CEILING LINE FOR INTERIOR ELEVATIONS.

STEP 3. COMPLETE FLOOR PLAN AND INTERIOR ELEVATIONS.

NORTH

WEST

EAST

SOUTH

KIT
8'×10'

FIGURE 16.27 ■ Steps in drawing interior elevations.

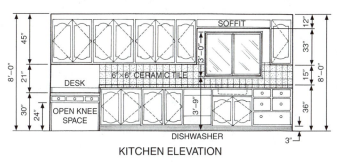

KITCHEN ELEVATION

FIGURE 16.28 ■ Kitchen wall interior elevation.

7. Windows and doors may be indexed by a code or symbol to a door or window schedule, if the style of the windows and doors are not shown on the elevation drawing. See also Figure 16.31.

8. The slope of the roof is shown by indicating a slope diagram.

9. Dimensions for small, complex, or obscure areas should be indexed to a separate detail.

10. Ground-line elevations are expressed as heights above a datum point (for example, sea level).

11. Heights of chimneys above the ridge line are dimensioned.

12. Floor and ceiling lines are shown with hidden lines.

13. Heights of planters, fences, and walls are dimensioned from the ground line.

14. Thicknesses of slabs are dimensioned.

15. Overall height dimensions are placed on the outside of subdimensions.

16. Thicknesses of footings are dimensioned.

17. Refer to Figure 16.32. When the level to be dimensioned is obscure or extremely close to other dimensions, use an elevation line symbol and label the level line.

18. Datum must be identified with a note, if not part of the elevation drawing.

Types of Dimensions

Two types of elevation dimensions are used: framing dimensions and finished dimensions. **Framing dimensions** show the actual distances between framing members. This is the most common method and is preferred by most builders. To avoid an accumulation of measuring errors, framing member dimensions are often dimensioned to their centers.

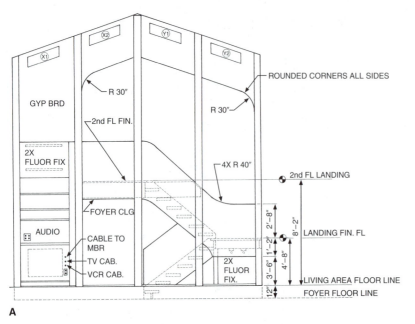

FIGURE 16.29 ■ (A) Two-level interior elevation; (B) Elevation view under construction. (*James Eismont, Photographer*); (C) Completed wall shown in elevation drawing of Figure 16.29B. (*James Eismont, Photographer*)

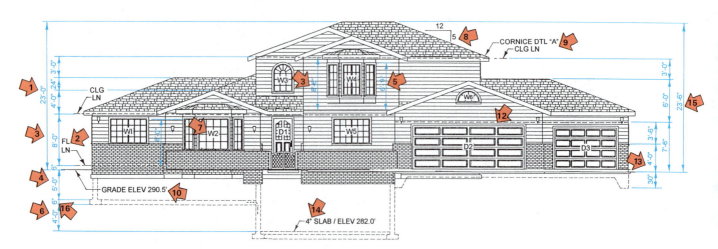

FIGURE 16.30 ■ Numbers indicate specific guidelines for dimensioning.

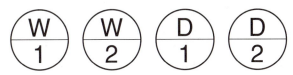

FIGURE 16.31 ■ Elevation codes indexed to door and window schedules.

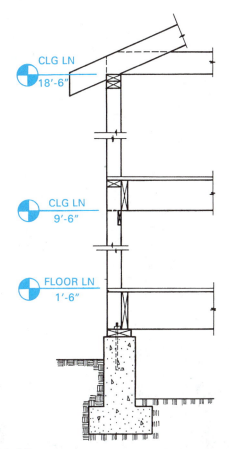

FIGURE 16.32 ■ Elevation height symbols.

Finished dimensions show the actual vertical distances between finished features, such as from the finished floor to finished ceiling levels. Interior dimensions may be added to a full section drawing as done in Figure 16.33. Placing dimensions on a separate interior elevation drawing as done in Figure 16.34 is preferred where possible. These two types of dimensions should not be alternately used on the same drawing, unless the exception is clearly noted. A note on each drawing or set of drawings should indicate which method was used.

Features that are shared by an interior and exterior wall, such as doors and windows, are usually shown on both interior and exterior elevations, but dimensioned only on the exterior elevation drawings. Many other features or components—such as cabinets, shelving, counters, ledges, railings, wall lamps, switches and receptacles—can only be dimensioned on interior elevations.

PRESENTATION ELEVATION DRAWINGS

Dimensioned elevation drawings, prepared to reveal every line and detail, lack realism. To add a more natural look to elevation drawings, landscape features are added to exterior elevations and wall textures and color are added to interior elevation drawings to create **presentation drawings**.

Exterior Presentation Elevations

Adding landscape features to an elevation usually hides some key lines and features. Therefore, trees, plants,

FIGURE 16.33 ■ Method of dimensioning both interior and exterior elevations.

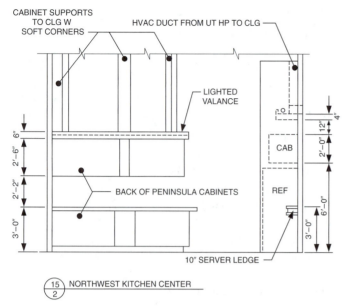

FIGURE 16.34 ■ Typical interior elevation dimensions.

and texture to siding, doors, windows, and roofs also gives realism to the drawing. Heavy foliage around the elevation shown in Figure 16.36 adds depth and texture to the drawing without hiding the main features of the design. Figure 16.37 shows some of the methods of drawing landscape features on an elevation. Figure 16.38 shows the use of a combination of water colors and dry markers to add texture and color to siding and windows. These can be added in color or black depending on the duplicating method to be used. Hard line drawings are recommended if the presentation drawing is to be reproduced with the entire set of plans.

Interior Presentation Elevations

Converting interior elevation working drawings into presentation elevations involves adding texture, color and shadows to wall surfaces, doors, windows, appliances, built-ins, and sometimes furniture. Figure 16.39 shows a rendered interior elevation used for presentation purposes. The rendering methods for pictorial drawings presented in Chapter 20 also apply to presentation elevation drawings. Rendered elevations are more accurate and easier to prepare than pictorial drawings, but pictorial drawings appear more realistic.

and other landscape features should not be added to elevation working drawings. A separate elevation should be traced or plotted for this purpose. The realistic appearance of the elevation shown in Figure 16.35 is created by adding foliage, cars, people, and shadows. Adding color

FIGURE 16.35 ■ Exterior presentation elevation drawing.

FIGURE 16.36 ■ Rendered exterior elevation.

FIGURE 16.37 ■ Landscape features for elevation drawings.

FIGURE 16.38 ■ Siding and window rendering.

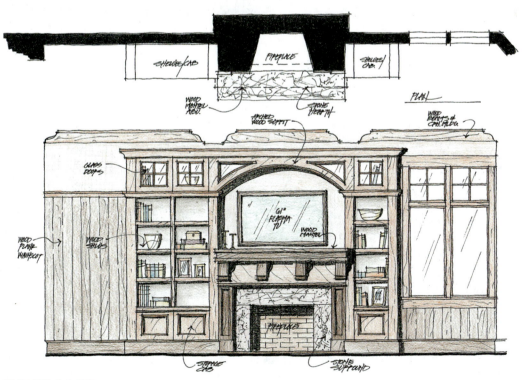

FIGURE 16.39 ■ Interior presentation elevation.

16

Drawing Elevations Exercises

1. Project the front, rear right, and left elevations of a floor plan of your own design. Add elevation symbols.

2. Draw a kitchen wall elevation of a kitchen of your design.

3. Sketch and dimension the front elevation of your home or another home with which you are familiar.

4. Using a CAD system develop a library of symbols for elevations to use in drawing your own design. Note or list which symbols are already included in the CAD program.

5. Project and sketch or draw the front elevation suggested in one of the pictorial drawings in Chapter 19.

6. Copy five of the trees, shrubs, or plants shown in Figure 16.37 as practice for creating a landscape rendering. Then create one with shadows to an elevation of your own design.

7. Create an exterior elevation drawing from the South elevation shown in Figure 16.6.

8. Draw a front elevation for the plan shown in Figure 14.27.

9. Draw an interior elevation of the living room fireplace wall shown in Figure 14.20.

10. Create an interior elevation presentation drawing from the elevation shown in Figure 16.6.

CHAPTER 17

Sectional, Detail, and Cabinetry Drawings

OBJECTIVES

In this chapter you will learn to:

- describe types of sectional drawings.
- communicate views of sections based on a cutting plane.
- draw sections, using correct codes and proper dimensioning.
- evaluate when a detail sectional drawing is needed.
- read and prepare detail drawings.
- design and prepare cabinet drawings.

TERMS

break-out sectional drawings
cabinet coding system
cutting plane
cutting-plane line

detail sections
full section
horizontal wall sections
longitudinal section

removed section
sectional drawings
transverse section
vertical wall sections

INTRODUCTION

Architectural sections are drawings that are important in the design process. They show details not visible on floor plans or elevations. Most architectural sections contain symbols, reference codes, and dimensions to indicate construction information. Other sections may only be sketches to compare heights, while still others may be pictorial drawings. Sections are used to show the exact details of construction. The ability to prepare technical architectural drawings depends on a thorough understanding of sectional drawings.

SECTIONAL DRAWINGS

Sectional drawings reveal the internal construction of an object. An architectural sectional drawing (or an architectural section) that is prepared for the entire structure is called a **full section**. A sectional drawing that shows only specific parts of a building is a **detail section**. The size and complexity of the parts usually determine whether a full section and/or detail sections are needed.

Because they provide all information needed for construction, architectural sections are used as working drawings. Sections are also useful as design concept drawings and presentation drawings.

Architectural designs that involve multiple levels and variations in height often require the preparation of full-section sketches (in addition to floor plans) during the design process. Sections are often needed to clarify the elevation relationships of ground, footing, floor, and roof lines.

When a presentation drawing is needed to show the general appearance of both the interior and exterior of a building in one drawing, a *pictorial section* is often completed. This is usually done using one-point perspective methods. See Figure 17.1.

FULL SECTIONS

In full-section drawings, entire buildings are drawn as if they were cut in half. The purpose of full sections is to convey how a building is constructed from the foundation through the roof.

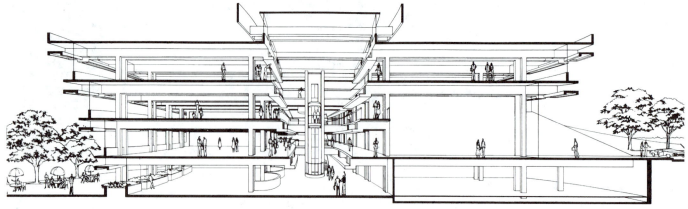

FIGURE 17.1 ■ Presentation section. (*Architectural Record, Morgan and McCrary Architects*)

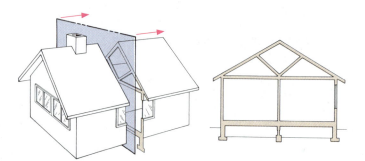

FIGURE 17.2 ■ Transverse section with cutting plane.

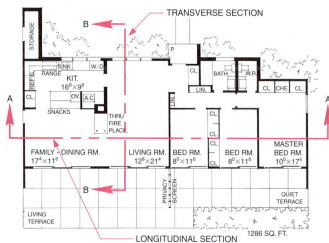

FIGURE 17.4 ■ Transverse and longitudinal cutting-plane lines on a floor plan.

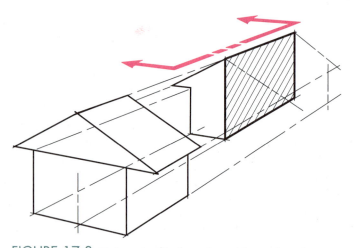

FIGURE 17.3 ■ Longitudinal section with cutting plane.

The Cutting Plane

A **cutting plane** is an imaginary plane that passes through a building. The position of a cutting plane is shown by the **cutting-plane line** that is drawn as a long heavy line with two dashes. Sectional drawings may show either transverse or longitudinal sections. A **transverse section** shows a cutting plane across the shorter or minor axis of a building, as shown in Figure 17.2. A **longitudinal section** shows a cutting plane along the length or major axis of a building. See Figure 17.3. Sections convey information about the inner construction. Both transverse and longitudinal full sections share the same external outlines as elevation drawings.

The cutting-plane line is usually placed on a floor plan to tell which part is drawn as a section. The arrows at the ends of the line tell the direction of the view. See Figure 17.4.

Because the cutting-plane line can easily interfere with dimensions, notes, and details, only the extreme ends of the cutting-plane line are indicated on most architectural drawings. The part of the line that is omitted on the drawing is then assumed to be a straight line between the ends of the cutting-plane line. When a cutting-plane line must be offset to show a different area, the offsetting corners of

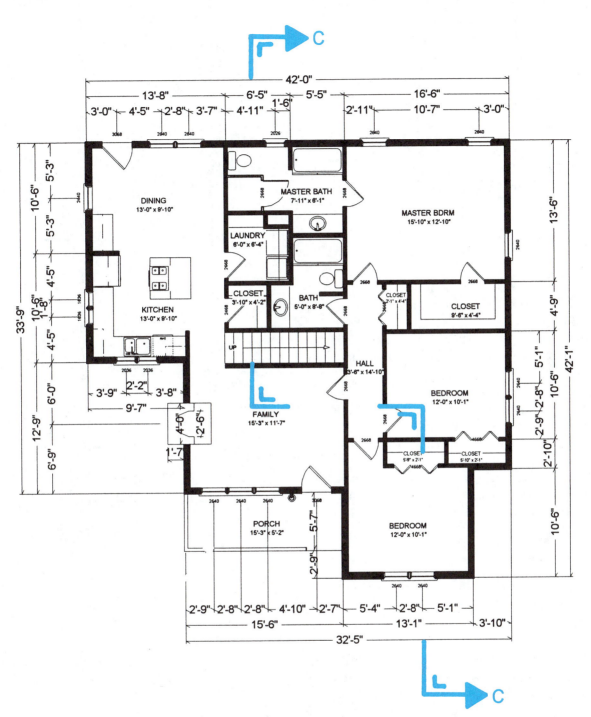

FIGURE 17.5 ■ Offset cutting-plane line.

the line are drawn as they appear in Figure 17.5. An offset cutting plane is often used to include sections of different walls on one sectional drawing.

Full sections provide builders with a total view of a structure from foundation through the roof. There is no limit to the number of cutting planes that can be drawn. Each section must be separately identified.

Symbols

When sections are referenced to another drawing, the symbol shown in Figure 17.6 should be used. This referencing method is the same method introduced in Chapter 4. Figure 17.7A shows a partial floor plan with coded cutting-plane lines. The cutting plan B4 is indexed to the

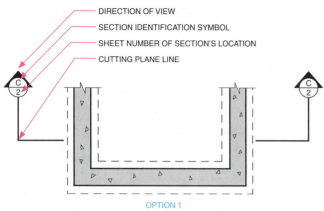

FIGURE 17.6 ■ Cutting-plane reference symbol.

longitudinal full section shown in Figure 17.7B. The transverse full section A5 shown in Figure 17.7C is also from the same floor plan.

A floor plan is a small-scale horizontal section. Therefore, many floor plan symbols are also used in plan detail drawings.

Many *section-lining symbols* appear realistic, as though materials were cut through. Others are simplified in order to save time on the drawing board.

A building material is only sectioned if the cutting-plane line passes through it. The outline of all other materials visible behind the plane of projection must also be drawn in the proper position and scale. Symbols for building materials are shown in Figure 17.8.

Scale

Because full sections show construction methods used in the entire building, they are drawn to a relatively small

USING CAD

CAD Section Symbol Library

Section symbols are used to represent a slice through a material. Symbols that represent a wide variety of materials are stored and accessed using the *Hatch* command. After the hatch pattern and boundary are selected, the material is added to a drawing by identifying the hatch border with a cursor. Hatch patterns should be applied to enable the pattern to be stretched or scaled without changing the pattern makeup.

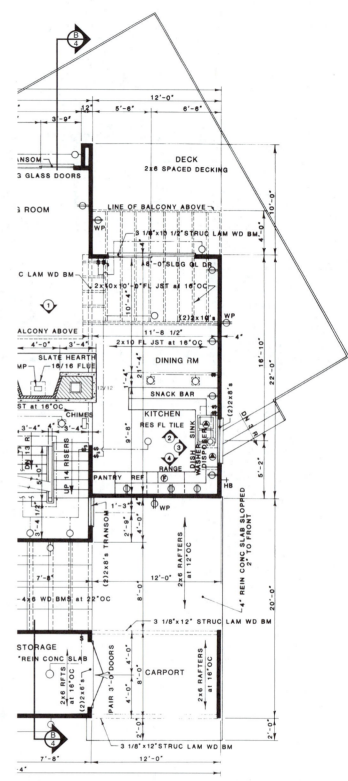

FIGURE 17.7A ■ Floor plan cutting plane referenced to the full section in Figure 17.7B.

scale (1/4″ = 1′-0″). However, a scale that is too small often makes the drawing and interpretation of smaller details extremely difficult. One method of maintaining a

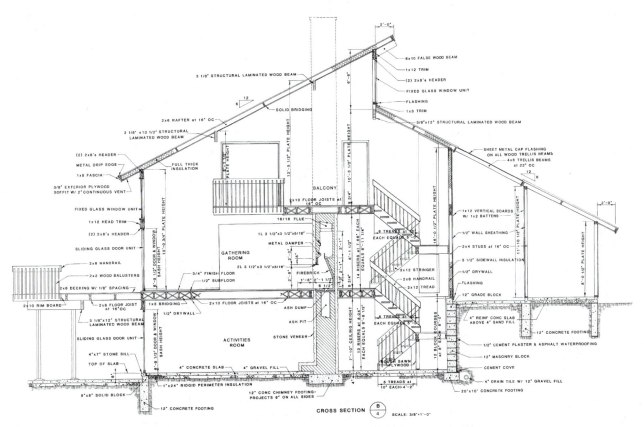

FIGURE 17.7B ▪ Longitudinal section referenced from the floor plan in Figure 17.7A.

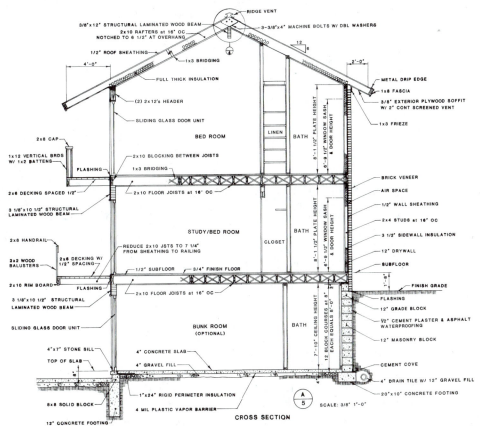

FIGURE 17.7C ▪ Transverse section referenced from floor plan in Figure 17.7A.

NAME	ABBRV	SECTION SYMBOL	ELEVATION	NAME	ABBRV	SECTION SYMBOL	ELEVATION
COMMON BRICK	COM BRK			WELDED WIRE MESH	WWM		
FACE BRICK	FC BRK			FABRIC	FAB		
FIREBRICK	FRB			LIQUID	LQD		
GLASS	GL			COMPOSITION SHINGLE	COMP SH		
GLASS BLOCK	GL BLK			RIDGID INSULATION SOLID	RDG INS		
STRUCTURAL GLASS	STRUC GL			LOOSE-FILL INSULATION	LF INS		
FROSTED GLASS	FRST GL			QUILT INSULATION	QLT INS		
STEEL	STL			SOUND INSULATION	SND INS		
CAST IRON	CST IR			CORK INSULATION	CRK INS		
BRASS & BRONZE	BRS BRZ			SHEET METAL (FLASHING)	SHT MTL FLASH		
ALUMINUM	AL			REINFORCING STEEL BARS	REBAR		

FIGURE 17.8 ■ Section symbols for building materials, Part 1.

larger scale and drawing area for large sections is to insert break lines. See Figure 17.9. Break lines indicate that much of the repetitive portion of the building was removed from the drawing. With break lines, a very large

area can be drawn at a larger and more readable scale and still fit on one sheet.

Another method is to draw detail sections of the removed portions of the building separate from the full-

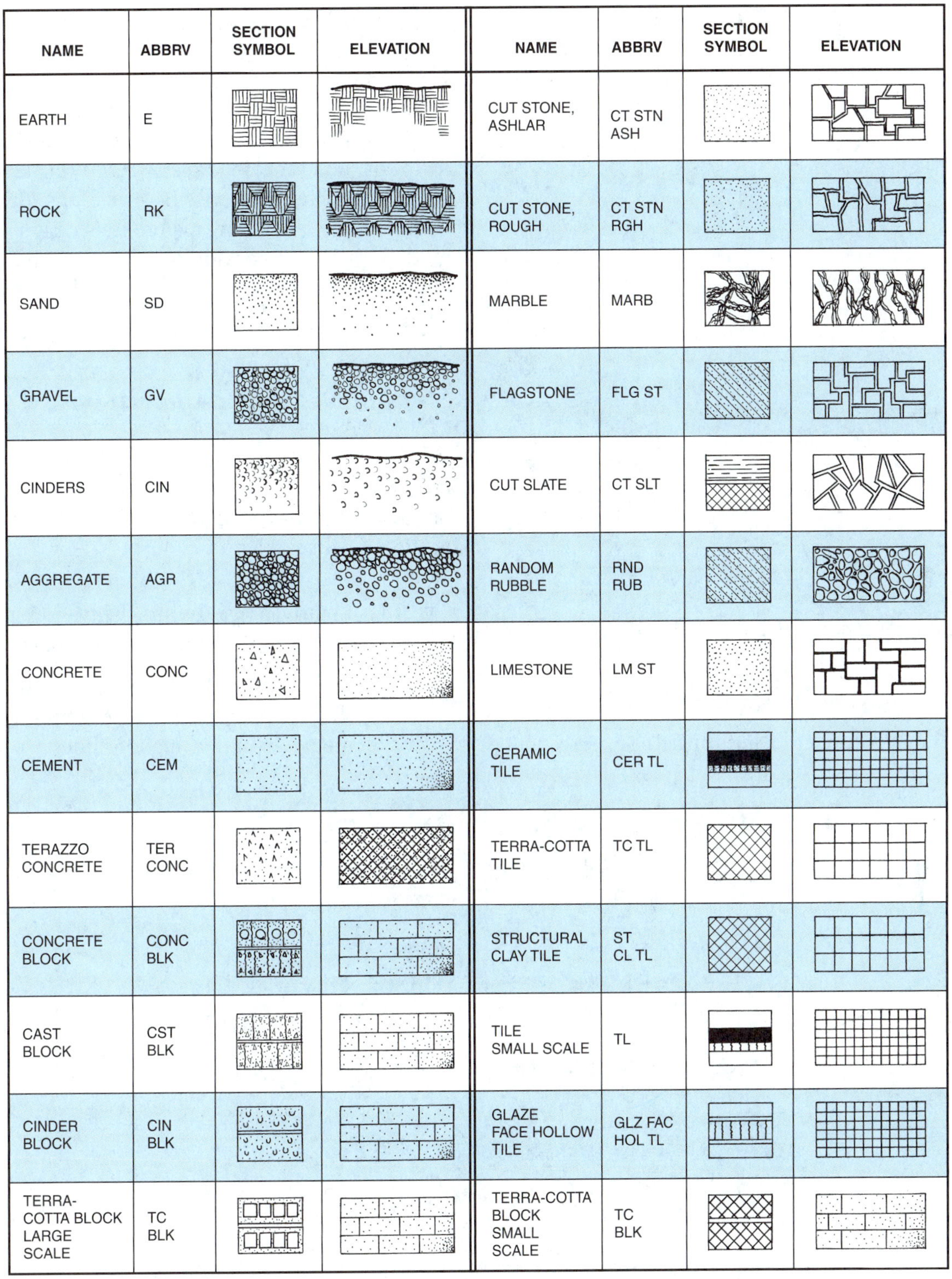

NAME	ABBRV	SECTION SYMBOL	ELEVATION	NAME	ABBRV	SECTION SYMBOL	ELEVATION
EARTH	E			CUT STONE, ASHLAR	CT STN ASH		
ROCK	RK			CUT STONE, ROUGH	CT STN RGH		
SAND	SD			MARBLE	MARB		
GRAVEL	GV			FLAGSTONE	FLG ST		
CINDERS	CIN			CUT SLATE	CT SLT		
AGGREGATE	AGR			RANDOM RUBBLE	RND RUB		
CONCRETE	CONC			LIMESTONE	LM ST		
CEMENT	CEM			CERAMIC TILE	CER TL		
TERAZZO CONCRETE	TER CONC			TERRA-COTTA TILE	TC TL		
CONCRETE BLOCK	CONC BLK			STRUCTURAL CLAY TILE	ST CL TL		
CAST BLOCK	CST BLK			TILE SMALL SCALE	TL		
CINDER BLOCK	CIN BLK			GLAZE FACE HOLLOW TILE	GLZ FAC HOL TL		
TERRA-COTTA BLOCK LARGE SCALE	TC BLK			TERRA-COTTA BLOCK SMALL SCALE	TC BLK		

FIGURE 17.8 (continued) ■ Section symbols for building materials, Part 2.

section drawing. **Removed sections** drawn at a larger scale are used to clarify small details.

Abbreviated full sections are often prepared as part of the design process or for presentation purposes. The section shown in Figure 17.10A is prepared to show only the outline and relationship of horizontal and vertical components, without dimensions or details. Figure 17.10B shows the elevation design concept study relating to this section.

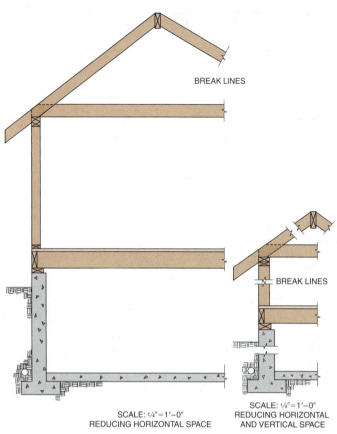

BREAK LINES

BREAK LINES

SCALE: ¼"=1'–0"
REDUCING HORIZONTAL SPACE

SCALE: ¼"=1'–0"
REDUCING HORIZONTAL
AND VERTICAL SPACE

FIGURE 17.9 ■ Use of break lines on small-scale sections.

Sectional Dimensions

Full sections expose the size and shape of building materials and components not revealed on floor plans and elevations. These sections are an excellent place on which to locate many detail dimensions. Full-section dimensions primarily show specific elevations, distances, and the exact size of building materials. See Figure 17.11. The guidelines for dimensioning elevation drawings apply also to full-elevation sections and to detail sections. (See Chapter 16.)

Steps in Drawing Full Sections

In drawing full sections, the architect "constructs" the framework of a house on paper. Figure 17.12 shows the progressive steps in the layout and drawing of a section of the side edge of a house.

STEP 1 Lightly draw the finished floor line approximately at the middle of the drawing sheet.

STEP 2 Measure the thickness of the subfloor and of the joist and draw lines representing these members under the floor line.

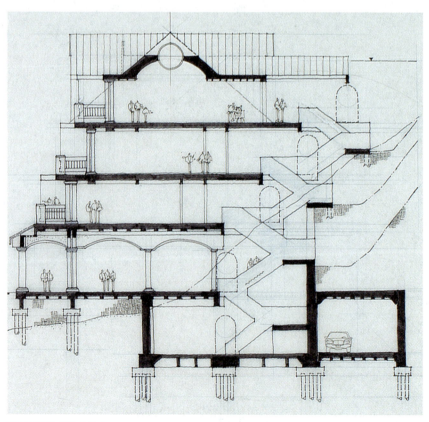

FIGURE 17.10A ■ Presentation section showing only major outlines.
(New York Institute of Technology)

FIGURE 17.10B ■ Elevation concept study relating to the section in Figure 17.10A. *(New York Institute of Technology)*

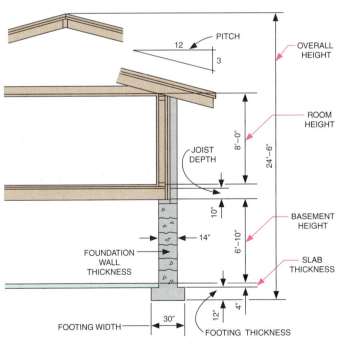

FIGURE 17.11 ■ Major dimensions used on a vertical section.

STEP 3 From the floor line, measure up and draw the ceiling line.

STEP 4 Measure down from the floor line to establish the top of the basement slab and footing line, and draw in the thickness of the footing.

STEP 5 Draw two vertical lines representing the thickness of the foundation and the footing.

STEP 6 Construct the sill detail and show the alignment of the studs and top plate.

STEP 7 Measure the overhang from the stud line and draw the roof pitch by projecting from the top plate on the angle determined by dividing the rise by the span.

STEP 8 Establish the ridge point by measuring the distance from the outside wall horizontally to the center of the ridge line. This is usually at the center of the structure.

STEP 9 Add details and symbols representing siding and interior finish.

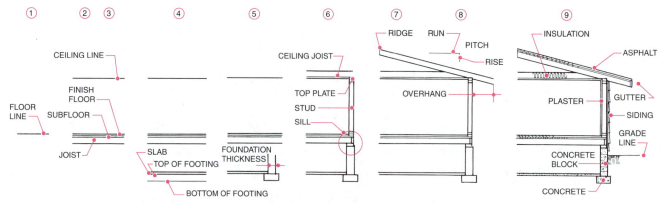

FIGURE 17.12 ■ Sequence of drawing an elevation section.

DETAIL SECTIONS

Because full sections are usually drawn to a small scale, many small parts are difficult to interpret. To reveal the exact position and size of many of these small parts, enlarged detailed sections are prepared. Detail sections clarify any construction feature that could not be described on the basic floor plans, elevations, or full sections. Detail sections may be prepared on a vertical (elevation) plane or a horizontal (plan) plane. Like full sections, detail sections are keyed to a plan or elevation view. See Figure 17.13.

Vertical Wall Sections

Vertical wall sections show exposed construction members on a vertical plane. They are prepared for exterior and interior walls.

Exterior Walls

Elevation drawings do not reveal construction details because of their small scale and because many details are hidden by siding materials. Several other methods are used to produce exterior wall section drawings large enough to show construction details and dimensions. These include the use of break lines and removed sections to reduce the length of the drawing.

Break Lines

Similar to full-section drawings, break lines are used on detail sections to reduce vertical distances. As you know, using break lines allows the area to be drawn

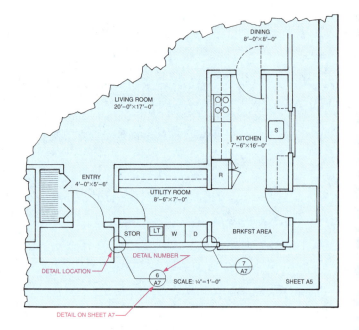

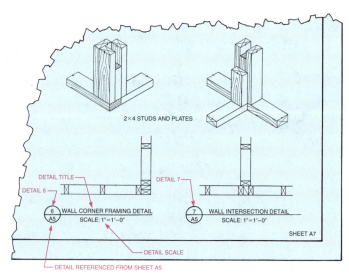

FIGURE 17.13 ■ Cross-referencing symbols used on detail drawings.

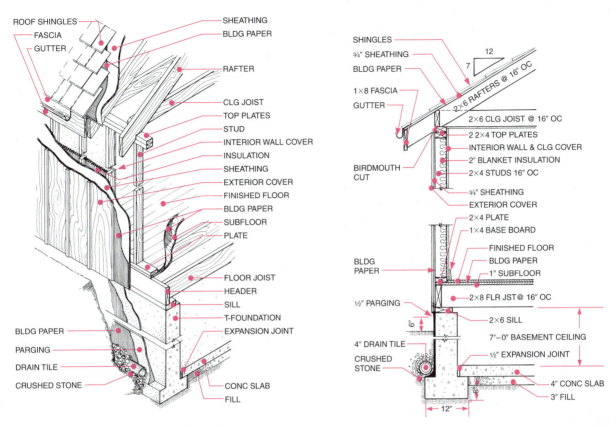

ROOF SHINGLES
FASCIA
GUTTER
SHEATHING
BLDG PAPER
RAFTER
CLG JOIST
TOP PLATES
STUD
INTERIOR WALL COVER
INSULATION
SHEATHING
EXTERIOR COVER
FINISHED FLOOR
BLDG PAPER
SUBFLOOR
PLATE
FLOOR JOIST
HEADER
SILL
T-FOUNDATION
EXPANSION JOINT
BLDG PAPER
PARGING
DRAIN TILE
CRUSHED STONE
CONC SLAB
FILL

SHINGLES
¾" SHEATHING
BLDG PAPER
1×8 FASCIA
GUTTER
BIRDMOUTH CUT
12
7
2×6 RAFTERS @ 16" OC
2×6 CLG JOIST @ 16" OC
2 2×4 TOP PLATES
INTERIOR WALL & CLG COVER
2" BLANKET INSULATION
2×4 STUDS 16" OC
¾" SHEATHING
EXTERIOR COVER
2×4 PLATE
1×4 BASE BOARD
FINISHED FLOOR
BLDG PAPER
1" SUBFLOOR
2×8 FLR JST @ 16" OC
2×6 SILL
7'–0" BASEMENT CEILING
½" EXPANSION JOINT
4" CONC SLAB
3" FILL
BLDG PAPER
½" PARGING
6"
4" DRAIN TILE
CRUSHED STONE
12"

FIGURE 17.14 ■ Use of break lines to reduce vertical drawing height.

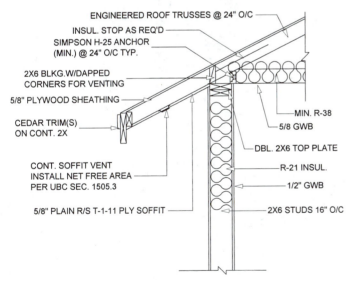

ENGINEERED ROOF TRUSSES @ 24" O/C
INSUL. STOP AS REQ'D
SIMPSON H-25 ANCHOR (MIN.) @ 24" O/C TYP.
2X6 BLKG.W/DAPPED CORNERS FOR VENTING
5/8" PLYWOOD SHEATHING
CEDAR TRIM(S) ON CONT. 2X
CONT. SOFFIT VENT INSTALL NET FREE AREA PER UBC SEC. 1505.3
5/8" PLAIN R/S T-1-11 PLY SOFFIT
MIN. R-38
5/8 GWB
DBL. 2X6 TOP PLATE
R-21 INSUL.
1/2" GWB
2X6 STUDS 16" O/C

FIGURE 17.15 ■ Cornice section detail.

larger than would be possible if the entire distance were included in the drawing. Break lines are used where the construction does not change over a long distance. Figure 17.14 shows the use of break lines to enlarge a frame-wall section.

Removed Sections

Sometimes it is impossible to draw an entire wall section to a large enough scale to show needed information, even when using break lines. In these cases, a removed section is drawn at a larger scale, separate from the original location. Removed sections are frequently drawn for the ridge, cornice, sill, footing, and beam areas.

Cornice sections are used to show the relationship between the outside wall, top plate, and rafter construction. See Figure 17.15. Some cornice sections show gutter details.

Sill sections, as shown in Figure 17.16, show how the foundation supports and intersects with the floor system and the outside wall.

A *footing section* is needed to show the width and height of the footing, the type of material used, and the position of the foundation wall on the footing. Figure 17.17 shows several footing details and the pictorial interpretation of each type.

Beam details are necessary to show how the joists are supported by beams and how the columns or foundation walls support the beams. As for all sections, the position of the cutting-plane line is extremely important. Figure 17.18 shows two possible positions of the cutting plane. If the cutting-plane line is placed parallel to the beam, you see a cross section of the joist, as shown in drawing A. If the cutting-plane line is placed perpendicular to the beam, you see a cross section of the beam, as shown in drawing B.

Interior Walls

To illustrate the methods of constructing inside partitions, sections are often drawn of interior walls at the base and at the ceiling. A base section shows how the wall-finishing materials are attached to the studs and how the intersection between the floor and wall is constructed. A crown section shows the intersection between the ceiling and the wall and how the finished construction materials of the wall and ceiling are related. Sections may also be prepared for stair and fireplace details.

Horizontal Wall Sections

Horizontal wall sections of exterior and interior walls are drawn to clarify how walls are constructed. Walls in these sections are similar to those on floor plans, but are drawn at a larger scale in order to show a horizontal sectional view of each construction member as shown in Figure 17.19.

Exterior Walls

Although a floor plan is a horizontal section, many construction details are omitted because of the small scale used. Therefore, larger horizontal sections are drawn to show the more exact construction details of corners,

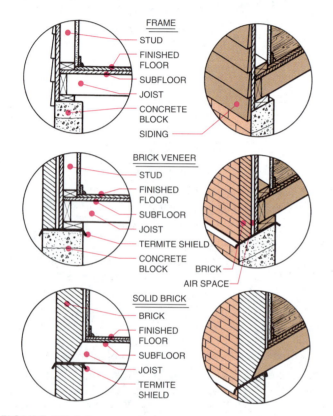

FIGURE 17.16 ■ Detailed sill sections.

FIGURE 17.18 ■ Two views of a section detail.

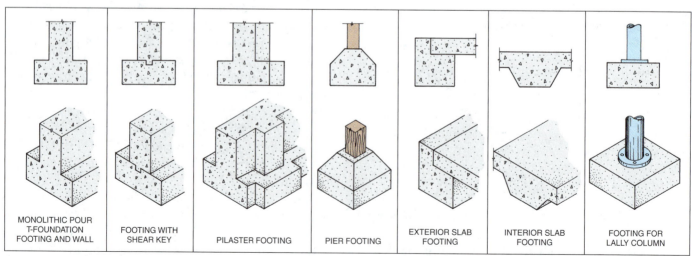

FIGURE 17.17 ■ Detailed footing sections.

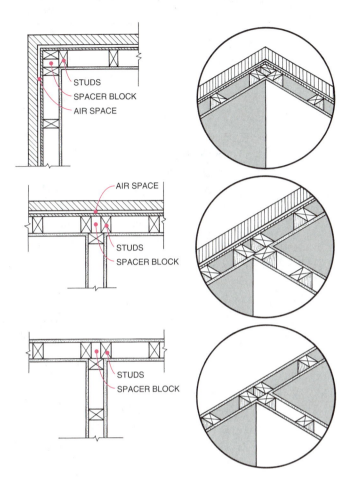

FIGURE 17.19 ■ Horizontal wall sections.

intersections, and window and door framing, as shown in Figure 17.20. This section represents the corner of the living room in Figure 14.24A.

Interior Walls

Typical horizontal sections of interior wall intersections indicate construction methods. For example, horizontal sections are needed to show the inside and outside corner construction of a paneled wall and how paneled joints and other building joints are constructed. See Figure 17.21. Figures 17.22A and 17.22B illustrate how crown molding detail sections and base molding detail sections are drawn. Figure 17.22C links these intersections to an exterior wall section. The cross section of standard molding shapes used on interior wall sectional drawings is shown in Figure 17.23. Molding and paneling is used to create paneled walls as shown in Figure 17.24. Note also the use of molding to provide a finished appearance to the beams in this illustration. A typical wall panel sectional drawing is shown in Figure 17.25. Horizontal sections are also very effective for illustrating various methods of attaching building materials.

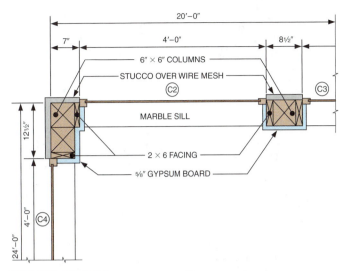

FIGURE 17.20A ■ Corner wall section detail.

FIGURE 17.20B ■ Corner under construction.

FIGURE 17.20C ■ Completed corner of living room shown in Figure 14.24A.

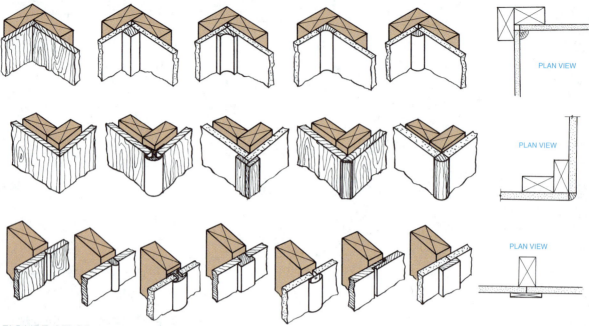

FIGURE 17.21 ■ Wall panel joints and corners.

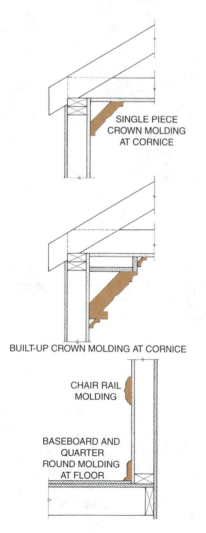

SINGLE PIECE
CROWN MOLDING
AT CORNICE

BUILT-UP CROWN MOLDING AT CORNICE

CHAIR RAIL
MOLDING

BASEBOARD AND
QUARTER
ROUND MOLDING
AT FLOOR

FIGURE 17.22A ■ Common molding and trim sections.

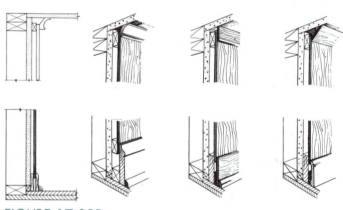

FIGURE 17.22B ■ Common crown and base molding sections.

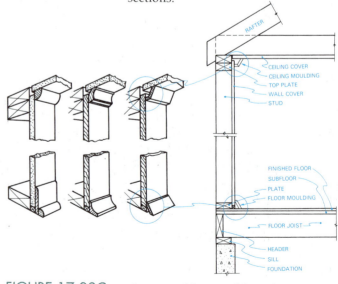

RAFTER

CEILING COVER
CEILING MOULDING
TOP PLATE
WALL COVER
STUD

FINISHED FLOOR
SUBFLOOR
PLATE
FLOOR MOULDING

FLOOR JOIST

HEADER
SILL
FOUNDATION

FIGURE 17.22C ■ Crown and base molding shown on a wall section.

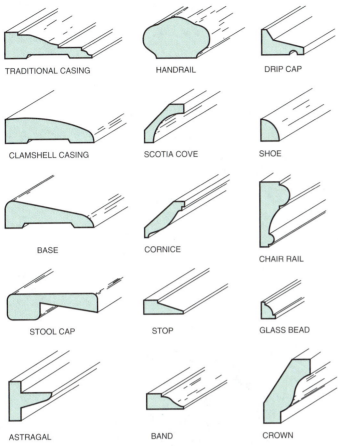

TRADITIONAL CASING	HANDRAIL	DRIP CAP
CLAMSHELL CASING	SCOTIA COVE	SHOE
BASE	CORNICE	CHAIR RAIL
STOOL CAP	STOP	GLASS BEAD
ASTRAGAL	BAND	CROWN

FIGURE 17.23 ■ Standard molding shapes.

Window Sections

Because window construction is hidden, sectional drawings are needed to show construction details. Figure 17.26 shows both horizontal and vertical window construction areas that are commonly sectioned. These include head, sill, and jamb construction. Window manufacturers generally use pictorial sections to illustrate key features and methods of installation. See Figure 17.27.

Head and sill sections are vertical sections. Preparing head and sill sections on the same drawing is possible only when a small scale is used. If a larger scale is needed, the head and sill must either be drawn independently or break lines must be used. Figure 17.28 shows the relationship between the cutting-plane line and the head and sill sections. The circled areas in Figure 17.29 show the areas that are removed when a separate head and sill section is prepared.

When a cutting-plane line is extended horizontally across the entire window, the resulting sections are known as jamb sections. Jamb details (the horizontal section) are projected from the window-elevation drawing. See Figure 17.30. The construction of both jambs is usually the same, with the right jamb drawing being the reverse of the left. Only one jamb detail is normally drawn. The builder then interprets one jamb as the reverse of the other.

FIGURE 17.24 ■ Use of molding on paneled walls, bookshelves, and beams. (*Hyde Park Lumber Co.*)

MULLION (VERTICAL RAIL)

SHEET ROCK
TOP RAIL
COVE MOLDING
PLYWOOD/FACE VENEER
STUD
PLATE
BASEBOARD
QUARTER ROUND
FINISHED FLOOR
SUBFLOOR
FLOOR JOIST

ELEVATION PLYWOOD PANEL

FIGURE 17.25 ■ Paneling elevation and section.

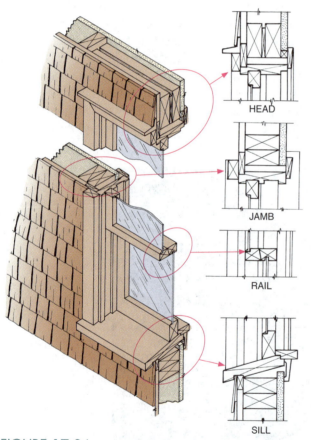

HEAD

JAMB

RAIL

SILL

FIGURE 17.26 ■ Sectioned window construction details.

FIGURE 17.27 ■ Pictorial section showing window construction. (*Kolbe and Kolbe*)

Door Sections

A horizontal section of all doors is indicated on a floor plan. However, a floor plan does not include sufficient detail for installation. Similar to window construction

sections, sill, head, and jamb sections are necessary to show door construction.

When a cutting-plane line is extended vertically through the sill and head, a section as shown in Figure 17.31 is revealed. These sections are often too small to show the desired degree of detail necessary for construction. A removed section is drawn at a larger scale to show head or sill sections.

Because doors are normally not as wide as they are high, an adequate jamb detail can be projected with-

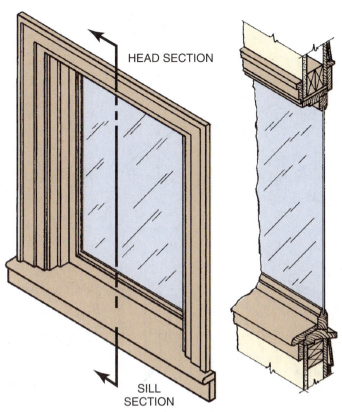

FIGURE 17.28 ■ Window head and sill sections in the vertical plane.

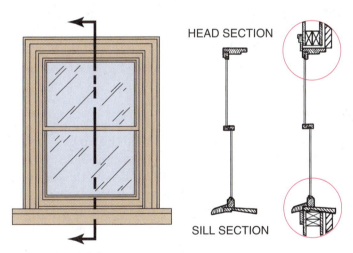

FIGURE 17.29 ■ Projection of window head and sill sections.

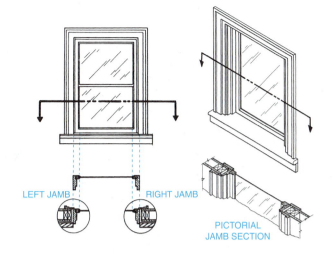

FIGURE 17.30 ■ Projection of window jamb sections.

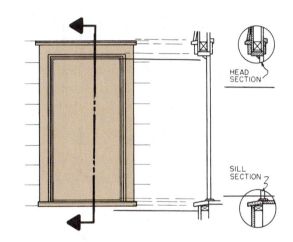

FIGURE 17.31 ■ Projection of door head and sill sections.

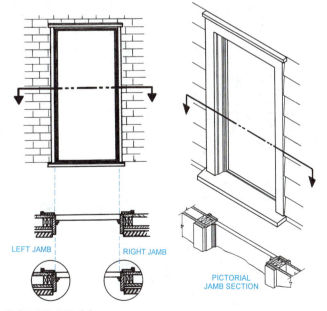

FIGURE 17.32 ■ Projecting left and right jamb sections.

out the use of break lines or removed sections. See Figure 17.32.

Sectional drawings of the framing details of the door sill, head, and jamb, exclusive of the door and door frame assembly, are often prepared for framing purposes. In such drawings, the framing section is drawn separately, with the door frame and door removed. Figure 17.33 shows the head and jamb construction used with common interior door types. Door manufacturers often

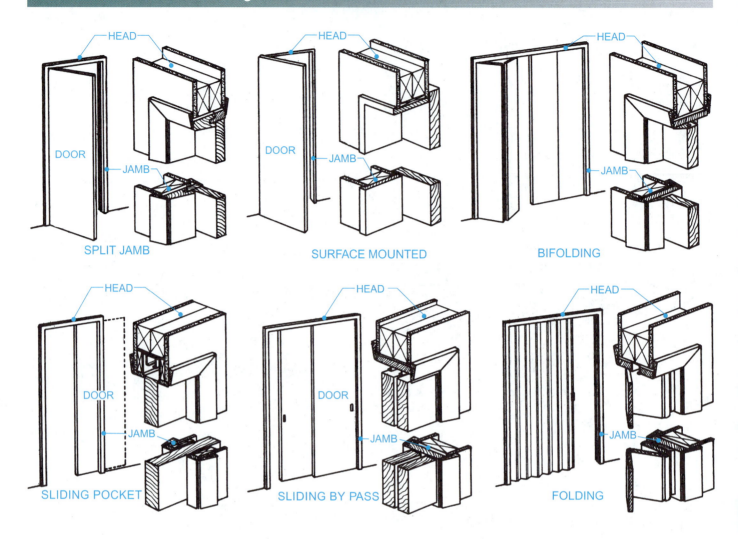

HEAD
DOOR
JAMB
SPLIT JAMB

HEAD
DOOR
JAMB
SURFACE MOUNTED

HEAD
JAMB
BIFOLDING

HEAD
DOOR
JAMB
SLIDING POCKET

HEAD
DOOR
JAMB
SLIDING BY PASS

HEAD
JAMB
FOLDING

FIGURE 17.33 ■ Framing for common interior door types.

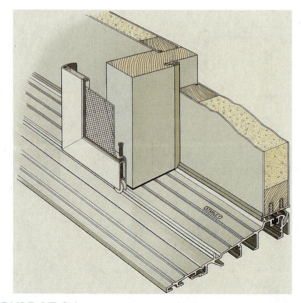

FIGURE 17.34 ■ Manufacturer's jamb section. (*Kolbe and Kolbe*)

prepare pictorial **break-out sectional drawings** to show the internal construction of components. These drawings, as shown in Figure 17.34, also show the framing, trim, and door in their proper locations. Drawings of garage doors and industrial-size doors are usually prepared with sections of the brackets and apparatus necessary to house the door assembly.

CABINETRY AND BUILT-IN COMPONENT DRAWINGS

Basic architectural plans are prepared at a scale too small to show the exact size and construction details of many cabinets and built-in components. On simple designs, details are often explained with a note or reference to a manufacturer's product. On larger, more complex designs, separate details are provided in pictorial, plan, elevation, and/or sectional drawings.

Cabinet Construction and Types

Cabinets are either custom-built (made to order) or manufactured using modular sizes. In either case, the quality of the finished product depends on the materials, joints, hardware (hinges, pulls, latches), finish, and the accuracy of construction and installation. The quality of components ranges widely from economy units, which use the least expensive materials and methods, to premium components, which resemble fine furniture.

Cabinets are either wall-hung or positioned on the floor (base cabinets). See Figure 17.35. Numerous styles and sizes are available, as shown in Figure 17.36. Most modular base cabinet sizes are standardized at 34 1/2″ in height and 24″ deep. Modular wall cabinets are usually 12″ deep, and floor cabinets are 24″ deep. Custom-built wall cabinets are usually 15″ deep and floor cabinets are 26″ deep. Cabinets are manufactured for baths, kitchens, and/or laundry rooms.

Materials used in the construction of cabinets and built-in components include hardwood, softwood, pressboard, stranded lumber, laminates, ceramic tile, marble, and synthetic materials. Though not always seen, the types of joints and fasteners used to attach parts have an important effect on the overall appearance and durability of cabinets. For example, simple butt joints are typically used on economy cabinets, whereas joints such as dovetails and mortise and tenons are used in premium units. See Figure 17.37.

Built-In Components

Because of the intricate details involved, only the outlines of most built-in components are drawn on floor

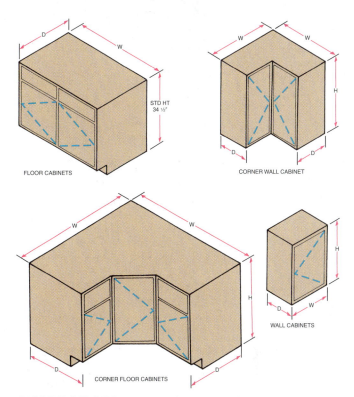

FIGURE 17.35A ■ Typical floor and wall cabinets.

SYM	AMT	DEPTH	HEIGHT	WIDTH
BASE CABINETS				
A	1	24"	34 1/2"	36"
B	2	24"	34 1/2"	36"
C	1	24"	34 1/2"	21"
D	1	24"	34 1/2"	48"
E	1	24"	34 1/2"	9"
WALL CABINETS				
1	1	12"	36"	36"
2	1	12"	18"	42"
3	2	12"	36"	12"
4	2	12"	36"	24"
5	1	12"	36"	30"
6	1	12"	36"	18"
7	1	12"	18"	30"
8	1	12"	30"	21"

FIGURE 17.35B ■ Common kitchen cabinet dimensions.

USING CAD

CAD Detail Blocks

Many architectural details are used repeatedly in otherwise original designs. These symbols can be created and stored as a *Block* by using the *Wblock* command and selecting the perimeter of the detail with the pick box.

WALL CABINETS   Wall Cabinets are 12" deep (excluding doors). Most wall cabinets available in 3" width increments from 9" to 48".

Single Door	Double Door	Wall End	45° Corner Glass Mullion Door	18" High Double Door	18" High Double Door

Available in 24", 30", 36", 42" heights.   *Available in 24", 30", 36", 42" heights.*   *Available in 30", 42" heights.*   *Available in 30" height.*   *Available in 30" width.*

BASE CABINETS   Base Cabinets are 24" deep (excluding doors) and 34½" high except where noted.

Base Tray	Single Door	Double Door	Single Drawer	Base Blind Corner	Sink Base Double Door

Available in 9" width Left or right hinging.   *Available in 12", 15", 18", 21", 24" widths. Left or right hinging.*   *Available in 27", 30", 33", 36", 39", 42", 45", 48" widths.*   *Available in 30", 36" widths.*   *Available in 36", 39", 42", 45", 48" widths.*   *Available in 24", 27", 30", 33", 36", 39", 42", 48" widths.*

TALL CABINETS   Tall Cabinets are 24" deep (excluding doors) except where noted.

VANITY CABINETS   Vanity Base Cabinets are 31½" high and 21" deep except where noted.

Single Oven	96" High Utility Cabinet	90" High Pantry Cabinet	Vanity Bowl	Vanity Bowl-Two Drawer	84" Vanity Linen

Available in 27", 30", 33" widths. Available in 84", 90", 96" heights.   *Available in 18", 24" widths. Available in 12" or 24" depths.*   *Available in 36" width.*   *Available in two door 24" to 42" widths in 3" increments. Three door available in 48" width. Four door available in 60" width.*   *Available in 24", 30", 36" widths. Available in 18" (space saver) depth.*   *Available 18" wide. Left or right hinging.*

FIGURE 17.36 ■ Standard modular cabinet dimensions. (*Merillat Industries, Inc.*)

plans and elevation drawings. Separate, large-scale dimensioned details are used to show construction and/or installation information.

Most built-in component designs include precise joinery, hidden hinges, roller sliding parts, and special hardware. Manufactured units for built-in products are usually prefabricated in modular units. The surrounding finished carpentry must be designed to blend with the unit. See Figure 17.38. This requires framing and detail drawings. When prefabricated units are not used, special framing drawings of the walls, shelves, fascia, and soffits are needed. See Figure 17.39. Built-ins of simpler design, such as shelves, mantels, and planters, are usually built on site. These may only require an interior elevation plus floor plan notes and dimensions. See Figure 17.40.

USING CAD

Manufacturers' Product Libraries

Many manufacturers supply customers with compact disks (CDs) that contain detail drawings of their products shown in various architectural applications. These fall primarily into two categories: DWG files or PDF files. CDs supplied in the DWG format can be stored, inserted into any drawing, and then altered. Details of this type are often altered to be consistent with the base drawing in line weight, dimensioning practice, or application. Imported PDF files cannot be significantly altered and must therefore be used as supplied.

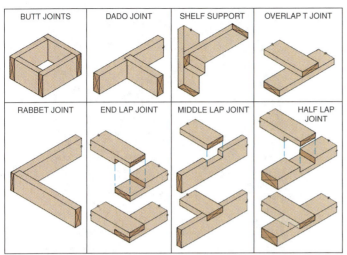

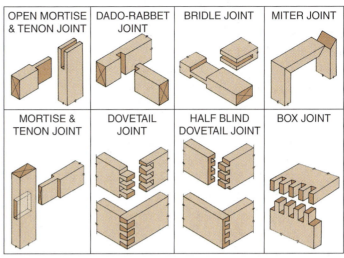

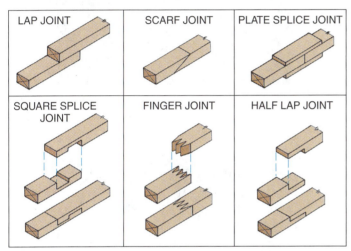

FIGURE 17.37 ■ Common cabinetry joints.

FIGURE 17.38 ■ Modular built-in components. (*Audio Tec Designs*)

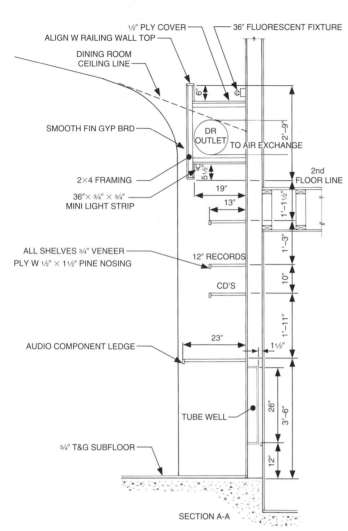

½" PLY COVER — 36" FLUORESCENT FIXTURE
ALIGN W RAILING WALL TOP
DINING ROOM
CEILING LINE
6"
SMOOTH FIN GYP BRD
DR OUTLET
TO AIR EXCHANGE
2'–9"
2×4 FRAMING
5½"
36" × ¾" × ¾"
MINI LIGHT STRIP
19"
2nd FLOOR LINE
13"
1'–1½"
ALL SHELVES ¾" VENEER
PLY W ½" × 1½" PINE NOSING
12" RECORDS
1'–3"
CD'S
10"
1'–11"
AUDIO COMPONENT LEDGE
23"
1½"
TUBE WELL
26"
3'–6"
¾" T&G SUBFLOOR
12"

SECTION A-A

FIGURE 17.39A ■ Detail drawing of a built-in entertainment center.

FIGURE 17.39B ■ Entertainment center under construction.

FIGURE 17.39C ■ Completed entertainment center.

Dimensioning Cabinetry and Built-Ins

When cabinets or built-ins are custom made or built on site, normal dimensioning practices are used. When factory-produced components are used, they must be precisely positioned with other components, cabinets, and/or framing. Appliances and plumbing fixtures are designed to stand alone, slide into a space between cabinets, or "drop into" (fit within) a countertop space. In all of these cases, adequate space must be dimensioned on plans and elevations or detailed to ensure proper fitting. Openings that are dimensioned to centerlines, such as doors and windows, need not be precise for construction purposes. Openings that are dimensioned between the sides denote a finished ("must hold") opening. This is necessary to ensure that appliances or prefabricated components will fit into the space. The word *finished* (or

"Fin") may also be added to a "must hold" dimension as in Figure 17.41. To ensure proper fitting of components all information also needs to be entered into the appropriate schedule. See Chapter 35.

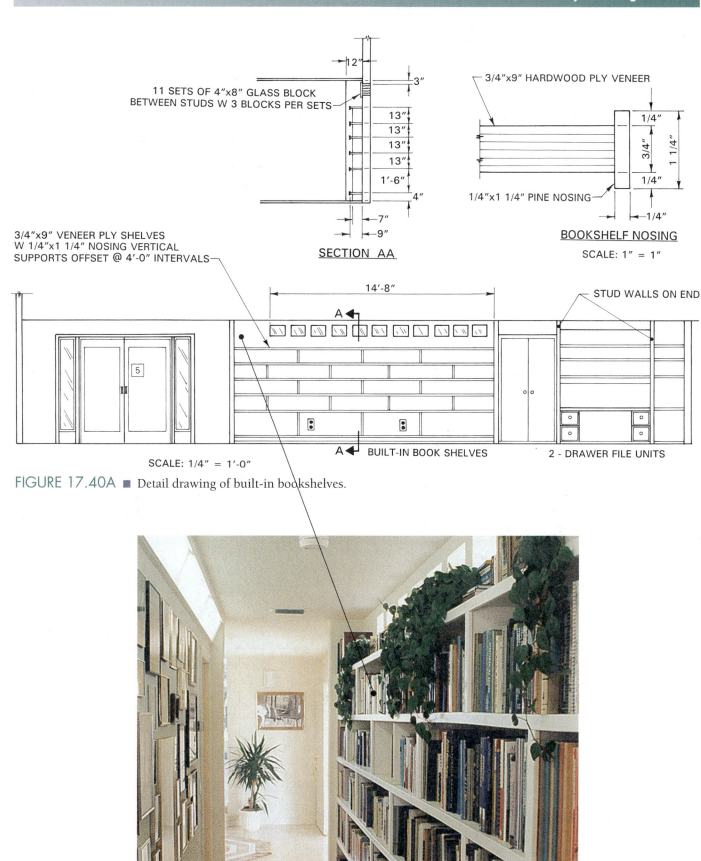

11 SETS OF 4"x8" GLASS BLOCK
BETWEEN STUDS W 3 BLOCKS PER SETS

12"

3"

13"
13"
13"
13"
1'-6"

4"

7"
9"

SECTION AA

3/4"x9" HARDWOOD PLY VENEER

1/4"
3/4"
1 1/4"
1/4"

1/4"x1 1/4" PINE NOSING

1/4"

BOOKSHELF NOSING
SCALE: 1" = 1"

3/4"x9" VENEER PLY SHELVES
W 1/4"x1 1/4" NOSING VERTICAL
SUPPORTS OFFSET @ 4'-0" INTERVALS

14'-8"

A

STUD WALLS ON END

5

A BUILT-IN BOOK SHELVES

2 - DRAWER FILE UNITS

SCALE: 1/4" = 1'-0"

FIGURE 17.40A ■ Detail drawing of built-in bookshelves.

FIGURE 17.40B ■ Completed built-in bookshelves.

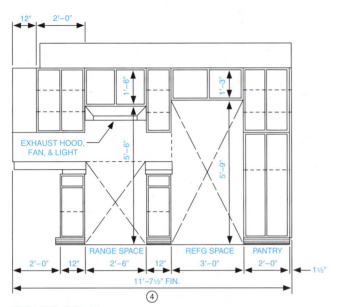

FIGURE 17.41 ■ Dimensioning practice for reserved appliance spaces.

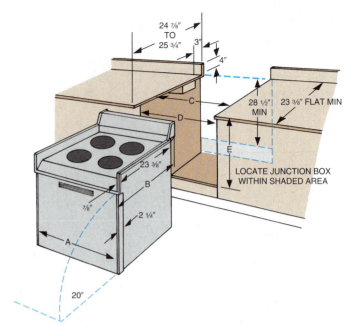

FIGURE 17.43A ■ Alphanumeric system of dimensioning cabinets.

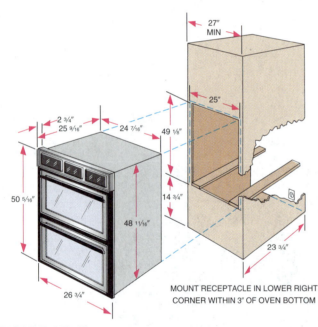

FIGURE 17.42 ■ Appliance positioning dimensions.

MODEL	A	B	C	D	E
1	26 3/4"	24 1/4"	26"	27"	26 1/16"
2	26 3/4"	24 1/2"	26"	27"	28 7/16"
3	24 3/8"	24 3/4"	22 1/2" min.	27"	34 1/2"
4	44 1/2"	24 3/4"	22 1/2" min.	27"	14 3/4"
5	28 1/8"	22"	24" min.	24"	32 1/2"
6	28 1/8"	24 3/4"	24" min.	27"	32 1/2"
7	49 5/8"	24 3/4"	24" min.	27"	13 1/4"
8	47 1/4"	24 3/4"	24" min.	27"	15 5/8"

FIGURE 17.43B ■ Cabinet and cutout demensions.

Some cabinetry must be built to accommodate all component dimensions (width, height, and depth). A drawing that shows the dimensions of the component and housing is then prepared for accurate positioning. See Figure 17.42. Positioning drawings are often included in the manufacturer's specifications.

Usually many sizes of models of a manufactured product are available. An alphabetical dimensioning system is used, like the one charted in Figure 17.43 to show model size differences.

When unusual design features are used, drawings of the surrounding framing must be detailed. See Figure 17.44. This is necessary whether the components are site built or factory built. Normally cabinets are factory built and assembled on site. Figure 17.45A is a dimensioned plan view of the completed cabinetry shown in Figure 17.45B. Many detailed drawings were prepared to complete this design.

A shortcut method of dimensioning cabinets is the use of a **cabinet coding system.** The manufacturer's code number for standard modular units is shown on the cabinet outlines drawn on a floor plan. See Figure 17.46. This system is also used for custom-made cabinets by referencing the code number to a detailed drawing of each cabinet.

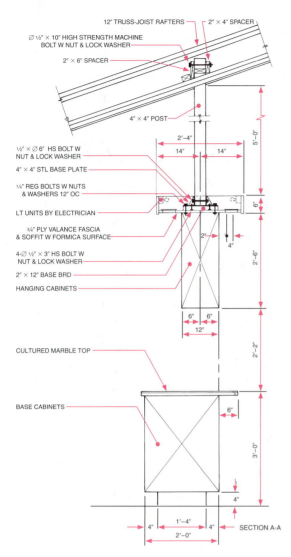

12" TRUSS-JOIST RAFTERS — 2" × 4" SPACER

∅ ½" × 10" HIGH STRENGTH MACHINE
BOLT W NUT & LOCK WASHER

2" × 6" SPACER

4" × 4" POST

½" × ∅ 6" HS BOLT W
NUT & LOCK WASHER

4" × 4" STL BASE PLATE

¼" REG BOLTS W NUTS
& WASHERS 12" OC

LT UNITS BY ELECTRICIAN

¾" PLY VALANCE FASCIA
& SOFFIT W FORMICA SURFACE

4-∅ ½" × 3" HS BOLT W
NUT & LOCK WASHER

2" × 12" BASE BRD

HANGING CABINETS

CULTURED MARBLE TOP

BASE CABINETS

2'-4"
14"   14"
5'-0"
6"
2"
4"
2'-6"
6"   6"
12"
2'-2"
6"
3'-0"
4"
4"   1'-4"   4"
2'-0"
4"

SECTION A-A

VALANCE & HANGING CABINET HANGER DETAILS
SCALE 1"=1'-0"

FIGURE 17.44A ■ Sectional drawing of a kitchen cabinet design.

FIGURE 17.44B ■ Completed kitchen area shown in Figure 17.44A.

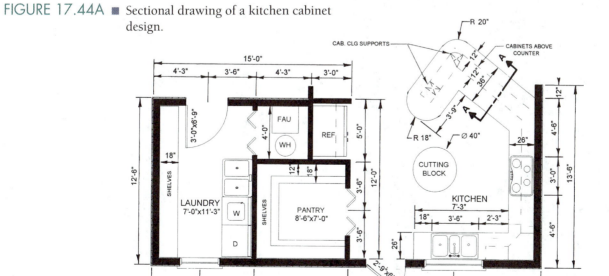

FIGURE 17.44C ■ Alternative floor plan design showing cabinetry detail dimensions.

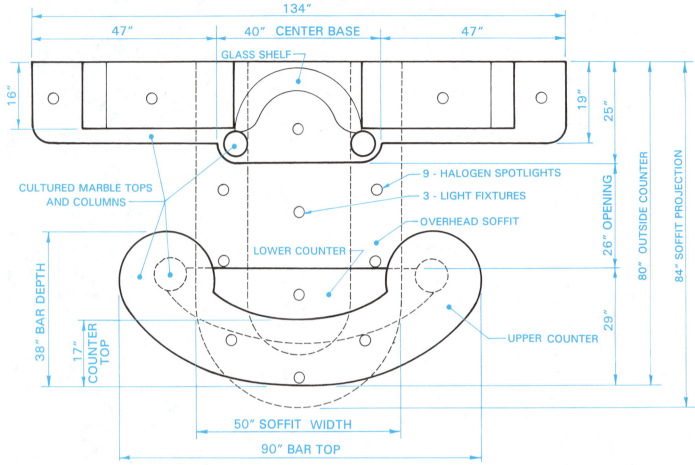

FIGURE 17.45A ■ Dimensioned cabinet plan view.

FIGURE 17.45B ■ Completed cabinetry shown in Figure 17.45A.

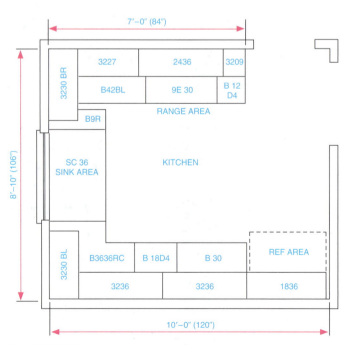

FIGURE 17.46 ■ Cabinet coding system.

Sectional, Detail, and Cabinetry Drawings Exercises

 1. Draw a full section of a house you have designed.

2. Draw a section through the view shown in Figure 17.5, revolving the cutting-plane line 90°.

 3. Draw a head, jamb, and sill section of a typical window and a typical door of the house you have designed.

 4. Draw a sill, cornice, and footing section of the house you have designed.

5. Name two methods of cabinet construction and describe what affects their quality.

6. List materials used in cabinet construction.

7. Draw a coded cabinet plan for the kitchen and bath of the house you are designing.

8. Cutting-plane line A5 is missing from the floor plan in Figure 17.7A. Find the location of A5 on this floor plan.

9. Select the best location and draw a longitudinal and transverse section through the plan shown in Figure 17.47.

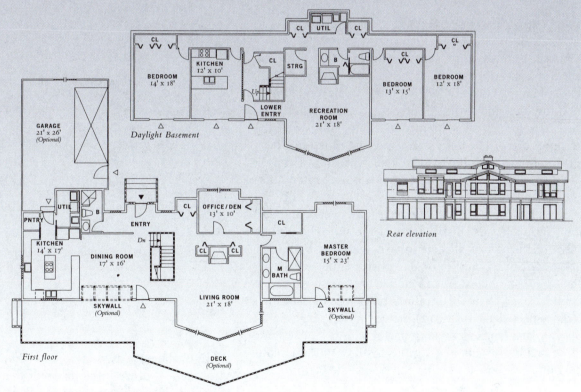

FIGURE 17.47 ■ Draw a longitudinal section A-A and a transverse section B-B through this plan. *(Lindal Cedar Homes)*

Site Development Plans

OBJECTIVES

In this chapter, you will learn to:

- identify the major elements used in site design.
- understand the role and uses of zoning ordinances in the design process.
- draw survey, plat, and plot plans.
- understand the polar coordinate system and its application to site plans.
- design, draw, and render landscape plans and elevations.

TERMS

building envelope
building permit
contour lines
density

landscape plans
phasing
planting schedules
plat

plot plans
setbacks
survey
zoning ordinance

INTRODUCTION

Landscape architecture is primarily concerned with the use of space and the integration of landform, site character, and architecture. Achieving this goes beyond simply planting trees and shrubs. It involves the development of the entire site. Site development is an integral part of the design process. The design sequence presented in Chapter 13 should be followed carefully when developing a site plan. A site design should provide proper orientation and use of natural features of the site.

Site plans describe the characteristics of the land and the relationship of all structures to the site. The outline and dimensions of all constructed features (buildings, driveways, etc.) and their exact position on the site are shown on site plans. Also included are the shape of the landform as well as the locations and types of plant material. Specialized site plans include survey plans, plot plans, plats, landscape plans, and renderings. These are discussed in this chapter. Various features of these plans are often combined into one composite site plan.

- Slope
- Soils
- Vegetation
- Wildlife and factors related to habitat, especially if rare and endangered species inhabit the site
- Hydrology:
  – Surface water
  – Flood hazard
  – Groundwater
  – Wetlands (FWW/Tidal)
- Climate (regional) and microclimate (specific to site)
- Geology
- Visual character

FIGURE 18.1 ■ Environmental factors affecting site design.

SITE ANALYSIS

Completing a site analysis is the first step in producing an acceptable site design. The design should meet the needs of the user, as well as protect and enhance the environment. Future inhabitants of the site must also be considered. Both environmental (Figure 18.1) and human-related elements (Figure 18.2) that influence development and design are analyzed in the site analysis. The surrounding area of a site should also be considered to determine future plans. For example, commercial zoning changes may affect plans for constructing a residence.

- Existing site, street layout, and topography
- Existing land use and zoning
- Historical significance and preservation
- Available utilities
- External factors
  - Noise
  - Site accessibility
- Demographics (population characteristics)
- Socioeconomic forecasts

FIGURE 18.2 ▪ Human factors affecting site design.

Suitability Levels

Environmental and human-related elements must be analyzed to determine the level of the site's suitability for development and building.

- *High suitability:* Many favorable conditions exist to make this area relatively inexpensive to develop with a minimum amount of environmental impact.

- *Moderate suitability:* Some special design and construction measures will be needed to modify this land and preserve the environment.

- *Low suitability:* Conditions exist that place serious restrictions on building in this area. For example, either environmental damage will result if the site is disturbed, or high construction costs will be needed to avoid damage to ecosystems or to protect the public.

- *Not suitable:* Disturbance or impact to this area will cause significant environmental damage or adversely impact the public safety and welfare. Costs to develop the area are excessive.

Suitability levels may apply to the entire site or to selected zones of a site. Refer back to the composite analysis shown in Chapter 13.

ZONING ORDINANCES

Before beginning the design concept, all local zoning ordinances must first be thoroughly checked. **Zoning ordinances** are laws or regulations designed to provide safety and convenience for the public and to preserve or improve the environment.

Local building codes must also be checked prior to beginning a design. Redesigning or redrawing plans, if they are in conflict with laws, is very time consuming and costly. Working drawings should not be started until the basic site plan is approved by the local zoning authorities.

Specific zoning ordinances may differ among communities. However, zoning categories are very similar. For example, zoning laws specify the type of occupancy, population density (number of persons in an area), land use, and the building type allowed in each zone of a community. Maps, usually called *tax maps*, are found in each community's building department.

Most codes divide municipalities (cities) into residential, commercial, and industrial zones. Residential zones are divided into single-family dwellings, multiple-family dwellings (duplex, triplex, and quadraplex), and apartments that include units for five or more families. Commercial zones include schools, offices, retail stores, and medical facilities. Industrial zones include factories, warehouses, or any facility requiring the movement and/or storage of large vehicles or equipment. Where appropriate, separate zones are established for hazardous areas, such as those where the danger of flooding exists.

Structural Types

The types of structures allowed in each community are related to the designated use of each zone. A community committee often regulates these ordinances. Zoning ordinances may be intended to maintain a degree of architectural consistency within a given area. Ordinances of this type may restrict or allow only specific styles, periods, materials, landscaping, heights, colors, or sizes.

Maximum and sometimes minimum building sizes are often specified to control the use of space, traffic, and the impact on the environment. For example, a single-family residential zone in one community may require a minimum size home of 2,000 sq. ft. and a maximum of 5,000 sq. ft. Another community may require a minimum of 3,000 sq. ft. with an unlimited maximum area. In addition to the amount of square footage allowed, some codes specify maximum building size with reference to street frontage and property lines.

All codes now include maximum building heights to allow neighbors maximum access to views, air circulation, and sunlight. A maximum height of 35 ft. is often used for residential zones. Many newer codes now include limits on the amount of space used in upper floors. See Figure 18.3. This pyramid principle, called the *daylight plane*, is required to allow more light to reach adjacent properties located on the north side. See Figure 18.4. The impact of shadowing is becoming more critical as building areas become more densely built. Using constructed models or computer-drawn models, shadow patterns are studied and checked during different seasons and times.

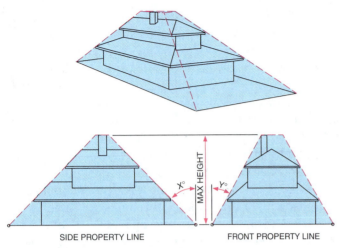

FIGURE 18.3 ■ The pyramid shape allows maximum sun penetration.

Land Coverage and Setbacks

Math Connection

Regardless of the square footage of a building, laws may also restrict the percentage of property allowed for building space. This is done to preserve as much *green space* as possible. Percentages vary from one community to another, usually from 25 percent to 40 percent. For example, a small site, 50′ × 120′, with a coverage of 40 percent can contain a maximum size building of 2,400 sq. ft. See Figure 18.5.

Zoning laws include codes that restrict the distance permitted from any building to the property lines. These distances are known as **setbacks**. Refer again to Figure 18.5. Some codes require that a structure be placed no closer than 5′ from a side property line. Other codes may require 10′ or more. Setback distances may be different between front, side, and rear property lines.

Setback lines are drawn within and parallel to property lines to indicate the acceptable building area for a lot. Setbacks and other zoning dimensions also vary for pools,

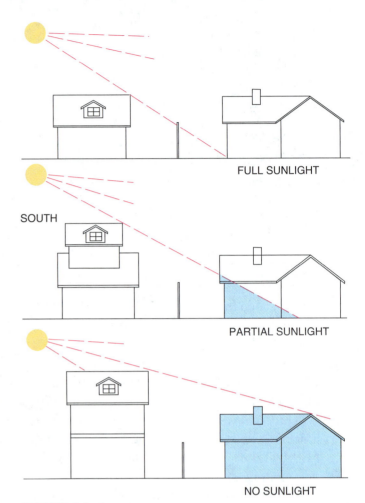

FIGURE 18.4 ■ Some codes restrict the blockage of sunlight on adjacent properties.

garages, corner lots, hillside sites, easements, parking areas, and walks, as shown in Figure 18.6. Figure 18.7 includes typical setback requirements for an interior lot, including adjacent roadway dimensions. Where lot shapes are irregular, setbacks must be held to the nearest building corner or tangent point to a property line, as dimensioned in Figure 18.8.

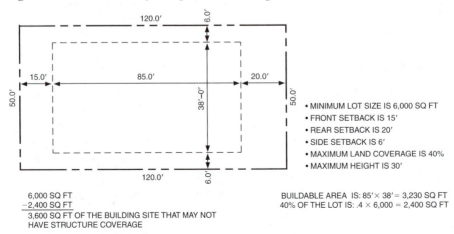

6,000 SQ FT
−2,400 SQ FT
3,600 SQ FT OF THE BUILDING SITE THAT MAY NOT HAVE STRUCTURE COVERAGE

• MINIMUM LOT SIZE IS 6,000 SQ FT
• FRONT SETBACK IS 15′
• REAR SETBACK IS 20′
• SIDE SETBACK IS 6′
• MAXIMUM LAND COVERAGE IS 40%
• MAXIMUM HEIGHT IS 30′

BUILDABLE AREA IS: 85′ × 38′ = 3,230 SQ FT
40% OF THE LOT IS: .4 × 6,000 = 2,400 SQ FT

FIGURE 18.5 ■ Typical setback ordinance requirements.

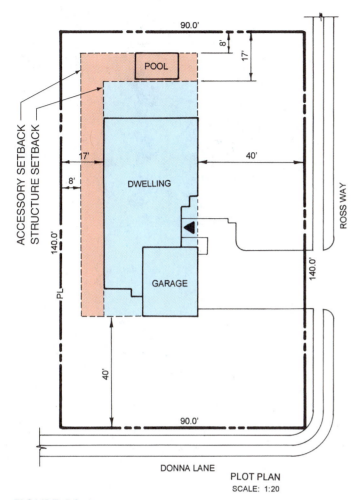

FIGURE 18.6 ■ Setback regulations applied to a corner lot.

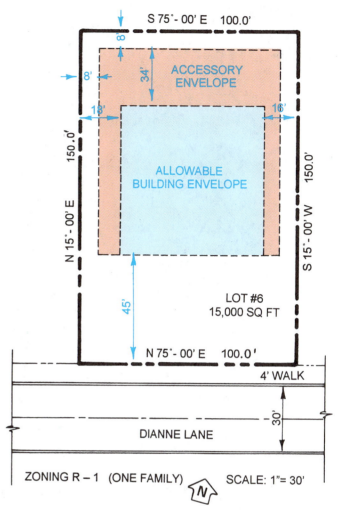

FIGURE 18.7 ■ Setback requirements applied to an interior lot.

Typical single-family-dwelling zoning ordinances (often called R-1) limit the front, rear, and side setbacks. They also limit the lot size and maximum land coverage. Figure 18.5 shows an example of how to calculate the **building envelope** (building area) for a 50′ × 120′ lot.

Figure 18.9 shows a lot with a building and accessory envelope that is determined as follows:

1. Draw setback lines for front yard, rear yard, and side yard setbacks. Also add accessory (other buildings and features) setbacks.

2. The remaining area (71′ × 66′) shows the allowable building envelope (4,686.0 sq. ft.).

3. A typical residential (R) code states the maximum gross floor area can be no more than 25 percent of the lot area. Thus 0.25 × 15,000 sq. ft. = 3,750 sq. ft. This is the maximum building envelope (or footprint) of the structure.

4. Note that the entire building envelope cannot be used. To calculate the amount that can be used:

Actual gross floor area = 3,750 sq. ft.

Allowable building envelope = 4,686 sq. ft.

4,680 − 3,750 = 936 must remain open in the building envelope.

5. Thus 80 percent (3,750 ÷ 4,686 = 0.80) of the allowable building envelope can be covered by the house. This does not include decks, patios, garages, or other buildings.

Setback, usage, land coverage, structural type, and height restrictions must all be combined to determine the location, size, and height of structure allowed for a particular lot. See Figure 18.10. It is important to check with the local zoning department to determine the exact requirements for setbacks on all floor levels. Setback dimensions are measured from the property line to the wall line, eave line, projecting fireplace, or cantilevered second story of the structure, depending on local setback codes.

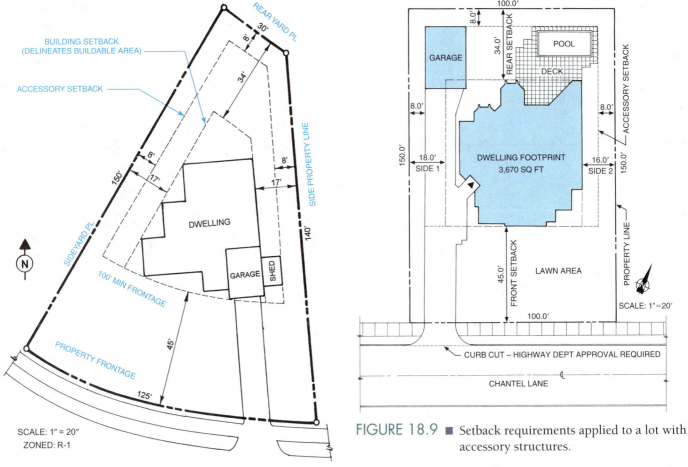

SCALE: 1″ = 20″
ZONED: R-1

FIGURE 18.8 ■ Setback requirements applied to an irregularly shaped lot.

FIGURE 18.9 ■ Setback requirements applied to a lot with accessory structures.

* MAXIMUM LENGTH OF SECOND STORY EQUALS 25% OF THE LOT'S DEPTH OR 35′ (WHICHEVER IS LESS)
* MAXIMUM LENGTH FOR THE SECOND STORY IS .25×120′ = 30′
* MINIMUM SIDE SETBACKS FOR THE SECOND STORY IS 50% OF THE STRUCTURE'S MAXIMUM HEIGHT. THE SECOND STORY SIDE SETBACK IS .5×25′ = 12.5′.

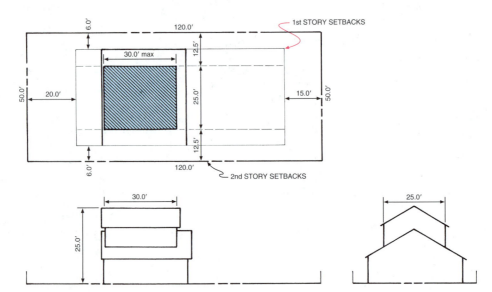

FIGURE 18.10 ■ Typical zoning ordinances, height restrictions, and setbacks for all levels.

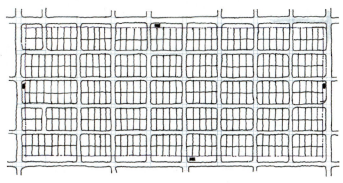

FIGURE 18.11A ■ Conventional housing development pattern.

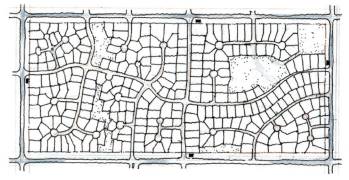

FIGURE 18.11B ■ Cluster housing pattern with limited open space.

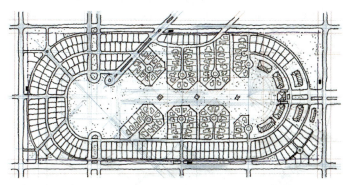

FIGURE 18.11C ■ Cluster housing development with expanded open space.

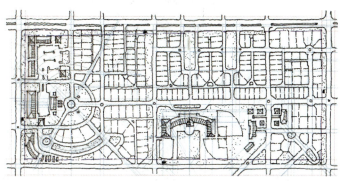

FIGURE 18.11D ■ Mixed-use community development.

Density Zoning

To create a good living environment and prevent over-building in an area, architects and builders who plan and build multiple-home developments must conform to density zoning laws. **Density**, in architectural terms, is the relationship of the number of residential structures and people to a given amount of space. The density of an area is the number of people or families per acre or square mile. For example, a town may have a density of 3 families per acre or 1,920 families per square mile.

The *average density* of an area is the ratio of all inhabitants to a specific geographic area. Density patterns may vary greatly within different parts of one area. Some parts of that area may be crowded, while other parts are less populated. In other patterns, the population may be evenly distributed. Density planning for an area must be based on the maximum number of people who will occupy the area, regardless of the patterns.

Figures 18.11A through 18.11D show the same area with different housing patterns designed to achieve different densities. Figure 18.11A shows a conventional single-family development pattern that yields greater numbers of houses but is visually monotonous and socially undesirable. Automobile use here is essential. Figure 18.11B shows a cluster development that contains single, quad, and duplex units to yield more open space. Curved streets add architectural interest by breaking roadway monotony. Automobiles are also essential here. The mixed-use development in Figure 18.11C includes single and multiple residential types with a commercial, retail, and school component. This plan yields better opportunity for open spaces and a walkable, livable, workable community. The need for automotive traffic is reduced. Athletic fields, lawn strips, and an organic road configuration, in Figure 18.11D create interest and a feeling of more open space. Increased density is achieved here with the use of zero-lot-line (ZLL) duplexes and townhouses. Auto traffic in this developed community is greatly reduced. Figure 18.11E shows an example of an area with the same average density. Only the density pattern has changed.

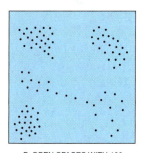

A. MOST PROPERTY PRIVATELY OWNED – 100 LIVING UNITS

B. OPEN SPACES WITH 100 CLUSTERED LIVING UNITS

FIGURE 18.11E ■ Comparison of areas with the same average density.

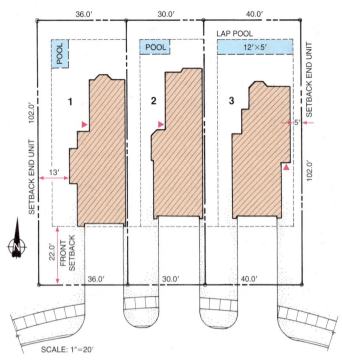

FIGURE 18.12 ■ Zero-lot-line properties.

To prevent overcrowding, local zoning ordinances may restrict the size of each building. This automatically restricts the number of families allowed to occupy a specific area. This approach also spreads the density patterns equally. As another approach, zoning laws may encourage clustering of many residents into fewer structures, such as high-rise apartments or townhouses (attached houses) with larger open public areas. In higher density developments, smaller size lots can be used more efficiently by eliminating one side yard (ZLL) and reducing the front, rear, and other side yard. See Figure 18.12.

A combination of plans involves zoning different parts of an area for single-family residences, townhouses, and high-rise apartments. The amount of space planned for each type of structure depends on the average density desired.

Building Permits

Building permits are required to ensure that sites and buildings conform to community standards for structural types, land coverage, setbacks, and density zoning. A well-planned community, such as the one shown in Figure 18.13, has a master plan for growth that includes code and zoning regulations. See Chapter 37. In addition to residential needs, community plans include provisions for schools, parks, traffic, shopping, police, fire protection, and often architectural style consistency.

Before a public structure or dwelling can be built, a building permit must be obtained from the local building department. Once the working drawings for a project are complete, a municipal building inspector checks each area of the design. The design must be in compliance with all existing codes and ordinances. If the drawings and specifications meet the code requirements, a building permit will be issued. See Figure 18.14. By carefully checking all local and regional code requirements before finalizing the design and preparing working drawings, revisions can be avoided.

In addition to local building departments, the administration of local codes may be co-regulated and/or controlled by other local or governmental agencies such as city planning commissions, air pollution control districts, fire departments, public health departments, water pollution control boards, and perhaps even art and design commissions or historical preservation societies. If a federal building is involved, the Department of Housing and Urban Development (HUD), Department of Health, Education and Welfare (HEW), Federal Housing Authority (FHA), or other agencies may also be involved in the approval process.

Zoning ordinances and some building codes do contain allowances for exceptions to the law. If a building cannot be designed or sited to conform to all local laws, builders can request a *variance* from the building department. In making this request, the builder must show that the exemption will not harm or inconvenience neighbors, the community, or the environment in any way. Variances are often requested for setbacks, styles, building sizes and types.

SURVEY PLANS

A **survey** is a drawing showing the exact size, shape, and levels of a property. When prepared by a licensed surveyor, a survey plan is used as a legal document to establish property rights. It is filed with the deed to the property. The lot survey includes the length of each boundary, tree locations, utility lines, corner elevations, contour of the land, and position of streams, rivers, roads, or streets. It also lists the owner's name and the owners or titles of adjacent lots. A survey drawing must be accurate and must also include a complete written description of the lot.

FIGURE 18.13 ■ Community development plan.

Establishing Dimensions

Math Connection

Surveying a site involves locating points (coordinates) on the earth's surface. To indicate and connect these points on a drawing, the *polar coordinate system* is used. In this system, a fixed, true north-south reference line, called a *meridian,* is established. The direction of a line on the survey drawing is given in relation to the meridian as the line's *bearing,* an angle toward east or west.

The exact plan and shape of a lot is shown by property lines. Property line dimensions are shown directly on the line by length and angle. Angles are dimensioned using either the American system or the azimuth system. See Figure 18.15. In the *American system* a compass is divided into four quadrants: NE, SE, SW, and NW. Angular dimensions are shown by noting the degrees, minutes, and seconds from either N, E, S, or W and toward another direction. There are 360 degrees (°) in a

PERMIT

Page 1 of 1

Permit #: 970012   Type: Residence   Issued: 1-3-- --   by: JP
Job Location: 603   Issue Loc:
Lot: 8   Subdiv: Lake Estates
Parcel: 8A
Owner: Kingston   Elev: 6'   Fl Map: A4
Project: 95 K   Seawall Datum
Job Description: Residence & Pool
Applicant Name: J R Smith   Type: Skeleton Frame
Applied Date: 1-2-- --   Appl Oper: 4
Contract Phone: 364 8752   Inspector Area: 6   Work w/o Permit Fee: 50
Contractor Name: T. Jones   Cert Nbr: 336
Business Name: Capitol Builders   Septic Tank: Dwg A1

Setbacks Front: 50   Left: 35   Right: 35   Rear: 35
FCC Code: 329
Square Footage: 3560   Rate: TBD   Job Value: 195,000
Number of Units: 1   Floors: 2   Buildings: 1 + Dock
ROW: 24' ℄ RD   Zoning: PUD   Map No: 00617
Minimum Floor Elevation: 6' Seawall Datum   ~~Residential~~/~~Commercial~~

H. Mitchell   1/3/-- --
——————————————————————————   ——————————————
Building Official or Authorized Signature   Date

FIGURE 18.14 ■ Sample building permit. (*Lake County, Florida, Building Department*)

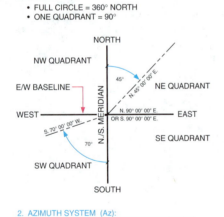

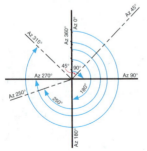

FIGURE 18.15 ■ American and azimuth property dimensioning systems.

circle, 60 minutes (') in a degree, and 60 seconds ('') in a minute. Thus a 45° line in the northeast quadrant is dimensioned N 45° 00' 00'' E. This means the line is 45° from north, heading toward east. Figure 18.16 illustrates how the Azimuth system is used to calculate the bearings of four property lines enclosing a site.

In the *azimuth system,* each line is dimensioned as an angle, reading clockwise from the north meridian, from 0° to 360°. These angular lines are drawn by aligning the 0° or 360° line on a protractor with the north meridian line. Place the center of the protractor at the intersection of the meridian and the east-west line. The degree is located on the circumference of the protractor and connected with a line to the protractor center.

Elevations (Heights)

The height of any point on a site is dimensioned from a fixed elevation point called a *datum*. The universal datum elevation is sea level. On small sites a fixed point on a road, sidewalk, or seawall may be used as datum. Datum is always zero and all elevation measurements are made up or down from the datum.

Several types of drawings and lines are used to show elevation distances and shapes on survey drawings. These include profile drawings (land sections), elevation point notations, and contour lines.

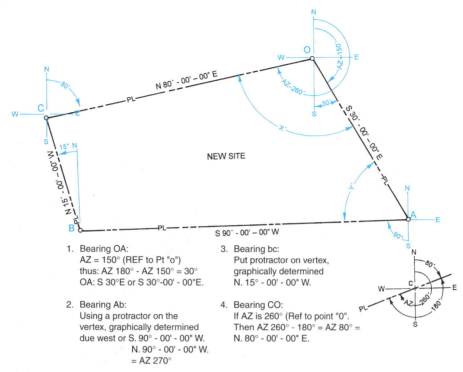

1. Bearing OA:
   AZ = 150° (REF to Pt "o")
   thus: AZ 180° - AZ 150° = 30°
   OA: S 30°E or S 30°-00' - 00"E.

2. Bearing Ab:
   Using a protractor on the
   vertex, graphically determined
   due west or S. 90° - 00' - 00" W.
   N. 90° - 00' - 00" W.
   = AZ 270°

3. Bearing bc:
   Put protractor on vertex,
   graphically determined
   N. 15° - 00' - 00" W.

4. Bearing CO:
   If AZ is 260° (Ref to point "0".
   Then AZ 260° - 180° = AZ 80° =
   N. 80° - 00' - 00" E.

FIGURE 18.16 ■ Steps in determining bearings using the azimuth system.

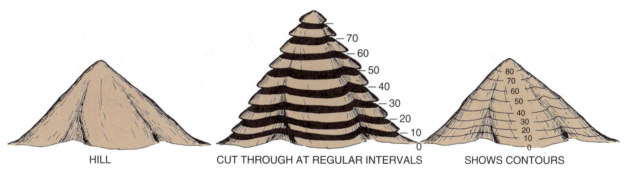

HILL   CUT THROUGH AT REGULAR INTERVALS   SHOWS CONTOURS

FIGURE 18.17 ■ Contour lines are imaginary cuts through the terrain.

Contour Lines

On maps or drawings that need to show terrain levels, **contour lines** connect points on the land surface that are the same elevation above datum. See Figure 18.17. Contour lines representing the same elevation are continuous, although some require a large distance to join. As shown in Figure 18.18, contour lines that are close together represent steep slopes. The closer the spacing, the steeper the slope.

The vertical distance between contour lines is called the *contour interval*. The interval can be any convenient distance, but is usually an increment of 5'. Contour intervals of 5', 10', 15', and 20' are common on large surveys. The use of smaller intervals (1' or 2') gives a more accurate description of the slope and shape of the terrain than does the use of larger intervals. Note the different levels of detail between the 2' and the 10' interval of the same terrain as shown in Figures 18.19 and 18.20. The size of the interval depends on the scale of the drawing and on the size of the area to be shown. Large contour intervals are normally used on large regional maps.

Figure 18.21 shows contour lines that are identical to those in Figure 18.20, except the contour numbers are in reverse order. The center elevation in Figure 18.20 represents a hill; the center elevation in Figure 18.21, however, represents a depression. Often the crest of a hill or the low point of a depression falls between contour intervals. In this case a plus sign (+) and the exact elevation level is noted if precision is required. The high point in Figure 18.20 would be referred to as hill 31.

Not every contour line includes an elevation number. The interval between numbers depends on the scale of the

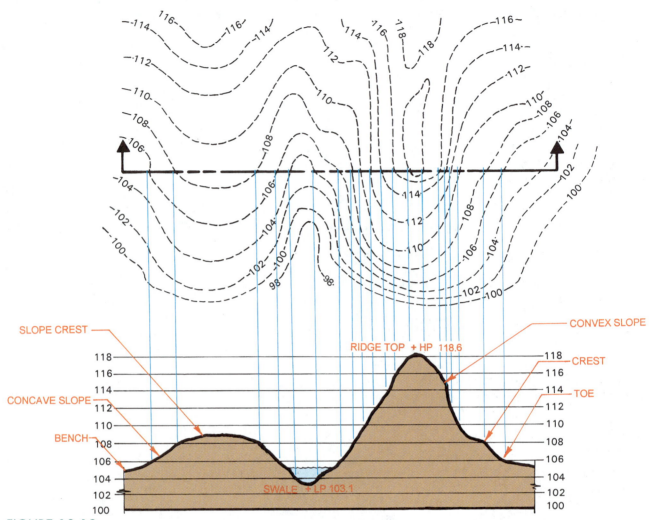

FIGURE 18.18 ■ Elevation profile (section) related to contour lines.

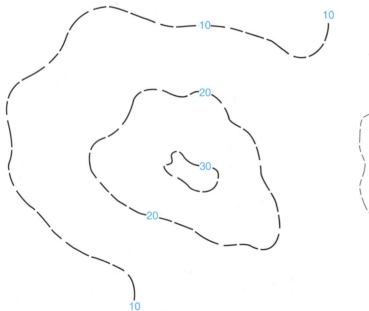

FIGURE 18.19 ■ Large intervals show only general slope direction.

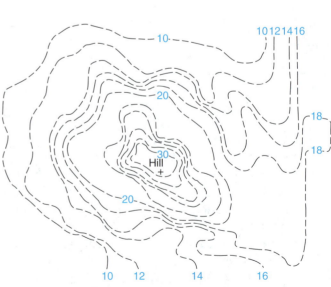

FIGURE 18.20 ■ Small intervals reveal slope details of a hill.

drawing. The interval noted in Figure 18.22 represents 10′ of elevation with heavy unnoted lines used every 50′.

Where land surfaces are to be altered during the construction process, proposed contour lines are added to the drawing. Land is either removed (*cut*) or added-to (*filled*). Figure 18.23 shows the addition of contour lines to a flat surface to denote the amounts of cut or fill. Figure 18.24A shows how cuts and fills are shown on a contour drawing. The original contour lines are drawn as dashed lines, whereas the proposed contour is shown with solid lines. This illustration also shows the projection of a profile elevation from these contour lines.

For large, flat projects such as parking lots and airport tarmacs (runway areas, etc.), contour lines may not be

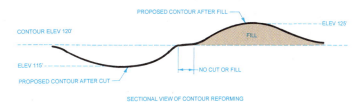

FIGURE 18.23 ■ Profiles showing cuts and fills on a flat surface.

USING CAD

Pictorial Contours on CAD

Computer programs are used to create pictorial contour drawings such as that shown in Figure 18.24B. This is done by combining X-Y coordinate input data and the corresponding datum level (Z) for each coordinate point. These points are then connected using surface modeling techniques to create the pictorial contour lines.

The individual data points are connected using the *Polylines* or *Splines* command to create the individual lines. The lines are then meshed together to create a surface model of the area being designed. Once the lines are created or a surface mesh is generated, the designer's viewpoint can be changed with the *Vpoint* command to get a true three-dimensional perspective. Use the *3D Orbit* command for the easiest control.

Digital graphics files are available from the U.S. Geological Survey (USGS) that contain information from USGS maps which can be used for large site plans. Local municipal planning departments can often supply digital map and/or aerial photographs with property lines superimposed.

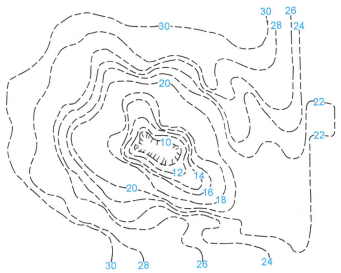

FIGURE 18.21 ■ Contour numbers reveal a depression.

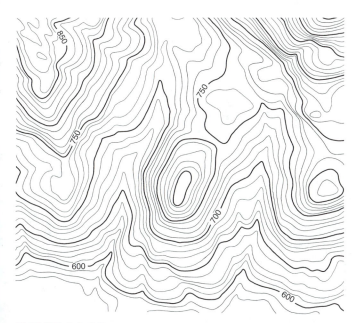

FIGURE 18.22 ■ Contour lines identifed every 10′.

used because the slope differential is less than a foot. Only the elevations of selected points are placed on the drawing. An arrow is drawn beside each point to show slope direction, as shown in Figure 18.25.

Symbols on Survey Drawings

Symbols are used extensively to describe the features of the terrain. Some symbols, such as tree symbols, resemble the appearance of a feature. Most survey symbols are graphic representations. The charts shown in Figure 18.26 include symbols and abbreviations used on survey

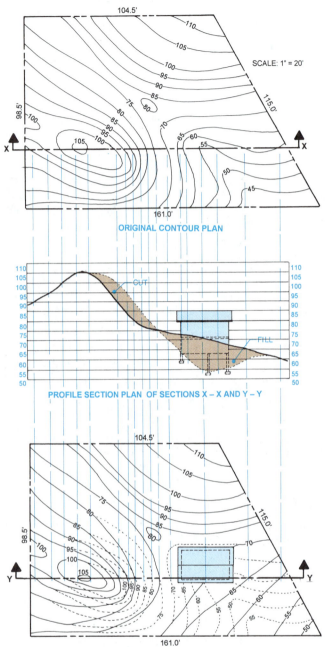

ORIGINAL CONTOUR PLAN

SCALE: 1" = 20'

PROFILE SECTION PLAN OF SECTIONS X – X AND Y – Y

PROPOSED CONTOUR PLAN (OLD CONTOURS ARE DOTTED)

FIGURE 18.24A ■ Dashed contour lines show original contour grades. Solid contour lines show new contours.

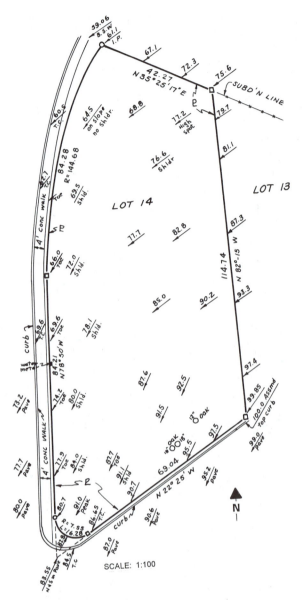

FIGURE 18.25 ■ Elevation notations showing slope direction.

SCALE: 1:100

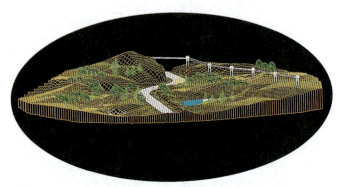

FIGURE 18.24B ■ Pictorial contour drawing. (*Autodesk, Inc.*)

USING CAD

Storing Site Details on CAD

Many site details are "stand-alone" details that are not an integral part of a site plan. Site details include removed sections, such as curb details and lighting fixtures, and planting instruction drawings, which may be included with landscape plans. It is very efficient to include these drawings in a site detail library, which eliminates having to redraw them on every drawing each time they are needed.

NAME	ABBREV	SYMBOL	NAME	ABBREV	SYMBOL	NAME	ABBREV	SYMBOL	NAME	ABBREV	SYMBOL	NAME	ABBREV	SYMBOL
TREES	TR		CULTIVATED AREA	CULT		BUILDINGS	BLDGS		LARGE RAPIDS	LRG RP				
GROUND COVER	GRD CV		WATER	WT		SCHOOL	SCH		WASH	WSH				
BUSHES SHRUBS	BSH SH		WELL	W		CHURCH	CH		LARGE WATERFALL	LRG WT FL				
OPEN WOODLAND	OP WDL		PROPERTY LINE	PR LN		CEMETARY	CEM		BOUNDARY, U.S. LAND SURVEY TOWNSHIP	BND US LD SUR TWN				
MARSH	MRS		SURVEYED CONTOUR LINE	SURV CON LN		POWER TRANSMISSION LINE	PW TR LN		BOUNDARY, TOWNSHIP APPROXIMATED	BND TWN				
DENSE FOREST	DN FR		ESTIMATED CONTOUR	EST CON		GENERAL LINE LABEL TYPE	GN LN		BOUNDARY, SECTION LINE U.S. LAND SURVEY	BND SEC LN US LD SUR				
SPACED TREES	SP TR		FENCE	FN		BOUNDARY, STATE	BND ST		BOUNDARY, SECTION LINE APPROXIMATED	BND SEC LN				
TALL GRASS	TL GRS		RAILROAD TRACKS	RR TRK		BOUNDARY, COUNTY	BND CNTY		BOUNDARY, TOWNSHIP NOT U.S. LAND SURVEY	BND TWN				
LARGE STONES	LRG ST		PAVED ROAD	PV RD		BOUNDARY, TOWN	BND TWN		INDICATION CORNER SECTION	COR SEC				
SAND	SND		UNPAVED ROAD	UNPV RD		BOUNDARY, CITY INCORPORATED	BND CTY		U.S. MINERAL OR LOCATION MONUMENT	U.S. MIN MON				
GRAVEL	GRV		POWER LINE	POW LN		BOUNDARY, NATIONAL OR STATE RESERVATION	BND NAT OR ST RES		DEPRESSION CONTOURS	DEP CONT				
WATER LINE	WT LN		HARD-SURFACE HEAVY DUTY ROAD – FOUR OR MORE LANES	HRD SUR HY DTY RD		BOUNDARY, SMALL AREAS: PARKS, AIRPORTS, ETC	BND		FILL	FL				
GAS LINE	G LN		HARD-SURFACE HEAVY DUTY ROAD – 2 OR 3 LANES	HRD SUR HY DTY RD		LEVEE	LEV		CUT	CT				
SANITARY SEWER	SAN SW		IMPROVED LIGHT DUTY ROAD	IMP LT DTY RD		RIVER	RV		LAKE, INTERMITTENT	LK INT				
SEWER TILE	SW TL		TRAIL UNIMPROVED DIRT ROAD	TRL UNIM DRT RD		STREAM PERENNIAL	ST PER		LAKE, DRY	LK DRY				
PROPERTY CORNER WITH ELEVATION	PROP CR EL	EL 70.5	ROAD UNDER CONSTRUCTION	RD CONST		STREAM INTERMITTENT	ST INT		SPRING	SP				
SPOT ELEVATION	SP EL	+ 78.8	BRIDGE OVER ROAD	BRG OV RD		STREAM DISAPPEARING	ST DIS		PILINGS	PLG				
WATER ELEVATION	WT EL	80	ROAD OVERPASS	RD OVP		SMALL RAPIDS	SM RP		SWAMP	SWP				
BENCH MARKS WITH ELEVATIONS	BM/EL	BM X 84.2 BM △ 84.2	ROAD UNDERPASS	RD UNP		SMALL WATERFALL	SM WT FL		SHORELINE	SH LN				

FIGURE 18.26 ■ Common survey symbols.

USING CAD

CAD Site Symbols

Site plans include symbols that represent land surface materials, plant materials, and site-related fixtures and devices. These are included in many CAD software programs but most are created, stored, and retrieved as *Blocks* using the *Block* command. Care must be taken to adjust the scale of these symbols to match the scale of the drawing. Land surface materials are available in blocks and are inserted into the drawing using the *Hatch* command. These symbols are also available for elevation drawings.

drawings. The survey plan in Figure 18.27 shows the application of the most common survey plan symbols.

Guidelines for Drawing Surveys

The numbered arrows in Figure 18.28 correspond to the following guidelines for preparing survey drawings:

1. Record the elevation above the datum of the lot at each corner.
2. Represent the size and location of streams and rivers by wavy lines (blue lines on geographical surveys).
3. Use a cross to show the position of existing trees. The elevation at the base of the trunk is shown on some drawings.
4. Indicate the compass direction of each property line by degrees, minutes, and seconds.

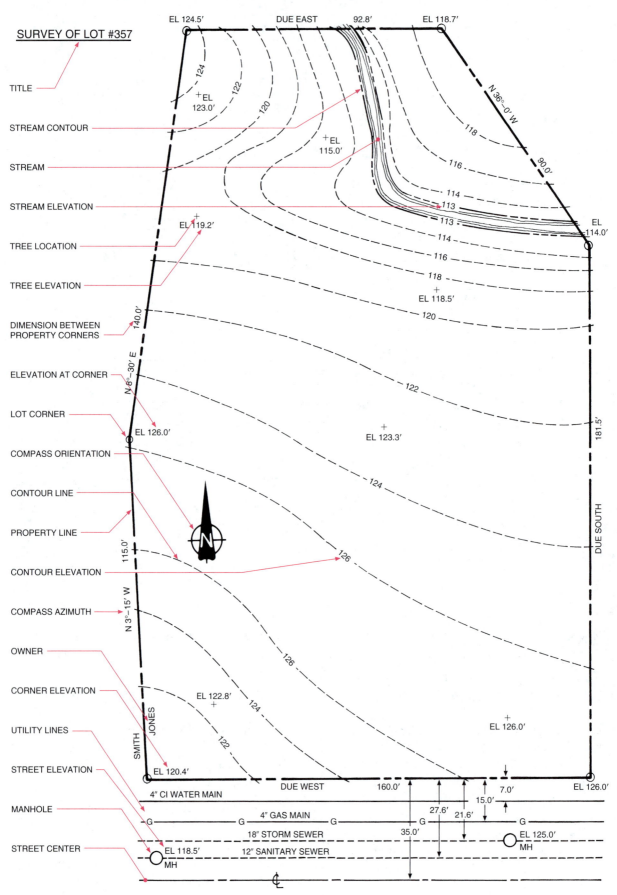

SURVEY OF LOT #357

TITLE

STREAM CONTOUR

STREAM

STREAM ELEVATION

TREE LOCATION

TREE ELEVATION

DIMENSION BETWEEN
PROPERTY CORNERS

ELEVATION AT CORNER

LOT CORNER

COMPASS ORIENTATION

CONTOUR LINE

PROPERTY LINE

CONTOUR ELEVATION

COMPASS AZIMUTH

OWNER

CORNER ELEVATION

UTILITY LINES

STREET ELEVATION

MANHOLE

STREET CENTER

EL 124.5′   DUE EAST   92.8′   EL 118.7′

+EL 123.0′

+EL 115.0′

N 36°–0′ W   90.0′

124   122   120   118   116   114   113   113   114   116   118   120

EL 114.0′

+EL 118.5′

EL 119.2′

140.0′

N 8°–30′ E

122

+EL 123.3′

EL 126.0′

124

181.5′

126

115.0′

N 3°–15′ W

126

DUE SOUTH

SMITH   JONES

EL 122.8′

124

122

+EL 126.0′

EL 120.4′

DUE WEST   160.0′   7.0′   EL 126.0′

4″ CI WATER MAIN   15.0′

G   4″ GAS MAIN   G   G   27.6′   21.6′   G   G

18″ STORM SEWER   35.0′   EL 125.0′   MH

EL 118.5′   12″ SANITARY SEWER
MH

FIGURE 18.27 ■ Application of survey plan symbols.

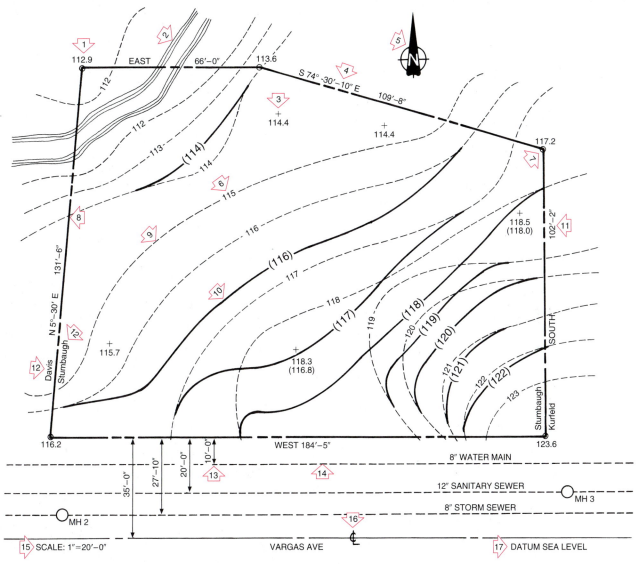

FIGURE 18.28 ■ Survey drawing keyed to survey guidelines.

USING CAD

Zoom and Pan on Site Plans

The *Zoom* and *Pan* commands are used constantly in preparing all types of architectural drawings. For example site plans often cover a D-size sheet (24″ × 36″) and are therefore over 10 times the size of a CAD monitor. The *Zoom* command is similar to the zoom function on a camera. It allows the entire drawing to be viewed and also enables small details or parts to be magnified or demagnified while drawing.

The area that is going to be magnified is selected by choosing the *View* pull-down menu, selecting *Zoom* and then selecting the *Window* option. A window is created around the area that will be magnified by picking opposite corners of the window. A specific magnification scale is required, which can be typed in as part of the *Zoom* command. Real-time zoom capabilities allow the drawing to be magnified only through wheel mouse controls.

The *Pan* command shifts the placement of the screen image on the screen. It does not change the magnification, just the position of the "viewing window."

5. Use a north arrow to show compass direction.

6. Break contour lines to insert the height of each contour above the datum.

7. Show lot corners by small circles or overlapping property lines.

8. Draw the property line symbol by using a heavy line with two dashes repeated throughout.

9. Show elevations above the datum by contour lines.

10. Show any proposed change in grade line by contour lines. Use dotted contour lines to show original grade and solid contour lines to show the new proposed grading levels.

11. Show lot dimensions directly on the property line. The dimension on each line indicates the distance between property corners.

12. Give the names of owners of adjacent lots outside the property line. The name of the owner of the site is shown inside the property line.

13. Dimension the distance from the property line to all utility lines.

14. Show the position of utility lines by dotted lines. Utility lines are labeled according to their function.

15. Draw surveys with an engineer's scale. Dimensions are shown as feet and decimal parts of a foot (for example, 6.5'). Common scales for surveys are 1" = 10' and 1" = 20'.

16. Show existing streets and roads either by centerlines or by curb or surface outlines. The centerline symbol is ℄ and is used for reference when an actual centerline is not drawn.

17. Indicate the datum level used as reference for the survey.

Geographical Survey Maps

U.S. Geographical Survey (USGS) maps are similar to property surveys except they cover extremely large areas. The entire world is divided into geographical survey regions. However, not all regions have been surveyed and mapped. Geographical survey maps show the general contour of the area, natural features of the terrain, and structures.

When large areas are covered, a small scale is used. When smaller areas are covered, as shown in Figure 18.29A, a

FIGURE 18.29A ■ Partial USGS map. (*U.S. Geological Survey*)

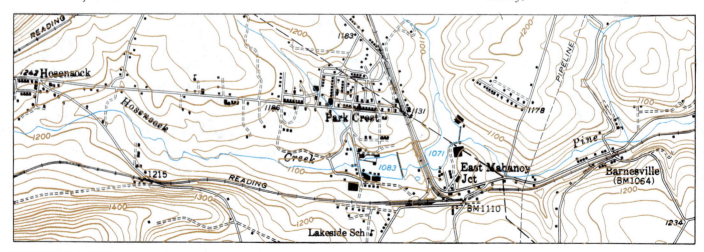

FIGURE 18.29B ■ Survey map related to aerial photograph in Figure 18.29C. (*U.S. Geological Survey*)

FIGURE 18.29C ■ Aerial photograph of area shown in Figure 18.29B. (*U.S. Geological Survey*)

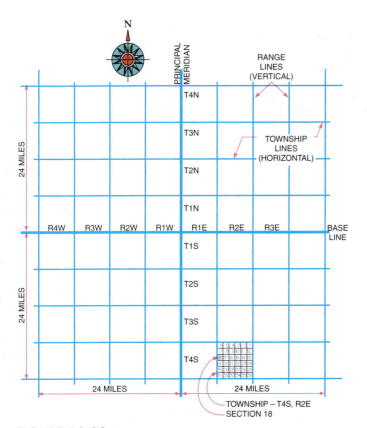

FIGURE 18.30 ■ Divisions of a 48-square-mile region.

larger scale such as 1:24000 is used. Figure 18.29B shows a portion of a geological survey map compared to an aerial photograph of the same area. The aerial photograph in Figure 18.29C represents the same area included in the map in Figure 18.29B.

Plats

A **plat** is a survey (map, chart, or plan) of multiple connected properties. Plats are legal descriptions of a land site and are identified by plat name, section, township, county, and state. A plat is part of a geographical survey region, which is divided into areas that contain further subdivisions. See Figure. 18.30.

Plats may include the compass bearing (direction) of the plat area, dimensions of each property line, and the position of all roads, utility lines, and easements, as shown in Figure 18.31. Some show only lot shapes, as in Figure 18.32. Others identify lots or buildings by numbers that refer to a more detailed survey. See Figure 18.33. Plats are prepared for residential developments, industrial parks, urban developments, and shopping complexes.

PLOT PLANS

Plot plans are used to show the size and shape of a building site and the location and size of all buildings on that site. The position and size of walks, drives, pools, streams, patios, and courts are also shown. Compass orientation of the lot is given, and contour lines are sometimes shown. Plot plans may also include details showing site construction features. See Figure 18.34. If a separate survey and/or a landscape plan is prepared, contour lines, utility lines, and planting details are usually omitted from a plot plan. Building foundations are

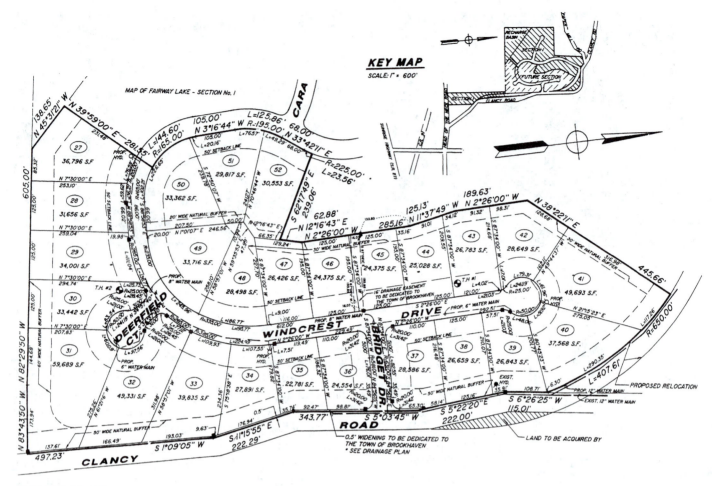

FIGURE 18.31 ■ Plat plan with property lines identified.

flat; this means the elevation is the same at all locations. Therefore, contour lines stop where they intersect building lines. Contour lines reappear outside at some point on the building perimeter. This keeps the contour planes continuous.

Guidelines for Drawing Plot Plans

When plot plans are prepared, the features indicated by the numbered arrows in Figure 18.35 should be drawn according to these guidelines:

1. Draw the outline of the main structure on the lot. Crosshatching is optional.

2. Draw the outlines of other buildings on the lot.

3. Show overall building dimensions.

4. Locate each building by dimensioning perpendicularly from the property line to the closest point on a building. On curved or slanted property lines,

dimension to points of tangency (touching but not intersecting). See Figure 18.37D. The property line shows the legal limits of the lot.

5. Show the position and size of driveways.

6. Show the location and size of walks.

7. Indicate elevations of key surfaces such as floors, ground line, patios, driveways, and courts.

8. Outline and show the symbol for surface material used on patios and terraces.

9. Label streets adjacent to the site.

10. Place overall lot dimensions either on extension lines outside the property line or near the property line.

11. Show the size and location of constructed recreation areas, such as tennis courts. (None on this plan.)

12. Show the size and location of pools, ponds, or other bodies of water.

FIGURE 18.32 ▪ Rendered plat plan showing lot shapes and positions.

FIGURE 18.33 ■ Plat plan showing numbered lots and vegetation.

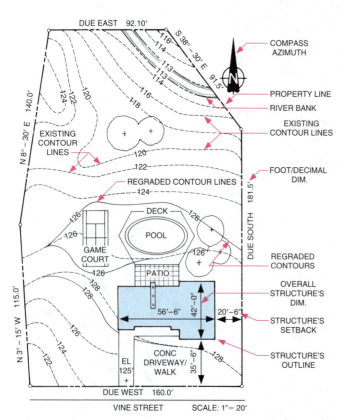

FIGURE 18.34 ■ Application of plot plan symbols.

13. Indicate the compass orientation of the lot with a north arrow.

14. Use a decimal (civil engineer's) scale, such as 1″ = 10′, or 1″ = 20′, for preparing plot plans.

15. Show the position of utility lines.

16. Include the compass direction with the perimeter dimensions for each property line.

17. Show trunk base location and coverage of all major trees. Trunk diameter may also be noted.

18. Label and dimension all landscape construction features and auxiliary structures.

19. Identify the location of entrances with symbols. Arrows are most often used. See Figure 18.36.

20. If a septic system is used, draw and dimension the location and minimum distance allowed from system components to the nearest building.

Variations

Although plot plans should be prepared according to the standards shown in Figure 18.35, many optional features may also be included in plot plans. For example, a plot

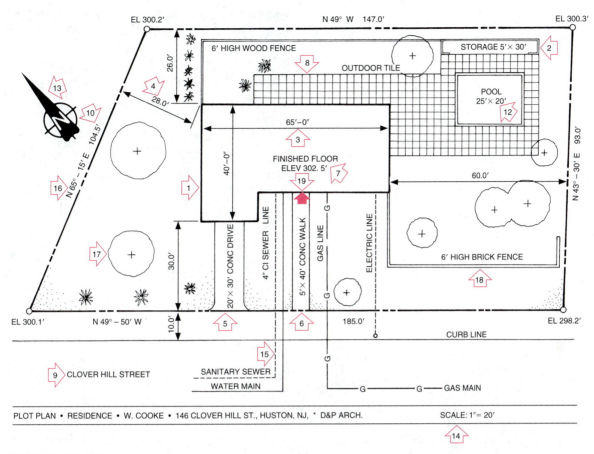

FIGURE 18.35 ■ Plan keyed to plot plan guidelines.

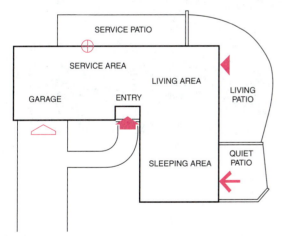

FIGURE 18.36 ■ Types of entrance symbols.

plan may show only the outline of the building, or it may include shading, outlines of the roof intersections, or crosshatching. Sometimes the interior partitions of a building are drawn to reveal connections between outside living areas and the inside rooms. Figures 18.37A through 18.37E shows a comparison between an undeveloped parcel of land and the completely developed

property. Also shown is the site plan, one of forty drawings, used to complete this project.

LANDSCAPE PLANS

Landscape plans are drawings that show the types and locations of vegetation. They may also show contour changes and the position of buildings. Such features are necessary to make the placement of the vegetation meaningful. Symbols are used on landscape plans to show the position of trees, shrubbery, flowers, vegetable gardens, hedges, and ground cover. See Figure 18.38.

Guidelines for Drawing Landscape Plans

The following guidelines for drawing landscape plans are illustrated by the numbered arrows in Figure 18.39.

1. Existing tree data including the existing topographic elevation, caliper (trunk diameter approximate; measured 1' above finish grade), type, and

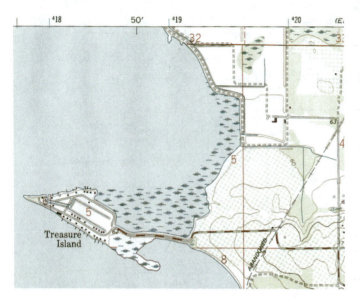

FIGURE 18.37A ■ USGS map of the site in Figure 18.37B.

FIGURE 18.37B ■ Site before development.

FIGURE 18.37C ■ Site after development.

the tree outline canopy are shown with a +. This data is usually provided as part of the survey, which the architect can utilize as a "base" for various site plans.

2. Existing topographic lines (dashed) are added, with numbers shown on or above the line.

3. Proposed topographic lines (solid) are also known as contour lines. Numbers are in brackets to differentiate existing and proposed data.

4. Property lines define site limits.

5. Landscape plans and site plans are prepared using an engineer's scale. This scale and sometimes a graphic scale as shown on the plan.

6. Buildings are outlined, crosshatched, or shaded depending on scale and use of the site plan document. A landscape plan at $1'' = 10'$ may show more detail (such as windows and door openings) than a $1'' = 100'$ scale plan.

7. Proposed trees and shrubs are shown by drawing an outline of the canopy (extent of branches).

FIGURE 18.37D ■ Dimensioned site plan.

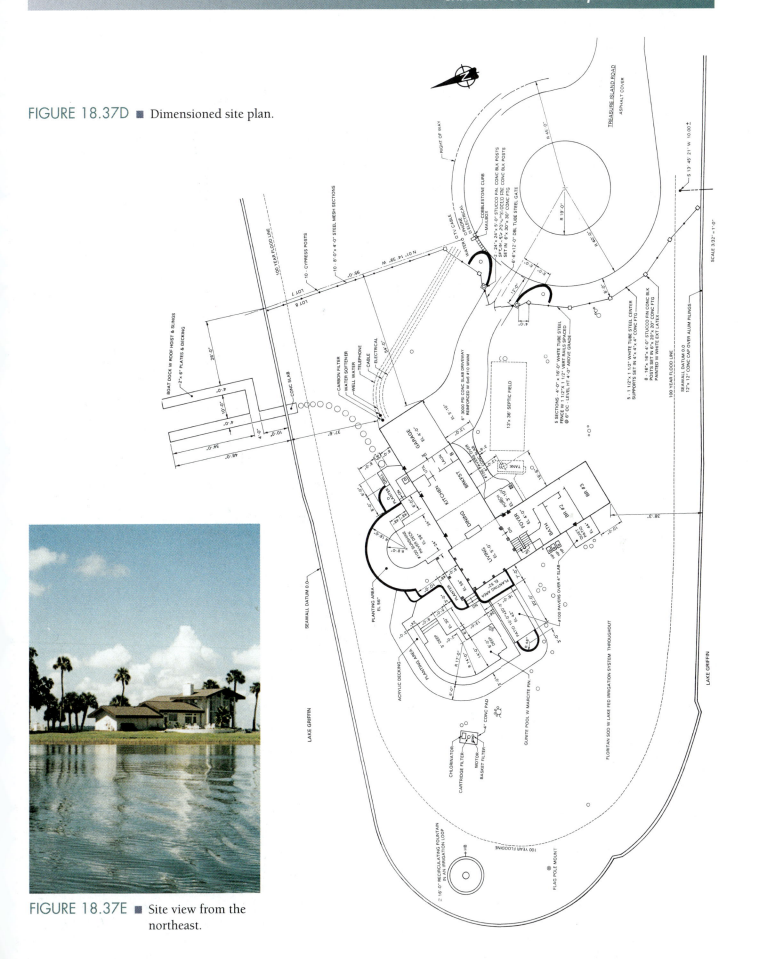

FIGURE 18.37E ■ Site view from the northeast.

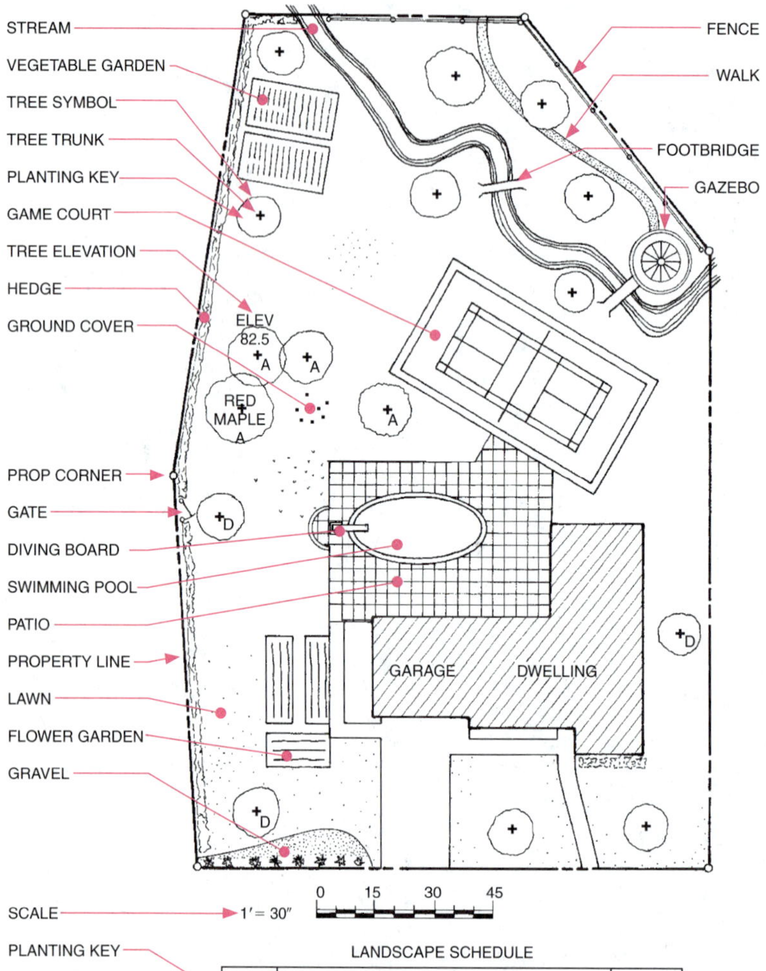
STREAM
VEGETABLE GARDEN
TREE SYMBOL
TREE TRUNK
PLANTING KEY
GAME COURT
TREE ELEVATION
HEDGE
GROUND COVER
PROP CORNER
GATE
DIVING BOARD
SWIMMING POOL
PATIO
PROPERTY LINE
LAWN
FLOWER GARDEN
GRAVEL
FENCE
WALK
FOOTBRIDGE
GAZEBO
ELEV
82.5
RED
MAPLE
GARAGE
DWELLING
SCALE
1′= 30″
0
15
30
45
PLANTING KEY

LANDSCAPE SCHEDULE

KEY	NAME	QUAN
A	RED MAPLE – ACER RUBRUM	4
B	AM. YELLOWWOOD – CLADRASTIS LUTEA	5
C	CHINESE ELM – ULMUS PARVIFOLIA	2
D	WHITE ELM – QUERCUS ALBA	3
E	SCOTCH PINE – PINUS STROBUS	2

FIGURE 18.38 ■ Landscape plan keyed to landscape plan guidelines. See Teacher's Resource Guide for the complete plant list.

8. Proposed trees and shrubs are identified by quantity and type, such as "(1 AMC)," which is labeled on or beside the tree symbol. The 1 refers to quantity, and the AMC is a key or an abbreviation that is indexed on the plant list.
9. Plant lists indicate quantity, key, botanical names, common name, size, type of rootball, and other pertinent data the landscape architect deems necessary for the proper identification.
10. Shrubs are sometimes shown as masses. The quantity of plants is determined by the mature size of the specified plant. In some cases some plants are drawn less than mature size to provide a denser planting pattern. This is done to solve a multitude of design problems such as accentuating a pleasant view or screening an unpleasant view.
11. Outlines of constructed recreation areas are shown.
12. Mass plantings of shrubs, ground cover perennials, and annuals are drawn as outlines. The complexity of the outline shapes is dependent on scale. In complex plantings, areas are crosshatched or "textured" to differentiate adjacent areas. The number and type are keyed in the same way trees are keyed. Mass perennial plantings may list numerous plants to be "placed in the field" by the

FIGURE 18.39 ■ Application of landscape symbols.

FIGURE 18.40 ■ Typical landscape plan.

landscape architect. Enlarged plans may be prepared at a larger scale for perennial beds. This is done to show individual plant locations.

13. Paved areas (hardscapes) are outlined and sometimes hatched or textured to indicate scale and enhance the "readability" of the drawing. See Figure 18.40.

14. Water edge is indicated by labeling WL (waterline). The body of water is also labeled.

15. Streams can be depicted by a double line or a labeled centerline in the case of intermittent (that is, seasonal) streams.

16. Orchards or allées are shown by outlining the tree canopy as one mass with a centerline connecting the trunks.

17. Vegetable gardens are shown by outlining the plant furrows.

18. Evergreen mass plantings, such as as trees or hedges, are used to provide privacy and winter windbreaks.

19. Trees provide shade and scale to the site. At times landscape architects specify very large trees with calipers (trunk) over 30″ to solve various design issues.

20. Shrubs are used to define spaces, outline walks, and conceal foundation walls. Shrubs can also accentuate or de-emphasize the landform/building relationship.

21. Flower gardens are shown by outlining their shapes.

22. Lawns are shown as outlines with or without "stippling" (sparsely placed dots).

23. Bridges and other structures are shown with a standard bridge symbol.

24. Landscape plantings should enhance the function and aesthetics of the site. Ideal landscape plans work in conjunction with the grading of the site to enhance the site opportunities and minimize the constraints.

25. Plantings in front of the house should address the major architectural "lines" of the structure. Generally landscape design provides for a mixture of plant types, forms, and colors to allow for the experience of *progressive realization* as one approaches the house. In other words, the entire house or site is not seen all at once.

26. A north arrow is always shown because building orientation can have a major effect on landscape design.

Phasing

The complete landscaping of a site may be prolonged over several years because of a lack of time or money. **Phasing** spreads the project over a larger period of time. Parts of the plan are completed at different times.

When a landscape plan is phased, the total plan is drawn. Then different shades or colors are used to identify the items that will be planted in the first month or year, and in successive months or years.

 ## SITE RENDERING

Landscape renderings are used by landscape architects, architects, planners, and environmental professionals to realistically communicate the scale and visual character of a project. They visually convey the integration of the architecture and the site. Proper landscape design is accomplished by using a variety of plants, pavings, water, topography, scale, form, color, and texture.

Rendering Media

Landscape drawings require the use of many different types of inks, markers, pens, pencils, paints, and papers. You will learn more about these and techniques for using them in Chapter 20.

If rendering techniques are to be added directly to a print, a heavy blackline or brownline print, not a blueline print, should be used. Blueline prints make a poor rendering medium if defined shapes and lines are to be maintained. Blue lines do not stand out when other colors are added.

Plan Rendering

Plan views are the most common form of site illustration because plans best define the overall scope of a project or development. A rendered landscape plan can be used to describe the landscaping of large parcels or small residential sites. A landscape rendering can instantly tell the viewer whether the site is a hilly, wooded area or a grassy knoll. A rendered landscape plan typically shows trees, shrubs, ground cover, grass, walks, driveways, curbs, steps, pools, patios, rock outcrops, walls, and bodies of water.

The amount of rendering and intensity of detail should be consistent with the scale of the drawing. To describe a large site development, a minimum amount of rendering and detail is often adequate because of the small scale used. See Figure 18.41.

A large site at the scale of $1'' = 200'$ can show trees and existing vegetation in simple groups or masses. A small residential plan at a scale of $1/4'' = 1'\text{-}0''$ can show a considerable amount of detail. Individual trees and even branches can be shown, as well as textured paving and ground cover. Elements such as pools, arbors, shrubs, and decks can also show some amount of texture.

Before rendering a full landscape plan, the individual elements should be mastered. The following illustrations show a progression of rendering techniques designed to build specific skills before rendering a complete plan. In these illustrations note how the image changes when color, value, texture, and shadow all come together.

Paving

Figure 18.42 shows three steps in plan rendering of different paving types. Practice these steps with successive sheets of tracing paper until you are satisfied.

Trees and Shrubs

In a plan view, tree branches cover and hide plant material, ground cover, and some structural areas located below. Because of their shape, the angle of the sun, and resulting shadows, treetops in a plan view contain a mixture of light, shade, and shadow. These are rendered by

FIGURE 18.41 ■ Rendered development plan.

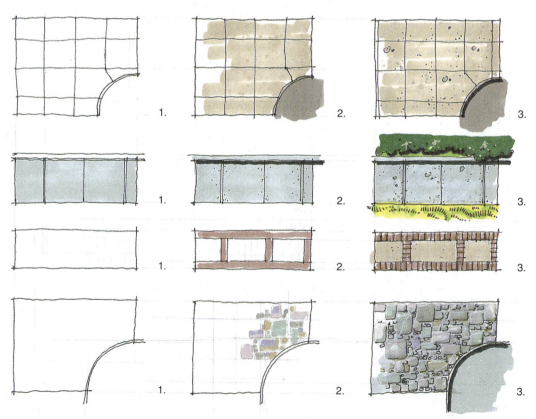

FIGURE 18.42 ■ Steps in rendering pavement surfaces in plan view.

FIGURE 18.43 ■ Plan view of rendered trees and shrubs.

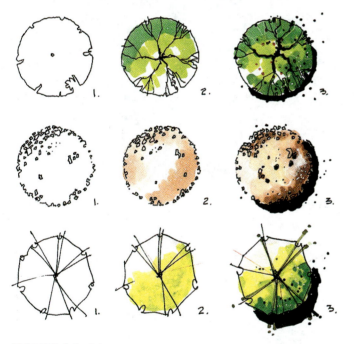

FIGURE 18.44 ■ Steps in rendering trees.

adding value, color, and texture. See Figure 18.43. Also note that larger trees cast larger shadows, and conifers (pine trees) cast shadows of a different shape. Deciduous trees cast different shadows in summer than in winter. Figure 18.44 illustrates the sequence of rendering several types of trees on a landscape plan.

Water

The rendering of water is challenging because water tends to be monochromatic (one color) and, hence, ap-

pears flat. Many aspects of the appearance of water result from its reflective qualities, which are difficult to render. Several techniques can be used to illustrate the reflections and movement of pool and stream water. See Figure 18.45.

People, Vehicles, and Other Objects

Including people, vehicles (especially automobiles), and other familiar objects in a rendering helps establish the size of the project and its various components. The key to effectively rendering people and objects is the proper use of scale and shadow. Everything in the rendering must be drawn at the same scale. In plan views, shadows are also very important. See Figure 18.46.

Rendering a Complete Landscape Plan

The techniques just described were applied in rendering the complete landscape and seascape plan shown in Figure 18.47. To prepare a complete landscape rendering, begin by determining the color scheme on a separate but similar print or tracing. Then render higher elements, such as trees, arbors, and trellises for plants. Next, render lower elements, such as shrubs, decks, and ground cover. Render people, vehicles, and other objects last. Finally, add texture and shadows.

Elevation Rendering

Plan views may be best for showing all the features of a site, but elevation drawings are also needed to show landscape features as realistically viewed from eye level. Landscape elevation renderings combine drawings of trees, shrubs, walls, topography, vehicles, and people with the structure to produce a realistic view of the site. These drawings are used by design professionals to reveal form and scale of the project. The addition of color, value, texture, shades, and shadows provides depth to elevation drawings. Rendered elevation drawings are effective tools for sales and client-approval presentations because they help viewers perceive spatial depth and the final appearance of the project. See Figure 18.48.

Trees and Shrubs

Vegetation sizes, shapes, colors, and textures vary greatly. These differences must be considered in rendering. See Figure 18.49A. Trees and shrubs are located and rendered to show them at full maturity even though they will be planted at much smaller sizes. Trees especially should

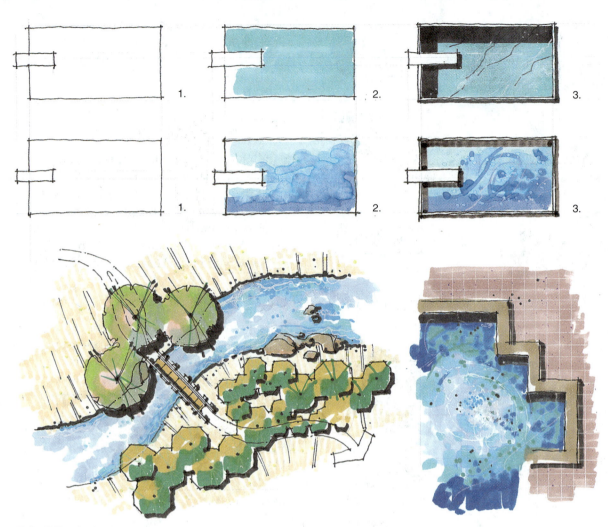

FIGURE 18.45 ■ Plan view techniques for rendering water surfaces.

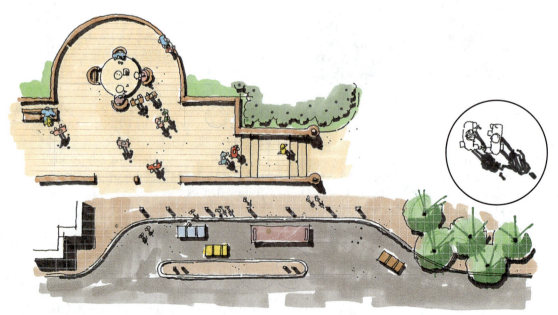

FIGURE 18.46 ■ People's shadows add realism to this plan.

FIGURE 18.47 ■ Rendered marina and park site plan.

help define space without totally blocking desirable natural views outside and eye-level views from inside structures. See Figure 18.49B. Where dense vegetation is used as a baffle, outlining or rendering the winter form of deciduous trees and shrubs is recommended.

FIGURE 18.48 ■ Rendered elevation of a natural waterfall.

Water

Water in elevations is found in the form of natural waterfalls, constructed waterfalls, and fountains. Water as a focal point in an elevation should be vibrant and alive. This is accomplished through the use of color and texture, with a proper mix of current flow and froth (bubbling). Figure 18.50 shows a waterfall as part of an elevation profile drawing. Figure 18.51 shows techniques for adding the illusion of water action to the elevation drawing.

People and Familiar Objects

As with complete plan renderings, adding people and familiar objects to an elevation rendering adds both realism and scale. The addition of both people and the automobile to Figures 18.46 and 18.50 provides much needed size comparisons in these illustrations.

Rendering Walls

Retaining walls are a common element in most multilevel landscape designs. These are constructed from wood timbers, brick, stone, and cast-concrete forms. Effective

FIGURE 18.49A ■ Rendering techniques for different trees in elevation.

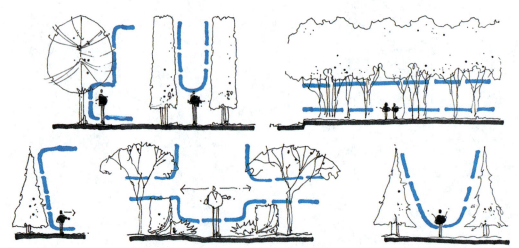

FIGURE 18.49B ■ Open space and views are considered when selecting and locating trees.

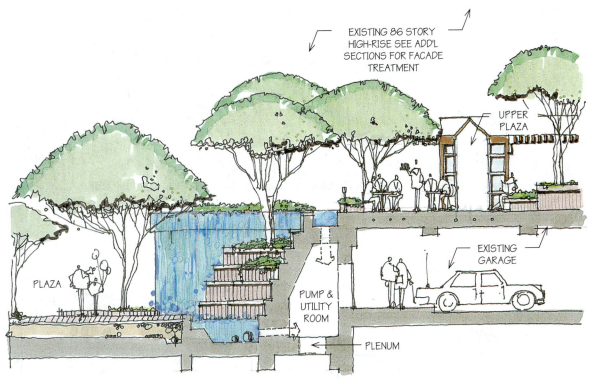

FIGURE 18.50 ■ Rendering of a waterfall in a profile drawing.

FIGURE 18.51 ■ Rendering technique to show water action.

FIGURE 18.52 ■ Techniques for rendering a masonry wall elevation.

the levels. Adding vegetation, people, and automobiles also provides a sense of scale and proportion. Refer back to Figure 18.50.

Rendering Site Structure Elevations

When adding landscape features to a structure elevation, care must be taken to not hide the major lines of the building. First add color, texture, and shadows to the building. Then begin rendering the foreground and work toward the background. Notice how this was accomplished in Figure 18.53. Here the vegetation and boulder

renderings duplicate the wall size, color, joints, texture, shades, and shadows. See Figure 18.52.

Rendering Elevation Sections

When a design contains multiple levels, a rendered section is often the best way to describe the relationship of

FIGURE 18.53 ■ Landscape features added to a fence and trellis.

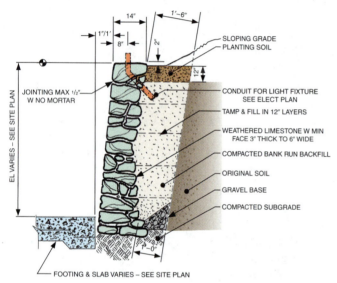

FIGURE 18.54 ■ Wall section detail prepared for landscape construction.

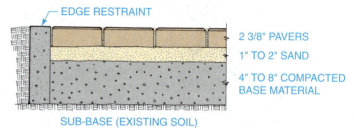

FIGURE 18.55 ■ Sectional detail prepared for paving construction.

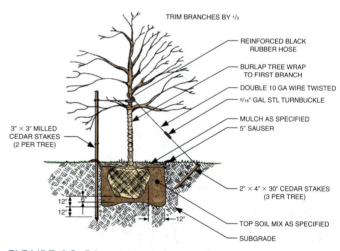

FIGURE 18.56 ■ Illustrated planting instructions.

SITE DETAILS AND SCHEDULES

Site development involves the design of landscape features, such as plants, with structural elements. Details, usually sections or profiles, are needed to ensure that construction is completed as designed. Figure 18.54 shows a landscape profile section of a rock wall to be used by landscape contractors. Figure 18.55 is a sectional drawing prepared for use by paving contractors. Planting instructions are detailed to ensure that the installation of

major plant materials meets horticultural planting standards. See Figure 18.56.

Planting schedules are prepared and indexed with numbers corresponding to a landscape plan. These schedules function as a guide for the purchase and placement of each size and species of plant material. See Chapter 35 for schedules.

placement are rendered without hiding the basic elements of the design.

CHAPTER

18

Site Development Plans Exercises

1. Identify and discuss the environmental and human-related influences that affect site design.

2. Describe the zoning daylight plane ordinances for second-story setbacks. List why these are important.

3. Draw the setback and building area for a lot 130′ × 65′ according to the zoning requirements in your community. Determine the maximum size building possible for a site with 35 percent land coverage. Determine the maximum size of a house for that site.

 4. Describe the zoning laws and the density pattern you would prefer for an area in which you wish to locate a house of your design.

5. Locate a house on a 60′ × 120′ lot. Setbacks are 5′ on the sides and 20′ front and rear.

 6. Prepare a survey plan for a home you are designing.

7. Determine the bearing of property lines for a lot in your area using both the azimuth and American systems. Estimate the contour lines and include them in a sketch of this property. Complete a profile view.

8. Study a plot plan in this chapter and identify the highest and lowest levels above datum. What changes and/or structures would you recommend for this site?

9. Sketch a plat of your neighborhood using roads or streets as the outer boundaries.

 10. Draw a plot plan of a house you are designing.

11. Draw a plot plan of a property in your area.

12. Draw and render a landscape plan for a house and property of your own design.

13. Redesign, draw, and render a landscape plan for a property in your area.

14. Find 10 drawings in this text that are part of the design shown in Figure 18.37.

15. Add landscape features to the sketches shown in Figure 18.57.

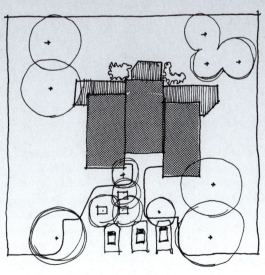 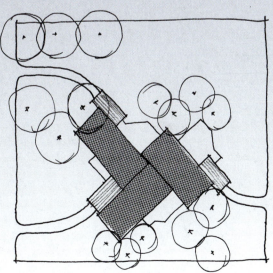

FIGURE 18.57 ■ Add symbols representing additional landscape features to these plans.

PART 5

5

Presentation Methods

19 CHAPTER
Pictorial Drawings

OBJECTIVES

In this chapter, you will learn:

- to differentiate between isometric, oblique, and perspective drawings.

- geometric principles involved in projecting lines (from a given point or at a constant angle) to create 3D images.

- to apply principles of perspective drawing to create interior and exterior pictorial drawings.

- projection methods for drawing pictorials.

TERMS

bird's-eye view
horizon line
isometric drawings
oblique drawings

one-point perspective
parallel angle projection
perspective drawings
picture plane

station point
three-point perspective
two-point perspective
vanishing point

INTRODUCTION

Pictorial drawings are picture-like drawings. Unlike elevation drawings that reveal only one side of an object, pictorials show several sides of an object in one drawing.

 ## TYPES OF PICTORIAL PROJECTION DRAWINGS

Three types of pictorial drawings are used in architecture—oblique, isometric, and perspective drawings, as shown in Figure 19.1. Oblique and isometric drawings are created by **parallel angle projection.** Perspective drawings are more complex to draw but appear more realistic.

Oblique Drawings

Oblique drawings are created by projecting parallel lines from an elevation drawing at an angle of 10° to 30° from the vertical plane. Because oblique drawings are prepared to scale, dimensions can be added, as shown in Figure 19.2. They are often used as a substitute or to clarify working drawings that may be difficult to understand.

Isometric Drawings

Isometric drawings are prepared with parallel receding lines drawn at 30° from the horizontal plane. Because of the visual distortion created by receding parallel lines, isometric drawings are usually used for details, as shown in Figure 19.3, or to describe small areas, as shown in Figure 19.4. Isometric drawings are sometimes dimensioned and used as working drawings because of their true dimensions.

If used to draw an entire building, parallel angle drawings appear distorted. Therefore, pictorial drawings of total structures are usually prepared in perspective form, which results in a more realistic view.

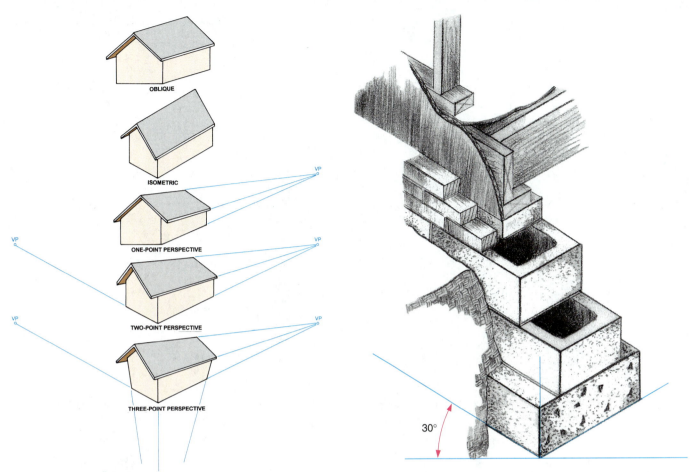

FIGURE 19.1 ■ Three types of architectural pictorial drawings.

FIGURE 19.3 ■ Isometric detail drawing.

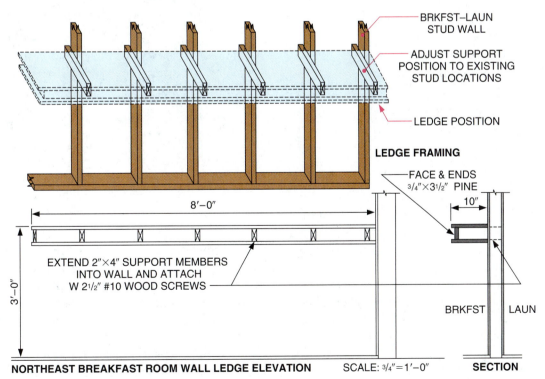

FIGURE 19.2 ■ Oblique construction detail drawing.

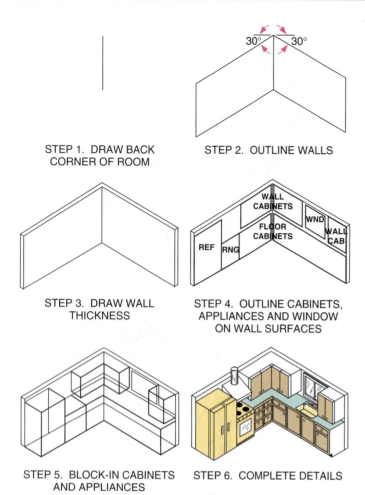

STEP 1. DRAW BACK CORNER OF ROOM

STEP 2. OUTLINE WALLS

STEP 3. DRAW WALL THICKNESS

STEP 4. OUTLINE CABINETS, APPLIANCES AND WINDOW ON WALL SURFACES

STEP 5. BLOCK-IN CABINETS AND APPLIANCES

STEP 6. COMPLETE DETAILS

FIGURE 19.4 ■ Interior isometric drawing.

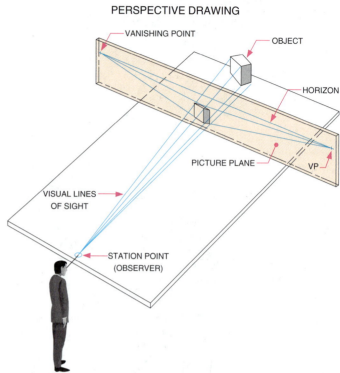

FIGURE 19.5 ■ Visualization of a two-point perspective shape on a picture plane.

PERSPECTIVE DRAWINGS

In **perspective drawings**, receding lines of a building appear to meet. They are not drawn parallel. A perspective drawing, more than any other kind of drawing, most closely resembles the way people actually see an image. If you look down railroad tracks, the parallel tracks appear to come together and vanish at a point on the distant horizon. Similarly, horizontal lines on a perspective drawing appear to meet at a distant point. The point at which these lines seem to meet and disappear is known as the **vanishing point**.

On two-point perspective drawings, a **horizon line** is established. This line is the observer's eye level. The location of the observer is the **station point**, as shown in Figure 19.5. A **picture plane** is an imaginary plane between the station point and the object on which a perspective view is observed. The vanishing points in a perspective drawing are always placed on the horizon

FIGURE 19.6 ■ Effect of horizon line placed through the center of a building. (*Jenkins & Chin Shue, Inc.*)

line. If the horizon line is placed through the center of a building, the building will appear to be at eye level (Figure 19.6). If the horizon line is placed low or below a building, the building will appear as if you were looking up at it (Figure 19.7). If the horizon line is placed above a building, it will appear to be below your line of sight (Figure 19.8). Objects, placed close to or on the horizon line are less distorted than objects placed a greater distance from the horizon.

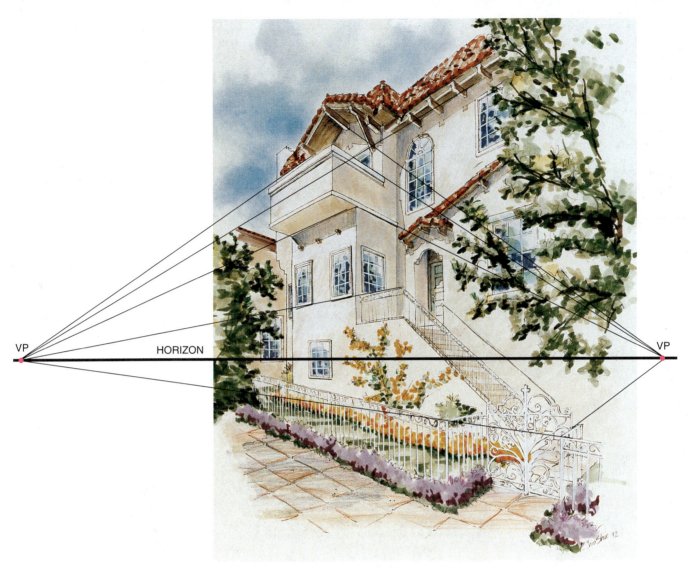

FIGURE 19.7 ■ Effect of horizon line placed low on a building. (*Jenkins & Chin Shue, Inc.*)

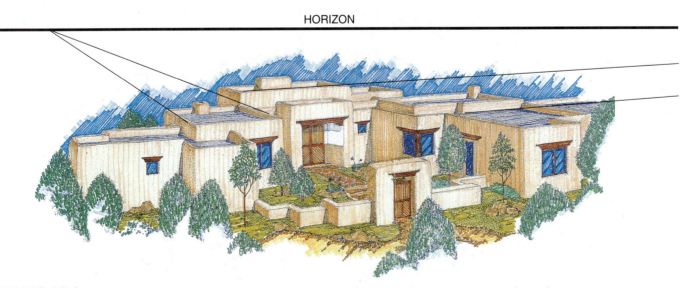

FIGURE 19.8 ■ Effect of horizon line placed above a building. (*Turner, Lechmer & Romero, Architects*)

We are accustomed to seeing areas decrease in depth as they recede from our point of vision. That is why the sides of an isometric drawing (prepared with the true dimensions of a building) appear distorted. To make perspective drawings appear more realistic, the actual lengths of the receding side lines of the building are shortened. Because perspective drawings do not reveal the true size of the building, they are not used as working drawings.

EXTERIOR PERSPECTIVE DRAWINGS

Perspective drawings of building exteriors are either one-, two-, or three-point perspective drawings depending on the size and proportion of the structure.

One-Point Perspective

A **one-point perspective** is a drawing in which the front view is drawn to its true scale and all receding sides are projected to a single vanishing point. If the vanishing point is placed directly to the right side of an interior or enclosed space as shown in Figure 19.9, the left side will show. If the vanishing point is placed on the left side of an interior or enclosed space, as shown in Figure 19.10, the right side will show. If the object is placed above the horizon line and vanishing point, the bottom of the object will show. If the object is placed below the horizon line and vanishing point, the top of the object will show. Vanishing points need not always fall outside the building outline. When vanishing points are located within the building outline only the frontal plane will show. Figure 19.11A shows an elevation drawing that has been converted to a one-point perspective by projecting elevation lines to a center vanishing point. Compare the photograph of the house shown in Figure 19.11B with the drawing in Figure 19.11A.

A one-point perspective drawing is relatively simple to create. The front view is drawn to the exact scale of the building. The corners of the front view are then projected to one vanishing point.

FIGURE 19.9 ■ Effect of the vanishing point placed on the right. (*Addison Estates, Boca Raton, FL*)

FIGURE 19.10 ■ Effect of the vanishing point placed on the left. (*Jenkins & Chin Shue, Inc.*)

FIGURE 19.11A ■ Elevation converted to a one-point perspective.

FIGURE 19.11B ■ Photograph of the drawing in Figure 19.7A. (*Lindal Cedar Homes*)

Follow these steps when drawing or sketching a one-point perspective. See Figure 19.12.

STEP 1   Draw the horizon line and mark the position of the vanishing point. If the vanishing point is to the left and outside the building, the left side of the building will show. If the vanishing point is to the right and outside the building, the right side of the building will show.

STEP 2   Draw the front view of the building to a convenient scale.

STEP 3   Project all visible corners of the front view to the vanishing point.

STEP 4   Estimate the length of the house. Draw lines parallel with the vertical lines of the front view to indicate the back of the building.

STEP 5   Make all object lines heavy, such as roof overhang. Erase the projection lines leading to the vanishing point.

Two-Point Perspective

A **two-point perspective** drawing is one in which the receding sides are projected to two vanishing points, one on opposite ends of the horizon line. See Figure 19.13. In a two-point perspective, no sides are drawn exactly to scale. All sides recede to vanishing points. Therefore, the only true-length line, the one that is to scale, on a two-point perspective may be the vertical line in the corner of the building from which both sides are projected.

When the vanishing points are placed close together on the horizon line, considerable distortion results because of the acute receding angles, as shown in Figure 19.14. When the vanishing points are placed farther apart, the drawing looks more realistic. Placing the drawn object closer to the horizon also helps create a more realistic appearance.

One vanishing point is often placed farther from the station point than the other vanishing point. This placement allows one side of the building to recede at a

HORIZON   VP
① DRAW THE HORIZON AND LOCATE ONE VANISHING POINT

      VP

② DRAW A FRONT VIEW OF THE HOUSE – ESTIMATE SIZES AND LOCATION

HORIZON   VP
③ DRAW CORNERS TO THE VANISHING POINT

HORIZON   VP
④ ESTIMATE HOUSE LENGTH

HORIZON   VP
⑤ ADD ROOF OVERHANG

FIGURE 19.12 ■ Sequence of projecting a one-point perspective drawing.

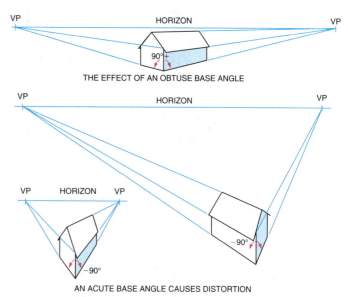

THE EFFECT OF AN OBTUSE BASE ANGLE

AN ACUTE BASE ANGLE CAUSES DISTORTION

FIGURE 19.14 ■ Effect of vanishing point and horizon line positioning.

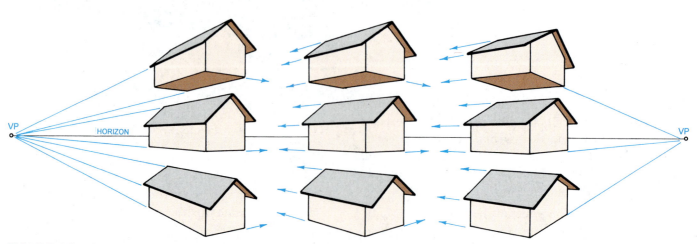

FIGURE 19.13 ■ Vertical and horizontal position options on a two-point perspective.

sharper angle than the other. The left vanishing point in Figure 19.15 is placed closer to the building than the right vanishing point. This exposes more of the right side of the structure. The reverse is true in Figure 19.16 where the right vanishing point is closer to the building. In drawing or sketching a simple two-point perspective, the steps outlined in Figure 19.17 can be followed.

Three-Point Perspective

Three-point perspective drawings are used to overcome the height distortion of tall buildings. In a one- or two-story building, the vertical lines recede so slightly that, for practical purposes, they are drawn parallel. However, the top or bottom of extremely tall buildings appears

FIGURE 19.15 ■ Effects of placing vanishing point closer to the left side of the building. (*Jenkins & Chin Shue, Inc.*)

FIGURE 19.16 ■ Effects of placing vanishing point closer to the right side of the building. (*Jenkins & Chin Shue, Inc.*)

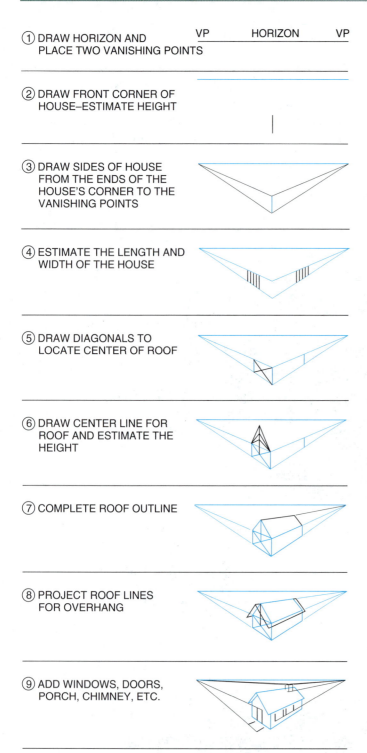

① DRAW HORIZON AND PLACE TWO VANISHING POINTS

② DRAW FRONT CORNER OF HOUSE–ESTIMATE HEIGHT

③ DRAW SIDES OF HOUSE FROM THE ENDS OF THE HOUSE'S CORNER TO THE VANISHING POINTS

④ ESTIMATE THE LENGTH AND WIDTH OF THE HOUSE

⑤ DRAW DIAGONALS TO LOCATE CENTER OF ROOF

⑥ DRAW CENTER LINE FOR ROOF AND ESTIMATE THE HEIGHT

⑦ COMPLETE ROOF OUTLINE

⑧ PROJECT ROOF LINES FOR OVERHANG

⑨ ADD WINDOWS, DOORS, PORCH, CHIMNEY, ETC.

FIGURE 19.17 ■ Steps in drawing a two-point perspective.

smaller than the area nearest the viewer. A third vanishing point may be added to provide the desired recession. See Figure 19.18. The farther away the third vanishing point is placed from the object, the less the distortion. If the lower vanishing point is placed so far below or above the horizon that the angles are hardly distinguishable, then the advantage of a three-point perspective may be lost because the vertical lines are almost parallel.

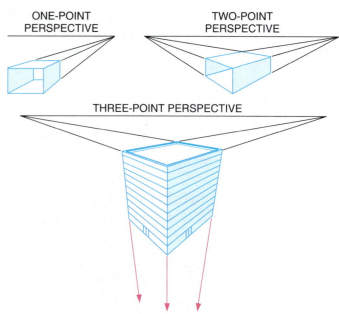

FIGURE 19.18 ■ Comparison of one-, two-, and three-point perspective drawings.

INTERIOR PERSPECTIVE DRAWINGS

A pictorial drawing of the interior of a building may be an isometric drawing, a one-point perspective, or a two-point perspective drawing. Pictorial drawings may be prepared for an entire floor plan. More commonly, however, a pictorial drawing is prepared for a partial view of a single room or to show a particular interior detail.

One-Point Perspective

A one-point perspective of the interior of a room is a drawing in which all the intersections between walls, floors, ceilings, and furniture are projected to one vanishing point. Drawing a one-point perspective of the interior of a room is similar to drawing the inside of a box with the front of the box removed. In a one-point interior perspective, walls perpendicular to the plane of projection, such as the back wall, are drawn to scale. The vanishing point on the horizon line is then placed somewhere on this wall. In the sketch shown in Figure 19.19A and B, the vanishing point is located vertically just above eye level and horizontally centered in the room. The points where this wall intersects the ceiling and floor are then projected from the vanishing point to form the intersection between the side walls and the ceiling and the side walls and the floor.

FIGURE 19.19A ■ Vanishing point lines projected to an interior single point. (*John Henry, Architect*)

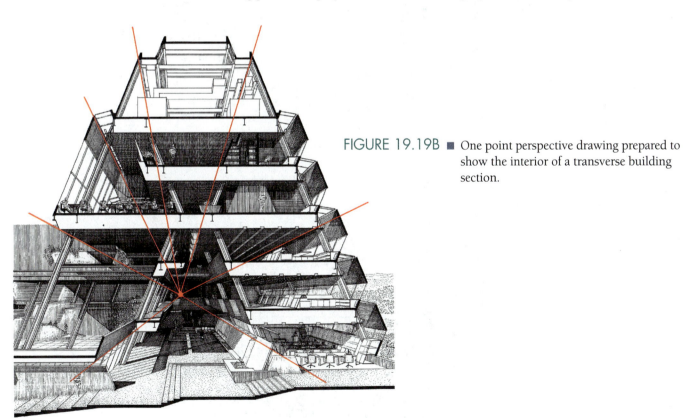

FIGURE 19.19B ■ One point perspective drawing prepared to show the interior of a transverse building section.

Vertical Placement

Vertical placement of the vanishing point affects the appearance of a drawing. If the vanishing point is placed high, as in Figure 19.20A, very little of the ceiling will show in the projection, but much of the floor area will be revealed. If the vanishing point is placed near the center

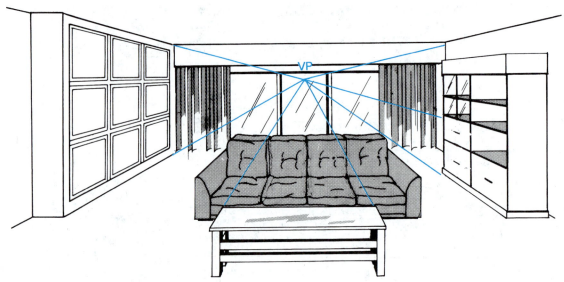

FIGURE 19.20A ■ Effects of high vanishing point height.

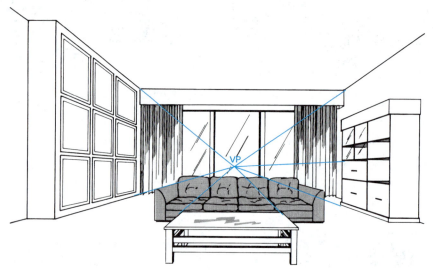

FIGURE 19.20B ■ Effect of a middle vanishing point height.

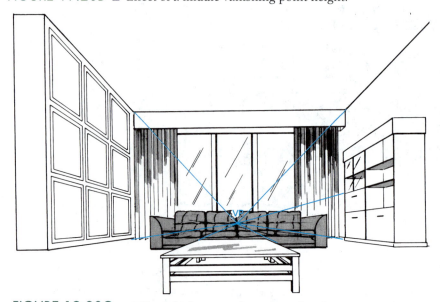

FIGURE 19.20C ■ Effect of a low vanishing point height.

of the back wall, as in Figure 19.20B, an equal amount of ceiling and floor will show. If the vanishing point is placed low on the wall, as in Figure 19.20C, much of the ceiling but very little of the floor will be shown. Because the horizon line and the vanishing point are at your eye level, you can see that the position of the vanishing point affects the angle from which you view the object.

Moving the vanishing point horizontally from right to left on the back wall has an effect on the view of the side walls. This effect is the same for interior walls and enclosed exterior walls. If the vanishing point is placed toward the left, more of the right wall will be revealed. See Figure 19.21. Conversely, if the vanishing point is placed near the right side, more of the left wall will be revealed in the projection. This was done in Figure 19.22 to expose the left wall. The arrow in the floor plan shows the viewer's direction and location. If the vanishing point is placed in the center, an equal amount of right wall and left wall will be shown. When one wall should dominate, place the vanishing point on the extreme end of the opposite wall.

When a vanishing point is placed in a plan drawing, a perspective view from above is created. These are often called a **bird's-eye view.** Unlike vertical perspective views, these horizontal views show all four walls of a room as in Figure 19.23.

When drawing wall offsets and furniture, always block in the overall size of the item to form a perspective view. The steps are shown in Figure 19.24. The details of furniture or closets or even of persons can then be completed within this blocked-in cube or series of cubes.

Two-Point Perspective

Two-point perspectives are normally prepared to show the final design and decor of two walls of a room. The vertical true-length or base line on an interior two-point perspective is similar to the base line on an exterior two-point perspective. On an interior drawing, this line may be a corner of a room, an article of furniture, or other vertical line. Not only are the walls projected to the vanishing points in a two-point perspective, but each object in the room is also projected to the vanishing points. The sequence of steps in drawing two-point interior perspectives is shown in Figures 19.25A through 19.25C.

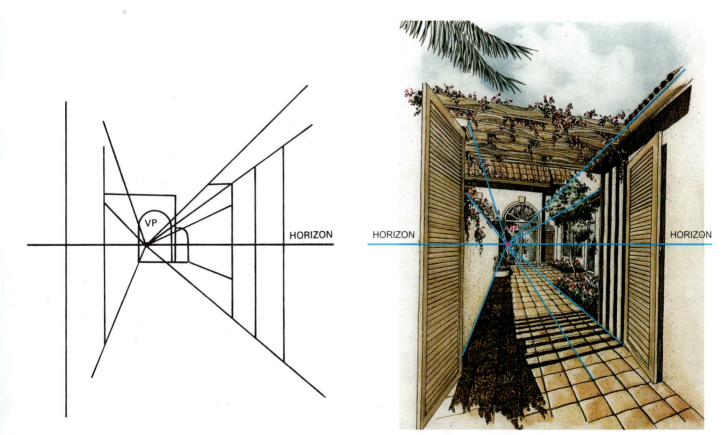

FIGURE 19.21 ■ Effect of left side vanishing point placement. (*Jenkins & Chin Shue, Inc.*)

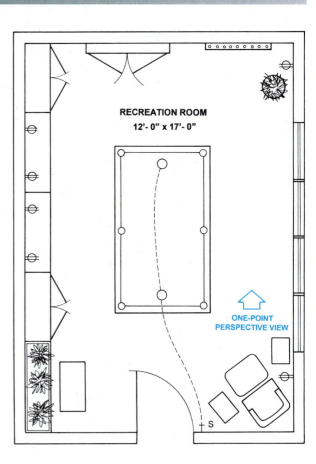

RECREATION ROOM
12'- 0" x 17'- 0"

ONE-POINT
PERSPECTIVE VIEW

FIGURE 19.22 ■ Effect of right side vanishing point placement.

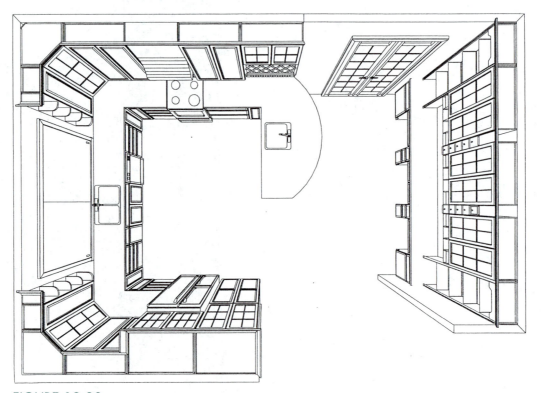

FIGURE 19.23 ■ One-point plan perspective; also called a bird's-eye view.

First step

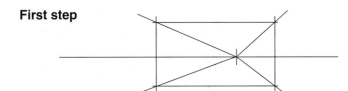

Second step

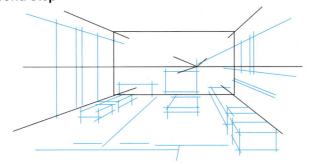

Final steps

FIGURE 19.24 ■ Steps in developing an interior one-point perspective. (*Home Planners, Inc.*)

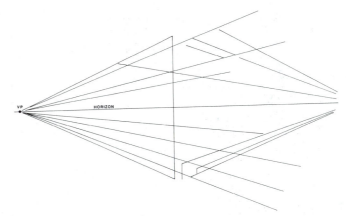

FIGURE 19.25A ■ Layout step for developing a two-point interior perspective. (*Jenkins & Chin Shue, Inc.*)

FIGURE 19.25B ■ Major outlines added. (*Jenkins & Chin Shue, Inc.*)

PROJECTION METHODS

Several methods can be used to project lines to vanishing points on perspective drawings. These include connecting lines with a straight edge, drawing over underlay perspective grid sheets (Figures 19.26A and 19.26B), or using CAD grid functions.

FIGURE 19.25C ■ Completed interior two-point perspective. (*Jenkins & Chin Shue, Inc.*)

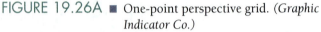

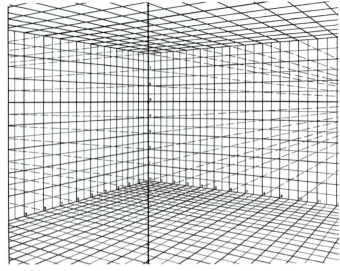

FIGURE 19.26A ■ One-point perspective grid. (*Graphic Indicator Co.*)

FIGURE 19.26B ■ Two-point perspective grid. (*Graphic Indicator Co.*)

CHAPTER

19

Pictorial Drawings Exercises

1. Draw a two-point perspective of a building of your own design.

2. Draw a one-point and a two-point perspective of your own home.

3. Sketch a three-point perspective of the tallest building in your community.

4. Prepare a one-point interior perspective of your own room.

5. Prepare a one-point perspective of a room in the house of your own design.

6. Draw a one-point interior perspective of a classroom. Prepare one drawing to show more of the ceiling and left wall. Prepare another drawing to show more of the floor and right wall.

CHAPTER 20

Architectural Renderings

OBJECTIVES

In this chapter, you will learn to:

- recognize the wide selection of media available for renderings.

- evaluate when to use which media to achieve an artistic effect.

- add realism to drawings by the use of shading, shadows, texture, entourage, and landscapes.

- follow the correct sequence for preparing a rendering.

TERMS

acrylics
entourage

pastels
render

wash drawing

INTRODUCTION

Because pictorial drawings are three dimensional, their shape resembles a realistic view of a building. Our eyes see more than shape though. We see color, texture, shades, shadows, people, and landscape features. In a rendering, these features are added to a pictorial drawing. To **render** a drawing is to make the drawing appear more realistic—whether it is a plan, an elevation drawing, or a perspective drawing. Drawings are rendered by adding realistic texture and establishing shade and shadow patterns. This may be done using a variety of media.

RENDERING MEDIA

A rendering may utilize only one medium, or several media may be combined to create various images. Media used to render drawings include pencils, charcoal, ink, watercolors, felt markers, **pastels** (light-colored, water-based chalk), oil paint, and **acrylics** (water-based permanent paints).

Pencil Renderings

Soft (B) pencils are effective media for rendering architectural pictorials. Changes in the weight and density of

lines create many tones. See Figure 20.1. Variations in the spacing of pencil lines and in the pressure of the pencil can create values and different contrasting effects. Smudge blending to add tone is accomplished by rubbing a finger over soft penciled areas. To add surface realism, extremely soft charcoal pencils can be used over a textured surface, as done in Figure 20.2.

Pencil renderings are popular because shading and texture can easily be added to penciled pictorial outlines. Colored or pastel pencils can be used to make the various colors and values of building surfaces appear very realistic.

Ink Renderings

Because ink lines cannot be blended, the distance between lines is controlled to create the appearance of texture, light, shade, and density. Figure 20.3 shows ink line patterns as rendering techniques. Ink lines and strokes placed close together produce dark effects. Farther apart, they create lighter effects. See Figure 20.4.

Watercolor and Wash Drawings

The use of wash drawings or watercolors is a fast and effective method of adding realism to pictorial drawings. When only black and gray tones are used, this type of rendering is called a **wash drawing**. When color is added, these drawings are known as *watercolors*. Watercolor

FIGURE 20.1 ■ Pencil rendering techniques.

FIGURE 20.2 ■ Charcoal rendering. (*Avery Architectural and Fine Arts Library*, Hugh Ferris, Illustrator)

FIGURE 20.3 ■ Common ink strokes.

FIGURE 20.4 ■ Ink rendering. *(Home Planners, Inc.)*

FIGURE 20.5 ■ Watercolor rendering. *(Amelia Island Estates)*

paints blend to create a variety of attractive color and gradation effects. Therefore, watercolors are used extensively for presentations and for advertising. Perhaps the most effective use of watercolor techniques is for the pictorial combination of landscape settings with structures. See Figure 20.5.

Oil and Acrylic Renderings

Architectural renderings in oil paints or acrylics are more time consuming and expensive than any other medium. For this reason, they are rarely prepared, except as works of art for display.

Felt-Marker Renderings

The use of felt (or felt-tip) markers is a fast way of adding color to pictorial drawings. Stroke lines do not blend easily, so this method is usually restricted to adding patches of color to existing drawings.

Pressure-Sensitive Overlays

A popular technique that is used to convert a perspective drawing into a rendering is to apply a pressure-sensitive overlay to the drawing. Preprinted pressure-sensitive screens are used to add tones, texture, and shadow. These

USING CAD

Entourage CAD Libraries

CAD entourage libraries or blocks contain people, animals, and vehicles in a wide variety of sizes, shapes and positions. In using entourage symbols, their size can be manipulated using the *Stretch* command and can be located using the *Move* command. The *Stretch* command can be used to either elongate or condense a feature. The *Move* command allows entities to be moved to a new location by pointing the cursor to the object, then dragging it to the new position. If more than one entity is needed in a pattern the *Copy-Multiple* command can be used by inputting the location, spacing, and number of entries.

effects are created by variations in the distance between lines or dots, in the width of lines, and in the blending of lines. Pressure-sensitive overlays can be used to create gray tones or solid black areas for contrast or for light and shadow patterns.

Media Combinations

Most architectural renderings include a variety of media, depending on the cost and the emphasis desired. Archi-

tectural illustrator Mark Englund (Figure 20.6) is shown preparing a magazine rendering using layers of ink and water colors. The illustration shown in Figure 20.7 was prepared using fine ink line work. This same view is rendered in Figure 20.8 using softer wash techniques, which reveals fewer fine details. Note the difference in appearance and in the amount of detail presented in the two renderings.

FIGURE 20.6 ■ Architectural illustrator at work. (*Photo by Mark Englund*)

FIGURE 20.7 ■ Detailed rendering in ink and watercolor. (*Scholz Design, Inc.*)

Interior design features are frequently rendered using watercolor over line work to reveal color and texture of materials. See Figure 20.9. Other media normally used in combination with pencil or ink line drawings include airbrush, pastels, and felt markers.

Another method of preparing pictorial renderings is to use colored or gray illustration board. When this method is used, all lines, shades, or tints are added with a variety of white watercolor and/or acrylic paint. Notice how the black, white, and gray tones in Figure 20.10 emphasize texture and depth.

The effectiveness of any medium is related to how skillfully it is used to create realistic textures, shades, shadows, and landscape features. Figure 20.11 illustrates the artful blending of these elements to produce a realistic and dramatic rendering.

FIGURE 20.8 ■ Watercolor rendering of house in Figure 20.8A. (*Scholz Design, Inc.*)

FIGURE 20.9 ■ Rendering with pastels and watercolors over a line drawing. (*Jenkins & Chin Shue, Inc.*)

FIGURE 20.10 ■ Rendering over gray stock. (*John Henry, Architect*)

EFFECTS OF LIGHT

Light Source and Shade

When shading a building, consider the location of the sun or other light sources. Areas exposed to the light source should appear lighter. Areas not exposed to a light source should be shaded or darkened. See Figure 20.12. When an object with sharp corners is exposed to a light source, one side may be extremely light and the other

side of the object extremely dark. However, objects and buildings often have areas that are round (cylindrical). These areas change gradually from dark to light. A gradual shading from extremely dark to extremely light must be made.

Shadow

To determine which areas of a building should be drawn darker to indicate shadowing, the angle of the light source must first be established. Once the angle of the light source is established, then all shading should be consistent with the direction and angle of the shadows, as shown in Figures 20.13A and 20.13B. In addition to the light source, consider the building outline and site contours when drawing shadows on buildings. Note the connection between the light source angle, building outlines, and site

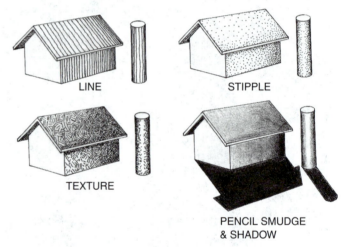

FIGURE 20.12 ■ Shading methods.

FIGURE 20.11 ■ Dramatic multimedia rendering. (*John Henry, Architect*)

FIGURE 20.13A ■ Shadow effect from a low light source.

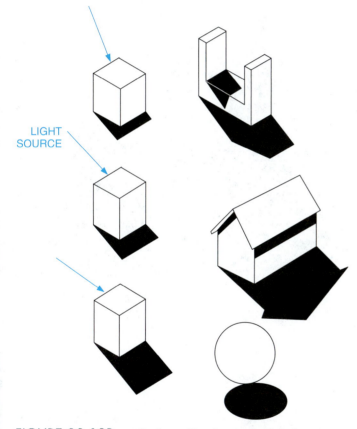

LIGHT SOURCE

FIGURE 20.13B ■ Shadow effect from a high light source.

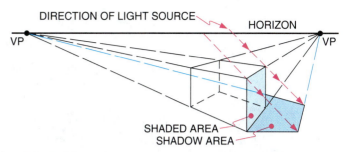

DIRECTION OF LIGHT SOURCE
HORIZON
VP
VP
SHADED AREA
SHADOW AREA

FIGURE 20.14A ■ Shadow pattern related to vanishing point (VP) and light source.

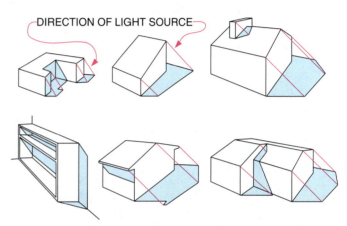

DIRECTION OF LIGHT SOURCE

FIGURE 20.14B ■ Shadows reflect building shapes.

contours. Also, shadowing is often used to reveal hidden features such as overhang depth, building depressions, offsets, and extensions. See Figures 20.14A and 20.14B.

Various techniques can be used to indicate shadows. Figure 20.15 shows how depth and shadows are rendered on windows. Some windows show reflected light, and others are drawn to reveal the room behind, as though the window were open. Keep in mind that most windows look dark during the day because the inside of the building is darker than the outside. To produce realistic-looking windows, dark colors or even black surfaces are often used, as shown earlier in Figure 20.4. Foliage and blue sky reflections on windows are often used to depict a very sunny site, as done in Figure 20.16.

TEXTURE

Giving texture to an architectural drawing means making building materials appear as smooth or as rough as they actually are. Smooth surfaces are very reflective and hence are very light. Only a few reflection lines are usually necessary to illustrate smoothness of surfaces such as aluminum, glass, and painted surfaces. For rough surfaces, the material can often be shown by shading and texturing. A variety of texturing methods and rendering techniques for the exterior materials used on siding and roofs can be used. The building shown in Figure 20.17A is rendered with black ink only. The same building is shown rendered in color in Figure 20.17B. Figure 20.18 is a pencil rendering that emphasizes the different textures of the siding and roof. Notice how the shading of the texture depends on the light source. Also note how the shadows from trees and overhangs are incorporated into the texture and values to add more realism to the drawing.

FIGURE 20.15 ■ Shadows used to render windows.

FIGURE 20.16 ■ Window reflections used to denote brightness. (*Jenkins & Chin Shue, Inc.*)

FIGURE 20.17B ■ Color rendering of the house shown in Figure 20.17A. (*Jenkins & Chin Shue, Inc.*)

FIGURE 20.17A ■ Black ink rendering. (*Jenkins & Chin Shue, Inc.*)

FIGURE 20.18 ■ Pencil rendered textures.

CHIMNEYS

FENCES AND WALLS

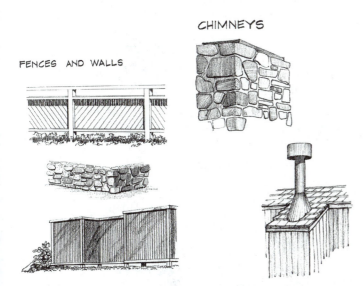

FIGURE 20.19 ■ Rendering of stand-alone features.

Rendering "stand-alone" features such as fences, walls, and chimneys requires close attention to independent shadow and shade effect on texture. See Figure 20.19.

 ENTOURAGE

The term **entourage** refers to the people or objects that are part of a building's surroundings and are used to enhance the size, distance, and reality of renderings. Sketches of people—sitting, standing, and walking—are often necessary to show the relative size of a building and to put the total drawing in proper perspective. Because people should not interfere with the view of the building, architects frequently draw people in outline or in extremely simple form. In a drawing, people may indicate pedestrian traffic patterns, as well as provide a feeling of perspective and depth. Automobiles are also added to architectural renderings to indicate relative size and to give a greater feeling for external traffic patterns.

People, boats, and automobile outlines in a variety of settings, angles, and scales are available on pressure-sensitive sheets or in traceable entourage publications. See Figure 20.20. Entourage figures are often photo reduced or enlarged to fit within a drawing. Figures closer to the vanishing point are progressively smaller than figures near the station point. Figure 20.21 illustrates this effect and also shows the difference between a black line multiple-building drawing and a full-color rendering of the same site.

Computer 3D CAD programs now offer libraries on disk. These libraries are similar to entourage publications and provide a wide variety of drawings, including people in walking, sitting, and standing positions.

 LANDSCAPE

In rendering, adding landscape features—whether to pictorial or elevation drawings—involves drawing trees, shrubs, ground cover, drive ways, and walkway surfaces. Trees should be placed so as not to block out the view of the buildings. See Figure 20.22.

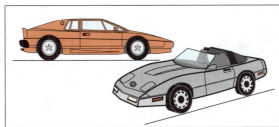

FIGURE 20.20 ■ Entourage sample drawings.

FIGURE 20.21 ■ Relationship of people to the vanishing point (VP) position in black and white and color. (*Jenkins & Chin Shue, Inc.*)

FIGURE 20.22 ■ Pictorial rendering techniques used for landscape features.

FIGURE 20.23 ■ Types of landscape feature renderings. *(Tangerine Bay Club, Tangerine Development Corp.)*

Rendering landscape pictorial drawings also involves using the same techniques to render accessory structures, water features, retaining walls, trellises for vines or other plants, and many other features. See Figure 20.23. Also, refer back to Chapter 18.

USING CAD

Rendering Pictorial Drawings

Surface materials can be added to pictorial drawings as shown in Figure 20.24, using the Photoshop color library and the *Hatch* command. Trees and other landscape features are also available in 3D form but must be adjusted to fit the scale of the drawing.

FIGURE 20.24 ■ Pictorial drawing using CAD material and landscape library input. *(Kurt Alan Williams)*

CHAPTER

20

Architectural Renderings Exercises

1. Render a perspective drawing of your own house.
2. Render a perspective drawing of a house of your own design.
3. Render a perspective sketch of your school. Choose your own medium: pencil, pen and ink, watercolors, pastels, or airbrush.
4. Collect illustrations that could be adapted for use on renderings: drawings of people, cars in different sizes and positions, landscapes, plants, etc.

CHAPTER 21

Architectural Models

OBJECTIVES

In this chapter, you will learn to:

- describe architectural models made for design study purposes.
- explain the differences between presentation and design study models.
- tell what input is needed to create a computer model.
- construct an architectural model.

TERMS

basic layout model
design study model
detailed model

interior design model
landform model
presentation model

solid form model
structural model

INTRODUCTION

Architectural models are three-dimensional representations of a complete design. Models are made to scale and can be viewed from any angle or distance. They may be a constructed replica of a building, or they may be computer images.

The two basic types of models are design study models and presentation models. To check basic design ideas or construction methods during the design process, a **design study model** is made. A **presentation model** is used for sales purposes because people can understand a design more easily by looking at a model than they can by looking at drawings.

 DESIGN STUDY MODELS

Architectural models have been used for centuries to visualize the final appearance, function, and construction of a design (Figure 21.1).

Design study models are helpful during the design process. They are used to check the form of a structure, verify the basic layout, clarify construction methods, or show interior design options. They can also be used to finalize the orientation of a structure on a site. Models of this type are used to study sun angles and show shadow patterns at different times and on different days. See Figure 21.2.

FIGURE 21.1 ■ Ancient use of an architectural model. (*Celotex Corp.*)

Solid Form Models

Before final dimensions are applied to a design, a **solid form model** like the one shown in Figure 21.3 is often made. It is used to check the overall proportions of a building. Solid form models that contain no details can help to study the size and relationship of building clusters. Solid form models are made from Styrofoam™, balsa, clay, or soap.

For buildings designed for high-density areas, solid form models of adjacent buildings are often needed to show the relative size and position of all buildings. Solid form models are also used in community and city planning. Figure 21.4A shows a model of this type that was used to study a renewal proposal for a portion of Manhattan. Figure 21.4B shows the related plan of this area.

Basic Layout Models

Design study models also provide a means to check the overall layout and function of a design. First the preliminary floor plan and elevations must be available. Then **basic layout models** are made. These models are constructed to the same scale as the floor plans and elevations, usually 1/8″ = 1′-0″ or 1/4″ = 1′-0″. They are not finely detailed because they will probably be revised and altered many times. The layout model shown in Figure 21.5A is shown in front of the completed house it represents. This model was used by contractors to help explain design details to subcontractors during construction (Figure 21.5B).

FIGURE 21.2 ■ Design study model.

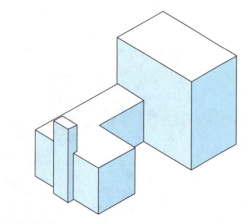

FIGURE 21.3 ■ Solid form model.

FIGURE 21.4A ■ City planning model.

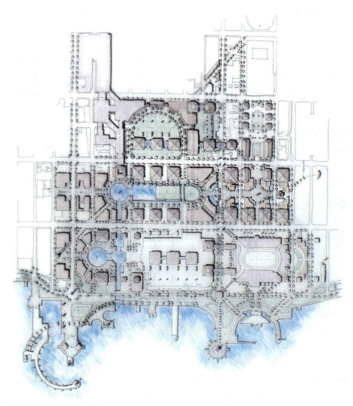

FIGURE 21.4B ■ Concept drawing of the area shown in Figure 21.4A.

FIGURE 21.5A ■ Basic layout model with completed structure. (*James Eismont, Photographer*)

FIGURE 21.5B ■ Model used to check progress on site.

Structural Models

Only the structural members of a building are shown on a **structural model**. Builders use these models to check unique structural methods or to study framing options. These checks are especially important if many houses with the same structural design are to be built.

Building a model of this type is also a good way to learn framing methods.

Structural models are usually built to a scale of 1″ = 1′-0″, 3/4″ = 1′-0″, or 1/2″ = 1′-0″. Smaller scales 1/8″ = 1′-0″ or 1/4″ = 1′-0″ are difficult to construct.

Structural members are cut to scale or purchased from model stock. Some parts are assembled into panels by

FIGURE 21.6 ■ Structural model used for client orientation. (*Lindal Cedar Homes*)

placing them directly over a wall-framing drawing or over elevation drawings. The panels and other components are then assembled with glue and pins as shown in Figure 21.6. A structural model is built in the same sequence as a full-size house.

Interior Design Models

An **interior design model** is used to show individual room designs. These can be shown effectively with one-room models. One-room models usually include the floor, ceiling, and three walls. One wall remains open for viewing, like a dollhouse. One-room models are usually built to a scale of 1″ = 1′-0″ or 1 1/2″ = 1′-0″. Scaled human figures, decor, and furniture add to the realism of interior design models. See Figure 21.7.

PRESENTATION MODELS

Most presentation models are used to promote the sale of a building, land parcels, or community development projects. **Presentation models** replicate (copy) the actual appearance of the real project in as much detail as the scale allows. A presentation model, such as that shown in Figure 21.8, often includes the building site.

FIGURE 21.7 ■ Interior room model. (*Helene Norman*)

FIGURE 21.8 ■ Presentation model. (*NYIT*)

Landform Models

Presentation models are frequently used to show the landform around a building. A **landform model** represents the shape and slope of a site. In this model, the thickness of each layer is equal (in scale) to the different levels of the land. The shape of each layer is the same shape as the contour lines on a survey drawing. Where contour intervals are large, landform layers are added to create a smoother contour, especially on steep slopes. See Figure 21.9.

To develop smooth contours, small posts are used to represent the contour height. These posts are spaced throughout the model site. Flexible screens are then laid on the posts and covered with paper-mâché, which can then be smoothed. See Figure 21.10.

Detailed Models

City and housing tract developers are the largest users of presentation models. To show building relationships, developers often surround a **detailed model** with solid form models of adjacent and nearby buildings, as shown

FIGURE 21.9 ■ Landform model using contour interval construction. (*NYIT*)

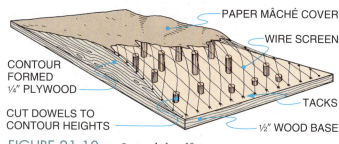

FIGURE 21.10 ■ Smooth landform construction.

in Figure 21.11. Constructed models of this type are often used to create a subject for photographs. A photograph can be retouched later to add or eliminate details and features.

Housing developers use landform models with small-scale solid form models to show specific lot locations and their relationship to other features. The placement of people, cars, and trees on these models adds realism and better defines the scale of the project. See Figure 21.12.

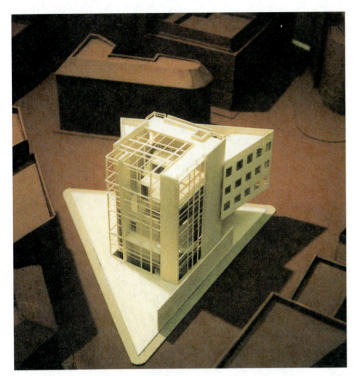

FIGURE 21.11 ■ Detailed model with surrounding buildings in solid form. *(NYIT)*

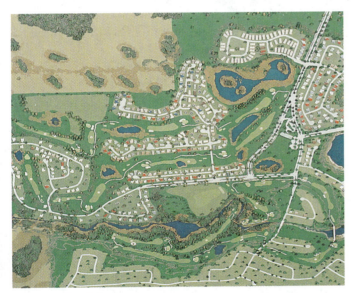

FIGURE 21.12 ■ Housing development model. *(The Plantation at Leesburg)*

STEPS IN CONSTRUCTING A MODEL

An accurate and realistic model is constructed to a precise scale with careful attention to detail. Many materials and items used to build models, such as those shown in Figure 21.13A and 21.13B, may not be available in the

PART	MODEL MATERIALS	METHODS OF CONSTRUCTION
■ STRUCTURE		
base	plywood; particleboard hollow core panel	Cut to maximum, but convenient, size for structure and site.
walls	softwood; cardboard; acrylic; matboard; foamboard; Styrofoam™; wallpaper; fabric; plywood	Cut walls to exact dimensions of elevations. Allow for overlapping of joints at corners. Wall thickness must be to scale.
floors	flocking for carpet; printed paper of floor type; thin wood veneer; vinyl scraps	Paint floor area with slow-drying colored enamel and apply flock. Remove excess when dry. With paper, glue in place. With wood veneer, rule black lines for strip effect and glue in place.
windows and doors	purchased, premade strips of wood or plastic; thin, clear acetate; acrylic	Glue premade windows and doors in place. Cut strips for casing, sill, and window frames, and glue in place. Draw the windows and doors directly on the walls.
roofs	thin, stiff cardboard or wood; paint; colored sand; wood pieces for shingles; premade/printed roof coverings; sandpaper	Cut out roof patterns and assemble. For roof coverings, glue on sand, wood pieces for shingles, or preprinted roof coverings.
■ BUILDING MATERIALS		
siding materials	scored sheets of balsa wood; foamboard; preprinted paper patterns; wood strips for board and batten and horizontal siding	Glue or paint siding materials to model walls.
stucco	spackle; plaster of paris; sandpaper; sand and thick paint	Mix and dab on with a brush leaving a rough texture.
brick and stone	printed paper; embossed plastic sheets; thin softwood	Glue paper in place. Cut grooves in wood, and paint color of bricks or stone.
wood paneling	printed paper; thin veneer wood; molded plastic sheets	Glue paper or plastic sheet in place. With veneer wood, rule lines for strip effect.

FIGURE 21.13A ■ Model materials for structure and building surfaces.

PART	MODEL MATERIALS	METHODS OF CONSTRUCTION
■ FURNISHINGS furniture, appliances, fixtures	commercial models; doll furniture (to scale); cardboard; softwood; clay; Styrofoam™; soap; fabric; paint	Purchase commercial model furniture or carve/sculpt to shape. Paint or flock for a finish.
fireplaces	(Refer to brick and stone materials.)	Carve fireplace, and simulate finish.
■ BUILDING SITE topography	wood base; wire screen; papier-mâché; fine gravel	Build up sloped areas with sticks and wire screen. Place papier-mâché over wire, and glue gravel for soil effect.
geologic features, terrain, water	stones; sand; colored gravel and sand	Glue small rocks for boulders. Paint high-gloss blue paint for water.
swimming pools	wood strips; blue paint or paper; sheet glass; acrylic	Outline pool with glued wood strips. Paint or glue paper for water and attach clear glass. Ripple acrylic surface while drying for ripple effect.
■ LANDSCAPE grass	green paint and flock	Paint grass area and apply flock. Remove excess when dry.
trees and bushes	sponges; lichen; small twigs	Grind up sponges and paint shades of green. Glue pieces to twigs for trees and bushes. Lichen may be purchased in model stores in bulk or as model trees.
wood fences	wood strips	Glue together to form a fence.
masonry fences	(Use same materials as for masonry walls.)	Form the same as walls to fence size.
gazebo	wood strips	Assemble from a working drawing.
■ MISCELLANEOUS automobiles	commercial models; toys; clay; soap; Styrofoam™	Purchase or shape from soft materials and paint.
people	commercial models; toys; clay; soap; Styrofoam™	Purchase or shape from soft materials and paint.
■ DRIVEWAYS	sandpaper, sand over glue, paint	

FIGURE 21.13B ■ Model materials for furnishings, site, landscape, and movable items.

scale needed. These must then be constructed by the model maker. Model makers may use manufactured parts and/or fabricate materials.

Methods of model construction vary depending on the material and the amount of detail required. The following procedures represent a typical sequence for constructing solid wall models.

STEP 1 *Floor plan base.* Attach a print of all floor plans to a sheet of rigid foamboard or Styrofoam™ sheet with rubber cement or use a preglued sheet, as shown in Figure 21.14. Allow some space on all sides.

STEP 2 *Wall construction.* Glue a print of all exterior elevations to a sheet of foamboard or softwood. This foamboard should be the same scaled thickness as the outside walls indicated on the floor plan. Cut the foamboard to create a wall for each elevation as shown in Figure 21.15. Be sure to add sufficient material to allow exterior corners to overlap unless corners are mitered.

STEP 3 Repeat this procedure for all interior partitions. If interior elevations are not available, the outline of each partition must be drawn onto and cut from foamboard. Usually all windows and doors are cut out of interior and exterior walls, as shown in Figure 21.16. An alternative, particularly for design study models, is to simulate them using paint or a suitable material.

STEP 4 The next step is to attach window and door trim and acetate. Figure 21.17 shows one method of constructing model window trim. Figure 21.18 shows the application of acetate to an inside wall with transparent tape. Doors may be cut from thin, stiff cardboard and hinged with transparent tape. If windows and doors are not cut out, they could just be shaded black or gray for effect, as in Figure 21.19, or the selected material should be glued in place.

STEP 5 *Wall attachment.* Glue the exterior walls to the floor plan base and to each other at the corners. Fit the corners carefully so that the overall outside wall lengths align properly with the floor plan corners. Use a 90° triangle to check the plumb and squareness of the tops of corner intersections. Glue interior partitions to the appropriate partition lines indicated on the floor plans. See Figure 21.20. Glue intersecting corners to outside walls and to other partitions.

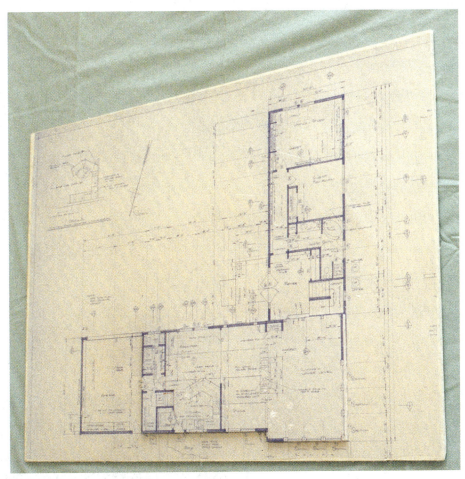

FIGURE 21.14 ■ Model floor plan base.

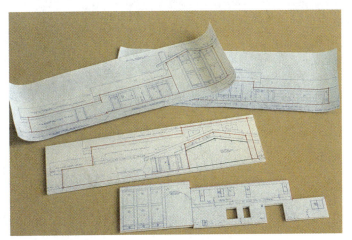

FIGURE 21.15 ■ Elevation drawings glued to foam board.

STEP 6 *Wall finishing.* Paint interior walls and floors with white or lightly tinted tempera paint. Apply texture or simulated (imitation) coverings to represent floor tiles, masonry surfaces, fireplaces, and chimneys. Add siding materials or textures to exterior surfaces. See Figure 21.21.

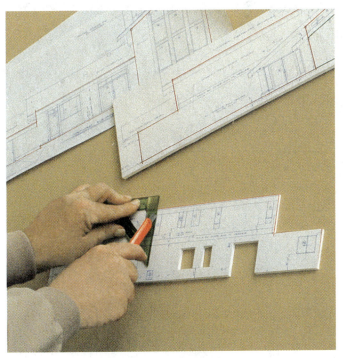

FIGURE 21.16 ■ Cutting windows and doors in elevation panel.

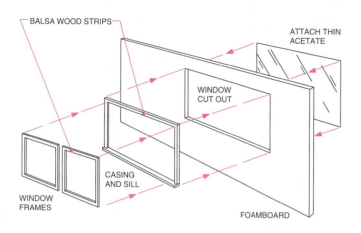

FIGURE 21.17 ■ Window trim assembly.

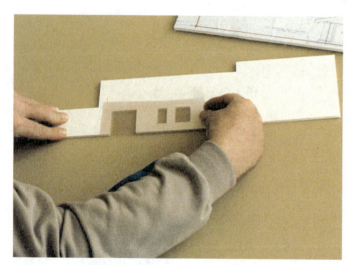

FIGURE 21.18 ■ Attaching acetate windows.

Spraying the finished surfaces with fixative will help surface treatments adhere better.

STEP 7 *Cabinetry and fixtures.* Make solid 3D forms to represent built-ins. These may include kitchen and bath cabinet fixtures, fireplaces, chimneys, and bookshelves. If desired, add color to cabinets, countertops, or fixtures.

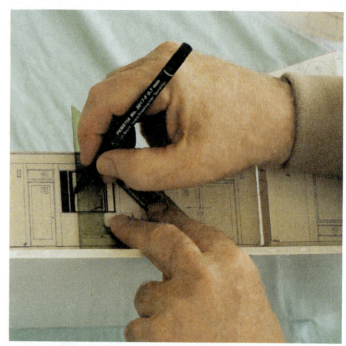

FIGURE 21.19 ■ Darkening windows on a solid panel.

FIGURE 21.20 ■ Attaching walls to base and adjoining walls.

STEP 8 *Roof construction.* Roofs should be constructed separately so that they can easily be removed to reveal the interior. Flat roofs can be constructed directly from a roof plan. The pitched roof in Figure 21.22 must be cut to align with the pitch indicated on an elevation drawing. Construct the roof in panels to represent continuous flat surfaces. Then glue the panels together. Other roof components and simulated roof coverings can then be added. Construct the chimney and glue it to the top of the roof to align with the interior chimney. This is done to

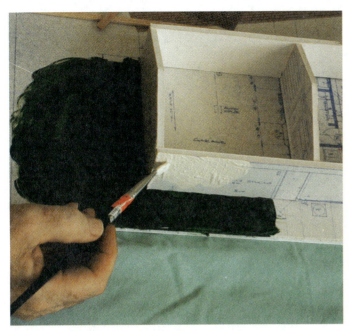

FIGURE 21.21 ■ Adding surface texture to walls.

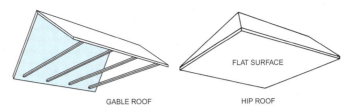

FIGURE 21.22 ■ Model roof construction.

taining walls might need to be added to separate parts of the site. Add landscape features such as trees, fountains, shrubbery, and ground cover. Cars, people, and furniture can be included to help define the scale and the areas.

allow the part of the chimney above the roof to be removed along with the roof.

STEP 9 *Outdoor areas.* After the structural part of the model is complete, remove the extra base material from the floor plan. Proceeding outward from the house, add outdoor areas, such as patios, pools, ponds, decks, or lanais. Include all the features that are within the property's perimeter.

STEP 10 *Landscape features.* A complete architectural model includes landscaping and details of the site. Construct walkways, driveways, and steps to connect different levels of the property. Re-

USING CAD

CAD Modeling

Many types of 3D computer models are used in the design process as described in Chapter 5. Computer-generated models can be used to check the structural stability, orientation, and pictorial appearance of the design. Some elements of the design are often changed after the model is studied.

A 3D CAD model can eliminate the need for a physically constructed model. The 3D construction and shading capabilities provide a depth to imagery on an otherwise flat computer screen. The main advantage of working in 3D is the accuracy and special relationships that can be created among objects.

CHAPTER
21 Architectural Models Exercises

1. What are the basic purposes of architectural models?
2. Describe the types of design study models and explain their functions.
3. What features does a presentation model usually include?
4. List the steps for constructing a model.

5. Construct a basic layout model of a house shown in Chapter 1.
6. Construct a model of the house you designed.
7. Complete a computer model of the house you designed.

 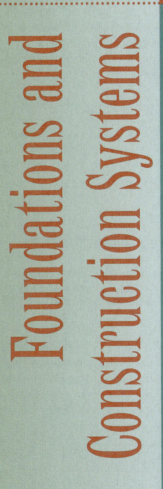
Foundations and Construction Systems

22 CHAPTER

Principles of Construction

OBJECTIVES

In this chapter, you will learn to:

- name and define physical forces that act on a building.

- describe the factors that determine the strength of structural components.

- draw a modular floor plan, elevation, and detail drawing.

TERMS

bearing-wall structures
building load
cantilever
compression force
dead load
deflection

lateral (horizontal) load
live load
modular components
module
prebuilt home
prebuilt module

prefabricated home
shear force
skeleton-frame structures
tension force
torsion force

INTRODUCTION

New construction materials and new methods of using conventional materials provide designers with great flexibility in construction design. Stronger buildings can now be erected with lighter and fewer materials.

Although the basic principles for preparing construction drawings are the same for all types of construction, the use of symbols, conventions, and terms changes from system to system. Construction systems are broadly divided into four material groups: wood, steel, masonry, and concrete. Specific drawings for these groups will be discussed in detail in subsequent chapters. Most contemporary designs include a combination of these systems and materials. Sometimes parts are preassembled and brought to the building site. Information about construction using prefabricated components is included in this chapter.

FIGURE 22.1 ■ Bearing-wall construction. (*GAF Corp.*)

STRUCTURAL DESIGN

Structurally, buildings are divided into two types: bearing-wall structures and skeleton-frame structures. **Bearing-wall structures** have solid walls that support the weight of the walls, floors, and roof.

As covered in Chapter 1, the early buildings of antiquity and, later, log cabins were bearing-wall structures as shown in Figure 22.1. With the development of post-lintel construction and the arch, larger openings could be designed into solid walls. Later, with the development of lighter and stronger materials such as steel and structural lumber, **skeleton-frame structures** could be constructed. Skeleton-frame structures have an open, self-supporting

FRAMEWORK

STRUCTURAL TIES

PROTECTIVE COVER

FIGURE 22.2 ■ Skeleton frame compared to vertebrate structure.

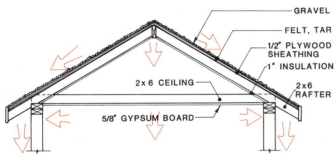

GRAVEL
FELT, TAR
1/2" PLYWOOD SHEATHING
1" INSULATION
2 x 6 CEILING
2 x 6 RAFTER
5/8" GYPSUM BOARD

FIGURE 22.3 ■ Major lines of force.

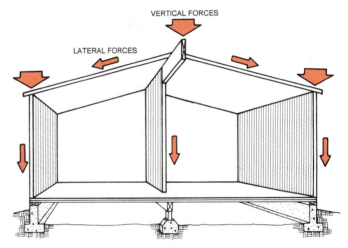

VERTICAL FORCES

LATERAL FORCES

FIGURE 22.4 ■ Interior partitions help support roof loads.

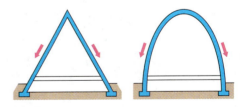

FIGURE 22.5 ■ Roof load forces transferred directly to the foundation.

framework covered by an outer, nonbearing surface. Most contemporary buildings are of the skeleton-frame type. Figure 22.2 illustrates the concept of skeleton-frame construction compared to the human anatomy. Skeleton-frame structures using steel are known as *steel-cage* or *curtain-wall construction.*

In both types of construction, structural stability is based on the strength and placement of the structural members.

The design of structures involves the laws of physics. Not only must a foundation be designed to support the loads (weight, or mass) of a structure, but the structure itself must also withstand the forces acting on it. Figure 22.3 shows a building's major lines of force supported by a foundation through the exterior walls. Figure 22.4 illustrates how an interior bearing partition helps support roof loads.

The walls of a structure are given stability by their attachment to the ground and to the roof. The roof is supported by wall framework, interior partitions, or columns. Each exterior wall and bearing partition is supported by the foundation, which in turn is supported by footings. Footings distribute building loads over a wide area of load-bearing soil and thus tie the entire structural system to the ground. In A-frame or continuous-arch construction, the roof is supported by direct connection with the foundation, as shown in Figure 22.5.

Every architectural designer needs to understand the relationships among loads, forces, and strength of materials. The proper selection and use of materials depends on understanding the loads and forces acting on these materials.

Regardless of the materials and methods used, the physical principles of structural design remain constant.

Structural Forces

Science Connection

Four types of force that exert stress on building materials are compression, tension, shear, and torsion. See Figure 22.6. **Compression forces** push on objects. They tend to flatten materials. **Tension forces** pull on objects. They stretch materials. A supporting chain on a hanging light fixture is in tension. **Shear forces** tend to make one part of an object slide past another part. Excessive shear loads may cause material fractures by abrupt action, as scissors cut a piece of paper. **Torsion force** twists an object. It can twist a member out of shape or fracture it completely by overloading an end or by movement of a connecting member. In wood construction the direction of grain is important because compression and tension forces are minimized when these forces are parallel to the grain. The shear strength of wood is greatest across the grain.

Loads

Loads supported by buildings include live loads and dead loads. **Live loads** are the weight of all movable objects, such as people and furniture. Live loads also include the

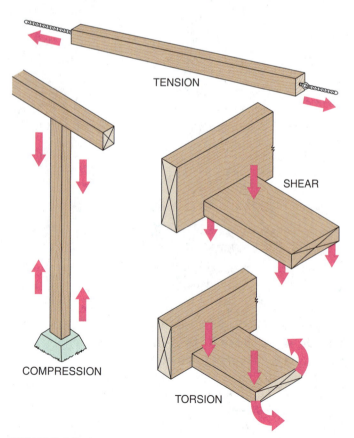

FIGURE 22.6 ■ Types of forces.

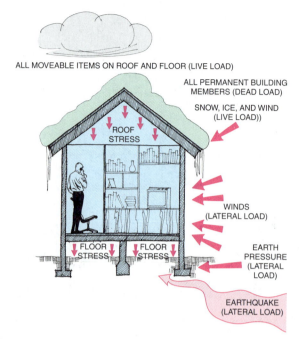

FIGURE 22.7 ■ Types of live and dead loads.

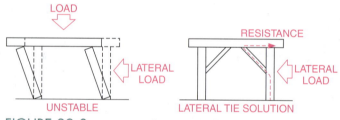

FIGURE 22.8 ■ Diagonal ties provide rigidity.

weight of snow and the force of wind. **Dead loads** are the weight of building materials and permanently installed components. Every piece of lumber, brick, glass, and nail adds to the dead load of a structure. The total weight or mass of all live and dead loads is known as the **building load.** See Figure 22.7.

Most loads follow lines of gravity. However, wind, earth (next to the foundation), and earthquakes also act on a building. These can exert **lateral (horizontal) loads.** There are several ways to counteract the force of these lateral loads. The early Egyptians recognized that the triangle provided the most rigidity with the fewest number of members. Likewise, creating triangular support for right-angle intersections stabilizes a structure, as shown in Figure 22.8.

Figure 22.9 shows how a roof acts to stabilize exterior walls, and Figure 22.10 describes how reinforcement provides protection from compression and tension forces.

Roof loads are measured in pounds per square foot (PSF). See Figure 22.11. For all practical purposes, snow and wind loads are combined and considered as one total live load. For example, the combined wind and snow loads in the South Pacific are 20 pounds per square foot (PSF)

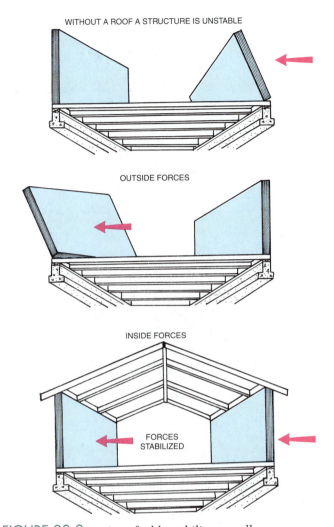

WITHOUT A ROOF A STRUCTURE IS UNSTABLE

OUTSIDE FORCES

INSIDE FORCES

FORCES STABILIZED

FIGURE 22.9 ■ A roof adds stability to walls.

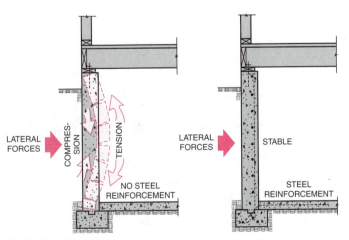

LATERAL FORCES

COMPRES-SION

TENSION

NO STEEL REINFORCEMENT

LATERAL FORCES

STABLE

STEEL REINFORCEMENT

FIGURE 22.10 ■ Reinforcement adds to wall stability.

compared to 30 PSF in the central/western parts of the United States and 40 PSF in the northern parts of the United States.

The design of a structure affects its ability to withstand loads. For example, the pitch of a roof helps determine

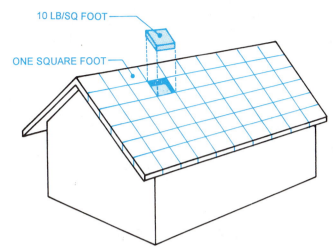

10 LB/SQ FOOT

ONE SQUARE FOOT

FIGURE 22.11 ■ This roof load is 10 pounds per square foot.

its ability to withstand snow and wind loads. Snow loads are exerted in a vertical direction. Wind loads are exerted in a horizontal direction. Therefore, a high-pitched roof will withstand snow loads better than a low-pitched roof (see Figure 22.12), but the reverse is true of wind loads (see Figure 22.13).

Buildings must support their own weight. Dead loads can be calculated for each building or specified in pre-calculated building codes. Material types, their size, and the spacing of members are also calculated or specified by code to prevent the building of structures that are unstable due to excess loading.

Roof loads include live loads (snow and wind) as well as the dead loads of shingles, sheathing, and rafters. All loads are computed on the basis of pounds per square foot, or kilograms per square meter if metric measurements are used. The typical asphalt- or composition-shingle roof weighs approximately 10 to 12 PSF. Thus a 40′ × 20′ (800 sq.ft.) asphalt-shingle roof should be designed to carry a dead load of 8,000 to 9,600 pounds (800 sq.ft. × 12 PSF = 9,600 lbs.). A Spanish tile roof weighs 17 PSF.

Resistance

Compression, tension, shear, and torsion forces create enormous stress on building materials. If a material is sufficiently strong, this stress will create little or no damage. However, if the material is weaker than the forces applied, the resulting stress can compress, stretch, slice, twist, or completely fracture a member. Therefore, the resistance of each structural member must always be equal to or greater than the force applied. see Figure 22.14.

To ensure that every structural member can resist the forces and stresses created by building loads, the

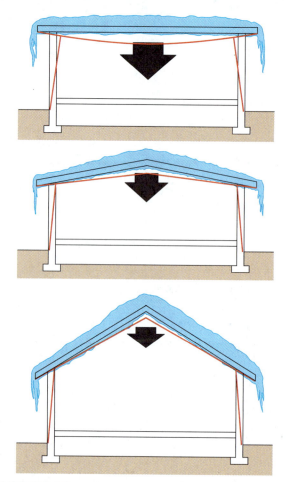

FIGURE 22.12 ■ Snow loads related to roof pitch.

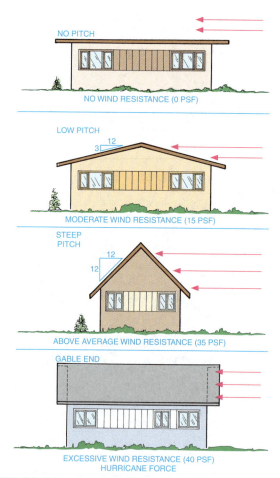

FIGURE 22.13 ■ High-pitched roofs create high wind resistance.

material, size, shape, placement, and spacing must be carefully planned. If any of these factors change during the building process, another design element may need to be changed to maintain the structural integrity of the building. For example, if a load is increased, stronger materials or a shorter distance between supporting members would be needed.

Strength of Materials

The *strength* of a construction material is the material's capacity to support loads by resisting compression, tension, shear, and torsion forces or stresses. The structural strength of a member, or a construction component, depends on the type, size, and shape of the material. Figure 22.15 illustrates how compression and tension forces act on a member to produce deflection. Different structural materials have varying capacities to resist stress and to support building loads. For example, a steel member can support more weight than a wood member of the same size. The load-bearing capacity will also vary among species of wood because different fibers have different stress levels.

Science Connection

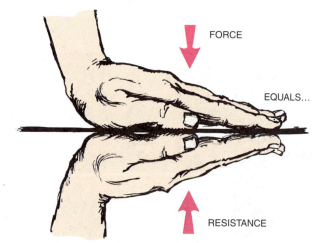

FIGURE 22.14 ■ Resistance must be equal to or greater than the force to provide stability.

Deflection, or bending, stress results from both compression and tension forces acting on a member at the same time. See Figure 22.16. Reinforcing the bottom half of a member balances this stress.

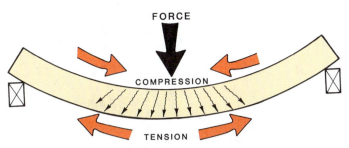

FORCE

COMPRESSION

TENSION

FIGURE 22.15 ■ The effect of force on compression and tension.

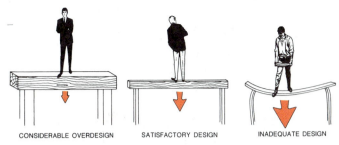

CONSIDERABLE OVERDESIGN   SATISFACTORY DESIGN   INADEQUATE DESIGN

FIGURE 22.17 ■ Relationship of member size to resistance.

STRUCTURAL MEMBER 10'-0" SPAN	APPROX DEFLECTION	STRUCTURAL MEMBER 10'-0" SPAN	APPROX DEFLECTION
500 LB FORCE — 10'-0" — 2" × 6" LAID FLAT	10"	500 LB FORCE — 4" × 8" LAMINATED WOOD BEAM	.07"
500 LB FORCE — 2" × 6" ON EDGE	.75"	500 LB FORCE — 4" × 6" REINFORCED CONCRETE BEAM	.05"
500 LB FORCE — 2" × 12" ON EDGE	.10"	500 LB FORCE — STEEL S BEAM	.02"
500 LB FORCE — 6" × 8" ON EDGE	.10"	500 LB FORCE — 24" FLAT ROOF TRUSS	.02"
500 LB FORCE — 8" DIAM LOG	.08"	500 LB FORCE — 48" HOWE TRUSS	.01"

FIGURE 22.16 ■ Deflection differences among structural members.

Larger members can obviously support greater loads than smaller members of the same material. Material sizes greatly affect load resistance. Inadequate size selection can result in material failure. However, choosing oversized materials can result in extreme material waste as shown in Figure 22.17. Builders must avoid excess notching or drilling of structural members, such as for inserting pipes, because notching and drilling reduce the structural strength of the member.

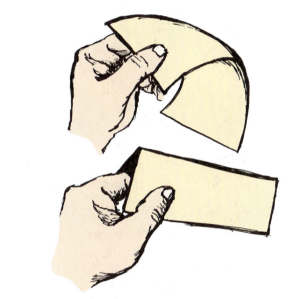

FIGURE 22.18 ■ Member stability related to its shape.

Shape

The different shape of a material influences its ability to support loads. For example, a sheet of thin paper will not stand up by itself. When the paper is folded in half, however, it can support itself plus a light object. See Figure 22.18. Folded several times, this same piece of paper will support a heavier object. Figure 22.19 shows how this principle relates to deflection, rigidity, and the load-bearing capacity of a structural member.

Placement

The strength of a building material is significant only when it becomes an integral part of a structure. Most materials are somewhat flexible until tied in to a structure. Orientation, or position, is also an important factor. See Figure 22.20. A structural member placed on its narrower edge will increase the horizontal deflection but the vertical deflection will be unaffected. Members with their widest dimension positioned parallel to the load direction will resist greater loads than members placed with their smallest dimension in the load direction. Combining the

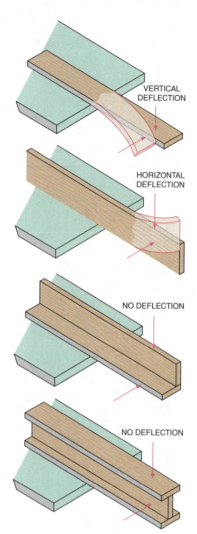

FIGURE 22.19 ■ Member shape related to deflection.

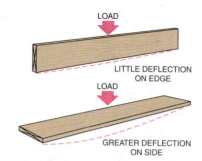

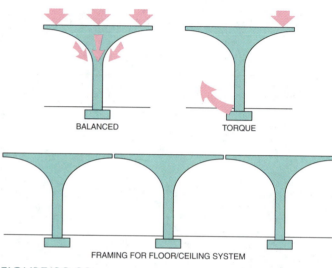

FIGURE 22.20 ■ Member position related to deflection.

FIGURE 22.21 ■ Center-supported cantilevered members.

horizontal and vertical components of the member in the form of a channel or I beam reduces both the vertical and the horizontal deflection.

Spans and Spacing

Spacing is the distance between parallel structural members. *Span* is the distance a member extends between vertical supports. The maximum allowable span of a member is directly related to the loads applied to it and the strength of the structural member. Obviously, stronger materials can support greater loads at greater distances with less deflection. Decreasing the span while using the same structural members can increase the load-bearing capacity of each member.

Cantilever is the term used when only one end of a horizontal structural member is supported. Cantilevered members can be center supported (Figure 22.21) or eccentric (off center). When the center supports equal dead loads on all sides, the member has equilibrium. Supporting on one side only would cause a torque (torsion stress). Eccentric cantilevered members are supported on one side opposite an unsupported end. Eccentric cantilevering requires stronger materials and stronger anchorage on the supported side. Deflection increases as the distance from the support and/or the amount of load increases as shown in Figure 22.22. Roof overhangs, balconies, and decks are structural features that are often cantilevered (Figure 22.23).

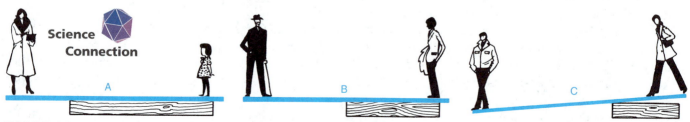

FIGURE 22.22 ■ Effect of loads and overhang length on cantilevered member.

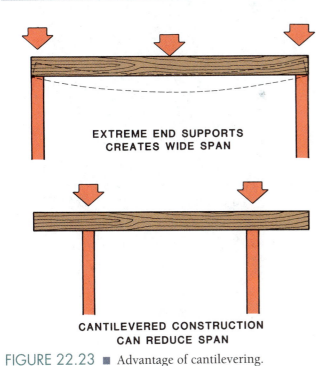

EXTREME END SUPPORTS
CREATES WIDE SPAN

CANTILEVERED CONSTRUCTION
CAN REDUCE SPAN

FIGURE 22.23 ■ Advantage of cantilevering.

MODULAR CONSTRUCTION

One of the most significant advances in structural design and construction is the manufacture of preconstructed parts, or components, of a building. The construction business is one of the last segments of industry to fully use standardized, interchangeable parts. Most basic materials—wood, steel, and masonry—are now available in standardized sizes. The sizes may vary, but modular materials and components are designed to fit together with precision.

Modular Components

Math Connection

Modular components are designed as parts or sections to be constructed away from the building site. This eliminates much on-the-job construction work. Typical components may range from preassembled wall sections and windows to molded bathrooms.

When more components are used, on-site construction work changes from piece-by-piece building to the assembling of components.

Size Standardization

Designing with modular components means the designer must adhere strictly to standard sizes in creating an ar-

chitectural plan. Sizes of the components are uniform, with many different interchangeable parts.

Modular building design is based on measurements that are divisible by the same base unit. The base unit is known as a **module**. For example, a system based on a module of 4″, 16″, 24″, and 48″ is used for U.S. customary measurements. The metric modular system is based on 100, 300, 600, and 1200 millimeters (mm).

Modular design and construction involves all three dimensions: length, width, and height. The overall width and length dimensions are the most critical in the planning process. The modular planning grid shown in Figure 22.24 is a horizontal plane divided into equal spaces in length and width. It provides the basic control for the architectural modular coordination system. The entire grid is divided into equal spaces of 4″, 16″, 24″, and 48″. See Figure 22.25. All module sizes divide equally into 4′ × 8′ panels. The 16″ unit is used in multiples for wall, window, and door panels to provide an increment small enough for flexible planning. Increments of 24″ and 48″ are used for overall dimensions. The 24″ module is called the *minor module*. The 48″ module is called the *major module*. See Figure 22.26.

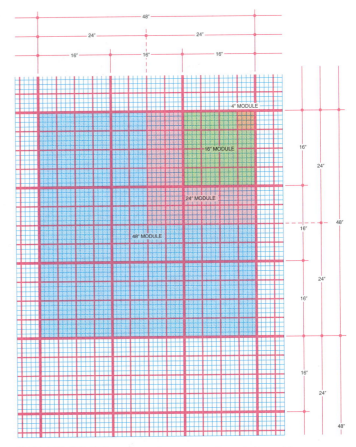

FIGURE 22.24 ■ Two-dimensional modular grid.

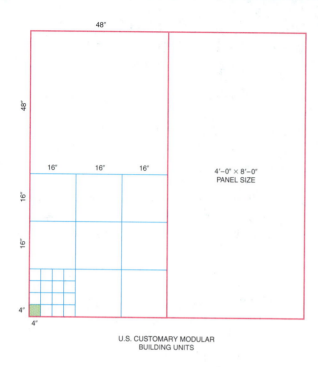

U.S. CUSTOMARY MODULAR
BUILDING UNITS

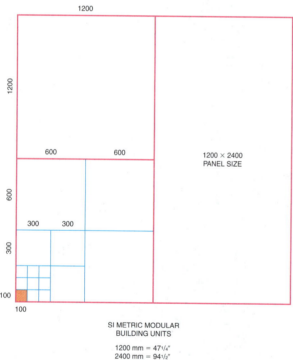

SI METRIC MODULAR
BUILDING UNITS

1200 mm = 47 1/4"
2400 mm = 94 1/2"

FIGURE 22.25 ■ U.S. customary and S.I. metric modular building units.

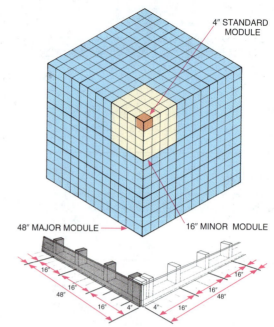

FIGURE 22.26 ■ Three-dimensional view of the standard, major, and minor modules.

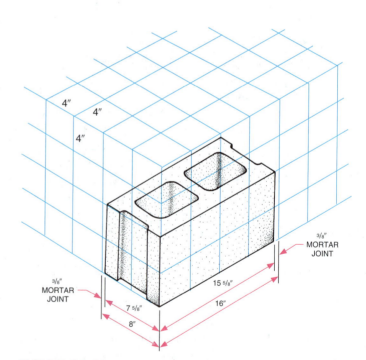

FIGURE 22.27 ■ Concrete block dimensions based on a 4″ modular unit.

Many building materials, such as lumber, plywood, brick, tile, and concrete block are available in modular units. See Figure 22.27. The use of small modular materials, usually based on a 4″ module, saves time and expense. Large modular units, such as wall panels, door assemblies, and window assemblies, can save much more time and resources. See Figure 22.28. To ensure that all components fit as planned, components must be designed to align with modular grid lines. This requires that architectural drawings be prepared such that building dimensions align with established modular grids.

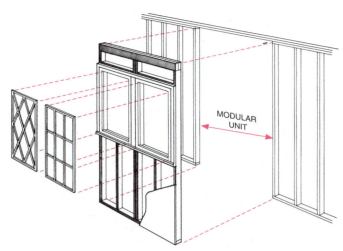

FIGURE 22.28 ■ Components of a manufactured window assembly align with the same modular standard.

Modular Drawings

Modular drawings are developed just like other architectural drawings, except both horizontal and vertical members are designed to align with grid lines. When drawing a modular design, space must still be provided for such items as doors and windows, plumbing runs, medicine cabinets, closets, and fireplaces.

Conventional (nonmodular) and modular components may be combined in the same plan. Framing for the modular components is then extremely critical. Any variation in the framing opening will result in a misfit when the modular units are positioned into the structure.

Modular Floor Plans

To build a modular structure, all drawings must conform to modular standards. For example, basic floor plans should be drawn on a modular grid. All nonmodular dimensions need to be aligned or converted to the nearest modular grid line. Figure 22.29 shows a floor plan aligned with a 16″ modular grid. Likewise, stud layouts are also drawn on a modular grid, as are the studs in Figure 22.30. If a window or door assembly is not available in a modular width, align one side on a 16″ grid to reduce the number of studs required.

Dimensions required by existing building laws or from built-in equipment must still be incorporated and

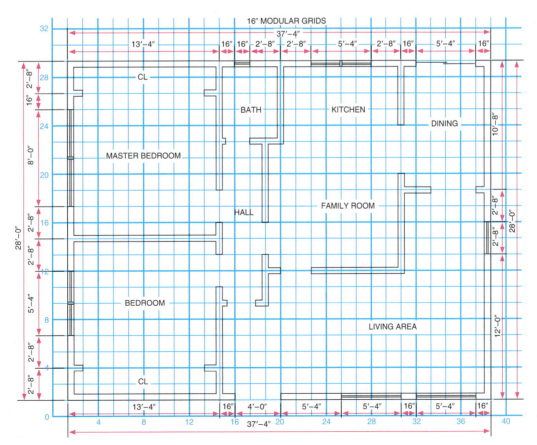

FIGURE 22.29 ■ Floor plan aligned with a 16″ modular grid.

coordinated with any prebuilt modular system. This is accomplished by conventional dimensioning. Remember, all dimensions are located either from surface to surface or from centerline to centerline.

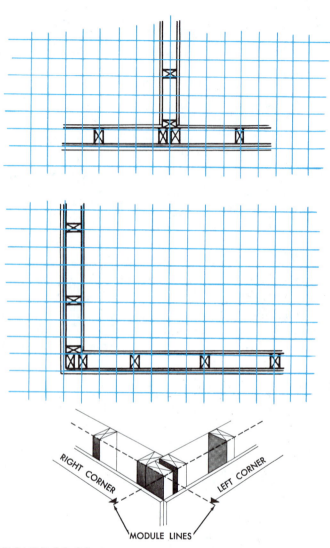

FIGURE 22.30 ■ Stud layout aligned on a 4″ modular grid with studs at 16″ on center.

Modular Elevations

As with modular-component floor plans, the modular-component elevation drawings are prepared using a 16″ grid. Elevation drawings closely resemble conventional elevations, except the components are identified and shown in their proper relationship with modular alignments. See Figure 22.31. Likewise, elevation framing drawings must be aligned with established modular grid lines, as shown in Figure 22.32.

Modular Details

Many modular drawings are needed to describe every detail of a modular plan. As in a conventional design, more detailed drawings improve the chances of achieving the desired outcomes. Details may be required for standard floor and roof components, as well as for many nonstandard components.

To ensure the correct alignment of key horizontal and vertical surfaces, such as walls and floors, detail drawings are often prepared using a modular grid. See Figure 22.33. An elevation detail section is used primarily to provide exact data for the establishment of critical height dimensions, such as for foundations and floors. Plan detail drawings are used to provide exact alignment data for width and length dimensions.

Modular Dimensioning

Dimensions that align with a module are known as *grid dimensions*. Dimensions that do not align with a module are known as *nongrid dimensions*. Figure 22.34 shows the two methods of indicating grid dimensions and nongrid dimensions. Grid dimensions may be shown by conventional arrowheads, and nongrid dimensions are shown by dots (small circles) instead of arrowheads.

Overall dimensions are first established. Next, nonmodular dimensions are incorporated into the plan.

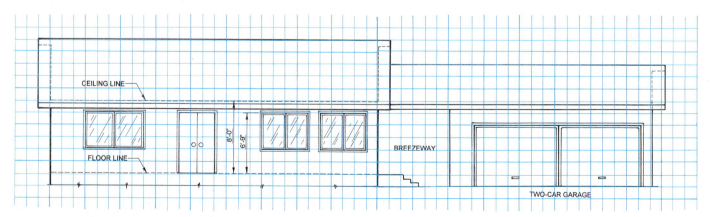

FIGURE 22.31 ■ Elevation drawing on a vertical modular grid.

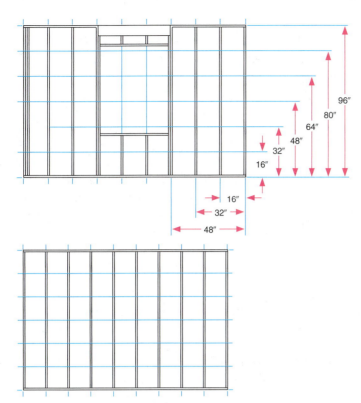

96"
80"
64"
48"
32"
16"

16"
32"
48"

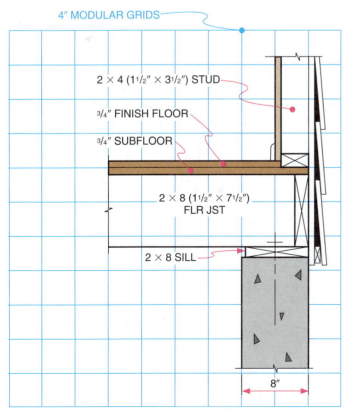

FIGURE 22.32 ■ Vertical modular size increments for elevation framing drawings.

GRID DIMENSIONS
NONGRID DIMENSIONS

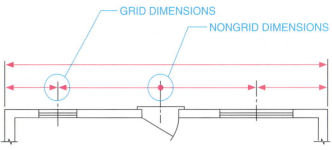

FIGURE 22.34 ■ Modular and nonmodular dimension methods.

4" MODULAR GRIDS

2 × 4 (1½" × 3½") STUD

¾" FINISH FLOOR

¾" SUBFLOOR

2 × 8 (1½" × 7½") FLR JST

2 × 8 SILL

8"

FIGURE 22.33 ■ Sill detail on a modular grid.

Then panels for exterior doors, windows, and exterior walls are located. The floor plan in Figure 22.29 has been properly fitted on the modular grid. It is important that the foundation, floors, walls, windows, doors, partitions, and roof are properly aligned. Establishing all the dimensions and components precisely on the 16", 24", and 48" spaces of the modular grid ensures the accurate fitting of the components.

In many construction detail drawings, it is possible to eliminate the placement of some dimensions by placing the grid lines directly over the drawing. When the grid lines coincide exactly with the material lines, no dimensions are needed. For example, using a 4" module, each line represents 4". Any building material that is an increment of 4" will align exactly with a line on this grid.

When using a CAD system, grid dimensions are created on the monitor using the *Grid* command. All lines can be snapped to align with grid lines. Nongrid dimensions are drawn in the free pick mode.

USING CAD

CAD Grids and Modules

Grid spacing and points can be selected to align with any modular scale and spacing. Grids should be selected to represent the smallest grid unit to be used on the drawing. Typical grid sizes for floor plans and elevations are 12" and 4" or 6" for details. Most modular grids are available with major and minor module lines in different weights and/or colors. To make lines conform to a vertical or horizontal axis with or without the use of grids, the Ortho command can be used.

CAD grid lines can also be placed on a different layer than object lines. Later these grid lines can be plotted either in a second color or with a fine line to show the modular alignment. Nonmodular lines must be drawn in a nongrid mode so that they don't automatically snap to the grid lines.

Manufactured Buildings

To some degree all contemporary buildings are manufactured. Not every component is totally built on site, even for a completely custom home. The amount of manufactured components used in buildings varies. Most companies combine mass-production methods with other construction design to minimize custom-job work without sacrificing quality.

Precut Structures

Precut structures are built from materials that are cut to specification at a factory and then assembled on site by conventional ("stick-built") methods. Designing precut buildings involves using standard sizes of materials and components to eliminate waste and on-site labor time.

Prefabricated Homes

In the most common type of **prefabricated homes**, the major components, such as the walls, decks, and partitions, are assembled at a factory. The utility work, such as installation of electrical, plumbing, and heating systems, is completed on site. The final finishing work, such as installation of prehung doors, prefinished roof coverings, and prefinished walls, is also done on site.

Prebuilt Homes

Prebuilt homes are manufactured houses that are totally built in a factory. The first completely factory-built homes were mobile homes. The mobile home buyer, like the buyer of a conventional factory-built home, has little option to adjust or customize the basic design. Nonetheless, there are opportunities to select different sizes, models, and interiors. All electrical, plumbing, and HVAC units are built into the structure.

Prebuilt homes are made in widths up to 28′. Most states allow only 14′ widths on roadways. A manufactured house can be produced in two 14′ halves for shipping. Because most trucks can only haul a 48′ length, a 28′ × 48′ building represents two 14′ × 28′ prefab units. To overcome these size restrictions, multiple prefab units are often combined on the job site to create large structures.

Prebuilt Modules

When larger structures or different configurations are needed, manufactured **prebuilt modules** can be combined to form a variety of floor plan shapes. Construction modules can be manufactured as complete rooms or as independent functional areas. In either case, they contain complete built-in components such as kitchen and bath cabinets, major appliances, and fixtures.

Area modules contain room clusters that can be combined with other room or room cluster modules.

Modules can be used as elements of an expandable plan that may take years to complete. Combinations of modules can be designed to attach horizontally side to side or end to end or vertically bottom to top. They may also be located separately in a campus plan and connected by walkways or lanais.

Designing prebuilt houses or modules involves conforming to factory production standards and sizes. Nevertheless, a designer can develop optional housing models and a wide variety of modular configurations for customer consideration.

In summary the standardization of materials and building components into modular sizes has created many opportunities for architectural innovation while controlling the costs normally associated with creative designs.

Principles of Construction Exercises

1. Describe four structural forces and give an example of how each can be counteracted.

2. Explain what makes up a building load. How does it relate to construction?

3. Sketch an elevation that minimizes live loads from snow and wind. Explain why you chose that design.

4. Explain how size, shape, placement, and spacing of structural members affect the strength of a building.

5. After selecting the size of a roof and materials you would use, calculate the weight of the roof.

6. Use the grid snap to design a modular floor plan and elevation.

7. Draw a floor plan of your own design based on the grid system in this chapter. Include modular dimensions.

8. Sketch an elevation that combines modular and nonmodular components. Use grid and nongrid dimensions.

9. Draw a construction detail using a 4″ grid system.

10. Design a small commercial building using a modular grid of 8′.

11. Trace the modular plan shown in Figure 22.35 and draw modular grids on an overlay.

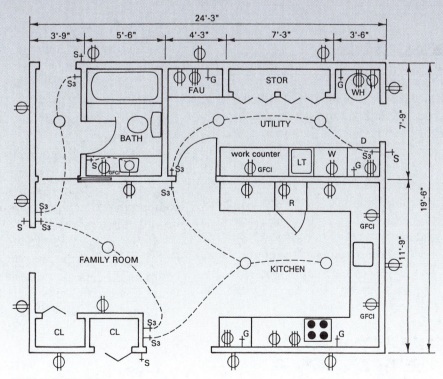

FIGURE 22.35 ■ Trace this drawing and draw modular grids on an overlay.

23 CHAPTER

Foundations and Fireplace Structures

OBJECTIVES

In this chapter you will learn to:

- describe the types of foundations.
- identify the components and materials used in foundations.
- design a fireplace with sufficient structural support and appropriate safety components.

- draw foundation plans.
- relate the layout and excavations for a building to the type of foundation it will have.

TERMS

damper
firebox (fire chamber)
flue
footings (footers)

foundation sills
permanent wood foundations
pier-and-column foundations

reinforcing bars (rebars)
slab foundations
T-foundations

INTRODUCTION

Every structure needs a foundation. The function of a foundation is to provide a level and uniformly distributed support for the structure. Foundations must be strong enough to support and distribute the load of the structure. Foundations must remain level to prevent the walls from cracking and the doors and windows from sticking. They also fulfill other functions, such as to help prevent cold air and dampness from entering buildings, to waterproof basements, and to form the supporting walls for basements.

The methods and materials used in constructing foundations vary greatly in different parts of the country and are continually changing. The basic principles of foundation construction are the same, though, regardless of the application.

FOUNDATION MATERIALS AND COMPONENTS

The components and materials in foundations vary, depending on the foundation type, size, and design. Mate-

rials used include concrete, concrete block, steel reinforcement bars, welded wire mesh, and a variety of wood and composite form materials.

Bearing Surface

The area under a foundation—the soil—must be capable of bearing the load, or weight, of the foundation and the structure. This bearing surface, or bearing soil, must be compactable, contain no clays or organic matter, drain easily, and be freeze resistant in cold climates. See Figure 23.1.

Minimum amounts of settlement will occur in bearing soil regardless of the amount of compaction. Therefore, compaction must be uniformly distributed to eliminate voids. Compaction must also be level to ensure that the supported structure will not shift out of plumb.

Concrete

The basic material used to pour foundation bases, walls, and floors is concrete. Concrete is a combination of cement (clay and limestone), water, stone aggregate, and chemicals that improve strength or workability. See

TYPE OF SOIL	BEARING CAPACITY (POUNDS PER SQ. FT.)
Soft clay, loose dirt, loam	2,000
Dry sand and hard clay	4,000
Hard sand or gravel	6,000
Partially cemented sand or gravel	20,000

FIGURE 23.1 ■ Bearing capacity of typical bearing soils.

Chapter 26 for more information on concrete characteristics and related building systems.

Concrete Block

Economy and ease of construction have made concrete block a popular material for foundation walls. Foundation blocks are manufactured in lengths of 16″, heights of 8″, and in modular widths of 4″, 6″, 8″, 10″, and 12″. Each dimension is actually 3/8″ smaller than the listed size to allow for mortar. This keeps the finished dimensions in modular units.

Reinforcing Bars

Concrete resists compression very well. To resist tension forces, steel bars are added to concrete slabs, beams, and columns. These bars are known as **reinforcing bars** or **rebars.** Steel rebars are either smooth or deformed (grooved or embossed). Deformed bars create a stronger bond between bar and concrete because the concrete is held in place by the grooves or depressions. Rebars are sized by numbers (1 through 18) representing 1/8″ increments, up to 2 1/4″. See Figure 23.2. The bar size number, mill number, symbol of the steel type, and the grade are marked on each rebar.

Beam rebars are located horizontally near the bottom of beams to provide maximum tension resistance. Similarly, slab rebars are placed horizontally and in parallel rows close to the bottom of the slab. To prevent cracking due to temperature and moisture changes, some rebars are also placed perpendicular to the load-supporting bars. These rebars are known as temperature bars.

Because a minimum thickness (1″ to 3″) must be maintained between rebars and the concrete surface, fixtures known as *bolsters* (*saddles*) and *chairs* are used to hold the bars in place during slab pouring. U-shaped rods, known as *stirrups*, are used for this purpose in beam pouring. See Figure 23.3.

BAR SIZE	AREA– SQ. IN.	WEIGHT– LBS. PER FT.	DIAMETER– INCHES
3	.11	.376	3/8″
4	.20	.668	1/2″
5	.31	1.043	5/8″
6	.44	1.502	3/4″
7	.60	2.044	7/8″
8	.79	2.670	1″
9	1.00	3.400	1 1/8″
10	1.27	4.303	1 1/4″
11	1.56	5.313	1 3/8″
14	2.25	7.650	1 3/4″
18	4.00	13.600	2 1/4″

FIGURE 23.2 ■ Rebar sizes.

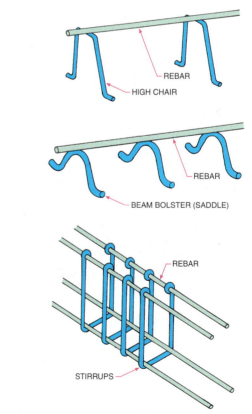

FIGURE 23.3 ■ Rebar support fixtures.

The exact position of rebars in a slab is shown on a foundation sectional drawing. The location of rebars in a wall is shown on a plan detail and/or elevation section.

SQUARE PATTERN AND GAUGE	RECTANGULAR PATTERN AND GAUGE
6 × 6–10/10	6 × 12–4/4
6 × 6–8/8	6 × 12–2/2
6 × 6–6/6	6 × 12–1/1
6 × 6–4/4	
	4 × 12–8/12
4 × 4–10/10	4 × 12–6/10
4 × 4–8/8	
4 × 4–6/6	4 × 16–8/12
4 × 4–4/4	4 × 16–6/10

FIGURE 23.4 ■ Wire mesh spacing and gauges.

Wire Mesh

Steel-welded wire mesh or wire fabric is often used in slabs in place of rebars. Square, rectangular, and triangular patterns are available. Square patterns are the most common. Wire mesh is specified on construction drawings by the spacing (in inches) between wire strands and by the gauge of the wires. The term *gauge* refers to the diameter of the wire. Be aware, however, that the larger the gauge number, the smaller the wire diameter. The two intersecting wires in a pattern are known as longitudinal (long direction) and transverse (short direction) wires. A rectangular pattern with transverse wires spaced 4″ apart and longitudinal wires spaced 12″ apart is labeled 4 × 12. This is followed by the gauge of each of the wires. For example, a wire mesh may consist of a #6 gauge for the transverse wire and a #10 gauge for the longitudinal wire. All this information is noted on the drawing as follows: 4 × 12—6/10 to indicate spacing and wire gauges. See Figure 23.4.

Footings

Footings, or **footers**, are the bases of foundations and foundation walls. There are two types: continuous and individual. Continuous footings extend under walls and around the perimeter of a foundation. The footings are often wider than the foundation walls in order to distribute the weight of the building over a larger area. See Figure 23.5. Individual footings (called piers) support vertical structural members.

Concrete is commonly used for footings because it can be poured to maintain a firm contact with the sup-

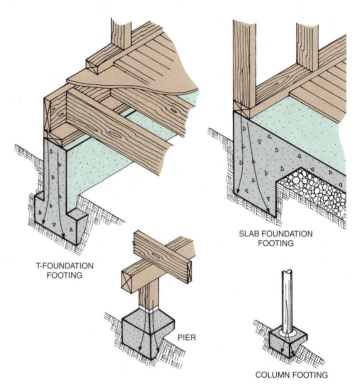

FIGURE 23.5 ■ Types of footings.

porting soil. Concrete is also effective because it can withstand heavy weights and is a decay-proof material. Steel reinforcements (rebars) are added to concrete footings to keep the concrete from cracking and to provide additional support. In some circumstances other masonry types and even treated wood can be used for footings.

Footings must be laid on solid ground to support the weight of the building effectively and evenly. In cold climates, footings must be placed below the frost line (the depth to which the soil freezes). Always consult the local building code for frost line foundation depth requirements before establishing footing depths. Stepped footings at different levels are used on sloping sites. See Figure 23.6.

TYPES OF FOUNDATIONS

The type of foundation depends on the nature of the soil, slope of the terrain, the size and weight of the structure, the climate, building laws, and the relationship of the floor line to the grade line (ground). See Figure 23.7. Foundations covered in this chapter are divided into four basic types: T-foundations, slab foundations, pier-and-column foundations, and permanent wood foundations. Figure 23.8 shows the relationship of these foundation

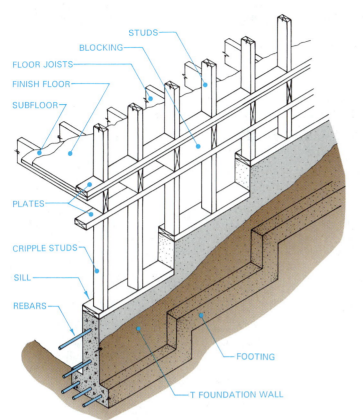

FIGURE 23.6 ■ Continuous stepped footing.

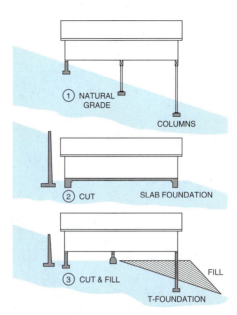

FIGURE 23.7A ■ Effect of grade contour on foundations.

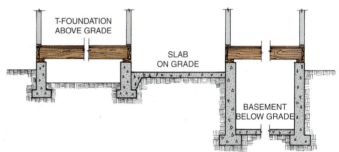

FIGURE 23.7B ■ Foundation position related to grade line.

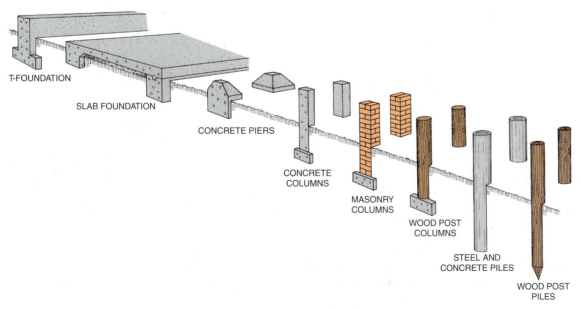

FIGURE 23.8 ■ Types of foundations.

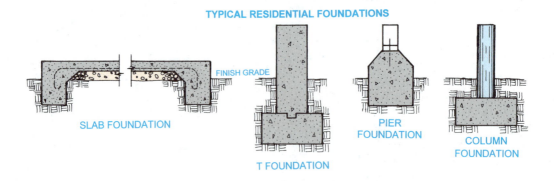

TYPICAL RESIDENTIAL FOUNDATIONS

SLAB FOUNDATION

FINISH GRADE

T FOUNDATION

PIER FOUNDATION

COLUMN FOUNDATION

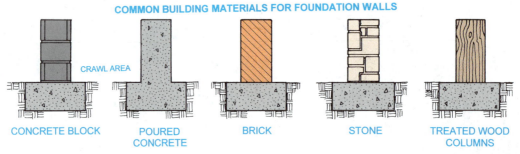

COMMON BUILDING MATERIALS FOR FOUNDATION WALLS

CRAWL AREA

CONCRETE BLOCK

POURED CONCRETE

BRICK

STONE

TREATED WOOD COLUMNS

FIGURE 23.9 ■ Sections of foundation types.

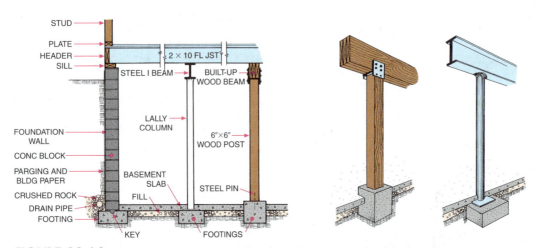

STUD
PLATE
HEADER
SILL

2 × 10 FL JST

STEEL I BEAM

BUILT-UP WOOD BEAM

FOUNDATION WALL
CONC BLOCK
PARGING AND BLDG PAPER
CRUSHED ROCK
DRAIN PIPE
FOOTING

LALLY COLUMN

6"×6" WOOD POST

BASEMENT SLAB

FILL

STEEL PIN

KEY

FOOTINGS

FIGURE 23.10 ■ T-foundations with different walls, posts, and columns.

types to the ground line, and Figure 23.9 shows how these foundation types are drawn as elevation sections.

T-Foundations

A **T-foundation** consists of a footing and a poured concrete or concrete block wall that forms an inverted T. T-foundations are necessary in structures with basements or when the underside of the first floor must be accessible. See Figure 23.10. The details of construction relating to several variations of a T-foundation plan are shown in Figure 23.11.

T-foundations are prepared by pouring the footing into an excavated trench, leveling the top of the footing, and erecting a concrete block or masonry wall on top of the footing. If poured concrete foundation walls are to be used, building forms are erected on top of the footing. Concrete is poured into these forms.

Foundation Walls

The function of foundation walls is to support the load of the structure and to transmit its weight to the footing.

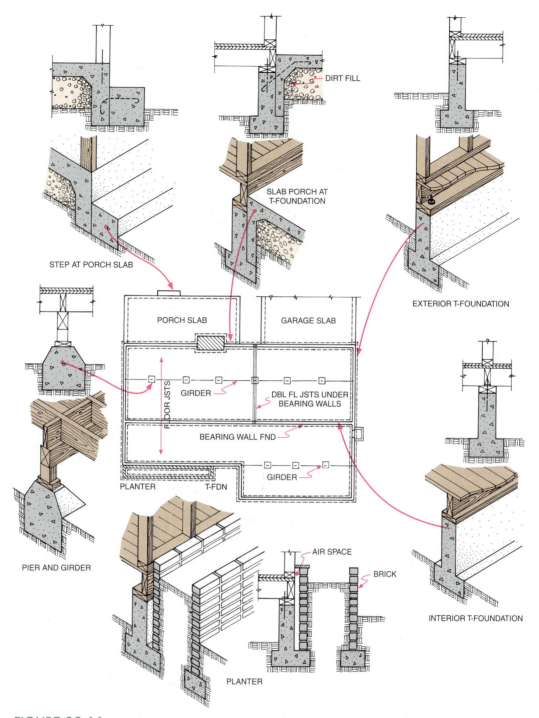

DIRT FILL

SLAB PORCH AT
T-FOUNDATION

STEP AT PORCH SLAB

EXTERIOR T-FOUNDATION

PORCH SLAB   GARAGE SLAB

FLOOR JSTS

GIRDER   DBL FL JSTS UNDER
BEARING WALLS

BEARING WALL FND

PLANTER   T-FDN   GIRDER

PIER AND GIRDER

INTERIOR T-FOUNDATION

AIR SPACE

BRICK

PLANTER

FIGURE 23.11 ■ T-foundation plan with different floor and wall types.

Foundation walls are normally made of poured concrete, concrete block, stone, or brick. See Figure 23.12. When a complete excavation is made for a basement, foundation walls also provide the walls of the basement, as shown in Figure 23.13. Figure 23.14 shows how foundation walls are drawn as plan and elevation sectional views.

Foundation walls can support heavy compression loads. However, lateral earth loads are often a problem. Imagine a sheet of paper standing on end. It will not stand unsupported. Fold it at right angles and it may stand, but wobble. Fold it again in the shape of a V and it gains great stability. See Figure 22.18 in Chapter 22. This same principle applies to foundation walls. Each

FIGURE 23.12 ■ Common foundation wall materials.

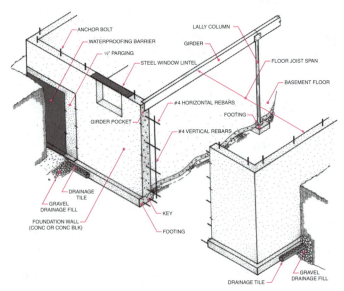

FIGURE 23.13 ■ Foundation walls may also serve as basement walls.

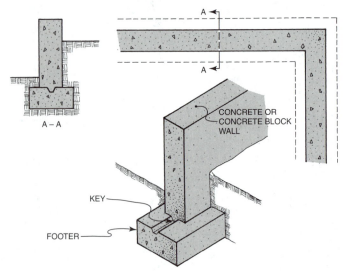

FIGURE 23.14 ■ T-foundation shown in plan and elevation section.

corner, offset, or pilaster adds strength, as shown in Figure 23.15.

Pilasters are reinforcements in a wall designed to provide more rigidity without increasing the width of the

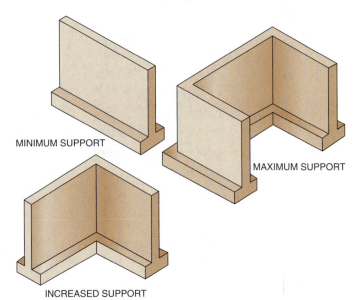

FIGURE 23.15 ■ Foundation shape affects lateral load resistance.

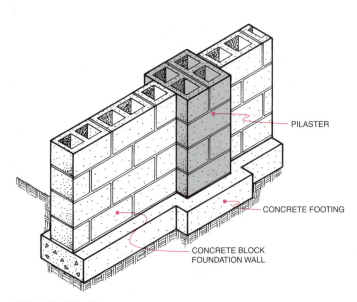

FIGURE 23.16 ■ Concrete wall pilaster.

entire wall. See Figure 23.16. Pilasters are also used to support girders (main horizontal members) instead of making girder pockets (inset spaces) in the foundation walls. Building codes normally specify the size and spacing of pilasters, depending on the wall width, height, and material.

Sills

Foundation sills are wood or steel members that are fastened to the top of foundation walls. Sills provide the base for attaching floor systems to foundations. See Figure 23.17. Wood sills must be pressure treated, or a

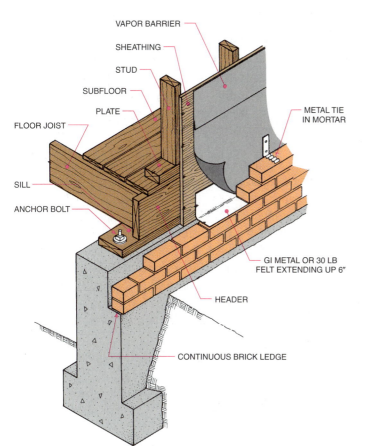

FIGURE 23.17 ■ Sills connect floors and walls to foundations.

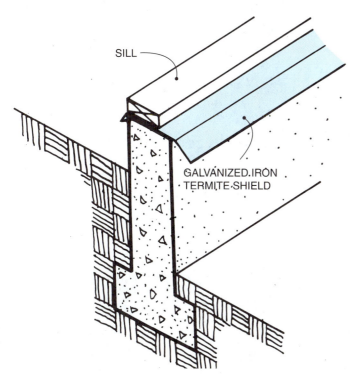

FIGURE 23.18 ■ Termite shield placed between the wood sill and concrete wall.

sheet-metal termite shield (Figure 23.18) must be placed between the wood and concrete. This applies to all wood in contact with masonry, concrete, or soil. Building laws specify the distance required from the bottom of the sill to the grade line inside and outside the foundation.

Anchor Devices

Anchor devices are embedded in the top of the foundation walls or piers. See Figure 23.19. The exposed part of the device is attached to the first wood member (the sill). Anchor bolts are used most often. Sizes of bolts typically used in residences are 1/2″ or 5/8″ in diameter and 10″ long. They are usually spaced 4′ apart, starting 1′ from each corner. Bolts may be embedded into drilled holes or shot into the concrete with low-caliber power activated fastening guns.

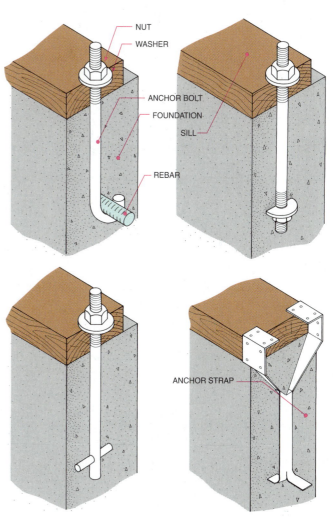

FIGURE 23.19 ■ Types of foundation-to-sill anchor devices.

Cripples

Cripples are used to raise floor levels without building a higher foundation wall. See Figure 23.20. Because the load of the structure must be transmitted through the cripples, these are usually heavy members, often four-by-fours (4 × 4's) spaced closer than the normal 16″ on center. "On Center" (OC) refers to the distance (spacing) measured from the center of one member to the center of the next. Shear stress plywood must be used over cripples to overcome lateral stresses.

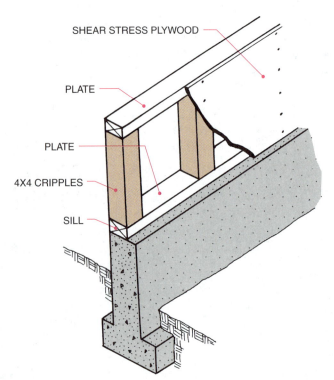

FIGURE 23.20 ■ Cripples extend foundation height.

Slab Foundations

Slab foundations, or "slabs," are made of reinforced concrete. They are either monolithic (one piece) or separate pieces. In a monolithic slab, the slab floor and footing are poured as one piece. See Figure 23.21. Rebars, wire mesh, plumbing line risers, waterproof membranes, electrical conduits (tubes for wires), and HVAC (heating, ventilating, and air conditioning) ducts (if in the floor) must all be securely in place on a compacted soil base before pouring. Figure 23.22 shows a sectional drawing of an HVAC duct embedded in a concrete slab. Figure 23.23 shows a chart of the number of cubic feet of concrete needed for different slab thicknesses.

In slab foundations that are not monolithic, the footings are poured separately from the floor slabs, as shown in Figure 23.24. Figure 23.25 shows a variety of slab floor, wall, and footing details used in different foundation situations.

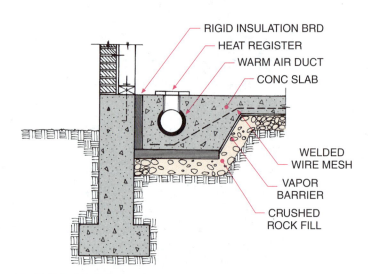

FIGURE 23.22 ■ HVAC duct embedded in a slab.

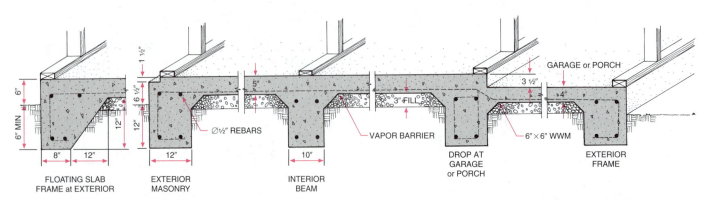

FIGURE 23.21 ■ Types of monolithic slab foundations.

CUBIC YARDS OF CONCRETE FOR SLABS

SLAB'S THICKNESS	10 SQ. FT.	25 SQ. FT.	50 SQ. FT.	100 SQ. FT.	200 SQ. FT.	300 SQ. FT.
4 in.	.12 cu. yd.	.31 cu. yd.	.62 cu. yd.	1.23 cu. yd.	2.47 cu. yd.	3.7 cu. yd.
5 in.	.15 cu. yd.	.39 cu. yd.	.77 cu. yd.	1.54 cu. yd.	3.09 cu. yd.	4.63 cu. yd.
6 in.	.19 cu. yd.	.46 cu. yd.	.93 cu. yd.	1.85 cu. yd.	3.7 cu. yd.	5.56 cu. yd.

FIGURE 23.23 ■ Data for calculating amounts of slab concrete.

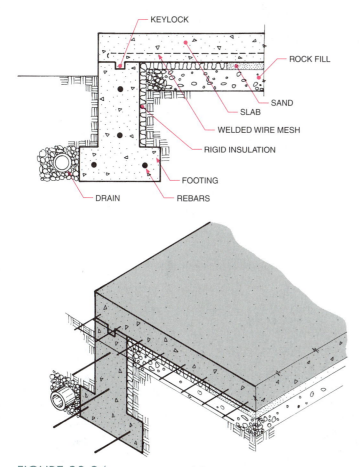

FIGURE 23.24 ■ Two-piece slab.

Figure 23.26 shows how a concrete slab floor relates to the fill, vapor barrier, and rigid insulation below and the flooring above. Figure 23.27A shows the use of rebars to connect a slab with a foundation wall. Figure 23.27B shows a slab detail for a monolithic (solid, one-piece) pour with rebars and welded wire mesh. Where concrete or concrete block is used separately for foundation walls,

the slab and footing must be secured to the foundation wall with rebars.

Building codes specify the minimum distance from the top of a slab to the grade line, usually 6″ to 8″. Because slabs lose heat around the perimeter, adequate insulation is important.

Pier-and-Column Foundations

Pier-and-column foundations consist of individual footings (piers) upon which posts and columns are placed. Posts and columns are vertical members used to support floor systems. See Figure 23.28, page 410.

Individual footings are known as *piers*. Piers may be sloped or stepped in order to spread the load of the structure on a wider base. See Figure 23.29, page 410.

Posts and columns are vertical members that support girders and beams and transmit their weight and the weight of the entire building to the footings. See Figure 23.30, page 410. The terms *post* and *column* are often used interchangeably. Generally, short vertical supports are called *posts* and longer vertical members are known as *columns*. Posts are usually made of wood, and columns are usually steel or masonry. Figure 23.31, page 411, illustrates how concrete column footings act as piers in supporting foundation walls.

Pole foundations are designed with enough vertical stability to function as extended piers. Figure 23.32, page 411, shows a pictorial plan and elevation drawing of a typical pole foundation. These drawings show the pole intersections with the floor and roof structure. The size, type, and spacing of members depends on the soil slope, soil conditions, and the weight of the structure.

Piers and posts or columns may be used as the sole support of the structure. They also may be used in conjunction with foundation walls to provide intermediate

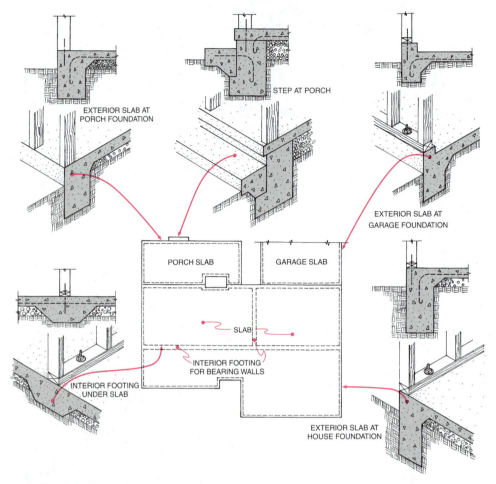

STEP AT PORCH

EXTERIOR SLAB AT
PORCH FOUNDATION

EXTERIOR SLAB AT
GARAGE FOUNDATION

PORCH SLAB

GARAGE SLAB

SLAB

INTERIOR FOOTING
FOR BEARING WALLS

INTERIOR FOOTING
UNDER SLAB

EXTERIOR SLAB AT
HOUSE FOUNDATION

FIGURE 23.25 ■ Slab foundation details.

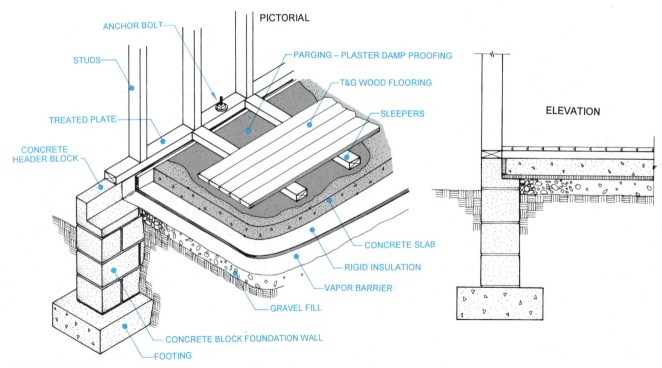

PICTORIAL

ANCHOR BOLT

STUDS

PARGING – PLASTER DAMP PROOFING

T&G WOOD FLOORING

SLEEPERS

ELEVATION

TREATED PLATE

CONCRETE
HEADER BLOCK

CONCRETE SLAB

RIGID INSULATION

VAPOR BARRIER

GRAVEL FILL

CONCRETE BLOCK FOUNDATION WALL

FOOTING

FIGURE 23.26 ■ Concrete slab floor with wood flooring system.

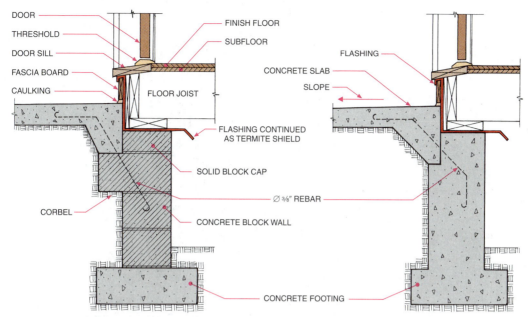

FIGURE 23.27A ■ Rebars connect the slab with the foundation.

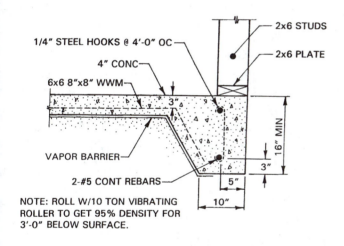

EXTERIOR SLAB FOOTING DETAIL - TYPICAL
SCALE: 1″ = 1′-0″

FIGURE 23.27B ■ Slab detail showing rebars and welded wire mesh placement.

support for horizontal members. See Figure 23.33, page 412. Fewer materials and less labor are needed for pier-and-column foundations, but they are seldom used in basements because they occupy needed open space.

Beams and Girders

Beams are horizontal structural members that support a load. Girders are large beams that are the major horizontal support members for a floor system. In com-

mon practice, the terms *beam* and *girder* are often used interchangeably. However, technically, girders are members that are supported by piers and columns and secured to foundation walls. Beams and girders may also be supported on foundation walls or indentations (pockets) in foundation walls as shown in Figure 23.34, page 412.

Girder sizes are closely regulated by building codes. The allowable span of the girder depends on the size of the girder. A decrease in the size of a girder means that the span must be decreased. This is done by adding additional pier-and-column supports underneath the girder.

Most wood girders for residential construction are built up from 2 × 6's, 2 × 8's, or 2 × 10's that are spiked or nailed together. Steel beams or girders can perform the same function as wood beams or girders, but steel members can span larger distances than wood members of the same size.

Joists

Joists are the parts of the floor system that are placed perpendicular to the girders. See Figure 23.35, page 412. Joists span either from girder to girder or from girder to the foundation wall. The ends of the joists butt against a header or extend to the end of the sill. Bridging is placed between joists. Bridging consists of smaller structural members fastened between the joists to add stability and keep spacing consistent.

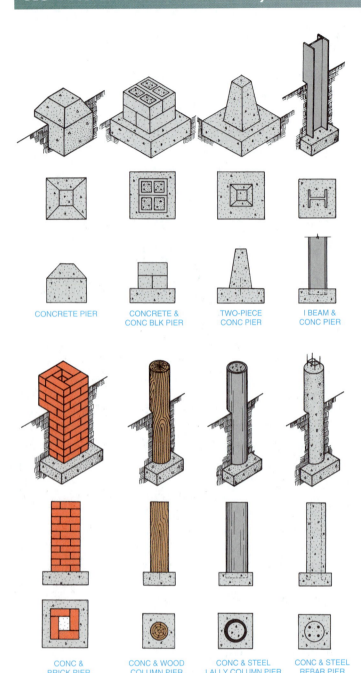

CONCRETE PIER   CONCRETE & CONC BLK PIER   TWO-PIECE CONC PIER   I BEAM & CONC PIER

CONC & BRICK PIER   CONC & WOOD COLUMN PIER   CONC & STEEL LALLY COLUMN PIER   CONC & STEEL REBAR PIER

FIGURE 23.28 ■ Materials used for piers and columns.

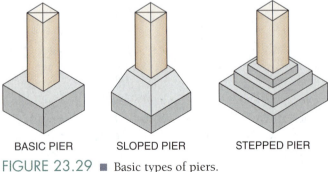

BASIC PIER   SLOPED PIER   STEPPED PIER

FIGURE 23.29 ■ Basic types of piers.

GIRDER

POST

PIER

FIGURE 23.30 ■ Posts transmit weight to footings.

Piles

When supports are driven into supporting soil or bedrock, without a separate footing, they are known as piles. Masonry, wood, steel, and concrete are used for piles, as shown in Figure 23.36. Piles are used to support large structures. They are driven deep into the soil to support structures on sites where stable soil conditions do not exist near the surface.

Piles support building loads in several ways: by friction with the soil, with self-contained footings as shown in Figure 23.37, or through contact with bedrock. Although bedrock may support the compression load of a building, sufficient soil bearing capacity and/or horizontal ties must be used to prevent piles from drifting out of position. Figure 23.38 shows the use of horizontal ties to prevent the lateral drifting of piles.

Permanent Wood Foundations

Permanent wood foundations may be constructed similarly to wood frame walls in other parts of a building. See Figure 23.39. There is one important difference: the plywood and lumber components of permanent wood foundation walls are pressure treated with wood preservatives that become chemically bonded in the wood. This process permanently protects the foundation from fungi, termites, and other causes of decay. The lumber species used must be highly stress resistant.

Permanent wood foundation walls are engineered to absorb and distribute loads and stresses that frequently crack and split other types of foundations. Another advantage of permanent wood foundation design is that it prevents the types of moisture problems that typically plague conventional basements. The design incorporates

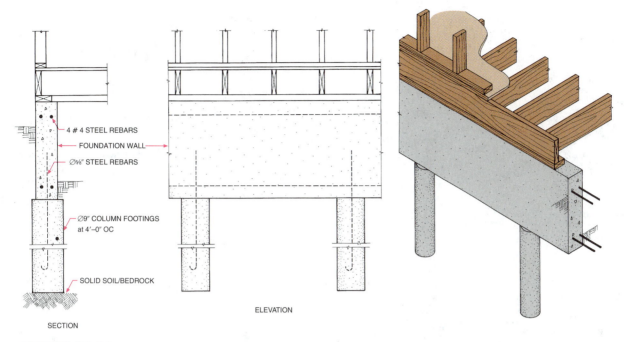

FIGURE 23.31 ■ Column footings.

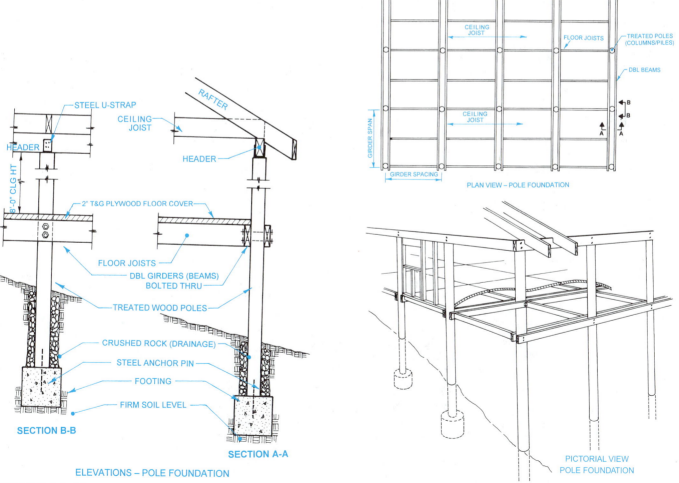

FIGURE 23.32 ■ Pole/column construction.

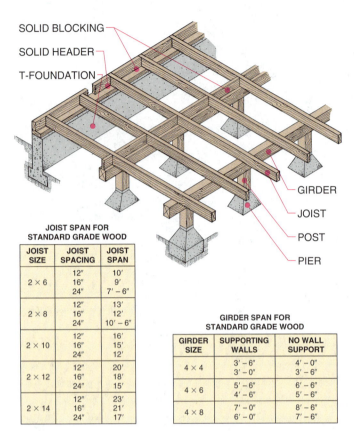

SOLID BLOCKING
SOLID HEADER
T-FOUNDATION

GIRDER
JOIST
POST
PIER

JOIST SIZE	JOIST SPACING	JOIST SPAN
2 × 6	12" 16" 24"	10' 9' 7' – 6"
2 × 8	12" 16" 24"	13' 12' 10' – 6"
2 × 10	12" 16" 24"	16' 15' 12'
2 × 12	12" 16" 24"	20' 18' 15'
2 × 14	12" 16" 24"	23' 21' 17'

JOIST SPAN FOR STANDARD GRADE WOOD

GIRDER SPAN FOR STANDARD GRADE WOOD

GIRDER SIZE	SUPPORTING WALLS	NO WALL SUPPORT
4 × 4	3' – 6" 3' – 0"	4' – 0" 3' – 6"
4 × 6	5' – 6" 4' – 6"	6' – 6" 5' – 6"
4 × 8	7' – 0" 6' – 0"	8' – 6" 7' – 6"

FIGURE 23.33 ■ Piers and posts used for intermediate support.

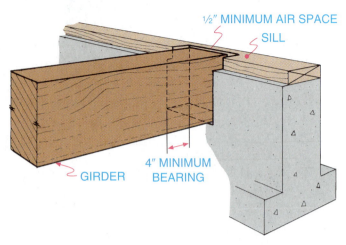

½" MINIMUM AIR SPACE
SILL
GIRDER
4" MINIMUM BEARING

FIGURE 23.34 ■ Girder pocket intersection.

FIREPLACE CONSTRUCTION

The construction of fireplaces is directly related to foundation design because structural support is essential for safe fireplace and chimney construction.

Today, fireplaces are rarely constructed completely on site. Most fireplaces consist of manufactured components around which framing or masonry walls are placed. Fireplace components are designed to produce fire and to provide for safety, convenience, and efficiency. Figure 23.42 reveals the components of a typical fireplace foundation and chimney structure.

moisture deflection and diversion features, such as vapor barriers, horizontally along the ground below the floor and vertically along the outside of the foundation. See Figure 23.40. Figure 23.41 shows a pictorial drawing of a wood foundation system.

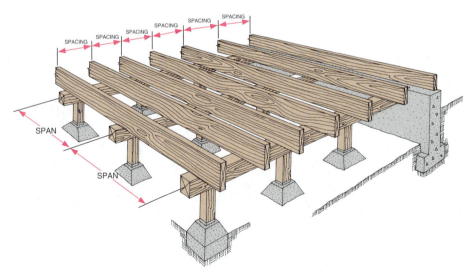

SPACING
SPAN

FIGURE 23.35 ■ Joists resting on girders.

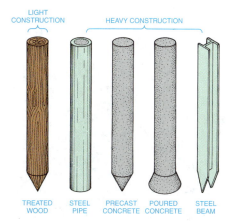

FIGURE 23.36 ■ Pile shapes and materials.

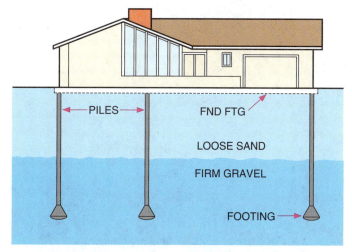

FIGURE 23.37 ■ Deep piles used in loose upper level soil.

FIGURE 23.38 ■ Horizontal ties keep piles plumb.

Fire-Producing Components

A firebox to support the fire, a damper to control air flow, and a flue system to exhaust fumes are necessary to create and maintain combustion in most fireplaces. A **firebox** or

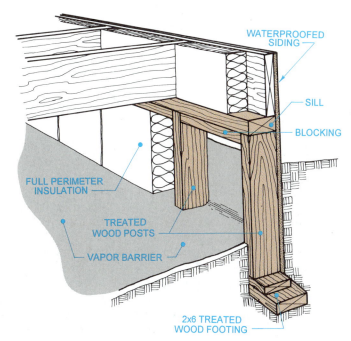

FIGURE 23.39 ■ Permanent wood foundation members.

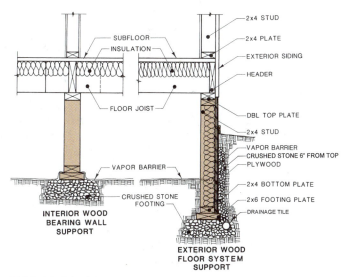

FIGURE 23.40 ■ Section of a wood foundation wall and posts.

fire chamber, contains and supports a fire while reflecting heat and exhausting smoke through a flue system. Fire chambers may be built on site, as shown in Figure 23.43, but most fireboxes are factory built (Figure 23.44) with the adjoining structure, foundation, hearth, and chimney built on site.

The floor of the firebox or fire chamber (also called the internal hearth) and the wall surfaces that are in direct contact with the fire are covered with firebrick, which is laid in fire-repellent mortar known as fireclay.

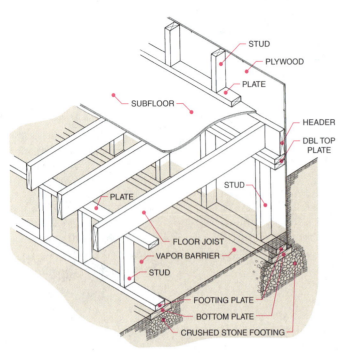

FIGURE 23.41 ■ Wood foundation details.

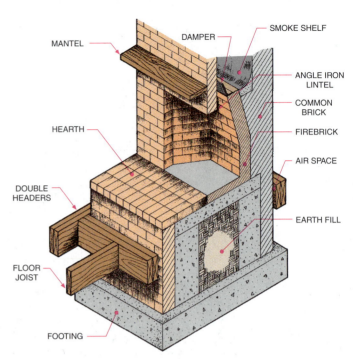

FIGURE 23.43 ■ Site-built firebox.

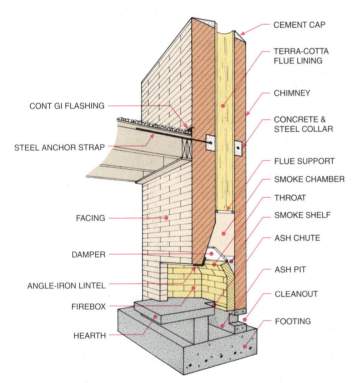

FIGURE 23.42 ■ Fireplace and chimney structure.

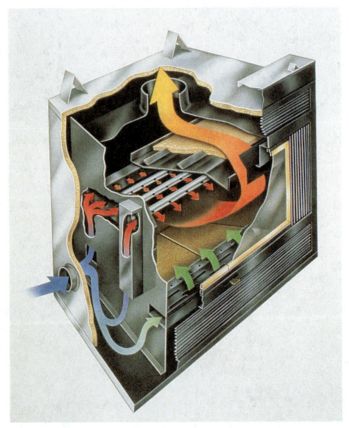

FIGURE 23.44 ■ Factory-built firebox. (*Majestic*)

If the depth of the firebox is excessive, only a small percentage of heat will be reflected into the room. Smoke may escape into the room if the depth is too shallow. The size and proportion of the firebox is critical to the effective operation of a fireplace. Therefore, care must be taken to use the most efficient ratio (usually 1:10) of flue size to firebox opening. These ratios are identified in Figure 23.45A as W, H, D, and T. The chart in Figure 23.45B lists the fireplace dimensions recommended for configuration. In this chart the symbol ⌀" (sq. in.) is used to represent square inches. The symbol ⌀' (sq. ft.) is also used to represent square feet or sq. ft.

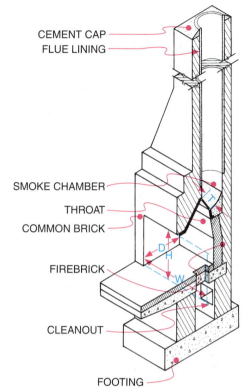

FIGURE 23.45A ■ Locations of ratios used to calculate fireplace dimensions.

The **flue** is the opening in a chimney through which smoke passes. A **damper** is a door that separates the firebox from the flue area. When the fireplace is in use, this door is opened to allow the upward flow of hot air to create a draft that expels smoke and gas from the firebox to the flue. When the fireplace is not in use, the damper is closed to prevent downdrafts from the flue to the firebox. Part of a damper system is a smoke shelf. When the damper is open, it deflects cold downdrafts into the rising warm air currents.

Designing for adequate warm-air rise (called *draw*) is critical for proper fireplace functioning. Inadequate draw, either from an undersized flue or from improper chimney placement, can result in smoke leaking into the room. The cross section area of a flue should be a minimum of one-tenth the area of the fireplace opening. One flue is necessary for each fireplace or furnace, but multiple flues can extend vertically through one chimney if properly offset, as shown in Figure 23.46. The recommended angle for flue offset ranges from 60 percent to a maximum of 45 percent. Often masonry caps or uneven flue projections prevent downdrafts. See Figure 23.47.

Chimneys

Chimneys extend from the footing through the roof of a house. The chimney extends above the roof line to provide a better draft for drawing the smoke and to eliminate the possibility of sparks igniting the roof.

FIREPLACE DIMENSIONS (*inches*)		FIREPLACE WIDTH W	RECTANGULAR FLUES (*inches*)			ROUND FLUES (*inches*)	
			Nominal or Outside Dimension	Inside Dimension	Effective Area	Inside Diameter	Effective Area
W	24 to 84						
H	2/3 to 3/4 W	24	8 1/2 × 8 1/2	7 1/4 × 7 1/4	41⌀"	8	50.3⌀"
D	1/2 to ⎱ 16 to 24 (Rec) for Coal 2/3 H ⎰ 18 to 24 (Rec) for Wood	30 to 34	8 1/2 × 13	7 × 11 1/2	70⌀"	10	78.54⌀"
FLUE (effective area)	1/8 WH for unlined flue 1/10 WH for rectangular lining 1/12 WH for circular lining	36 to 44 46 to 56	13 × 13 13 × 18	11 1/4 × 11 1/4 11 1/4 × 6 1/4	99⌀" 156⌀"	12 15	113.0⌀" 176.7⌀"
T (area)	5/4 to 3/2 flue area	58 to 68	18 × 18	15 3/4 × 5 3/4	195⌀"	18	254.4⌀"
T (width)	3" minimum to 4 1/2" minimum	70 to 84	20 × 24	17 × 21	278⌀"	22	380.13⌀"

FIGURE 23.45B ■ Key dimensions of efficient fireplaces.

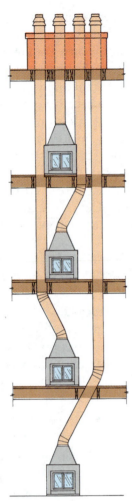

FIGURE 23.46 ■ Positioning of multiple flues in one chimney.

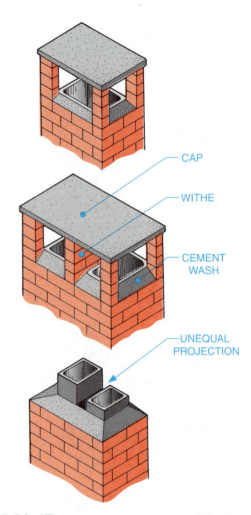

FIGURE 23.47 ■ Chimney tops designed for downdraft control.

Spark arrestors are required by many building codes. See Figure 23.48.

The minimum required height of a chimney above the roof line varies somewhat among local building codes. In most areas the minimum distance is 2′ (610 mm) if the chimney is closer than 15′ to the nearest ridge. Saddles, as seen in Figure 23.49, may be required to stabilize an excessively high and narrow chimney and to divert water. The fireplace footing must also be designed to support the weight of the entire fireplace and chimney.

Safety Components

Some components are not necessary for fire maintenance, but are necessary to prevent flames from spreading outside the firebox. These are covered specifically in building codes.

■ **Hearth.** Most codes require an extended hearth to cover a floor area 16″ beyond the firebox face and

6″ to each side of the firebox. Hearths may be elevated or flush with the floor level.

■ **Materials.** Noncombustible materials must be used in fireplace components that will either be in contact with flames or be excessively heated during operation. Firebrick and fire mortar must be used in the firebox. Generally all masonry materials such as concrete, brick, stone, tile, or marble are safe for use outside the firebox. Wood products are not acceptable. Flues must be made of heat-reflecting materials such as terra-cotta, or a 3″ minimum air space must be provided between metal flues and wood framework. An airspace or fire-resistant wallboard or insulating material must be placed between any wood members and a firebox.

■ **Structural support.** During the structural design phase, provisions must be made so that the heavy masonry loads of the fireplace and chimney assemblies rest on a solid footing. A solid reinforced masonry fireplace and chimney may require a footing depth of 12″ to 24″, depending on the size of the chimney structure. Footings should extend at least 6″ beyond the fireplace outline on all sides. Wood-burning freestanding wood stoves do not require foundations.

■ **Safety screens.** To prevent sparks from projecting outside the firebox, a glass or wire safety screen should cover the fireplace opening.

Convenience Components

Some items are not necessary for fire production or safety but do add to the convenience and efficiency of using a fireplace.

■ **Glass enclosures.** Glass enclosures allow the flames to be seen while preventing smoke or sparks from escaping. When glass enclosures are closed, most heat from the fire is transmitted to the room through vents in the firebox. Built in glass screens block heat from entering the room, but also stop warm air from escaping as the fire diminishes.

■ **Ash pit.** Except for fireplaces on a slab construction, a metal trap door can be placed on the inner hearth. Cold ashes are then dumped through this door to a metal container below. This container has a cleanout door for the removal of ashes.

■ **Blowers and Remote Outlets.** To project heat into a room beyond normal air movement, blowers can be used. They may direct air into the room or to remote outlets. See Figure 23.50.

■ **Air intake ducts.** To add more oxygen to a firebox, a cold air return or an air duct connected directly from the outside can be added. See Figure 23.51. Some systems use blowers to accelerate air flow through the duct. The use of an outside duct is especially helpful in tightly insulated houses.

■ **Freestanding fireplaces.** Freestanding metal fireplaces constructed of heavy-gauge steel are available in a variety of shapes, as shown in Chapter 7. They are relatively light wood-burning stoves and therefore need no concrete foundation for support. A stovepipe leading into the chimney provides the exhaust flue. Because metal units reflect more heat than masonry, metal fireplaces are much more heat efficient, especially if centrally located. For safety, fire-resistant materials such as concrete, brick, stone, or tile must be used beneath and around these fireplaces.

FIGURE 23.48 ■ Spark arrestor and enclosure prevents downdrafts.

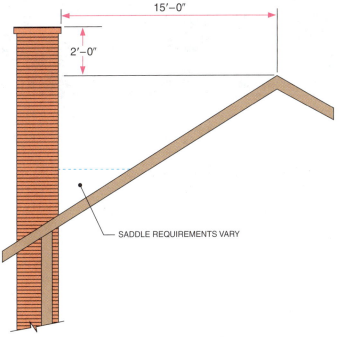

FIGURE 23.49 ■ Common chimney code requirements.

concrete footing is most often used for residential construction. Extra footing depth must be provided under the fireplace area. Wood-framed fireplaces with manufactured fireboxes and sheet-metal flues may require a shallow footing of 6″. A

FIGURE 23.50 ■ Remote outlets used to distribute heat. (*Olympic Fireplaces*)

FIGURE 23.51 ■ Air duct installed to bring oxygen to the firebox.

FOUNDATION DRAWINGS

Before learning to draw foundation plans, an understanding of foundation layout and excavation is necessary.

Layout

Math Connection

Establishing the exact position of a building on a lot requires locating each building corner. This is done by measuring the distance from property lines to building corners on the plot plan.

If the position of only one building corner is shown on the plot plan, all others can be plotted by turning angles with a transit. Right angles (90°) can be plotted by using the 3.4.5 unit method. See Figure 23.52. The Pythagorean theorem may be used to determine if the measurements are correct.

Pythagorean Theorem:

Square of the hypotenuse of a right triangle = the sum of the square of the two sides:

$$c^2 = a^2 + b^2$$

Example:

$$c^2 = a^2 + b^2$$
$$\sqrt{c^2} = \sqrt{a^2 + b^2}$$
$$c = \sqrt{32^2 + 24^2}$$
$$c = \sqrt{1{,}024 + 576}$$
$$c = \sqrt{1{,}600}$$
$$c = 40''$$

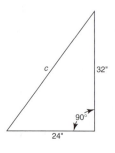

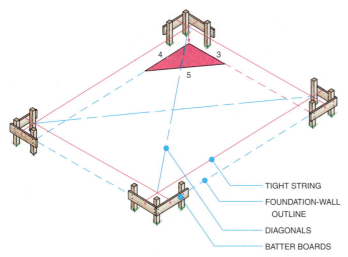

FIGURE 23.52 ■ Building layout using the 3.4.5 unit method.

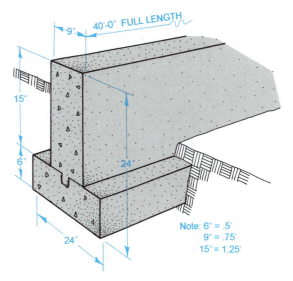

FIGURE 23.53 ■ Dimensions used to compute foundation and footing volume.

If only one point is dimensioned, the azimuth (compass heading) of one side needs to be known to accurately lay out the building. (You may wish to refer back to Chapter 18.) For rectangular areas the accuracy of the layout can be checked by measuring across the diagonals. The diagonal distances will be equal if the measurements are true. Once each corner is located, string attached to batter boards is used to identify the corners during excavation, as shown in Figure 23.52.

Excavations and Forming

Foundation plans should clearly show whether areas of a foundation are to be completely excavated for a basement, partly excavated for a crawl space, or unexcavated. The depth of the excavation should also be indicated on the elevation drawings. If a basement is planned, the entire excavation for the basement is dug before the footings are poured. If there is to be no basement, a trench excavation is made. To calculate the volume of an excavation or the amount of concrete needed, use the following formula:

$$\text{Formula: cu. yds.} = \frac{W' \times L' \times D'}{27}$$

Example (see Figure 23.53):

(T-foundation wall)

$$\text{cu. yds.} = \frac{.75' \times 40' \times 1.25'}{27} = \frac{37.5}{27} = 1.39 \text{ cu. yd.}$$

(footing)

$$\text{cu. yds.} = \frac{2' \times 40' \times .5'}{27} = \frac{40}{27} = 1.49 \text{ cu. yd.}$$

Total $= 2.88$ cu. yd.

Although slab foundations may only require trench footings, all organic soil material must be removed and replaced with nonorganic, compactable soil. Excavation must always be made at least 6″ below the frost line. All HVAC, electrical, or plumbing lines which are to be embedded in or pass through a poured area must be installed when the form for the foundation is constructed.

Some foundation plans include the position of plumbing lines if lines are to pass under or through a poured area. Usually, though, the position of plumbing lines is read from wall and fixture positions on either a floor plan or plumbing plan.

Foundation Plans

Foundation plan drawings are floor plans drawn at the foundation level. The same plan symbols, such as those for brick and concrete, show construction materials. Positions of footings and piers under the soil line are shown with hidden lines on a foundation plan.

T-Foundation Drawings

Figure 23.54 shows a T-foundation drawing. The sequence for drawing T-foundation or basement plans is similar to that for drawing floor plans. The outside perimeter line should be drawn first, then the interior partitions. Next, draw the parallel wall thickness lines, footing lines, and material symbols. Add the position of any piers, columns, footings, and beams. Include a floor joist directional label with dimensions. Figure 23.55 shows how different types of T-foundation walls and footing are drawn as plans and elevation sections. More detailed drawings, as shown

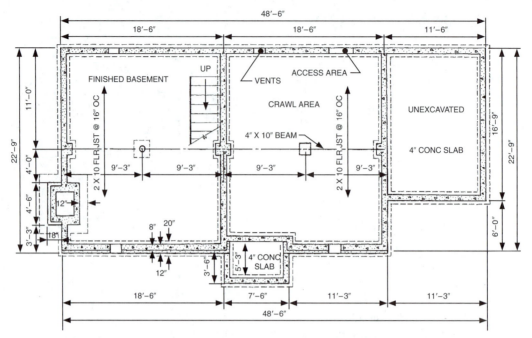

FIGURE 23.54 ■ T-foundation plan.

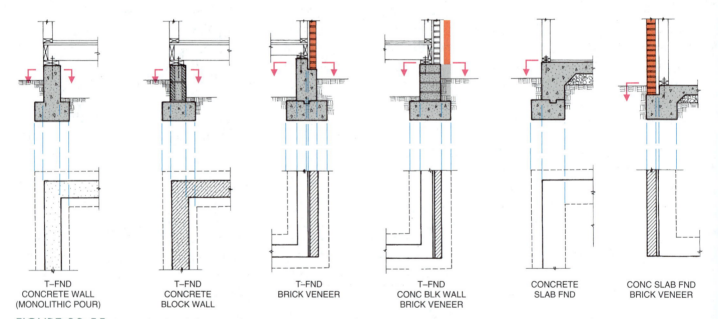

FIGURE 23.55 ■ Methods of drawing foundation walls and footings details.

in Figure 23.56, may include added dimensions and related components.

Slab Foundation Drawings

Slab drawings show only the outer perimeter line with hidden lines to represent footings. Slab drawings must note the thickness of the slab, size and spacing of reinforcing bars and wire mesh, concrete PSI (pounds per square inch), and thickness of waterproof membrane.

Footings for slab foundations are usually poured as part of the slab. The design of a monolithic (one-piece) pour and the location of rebars is indexed from the plan to detail drawings as shown in Figure 23.57.

Detail Drawings

Foundation plan views only show the location of all features on a horizontal plane. Most foundation details require elevation sections to convey vertical positions and

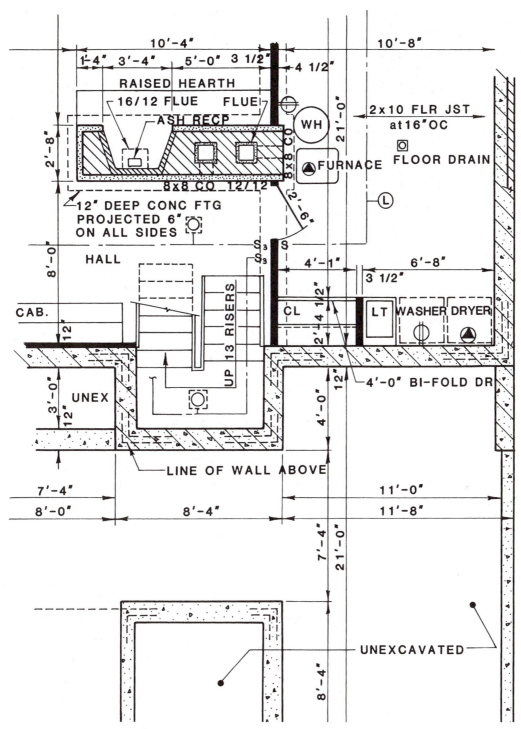

FIGURE 23.56 ■ Part of a detailed foundation plan.

dimensions. The most common types of foundation details are footing and sill details.

Footing detail drawings show a vertical section through the footing, foundation wall, or slab. These drawings show the size, shape, and material used in footings. See Figure 23.58. If many different footing types are used under partitions, an entire profile of the foundation may be drawn with the location of all footings, as shown in Figure 23.59.

Sill detail drawings show how the foundation, exterior wall, and floor system intersect. Sill details are viewed in elevation, pictorial, or plan sections. See Figure 23.60.

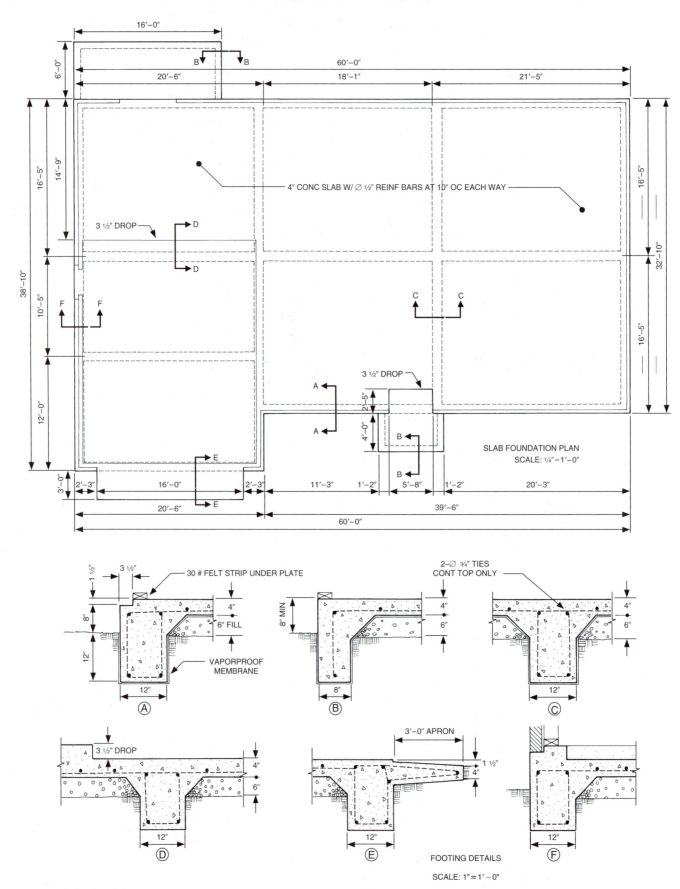

FIGURE 23.57 ■ Slab foundation plan and details. (*Vardy Vincent*)

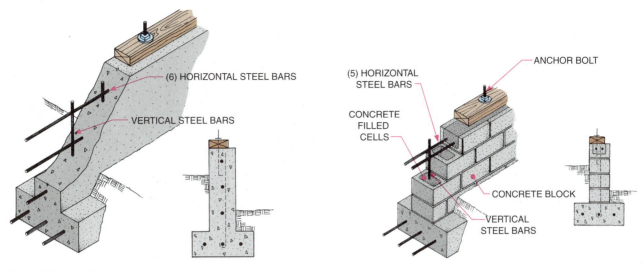

(6) HORIZONTAL STEEL BARS

VERTICAL STEEL BARS

(5) HORIZONTAL STEEL BARS

ANCHOR BOLT

CONCRETE FILLED CELLS

CONCRETE BLOCK

VERTICAL STEEL BARS

FIGURE 23.58 ■ Footing detail drawing.

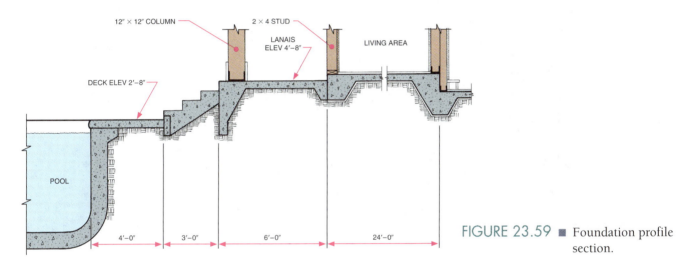

12" × 12" COLUMN

2 × 4 STUD

LANAIS ELEV 4'-8"

LIVING AREA

DECK ELEV 2'-8"

POOL

4'-0"   3'-0"   6'-0"   24'-0"

FIGURE 23.59 ■ Foundation profile section.

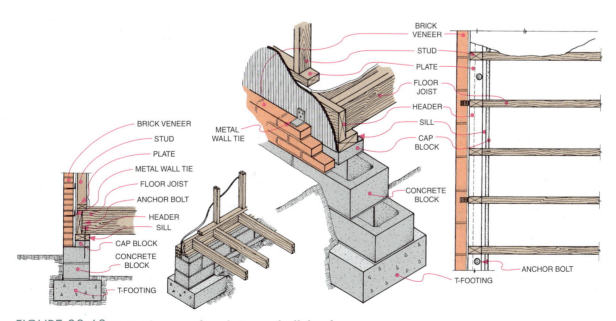

BRICK VENEER

STUD

PLATE

FLOOR JOIST

HEADER

SILL

CAP BLOCK

CONCRETE BLOCK

ANCHOR BOLT

T-FOOTING

BRICK VENEER

STUD

PLATE

METAL WALL TIE

FLOOR JOIST

ANCHOR BOLT

HEADER

SILL

CAP BLOCK

CONCRETE BLOCK

T-FOOTING

METAL WALL TIE

FIGURE 23.60 ■ Brick veneer foundation and sill details.

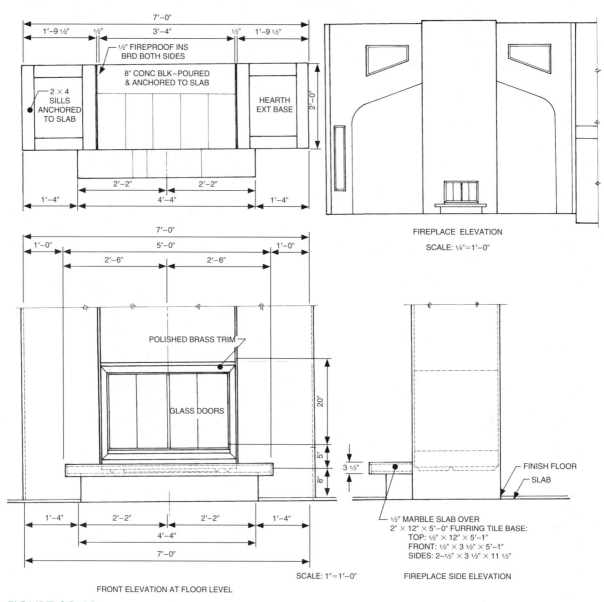

FIGURE 23.61 ■ Detailed design of the fireplace shown in Figure 23.51 and in Chapter 7.

Fireplace Drawings

A horizontal section through a fireplace firebox is drawn on floor plan drawings. The outline of the fireplace opening and the chimney design are shown on interior elevation drawings. However, floor plan and elevation scales are not large enough to show the amount of detail needed to build a fireplace system, so enlarged sections are used to show construction details for masonry fireplaces. For fireplaces that include manufactured components in framed walls, enlarged orthographic views and/or sectional framing drawings are necessary to show construction details. See Figure 23.61.

If the entire fireplace and chimney structure is to be built on site, a complete set of plans, elevations, and sectional drawings should be prepared, as in Figure 23.61. Figure 23.62 shows working drawings for a fireplace design.

Floor and roof framing drawings related to fireplaces and chimney construction are covered in more detail in Chapters 28 and 30.

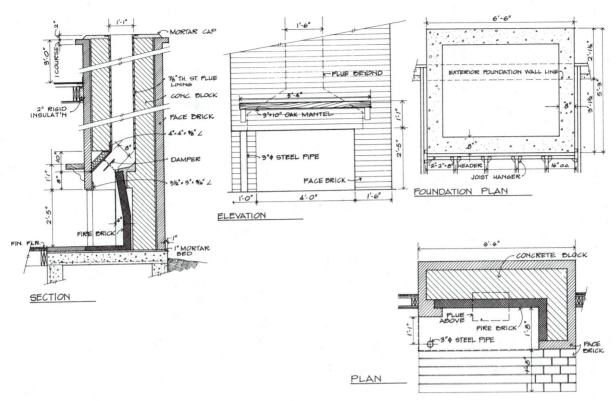

FIGURE 23.62 ■ Fireplace plan, elevation, and section details.

USING CAD

Foundation Drawing on CAD

Because foundation drawings align with the floor above, floor plans are used to establish the basic lines of the foundation plan. This is done by drawing the foundation plan on a separate layer over the floor plan layer, which can be shown with lighter lines or color. The floor plan layer can be periodically removed to check progress. The use of layers in this way also enables different drafters to accurately draw different layers for a complex design and then coordinate their work with ease.

CHAPTER 23

Foundations and Fireplace Structures Exercises

1. Draw the foundation plan for the house you are designing.

2. Draw or sketch a foundation plan for the design shown in Figure 14.22, using the scale 1/4" = 1'-0" for a T-foundation.

3. Draw or sketch a foundation plan for the design shown in Figure 14.22, using the scale 1/4″ = 1′-0″ for a slab foundation.

4. Draw or sketch a foundation plan for the design shown in Figure 14.18, using the scale 1/4″ = 1′-0″ for a pier foundation.

5. Draw a plan and elevation view of the fireplace in Figure 23.43, using the scale 1/2″ = 1′-0″. See Figure 23.45B for typical dimensions.

 6. Design a fireplace for the house you are designing.

 7. Draw a sill and footing detail for the house you are designing.

8. Design a T-foundation for a rectangular 25′ × 30′ cabin with a 48″ crawl space. Use a 24″ cripple wall for the top of the T-foundation's wall.

9. Draw the construction details for Exercise 8.

10. Design a foundation for a building site with a consistent 30-degree slope.

11. Refer to Figure 23.63; draw and dimension sections A-A, B-B, and C-C. Show a 5″ × 48″ × 76″ continuous concrete slab with a 24′ × 48′ section elevated 6″ above the 48′ × 52′ section.

    Perimeter footing to be 18″ deep and 12″ wide. Interior footing to be 18″ deep and 10″ wide. Include symbols and locations of rebars, welded wire mesh, base fill, insulation, and drain.

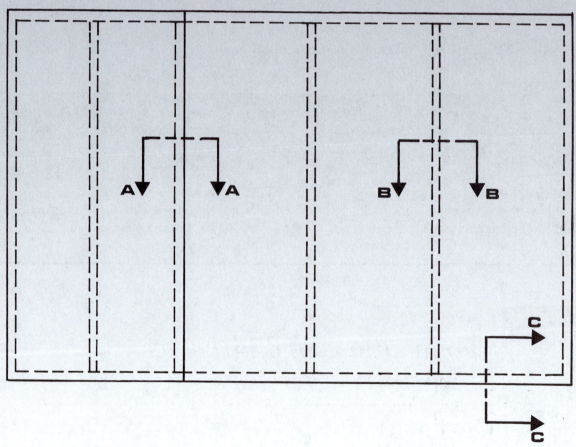

FIGURE 23.63 ■ Design and draw sections A-A, B-B, and C-C for the locations described by the cutting plane lines.

Wood-Frame Systems

OBJECTIVES

In this chapter, you will learn to:

- differentiate between skeleton-frame and post-and-beam construction.
- identify major characteristics of lumber, plywood, and structural timber.
- calculate the number of board feet in a piece of lumber.

TERMS

balloon framing
board foot
hardwoods
laminated timber

plank-and-beam construction
platform framing
plywood

post-and-beam construction
skeleton-frame construction
softwoods

INTRODUCTION

In wood-frame construction, wood structural members are joined to make an open framework for the structure. This open framework is then covered with layers of other construction materials to form the solid surfaces of floors, walls, and roofs.

There are several varieties of wood-frame construction. This chapter covers skeleton-frame construction and post-and-beam construction.

 ## SKELETON-FRAME CONSTRUCTION

In **skeleton-frame construction** (Figure 24.1), small structural members are joined in such a way that they share the loads of the structure. When the structural members are covered, they form complete walls, floors, and roofs. Because of the limited size of wood materials, the skeleton-frame method is considered light construction. See Figure 24.2.

Materials

Light construction materials include lumber, plywood, reconstituted wood, plastics, fasteners, and multimaterial components such as doors, windows, cabinets, plumbing, ductwork, and electrical fixtures.

FIGURE 24.1 ■ Skeleton-frame building under construction.

Lumber

Construction lumber is classified according to grade, species, size, and whether it has been treated. Lumber grades are determined by the number and location of defects, such as knots, checks, and splits, and by the degree of warp (deviation from a flat, even surface). For grading

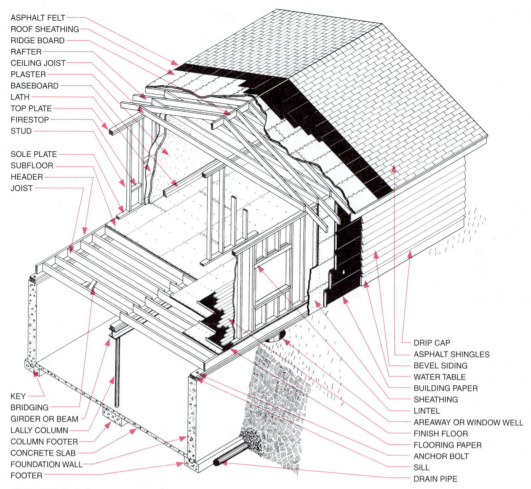

ASPHALT FELT
ROOF SHEATHING
RIDGE BOARD
RAFTER
CEILING JOIST
PLASTER
BASEBOARD
LATH
TOP PLATE
FIRESTOP
STUD

SOLE PLATE
SUBFLOOR
HEADER
JOIST

KEY
BRIDGING
GIRDER OR BEAM
LALLY COLUMN
COLUMN FOOTER
CONCRETE SLAB
FOUNDATION WALL
FOOTER

DRIP CAP
ASPHALT SHINGLES
BEVEL SIDING
WATER TABLE
BUILDING PAPER
SHEATHING
LINTEL
AREAWAY OR WINDOW WELL
FINISH FLOOR
FLOORING PAPER
ANCHOR BOLT
SILL
DRAIN PIPE

FIGURE 24.2 ■ Skeleton-frame building major components.

purposes, lumber is divided into two broad categories: hardwoods and softwoods. Small, young trees up to 12" in diameter are cut into 2 × 4 and 2 × 6 framing lumber. Large, older growth logs are cut to produce a wide range of lumber, from large beams to siding boards.

Hardwoods are used for surfaces that must withstand much wear, such as flooring and railings. Hardwoods are also used to make items that require a fine natural finish, such as cabinets and furniture. Hardwoods come from broad-leaved trees. The species most commonly used in construction include oak, walnut, birch, cherry, mahogany, and maple. Hardwood is graded from highest quality to lowest quality depending on the amount of usable material in each piece. See Figure 24.3. Hardwood lumber is available in lengths of 4' to 16', widths up to 12", and thicknesses up to 2".

Softwoods come from coniferous (needle-bearing) trees such as Douglas fir, pine, or cedar. They are used for structural members such as joists, rafters, studs, sheath-

WOOD GRADE	QUALITY LEVEL
FAS firsts	High Quality
FAS seconds	
Select	
Common #1	
Common #2	
Common #3A	Poor Quality
Common #3B	

FIGURE 24.3 ■ Hardwood lumber grades.

ing, and formwork. Most skeleton-frame lumber is softwood. Softwood lumber is divided into three grading classes: yard, structural, and factory (shop) lumber. *Yard lumber* is used for most light framing members, such as sheathing, bracing, subfloors, and casings. See Figure 24.4. *Structural lumber,* as the name implies, is used for

GRADE	USE	GRADE	USE
Selects and finish	Graded from the best side. Used for interior and exterior trim, molding, and woodwork where appearance is important.	No. 2 common (WWPA) Construction (WCLIB)	All sound tight knots with some defects, such as stains, streaks, and patches of pitch, checks, and splits. Used as paneling and shelving, subfloors, and sheathing.
B & BTR	Used where appearance is the major factor. Many pieces clear, but minor appearance defects allowed which do not detract from appearance.		
C Select	Used for all types of interior woodwork. Appearance and usability slightly less than B & BTR.	No. 3 common (WWPA) Standard (WCLIB)	Some unsound knots and other defects. Used for rough sheathing, shelving, fences, boxes, and crating.
D Select	Used where finishing requirements are less demanding. Many pieces have finish appearance on one side with larger defects on back.	No. 4 common (WWPA) Utility (WCLIB)	Loose knots and knotholes, up to 4" wide. Used for general construction purposes, such as sheathing, bracing, low-cost fencing, and crating.
Boards	Lumber with defects that detract from appearance but suitable for general construction.	No. 5 common (WWPA) Economy (WCLIB)	Large knots or holes, unsound wood, massed pitch, splits, and other defects. Used for low-grade sheathing, bracing, and temporary construction. Pieces of higher grade wood may be obtained by crosscutting or ripping boards without defects.
No. 1 common (WWPA) Select merchantable (WCLIB)	All sound tight knots, with use determined by size and placement of knots. Used for exposed interior and exterior locations where knots are not objectionable.		

FIGURE 24.4 ■ Yard lumber grades.

GRADE	USE
LF (Light Framing)	Used in thicknesses from 2" to 4" and widths from 3" and 4", for studs, joists, and rafters in light framing.
JP (Joints and Planks)	Used in thicknesses from 2" to 4" and widths over 2", for joists and rafters to be loaded on either side, or for planking when laid flat.
B&S (Beams and Stringers)	Used in thicknesses from 2" to 4". Widths over 2" must be loaded on narrow edge.
P&T (Posts and Timbers)	Used for posts or columns 5" × 5" and larger or where bending resistance is not critical.

FIGURE 24.5 ■ Structural lumber grades.

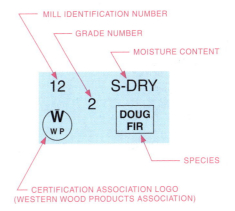

FIGURE 24.6 ■ Lumber grade marks.

load-bearing members and is classified according to use and grades. Grades are based on a lumber's stress resistance. See Figure 24.5. *Factory*, or *shop, lumber* consists of light members which are finished at a mill and used for trim, molding, and door and window sashes.

All lumber is graded by an authority at a lumber mill according to the American Lumber Standards. These standards ensure that lumber is grade marked with a variety of appropriate, accurate information. Lumber is labeled with a mill identification number and by a certification association logo. Other identifying information includes the grade number, moisture content, and species classification, as shown in Figure 24.6.

Structural lumber is defined as either rough or finished. Rough sizes represent the width and thickness of a piece of lumber as cut from a log. Rough lumber is also called *nominal size* lumber. Finished lumber sizes represent the actual dimensions of a member after final surfacing. See Figure 24.7. Finished lumber is also known as surfaced, *dressed,* dimensional, or actual size lumber.

When a drawing callout reads 2 × 4, builders know that the rough size is 2″ × 4″ but the actual size of the member is 1 1/2″ × 3 1/2″. Figure 24.8 shows the U.S. customary range of rough (nominal) and finished (actual) standard lumber sizes. Figure 24.9 shows the range of metric lumber sizes. Because lumber is not always surfaced on all sides, symbols (or codes) desig-

nate the number of sides that are surfaced. See Figure 24.10. Figure 24.11 shows the nominal thickness of lumber sizes from 1″ to 16″ and the resulting (dressed) thickness of each.

Plywood

Solid lumber is limited in width and has a tendency to warp, split, and check. **Plywood** can be made in wide sheets and is structurally stable. Plywood is manufactured from thin sheets (0.10″ to 1.25″) of wood laminated together with an adhesive under high pressure. The number of layers (plies) varies from three to seven. The grain of each ply is laid perpendicular to the grain of each adjacent layer. The grain of both outside sheets always faces the same direction. This layering process greatly reduces the tendency of plywood to warp, check, split, splinter, and shrink. Plywood is available in individual 4′ × 8′ sheets or in continuous panels up to 50′ in length. It is made in thicknesses of 1/8″, 3/16″, 1/4″, 5/16″, 1/2″, 5/8″, 3/4″, 1″, 1 1/8″, and 1 1/4″.

All plywood is divided into two broad categories: exterior (waterproof) or interior. Plywood for structural use is made with surfaces of softwood. Plywood for cabinets and furniture has hardwood surfaces. Because of the different uses, construction-grade plywood (softwood) and veneer-grade plywood (hardwood) are identified by two different quality-rating systems.

Construction-grade plywood is unsanded. It is identified by grade levels based on structural strength, as shown in Figure 24.12. Because of live-load differences, plywood panels used for structural purposes are marked

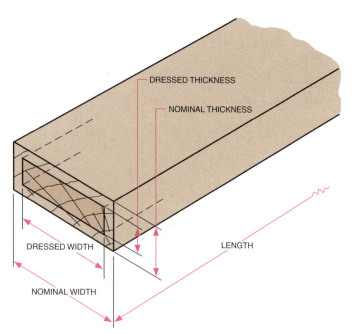

FIGURE 24.7 ■ Nominal and finished (dressed) lumber sizes.

LUMBER SIZES IN INCHES

Nominal Size	2 × 4	2 × 6	2 × 8	2 × 10	2 × 12	4 × 6	4 × 8	4 × 10	6 × 6	6 × 8	6 × 10	8 × 8	8 × 10
Dressed Size	1½ × 3½	1½ × 5½	1½ × 7½	1½ × 9½	1½ × 11½	3⁹⁄₁₆ × 5½	3⁹⁄₁₆ × 7½	3⁹⁄₁₆ × 9½	5½ × 5½	5½ × 7½	5½ × 9½	7½ × 7½	7½ × 9½

BOARD SIZES IN INCHES

	1 × 4	1 × 6	1 × 8	1 × 10	1 × 12
Nominal Size	1 × 4	1 × 6	1 × 8	1 × 10	1 × 12
Actual Size—Common	¾ × 3⁹⁄₁₆	¾ × 5⁹⁄₁₆	¾ × 7½	¾ × 9½	¾ × 11½
Actual Size—Shiplap	¾ × 3	¾ × 4¹⁵⁄₁₆	¾ × 6⁷⁄₈	¾ × 8⁷⁄₈	¾ × 10⁷⁄₈
Actual Size—T&G	¾ × 3¼	¾ × 5³⁄₁₆	¾ × 7⅛	¾ × 9⅛	¾ × 11⅛

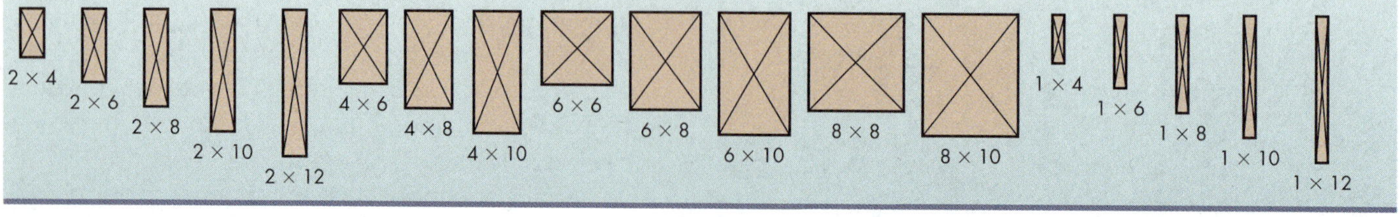

2 × 4   2 × 6   2 × 8   2 × 10   2 × 12   4 × 6   4 × 8   4 × 10   6 × 6   6 × 8   6 × 10   8 × 8   8 × 10   1 × 4   1 × 6   1 × 8   1 × 10   1 × 12

FIGURE 24.8 ■ U.S. customary lumber sizes.

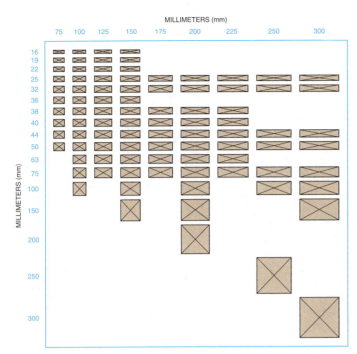

FIGURE 24.9 ■ Metric lumber sizes.

DESIGNATION	DESCRIPTION
S1S	Surfaced on one side
S2S	Surfaced on two sides
S1E	Surfaced on one edge
S2E	Surfaced on two edges
S1S1E	Surfaced on one side, one edge
S2S1E	Surfaced on two sides, one edge
S4S	Surfaced on four sides

FIGURE 24.10 ■ Lumber surfacing codes.

NOMINAL THICKNESS	DRESSED THICKNESS
1″	3/4″
2″	1 1/2″
3″	2 1/2″
4″	3 1/2″
5″	4 1/2″
6″	5 1/2″
8″	7 1/4″
10″	9 1/4″
12″	11 1/4″
14″	13 1/4″
16″	15 1/4″

FIGURE 24.11 ■ Standard lumber thicknesses.

GRADE	DESCRIPTION
Standard	For use as subflooring, roof sheathing, wall sheathing, and structural interior applications.
Structural Class I and II	For uses requiring resistance to tension, compression, and shear stress including box beams, stressed skin panels, and engineered diaphragms. High nail-holding quality and controlled grade and glue bonds.
CC Exterior	Meets all exterior plywood requirements.
BB Concrete-Form Panels, Class I and II	Edges sealed and oiled at the mill and used for concrete form panels.

FIGURE 24.12 ■ Plywood construction grades.

with two numbers indicating the structural rating. See Figure 24.13. The first number represents the maximum span (in inches) possible between supporting roof members. The second number represents the maximum allowable span when used for flooring.

Because hardwood plywood is used for making cabinets and furniture, veneer plywood grades (groups) are classified by a letter indicating the number of knots, checks, stains, and open sections in each panel, as shown in Figure 24.14. These letters are used to show only the group of the front and back plies (layers). When front and back plies are of different grades, letters are combined. For example, a grade of B-D means the front ply is B grade and the back ply is D grade. On Figure 24.15, notice the various kinds of information included in the hardwood plywood standard grade

marks stamped on each sheet. Wood species are classified in groups. See Figure 24.16.

When selecting hardwood plywood, several characteristics must be considered. Grain patterns, color and texture consistency, as well as specific species, smoothness, and finishability, should be matched for cabinets, paneling, and furniture.

Wood-Framing Methods

When multiple-level buildings are constructed using a wood skeleton frame, either the **platform framing**

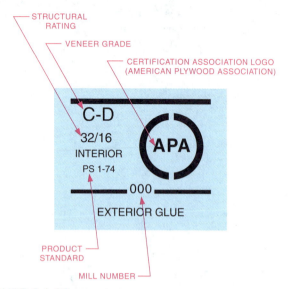

FIGURE 24.13 ■ Plywood construction grade marks.

GRADE	DESCRIPTION
Grade A	Paintable and smooth with no more than 18 neat boat, sled, or router type repairs made parallel with grain. Will accept natural finish.
Grade B	Solid surface with shims, circular repair plugs, or tight knots less than 1" wide permitted.
Grade C	Tight knots of less than 1 1/2", knotholes less than 1" wide, synthetic or wood repairs, limited splits, slices (gouges), discoloration, and sanding defects that do not impair strength permitted.
Grade C Plugged	Some broken grain, synthetic repairs, splits up to 1" wide, knotholes and bareholds (other holes) up to 1 1/4" × 1/2" permitted.
Grade D	Knots and knotholes up to 2 1/2" wide across grain or 3" wide if within limits permitted but restricted to interior use.

FIGURE 24.14 ■ Veneer plywood grades.

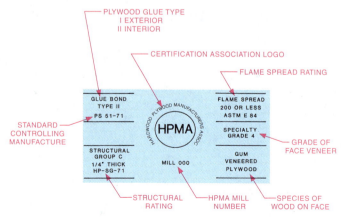

FIGURE 24.15 ■ Typical hardwood plywood grade mark.

GROUP 1	GROUP 2	GROUP 3	GROUP 4	GROUP 5
Beech	Cedar, port	Alder, red	Aspen	Basswood
Birch	Cypress	Birch, paper	Cedar	Poplar, balsam
Sweet	Douglas fir 2	Cedar, alaska	Incense	
Yellow	Fir	Fir, subalpine	Western	
Douglas fir 1	Balsam	Hemlock	red	
Maple, sugar	California red	Maple, bigleaf	Cottonwood	
Pine	White	Pine	Pine, Eastern	
Caribbean	Hemlock	Jack	white	
Ocote	Lauan	Ponderosa	Sugar	
Pine, south	Maple, black	Spruce		
Loblolly	Pine	Redwood		
Longleaf	Red	Spruce		
Shortleaf	Western			
Slash	white			
	Spruce			
	Yellow poplar			

FIGURE 24.16 ■ Wood species groups.

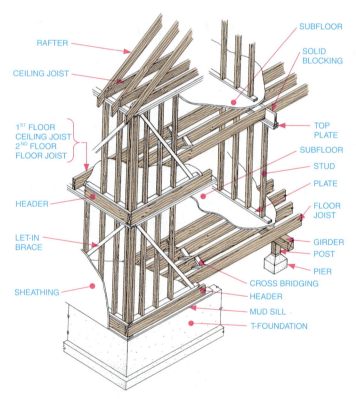

FIGURE 24.17 ■ Platform framing components.

method or the **balloon framing** method is used. In platform framing, the second floor rests directly on first-floor exterior walls. See Figure 24.17. In balloon-framed buildings, the first-floor joists rest directly on a sill plate, and the second-floor joists bear on *ribbon (ledger) strips* set into the studs. The studs are continuous for the full height of the building, and floor joists butt against the sides of the studs. See Figure 24.18.

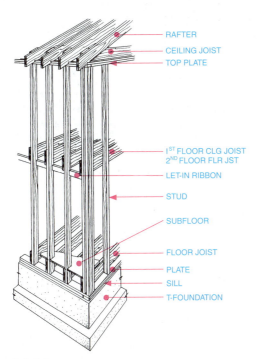

FIGURE 24.18 ■ Balloon framing.

Board-Foot Measure

Math Connection

Lumber is purchased in bulk by the board foot. One **board foot** is $1'' \times 12'' \times 12''$. To determine the number of board feet (BF) in a given piece of lumber, multiply the thickness (in inches) $\times$ width (in inches) $\times$ length (in feet) and divide the result by 12, as shown in Figure 24.19.

POST-AND-BEAM CONSTRUCTION

Large timbers have been used in construction for centuries, usually for floor and roof systems in buildings of bearing wall design. The development of large glass sheets and sheathing materials as well as improvements in manufacturing and transporting large wood members have made new uses possible. Today, heavy timbers may be used for walls as well as for floors and roofs. When heavy timbers are used in this manner, the construction method is called **post-and-beam construction**.

Post-and-beam construction uses larger structural members than skeleton-frame construction, and they are spaced farther apart. See Figure 24.20. The larger spacing means that fewer members are needed. Labor savings are also considerable because fewer members are handled and fewer intersections connected.

Many post-and-beam members remain exposed in the finished building. To create a more pleasing visual effect,

$$\text{Formula: BF} = \frac{T'' \times W'' \times L'}{12}$$

$$\text{Example: BF} = \frac{2'' \times 10'' \times 3'}{12} = 5 \text{ BF}$$

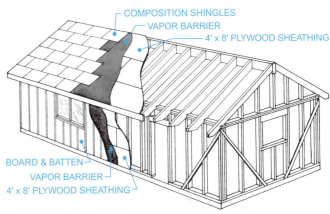

FIGURE 24.19 ■ Board feet measurement calculation example.

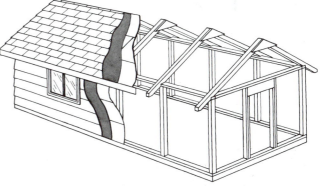

SKELETON-FRAME CONSTRUCTION

POST-AND-BEAM CONSTRUCTION

FIGURE 24.20 ■ Skeleton-frame and post-and-beam comparison.

a better grade of lumber is specified. Rigid insulation must be used above, instead of under, the roof planks to expose the natural plank ceiling. Plumbing and electrical lines must be passed through cavities in columns and/or beams. Bearing partitions and other heavy dead loads, such as bathtubs, must be located over beams, or additional support framing must be used.

Post-and-beam construction relies on the relationship of three basic components: *posts* or *columns, beams,* and *planks.* Vertical columns support horizontal beams. These beams support planks placed perpendicular to the beams, as in Figure 24.21. Floor and roof systems are supported by the beams, which, as stated, are supported by posts or columns, which transfer loads to footings. Member sizes and spacing vary depending on load requirements.

Floor Construction

Timber floor systems use heavy wood planks or T&G plywood placed over widely spaced beams, as shown in Figure 24.22. This type of floor system is called **plank-**

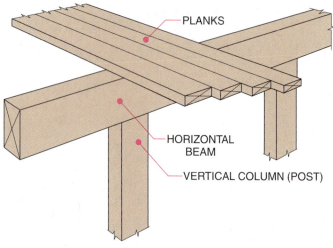

FIGURE 24.21 ■ Components of heavy timber, post-and-beam construction.

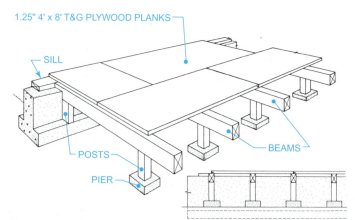

FIGURE 24.22 ■ Plank-and-beam floor construction using heavy plywood.

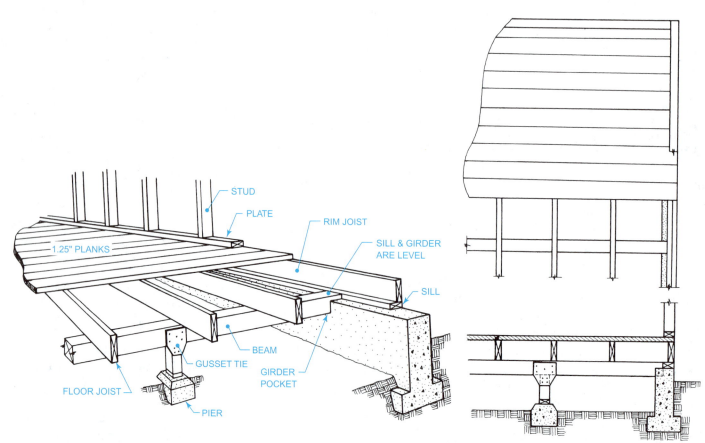

FIGURE 24.23 ■ Plank-and-beam floor construction using heavy lumber planks.

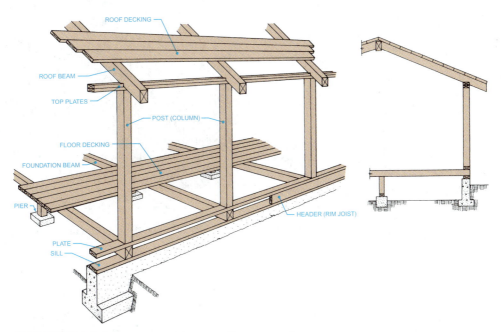

FIGURE 24.24 ■ Post-and-beam wall construction.

and-beam construction. Each evenly spaced beam replaces several of the intermediate joists that would be used in a skeleton-frame floor system. For example, a 24′ distance may require 19 conventional floor joists spaced at 16″ OC (on center). See Figure 24.23. However, only 7 joists placed 4′ OC may be needed to support the same loads in plank-and-beam construction. In this system, floor planks must be strong enough to avoid deflection at the middle or at midspan. In this design, joists support heavy lumber planks and beams support the joists. The beams, in turn, are supported by piers and posts in the center and by the foundation wall on the perimeter.

Wall Construction

Just as beams replace conventional joists in plank-and-beam floor systems, posts replace conventional studs in post-and-beam wall construction. The large open spans between the wall posts can be occupied by nonbearing material or components such as windows, doors, or insulating material. See Figure 24.24. For this reason nonstructural elements in a post-and-beam outside wall are known as *curtain walls*. An example of the construction of wall column intersections is detailed in Figure 24.25.

Roof Construction

In plank-and-beam roofs, beams replace conventional roof rafters and planks replace conventional roof sheath-

ing. There are two types of plank-and-beam roof systems: longitudinal and transverse.

In longitudinal systems, roof beams are aligned parallel with the long axis of the building. See Figure 24.26. In transverse systems, beams are aligned across the short width of the building, as illustrated in Figure 24.27. One end of the beam is supported by a post. The other end may rest on the top of a ridge beam, or it may be butted

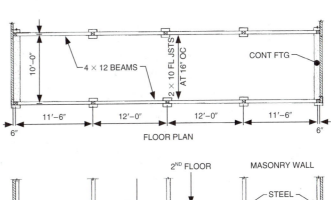

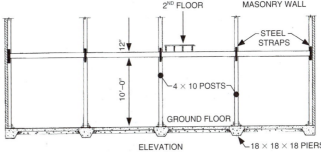

FIGURE 24.25 ■ Post-and-beam wall framing plan and elevation.

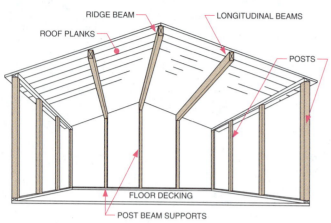

FIGURE 24.26 ■ Longitudinal roof framing.

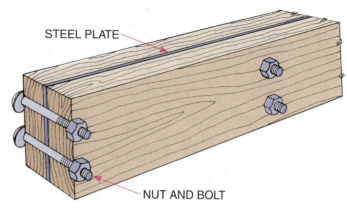

FIGURE 24.28 ■ Flitch beam.

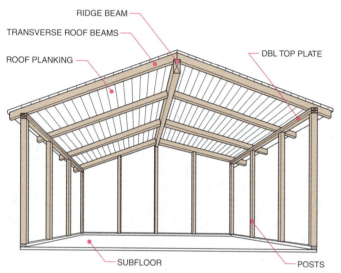

FIGURE 24.27 ■ Transverse roof framing.

and fastened against the side of the ridge beam. Transverse beams either intersect a ridge beam on pitched roofs or lie flat across the span on flat roofs.

Structural Timber Members

Three types of structural members are used in contemporary post-and-beam construction: *solid*, *laminated*, and *fabricated* components.

Solid Members

Solid wood timbers are available in thicknesses that range from 3″ to 12″. However, the use of sizes over 8″ is hampered by the tendency of large solid wood timbers to warp. One method of stabilizing larger solid wood members is to add steel plates to create *flitch beams,* which are stronger and remain straighter. See Figure 24.28.

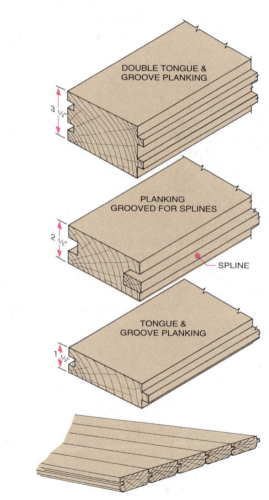

FIGURE 24.29 ■ Tongue-and-groove planking.

Solid planking in small widths (2″ and 4″) is more commonly used than plywood panel flooring. Because of the impact of live-load thrusts, solid planking is usually specified as tongue and groove (T&G). The T&G joint reduces deflection by tying the flooring planks together into one monolithic unit, as shown in Figure 24.29.

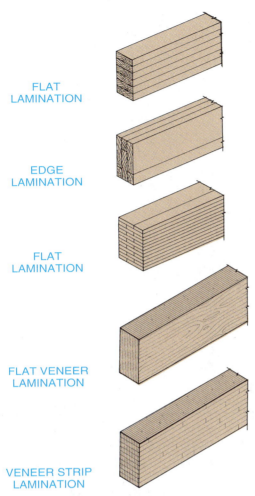

FLAT
LAMINATION

EDGE
LAMINATION

FLAT
LAMINATION

FLAT VENEER
LAMINATION

VENEER STRIP
LAMINATION

FIGURE 24.30 ■ Laminated beam types.

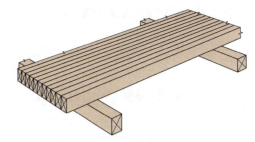

FIGURE 24.31 ■ Side-laminated decking.

Laminated Members

When larger timbers are needed to support heavier weights or greater spans, glue-laminated timbers are often used. **Laminated timbers** are made from thin layers (less than 2″) of wood, glued together either vertically or in patterns. See Figure 24.30. Laminated timbers are stronger than solid timbers because the grain direction is reversed (180°) in alternate layers. Because of its more consistent moisture content, there is less expansion in a laminated wood member than in solid wood. Glue-laminated members (or glulam for short) are manufactured in forms for columns, beams, and arches.

Laminated construction members used for plywood include *parallel stranded lumber* (PSL) and *laminated veneer lumber* (LVL). PSL is made from parallel strands of wood and are manufactured in large sizes suitable for heavy load-bearing posts and beams. LVL is made from thin veneer layers glued together with phenolic glue. The

joints are staggered over the length with the grain facing the same direction.

Oriented strand board (OSB) can be substituted for laminated plywood mainly for roof and wall sheathing. OSB is composed of shredded strands of wood that are glued and compressed into layers.

In addition to the laminated members for posts and beams, laminated decking is also available in 2″ thicknesses and in 6″ to 12″ widths. Laminated decking is specified in nominal sizes on construction drawings. When decking material is laminated, the layers may be offset to fit together. Another method of lamination is to align the wood grain vertically and laminate the sides for greater resistance to loads. See Figure 24.31.

Although lamination can create stronger, larger, and more structurally stable members, its most popular feature is its capability to be bent into a wide variety of structurally sound and aesthetically pleasing shapes, as shown in Figure 24.32. A variety of beam forms, including arches, are created by first bending thin, parallel layers of wood to a desired shape. Then the layers are glued and clamped together under pressure. When the glue dries, the member retains the new, bent form.

To indicate arches on architectural drawings, the base location of each arch is shown in the ground floor plan. The profile shape, including height and width dimensions, is drawn on elevation drawings and/or on elevation sectional drawings. See Figure 24.33.

Fabricated Members

Many fabricated products and materials are used in place of solid or laminated wood members. These include a variety of wood I-beams, truss joists, panels, box beams, strand lumber, and recycled plastic material.

Wood I-beams, and I-joists used for floor and ceiling joists and rafters, are constructed with a plywood or strand board web that is inserted into a groove in a laminated or machine-stressed wood flange. This construction produces a very straight, lightweight, strong, and stable member. Knockout areas can be used to install

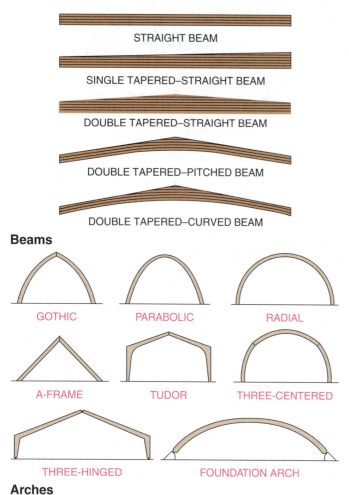

Beams

GOTHIC PARABOLIC RADIAL

A-FRAME TUDOR THREE-CENTERED

THREE-HINGED FOUNDATION ARCH

Arches

FIGURE 24.32 ■ Laminated beam forms.

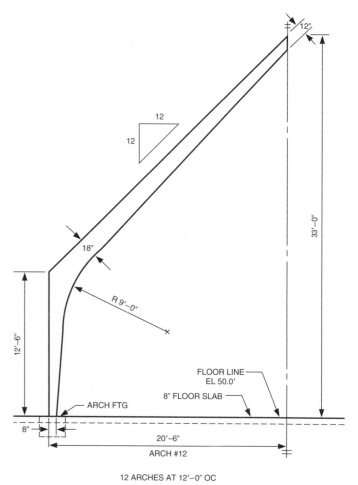

12 ARCHES AT 12'–0" OC

FIGURE 24.33 ■ Laminated arch dimensioning methods.

HVAC, electrical, or plumbing lines without sacrificing strength. See Figure 24.34. Wood I-beams can span much longer distances than comparably sized solid members.

Truss joists are constructed like a truss, but with parallel flanges usually made of 2 × 4's. See Figure 24.35. These members can support greater loads than the same size solid lumber. Figure 24.36 shows truss joists used as a substitute for solid roof rafters or conventional trusses.

Stressed-skin or *sandwich* panels are often used in place of solid or laminated members. These lightweight prebuilt or site-built panels are made by gluing and/or nailing plywood sheets to structural member frames. Because panels can easily be constructed using standard plywood sizes, they are often used for floor, wall, and roof panels. Stressed-skin panels can also be used to make box beams for spans up to 120' depending on load factors. Folded plate roofs and curved panels can also be either fabricated on the site or factory-built. The use

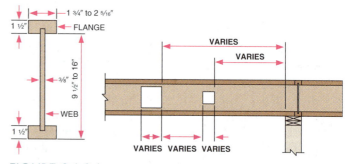

FIGURE 24.34 ■ Wood I-beam or truss details. (*TrusJoist Mac-Millan*)

of stressed-skin panels in roof construction is covered in Chapter 30.

Foam sandwich panels are made of two plywood or strand board sheets adhered to a core of polyurethane or polystyrene. These panels function as framing, insulation, and sheathing and may sometimes be used as finished interior wallboard.

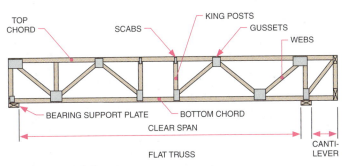

FIGURE 24.35 ■ Truss joist components.

FIGURE 24.36 ■ Truss joists used as roof rafters.

FIGURE 24.37 ■ Examples of heavy timber joint construction. (*American Olean Tile Company*)

Strand lumber is made of cellulose fiber strips. The fibers from poor quality or small trees, rice, rye, wheat, or straw are crushed into long strands. After being combined with formaldehyde and adhesives, the strands are woven, compressed, and heat treated. The finished shapes are very strong and can span long distances. They are used primarily for columns, beams, and large headers. In addition to PSL, LVL, and OSB lumber described earlier in the plywood section, *laminated strand lumber* (LSL) is frequently used in place of solid wood as core material for doors and short span headers.

Recycled thermoplastics are plastics that are ground, glued, and pressed into structural members. These thermoplastic members are capable of being sawed and glued, and they can accept and hold screws and nails. Contrary to wood, thermoplastic members will not rot, absorb moisture, expand, contract, or be infested by insects.

Timber Connectors

Because of heavy timber sizes, concentrated loads, and lateral thrust from winds and earthquakes, special joints and fasteners are required to attach post-and-beam members to the foundation and to other members. Nails are useful only as temporary holding devices, and lag screws can only be used in areas of limited stress. Figure 24.37 shows the application of heavy timber joint construction at the base, intermediate, and roof levels. Figure 24.38 shows several typical post-and-beam intersections that do not make use of brackets or straps. Figures 24.39A and 24.39B show common joints and brackets used in post-and-beam construction.

Base anchors and plates are used to attach the base of heavy timber posts to the foundation and to prevent wood deterioration. See Figure 24.40. Timber brackets can be embedded into concrete as shown in Figure 24.41, or timber can be placed into an impact-fastener base anchor. See Figure 24.42.

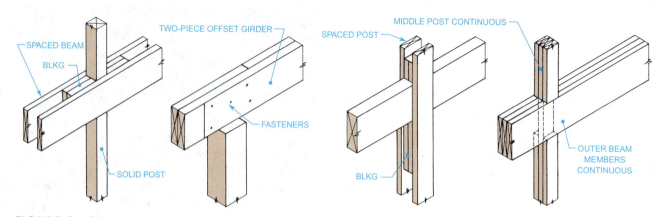

FIGURE 24.38 ■ Post-and-beam intersections without brackets.

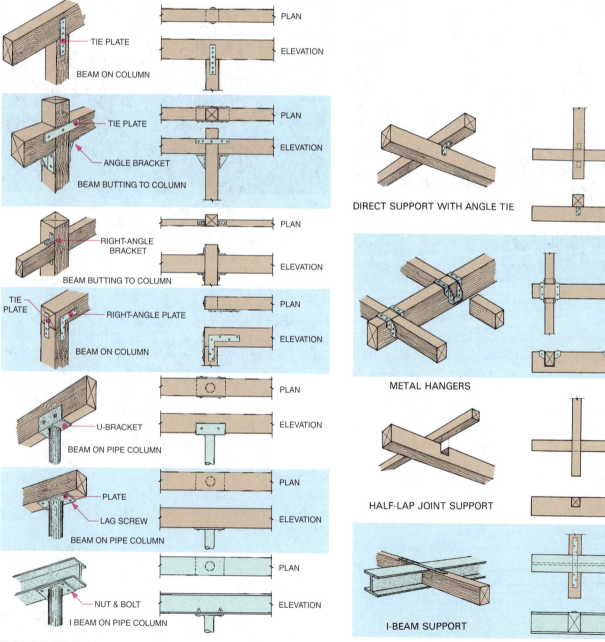

FIGURE 24.39A ■ Common post-and-beam joints.

FIGURE 24.39B ■ Common post-and-beam brackets.

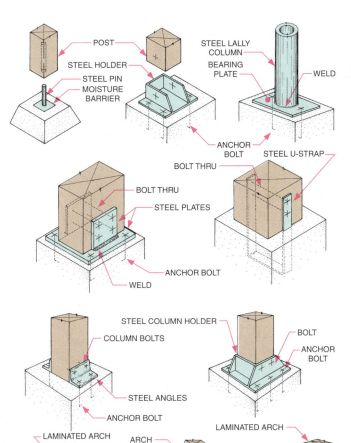

FIGURE 24.40 ■ Base anchors and plates.

Before installation.

After installation.

FIGURE 24.42 ■ Timber base anchors.

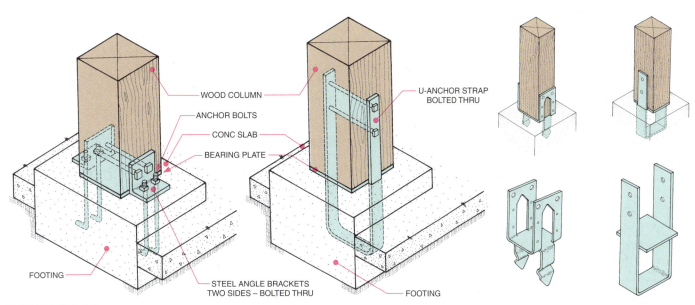

FIGURE 24.41 ■ Embedded timber brackets.

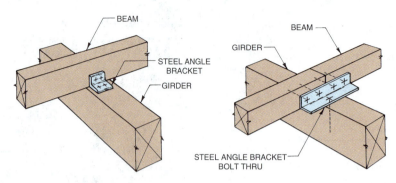

FIGURE 24.43 ■ Timber angle brackets.

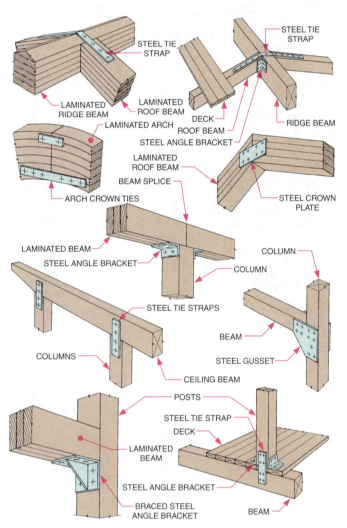

FIGURE 24.44 ■ Use of straps, gussets, and angle brackets.

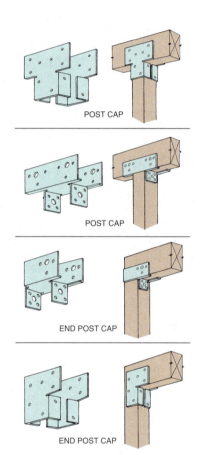

FIGURE 24.45 ■ Post caps.

Metal strap ties and gusset plates are used to attach posts to beams and to keep transverse roof beams aligned with ridge beams. Post-and-beam construction has excellent resistance to dead loads, which exert pressure directly downward. However, because of the large unsupported wall areas, lateral live loads can be a problem. Although diagonal ties or sheathing can help control the lateral thrust, angle brackets should be used to fasten perpendicular intersections to prevent lateral movement between members and help provide rigidity to joints.

Figure 24.43 shows the use of angle brackets to hold overlapping beams at right angles. Figure 24.44 shows the use of straps, gussets, and angle brackets to hold beams and columns together.

Post caps are used extensively when posts intersect beams at a beam joint, as shown in Figure 24.45. When

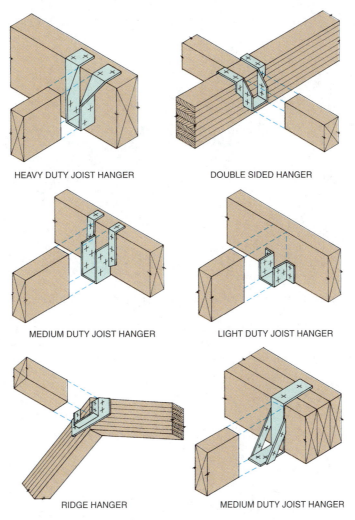

FIGURE 24.46 ■ Joist and beam hangers.

the end of a member intersects the side of another member, without resting on it, metal hangers are usually specified. See Figure 24.46. In some cases special truss clips may be used to prevent movement where two members intersect at angles other than 90°. See Figure 24.47. Figure 24.48 shows the types of mortise and tenon joints available for use with exposed joints.

Because interior timber connections are sometimes exposed, hidden joints and fasteners using dowels, rods, and half-lap joints, as shown in Figure 24.49, are often used. Split-ring connectors (Figure 24.50), which are extremely strong and easy to assemble, are also used for this purpose.

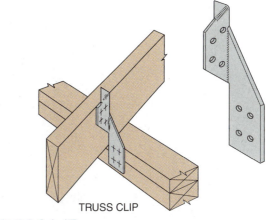

FIGURE 24.47 ■ Use of a truss clip.

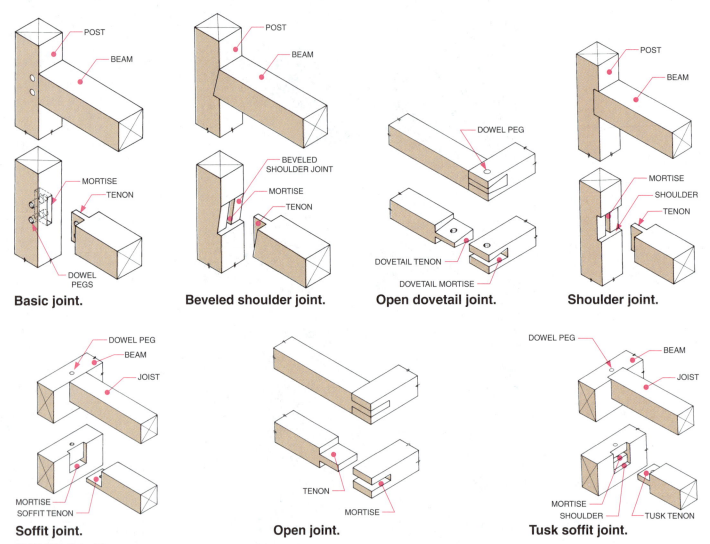

Basic joint.

Beveled shoulder joint.

Open dovetail joint.

Shoulder joint.

Soffit joint.

Open joint.

Tusk soffit joint.

FIGURE 24.48 ■ Types of mortise and tendon joints.

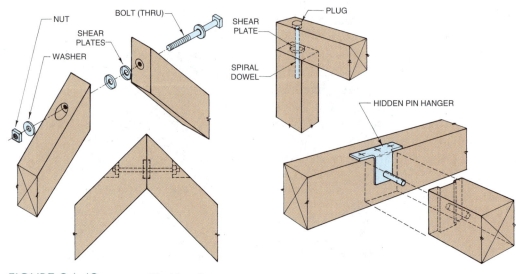

FIGURE 24.49 ■ Use of hidden fastners.

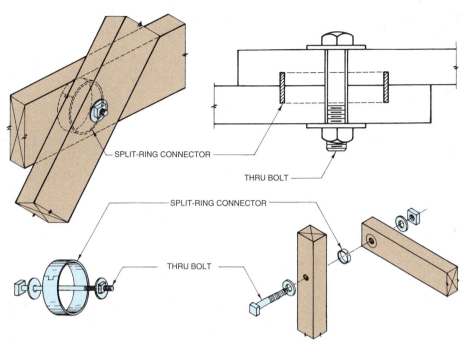

FIGURE 24.50 ■ Split-ring connector assembly.

CHAPTER 24

Wood-Frame Systems Exercises

1. Describe the major differences between skeleton-frame and post-and-beam construction.
2. Describe the differences between platform and balloon framing.
3. List six types of fabricated members used in light construction.
4. Select the lumber grade you will specify for the studs, rafters, sheathing, and joists for the house you are designing.
5. Describe the difference between yard, structural, and factory lumber.
6. How many board feet are in 400 wood members 2″ × 6″ × 12′?
7. What is the dressed size of a 2 × 4 and a 4 × 6 wood member?
8. What is the range of plywood thicknesses?
9. Name three uses for hardwoods and softwoods in light construction.
10. Check the types of plywood in your local lumber store and write a short description of each.
11. Check the types of construction grades of lumber in your local lumber store and write a short description of each.

25 CHAPTER

Masonry and Concrete Systems

OBJECTIVES

In this chapter, you will learn to:

- identify the types of masonry materials used in construction.
- describe four types of masonry walls.
- describe ways to strengthen concrete and prevent deflection.
- explain how concrete is used for slabs and other structural components.

TERMS

aggregate
lally column
masonry bond

one-way slab system
post-tensioning
prestressing

pretensioning
two-way slab system

INTRODUCTION

Masonry construction systems use brick, stone, concrete block, or clay tile products. Masonry units are arranged, usually row upon row (courses), to form structures such as walls. *Concrete* construction systems use structural members made of poured or precast concrete.

MASONRY CONSTRUCTION SYSTEMS

Construction systems that use masonry are usually combined with other systems, such as structural steel or skeleton-frame construction. Buildings usually are not constructed only with masonry materials because wood, steel, or reinforced concrete is needed for the large span floor and roof systems.

Masonry Materials

Masonry materials used for today's construction include a broad range of manufactured products. The many different types, sizes, shapes, and grades of masonry materials serve a variety of purposes.

Brick

Bricks are divided into two general categories: *common brick* and *face brick*. Color, texture, and dimensional tolerance are less consistent and critical for common brick than for face brick. Common brick is therefore less expensive and is generally used in unexposed construction areas. Common brick is graded according to structural characteristics. See Figure 25.1.

Face brick is used in exposed areas that require dimensional accuracy and absorption control, as well as consistent color and texture. Face brick is therefore graded according to these characteristics. See Figure 25.2. Many special types of face brick are available for specific con-

GRADE	USE
SW	Used for maximum exposure to heavy snow, rain, and/or continuous freezing conditions.
MW	Used for average exposure to rain, snow, and moderate freezing conditions.
NW	Used for minimum exposure to rain, snow, and freezing conditions.

FIGURE 25.1 ■ Grades of common brick.

TYPE	USE
FBX	Used where minimum size and color variations, and high mechanical standards are required.
FBS	Used where wide color variations and size variations are permissible or desired.
FBA	Used where wide variations in color, size, and texture are required or permissible.

FIGURE 25.2 ■ Grades of face brick.

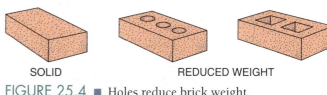

SOLID   REDUCED WEIGHT

FIGURE 25.4 ■ Holes reduce brick weight.

TYPE	SIZE
Standard	2 1/2″ × 3 7/8″ × 8 1/4″ 2 1/4″ × 3 3/4″ × 8″
Oversized	3 1/4″ × 3 1/4″ × 10″
Modular (1/4″ Joints)	2 1/2″ × 3 3/4″ × 7 3/4″ 2 5/16″ × 3 3/4″ × 7 3/4″ 2 1/2″ × 3 3/4″ × 11 3/4″
Modular (1/2″ Joints)	2 1/4″ × 3 1/2″ × 7 1/2″ 2 1/4″ × 3 1/2″ × 11 1/2″ 2 1/16″ × 3 1/2″ × 7 1/2″

FIGURE 25.5 ■ Common brick sizes.

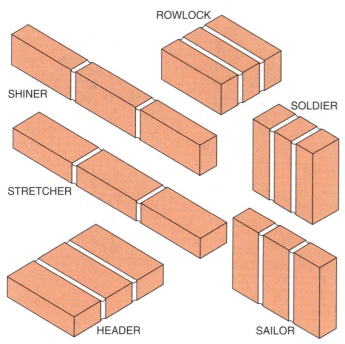

FIGURE 25.3 ■ Laid brick positions.

TYPE	SIZE
Standard	2 1/2″ × 3 1/2″ × 11 1/2″
Norman	2 3/16″ × 3 1/2″ × 11 1/2″ 2 1/4″ × 3″ × 11 11/16″
Roman	1 1/2″ × 3 1/2″ × 11 1/2″

FIGURE 25.6 ■ Face brick sizes.

Concrete Block

struction needs; for example, glazed brick, fire brick, cored brick, and paving brick.

Bricks are also classified by their positioning in construction. See Figure 25.3.

Most bricks are rectangular. Special shapes for sills, corners, and thresholds are also available or can be made to order. Bricks usually have holes, as shown in Figure 25.4. Holes reduce the weight of bricks and increase bonding. Masonry bonds are discussed later in this chapter.

Sizes differ among brick types. Common bricks come in standard, oversized, and modular sizes. See Figure 25.5. Face bricks come in standard, Norman, and Roman sizes. See Figure 25.6. Modular bricks are standardized to align on 4″ grids after mortar joint dimensions are added. Increments of 4″, 8″, 12″, and so forth, fit into established measured spaces.

Concrete blocks are made in many different shapes for a wide variety of construction purposes, as shown in Figure 25.7. Concrete blocks are either solid, hollow-core, or split-face for exposed surfaces. The weight, texture, and color of each block are determined by the types of aggregate used. **Aggregate** is a combination of sand and crushed rocks, slate, slag, or shale. When concrete masonry does not need to support heavy loads, lightweight masonry blocks are ideal. Lightweight blocks are molded by adding fly ash cinders to concrete.

Similar to modular bricks, concrete block is manufactured in modular sizes. That means the actual size of each block is 3/8″ smaller than the space to be filled. The difference allows for the thickness of the mortar joint. For example, the dimensions of an 8″ × 8″ × 16″ concrete block are actually 7 5/8″ × 7 5/8″ × 15 5/8″. Because their sizes are standardized, modular concrete

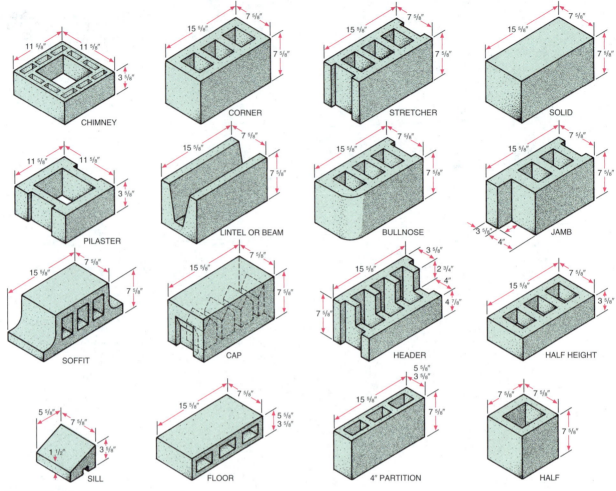

FIGURE 25.7 ■ Concrete block sizes and shapes.

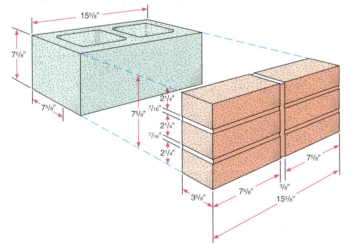

FIGURE 25.8 ■ Modular concrete block and brick dimensions.

blocks can be used in conjunction with modular bricks. See Figure 25.8.

Some lightweight masonry blocks are made from a combination of sawdust and cement.

Stone

For centuries, natural stones were used as a major structural material. Today stone is primarily used decoratively except for landscape construction. The stone wall shown in Figure 25.9 is decorative and functions as a privacy baffle. The stone chimney in the background is structural. Stone masonry is classified by the type of material, shape of cut, finish, and laying pattern. The most common types of stone used in construction are sandstone, limestone, granite, slate, and marble. These can be cut and arranged into a variety of patterns, as shown in Figure 25.10.

Structural Clay Tile

Hollow-core structural tile units are larger than bricks and can be either load-bearing or non-load-bearing structures. Structural tiles are used for partitions, fireproofing, surfacing, or furring. See Figure 25.11. Load-bearing tile is graded according to structural characteristics: LBX for tile exposed to weathering and LB for tile not exposed to weathering or frost. Non-load-bearing facing tile is graded

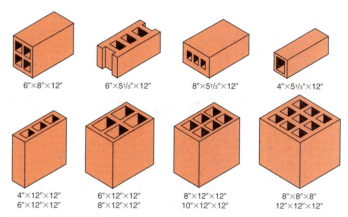

FIGURE 25.11 ■ Structural tile sizes and shapes.

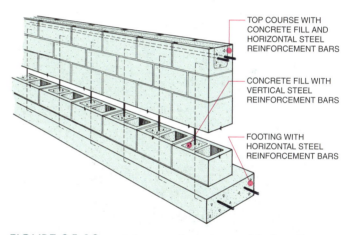

TOP COURSE WITH CONCRETE FILL AND HORIZONTAL STEEL REINFORCEMENT BARS

CONCRETE FILL WITH VERTICAL STEEL REINFORCEMENT BARS

FOOTING WITH HORIZONTAL STEEL REINFORCEMENT BARS

FIGURE 25.12 ■ Rebars used in concrete block walls.

FIGURE 25.9 ■ Decorative and structural uses of stone. (*Alfred Karram Design*)

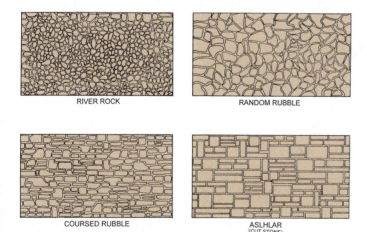

RIVER ROCK

RANDOM RUBBLE

COURSED RUBBLE

ASLHLAR
(CUT STONE)

FIGURE 25.10 ■ Stonework patterns.

by surface texture, its stain resistance, color consistency, and dimensional accuracy: FTX for high quality, FTS for low quality.

Masonry Walls

Four basic types of masonry wall construction are solid, cavity, facing, and veneer.

Solid Masonry Walls

Most masonry bearing-wall construction is solid. Solid masonry construction can utilize almost any masonry material if it is laid flat to support loads. However, the material used must be able to withstand the loads involved.

Concrete block is commonly used for solid load-bearing walls. For heavy loads and/or high walls, steel reinforcing rods (rebars) are added and the block cells are filled with concrete for structural stability. See Figure 25.12. When solid masonry walls are constructed with combinations of materials (such as concrete block and brick), steel reinforcement is mandatory, as shown in Figure 25.13A. Figure 25.13B shows several alternative steel ties and flashing used to join two parallel solid

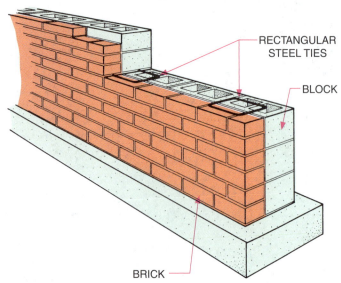

RECTANGULAR
STEEL TIES

BLOCK

BRICK

FIGURE 25.13A ■ Steel tie reinforcement between masonry walls.

HORIZONTAL STEEL
JOINT REINFORCEMENT

FIGURE 25.14 ■ Steel reinforcement between masonry courses.

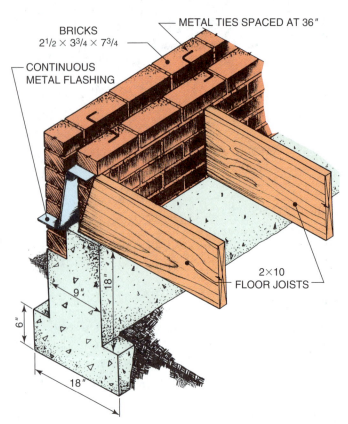

BRICKS
$2^1/2 \times 3^3/4 \times 7^3/4$

METAL TIES SPACED AT 36″

CONTINUOUS
METAL FLASHING

2×10
FLOOR JOISTS

18″

9″

6″

18″

FIGURE 25.13B ■ Steel ties and flashing on parallel brick walls.

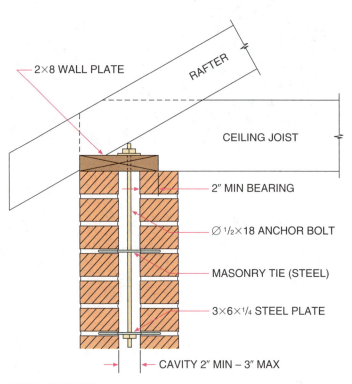

2×8 WALL PLATE

RAFTER

CEILING JOIST

2″ MIN BEARING

∅ $^1/_2 \times 18$ ANCHOR BOLT

MASONRY TIE (STEEL)

$3 \times 6 \times ^1/_4$ STEEL PLATE

CAVITY 2″ MIN – 3″ MAX

FIGURE 25.15 ■ Elevation section of a masonry-cavity wall.

Masonry-Cavity Walls

To reduce dead loads and to improve temperature and humidity insulation, cavity walls are preferred to solid masonry walls. In cavity wall construction, two separate and parallel walls are built several inches apart. A structural tie, usually metal, bonds the walls together. See Figure 25.15.

brick walls. Reinforcement between courses or between materials is particularly necessary for walls subject to earthquakes, heavy storms, wind, and lateral earth loads. See Figure 25.14.

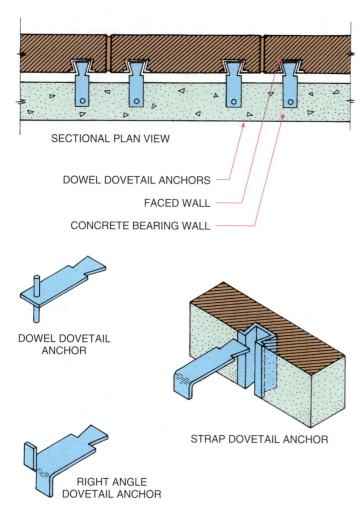

SECTIONAL PLAN VIEW

DOWEL DOVETAIL ANCHORS

FACED WALL

CONCRETE BEARING WALL

DOWEL DOVETAIL
ANCHOR

STRAP DOVETAIL ANCHOR

RIGHT ANGLE
DOVETAIL ANCHOR

FIGURE 25.16 ■ Metal ties bonding a masonry-faced wall.

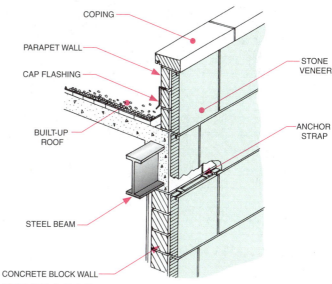

COPING

PARAPET WALL

CAP FLASHING

BUILT-UP
ROOF

STEEL BEAM

CONCRETE BLOCK WALL

STONE
VENEER

ANCHOR
STRAP

FIGURE 25.17 ■ Masonry-veneer wall.

Masonry-Faced Walls

Walls are often faced with different masonry materials. Any type masonry wall can be faced with another facing material. For example, a faced wall may consist of common bricks faced with structural tile or concrete block faced with brick. Regardless of the material, the two walls are always bonded so that they become one wall structurally. The bonding material can be metal ties as shown in Figure 25.16, steel reinforcing rods, or masonry units laid on end to intersect the opposite wall. Always remember that walls with different coefficients of expansion are never faced together. Their differing rates of expansion and contraction under extreme temperature-change conditions can cause cracks and damage the structural integrity of the wall.

Masonry-Veneer Walls

Veneer walls, like masonry-faced walls, include two separate walls constructed side by side. Unlike faced walls, the veneer wall is not tied to the other wall to form a single structural unit. The veneer wall is simply a non-load-bearing decorative facade, although the two walls may be connected with masonry ties or adhesives. See Figure 25.17.

A veneer wall may include two different masonry materials or include a skeleton-frame wall veneered with a masonry material. In the latter case, the space between the wood and masonry walls (usually 1″) may remain empty or may be filled with insulation, depending on climactic conditions. A wall detail or sectional drawing is usually prepared to provide this information.

Masonry Bonds

A **masonry bond** for walls is the pattern of arranging and attaching masonry units in courses (rows). Masonry can be placed in a variety of bond patterns, as shown in Figure 25.18. Different patterns can make the same size, shape, and material appear completely different. Various types of mortar joints are specified on construction drawings. See Figure 25.19.

CONCRETE CONSTRUCTION SYSTEMS

Early Romans crushed and processed rocks to create cement for bonding their structures. Today, cement continues to be an important material in construction. Cement is manufactured primarily from clay and limestone. *Concrete,* so widely used in today's construction, is basically made from cement combined with water and aggregate (such as sand and gravel).

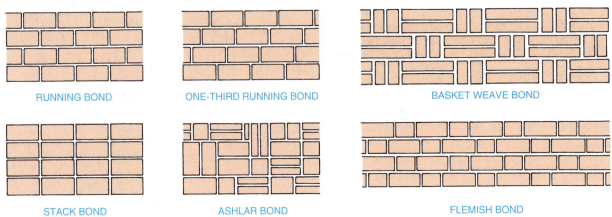

FIGURE 25.18 ■ Common types of masonry bonds.

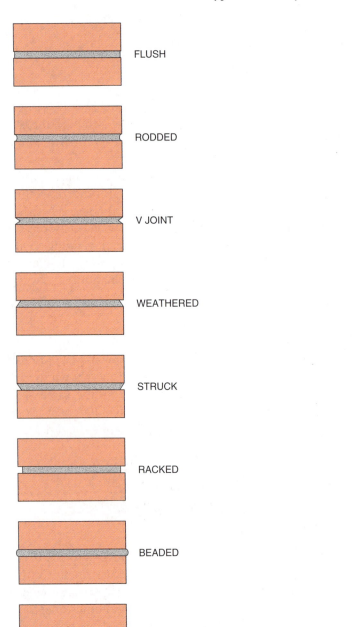

FIGURE 25.19 ■
Types of mortar joints.

A typical concrete mix consists of 41% crushed rock, 26% sand, 16% water, 11% portland cement, and 6% air. Engineering researchers are currently developing inexpensive and lightweight concrete made from recycled glass and dredge materials. The new material combines the advantages of thermal and sound insulation properties with adequate strength, durability, and reduced weight. Fiber-reinforced concrete provides greater fracture-resistant properties and energy absorption capacity than conventional concrete mixes.

Another development in concrete is autoclaved aerated concrete (AAC), which is produced from lime, sand, gypsum, cement, aluminum powder, and water. ACC has low heat conductivity (U) values and high thermal (R) values. It is also noncombustible, lightweight, insect proof, fire resistant, and can be sawed or drilled with common hand tools. See chapter 32 for a detailed description of R and U values.

Different types of concrete are identified by their compressive strength, measured in pounds per square inch (PSI). Concrete strengths range from 2500 PSI to 4000 PSI for most residences and up to 8000 PSI for large industrial buildings. Concrete volume is measured in cubic yards.

Concrete can either be cast in place on site or precast off site and shipped to the site as finished girders, beams, slabs, columns, or other components. Either way, concrete may be reinforced, prestressed, or poured plain depending on the construction application and/or site conditions.

Reinforced Concrete

Because concrete is weak in tensile strength but has a strong resistance to compression stress, it was previously used only for nontension applications such as ground-level slabs, walks, or roadways. However, when materials with high tensile strength are added to the concrete, the tensile strength of the concrete is greatly increased, and its com-

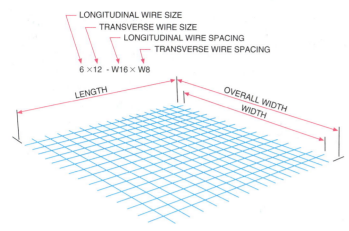

FIGURE 25.20 ■ Welded wire mesh specifications.

⌀ IN INCHES		GAUGE NUMBER	ACTUAL SIZE OF WIRE	WOVEN WIRE MESH SPACING
DECIMAL	FRACTION			
.2437	1/4″	3	●	–
.2253	7/32″	4	●	–
.2070	13/64″	5	●	–
.1920	3/16″	6	●	2 1/2″
.1770	11/64″	7	●	2 1/4″
.1620	5/32″+	8	●	2″
.1483	5/32″–	9	●	1 3/4″
.1350	9/64″	10	●	1 1/2″
.1205	1/8″	11	●	1 1/4″
.1055	7/64″	12	●	1″
.0915	3/32″	13	●	–
.0800	5/64″	14	●	3/4″
.0625	1/16″	16	●	3/8″ or 1/2″

FIGURE 25.21 ■ Welded wire mesh sizes and spacing.

pression strength is doubled. Improvements in the reinforcing of concrete are mainly responsible for the increased use of concrete in all types of building construction.

Reinforcement Materials

Reinforcement bars (rebars) and welded wire mesh provide the steel that converts concrete into reinforced concrete. The use of rebars and welded wire mesh was described in relation to foundations and slabs in Chapter 23.

Welded wire mesh is designed to evenly distribute stress forces and prevent cracking. The spacing of the longitudinal and transverse wires and the size of each are noted in Figures 25.20 and 25.21.

The use of rebars for reinforcing slabs was covered in Chapter 23. Rebars are also used in the formation of concrete beams, columns, walls, and suspended decks. See Figure 25.22. A marking system identifies rebar manufacturer, size, steel type, and PSI grade, as shown in Figure 25.23.

Prestressed Concrete

When loads are added to concrete members, some deflection (sag) occurs in the center of the member. This happens to all materials under load. However, because concrete has very low tensile strength, excessive deflection can result in tension cracking or complete member failure. This is caused by compression of the upper side and tensioning (stretching) of the lower side. See Figure 25.24. To counteract these unstable compression and tension stresses, concrete is often prestressed.

Prestressing is a method of compressing concrete so that both the upper and lower sides of a member remain in compression during loading. Prestressing can be accomplished either by pretensioning or posttensioning.

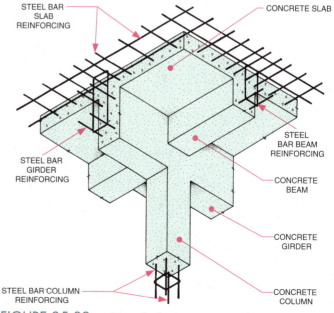

FIGURE 25.22 ■ Use of rebars in structural concrete.

Pretensioning

In **pretensioning**, deformed steel bars called *tendons* are stretched (tensioned) between anchors and the concrete is poured around the bars. Once the concrete has cured, the tension is released and the bars attempt to return to their original, shorter length. However, the concrete that has hardened around the bar grooves holds the deformed bars at nearly their stretched length. This creates a continual state of compressive stress that can be compared to pressing a row of blocks together. See Figure 25.25.

As a further aid to prevent bending, concrete members are prestressed by draping tendons near the bottom of the member, as shown in Figure 25.26. This bottom tension buckles the member upward so that when the anticipated loads are added, the beam straightens to a level position.

Post-Tensioning

Post-tensioning is done after concrete has cured. In post-tensioning, tendons are either placed inside tubes embedded in the concrete or the tendons have been greased to allow slippage. The tendons are then stretched with

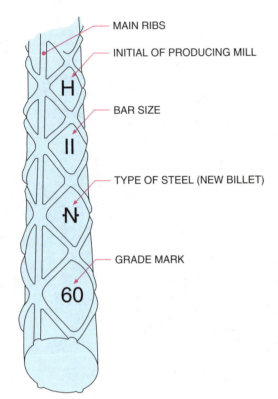

FIGURE 25.23 ■ Rebar marking system.

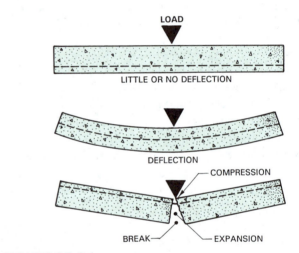

FIGURE 25.24 ■ Loading effect on concrete.

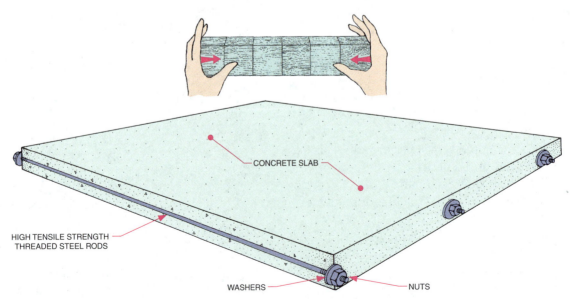

FIGURE 25.25 ■ Principle of pretensioning.

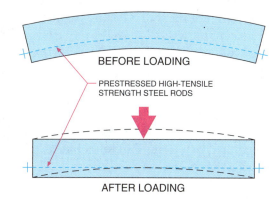

FIGURE 25.26 ■ Prestressing with draped tendons.

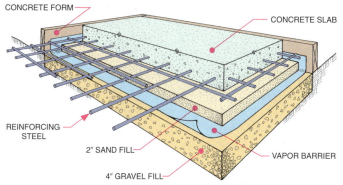

FIGURE 25.27 ■ Grade-level slab construction.

hydraulic jacks and the ends anchored. This creates compressive stress because the ends of the tendons pull toward the center. Post-tensioning can be done at a factory or on site to reduce shipping weight, especially for large members.

Concrete Structural Members

Concrete has been used for centuries for foundations, walls, and ground-supported slabs. Not until low-tensile-strength concrete was reinforced with high-tensile-strength steel, however, could concrete be used for structural components such as columns, beams, girders, and suspended-slab floor and roof systems.

Concrete Slabs

Once slabs could only be poured in place at grade level, as shown in Figure 25.27. However, with advances in steel reinforcement methods, structural slab members can now be manufactured off site. Elevated slabs can also be poured using parallel rebars aligned with the beam direction. To prevent cracking due to temperature and moisture changes, rebars are also placed perpendicular to the load-supporting rebars. These rebars are known as temperature rebars (or bars). See Figure 25.28.

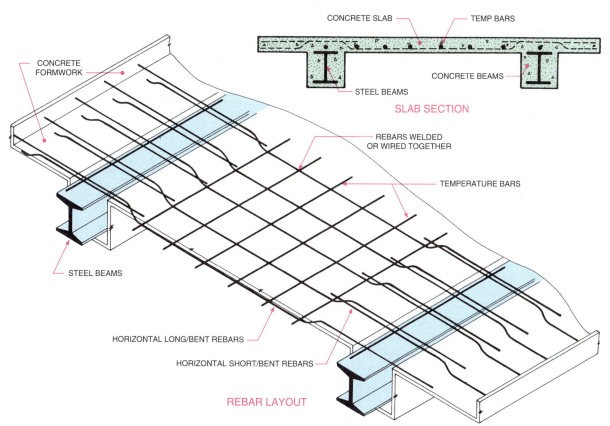

FIGURE 25.28 ■ Temperature and load-supported bar placement.

Columns

Concrete columns are vertical members that support weights transferred from horizontal beams and girders. Concrete columns are made structurally sound by the addition of rebars. Another method is to fill a hollow steel column with concrete. Such a member is called a **lally column.** A large concrete column like that shown in Figure 25.29 is called a *caisson.*

Sectional drawings convey the exact relationship of column, beam, and reinforcement material, as shown in Figure 25.30. If the exact position of each column is not dimensioned on a floor plan, column schedules are prepared. Column schedules include coding that is indexed to a column plan.

Beams and Girders

Concrete girders are major horizontal members that rest on columns. Beams are horizontal members supported by girders or columns. Concrete beams and girders are reinforced with steel rebars to increase tensile strength. Some reinforced concrete beams are rectangular, but most are wider at the top.

Rebars or WWM (welded wire mesh) placed low in a beam will prevent cracking. Rebars or WWM placed too high in a beam may bend and result in cracking. For very heavy loads, a top and bottom row may be needed. This practice is called *draped reinforcement.* See Figure 25.31.

The position of girders and beams is shown on floor plans with dotted lines or indexed to a beam schedule, similar to a column schedule.

Lintels

Short horizontal members that span the top of openings in a wall are known as lintels (headers). Concrete lintels are either poured in a form, into lintel blocks, or precast as a small concrete beam.

Cast-in-Place Concrete

Forms for pouring concrete for footings, foundations, slabs, and walls have been used for a long time. However, new developments in reinforced and prestressed concrete

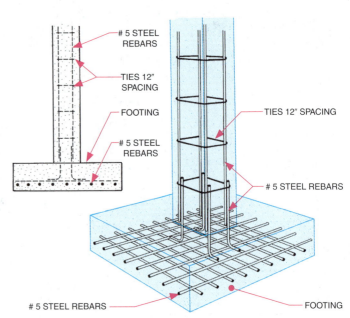

FIGURE 25.30 ■ Column and footing rebar placement.

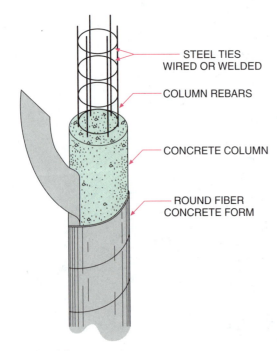

FIGURE 25.29 ■ Steel-reinforced concrete column.

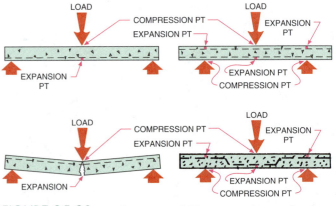

FIGURE 25.31 ■ Placement of WWM or rebars in beams or suspended slabs to prevent cracking.

enable builders to erect structures with extremely complex contours. Concrete shells are a type of concrete system that uses poured reinforced concrete. A light steel structure is erected. Then concrete is poured or sprayed over the steel frame. The concrete holds the steel in place after hardening.

Drawings for cast-in-place concrete systems need to include the outline and dimensions of the finished job, including the position of rebars and joints.

Precast Concrete

Precast concrete is the opposite of cast-in-place concrete. Precasting of concrete involves pouring the concrete into wood, metal, or plastic molds. Precast concrete is usually reinforced. Once set, the molded concrete is placed in position in its hardened form.

Although concrete block is the most commonly used precast concrete material, it is considered a masonry material, like brick. Precast concrete structural members include wall panels, girders, beams, and a variety of slabs. See Figure 25.32. Available in solid, hollow-core, and single and double tee shapes, precast slabs are used for

walls, floors, and roof decks. Wall panels are solid precast units used either for bearing or non-bearing walls, depending on the amount of reinforcement. Because the exterior sides of concrete wall panels are usually exposed, special textured finishes are often applied during the casting process. These panels may be combined with layers of insulation to form a complete monolithic wall unit.

Concrete floor system drawings show only the dimensions of slabs that are poured in place. For precast systems, the locations of ribs and/or support beams are shown with dotted lines. See Figure 25.33. All other information is found on detail and/or sectional drawings keyed to the general floor framing plan.

Slab Component Systems

Precast slab components or cast-in-place slabs are divided into two types: one-way systems and two-way systems.

One-Way Systems

In **one-way slab systems** the rebars are all parallel. One-way system girders, which rest on columns, are parallel to the rebar alignment. One-way solid slabs are

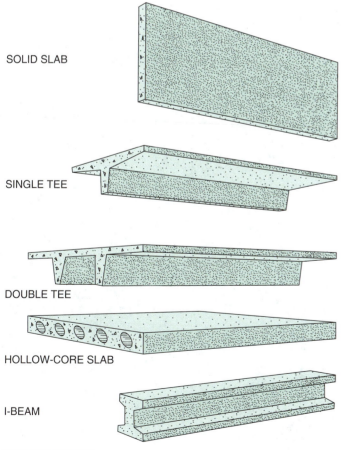

SOLID SLAB

SINGLE TEE

DOUBLE TEE

HOLLOW-CORE SLAB

I-BEAM

FIGURE 25.32 ▪ Precast concrete members.

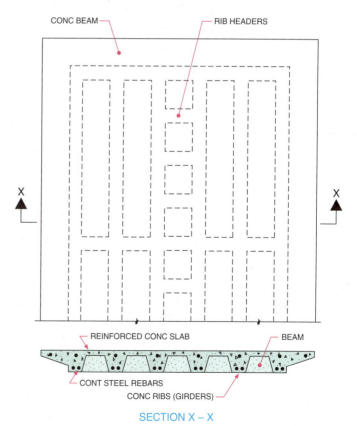

CONC BEAM — RIB HEADERS

X   X

REINFORCED CONC SLAB — BEAM

CONT STEEL REBARS

CONC RIBS (GIRDERS)

SECTION X – X

FIGURE 25.33 ▪ Method of drawing ribs and beams in a precast slab.

extremely heavy and are therefore impractical for most spans over 12 feet. To lighten the dead load, ribbed one-way slabs are often used. See Figure 25.34. The ribbed slab is a thin slab (2″ to 3″ thick), supported by cast ribs. These units are constructed of precast slab tees or are cast in place. When ribbed slabs are to be poured in place, a ribbed slab plan shows the horizontal rib positions with dotted lines. Dotted lines are also used on detail drawings to show the position of rebars in the slab and in the ribs.

Two-Way Systems

In **two-way slab systems** the rebars, girders, and beams are placed in perpendicular directions. See Figure 25.35.

When ribs extend in both directions, the system is known as a waffle slab.

Pan and *waffle* slabs are used to cast suspended floor and roof systems for spans up to 60 feet. A pan slab is created by pouring concrete into molded fiberglass forms on site. Waffle slabs may be poured on site or prefabricated off site. Temporary fiberglass or metal pans (domes) are placed, open side down, 4″ to 7″ apart on a temporary floor. No pans are placed around columns. Rebars are added and concrete is then poured to a depth of several inches over the pans. After the concrete has cured, the pans and temporary flooring are removed, and a suspended waffled floor (or roof) results. See Figure 25.36. This type of cast-in-place system is lightweight (for concrete), sound resistant, fireproof, and economical.

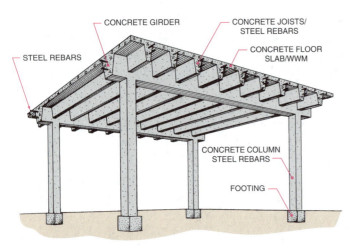

FIGURE 25.34 ■ Ribbed one-way concrete slab system.

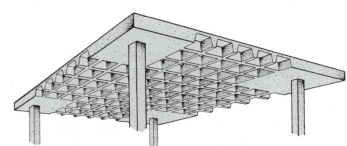

FIGURE 25.36 ■ Waffle concrete slab system.

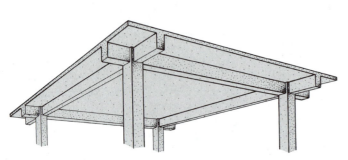

FIGURE 25.35 ■ Two-way concrete slab system.

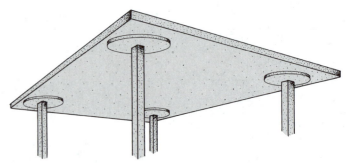

FIGURE 25.37 ■ Flat concrete slab system.

A *flat slab* is a two-way slab unit that rests directly on columns without a girder or beam support. A flat slab floor (or roof) system is actually a series of individual slabs, with the center of each slab resting independently on a column. All of the slab's weight is directed through the columns to footings. When the slabs are joined, a unified floor is created. In flat slab systems the supporting columns are strengthened by the addition of a thicker slab area (*drop panel*) around columns. See Figure 25.37. A column capitol, or flared head, also helps spread the slab loads onto a column in this type of construction.

Concrete Joints

Because concrete expands and contracts with changes in moisture levels and temperature, relief joints (expansion joints) are required to allow for these fluctuations. Some construction drawings and/or specifications indicate minimum dimensions for placement of expansion joints. Some drawings show specifically where joints are required or must be avoided.

Concrete Wall Systems

Conventional Concrete Walls

Most concrete walls are poured between wood or metal forms. Most walls of this type are used with wood-frame construction. Reinforced concrete walls are also the main structural material for both walls and roofs in an earth-sheltered construction, as shown in Figure 25.38.

Insulated Concrete Walls

There are three types of preinsulated concrete walls. The first is an insulated wall form system that is composed of two polystyrene panels separated by plastic ties that provide the wall thickness needed, as shown in Figure 25.39. Once concrete is poured into the cavity, the forms remain as part of the finished wall and act as insulation. The second type is a polystyrene block wall system in which interlocking rigid foam blocks are stacked to form a wall, as shown in Figure 25.40. Concrete is poured into the hollow core of the blocks creating a rigid wall. The third type is a conventionally formed (with wood or metal) wall. A plastic foam panel is suspended in the center of the wall while concrete is poured, creating a foam core sandwich with concrete on the outside and insulation on the inside.

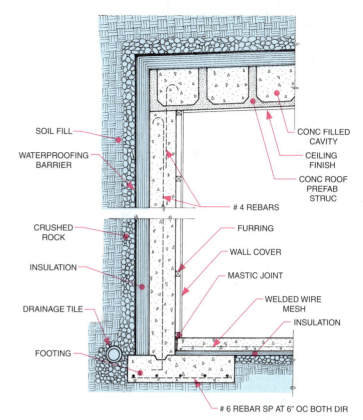

FIGURE 25.38 ■ Reinforced concrete walls in an earth-sheltered building.

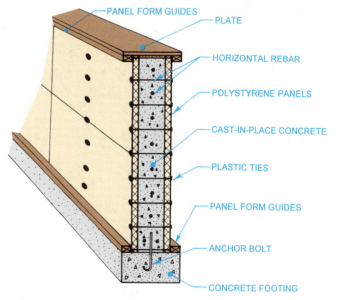

FIGURE 25.39 ■ Section of a preinsulated wall form system.

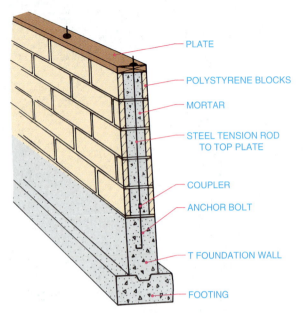

PLATE

POLYSTYRENE BLOCKS

MORTAR

STEEL TENSION ROD
TO TOP PLATE

COUPLER

ANCHOR BOLT

T FOUNDATION WALL

FOOTING

FIGURE 25.40 ■ Preinsulated block wall system.

All preinsulated systems are strong, durable, fire resistant, and termite proof. They also provide excellent insulation and create an effective vapor barrier. In preparing floor plan drawings for preinsulated walls, the thickness of the insulation must be added to the thickness of the concrete, interior wall covering, and external siding to determine the total wall thickness.

Rammed Earth Walls

Rammed earth walls are created by compacting a moistened mixture of soil and cement in an open-bottom wall form. By means of pneumatic tampers, the mixture is compressed into distinctive strata. To ensure stability and conform to code, this method is usually combined with post-and-beam construction, which provides the necessary support.

Masonry and Concrete Systems Exercises

1. Name the types of bricks and their uses.

2. Describe the types of masonry walls.

3. Describe the types of concrete construction systems.

 4. Draw a wall section for the house you are designing using one of the masonry systems described in this chapter.

CHAPTER 26

Steel and Reinforced-Concrete Systems

OBJECTIVES

In this chapter, you will learn to:

- describe three types of steel construction and explain the basic purpose of each.

- identify manufactured steel forms and their function as structural members.

- read and interpret steel symbols, weld notations, identification, and measurements for working drawings.

- relate the types of fasteners and intersections of steel members to construction methods.

TERMS

channel
large-span construction
L-shape (angle)

rolled steel
S-shape
steel cage construction

tee
W-shape

INTRODUCTION

Steel can span greater distances and support greater loads than any other conventional building material. However, steel is extremely heavy and thus creates heavy dead loads. This chapter describes structural steel members and explains how they are fastened together to form building components.

STEEL BUILDING CONSTRUCTION

In steel construction, plates, bars, tubing, and rolled shapes are used for columns, girders, beams, and bases. There are three general types of steel construction systems: steel cage, large span, and cable supported.

When steel members are used in a manner similar to skeleton-frame wood members, the system is known as **steel cage construction**. The terms *steel skeleton-frame* and *steel cage* construction are often used interchangeably. The major structural members used in steel cage construction are columns, girders, and beams. The definitions of structural members remain the same for steel construction as for other types of construction. Columns are vertical members that rest on footings or piers. Girders are horizontal mem-

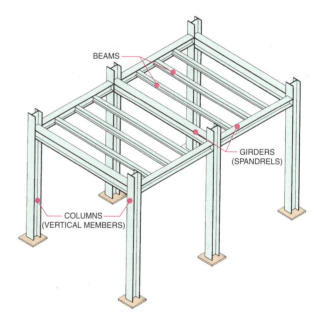

FIGURE 26.1 ■ Spandrels (girders) are connected to columns and beams are connected to spandrels.

bers that extend between columns. They are sometimes called *spandrel beams* if they connect to columns erected on the perimeter. See Figure 26.1. Beams are horizontal members placed on or between girders. *Purlins* are beams that connect roof trusses or rafters.

Beams are supported by girders, which in turn are rigidly attached to columns, through which all loads are transmitted through bearing plates to footings. Because all live and dead loads are transmitted through the columns, there is no need for additional exterior or interior bearing walls in steel cage construction. This enables buildings to be built extremely high with a minimum of interior obstruction. It also allows for the use of large exterior wall panels that have no structural value. These walls, called curtain walls, are often constructed of glass.

Even steel cage construction cannot provide the enormous amount of unobstructed space needed in structures such as aircraft hangars, sports stadiums, and convention centers. For these structures, large trusses or arches are necessary to span long distances. Such construction is called **large-span construction.**

Structural steel is manufactured in many forms and types. Structural steel types are specified by their metallurgical characteristics and minimum stress yields, as designated by the American Society for Testing Materials (ASTM). See Figure 26.2.

Preparing structural steel drawings requires a working knowledge and understanding of steel symbols, weld notations, identifications, drawing conventions, and measurement, fastening, and intersection methods.

ASTM TYPE	MIN YIELD* STRESS POINT	MANUFACTURED FORMS	DESCRIPTION
A36	36,000 PSI	Sheets Plates Bars Shapes Rivets Nuts Bolts	A medium carbon steel that is the most commonly used structural steel. Suitable for buildings and general structures, and capable of welding and bolting.
A440	42,000 PSI	Plates Bars Shapes	A high-strength, low-alloy steel suitable for bolting and riveting, but not welding. Used for lightweight structures—high resistance to corrosion.
A441	40,000 PSI	Plates Bars Shapes	A high-strength, low-alloy steel modified to improve welding capabilities in lightweight buildings and bridges.
A572	41,000 PSI	Limited types of shapes Bars & plates	A high-strength, low-alloy economical steel suitable for boltings, riveting, and welding with lightweight high toughness for buildings and bridges.
A242	42,000 PSI	Plates Bars Shapes	A durable, corrosion-resistant, high-strength, low-alloy steel that is lightweight and used for buildings and bridges exposed to weather. Can be welded with special electrodes.
A588	42,000 PSI	Plates Bars Shapes	A lightweight, corrosion-resistant, high-strength, low-alloy steel with high durability in high thicknesses used for exposed steel.
A514	90,000 PSI	Limited shapes & plates	A quenched and tempered alloy steel with varying strength, width, thickness, and type.
A570	25,000 PSI	Plates Light shapes	A light gauge steel used primarily for decking, siding, and light structural members.
A606	45,000 PSI	Plates	A high-strength, low-alloy sheet and strip steel with high atmospheric corrosion resistance.

*American Society for Testing Materials

FIGURE 26.2 ■ Structural steel types.

STEEL STRUCTURAL MEMBERS

Steel used for structural purposes is manufactured in plates, bars, pipes, tubing, and a variety of other shapes. Many of these items are formed by passing steel between a series of rollers. Steel that has been shaped by this method is called **rolled steel**.

Plates

Structural plates are flat sheets of rolled steel. These sheets range in thickness from 1/8″ to 3″ and in width from 8″ to 60″. Plates are specified by thickness, width, and length in that order. See Figure 26.3.

Plates are used as webs in built-up girders and columns, as shown in Figure 26.4, and to reinforce other webs or flanges of structural steel shapes. Bearing plates provide bearing surfaces between columns and concrete footings. See Figure 26.5.

Bars

Steel bars used for structural purposes are available in round, square, hexagonal, and flat (rectangular) cross-section shapes. See Figure 26.6. Square, hexagonal, and round bars are manufactured in 1/16″ increments from 1/16″ to 12″. Flat bars are manufactured in 1/4″ increments up to 8″. Round bars are specified by diameter; square bars by width or gauge number; flat bars by width and thickness; and hexagonal bars by the distance across the flats (AF). Steel bars are used primarily for bracing other structural components and for concrete reinforcement.

Steel Pipe and Structural Tubing

Steel pipe and tubing are used extensively in exposed areas because of their clean, pleasing lines. Structural pipe and tubing are available in round, square, and rectangular cross-section shapes, as shown in Figure 26.7.

Hollow steel pipe is manufactured in sizes from 1/2″ to 12″ (inside diameter) and in three strength classes. Strength classes relate to wall thicknesses and are either standard weight (STO), extra strong (x-strong), or double extra strong (xx-strong). On structural drawings, pipe is specified by diameter and strength. Thus, a 4″ double extra-strong steel pipe is labeled: pipe 4 xx-strong. See Figure 26.8. When hollow steel pipe is used as a vertical structural support, it is called a lally column.

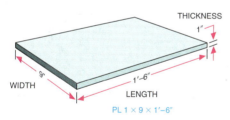

FIGURE 26.3 ■ Steel plate specifications.

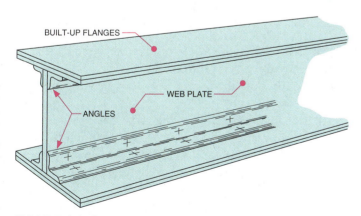

FIGURE 26.4 ■ Built-up steel plate girder.

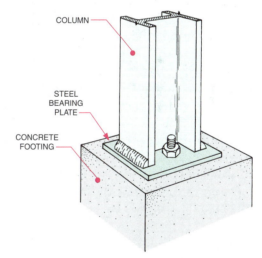

FIGURE 26.5 ■ Steel bearing plate.

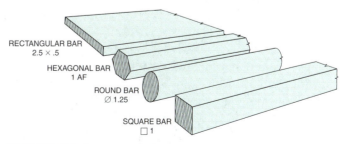

FIGURE 26.6 ■ Steel bar shapes.

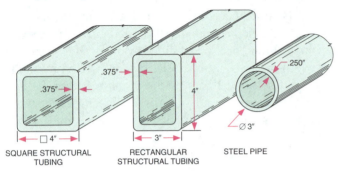

FIGURE 26.7 ■ Structural steel pipe and tube shapes.

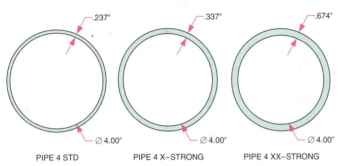

FIGURE 26.8 ■ Steel pipe thicknesses.

Square structural tubing is specified by cross-section width and thickness and is available in sizes from 2″ × 2″ to 10″ × 10″ outside dimension (OD). Rectangular tubing sizes range from 3″ × 2″ to 12″ × 8″ OD. Structural tubing is specified by the symbol TS followed by the width, thickness, and wall thickness. A rectangular structural tube 4″ wide and 3″ thick with a wall thickness of 1/4″ is therefore labeled TS 4 × 3 × .25. Round structural tubing is specified by outside diameter and wall thickness. For example, a 4″ diameter tube with a 1/4″ wall thickness is labeled 4 OD × .25. The length dimension of tubing is placed at the end of the note and/or on the working drawing.

Other Structural Steel Shapes

In addition to plates and bars, steel is rolled into channels, tee sections, and a number of other shapes. Their designations are shown in Figure 26.9. Steel shapes are designated on construction drawings by shape symbol, depth in inches, and weight in pounds per foot. Figure 26.10 shows the most common structural steel shapes and the related drawing symbols.

L-shapes (angles) are structural steel members rolled in the (cross-section) shape of the letter L with legs of equal or unequal length. (Equal leg lengths are available in sizes of 1″ to 8″. Unequal leg lengths range from 1 3/4″ to 9″.) Whether equal or unequal, the thickness of each

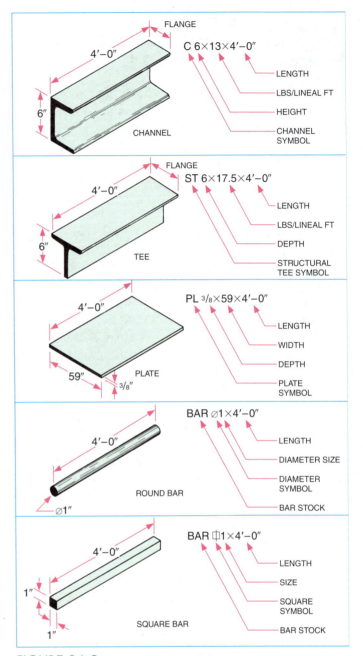

FIGURE 26.9 ■ Structural steel designations.

leg (called wall thickness) is always the same. L-shapes (angles) are specified on drawings by the symbol L followed by the length of each leg, followed by the wall thickness and length.

All inch marks are omitted on shape notations used on structural drawings since all sizes are assumed to be in inches unless otherwise specified. For example, an L-shape member with one 2″ and one 3″ leg and a wall thickness of 1/2″ is specified: L2 × 3 × .5. L-shape members are used as components in built-up beams, columns, and trusses. They are also used for connectors and as lintels in light- or short-span construction.

NAME	SECTIONAL FORM	SYMBOL	PICTORIAL
WIDE FLANGE		W	
AMERICAN STANDARD BEAM		S	
TEE		T	
ANGLE		L	
ZEE		Z	
AMERICAN STANDARD CHANNEL		C	
BULB ANGLE		BL	
LALLY COLUMN		◎	
SQUARE BAR		⌗	
ROUND BAR		φ	
PLATE		℔	

FIGURE 26.10 ■ Structural steel shapes.

Channels are rolled into a cross-section shape resembling the letter U, with the inner faces of flanges shaped with a 2/12 pitch. Channels are classified by depth, from 3″ to 15″. Two types of channels are specified for structural use: American Standard channels (C) and Miscellaneous channels (MC). Channels are specified by symbol (C or MC), followed by the depth times the weight per foot. An 8″-deep Standard channel that weighs 11.2 lb/ft is labeled C 8 × 11.2. Channels are used for roof purlins, lintels, and truss chords and to frame-in floor and roof openings.

S-shapes (formerly I-beams) are rolled in the shape of a capital letter I. American Standard shapes have narrow flanges with a 2/12 inside pitch. S-shapes are classified by the depth of the web and the weight per foot. The web is the portion between the flanges. Web depths

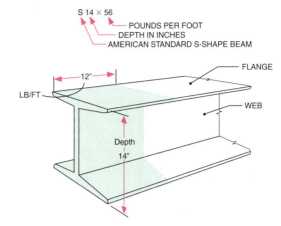

FIGURE 26.11 ■ S-shape designations.

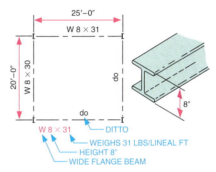

FIGURE 26.12 ■ W-shape designations.

range from 3″ to 24″. S-shapes are designated by their symbol (S) followed by the web depth and the weight per lineal foot. See Figure 26.11. For example, an S-shape member with a 14″-deep web that weighs 56 lb/ft is labeled S 14 × 56. On some drawings, the length may be added to the designation rather than as a dimension on the drawing. S-shapes are used extensively as columns because of their symmetry. Their narrow flanges are applicable to many designs where size restrictions are a problem.

W-shapes (formerly wide-flange or H beams) are similar to S-shapes but with wider flanges and comparatively thinner webs. Their capacity to resist bending is greater than that of S-shapes. W-shapes are designated in the same manner as S-shapes. See Figure 26.12. For example, W 18 × 62 describes a W-shape member with an 18″-deep web weighing 62 lb/ft. W-shapes are available in depths from 4″ to 36″. Lighter weight versions of W-shape members are known as M-shapes.

Structural **tees** are made by cutting through the web of an S or W shape, although some tees are rolled to order. If the web is cut exactly through the center, two identical tees result. The symbol for a tee is the capital letter T. On structural drawings the tee symbol includes the shape

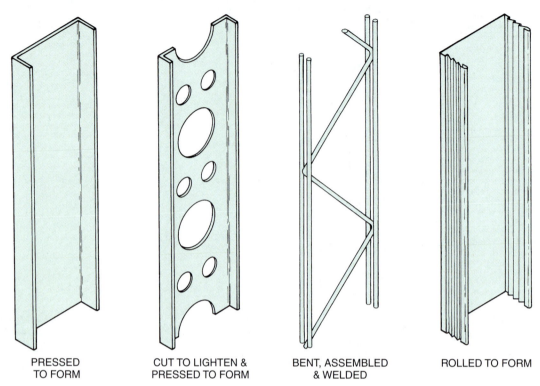

PRESSED
TO FORM

CUT TO LIGHTEN &
PRESSED TO FORM

BENT, ASSEMBLED
& WELDED

ROLLED TO FORM

FIGURE 26.13 ■ Types of metal joists and studs.

from which the tee was cut (S, W, or M) followed by the letter T, the depth of cut (from web to flange), and the weight per foot. Therefore, a tee-shape member cut in half from a W 12 × 50 would be specified WT 6 × 25. (6 is half the depth and 25 is half the weight per foot.) Tees are most commonly used for truss chords and to support concrete reinforcement rods.

Metal studs, as shown in Figure 26.13, are manufactured in different forms. Stud thicknesses vary depending on the loads to be supported. Widths are manufactured to be identical with conventional wood stud sizes. Fireproof steel framing can be substituted for conventional wood framing. It is insect-proof and will not decay, expand, contract, warp, or split.

STEEL FASTENERS AND INTERSECTIONS

Major structural steel members depend on a wide variety of joining methods and devices to function as a structurally stable frame. This includes the use of brackets, rivets, bolts, and welds to attach members to each other and to foundation piers and footings. See Figure 26.14. Some steel components can be assembled before shipping to a site. These are usually assembled

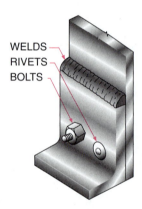

WELDS
RIVETS
BOLTS

FIGURE 26.14 ■ Methods of joining steel members.

and welded at a fabrication shop. All other members are assembled and permanently fastened at the building site. For example, some brackets may be welded to a girder at a shop, then bolted or welded to a column at the site during construction.

Brackets

Most structural steel members intersect at right angles. Many different types of brackets are used to provide a perpendicular surface for bolting, riveting, or welding. Figure 26.15 shows the use of brackets and welds to as-

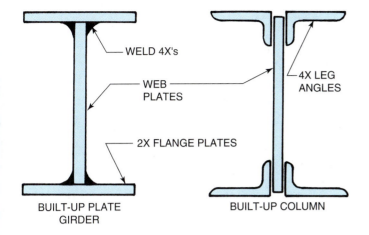

WELD 4X's

WEB PLATES

2X FLANGE PLATES

BUILT-UP PLATE GIRDER

4X LEG ANGLES

BUILT-UP COLUMN

WIDE FLANGE BEAM

WEB PLATES

2X COVER PLATES

COVER-PLATED GIRDER

4X LEG ANGLES

COVER-PLATED BUILT-UP COLUMN

Bracket information on a structural drawing shows the size of the bracket legs followed by the thickness, width, shape symbol, and fastening device information. See Figure 26.16. If brackets are to be welded to a member at a fabrication shop, a detail drawing is not provided in the field. Only the assembled intersection of the joint is drawn for field reference. Only shop fabricators are provided with a complete set of details.

Rivets

Rivets are used to connect steel members. Figure 26.17 shows the most common types of rivets use for shop fabrication and job-site use. Figure 26.18 shows the four basic types of rivet heads. The type of rivet specified is shown at the end of the drawing notation. Rivets are made of soft steel and, when heated rivets are cooled, they tend to shrink. The shrinking decreases the tightness of the joint. Consequently, bolts are now used more extensively than rivets in the erection of structural steel.

Bolts

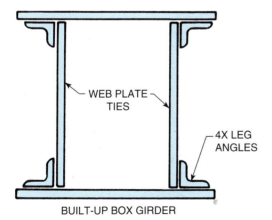

WEB PLATE TIES

4X LEG ANGLES

BUILT-UP BOX GIRDER

FIGURE 26.15 ■ Methods of joining built-up girders.

High-strength bolts and nuts can carry loads equal to rivets of the same size, but they can be turned tighter because of their high tensile strength. Bolts used in steel construction are either high-strength or unfinished bolts. High-strength bolts are used to connect extremely heavy load-bearing members such as girders, beams, and columns. See Figure 26.19. They are also used to attach members where shear loads are transmitted through the bolts.

Unfinished bolts are used for lighter connections, where loads are transmitted directly from member to

semble built up girders. Angles, L-shapes, and bent or welded plates are used for this purpose. Angles, nuts, and bolts help join a girder to the top of a column.

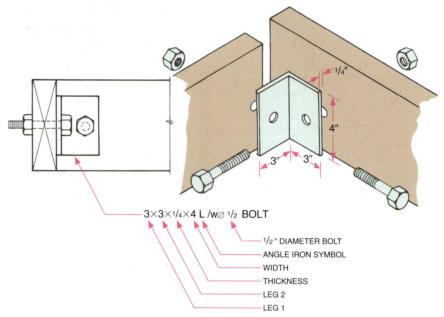

$3 \times 3 \times 1/4 \times 4$ L /wØ $1/2$ BOLT

$1/2$" DIAMETER BOLT
ANGLE IRON SYMBOL
WIDTH
THICKNESS
LEG 2
LEG 1

FIGURE 26.16 ■ Angle bracket and bolt notations.

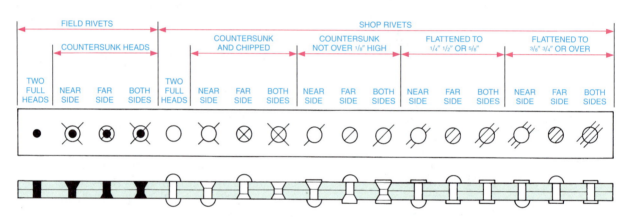

FIGURE 26.17 ■ Types of field and shop rivets.

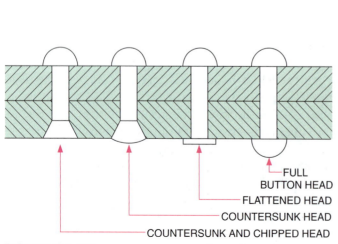

FULL
BUTTON HEAD
FLATTENED HEAD
COUNTERSUNK HEAD
COUNTERSUNK AND CHIPPED HEAD

FIGURE 26.18 ■ Common types of rivet heads.

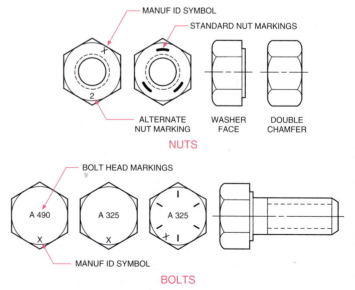

MANUF ID SYMBOL
STANDARD NUT MARKINGS
ALTERNATE NUT MARKING
WASHER FACE
DOUBLE CHAMFER
NUTS

BOLT HEAD MARKINGS
A 490
A 325
A 325
MANUF ID SYMBOL
BOLTS

FIGURE 26.19 ■ High-strength nut and bolt notations.

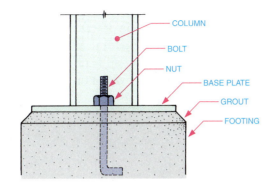

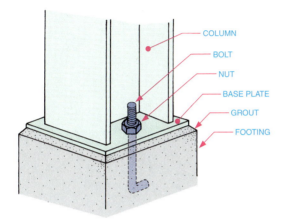

FIGURE 26.20 ■ Unfinished bolts used on a column base.

WELD SYM	WELD NAME
	FILLET
	PLUG/SLOT
	SPOT/PROJECTION
	SEAM
	BACK/BACKING
	SURFACING
//	SCARF (BRAZING)
	FLANGE – EDGE
	FLANGE – CORNER
	GROOVE – SQUARE
V	GROOVE – V
	GROOVE – BEVEL
Y	GROOVE – U
	GROOVE – J
	GROOVE – FLARE V
	GROOVE – FLARE BEVEL

FIGURE 26.21 ■ Types of welds.

member. For example, unfinished bolts could be used when a beam rests directly on a girder because there is no vertical shear load on the bolts holding these two members together. Unfinished bolts are also used to anchor column base plates to footings, as shown in Figure 26.20, since there is also no shear stress at this location.

Welds

Welding is a popular method of connecting structural steel members. It has some advantage over bolting. For example, fabrication is simplified by reducing the number of individual parts to be cut, punched with holes, handled, and installed. The major types of welds and their symbols are illustrated in Figure 26.21.

The convention used to locate welding information on drawings is a horizontal reference line with a sloping arrow directed to the joint. See Figure 26.22. The arrow may be directed right or left, upward or downward, but always at an angle to the reference line. If no extra marking is shown, a shop weld is assumed. Other drawing symbols give additional instructions. A triangular flag indicates a field weld. An open circle means weld all around the member.

WELD NOTATION:

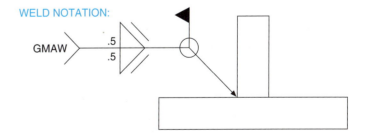

MEANS:
FILLET WELD BOTH SIDES
GAS METAL ARC WELDING
FLUSH CONTOUR
WELD ALL AROUND
WELD IN FIELD

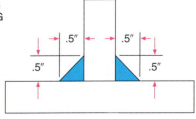

FIGURE 26.22 ■ Welding symbol used on a detail drawing.

The basic weld symbols or supplementary weld symbols are located midway on the horizontal reference line. See Figure 26.23. The symbol is located below the line if the weld is to be placed on the near side of the workpiece where the arrow points. The symbol is placed above the line if the weld is to be placed on the far side. The symbol is placed above and below if both sides are to be welded. The side of the weld (or its depth) is indicated to the left of the basic symbol. The length of the weld is shown to the right of the symbol. When long joints are used, intermittent welds are often specified. These are indicated by the length of weld followed by the center-to-center spacing (pitch). Such welds are usually staggered on either side of a joint. Figure 26.24 shows the position of supplementary weld symbols on the horizontal line.

The tail of the reference line may contain information about the kind of material or process required. This feature is not often used on structural steel details. When no information is required, the tail is omitted. Figure 26.25 shows the position of information on a welding symbol.

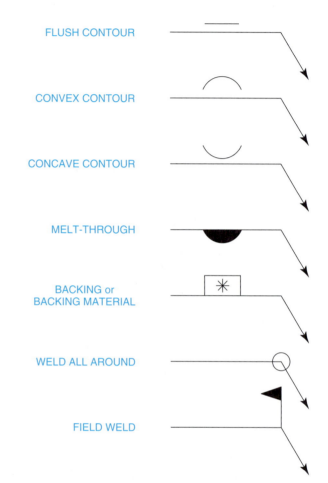

FIGURE 26.24 ■ Supplementary weld symbols.

F – Finish symbol
⌒ – Contour symbol
A – Groove angle: included angle of countersink for plug welds
R – Root opening: depth of filling for plug and slot welds
S – Depth of preparation
– Size or strength for specific welds
– Height of weld reinforcement
– Radii of flare-bevel grooves
– Radii of flare-V grooves
– Angle of joint (brazed welds)
(E) – Effective throat
T – Specific process or reference
L – Length of weld
– Length of overlap (brazed joints)
P – Pitch of welds (center-to-center spacing)
1 – Weld located on opposite side of arrow
2 – Weld located on same side of arrow
(N) – Number of spot or projection welds
⌐ – Weld made in field
o – Weld all around

FIGURE 26.23 ■ Common weld symbols.

USING CAD

CAD Welding Symbol Library

Welding symbol libraries contain symbols for each type of weld. They also include the symbol line on which the welding symbols can be located with the cursor. These are stored as base symbol *Blocks*. The numerals related to each weld are then added using the *Text* task.

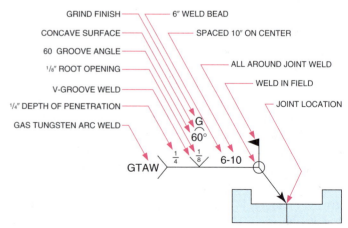

FIGURE 26.25 ■ Position of information on a weld symbol.

Although steel construction is made extremely rigid through the use of regular fasteners, special cross-bracing is often required to counteract lateral wind loads.

STRUCTURAL STEEL DRAWING CONVENTIONS

Structural steel drawings are of several types: design (schematic) drawings, working (shop) drawings, and erection drawings. *Design (schematic) drawings* are very symbolic and show only the position of each structural member with a single line. Notations describing each member's shape, size, and weight are included on each line. When several members with identical characteristics are aligned, the successive lines are labeled with a ditto symbol (DO) indicating that the shape, size, and weight of the member are identical to the previous member's. See Figure 26.26.

Working (shop) drawings are complete orthographic engineering drawings showing the exact size and shape of each member, including every cut, hole, and method of fastening. Figure 26.27 shows a structural steel detail drawing that is part of a series of detail drawings indexed to an erection set of plans. Figure 26.28 is an example of an erection plan. *Erection drawings* show the method and order of assembling each member, which is coded for easy field identification. The specific methods used to prepare structural steel floor, wall, and roof framing drawings are covered in Chapters 28, 29, and 30.

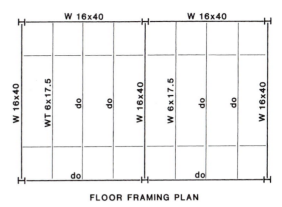

FIGURE 26.26 ■ Structural steel schematic drawing.

Lightweight Steel Framing

Lightweight steel structures resemble wood skeleton-frame buildings. Light-gauge zinc-coated steel members (C-studs, U-shaped tracks, and angles) are manufactured in structural wood sizes such as 2 × 4, 2 × 6, 2 × 8, etc. Using screws, not nails, contractors can build structures in a manner similar to that used for building wood structures. Steel members can also be joined with bolts, welds, or power-driven fasteners. Figure 26.29 shows how C-studs intersect with tracks to form a stud wall. Lightweight steel floor systems also resemble conventional wood floor framing, as shown in Figure 26.30. Lightweight steel framing provides

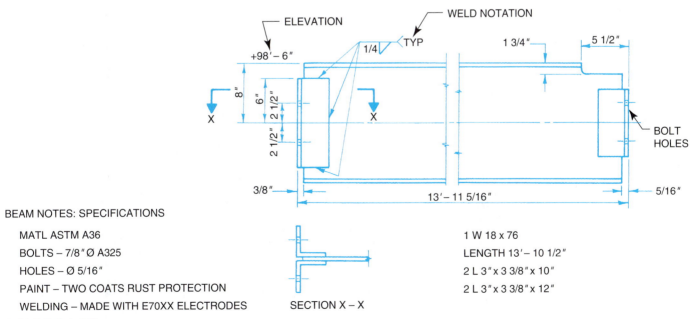

BEAM NOTES: SPECIFICATIONS

MATL ASTM A36

BOLTS – 7/8″ Ø A325

HOLES – Ø 5/16″

PAINT – TWO COATS RUST PROTECTION

WELDING – MADE WITH E70XX ELECTRODES

SECTION X – X

1 W 18 x 76

LENGTH 13′– 10 1/2″

2 L 3″ x 3 3/8″ x 10″

2 L 3″ x 3 3/8″ x 12″

FIGURE 26.27 ■ Steel detail working drawing.

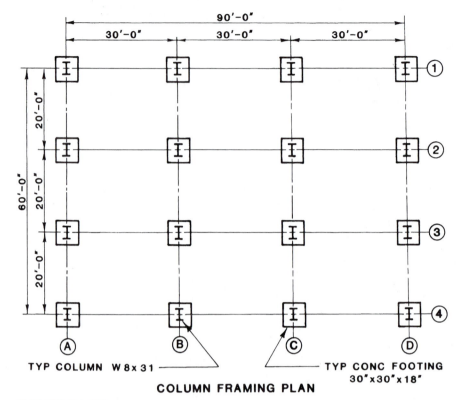

TYP COLUMN W 8 x 31

TYP CONC FOOTING
30"x30"x18"

COLUMN FRAMING PLAN

FIGURE 26.28 ■ Erection drawing showing column, footing, and girder locations.

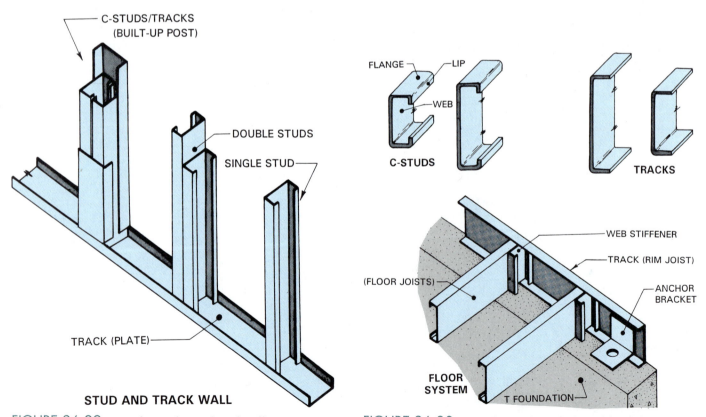

C-STUDS/TRACKS
(BUILT-UP POST)

DOUBLE STUDS

SINGLE STUD

TRACK (PLATE)

STUD AND TRACK WALL

FIGURE 26.29 ■ Lightweight steel stud wall.

FLANGE   LIP

WEB

C-STUDS

TRACKS

WEB STIFFENER

TRACK (RIM JOIST)

(FLOOR JOISTS)

ANCHOR BRACKET

**FLOOR
SYSTEM**

T FOUNDATION

FIGURE 26.30 ■ Lightweight steel floor framing.

structurally stable door lintels, jambs, and wall intersections, (Figure 26.31).

Lightweight steel members can be used with wood or concrete materials (Figure 26.32); however, special intersections must be designed in these cases. The most effective use of lightweight steel construction involves the use of steel for the entire structure. Figure 26.33 shows an example of an all-steel wall. This elevation section through an outside steel wall also shows the intersecting steel roof and floor members.

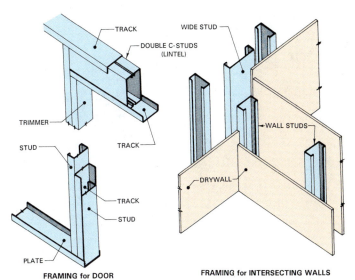

FRAMING for DOOR

FRAMING for INTERSECTING WALLS

FIGURE 26.31 ▪ Lightweight steel door and wall intersection framing.

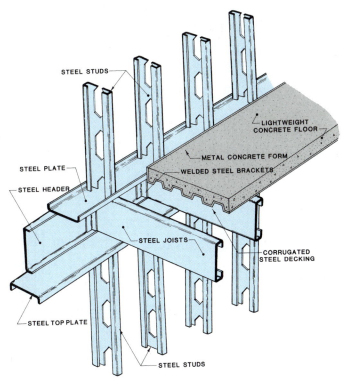

FIGURE 26.32 ▪ Lightweight steel wall intersecting concrete floor.

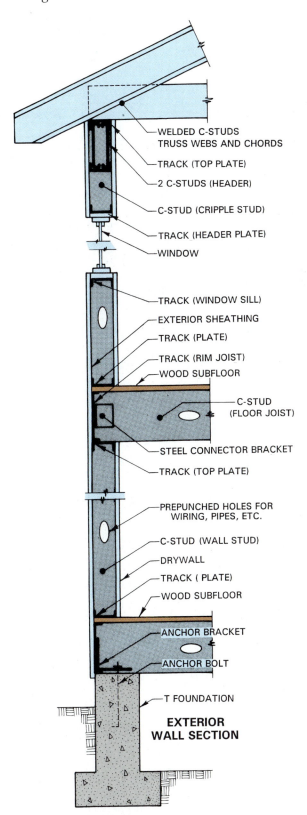

EXTERIOR WALL SECTION

FIGURE 26.33 ▪ Lightweight steel exterior wall section.

CHAPTER

26

Steel and Reinforced-Concrete Systems Exercises

1. Describe the differences between a design drawing, shop drawing, and erection drawing.

2. List the advantages and disadvantages of steel construction.

3. Describe the difference between beams, girders, columns, and purlins.

4. Name the types of structural steel shapes.

5. Draw the symbols for these steel shapes: wide flange beam, tee, angle, channel, and square bar.

6. Draw a structural steel framing plan and elevation drawing for the house you are designing.

7. Name the types of fastening methods used in steel construction.

8. Draw the symbols for these welds: fillet, spot, seam, edge flange, V groove, and backing.

9. Design, draw, and dimension a steel erection plan for the 28 × 56′ concrete block storage building shown in Figure 26.34. Show columns, footings, and girder locations. Allow 4′–0″ RO for window and 16′–0″ RO for the overhead door. Minimize interior columns and hold spans to a maximum of 18′–0″.

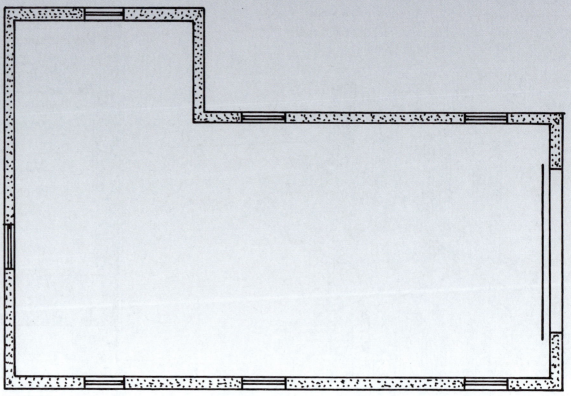

FIGURE 26.34 ■ Design, draw, and dimension a steel erection plan for this building.

CHAPTER 27

Disaster Prevention Design

OBJECTIVES

In this chapter, you will learn to:

- describe the measures that can be taken during construction to minimize potential damage from natural disasters.

- describe how to prevent gas leaks.

- name ways to provide fire protection for a structure and its residents.

- discuss methods for ensuring clean air and water in a building.

TERMS

cannon test
carbon monoxide
Environmental Protection Agency
   (EPA)

radon
safe room

toxic material
Underwriters Laboratories (UL)

INTRODUCTION

Hurricanes, tornadoes, high winds, floods, blizzards, gas leaks, wildfires—these disasters are not preventable. However, precautions can be designed into a structure to prevent or minimize the damage they cause.

PREVENTING WIND DAMAGE

Winds do not need to reach hurricane or tornado velocity to seriously damage or demolish a structure. To design a structure with maximum wind resistance, a continuous and strong structural link must be made from the foundation to the roof. Reinforced concrete construction resists wind loads better than does concrete block; and concrete block resists wind loads better than wood construction. However, these differences can be minimized through the use of structural ties and reinforcements (Figure 27.1). Many methods are employed to achieve a strong structure—depending on the type of construction.

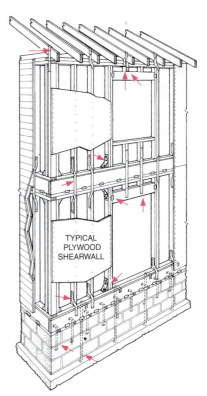

FIGURE 27.1 ■ Continuous load transfer design using structural link connectors. (*Simpson Strong-Tie Co., Inc.*)

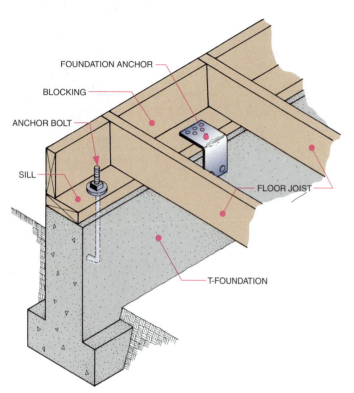

FIGURE 27.2 ■ Fastening sill to foundation with anchors.

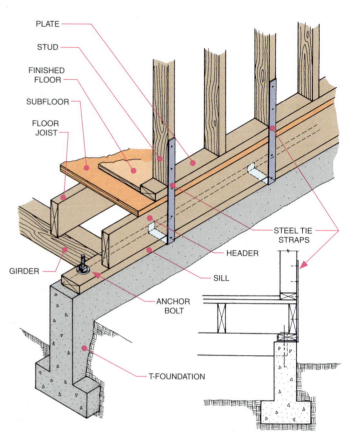

FIGURE 27.4 ■ Sill tie-down straps.

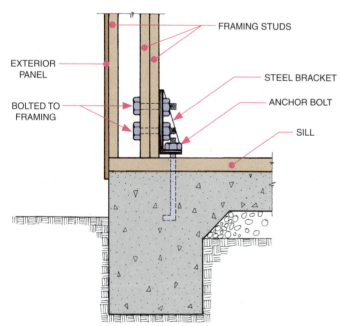

FIGURE 27.3 ■ Fastening sill to a slab foundation with a steel bracket.

To maximize structural strength, vertical rebars and continuous pours are required during concrete construction. In wood construction, the effective placement of timber connectors is vital.

Sills must be fastened firmly to the foundation using foundation anchors. Figure 27.2 shows how this is done using anchor bolts and L-brackets. Figure 27.3 shows the use of a steel bracket bolted to studs and a slab foundation anchor. Long anchors provide additional control by connecting the foundation and sill to the wall framing.

Tie straps help hold the wall framing to the sill, as shown in Figure 27.4. Additional blocking on the sill and between vertical members also helps anchorage and provides added rigidity to the structural frame. Steel ties between joists and the foundation provide protection against uplift. See Figure 27.5.

Nonmasonry wall surfaces should be covered with shear stress plywood or a material with equal wind velocity rating. See Figure 27.6. Windows must also be rated to withstand the same wind velocity as the structural walls.

On bilevel platform framed structures, the two levels must be connected in order to transfer the uplift forces from the upper studs to the lower studs. Figure 27.7 shows how metal straps are used, and Figure 27.8 shows bolts and brackets connecting two levels. Bolt hold-downs between joists and studs on both levels will also

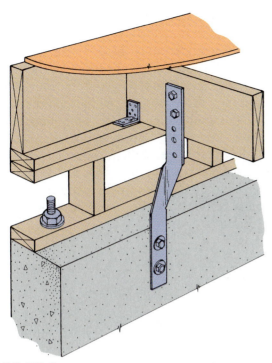

FIGURE 27.5 ■ Ties used to prevent uplift.

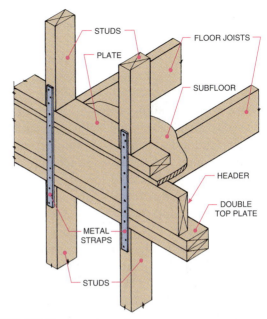

FIGURE 27.7 ■ Platform framing connectors that use metal straps.

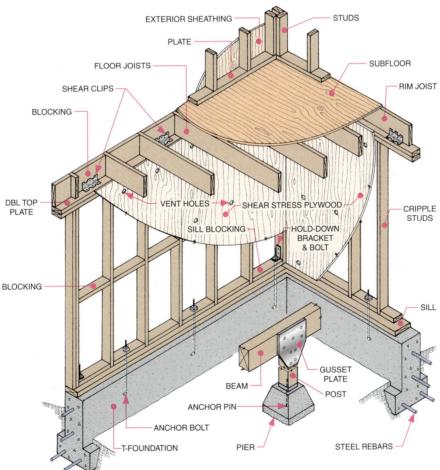

FIGURE 27.6 ■ Methods used to protect against wind and seismic (earthquake) damage.

provide this connection. When bolts are used, they must be firmly fastened with nuts.

Connections between wall framing and roof framing are also critical. Winds trapped under overhangs can damage or break away soffits and fascia boards. Any area where wind can be trapped should be ventilated to allow the air to escape. Roof sheathing should be connected with hurricane clips. Roofing screws or nails, not staples, must be used to attach sheathing, soffit, and fascia material. In high-wind areas, wind shear shingles should be specified and connectors used to attach top plates to studs, floor joist to girders, and top plates to rafters as shown in Figure 27.9. Structural straps as shown in Figure 27.10 hold studs and ceiling joists together.

In addition to structural reinforcements, other design features must be considered to minimize wind damage. Hip roofs withstand more wind force than gable roofs, so gable roof trusses should be cross braced to add more resistance. Large roof overhangs should be vented to prevent uplift. Specifying #1 grade lumber for roof

trusses also helps stabilize roof structures during high wind gusts.

Doors and windows must be designed to withstand wind gusts that could expose the inside of the structure to severe damage. Window glass, in hurricane- or tornado-prone areas, should withstand the **cannon test**—a 2 × 4 stud hurled 34 feet before impacting the window. Windows in these areas should also be equipped with roll-down hurricane shutters or fitted with supports to which plywood covers can be attached. Doors in hurricane areas should open out and have top and bottom reinforcement bolts with welded hinge pins.

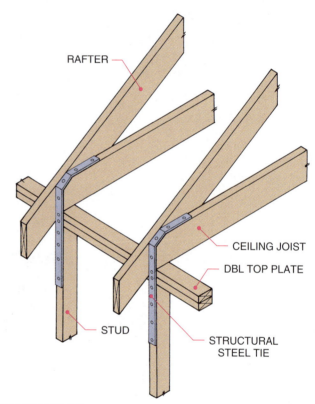

FIGURE 27.10 ■ Ceiling joist-stud structural tie.

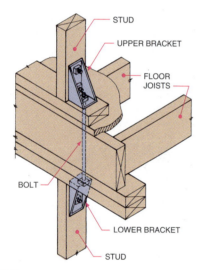

FIGURE 27.8 ■ Platform framing connectors that use bolts and brackets.

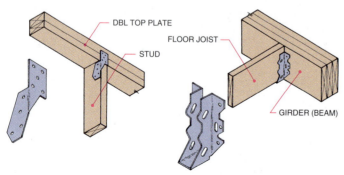
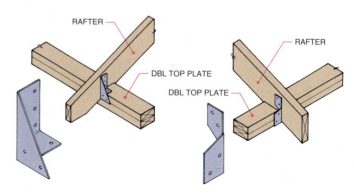

FIGURE 27.9 ■ Rafter-plate and hanger connectors.

PREVENTING EARTHQUAKE DAMAGE

The structural features designed to reduce wind damage also apply to earthquake damage reduction. Figure 27.11 illustrates many of these features. However, special attention is needed for earthquake protection as follows:

1. Design a continuous structural link from foundation to roof according to the same principles as those described earlier for preventing wind damage.

2. Specify that all gas appliances are to be connected with flexible lines.

3. Fasten built-in appliances to structural members, not to wallboard or trim. ✐

4. Don't use masonry chimney materials for above the roof line.

5. Specify push-type cabinet latches (touch latch).

6. Specify hook and eye latches on workshop cabinets.

7. Specify security film for windows—to prevent shattering.

8. Use ball-bearing supports to prevent columns from flexing or bending. See Figure 27.12.

9. Footings should cover the largest horizontal area possible.

10. Avoid using suspended floor systems, which could collapse as one unit on the floors below.

11. Ensure good site drainage.

12. Strap all gas, water, and vent pipes to framing members.

13. Bolt all tall, heavy wall furniture into wall studs.

14. Ensure that footings are poured on highly compacted, nonorganic soil or preferably on bedrock.

On large buildings, extreme measures are taken to help maintain a building's earthquake tolerance. These include X- or diamond-shaped wall reinforcements with large shock absorbers. Mexico's tallest building, the 57-story Torre Major Tower is to absorb and cushion tremor effect through the use of these latest earthquake technologies. Figure 27.13 illustrates the placement of the X- and diamond-shaped earthquake reinforcements in Torre Major Tower.

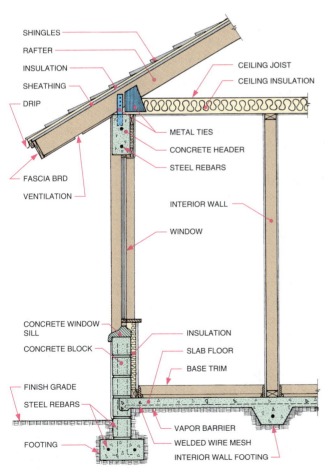

FIGURE 27.11 ■ Hurricane- and earthquake-resistant design components.

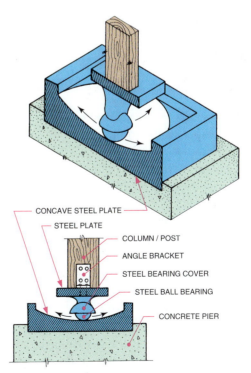

FIGURE 27.12 ■ Ball-bearing supports help absorb ground movement during an earthquake.

FIGURE 27.13 ■ Torre Major Tower is designed to absorb and cushion tremor effect.

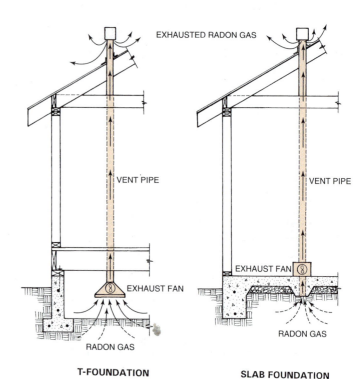

EXHAUSTED RADON GAS

VENT PIPE

VENT PIPE

EXHAUST FAN

EXHAUST FAN

RADON GAS

RADON GAS

T-FOUNDATION

SLAB FOUNDATION

FIGURE 27.14 ■ Radon prevention methods.

PREVENTING GAS LEAKS

The nature of gas is to expand and fill all available space. When gas escapes from containers, pipes, or the soil, it may be trapped in a sealed building. These trapped gases can seriously injure or kill people who inhale them. If ignited with a spark or flame, some gases can explode and cause great damage.

Natural gas, propane, carbon monoxide, and radon are potentially the most dangerous gases. Gas fumes created by many synthetic building materials and interior furnishings are suspected of causing long-range health problems.

Natural and Propane Gas

To minimize the potential for gas leakage, specify that pipes should be strapped to structural members and use flexible lines to all appliances. Locate propane containers as far from structures as possible. Specify electronic pilots for all gas appliances.

Carbon Monoxide (CO)

Carbon monoxide is a colorless, odorless gas produced by combustion. Poisoning from CO gas causes brain damage at low levels and death at high levels. Carbon monoxide is produced by all combustion appliances including wood fireplaces, stoves, grills, gas or oil furnaces, clothes dryers, and water heaters. Designers must ensure that gas emissions from these devices are adequately vented to the exterior and cannot escape into the interior. An inspection should be specified and CO detectors installed during construction. Detectors should be mounted on ceilings on each story.

Radon

Radon is a colorless gas produced by the natural decay of the element radium. This radioactive gas enters a structure from the soil. To minimize risk, specify polyethylene membranes under floors and slabs. Specify total parging (protective plasterwork) on all walls below the grade line. Specify a separate outside air source for fireplaces to avoid drawing radon in through small cracks in the floor system. Ventilate crawl spaces to the outside. Provide a fan-driven air escape vent to avoid gas buildup. Test for radon before construction begins. In areas with high levels of radon, a central vent from under the foundation through the roof can divert gas from entering the house. See Figure 27.14.

FIRE PREVENTION AND CONTROL

The best fire protection is to observe building codes and to use electrical and gas devices approved by **Underwriters Laboratories (UL)**, a nonprofit agency that tests products for safety. All electrical wiring and devices must also be grounded according to UL standards. Fire control measures include specifying smoke detectors on each level of the house. Fire extinguishers and water hoses should be specified to reach the kitchen, laundry, garage, and workshops. Designing buildings that are "firesafe" also involves following fire code structural design such as specifying firewalls in hazardous areas, firestop wall construction, and roof venting construction.

AIR PURIFICATION

To provide the maximum pure air, specify electronic air filters for all furnaces, air conditioners, and the air exchangers inside heat pumps. Air cleaners may be installed to quickly purify contaminated air.

WATER CONTROL

Water quality varies greatly from one community to another. To maximize water purity, specify a water system that may include a water softener, carbon filter, ultraviolet filter, and/or reverse osmosis unit. You may specify a dedicated faucet for pure drinking water and for the refrigerator ice-maker.

In addition to purifying consumable water, unwanted water in the form of moisture (humidity) must be controlled to eliminate the formation of fungi, mold, dry rot, and termite attacks. Diverting water from foundation walls through the use of swales, trenches, berms, or drains is the first line of defense. See Figure 27.15. The

use of vents, fans, cross ventilation, and dehumidifiers is most effective inside a structure.

CONTROLLING TOXIC POLLUTANTS

Although the effects of synthetic material vapors may not be immediate, the long-range health effects may be serious. Avoid the interior use of exposed and unsealed building materials that contain toxic materials as defined by the **Environmental Protection Agency (EPA)**, a federal regulatory agency. Use solid wood products where possible.

Toxic materials such as arsenic, asbestos, and formaldehyde are found in many building materials including structural members, coatings, insulation, and wall and floor coverings. Where possible designers should specify nontoxic materials recommended by the EPA.

PEST CONTROL

Wood-boring insects and wall-dwelling rodents can be controlled by specifying treatment barriers under the slab and footing. Ensuring that the foundation connections to the upper structure have no openings for insect penetration must be part of the structural design.

Household pests can be controlled in the design process by specifying the installation of wall cavity tubes. These tubes safely carry pesticides into the walls without detectable levels entering living spaces.

ACCIDENT PREVENTION

Eliminating electrical shock and fire hazards involves following the electrical design practices outlined in Chapter 31. This means designing circuits that won't overload (GFCI), outlets that minimize the use of long extension cords, and switches and wires with the capacity to carry the maximum load as designed.

Designers can prevent accidents by following the national code and Occupational Safety and Health Administration requirements that are appropriate to the building type. Special consideration should be given to the design of stairs, kitchens, baths, children's bedrooms, and swimming pools where most accidents occur.

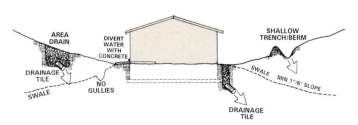

FIGURE 27.15 ■ Exterior water control methods.

SAFE ROOMS

Safe rooms are designed to provide personal and/or property safety from intruders, fire, hurricanes, tornadoes, earthquakes, or air pollution (poisonous gas, smoke, or other pollutant fallout). Safe rooms should be designed to be structurally stable, bulletproof, flame retardent, and able to withstand winds up to 260 MPH.

Most safe rooms are constructed of preengineered concrete panels (Figure 27.16). Some use special bulletproof film with wood construction if intruder protection is the only concern. A fireproof steel door should open inward. Safe rooms should be located in the center of the building with no windows or with an escape window that is bulletproof. Some safe rooms double as closets; they are often located in a basement.

Equipment for a safe room should include a cell phone, air purifier (scrubber), closed-circuit TV for the outside, battery radio, and generator with connecting lights. Furniture should include chairs, a table, and bed. Supplies should include water, nonperishable food, flashlights, and sleeping bags.

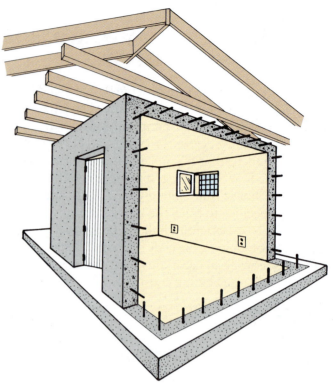

FIGURE 27.16 ■ Safe room design.

CHAPTER

27

Disaster Prevention Design Exercises

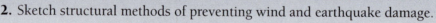

 1. List the design features you will use in the house you are designing to prevent wind, earthquake, gas, and fire damage.

2. Sketch structural methods of preventing wind and earthquake damage.

 3. Explain how you will provide for water and air purification in the house you are designing.

4. Sketch the ventilating system you will use to provide adequate ventilation in the house you are designing.

5. Locate and design a safe room for your house.

PART 7

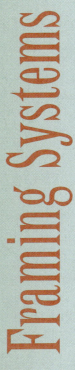

Framing Systems

28 CHAPTER
Floor Framing Drawings

OBJECTIVES

In this chapter, you will learn to:

- identify the components of floor systems.
- draw a floor framing plan that shows all structural parts.
- draw details of sills, supports, and stairwells.

TERMS

bays	grid column identification system	sequential column identification system
blocking	I-joists	
decking	lookout joists	truss joists

INTRODUCTION

The design of floor framing systems demands careful calculation of the live and dead loads acting on a floor. The exact size and spacing of floor framing members, plus the most appropriate materials, must be selected.

Some floor framing drawings are prepared to show only the structural support for the floor platform. Others may illustrate details of construction, such as the attachment of the floor frame to the foundation.

 ## TYPES OF PLATFORM FLOOR SYSTEMS

Systems that are supported by foundation walls and/or beams or girders are called platform floor systems. These differ from ground-level slab floors that are structurally part of the foundation. (Refer back to Chapter 23.) Platform floor systems are divided into three types: conventional, heavy timber (plank-and-beam) and panelized floor systems.

- Conventional systems. Conventionally framed platform systems provide a flexible method of floor framing for a wide variety of designs. Floor joists are usually spaced 16″ (406 mm) apart and are supported by side walls of the foundation and/or by girders. See Figure 28.1.

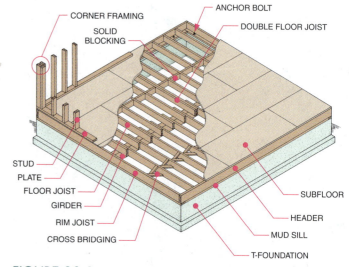

FIGURE 28.1 ■ Conventional platform floor system.

- Plank-and-beam systems. The plank-and-beam method of floor framing uses fewer and larger members than conventional framing. Because of the increased size and the rigidity of the larger members, longer distances can be spanned. Unlike conventional framing, no cross bridging is needed between joists. Extended blocking provides the needed rigidity. See Figure 28.2.
- Panelized systems. Panelized floor systems are composed of preassembled sandwich panels. The panels

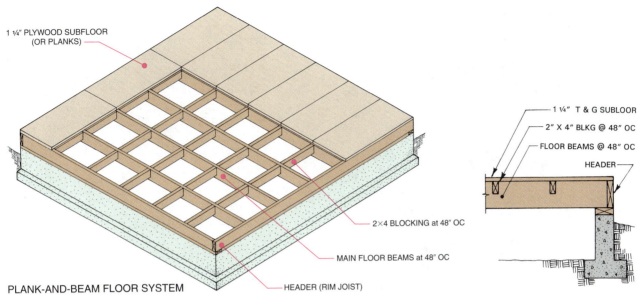

PLANK-AND-BEAM FLOOR SYSTEM

FIGURE 28.2 ■ Plank-and-beam platform floor system.

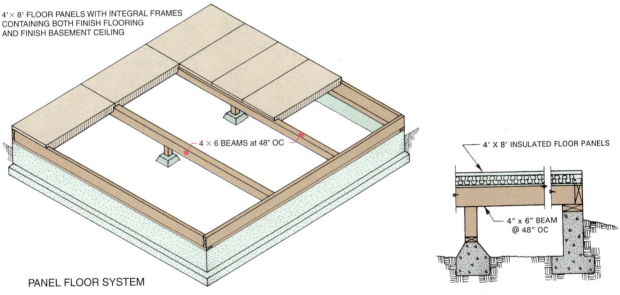

PANEL FLOOR SYSTEM

FIGURE 28.3 ■ Panelized platform floor system.

are made from a variety of skin and core materials. Panelized systems are used for long, clear spans. The main advantage of panelized systems is the reduction of on-site construction costs. See Figure 28.3.

FLOOR FRAMING MEMBERS

Floor framing members for platform floors consist of decking, joists, and girders or beams. Supporting walls and columns also are part of the floor framing system.

Decking

Decking is the surface of a floor system. Decking usually consists of a subfloor and a finished floor, although in some plank systems these are combined. Subfloor decking materials range from wood boards and plywood sheets to concrete slabs or corrugated steel sheets. The finished floor may be wood, ceramic, vinyl, concrete, or carpeting.

Wood Decking

Boards or plywood sheets used as subflooring are placed directly over the joists. See Figure 28.4. Unless the edges of

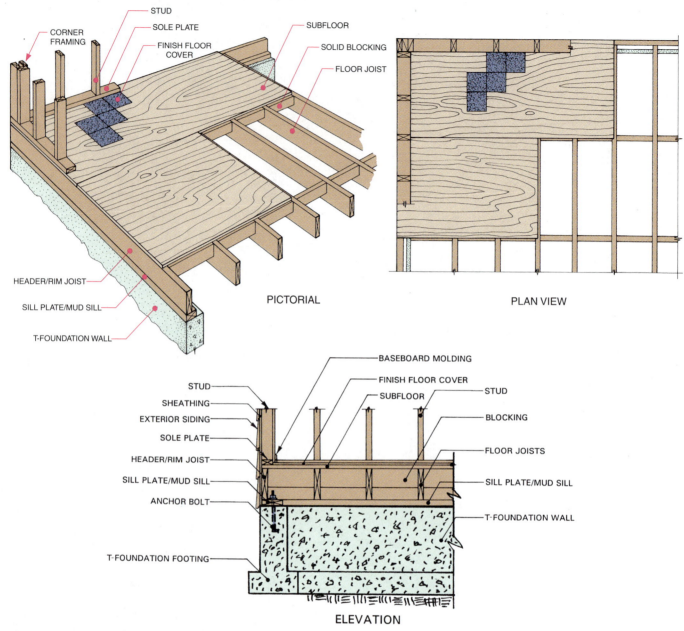

FIGURE 28.4 ■ Plywood subfloor system.

plywood sheets are tongue and grooved, blocking may be needed beneath the joints. **Blocking** consists of short pieces of lumber nailed between the joists. The blocking provides additional support for the subfloor joints. Sole plates for exterior and interior walls are laid directly on the subflooring.

The functions of the subfloor are to:

■ Increase the strength of the floor and provide a surface for laying a finished floor.

■ Help to stiffen the position of floor joists.

■ Serve as a working surface during construction.

■ Help to deaden sound.

■ Prevent dust from rising through the floor.

■ Help insulate.

■ Act as a buffer to soften and reduce the hard impact of slab floor construction. See Figure 28.5.

Finished flooring is installed over the subfloor. The finished floor provides a wearing surface over the subfloor. If there is no subfloor, the finished floor must be tongue-and-groove boards 1 1/2″ to 2″ thick. Hardwood such as oak, maple, beech, and birch is used for finished floors over wood subfloors. Vinyl, ceramic tile, marble, and carpeting are also used.

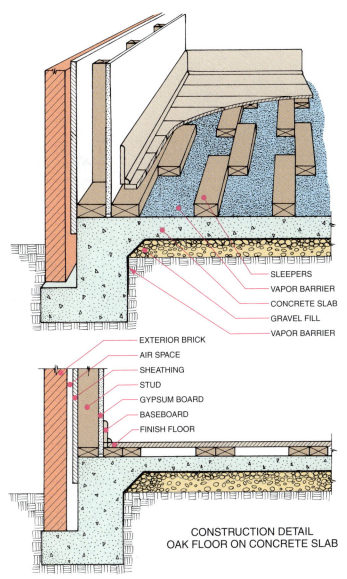

SLEEPERS
VAPOR BARRIER
CONCRETE SLAB
GRAVEL FILL
VAPOR BARRIER

EXTERIOR BRICK
AIR SPACE
SHEATHING
STUD
GYPSUM BOARD
BASEBOARD
FINISH FLOOR

CONSTRUCTION DETAIL
OAK FLOOR ON CONCRETE SLAB

FIGURE 28.5 ■ Subfloor sleepers on a concrete slab.

Steel and Concrete Decking

Steel decks for floors (and roofs) use corrugated sheets, interlocking galvanized steel panels, or cellular units over steel beams. As shown in Figure 28.6, steel deck details or sectional drawings are usually prepared to show the relationship of the decking to the structural support members.

When steel subfloors are used, they are usually constructed of corrugated sheet steel. These subfloors act as platform surfaces during construction and also provide the necessary subfloor surface for a concrete slab floor. Steel subfloors are not always needed with concrete slabs. The concrete slabs can function as subfloors if the concrete floor is precast with reinforcement bars.

Joists

Floor joists are horizontal members that rest on a foundation wall and/or girder (beam) and support the floor decking. Floor joists must support the maximum live load of the floor. Many variations of wood and steel joists are manufactured.

Solid Lumber Joists

Conventional residential framing normally uses solid lumber joists. Solid lumber joists are most commonly made from Douglas fir, pine, spruce, or hemlock. All joists must rest directly on a girder, beam, wall, or foundation wall. See Figure 28.7.

Double joists, known as *trimmers*, are used under partitions (interior walls) to provide added support. Sometimes a small space is left between these joists to provide a channel for electrical or piping access. See Figure 28.8. Where joists are to be level with the top of a girder, as shown in Figure 28.9, they should rest on a ledger board attached to the girder or hung on joist hangers.

Joists are often intersected at right angles to reduce spans. At these intersections, double joists and joist hangers are used to attach joists to trimmers or beams. See Figure 28.10.

To prevent drift and warp and to distribute loads more evenly, bridging is used between joists. See Figure 28.11.

Wherever joists need to be cut for an opening, such as a stairwell, chimney, or hearth, it is necessary to provide trimmers and auxiliary joists called *headers*. Headers are placed at right angles to trimmers to support the ends of joists that are cut. A header cannot be of greater depth than joists. Therefore, headers are usually doubled (placed side by side) to compensate for additional loads. See Figure 28.12, page 490. Double headers and trimmers are also used for floor, ceiling, and roof openings around fireplaces, chimneys, stairwells, and skylights. Figure 28.13A, page 490, shows a typical chimney opening through a floor structure. Note that all openings are double framed on all sides.

The size, spacing, and strength needed for joists depend on the loads acting on a floor. (Review Chapter 22 and see Appendix B for the physical effects of loads, material strength, member size, and spacing.) Standard joist sizes for most residential construction range from 2 × 6 to 2 × 14. Normal residential spacing of wood joists is from 12″ to 24″ OC (on center). The most common spacing is 16″ OC.

Floor framing systems are horizontal planes, and floor loads must be transferred horizontally to vertical supports

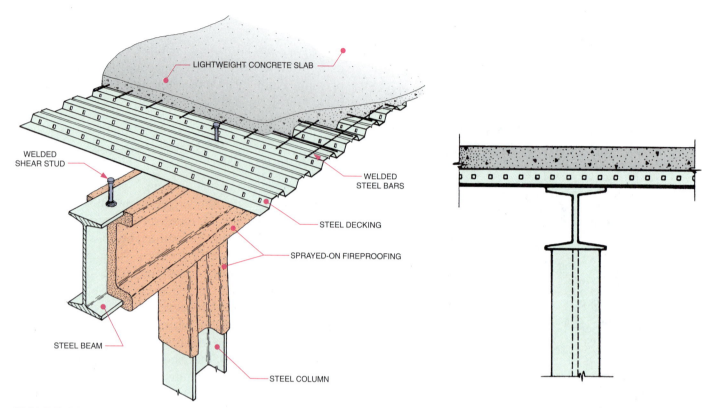

LIGHTWEIGHT CONCRETE SLAB

WELDED
SHEAR STUD

WELDED
STEEL BARS

STEEL DECKING

SPRAYED-ON FIREPROOFING

STEEL BEAM

STEEL COLUMN

FIGURE 28.6 ■ Steel deck details.

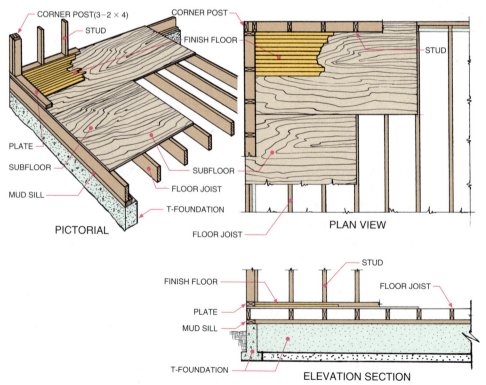

CORNER POST(3–2 × 4)

STUD

CORNER POST

FINISH FLOOR

STUD

PLATE

SUBFLOOR

MUD SILL

T-FOUNDATION

SUBFLOOR

FLOOR JOIST

PICTORIAL

FLOOR JOIST

PLAN VIEW

STUD

FINISH FLOOR

FLOOR JOIST

PLATE

MUD SILL

T-FOUNDATION

ELEVATION SECTION

FIGURE 28.7 ■ Solid lumber joists in a conventionally framed floor system.

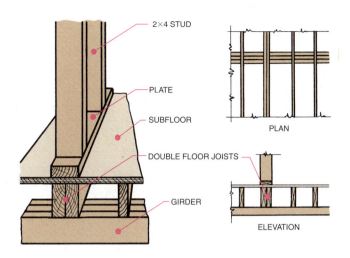

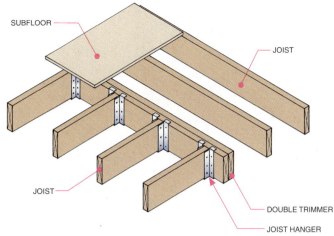

FIGURE 28.10 ■ Joist hangers used to change joist direction.

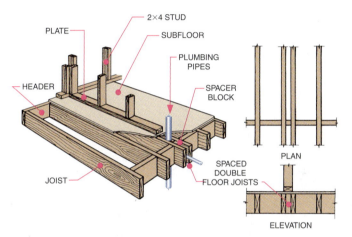

FIGURE 28.8 ■ Trimmer detail drawings.

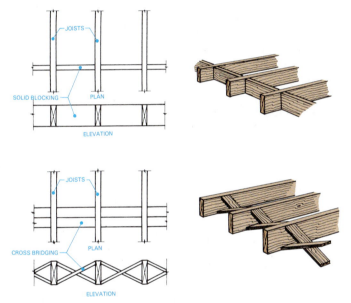

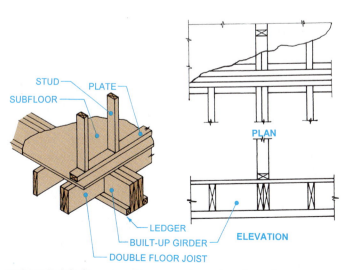

FIGURE 28.9 ■ Double joists aligned with girder top.

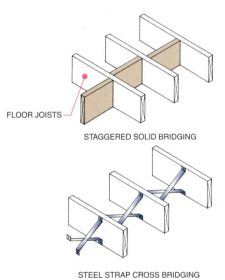

FIGURE 28.11 ■ Types of joist bridging.

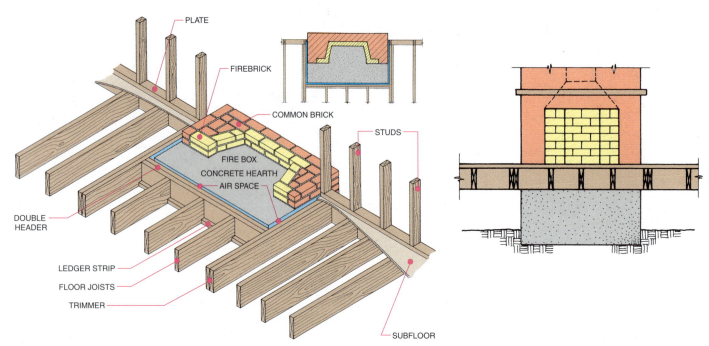

FIGURE 28.12 ■ Trimmers and headers around a fireplace structure.

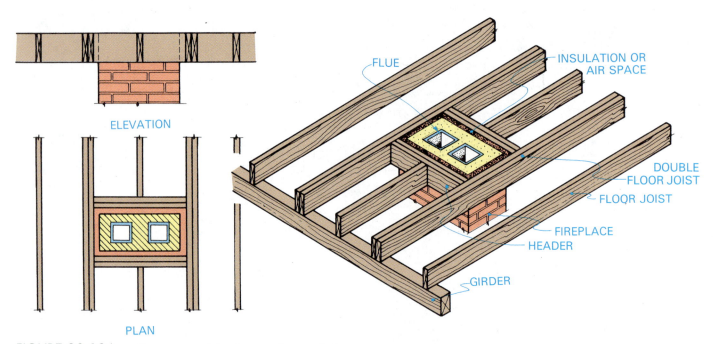

FIGURE 28.13A ■ Trimmers and headers used around chimney opening.

**Math
Connection**

such as columns or bearing walls. To support both fixed (dead) loads and moving (live) loads, a floor system must be rigid yet elastic enough to absorb bending stresses.

Live loads bear directly on the decking and joists. Therefore, the total live load for the room containing the heaviest furniture and the heaviest traffic should be used to compute the total load for the entire floor.

To find the live load in pounds per square foot, divide the total room load in pounds by the number of square feet supporting the load. The average live load for most residences is between 40 and 50 pounds per square foot. See Appendix B, "Mathematical Calculations," for calculating loads.

Engineering tables are used to select sizes and spacing for structural members of standard grade wood. Figure

ALLOWABLE SPANS FOR FLOOR JOISTS USING NONSTRESS-GRADED LUMBER

SIZE OF FLOOR JOISTS (INCHES)	SPACING OF FLOOR JOISTS (INCHES)	MAXIMUM ALLOWABLE SPAN (Feet and Inches)							
		GROUP I		GROUP II		GROUP III		GROUP IV	
		PLASTERED CEILING BELOW	WITHOUT PLASTERED CEILING BELOW	PLASTERED CEILING BELOW	WITHOUT PLASTERED CEILING BELOW	PLASTERED CEILING BELOW	WITHOUT PLASTERED CEILING BELOW	PLASTERED CEILING BELOW	WITHOUT PLASTERED CEILING BELOW
2 × 6	12	10-6	11-6	9-0	10-0	7-6	8-0	5-6	6-0
	16	9-6	10-0	8-0	8-6	6-6	7-0	5-0	5-0
	24	7-6	8-0	6-6	7-0	5-6	6-0	4-0	4-0
2 × 8	12	14-0	15-0	12-6	13-6	10-6	11-6	8-0	8-6
	16	12-6	13-6	11-0	11-6	9-0	10-0	7-0	7-6
	24	10-0	11-0	9-0	9-6	7-6	8-0	6-0	6-6
2 × 10	12	17-6	19-0	16-6	17-6	13-6	14-6	10-6	11-6
	16	15-6	16-6	14-6	15-6	12-0	13-0	9-6	10-0
	24	13-0	14-0	12-0	13-0	10-0	10-6	7-6	8-6
2 × 12	12	21-0	23-0	21-0	21-6	17-6	19-0	13-6	14-6
	16	18-0	20-0	18-0	19-6	15-6	16-6	12-0	13-0
	24	15-0	16-6	15-0	16-6	12-6	13-6	10-0	16-6

FIGURE 28.13B ■ Engineering tables used to select size and spacing of floor members.

28.13B is a table used to select joist sizes and spacing. To use this table, first select the wood group. Numbers indicate the quality of lumber. Number 1 is top quality, and number 4 is lowest quality. Next select the shortest span the joists must cross. The joist size and spacing (distance between members) is shown on the left. For example, for group 1, if the shortest span is 12'-0", then the smallest joist that can be used is a 2 × 8 at 16" OC, which has a maximum span of 12'-6". Figure 28.14 shows the normal girder spans for standard grade wood, with and without supporting walls. Always remember: As the size, spacing, and load vary, the spans must vary accordingly. If the span is changed, the joist and girder spacing must also change. You should not overdesign but you MUST NOT underdesign.

Floor Truss Joists

Truss joists have long, usually parallel, top and bottom chords connected by shorter pieces called *webs*. Triangular web patterns give truss joists the ability to span long distances, and they weigh less than solid lumber. Truss joists have open spaces through which plumbing and electrical lines can be run. They can also be designed to

GIRDER SIZE	SUPPORTING WALLS	NO WALL SUPPORT
4 × 4	3'-6"	4'-0"
	3'-0"	3'-6"
4 × 6	5'-6"	6'-6"
	4'-6"	5'-6"
4 × 8	7'-0"	8'-6"
	6'-0"	7'-6"

FIGURE 28.14 ■ Allowable girder spans in light construction.

accommodate plumbing, heating and air-conditioning ducts. Truss joists are manufactured using stress grade lumber or a combination of wood, metal, and/or composite materials. See Figures 28.15 and 28.16.

Floor I-Joists

Structurally, **I-joists** are similar to steel I-beams (S-beams). I-joist weight is reduced through the use of thin webs without sacrificing the strength and stability provided by the

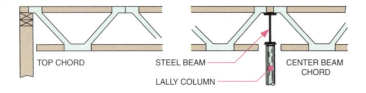
TOP CHORD

STEEL BEAM

LALLY COLUMN

CENTER BEAM CHORD

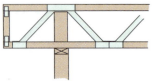
CANTILEVER BOTTOM CHORD

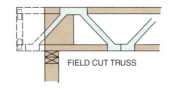
FIELD CUT TRUSS

FIGURE 28.15 ■ Fabricated truss joists with lumber flanges (chords) and steel webs.

For second-floor framing.

For larger cantilevered extensions.

FIGURE 28.16 ■ Use of truss joists.

flanges. See Figure 28.17. I-joist webs are usually made of stranded wood. The flanges are made of laminated or solid lumber. The web area can be cut, within limits, to receive HVAC, plumbing, or electrical lines without sacrificing strength. Figure 28.18 shows the use of I-joists with laminated headers and beams in a platform floor system.

Laminated Joists

Laminated members are made by bonding layers of material together. Solid and parallel strand lumber are used to manufacture laminated joists, headers, and beams that can span up to 60'. Laminated joists are extremely straight, dimensionally stable, and without checks, cracks, or twists.

FIGURE 28.17 ■ I-joists compared to solid joists. (*TrusJoist MacMillan*)

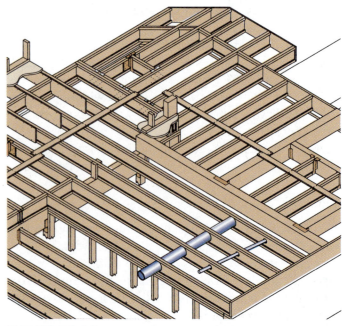
FIGURE 28.18 ■ I-joists in a platform system.

Steel Joists

Joists made of steel are either *bent sheet steel* or *open web* joists. See Figure 28.19. Open web steel joists are more common and consist of angles and bars welded into truss shapes.

There are three types of steel open web joists: short-span, long-span, and deep long-span. Because joists are closely spaced, one note is used on drawings to give the number, classification, spacing, and length of all joists in a series. Only the first few joists in a series are usually noted. For example, if there are 8 short-span joists spaced at 3' intervals over a 24' distance, the note should read 8SP @ 3'-0" = 24'-0". In addition, a notation is placed on a line representing the joists' direction and includes the length, class of the joist, and load range.

Girders and Beams

As explained previously, girders are the largest horizontal support members and rest on columns, posts, or exterior walls. In heavy construction, beams are the members that span the distances between girders. In residential timber construction, the terms beam and girder are often used interchangeably.

Wood Girders and Beams

Girders (or beams) used in wood construction are either built-up from solid lumber, laminated, or fabricated as shown in Figure 28.20. Girders are connected to joists with a varity of connectors. Figure 28.21 shows how wood girders (beams) are used to support joists. Figure

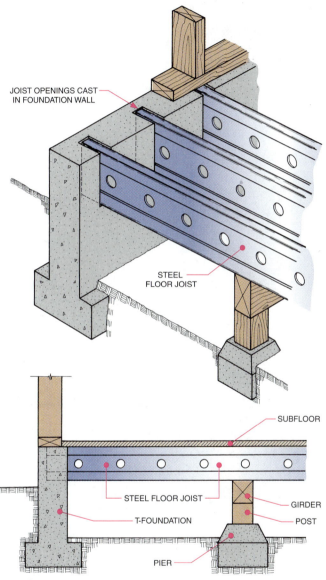

Bent sheet.

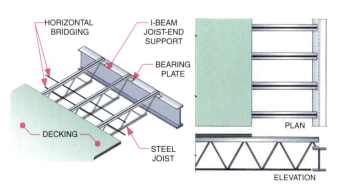

Open web.
FIGURE 28.19 ■ Types of steel joists.

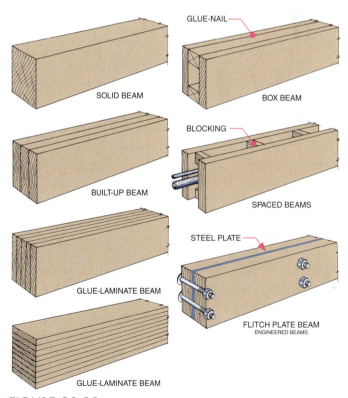

FIGURE 28.20 ■ Types of beams or girders and connectors.

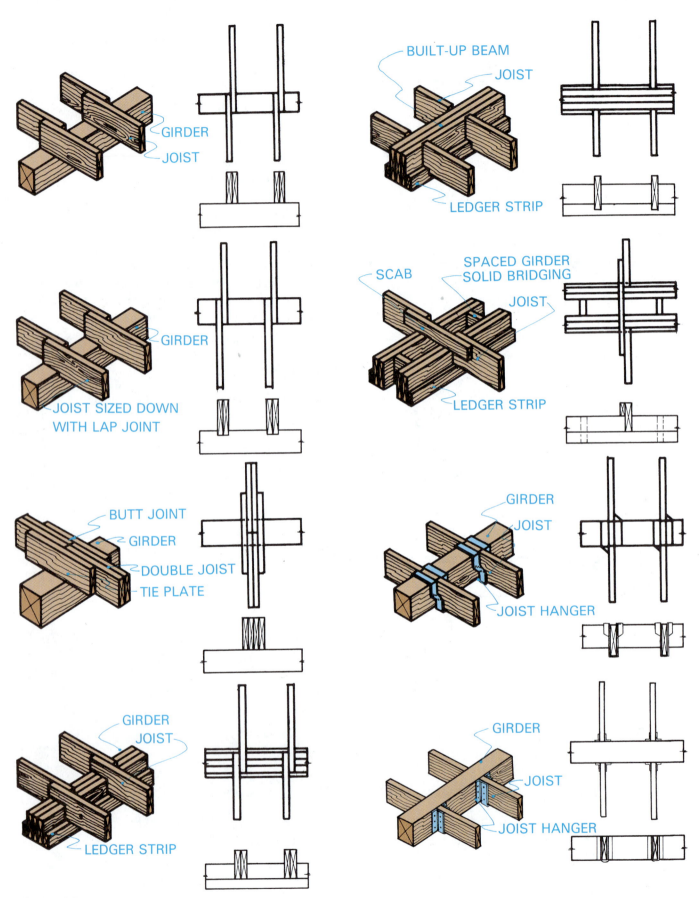

FIGURE 28.21 ■ Use of wood girders and connectors to support joists.

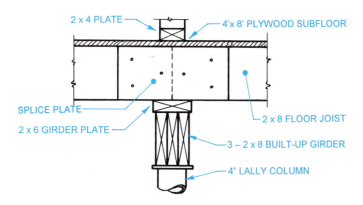

SECTION THROUGH GIRDER

FIGURE 28.22 ■ Built-up wood girder intersection with joist and column.

28.22 shows the use of a lally (pipe) column, built-up wood girder, and plate to support floor joists. Floor joists may be positioned on top of a girder or hung from the side of a girder with joist hangers.

When two or more girders are placed end to end to span the distance between outside supports, the joints between the girders must be placed directly over supporting columns or posts. Built-up girder members must be overlapped. Heavy timber girders should be half-lapped over columns. Members can be spliced to reduce compression, tension, bending, and torque forces. See Figure 28.23. Second-floor and higher level girders are supported by bearing partitions or by columns aligned with lower level columns. See Figure 28.24.

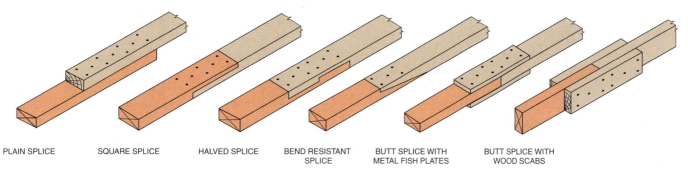

PLAIN SPLICE   SQUARE SPLICE   HALVED SPLICE   BEND RESISTANT SPLICE   BUTT SPLICE WITH METAL FISH PLATES   BUTT SPLICE WITH WOOD SCABS

FIGURE 28.23 ■ Splices used to reduce structural forces.

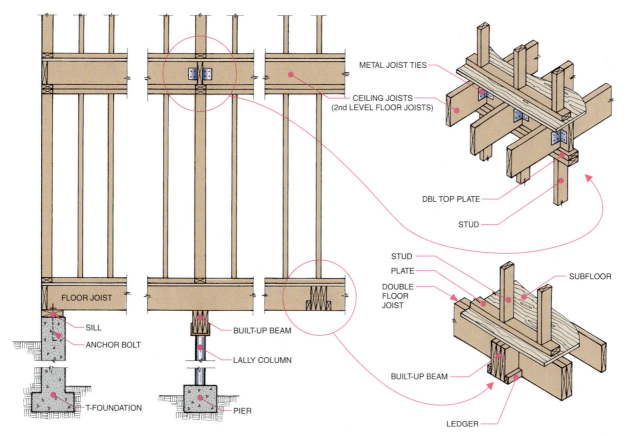

FIGURE 28.24 ■ Beam and bearing wall support of first- and second-floor framing systems.

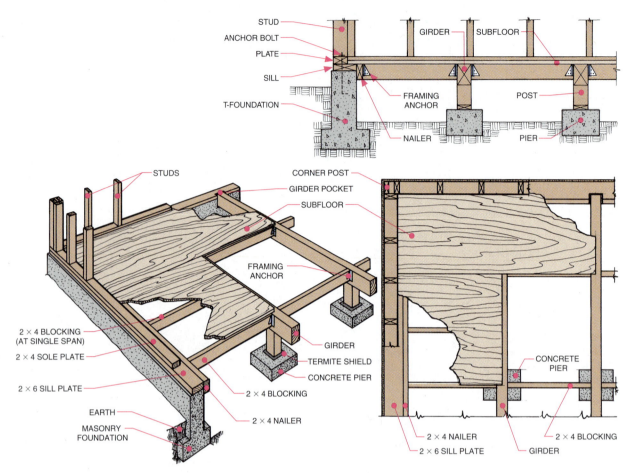

FIGURE 28.25 ■ Post-and-beam floor support system.

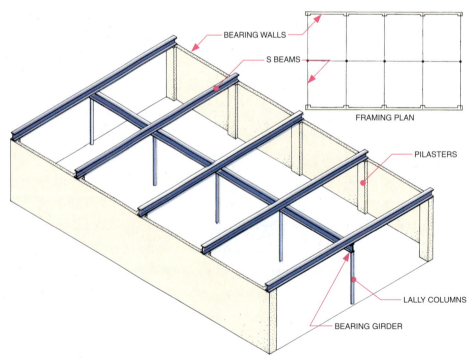

FIGURE 28.26 ■ Girders, beams, and columns used in steel cage construction.

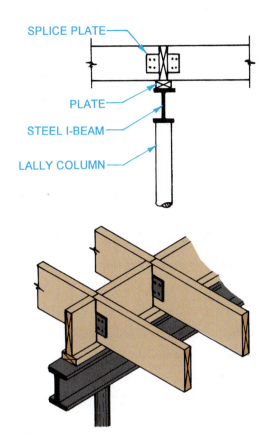

FIGURE 28.27 ■ Steel I-beam support detail.

SPLICE PLATE

PLATE

STEEL I-BEAM

LALLY COLUMN

Post-and-beam floor systems have no joists. In this system, the girders and blocking perform the function of joists. Girders rest directly on posts, and the subflooring rests directly on girders. In this type of construction, girders need to be spaced more closely together than girders that support joists. See Figure 28.25.

Steel Girders and Beams

Steel girders and beams may be solid steel (S-shape or W-shape) or built-up steel assemblies. They are bolted or welded together to form the major structural element of steel cage construction. See Figure 28.26. In wood construction, steel girders may be used in addition to or in place of wood girders. Figure 28.27 shows the use of a steel I-beam and lally column to support floor joists.

FLOOR FRAMING PLANS

If a set of architectural drawings does not include a separate floor framing plan, the builder, not the designer, de-

termines the framing design. Floor framing plans for wood framing and steel framing use different conventions and symbols.

Floor Framing Plans for Wood

Floor framing plans for wood structures range from simple to very detailed, as shown in Figure 28.28. In some cases, the direction of joists and girders may simply be shown on the floor plan (Figure 28.28A). In the most detailed floor framing plans, each structural member is represented by a double line to show exact thicknesses (Figure 28.28B).

The more simplified plan (Figure 28.28C) is a shortcut method of drawing floor framing plans. A single line is used to designate each member. Chimney and stair openings are shown by diagonals. Only the outline of the foundation and post locations is shown.

The abbreviated floor framing plan (Figure 28.28D) simply shows the entire area where the uniformly distributed joists are placed. The direction of joists is shown with arrows, and notes indicate the size and spacing of the joists. This type of framing plan is usually accompanied by numerous detail drawings.

While floor framing may be shown on drawings, the method of cutting and fitting the subfloor and finished floor panels is usually left up to the builder. When off-site or mass-produced floor systems are to be installed, a floor panel layout may be prepared. Its purpose is to ensure maximum use of materials and minimum waste.

Second-floor Framing Plans

Floor framing details for second floors or above are usually shown with a full section through an exterior wall. In balloon framing the studs are continuous from the foundation to the eave. See Figure 28.29. The second-level joists are supported by a ribbon board (ledger) that is recessed and nailed or screwed directly to the studs. In construction for platform framing, second-floor joists are placed on a top plate that rests on the platform (subfloor), as shown in Figure 28.30.

When a combination of exterior covering materials is used, the relationship between the floor system and the exterior wall is shown on an elevation section. See Figure 28.31.

If an upper-level floor is cantilevered over the first floor, the second-floor joists are either parallel or perpendicular to the first-floor top plate that supports the

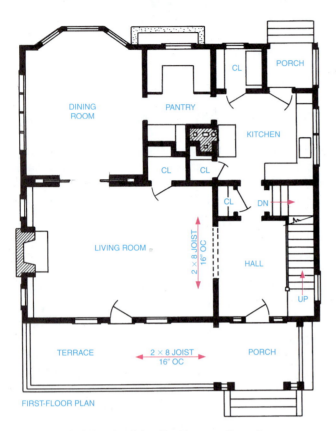

A. A method of showing joist direction on a floor plan.

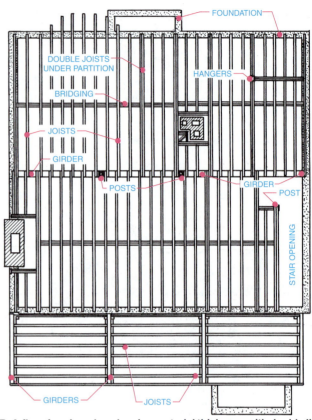

B. A floor framing plan showing material thickness with double lines.

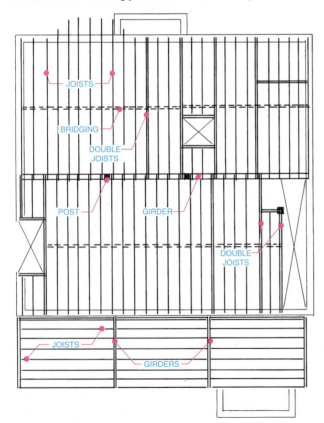

C. A simplified method of drawing floor framing plans with single lines.

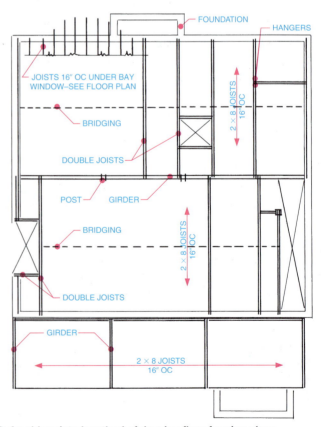

D. An abbreviated method of drawing floor framing plans.

FIGURE 28.28 ■ Types of floor framing plans.

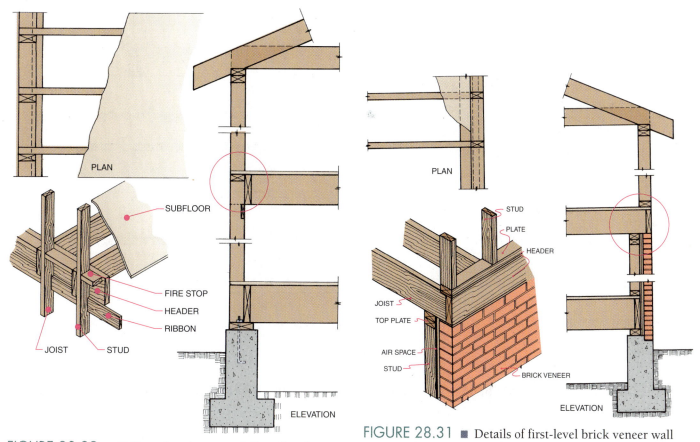

FIGURE 28.29 ■ Balloon framing second-floor details.

FIGURE 28.31 ■ Details of first-level brick veneer wall under second-level stud wall.

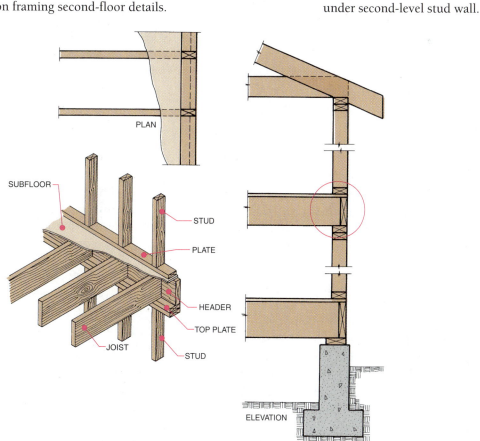

FIGURE 28.30 ■ Platform framing second-floor details.

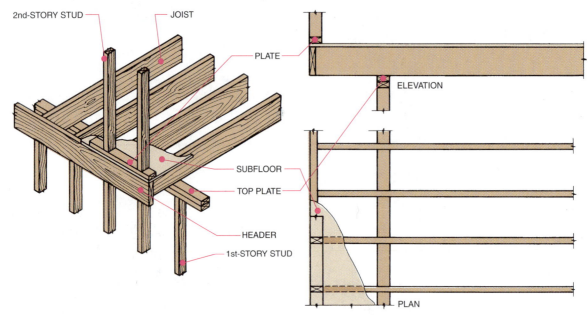

2nd-STORY STUD JOIST PLATE ELEVATION SUBFLOOR TOP PLATE HEADER 1st-STORY STUD PLAN

FIGURE 28.32 ■ Cantilever framing with joists perpendicular to an outside wall.

second floor. See Figure 28.32. When the joists are perpendicular to the wall, the construction is simple. When the joists are parallel to the wall, **lookout joists** must be used to support the cantilevered second-floor extension.

Details

Most floor framing plans are easily interpreted by experienced builders. Some plans may require additional details to explain a construction method. Details are drawn to eliminate the possibility of error in interpretation or to explain a unique condition. Details may be enlargements of what is already on the floor framing plan. They may be prepared for dimensioning purposes, or they may show a view from a different angle for better interpretation. On the floor framing plan in Figure 28.33, the circles indicate areas for which detail drawings have been made.

■ Sill support. The sill is the transition between the foundation and the exterior walls of a structure. Drawings that show sill construction details reveal not only the construction of the sill, but also the method of attaching the sill to the foundation. A sill detail is included in most sets of architectural plans. Some sill details are drawn in pictorial form. Pictorial drawings are easy to interpret but are difficult and time consuming to draw and dimension, and they are not orthographically accurate. Therefore, most sill details are prepared in two-dimensional

sectional form. Figure 28.34 shows a detail drawing in plan, elevation, and pictorial form.

The floor area in a sill detail sectional drawing usually shows at least one joist. This is done to show the direction of the joist and its size and placement in relation to the placement of the subfloor and finished floor. For example, the floor framing plan in Figure 28.34 is needed to indicate the spacing of joists and studs. The elevation section shows the intersections between the foundation sill and exterior wall. For this reason, both a plan and an elevation are needed to more fully describe this type of construction.

Finished detail drawings include all dimensions and notations necessary for construction.

Sill details are also required to show how materials are joined, such as masonry, wood, precast concrete, and structural steel. Special design features are shown on details as well. For example, firecuts are necessary in masonry walls. See Figures 28.35 and 28.36. Sill details are also needed to show the intersection between girders and foundation walls. Figure 28.37 shows a box sill used to support a girder. Some details may be expanded to show additional details. See Figure 28.38.

■ Intermediate support. If girders cannot safely span the distance between exterior supports, intermediate vertical supports (such as wood or steel columns, piers, or bearing walls) must be used to reduce the span. Detail drawings for intermediate supports consist of sections, elevations, or plan

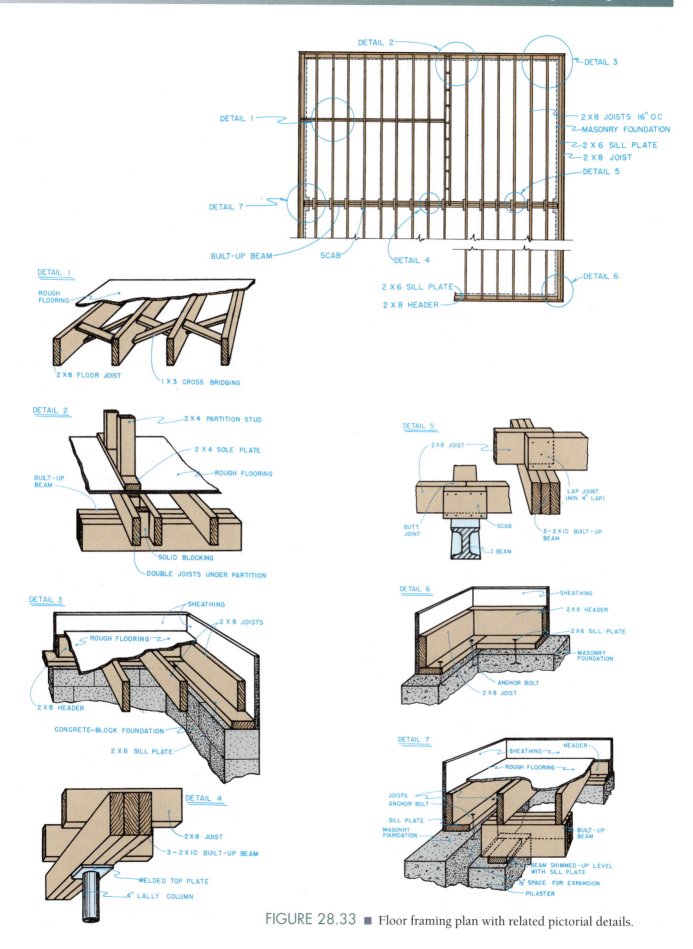

DETAIL 2

DETAIL 3

DETAIL 1

2 X 8 JOISTS 16" O C
MASONRY FOUNDATION
2 X 6 SILL PLATE
2 X 8 JOIST
DETAIL 5

DETAIL 7

DETAIL 4

BUILT-UP BEAM   SCAB

2 X 6 SILL PLATE
2 X 8 HEADER

DETAIL 6

DETAIL 1

ROUGH FLOORING

2 X 8 FLOOR JOIST   I X 3 CROSS BRIDGING

DETAIL 2

2 X 4 PARTITION STUD
2 X 4 SOLE PLATE
BUILT-UP BEAM
ROUGH FLOORING
SOLID BLOCKING
DOUBLE JOISTS UNDER PARTITION

DETAIL 5

2 X 8 JOIST
LAP JOINT (MIN 4" LAP)
BUTT JOINT   SCAB
3 - 2 X 10 BUILT-UP BEAM
I BEAM

DETAIL 3

SHEATHING
2 X 8 JOISTS
ROUGH FLOORING
2 X 8 HEADER
CONCRETE-BLOCK FOUNDATION
2 X 6 SILL PLATE

DETAIL 6

SHEATHING
2 X 8 HEADER
2 X 6 SILL PLATE
MASONRY FOUNDATION
ANCHOR BOLT
2 X 8 JOIST

DETAIL 4

2 X 8 JOIST
3 - 2 X 10 BUILT-UP BEAM
WELDED TOP PLATE
4" LALLY COLUMN

DETAIL 7

SHEATHING   HEADER
ROUGH FLOORING
JOISTS
ANCHOR BOLT
SILL PLATE
MASONRY FOUNDATION
BUILT-UP BEAM
BEAM SHIMMED-UP LEVEL WITH SILL PLATE
½" SPACE FOR EXPANSION
PILASTER

FIGURE 28.33 ■ Floor framing plan with related pictorial details.

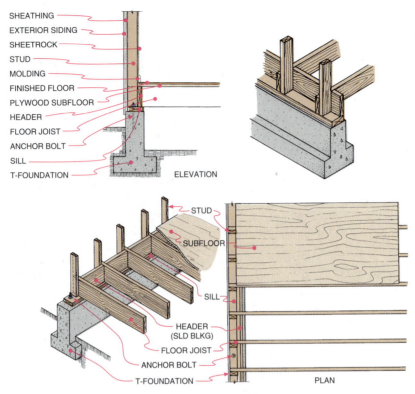

SHEATHING
EXTERIOR SIDING
SHEETROCK
STUD
MOLDING
FINISHED FLOOR
PLYWOOD SUBFLOOR
HEADER
FLOOR JOIST
ANCHOR BOLT
SILL
T-FOUNDATION
ELEVATION

STUD
SUBFLOOR
SILL
HEADER
(SLD BLKG)
FLOOR JOIST
ANCHOR BOLT
T-FOUNDATION
PLAN

FIGURE 28.34 ■ Balloon framing sill details.

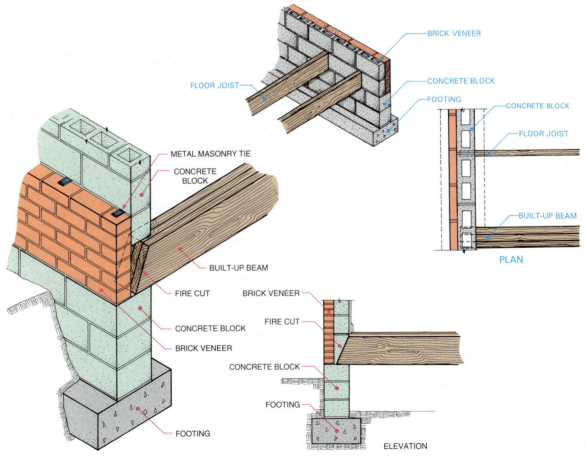

BRICK VENEER
CONCRETE BLOCK
FOOTING
FLOOR JOIST
CONCRETE BLOCK
FLOOR JOIST
BUILT-UP BEAM
PLAN

METAL MASONRY TIE
CONCRETE BLOCK
BUILT-UP BEAM
FIRE CUT
CONCRETE BLOCK
BRICK VENEER
FOOTING

BRICK VENEER
FIRE CUT
CONCRETE BLOCK
FOOTING
ELEVATION

FIGURE 28.35 ■ Brick veneer sill details showing firecut beams.

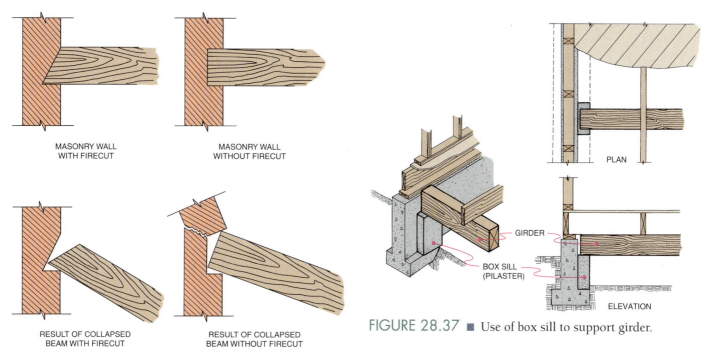

MASONRY WALL
WITH FIRECUT

MASONRY WALL
WITHOUT FIRECUT

RESULT OF COLLAPSED
BEAM WITH FIRECUT

RESULT OF COLLAPSED
BEAM WITHOUT FIRECUT

FIGURE 28.36 ■ Firecut used to prevent masonry
collapse.

PLAN

GIRDER

BOX SILL
(PILASTER)

ELEVATION

FIGURE 28.37 ■ Use of box sill to support girder.

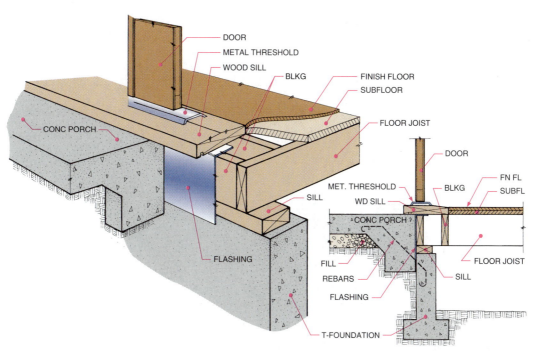

DOOR
METAL THRESHOLD
WOOD SILL
BLKG
FINISH FLOOR
SUBFLOOR
FLOOR JOIST
CONC PORCH

SILL
FLASHING

DOOR
MET. THRESHOLD
WD SILL
CONC PORCH
FILL
REBARS
FLASHING
T-FOUNDATION
FN FL
BLKG
SUBFL
FLOOR JOIST
SILL

FIGURE 28.38 ■ Relationship of door sill to interior wood floor and exterior concrete porch
slab.

views to show the intersections between footings,
vertical members, and horizontal girders. Intermediate framing details are needed, for example,
where level changes in floor level require special
support. See Figure 28.39.

■ Headers and trimmers. Building codes require
headers and trimmers around stairwells, chimneys,
and hearths and under heavy dead loads such as
bathtubs, waterbeds, and masonry furniture. To ensure headers and trimmers are used correctly, detail

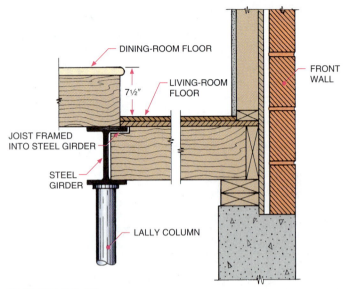

FIGURE 28.39 ■ Method of creating a one-step floor-level change.

drawings are prepared. For example, stairwell floor framing details show the position of joists, headers, and trimmers. See Figure 28.40.

Stairwell Framing

Because stairwell openings must be precisely shown on the floor framing plan, the stair system must be designed before the floor framing plan is drawn. Figure 28.41 shows the floor framing plan and a matching pictorial view of a stair system. Elevation views of stairwell systems are covered in Chapter 29, "Wall Framing Drawings."

Floor Framing Plans for Steel

Steel floor framing plans are similar to other plans, except the exact position of every column, girder, and beam

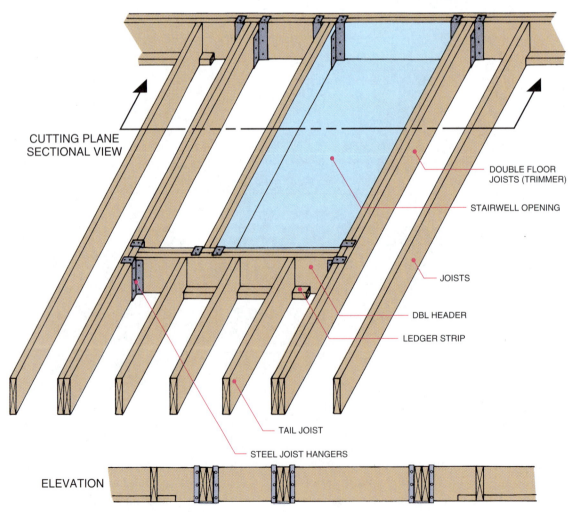

FIGURE 28.40 ■ Stairwell opening detail.

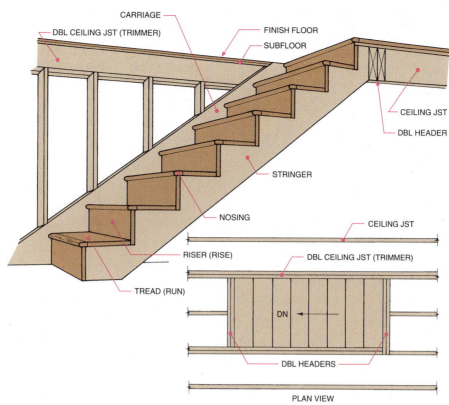

FIGURE 28.41 ■ Stairwell pictorial and plan view.

is classified and dimensioned. Grid systems are used to identify the position of each member.

The **sequential column identification system** is shown in the top drawing in Figure 28.42. In this system steel columns are identified by numbering them in sequence from left to right and from the rear to the front.

The **grid column identification system** is shown in the bottom drawing in Figure 28.42. In this system column rows are numbered on the horizontal perimeter from left to right. Letters are used to identify column rows on the vertical perimeter from front to rear.

Floor framing systems for steel construction include girders, beams, joists, and decking materials. A separate floor framing plan is prepared for each floor of a multilevel building. Although the framing for many floors may be nearly identical, this cannot be assumed unless specified. Usually there are slight differences on each floor plan. For this reason CAD layering is especially ideal for preparing highrise structural steel floor framing plans.

Layering (or pin graphics) as used in conventional drafting allows the drafter to draw a base floor plan and make specific floor changes without redrawing each floor separately. This is done by drawing each floor plan on

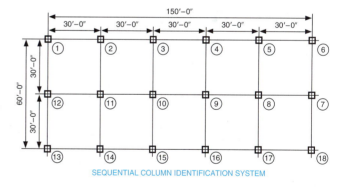

SEQUENTIAL COLUMN IDENTIFICATION SYSTEM

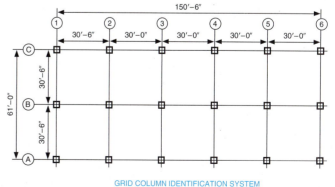

GRID COLUMN IDENTIFICATION SYSTEM

FIGURE 28.42 ■ The sequential column identification system (top) and the grid column identification system (bottom).

USING CAD

Floor Framing Plans on CAD

Floor framing plans are prepared on a separate layer using the corresponding floor plan layer as the base drawing. The joist framing can be represented as either a single-line or a double-line drawing. Use the *Line* command to create both, but note that the *Offset* command is used to create the second line of the joist. The *Copy-Multiple* command can be used to space the joists evenly. Interruptions for such features as stairwells and chimney openings can be deleted by identifying the segment to be removed and using the *Trim* command.

In drawing steel floor framing plans, major members, such as girders and beams, are shown with a solid heavy line with the identifying notation placed directly on or under the line. The length of each line represents the length of each member. If a continuous beam passes over a girder, a solid unbroken line is drawn through the girder line. However, if the beam stops and is connected to the girder, the beam line is broken. See Figure 28.43. Remember that solid lines represent continuous members. Broken lines indicate that the member intersects or is under a continuous member.

Spaces created between rows of members in two directions are known as **bays**. See Figure 28.44.

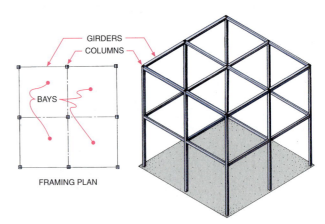

Bays with framing plan.

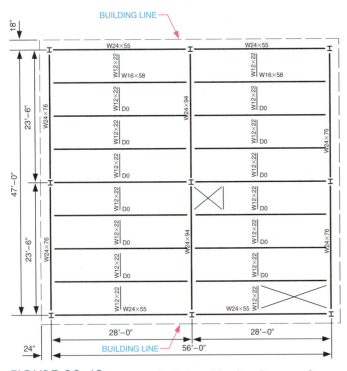

FIGURE 28.43 ■ Use of solid and broken lines to show locations and intersections of steel members.

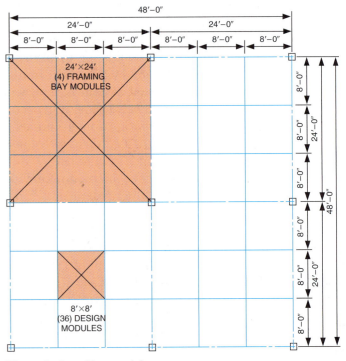

Dimensioning of bay modules.

FIGURE 28.44 ■ Use of construction bays in steel floor framing plans.

clear acetate. Drawings are aligned in layers with registration pins. The same spacing must be used between grid lines to ensure alignment of columns and other vertically oriented features, such as stairwells, plumbing lines, HVAC ducts, and electrical conduits. Each floor framing plan shows the position of each column that passes through the floor.

Three methods of dimensioning are used on structural steel drawings. In the first method, a description of each member includes the length placed directly on each schematic line. The second method uses notations to show only the shop size (width) and weight. Dimension lines are used to show the position and length of each member. The third method uses a coding system that relates each member to a schedule containing all pertinent information. See Figure 28.45.

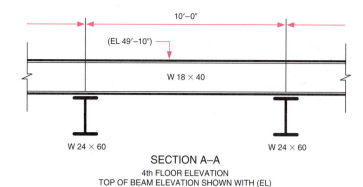

SECTION A–A
4th FLOOR ELEVATION
TOP OF BEAM ELEVATION SHOWN WITH (EL)

FIGURE 28.45 ■ Method of identifying steel beam size, type, and weight.

CHAPTER 28

Floor Framing Drawings Exercises

1. Draw a simplified plan view of a floor system shown in this chapter.

2. Redesign a floor framing plan shown in this chapter with the joists aligned in the opposite direction.

3. Add a 12′ × 14′ loft to the upper left corner of a 30′ × 35′ room. Draw the outlines and show the position of joists and girders.

 4. Develop a complete floor framing plan for one floor or as many floors as you are designing in your house plans.

5. Complete an abbreviated floor framing plan.

6. Draw the detail of a sill support in this chapter. Include callouts for all the parts shown.

7. Draw a slab foundation for a 30′ × 40′ house. Install sleepers for a wood finish floor.

8. Draw the construction details for a concrete porch adjacent to a T-foundation.

9. Draw two different construction details showing the end support of a girder at a T-foundation wall.

29 CHAPTER
Wall Framing Drawings

OBJECTIVES

In this chapter, you will learn to:

- draw an exterior wall framing elevation and plan.
- draw an interior wall framing elevation and plan.
- draw details and sections of walls.
- draw wall intersections.

TERMS

bracing
column schedule
curtain walls

drywall construction
framing elevation drawings
rammed earth

rough opening
siding
stud layout

INTRODUCTION

Wall framing provides the base to which coverings, such as siding and drywall, are attached. Typically, exterior wall framing supports the roof and ceiling loads. In most designs, the interior walls also help support these loads.

Wall framing drawings may consist of exterior and interior elevations, column and stud layouts, and details.

EXTERIOR WALLS

Exterior walls for most wood-frame residential buildings use either skeleton-frame or post-and-beam construction. The typical method of erecting walls for most buildings follows a conventional braced-frame system. Figure 29.1 shows the skeleton framing of a residence and the finished building. Prefabricated components have led to variations in the methods of erecting walls. Manufactured panels range from a basic wall-frame panel to a completed wall that includes plumbing, electrical work, doors, and windows. See Figure 29.2. Whether a structure is prefabricated or field constructed, the preparation of exterior framing drawings is the same.

Framing Elevations

Exterior walls are best constructed when a framing elevation drawing is used as a guide. A wall framing elevation

A. Skeleton frame of a house.

B. The completed house.

FIGURE 29.1 ■ A skeleton frame structure (A) under construction and (B) and completed.

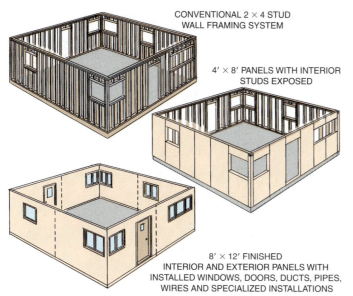

CONVENTIONAL 2 × 4 STUD
WALL FRAMING SYSTEM

4′ × 8′ PANELS WITH INTERIOR
STUDS EXPOSED

8′ × 12′ FINISHED
INTERIOR AND EXTERIOR PANELS WITH
INSTALLED WINDOWS, DOORS, DUCTS, PIPES,
WIRES AND SPECIALIZED INSTALLATIONS

FIGURE 29.2 ■ Basic types of wall construction.

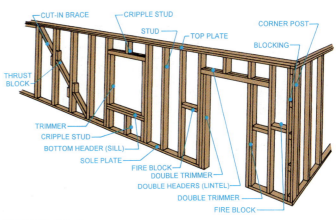

FIGURE 29.3 ■ Wall framing members.

drawing is the same as an elevation of a building with the siding materials removed. Figure 29.3 shows the basic wood framing members included in framing elevations.

To draw framing elevations, project lines from floor plans and elevations, as shown in Figure 29.4. The elevation supplies all the projection points for the horizontal framing members. The floor plan provides all the points of projection for locating the vertical members. Because floor plan wall thicknesses normally in-

clude the thickness of siding materials, care should be taken to project the outside of the *framing* line to the framing elevation drawing and not the outside of the siding line.

When drawing door and window **rough openings** (framing openings), first check the sizes listed in manufacturing specifications or door and window schedules. Rough openings are slightly larger than the size of doors or windows. Then project the position of the top, bottom, and sides of the door and window openings from the floor plan and elevation. Figure 29.5 shows examples of exterior wall framing elevations.

In drawing steel-framed elevations, only the locations of structural members are included. See Figure 29.6.

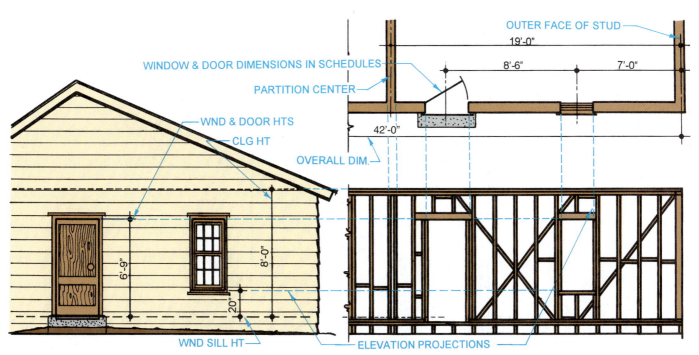

FIGURE 29.4 ■ Projections of a wall framing elevation.

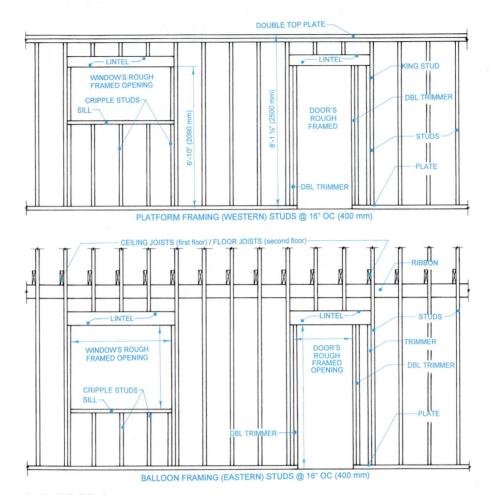

FIGURE 29.5 ■ Platform and balloon wall framing elevations.

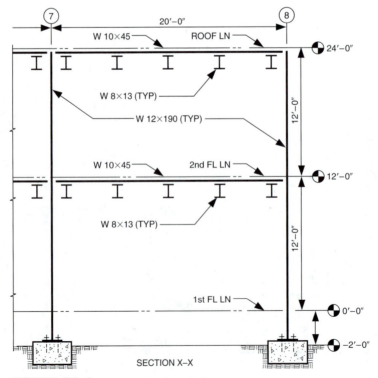

FIGURE 29.6 ■ Structural steel elevation.

Information about coverings for curtain walls is shown on wall elevations and wall details.

In Figure 29.7, the framing elevation is incorporated into a sectional drawing of the structure. This drawing shows an elevation section of the framing from the foundation-floor system to the roof construction. In a sectional drawing, any members intersected by a cutting-plane line, such as the joists, are indicated by crossed diagonals.

Bracing

To make wall frames rigid, structural lumber members are attached at an inclined angle. Members used for this purpose are called **bracing**. The bracing may be placed on the inside or outside of the wall, or between the studs. See Figure 29.8. Steel straps are often used for corner bracing, as shown in Figure 29.9.

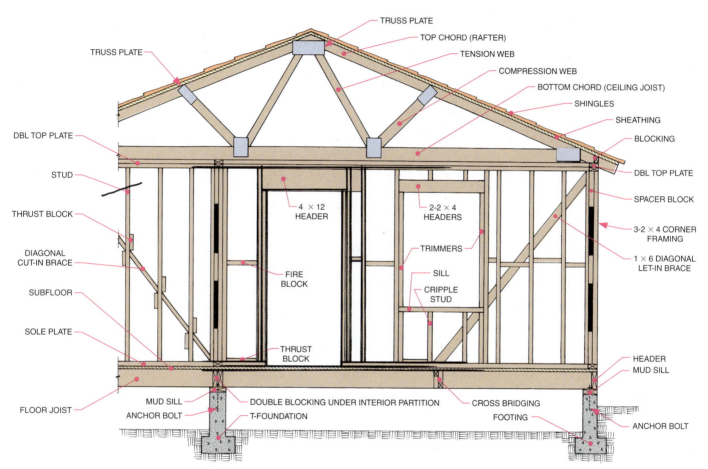

FIGURE 29.7 ■ Complete wall framing elevation.

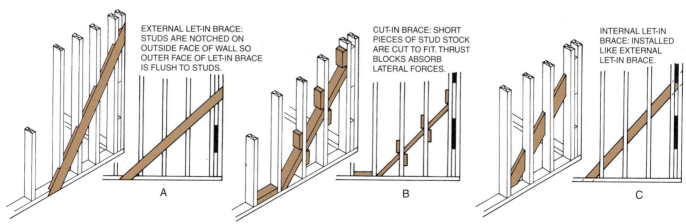

EXTERNAL LET-IN BRACE: STUDS ARE NOTCHED ON OUTSIDE FACE OF WALL SO OUTER FACE OF LET-IN BRACE IS FLUSH TO STUDS.

CUT-IN BRACE: SHORT PIECES OF STUD STOCK ARE CUT TO FIT. THRUST BLOCKS ABSORB LATERAL FORCES.

INTERNAL LET-IN BRACE: INSTALLED LIKE EXTERNAL LET-IN BRACE.

A   B   C

FIGURE 29.8 ■ Diagonal bracing on wall framing elevation.

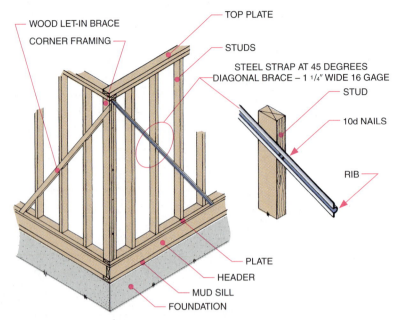

WOOD LET-IN BRACE
CORNER FRAMING
TOP PLATE
STUDS
STEEL STRAP AT 45 DEGREES
DIAGONAL BRACE – 1 1/4″ WIDE 16 GAGE
STUD
10d NAILS
RIB
PLATE
HEADER
MUD SILL
FOUNDATION

FIGURE 29.9 ■ Use of steel strap corner brace.

Difficulties may occur in interpreting the true position of headers, cripple studs, plates, and trimmers. Figure 29.10 shows how to illustrate the position of these members on framing elevation drawings to ensure proper interpretation.

Panels

Panel elevations show the attachment of the sheathing panels to the framing. To show the relationship between the panel layout and the framing, panel drawings and framing drawings are usually combined in one drawing. Diagonal dotted lines indicate the position of the panels. When only the panel layouts are shown, the outline of the panels and the diagonals are drawn solid. See Figure 29.11.

Dimensions

Dimensions on framing elevations include overall widths, heights, and spacing of all studs. See Figures 29.12 and 29.13. Control dimensions for the sizes of all rough openings (RO) for windows are also indicated. If the spacing of studs does not automatically provide the rough opening necessary for windows, the RO width of windows must also be dimensioned. Framing dimensions for a standard 8′ ceiling height using standard 93″ studs are shown in Figure 29.14.

The finished ceiling height will vary depending on the thickness of the finished floor and ceiling covering materials. A typical calculation of ceiling height is shown here:

Math Connection

Typical Standard Stud Height	....	93″
Double Top Plate	....	3″
Bottom Plate	....	1 1/2″
	Total	97 1/2″
Less the Finish Floor (3/4″) and Ceiling Cover (1/2″)	....	−1 1/4″
Typical Finished Ceiling Ht	....	96 1/4″
		(8′-0 1/4″)

Detail Drawings

Not all the information needed to frame an exterior wall can be shown on a framing elevation. Many details must be shown through the use of sectional drawings, exploded views, or pictorials.

Sections

Sectional wall framing drawings are either complete sections, partial sections, or removed sections indexed from a plan or elevation drawing.

(*continued page 516*)

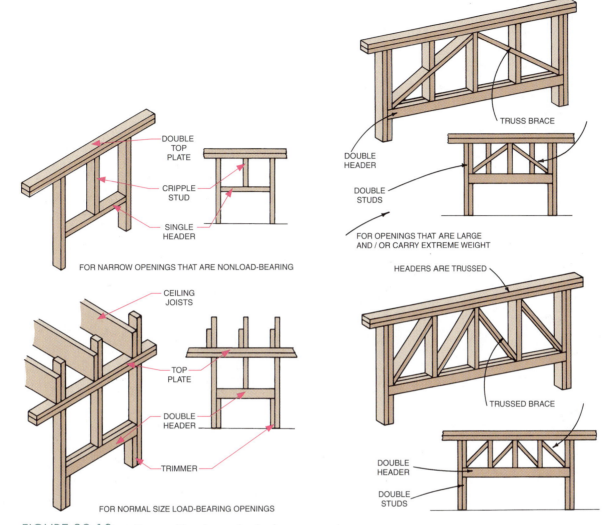

FIGURE 29.10 ■ Types of header and cripple construction.

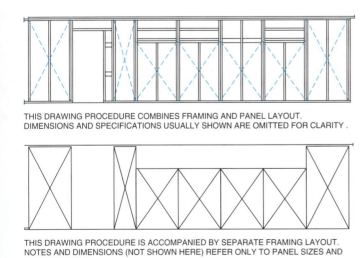

THIS DRAWING PROCEDURE COMBINES FRAMING AND PANEL LAYOUT.
DIMENSIONS AND SPECIFICATIONS USUALLY SHOWN ARE OMITTED FOR CLARITY .

THIS DRAWING PROCEDURE IS ACCOMPANIED BY SEPARATE FRAMING LAYOUT.
NOTES AND DIMENSIONS (NOT SHOWN HERE) REFER ONLY TO PANEL SIZES AND
SPECIFICATIONS.

FIGURE 29.11 ■ Use of diagonal lines to show panel positions.

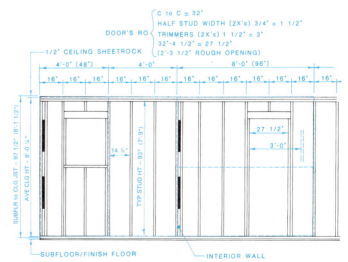

FIGURE 29.12 ■ Fully dimensioned wall framing elevation.

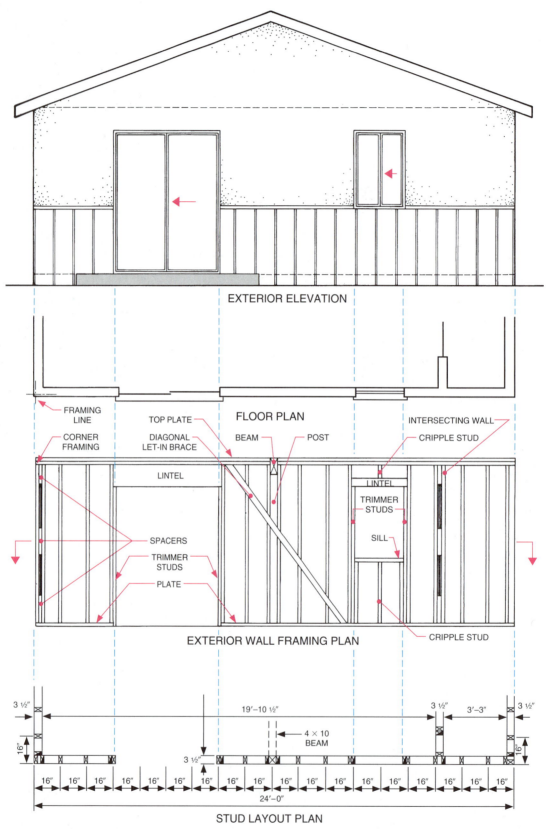

EXTERIOR ELEVATION

FLOOR PLAN

FRAMING
LINE

CORNER
FRAMING

TOP PLATE

DIAGONAL
LET-IN BRACE

BEAM

POST

INTERSECTING WALL

CRIPPLE STUD

LINTEL

LINTEL

TRIMMER
STUDS

SILL

SPACERS

TRIMMER
STUDS

PLATE

CRIPPLE STUD

EXTERIOR WALL FRAMING PLAN

3 ½″

19′–10 ½″

3 ½″   3′–3″   3 ½″

16″

4 × 10
BEAM

3 ½″

16″

16″ 16″ 16″ 16″ 16″ 16″ 16″ 16″ 16″ 16″ 16″ 16″ 16″ 16″ 16″ 16″ 16″ 16″

24′–0″

STUD LAYOUT PLAN

FIGURE 29.13A ■ Related exterior elevation, floor plan, framing elevation, and stud layout.

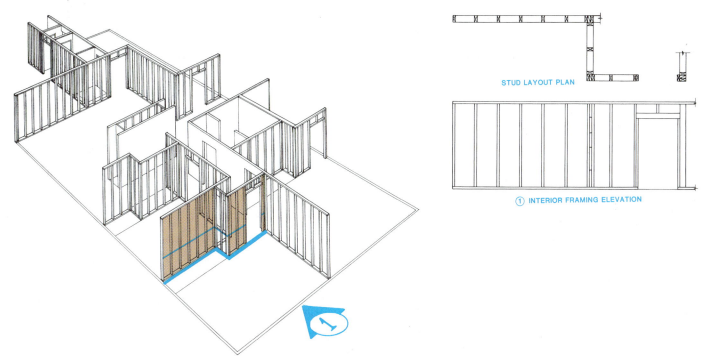

FIGURE 29.13B ■ Related interior framing elevation and stud layout.

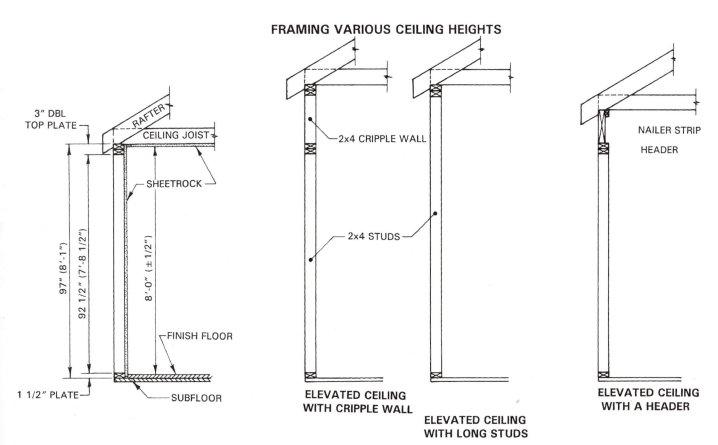

FIGURE 29.14 ■ Framing elevation for standard ceiling heights.

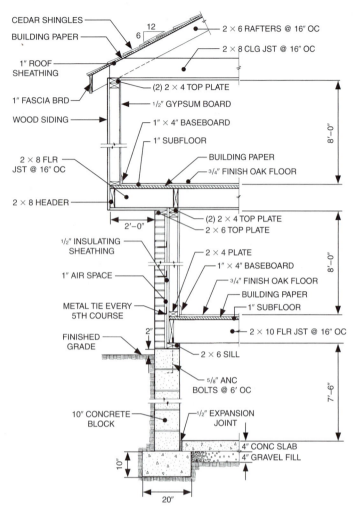

FIGURE 29.15 ■ Complete exterior wall section.

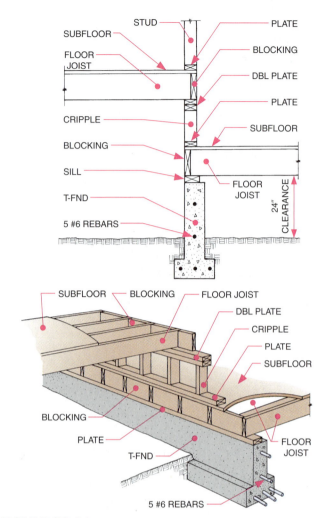

FIGURE 29.16 ■ Method of drawing a floor-level change.

Figure 29.15 shows a section through an outside wall from foundation to the roof. This is the most used wall section type because it often represents a typical section through most outside walls of a structure. When only a portion of a wall is unique, a partial section, as shown in Figure 29.16, is often used. Framing drawings do not include assembly devices such as nails, screws, glue, or sealers unless unique to the building process. These are included in the set of specifications.

With sectional breaks, such as on a full wall section, a larger scale can be used to allow more detail. Elevation drawings may also be partially sectioned to reveal construction details, as shown in Figure 29.17.

Exploded Views

Occasionally, exploded views are drawn to show internal wall framing construction if it is hidden when the total assembly is drawn. Figure 29.18 shows an ex-

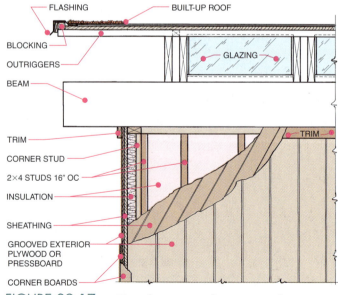

FIGURE 29.17 ■ Use of an exposed section to show framing details.

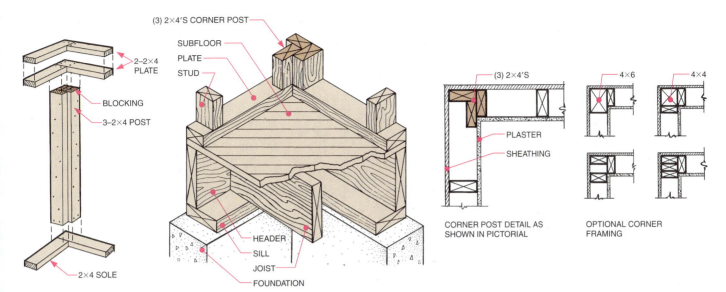

FIGURE 29.18 ■ Use of exploded views.

ploded view of a corner post construction, and Figure 29.19 shows an exploded view of the wall intersection of two stud partitions. Some drawings simply remove individual members of an assembly to reveal more detail, as was done in Figure 29.20. Exploded views are generally used in cabinet work.

Pictorials

Pictorial framing drawings help eliminate construction errors due to misreading of plans, elevations, and sectional drawings. Full wall framing pictorial drawings may be used for this purpose. However, most pictorial details are limited to a single intersection detail or a complex construction feature.

Finished Wall and Siding Details

The covering for an exterior wall is called **siding**. Siding material details are most often shown on elevation views. Plan and pictorial sections help builders to interpret drawings.

- *Lap siding* is horizontal siding applied over sheathing. Each piece covers (overlaps) part of the piece below it. Lap siding is available in solid redwood, cedar, and pine. Other materials are also used extensively, such as aluminum, steel, and fabricated boards. See Figure 29.21.

- *Board and batten siding* is vertical siding that originally consisted of a vertical board placed over studs, with the joints covered with vertical batten boards. Today, many board and batten sidings are made of

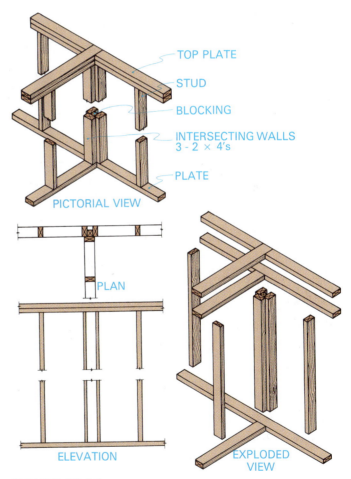

FIGURE 29.19 ■ Exploded views of wall intersections.

plywood or strandboard sheets with battens over the 48″ joints. Additional battens at 16″ intervals add consistency and help stabilize the vertical sheets. See Figure 29.22.

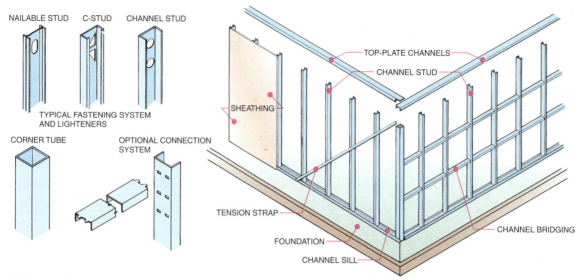

FIGURE 29.20 ■ Steel wall framing drawing with removed members.

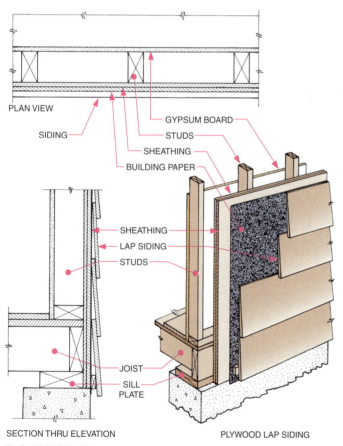

FIGURE 29.21 ■ Lap siding framing plan and elevation.

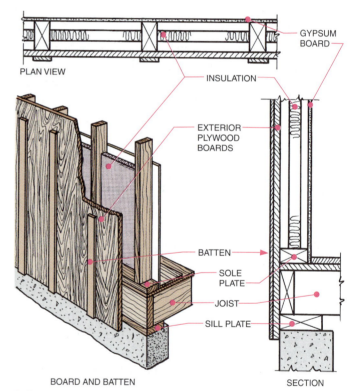

FIGURE 29.22 ■ Board and batten wall framing plan and elevation.

■ *Grooved panel boards* are a variation of vertical sheet siding. These are available with a wide variety of surface textures, grades, and groove sizes and shapes. Panels are also available that simulate the appearance of bevel or lap siding or shingles.

■ *Shingle siding* is applied as individual overlapped shingle boards, like roofing shingles, or as prefabricated shingle siding sheets. In either case they are applied over insulated sheathing.

■ *Stucco siding* (Figure 29.23), is applied with a trowel or sprayed. In wood construction, steel mesh must be applied over insulated sheathing to provide a base for the stucco. Stucco can be applied directly to concrete block without the use of mesh. Stucco is available in traditional three-coat portland ce-ment formula or in synthetic or insulated-finish system (EIFS) form.

■ *Solid masonry wall facing* materials include stucco, brick, stone, aluminum, steel, vinyl, or polyurethane siding. These are applied over solid concrete or concrete block. Brick is a material that can be used both structurally and as a finished wall face. Figure 29.24 shows a solid brick wall that performs both functions. Some brick walls include cavities (spaces between) to provide insulation, reduce dead loads, and lower material costs. Figure 29.25 shows typical wall sections of different types of brick cavity, solid brick, and brick veneer walls. Concrete block can be used in place of brick in these types of construction.

■ *Masonry veneer walls* provide brick or stone facing to wood-framed walls. See Figure 29.26. An entire wall may be a brick veneer wall. Where added support is required, a concrete block veneer wall may be used.

■ *Curtain walls* are used where structural steel or post-and-beam construction provides large wall spaces that are not part of the load-bearing structure. See Figure 29.27. One of the greatest advantages of steel cage construction is the unobstructed space provided by curtain walls. Building loads are transmitted through columns, so the remaining

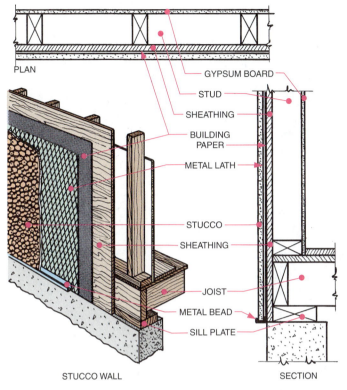

FIGURE 29.23 ■ Stucco wall framing plan and elevation.

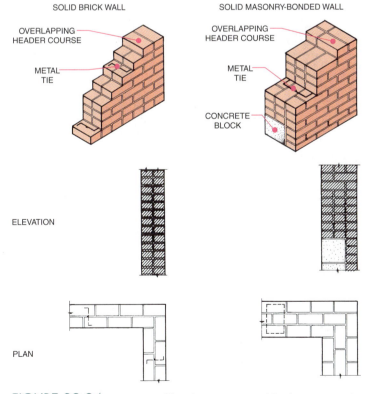

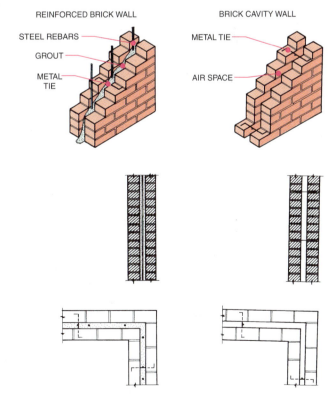

FIGURE 29.24 ■ Types of brick or concrete block construction.

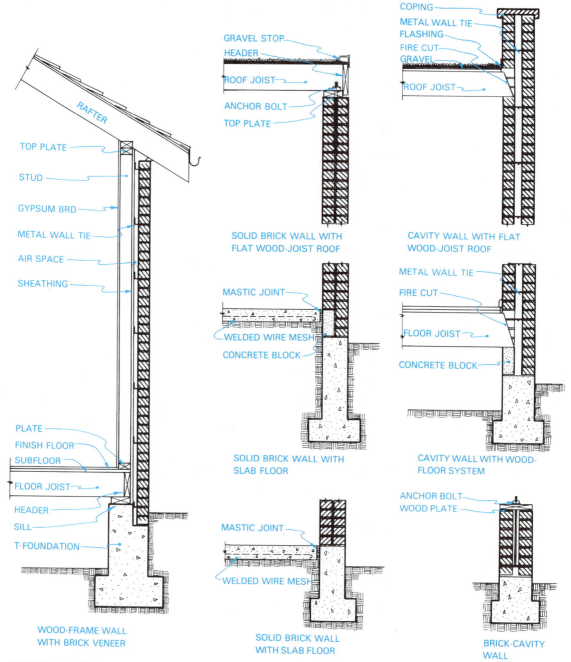

FIGURE 29.25 ■ Types of masonry construction with related roof and foundation intersections.

open wall space can be filled with any type of non-bearing (curtain) panels. These panels are usually prefabricated in modular units. Therefore, wall framing plans show only the position of modular units and not the construction details.

■ *Rammed earth* designs are gaining popularity in mild climates. Rammed earth walls are created by tamping a soil-cement mixture into two-foot thick forms. Some interior walls are formed between load bearing post-and-beam members. Others use rebar grids embedded into foundation footings.

■ *Straw bale* construction uses blocks manufactured from heat-compressed straw fibers. The straw surfaces are usually covered with stress.

■ *Log wall* construction is shown in Figure 29.28. To further insulate and finish interior walls, sheathing and drywall panels can be attached to plumbed furring strips.

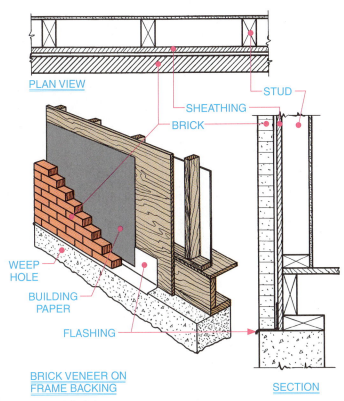

FIGURE 29.26 ■ Brick veneer wall plan and elevation.

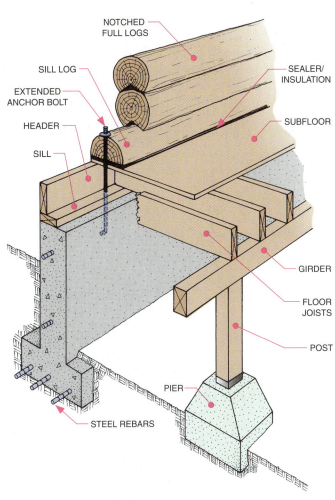

FIGURE 29.28 ■ Log wall construction.

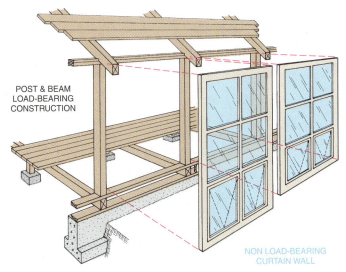

FIGURE 29.27 ■ Curtain wall construction.

Window Framing Drawings

Framing members around a window must not transfer structural movement or thermal stress to the glass. To design the framing needed to support each window, an understanding of the major components of a window is necessary.

USING CAD

Wall Framing Drawings on CAD

Using an exterior or interior wall elevation layer as a guide, repetitive vertical wall framing members such as studs, can be drawn on a superimposed layer using the *Copy-Multiple* command. This is done using the *Single Line* or *Polyline* task as with floor framing plans. The *Trim* command can be used to eliminate framing members from openings for doors, windows, or chimneys. Lintels, jambs, and plates can then be added individually.

Figure 29.29 shows the relationship between the rough opening for a window and the window frame, sash, and trim. Typical rough opening framing is shown in Figure 29.30. Rough opening dimensions are usually 1″ to 3″

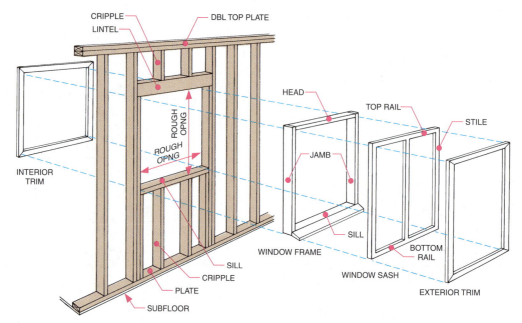

FIGURE 29.29 ■ Basic elements of a window assembly.

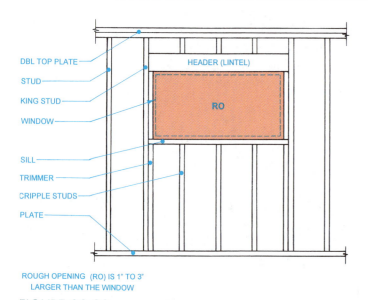

ROUGH OPENING (RO) IS 1" TO 3"
LARGER THAN THE WINDOW

FIGURE 29.30 ■ Typical rough opening framing.

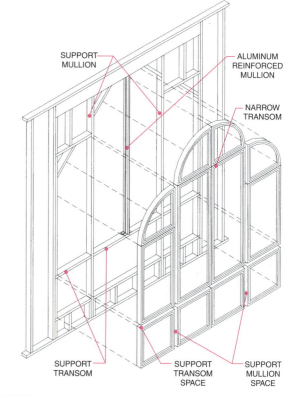

FIGURE 29.31 ■ Exploded view of manufactured window
components. (*Anderson Windows, Inc.*)

larger than the window dimensions. Window manufacturers often use exploded views to show these relationships, as seen in Figure 29.31.

Window construction details are usually shown on a head, jamb, rail, or sill section. See Figure 29.32. The positions of trimmers, headers, and sill-support members may also be shown on these details. The method of weatherproofing between window components and panel framing is illustrated in Figure 29.33.

There are hundreds of window manufacturers, and each has hundreds of window styles and sizes. It is therefore necessary to refer to the manufacturer to determine the exact dimensions and rough opening for each win-

dow. This information is usually included in schedules and specifications. (See Chapter 35.)

Rough opening dimensions vary among window manufacturers for each window style. Figures 29.34 and 29.35 show a common chart used by manufacturers to summarize rough opening requirements.

FIGURE 29.32 ■ Sill and head section of a manufactured window. (*Anderson Windows, Inc.*)

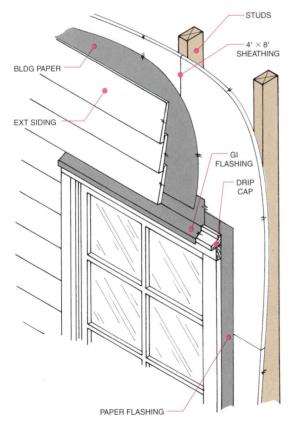

FIGURE 29.33 ■ Typical window waterproofing installation.

When fixed windows or unusual window shapes or sizes are to be constructed in the field—or even at a factory—complete framing details must be drawn. Unusual

ADD 3½″ FOR MASONRY ROUGH OPENING				
ROUGH OPENING →22½″	3′–6⅛″	5′–1¾″	6′–9⅜″	8′–5″
SASH OPENING → 19″	3′–2⅝″	4′–10¼″	6′–5⅞″	8′–1½″

FIGURE 29.34 ■ Common rough opening dimensions for windows.

and nonstandard sizes and components require more complete detail drawings. *Fixed sheet glass* thicknesses range from 3/32″ to 7/16″. Widths range from 40″ to 60″ and lengths range from 50″ to 120″. *Plate glass 1/8″* to 1/2″ thick ranges from 80″ × 130″ to 125″ × 280″. *Glass block* windows also require a head, jamb, and sill detail. See Figure 29.36. Glass block rough openings must allow space for the block size, plus mortar or channel space as prescribed by the manufacturer.

Door Framing Drawings

Component drawings of a door assembly include the wall framing with the rough opening, the door frame (head and side jambs), the door, and sometimes the sill (threshold). Doors are usually prehung (attached with hinges) to the jamb, and the complete assembly is fit into a rough opening in a framed wall. See Figure 29.37. A lintel must be placed above the head jamb to prevent the jamb from sagging. Lintels distribute building loads to vertical support members such as studs, trimmers, or posts.

Figure 29.38 shows a large solid lintel with future trimmer positions noted. Solid wood lintel sizes and makeup vary greatly. Figure 29.39 illustrates the typical lintel sizes used in light wood construction. Lighter weight headers using strand board and insulation cores, as shown in Figure 29.40, are more energy efficient (R20) and dimensionally stable.

Modular door units may include multiple doors and windows that must also fit into a rough opening. See

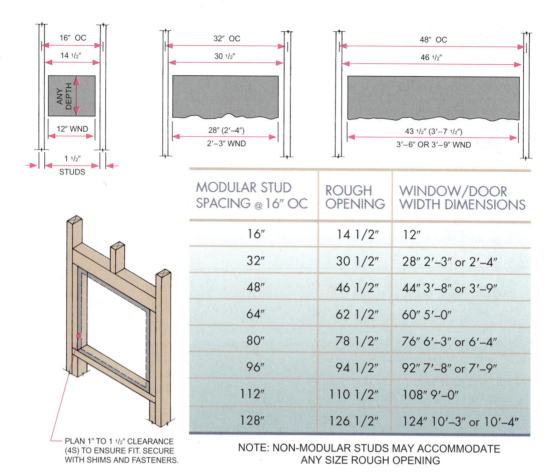

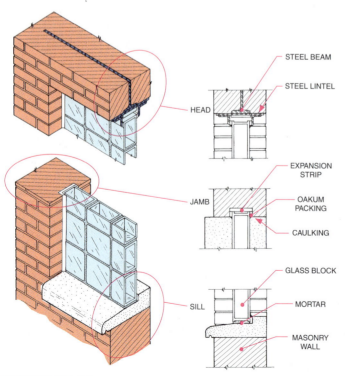

PLAN 1" TO 1 1/2" CLEARANCE
(4S) TO ENSURE FIT. SECURE
WITH SHIMS AND FASTENERS.

MODULAR STUD SPACING @ 16" OC	ROUGH OPENING	WINDOW/DOOR WIDTH DIMENSIONS
16"	14 1/2"	12"
32"	30 1/2"	28" 2'–3" or 2'–4"
48"	46 1/2"	44" 3'–8" or 3'–9"
64"	62 1/2"	60" 5'–0"
80"	78 1/2"	76" 6'–3" or 6'–4"
96"	94 1/2"	92" 7'–8" or 7'–9"
112"	110 1/2"	108" 9'–0"
128"	126 1/2"	124" 10'–3" or 10'–4"

NOTE: NON-MODULAR STUDS MAY ACCOMMODATE
ANY SIZE ROUGH OPENING

FIGURE 29.35 ■ Rough openings and modular sizes for windows and doors.

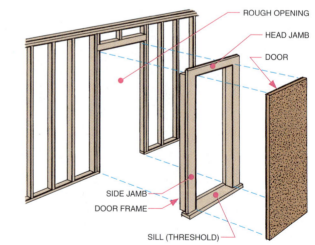

FIGURE 29.37 ■ Relationship of door jamb and rough opening.

Figure 29.41. Specifying, dimensioning, and constructing accurate rough openings for these units is critical to successful door functioning. See Figure 29.42. Like windows, rough openings for doors are found in manufacturer's specifications, which are included in a door

FIGURE 29.36 ■ Glass block wall section.

schedule. (See Chapter 35.) Add 3 1/2″ to the width and 1 1/2″ to the height of a door if the rough opening is not specified by a manufacturer.

Head, sill (threshold), and jamb sections are just as effective in describing door framing construction as they are in showing window framing details (Figure 29.43). Because doors extend to the floor, the relationship of the floor framing system to the position of the door is

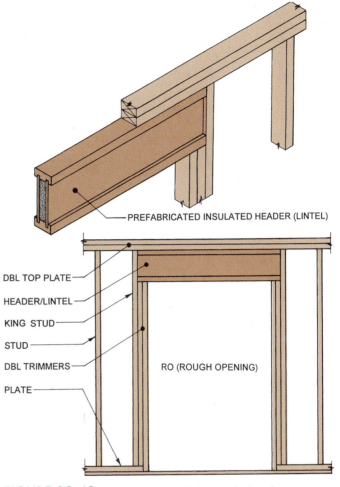

PREFABRICATED INSULATED HEADER (LINTEL)

DBL TOP PLATE

HEADER/LINTEL

KING STUD

STUD

DBL TRIMMERS

PLATE

RO (ROUGH OPENING)

FIGURE 29.40 ■ Preinsulated lightweight header. (*Superior Wood Systems, Inc.*)

FIGURE 29.38 ■ Solid door lintel.

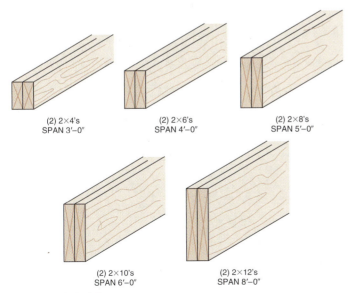

(2) 2×4's
SPAN 3′–0″

(2) 2×6's
SPAN 4′–0″

(2) 2×8's
SPAN 5′–0″

(2) 2×10's
SPAN 6′–0″

(2) 2×12's
SPAN 8′–0″

FIGURE 29.39 ■ Typical wood lintel sizes and types.

FIGURE 29.41 ■ Prehung double French doors. (*Marvin Windows & Doors*)

critical. The method of intersecting the door and hinge with the wall framing is also important.

Framing for sliding doors (Figure 29.44) also requires careful detailing of intersections. See Figure 29.45. If thresholds are not part of the exterior door assembly, a separate detail is necessary to show the exact type and alignment of the door and the threshold. See Figure 29.46.

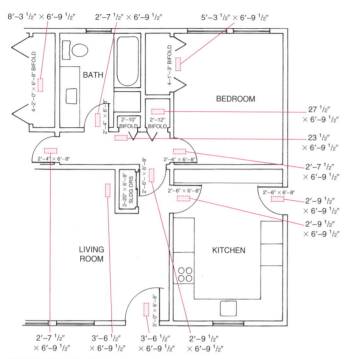

8'-3 1/2" × 6'-9 1/2"   2'-7 1/2" × 6'-9 1/2"   5'-3 1/2" × 6'-9 1/2"

4'-2'-0" × 6'-8" BIFOLD

4'-1"-3" BIFOLD

BATH

BEDROOM

27 1/2"
× 6'-9 1/2"

2'-4" × 6'-8"

2'-10"
BIFOLD

2'-12"
BIFOLD

23 1/2"
× 6'-9 1/2"

2'-4" × 6'-8"

2'-4" × 6'-8"

2'-7 1/2"
× 6'-9 1/2"

2-20" × 6'-8"
SLDG DRS

2'-6" × 6'-8"

2'-6" × 6'-8"

2'-9 1/2"
× 6'-9 1/2"

2'-6" × 6'-8"

2'-9 1/2"
× 6'-9 1/2"

LIVING
ROOM

KITCHEN

3'-0" × 6'-8"

2'-7 1/2"
× 6'-9 1/2"

3'-6 1/2"
× 6'-9 1/2"

3'-6 1/2"
× 6'-9 1/2"

2'-9 1/2"
× 6'-9 1/2"

FIGURE 29.42 ■ Rough opening dimensions for standard door sizes.

FIGURE 29.44 ■ Typical sliding glass door components.

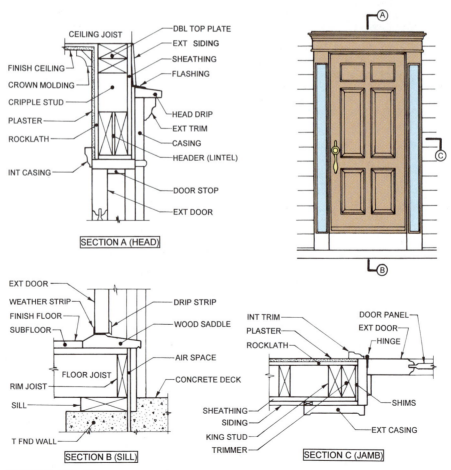

CEILING JOIST

DBL TOP PLATE

EXT SIDING

SHEATHING

FLASHING

FINISH CEILING

CROWN MOLDING

CRIPPLE STUD

HEAD DRIP

PLASTER

EXT TRIM

ROCKLATH

CASING

HEADER (LINTEL)

INT CASING

DOOR STOP

EXT DOOR

SECTION A (HEAD)

EXT DOOR

WEATHER STRIP

FINISH FLOOR

DRIP STRIP

SUBFLOOR

WOOD SADDLE

FLOOR JOIST

AIR SPACE

RIM JOIST

CONCRETE DECK

SILL

T FND WALL

SECTION B (SILL)

INT TRIM

PLASTER

ROCKLATH

DOOR PANEL

EXT DOOR

HINGE

SHEATHING

SIDING

KING STUD

TRIMMER

SHIMS

EXT CASING

SECTION C (JAMB)

FIGURE 29.43 ■ Wall framing related to door assembly.

INTERIOR WALLS

Interior framing drawings include plan, elevation, and pictorial drawings of partitions and wall coverings. Detail drawings of interior partitions may also show intersections between walls and ceilings, floors, windows, and doors. See Figure 29.47.

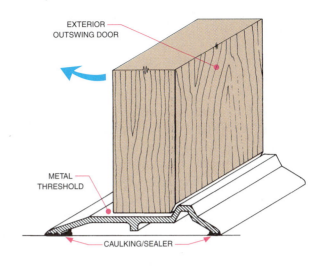

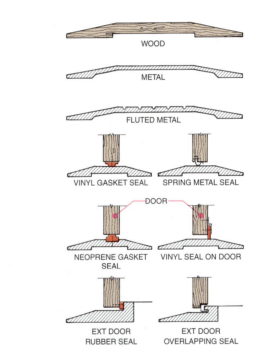

FIGURE 29.46 ■ Common types of thresholds and weatherstripping.

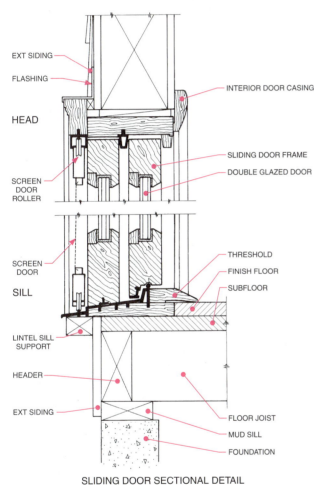

SLIDING DOOR SECTIONAL DETAIL

FIGURE 29.45 ■ Typical sliding door framing.

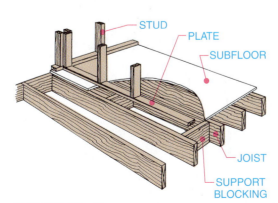

FIGURE 29.47 ■ Wall framing intersection detail.

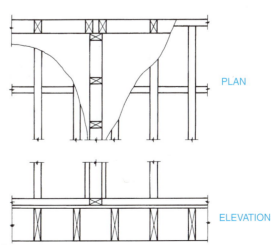

Framing Elevations

Framing elevation drawings show direct two-dimensional views of the framing. They are most effective for showing the construction of interior partitions. Interior partitions are projected from a floor plan, as viewed from the center of the room. To ensure the correct interpretation of the partition, each interior elevation drawing should include a label that indicates the room, a reference number, and/or the compass orientation of the wall. If either the room name, compass direction, or reference is omitted, the elevation may easily be misinterpreted and confused with a similar wall in another room. Figure 29.48 shows (A) an interior elevation drawing with a top and side view, (B) the same area under construction, and (C) the completed wall. Note the drawing is labeled Southeast Foyer Wall Section A-A. The A-A reference is found on the floor plan.

A complete study of the floor plan, elevation, plumbing diagrams, and electrical plans should be made prior to the preparation of interior wall framing drawings. Provisions must be made in framing drawings for special needs and to allow openings for electrical, plumbing, and HVAC installations. Figure 29.49 shows a framing drawing used to accommodate plumbing pipes. Figure 29.50 shows how wall framing is adjusted to allow space for heating registers.

When a stud must be broken to accommodate various items, the framing drawing must show the recommended construction. A structural stud should never have more than half its thickness removed.

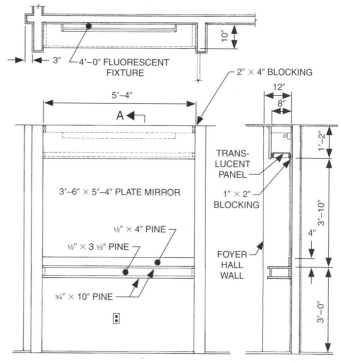

A. Special partition framing sections.

B. Same area under construction. (James Eismont, Photographer)

C. Completed area. (James Eismont, Photographer)

FIGURE 29.48 ■ Relationship of (A) framing drawings, (B) construction, and (C) the finished job.

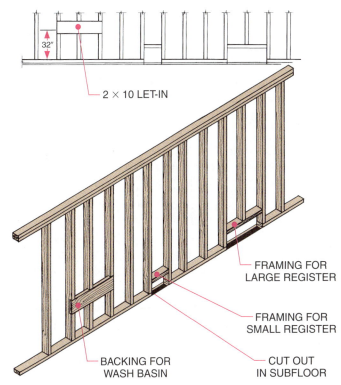

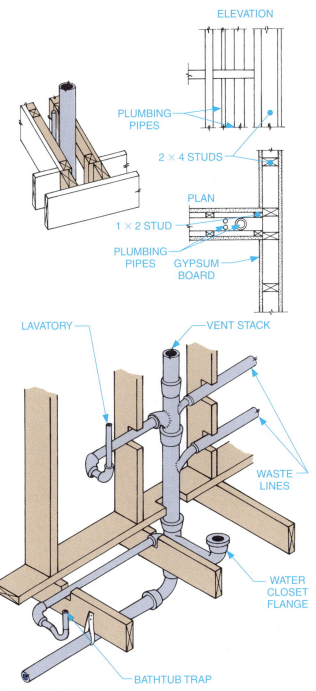

ELEVATION

PLUMBING PIPES

2 × 4 STUDS

PLAN

1 × 2 STUD

PLUMBING PIPES

GYPSUM BOARD

LAVATORY

VENT STACK

WASTE LINES

WATER CLOSET FLANGE

BATHTUB TRAP

FIGURE 29.49 ■ Framing adjustments for plumbing lines.

2 × 10 LET-IN

FRAMING FOR LARGE REGISTER

FRAMING FOR SMALL REGISTER

CUT OUT IN SUBFLOOR

BACKING FOR WASH BASIN

FIGURE 29.50 ■ Framing adjustments for heating outlets.

Built-in wall items may only require a partial framing elevation if location dimensions are included in an interior wall elevation or the heights are noted on a floor plan. See Figure 29.51.

Columns

Steel, concrete, masonry, and wood columns or posts perform the function of load-bearing partitions. They support

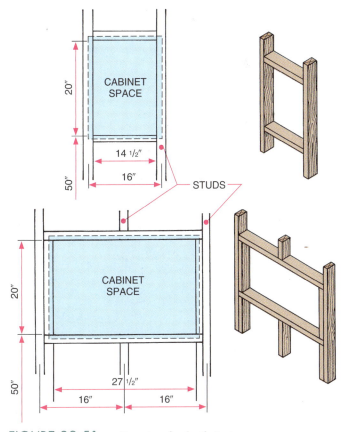

CABINET SPACE

20"

14 1/2"

16"

50"

STUDS

CABINET SPACE

20"

27 1/2"

16"   16"

50"

FIGURE 29.51 ■ Framing for built-in items.

girders and beams upon which floor decking and ceiling systems rest. Horizontal members can also be supported at the same height as the post through the use of hangers or blocking. See Figure 29.52.

Steel column positions may be shown on floor plans or on elevation drawings. On a floor plan, the style, size, and weight may be noted on each outline of the column, such as A1, B2. This information may also be shown on a column schedule. **Column schedules** are schematic elevation drawings showing the entire height of a building and the elevation of each floor, base plate, and column splice. The type, depth, weight, and length of columns with common characteristics are shown under the column mark for each column. Figure 29.53 shows a pictorial floor plan with column locations related to the column schedule shown in Figure 29.54. There are 13 columns in this plan.

Large structures, are designed with curtain walls (see Figure 29.27) on the exterior. In this type of construction, nonstructural wall panels cover the structural steel frame. Figure 29.55 shows the common types of steel-frame structures and their relationship to building height.

Columns with common specifications are grouped together at the top of the schedule. Under each grouping a heavy vertical line represents the height of each column, with the type, size, and weight noted on each. For example, columns with marks A1, C1, A3, and C3 are all

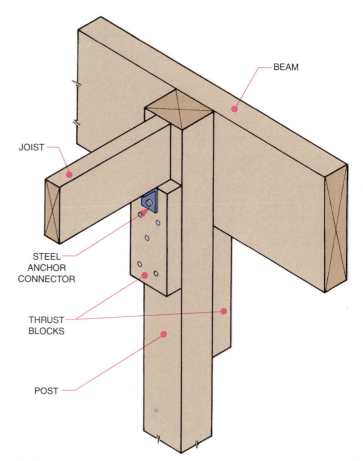

FIGURE 29.52 ■ Post support for joists and beams.

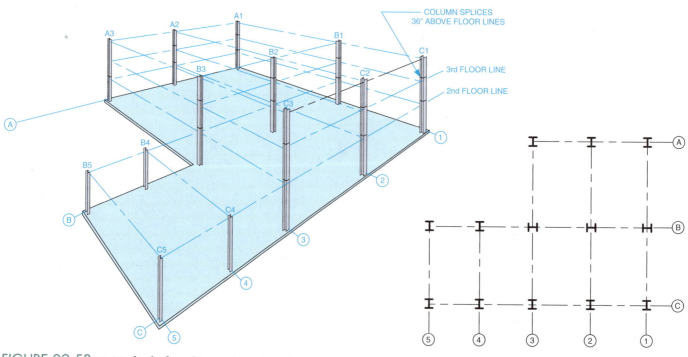

FIGURE 29.53 ■ Method of marking column locations.

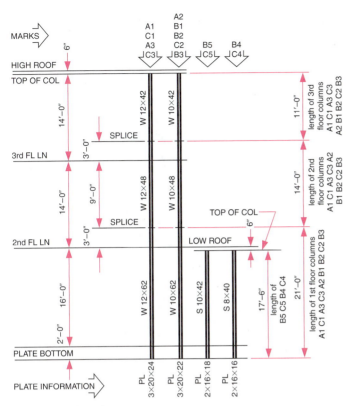

FIGURE 29.54 ■ Column schedule showing height, size, and column type.

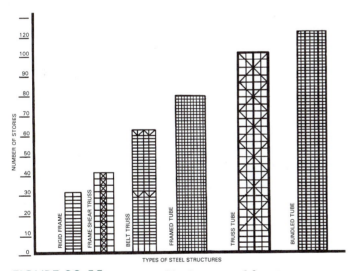

FIGURE 29.55 ■ Types of high-rise steel framing.

12″-wide flange shapes and extend vertically 46′-0″ from base to top (2′ + 16′ + 14′ + 14′). Three individual columns comprise each of these. The bottom length is 21′-0″ from plate to first splice (3′ above the floor line), 14′-0″ from the first splice to the second splice, and 11′-0″ from the second splice to the top. The second row of column marks (A2, B1, B2, C2, and B3) have the same

lengths as the first row but are 10″-wide flange shapes. The third-row and fourth-row columns are continuous 17′-6″, without splices; row three columns (B5, C5) are 10″ S-shapes and row four columns (B4, C4) are 8″ S-shapes.

Details

Additional drawings of wall construction, besides the basic structural framing, are often needed. Such drawings include molding and trim, detail intersections, interior doors, stair elevations, and wall coverings.

Molding and Trim

For finished interior walls, moldings and trim are used at intersections. Small-scaled drawings cannot accurately show the size, shape, position, and material used for molding and trim. Thus, detail sections are prepared, such as for crown (ceiling) and base (floor) moldings. Where more intricate designs are used, a detail drawing that shows the different molding segments is prepared. See Chapter 17.

Interior Doors

Pictorial or orthographic section drawings of the jamb, sill, and head should be prepared to illustrate the framing around interior doors. See Figure 29.56. A detailed drawing need not be prepared for each door, but one should be prepared for each *type* of door used. The position of headers, particularly to support closet sliding doors and pocket doors, is critical to the vertical fit and horizontal matching of doors. Drawings should also be keyed to the door schedule for identification. A pictorial section of an interior-door jamb may be needed to show variations because of different wall coverings, such as plaster, gypsum board, or paneling. See Figure 29.57.

Stairs

Just as rough openings must be allowed for windows and doors, floor openings must be planned for stairwell framing. A sectional wall elevation drawing is often used to describe the vertical distances that are not found on floor framing plans.

Abbreviated elevation sections are adequate to show only the rise, run, stairwell opening, and headroom clearance. Figure 29.58 shows an abbreviated elevation for a straight-run stair system. Figure 29.59 shows the same type of drawing for an L-shaped system. Where more

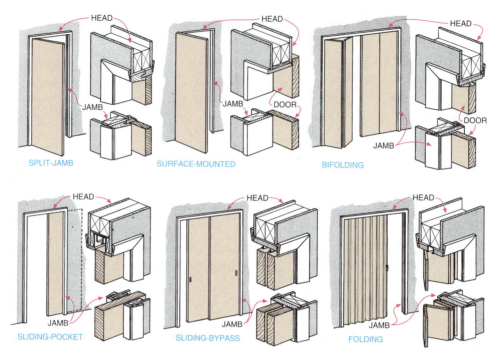

FIGURE 29.56 ■ Framing and trim details for interior doors.

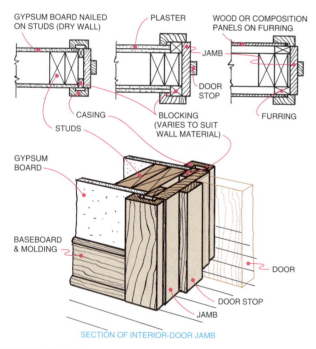

FIGURE 29.57 ■ Types of door trim for different wall coverings.

detailed information is needed, a completely dimensioned elevation drawing is used, as shown in Figure 29.60.

Tread width is shown on floor plans and floor framing plans. Riser height may be shown on interior elevations and on stair framing drawings. To calculate the number of risers use the following formula:

FORMULA:

$$\frac{\text{height of stairs}}{\text{height of each riser}} = \text{riser number}$$

EXAMPLE: 9'-4" = 108" + 4" = 112"

$$\frac{112''}{7''} = 16 \text{ risers}$$

$$\text{OR } \frac{115''}{7''} = 16.4 \text{ risers}$$

(adjust 16 risers at 7.15" each)

Math Connection

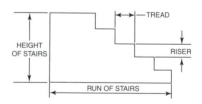

See Chapter 14 for more information about calculating stair dimensions.

Wall Coverings

Basic types of wall-covering materials used for finished interior walls include plaster, drywall, paneling, tile, and

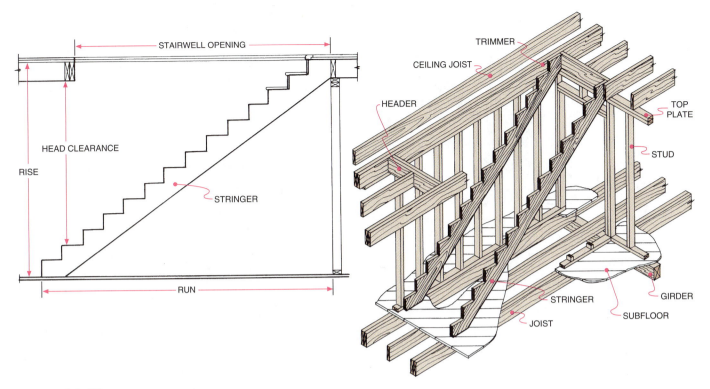

FIGURE 29.58 ▪ Straight-run stair system.

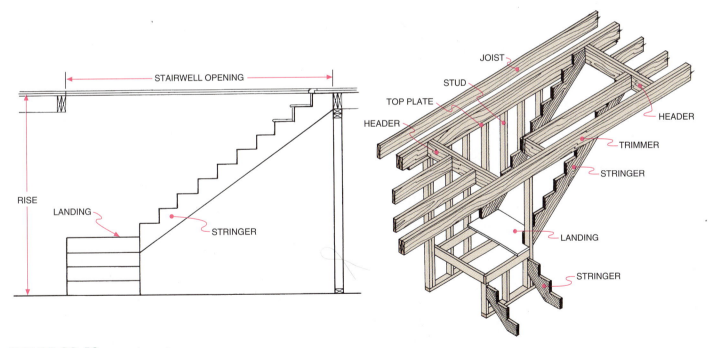

FIGURE 29.59 ▪ L-shaped stair system.

masonry. Each type requires a different method of attachment to the wall.

- ▪ *Plaster* is applied to interior walls over wire lath or gypsum sheet lath. Plaster walls are very strong and sound-absorbing. Plaster is also decay-proof and

termite-proof. However, plaster walls crack easily, they take months to dry, and the installation costs are high.

- ▪ **Drywall construction** is a system of interior wall finishing using prefabricated sheets of materials. A variety of manufactured materials may be used,

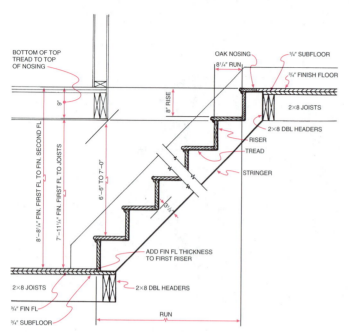

FIGURE 29.60 ■ Fully dimensioned stair assembly drawing.

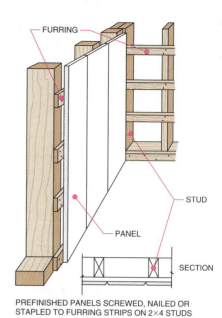

PREFINISHED PANELS SCREWED, NAILED OR
STAPLED TO FURRING STRIPS ON 2×4 STUDS

FIGURE 29.62 ■ Furring strips provide a horizontal surface for attaching paneling.

drywall, called *greenboard,* is specified where walls are to be exposed to excess moisture such as on baths and kitchens. *Blueboards* are specified where plaster is to be applied.

■ When *paneling* is used as an interior finish, horizontal furring strips should be placed on the studs to provide a nailing or gluing surface for the paneling. See Figure 29.62. The type of joint used between panels should be determined by developing a separate detail. The method of intersecting the outside corners of paneling should also be detailed. Outside corners can be intersected by mitering, overlapping, or exposing the paneling. Corner boards, metal strips, or molding may be used on the intersections. Inside corner intersections can be constructed by butting the wall coverings or by using corner moldings.

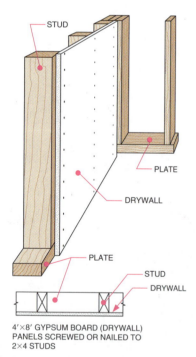

FIGURE 29.61 ■ Drywall construction.

STUD LAYOUTS

A horizontal framing section called a **stud layout** is a plan similar to a floor plan, except it shows the position of each wall framing member, exterior and interior. Stud layouts are used to show how studs are spaced on the plan and how interior partitions fit together.

A stud layout is a horizontal section through walls and partitions. The cutting-plane line is placed approxi-

such as fiberboard, gypsum wallboard, stranded lumber sheets, sheet-rock, and plywood. Drywall is nailed or screwed directly to studs. Then the drywall joints are finished. See Figure 29.61. Drywall thicknesses range from 1/4″ to 1″. Width and length range from 2′ × 8′ to 4′ × 14′. Moisture-resistant

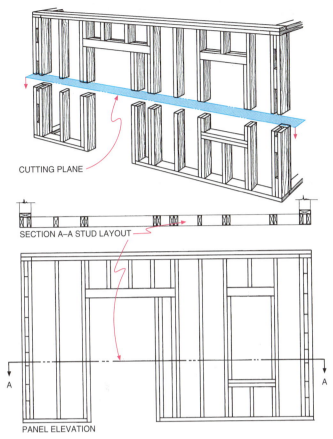

CUTTING PLANE

SECTION A–A STUD LAYOUT

A   A

PANEL ELEVATION

FIGURE 29.63 ■ A stud layout is a section through wall framing.

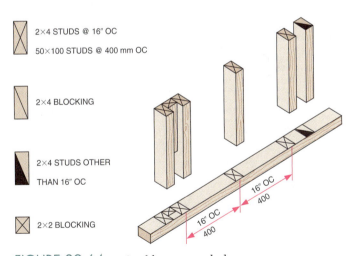

2×4 STUDS @ 16″ OC

50×100 STUDS @ 400 mm OC

2×4 BLOCKING

2×4 STUDS OTHER THAN 16″ OC

2×2 BLOCKING

16″ OC 400

16″ OC 400

16″ OC 400

FIGURE 29.64 ■ Stud layout symbols.

mately at the midpoint of the panel elevation. See Figure 29.63. The exact position of each stud that falls on an established modular center (16″, 32″, or 48″) is usually shown by diagonal crossed lines. Studs other than those on 16″ centers are shown with different symbols. Different symbols also identify studs that are different in size,

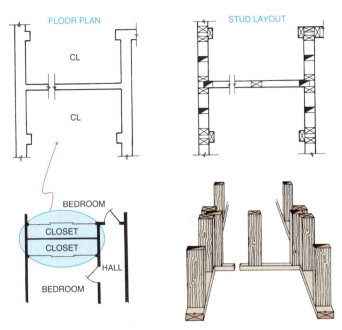

FLOOR PLAN   STUD LAYOUT

CL

CL

BEDROOM

CLOSET

CLOSET

HALL

BEDROOM

FIGURE 29.65 ■ Stud position shown on a stud layout.

such as blocking and short pieces of stud stock. See Figure 29.64. A coding system of this type eliminates the need for dimensioning the position of each stud. The practice of coding studs and other members in the stud plan also eliminates the need for repeating dimensions of each stud.

To conserve space, nonbearing studs are sometimes turned so they are flat. This rotation should be reflected in the stud layout. See Figure 29.65.

Dimensions

Detailed dimensions are normally shown on a stud symbol key or on a separate enlarged detail. Distances that are typically dimensioned on a stud layout include the following:

■ Inside framing dimensions of each room (stud to stud)

■ Framing width of the halls

■ Rough opening for doors and arches

■ Length of each partition

■ Width of partition where dimension lines pass through from room to room. This provides a double check to ensure that the room dimensions plus the partitioned dimensions add up to the overall dimension. When a stud layout is available, it may be used on the job to establish partition positions.

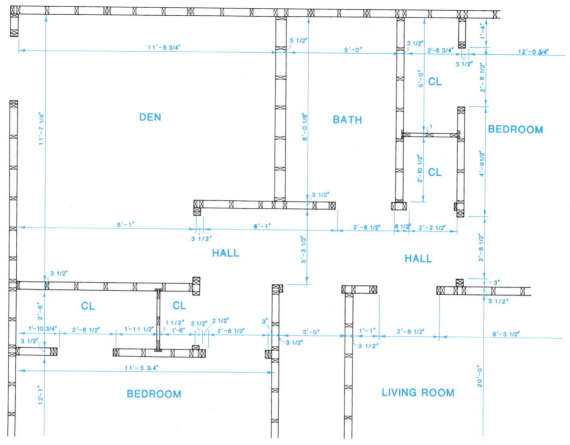

FIGURE 29.66 ■ Dimensioned stud layout.

Stud Details

Stud layouts are of two types: *complete plans* (Figure 29.66), which show the position of all framing members on the floor plan, and *stud details,* which show only the position and relationship of some studs or framing intersections. For example, the position of each stud in a corner-post layout is frequently detailed in a plan view. See Figure 29.67.

Occasionally, siding and inside-wall covering materials are included on this plan. Preparing this type of drawing without showing the covering materials, however, is the quickest way to show corner-post construction. If wall coverings are included on the detail, the complete wall thickness can be drawn. In a plan section, care should be taken to show the exact position of blocking because it may not pass through the cutting-plane line. Blocking should be labeled or a symbol used to prevent the possibility of mistaking it for a full-length stud. See Figure 29.68.

When laying out the position of all studs, remember that the finished dimensions of a 2 × 4 stud are actually 1 1/2″ × 3 1/2″. For the exact dressed sizes of other rough stock,

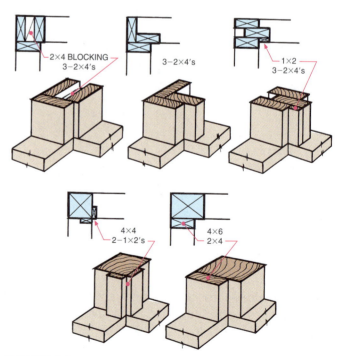

FIGURE 29.67 ■ Corner-post layout shown with stud layouts.

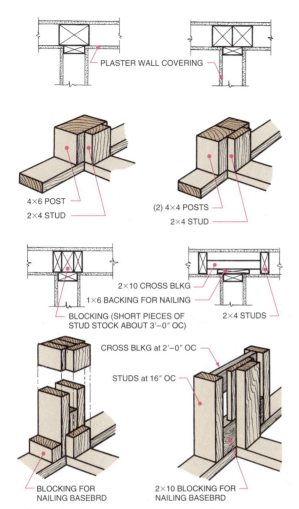

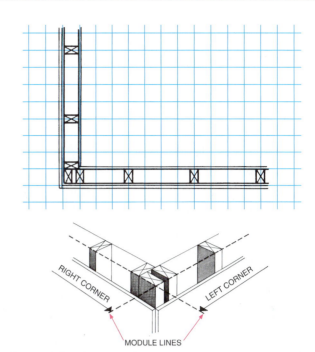

FIGURE 29.68 ■ Wall covering relationship with stud positions.

FIGURE 29.69 ■ Stud layout applied to a modular grid.

FIGURE 29.70 ■ Nonbearing steel stud wall.

refer back to Chapter 24. Stud lengths precut to 7′-7 1/2″ and combined with a double top plate (3″) and a bottom plate (1 1/2″) yield a floor-to-ceiling length of 8′-0″.

Modular Plans

In drawing modular framing plans, partitions should fit in relation to modular grid lines. Allowances for exterior wall thicknesses need to be indicated. See Figure 29.69. Space must also be provided for such items as door and window placement, medicine cabinets, closets, fireplaces, and plumbing runs.

Steel Studs

Steel studs are one-half the weight of wood studs. They are stronger, won't warp or split, and are moisture and insect resistant. Some nonbearing steel studs are prepunched for electrical or plumbing lines or for attachment to wood

sills, plates, or other studs. See Figure 29.70. A steel symbol should be added to stud layouts where steel studs are specified.

CHAPTER

29

Wall Framing Drawings Exercises

1. Prepare an exterior wall framing plan for a home of your own design.

2. Draw a stud layout (16″ OC) for a floor plan. Indicate the rough openings for doors and windows. Show a corner post and intersection detail.

3. Draw a floor plan and a wall framing elevation using post-and-beam construction for a 15′ × 10′ storage shed.

4. Draw a window detail of a window component: a head, jamb, or sill.

5. Draw a framing elevation for a double swinging door. Include the dimensions.

6. Project a framing elevation drawing of a kitchen wall in Figure 16.24.

7. Draw an interior wall framing plan for a kitchen, bath, and living area wall of the house you are designing.

8. Prepare a stud layout (1/4″ = 1′-0″). for a home of your own design.

9. Complete a stud detail of a corner post and another intersection.

10. Draw the construction details for a standard T-foundation with a brick veneer wall.

11. Draw a wall framing elevation of the four exterior walls shown in Figure 29.71.

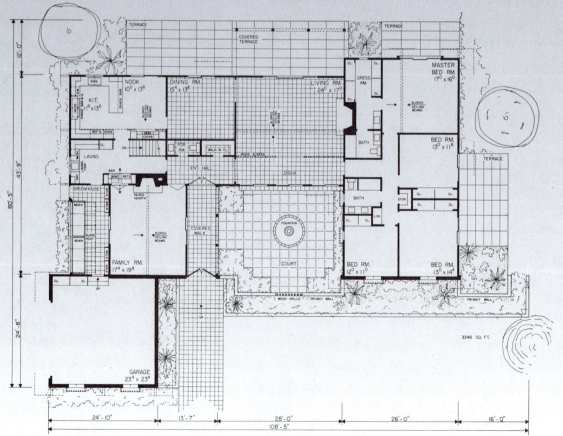

FIGURE 29.71 ■ Draw exterior wall framing elevations of all exterior walls shown on this plan.

Roof Framing Drawings

OBJECTIVES

In this chapter, you will learn to:

- describe roof framing members, components, and methods.
- calculate roof pitch.
- draw a roof framing plan showing structural members, sizes, pitch, and spacing.
- draw roof framing details and elevations.

TERMS

cornice	rafter	slope diagram
fascia	ridge board	soffit
flashing	rise	span
pitch	run	

INTRODUCTION

Roof styles and construction methods developed through the centuries. Pitches (slopes) of roofs were changed, gutters and downspouts were added for better drainage, and overhangs were extended to provide more protection from the sun. The size of roofs and the types of materials changed accordingly. The structure of a roof affects the choice of roof framing and roof covering materials, as well as the interior ceiling systems.

ROOF FUNCTION

The main function of a roof is to provide protection from rain, snow, sun, and hot or cold temperatures. In a cold climate, a roof is designed to withstand heavy snow loads. In a tropical climate, a roof provides protection mainly from sun and rain.

The walls of a structure are given stability by their attachment to the ground and to the roof. Buildings are not structurally sound without roofs. As explained in Chapter 22, live loads that act on the roof include wind loads and snow loads, which vary greatly from one geographical area to another. Dead loads that bear on roofs include the weight of the structural members and coverings. All loads are computed by pounds per square foot, or kilograms per square meter if metric measurements are used. (For computation of loads for an entire structure, see Appendix B, "Mathematical Calculations.")

ROOF FRAMING MEMBERS

The structural members of a roof must be strong enough to withstand many types of loads.

Wood Roof Members

Beams used for roof construction may be made from the same materials used for floor framing. A ridge beam, also called a ridge or **ridge board,** is the top member in the roof assembly. See Figure 30.1. In post-and-beam construction, the ridge beam or board may be exposed on the inside of a building. Figures 30.2 and 30.3 show the intersection of a ridge board and roof sheathing.

Roof **rafters** intersect ridge boards and rest on the tops of outside walls. Rafters may be selected from the same materials as floor joists. They may be solid lumber, truss joists, I-joists, laminated, or stranded lumber.

Steel Roof Members

Roofs framed with structural steel use steel girders, beams, and joists in the same manner as steel-framed floors. Most steel-framed roofs are flat, but steel ridge beams and rafters are also used on pitched roofs with large spans.

Concrete Roof Members

At one time concrete was never used in roof construction because normal spans could not be achieved. Technological developments have dramatically increased the use of concrete for roofs.

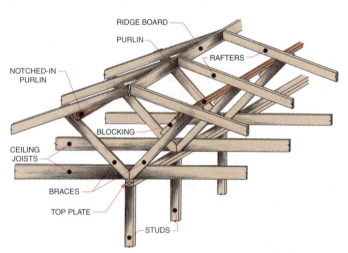

FIGURE 30.1 ▪ Ridge board related to roof framing members.

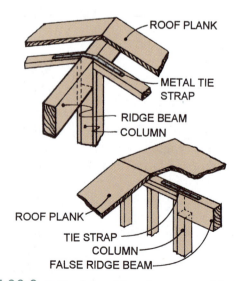

FIGURE 30.2 ▪ Roof sheathing shown over ridge beam.

Reinforced Concrete

Reinforced concrete is either precast or poured-on-site concrete with steel reinforcing rods inserted for stability

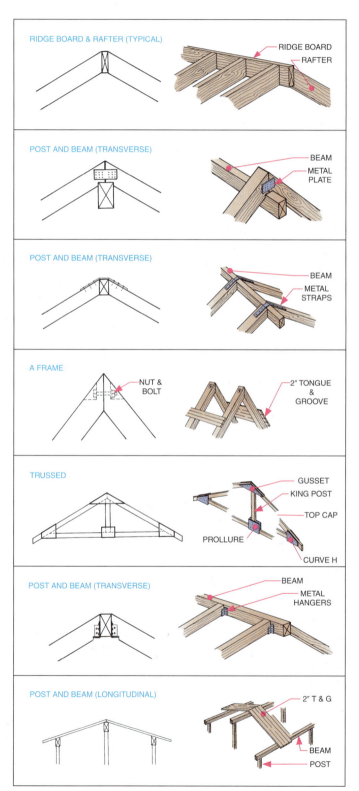

FIGURE 30.3 ▪ Typical ridge board intersections.

and rigidity. Reinforced concrete slabs are used extensively for roof systems where short spans make prestressing unnecessary. A wood-framed flat roof and a reinforced concrete roof may have the same covering. See Figure 30.4.

Prestressed Concrete

A prestressed concrete roof, beam, or slab is made by stretching steel rods and pouring concrete around them. When the stretched rods are released, the rods try to return to their original shape, and this creates tension on the concrete. Prestressing strengthens the concrete and allows beams to have a lower ratio of concrete depth to span.

Precast Concrete

Precast concrete roof members may be poured off site into molds, with steel reinforcement. When high stresses will not be incurred, precasting without prestressing is acceptable for most short span construction.

Concrete Shells

Concrete shells are curved sheets of lightweight concrete poured or sprayed onto steel mesh and bars that provide a temporary form until the concrete hardens. Once the concrete hardens, the steel is locked in place and the structure becomes rigid and stable.

ROOF FRAMING COMPONENTS

Trusses

Trusses for roofs are prefabricated components that perform all the functions of rafters, collar beams, and ceiling joists. Trusses consist of a top chord and bottom chord, joined with diagonal and vertical members called *webs*. Figure 30.5 shows the members of a truss, and Figure 30.6 shows examples of common lightweight wood trusses. These truss members are held rigidly in place with bolts, ring connectors, and/or gussets. Plywood gussets or sheet-metal gussets are used in light construction. Figure 30.7 shows the common locations of gusset plate connectors on a typical wood truss. Figure 30.8 illustrates the application of gussets on truss joists. Heavy timber trusses require heavy-duty

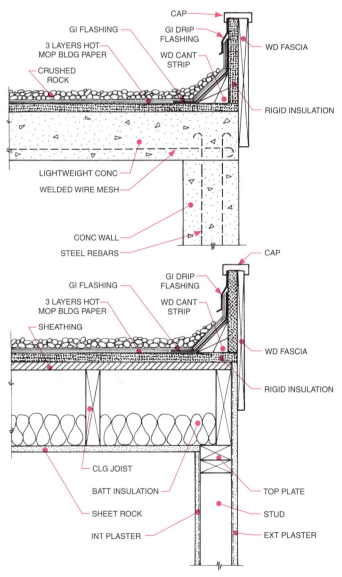

FIGURE 30.4 ■ Wood-framed roof compared to reinforced concrete roof construction.

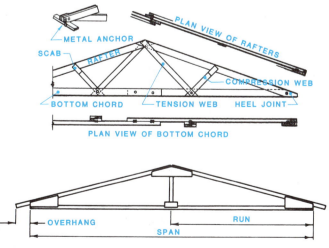

FIGURE 30.5 ■ Common truss members.

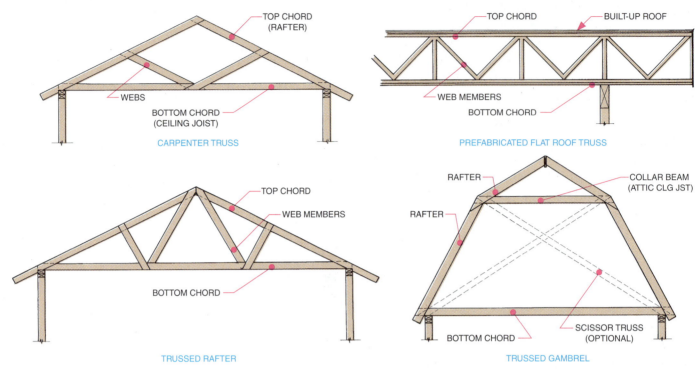

FIGURE 30.6 ■ Common lightweight wood trusses.

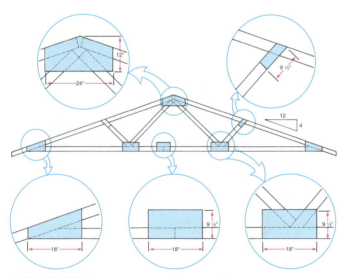

FIGURE 30.7 ■ Location of gusset plate connectors on a truss.

FIGURE 30.8 ■ Gusset plate applications on trusses.

steel gusset plates that are welded together and bolted through the top and bottom chords.

Widely spaced trusses are tied together horizontally to make the roof system structurally stable. Horizontal members known as *purlins* are used for this purpose. See Figure 30.9. On a trussed roof, purlins perform the same function as joist bridging in conventional construction.

Trusses prevent sags and cracks in the roof system because they are structurally independent and resist both compression and tension forces. See Figure 30.10. Because trusses can span larger distances than conventionally framed roofs, the spacing of interior partitions is more flexible. Fewer or no bearing partitions may be needed. Trusses may rest on steel beams columns, masonry walls, or on exterior wood-framed walls. See Figure 30.11.

Truss type and design depend on the length of span, room height requirements, spacing, roof pitch, live and dead loads, and cost factors. Several types of trusses are manufactured for structural steel construction. See Figure 30.12. A variety of truss designs are used in light res-

idential construction. See Figure 30.13. More complete truss specifications are found in Appendix B, "Mathematical Calculations."

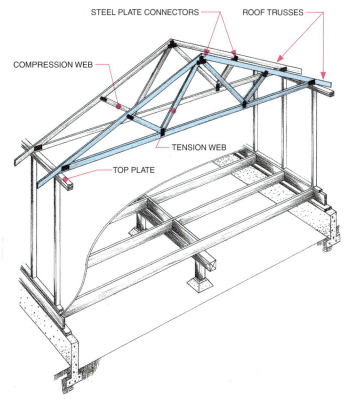

FIGURE 30.11 ■ Trusses bearing on exterior skeleton-frame walls.

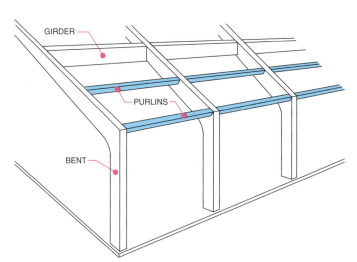

FIGURE 30.9 ■ Purlins provide stability between bents and girders.

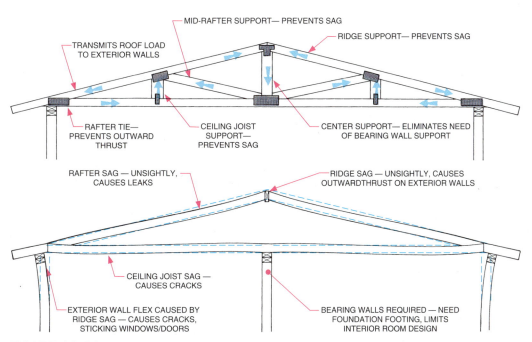

FIGURE 30.10 ■ Trusses resist tension and compression forces.

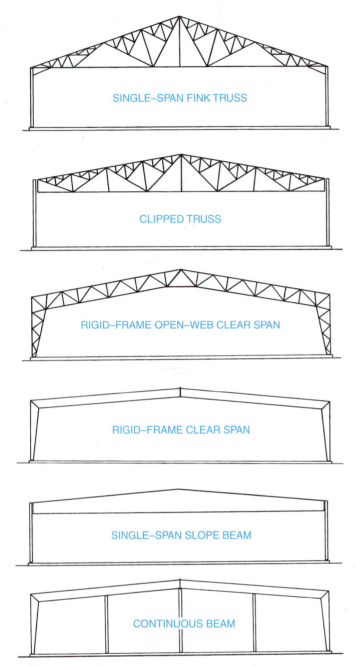

FIGURE 30.12 ■ Common types of steel trusses.

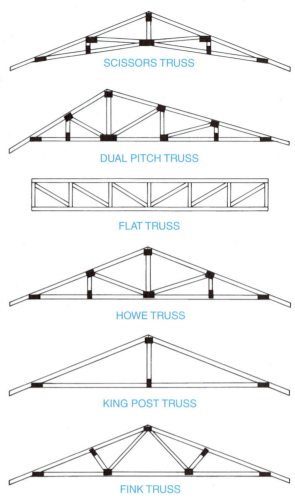

FIGURE 30.13 ■ Light residential construction truss types.

Roof Panels

There are many forms of lightweight, prefabricated roof panels. See Figure 30.14. These units can be designed to resist loads, span great distances, or eliminate the need for trusses.

Stressed Skin Roof Panels

These panels are constructed of plywood or stranded panels and seasoned lumber. The framing plywood skin acts as a unit to resist loads. Glued joints transmit the shear stresses, making it possible for the structure to act as one piece.

Curved Panels

The three types of curved panels are the sandwich (or honeycomb paper-core) panel, the hollow-stressed end panel, and the solid-core panel. The arching action of these panels permits spanning across great distances with a relatively thin cross section, as illustrated by the curved ceiling in Figure 30.15.

Folded Plate Roofs

These roofs are thin skins of plywood reinforced by rafters to form shell structures that can utilize the strength of plywood. The use of folded plate roofs eliminates trusses and other roof members. The tilted plates lean against one an-

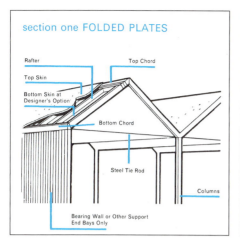

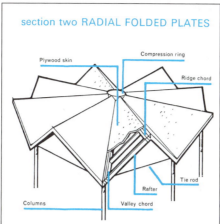

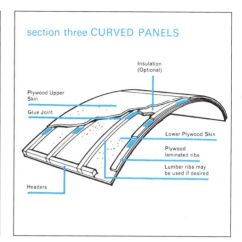

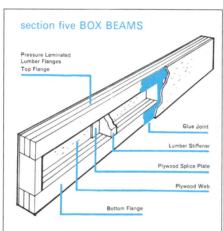

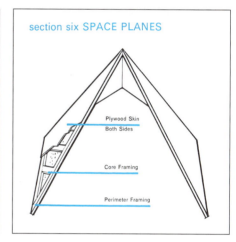

FIGURE 30.14 ■ Prefabricated roof units.

FIGURE 30.15 ■ Curved ceiling.

other, acting as giant V-shaped beams that are supported by walls or columns.

Cornices

The area of the roof that intersects with the outside walls and extends to the end of the roof overhang is the **cornice**. Detail drawings are necessary to show cornice areas. These detail drawings include part of the wall framing, roof framing, and methods of attaching the roof structure to the wall. Figure 30.16 shows several methods used to draw cornice details.

Several types of rafter cuts are used at intersections to help hold rafters onto top wall plates. For example, the *bird's mouth cut* shown in Figure 30.17 provides a level surface for the intersection of the rafters and top plates. The area bearing on the plate should not be less than 3″ (76 mm).

The outer vertical edge of an overhang is the **fascia**, and the horizontal bottom of an overhang structure is the **soffit**. Figure 30.18 shows four types of soffit design. Soffits

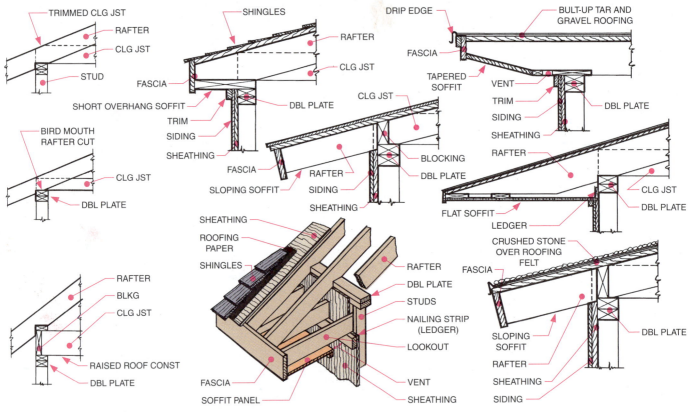

FIGURE 30.16 ■ Typical cornice framing details.

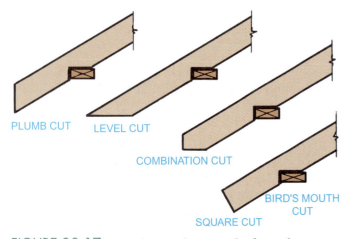

FIGURE 30.17 ■ Bird's-mouth cuts and rafter tail cuts.

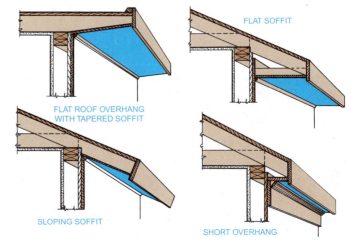

FIGURE 30.18 ■ Types of soffit design.

are made from plywood or sheet-metal panels. Soffit panels should contain screened openings to allow air to pass between rafters or to circulate through crawl spaces if rafters are exposed. Sheet-metal soffits are available in solid or ventilated designs to allow air flow. Regardless of the material used, some air flow must be designed into the cornice, as shown in Figure 30.19, to provide ventilation and prevent dry rot. Figure 30.20 outlines the steps in laying out and drawing a typical cornice elevation detail.

Cantilevered *lookout rafters* are used where common rafter extensions cannot create an overhang. See Figure 30.21. Lookout rafters are placed perpendicular to the common rafter direction.

Collar Beams and Knee Walls

Collar beams provide a tie between opposing rafters. They are usually placed on every rafter or every second

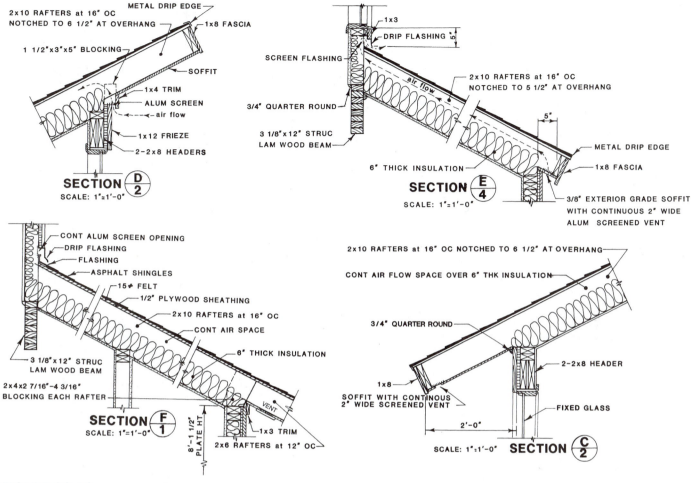

FIGURE 30.19 ■ Cornice design incorporating insulation and air flow.

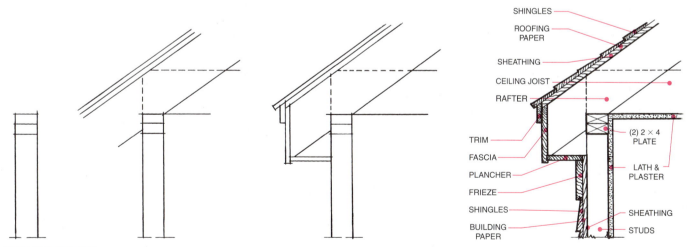

FIGURE 30.20 ■ Steps in drawing cornice details.

or third rafter. Collar beams are used to reduce the rafter stress that occurs between the top plate and the top of the rafter. They lock rafters into position. They may also act as ceiling joists for finished attics. On low-pitched roofs, collar beams may be required to counteract the lateral (outward) thrust of joists. See Figure 30.22.

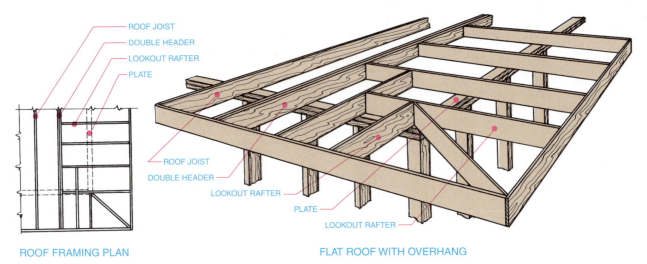

ROOF FRAMING PLAN

FLAT ROOF WITH OVERHANG

FIGURE 30.21 ■ Lookout rafters used to extend overhang.

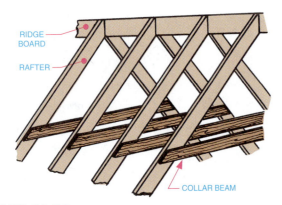

FIGURE 30.22 ■ Collar beams are used to reduce stress on rafters.

Knee walls are vertical studs that project from ceiling joists or attic floors to roof rafters. See Figure 30.23. Knee walls add rigidity to the rafters and may also provide half-wall framing for finished attics.

Dormers

As covered in Chapter 15, there are two basic types of dormers. These have different framing requirements. Figure 30.24 illustrates the methods of drawing a related plan view and front elevation of a gable end dormer. Figure 30.25 shows a side elevation of a different gable end dormer, and Figure 30.26 shows a framing elevation for a typical individual shed dormer.

Parts of the roof framing often extend above the roof line. Dormer rafters are one example. These are sometimes drawn with dotted lines. The details of intersecting

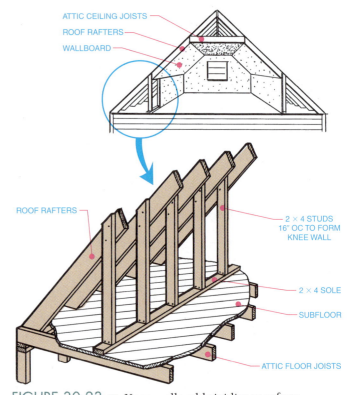

FIGURE 30.23 ■ Knee walls add rigidity to rafters.

dormer roof framing with dormer walls is shown in the framing plan. See Figure 30.25.

Dormer rafters and walls do not lie in the same plane as the remainder of the roof rafters. A framing elevation drawing is therefore needed to show the exact position of the dormer members and their tie-in with the common roof rafters and with other roof framing members.

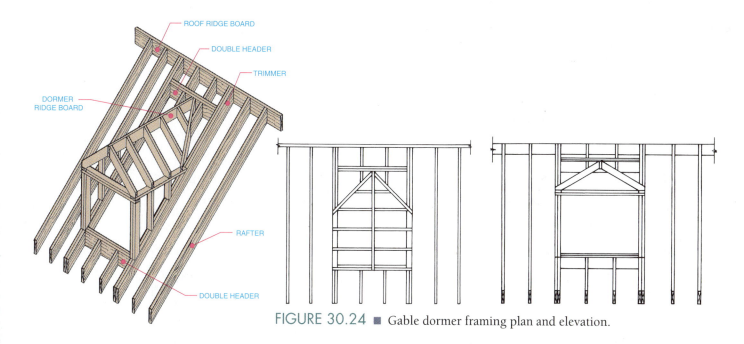

FIGURE 30.24 ■ Gable dormer framing plan and elevation.

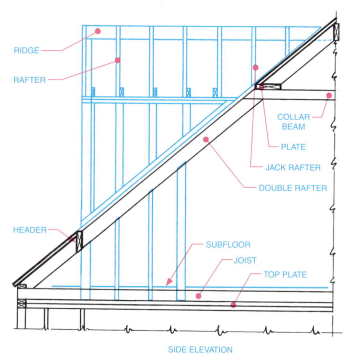

FIGURE 30.25 ■ Gable dormer side elevation.

USING CAD

Roof Framing on CAD

The top-level floor plan layer is used as a guide in preparing the roof framing plan. The *Line* command or the *Polyline* command is then used to draw rafters. The *Trim* or *Break* command is then used to remove portions of rafters for openings such as chimneys, skylights, and dormers. Double rafters and special framing for intersections are then added using the *Offset* command.

ROOF FRAMING DRAWINGS

Plans

To convey the structural design of a roof to a builder, plans must be accurate and complete. A *roof plan* shows only the outline and the major object lines of the roof. A roof plan is not a framing plan but a plan view of the roof. A roof plan can be used, however, as the basic outline to develop a roof framing plan. A *roof framing plan* exposes the exact position and spacing of each structural member. See Figure 30.28.

Chimney Details

If a detailed roof framing plan is not prepared, chimney framing roof details are often provided on separate detail drawings. Chimney details include the position of ceiling joists, roof rafters, and type of construction. See Figure 30.27.

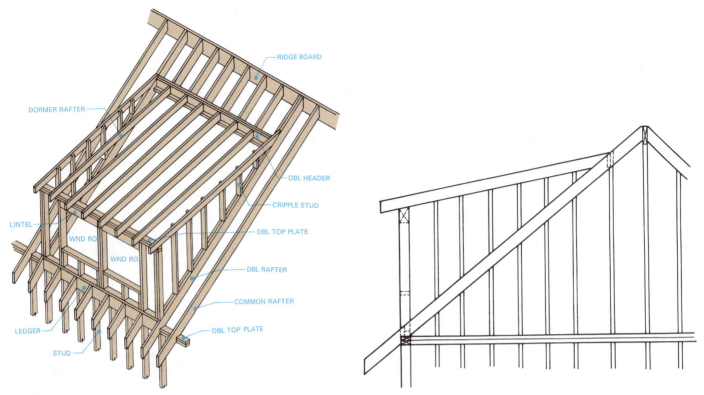

FIGURE 30.26 ■ Shed dormer framing elevation.

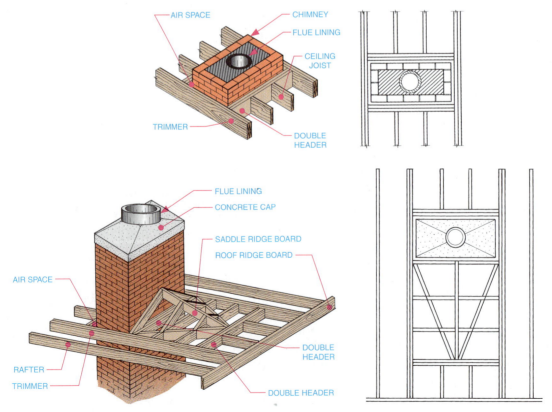

FIGURE 30.27 ■ Chimney framing details showing flat roof and saddle intersections.

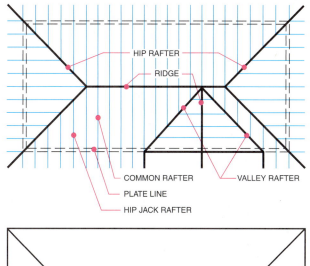

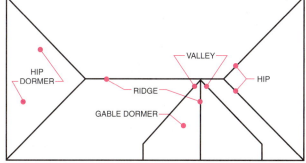

FIGURE 30.28 ■ Roof framing plan compared to a roof plan.

Roof framing plans may be either simplified, single-line plans or detailed, double-line plans. A single-line plan is acceptable only to show the general relationship and spacing of the structural members. When more details concerning the exact construction of intersections and joints are needed, a double-line plan showing the thickness of each member should be prepared. See Figure 30.29. This type of plan is necessary to indicate the relative placement of one member compared with another; that is, to indicate whether one member passes over or under another. In a double-line plan, the width of ridge boards, rafters, headers, and plates should be drawn to the exact scale. If a complete roof framing plan of this type cannot fully describe construction framing details, then additional removed pictorial or elevation drawings should be prepared. See Figure 30.30. These separate drawings can also be enlarged, for example, to better show detail dimensions.

In contrast to a true orthographic projection, only the outline of the top of the rafters is shown on roof framing plans. Areas underneath the rafters are shown by dotted lines. Roof framing plans show only horizontal relationships of members, such as thickness, length, and horizontal spacing. See Figure 30.31. In a top view (plan view), you cannot show vertical dimensions such as structural heights and pitches.

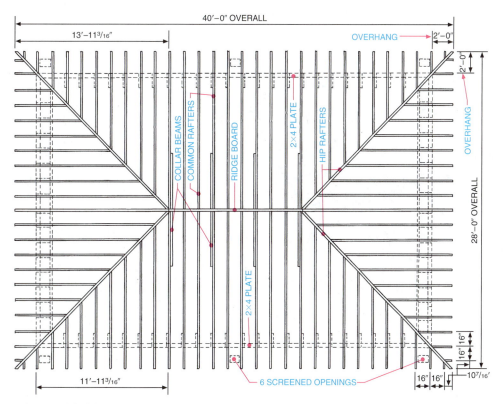

FIGURE 30.29 ■ Double-line roof framing plan.

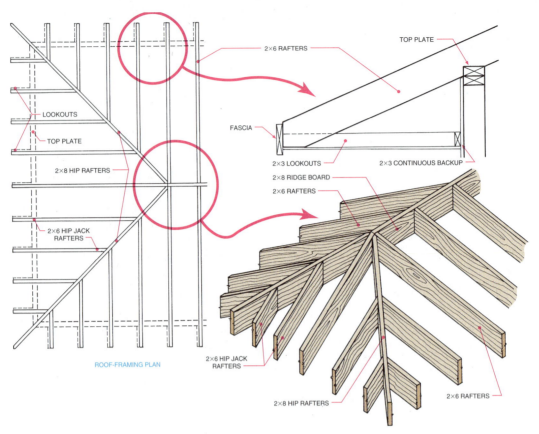

FIGURE 30.30 ■ Roof framing detail plan and elevation.

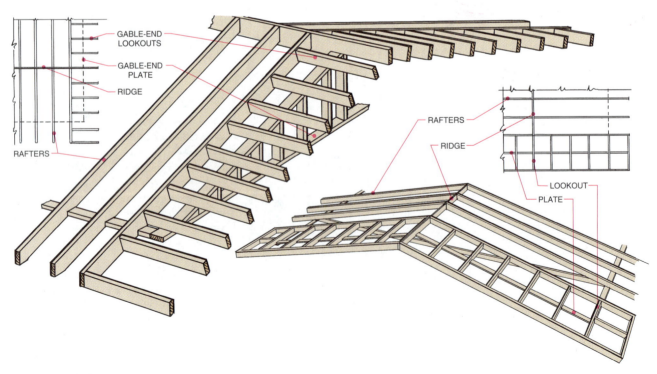

FIGURE 30.31 ■ Roof framing plans show only tops of members.

Elevations

The angle or vertical position of any roof framing member should be shown on a roof framing elevation. If a comparison of different heights and pitches is desired, a *composite framing-elevation* drawing should be prepared. This elevation is one that can be projected from the roof framing plan or from corresponding lines on the elevation drawings. Figure 30.32 shows the projection of an elevation framing drawing from a framing elevation plan. In Figure 30.33 the roof framing details are abbreviated on a sectional elevation drawing of the entire structure.

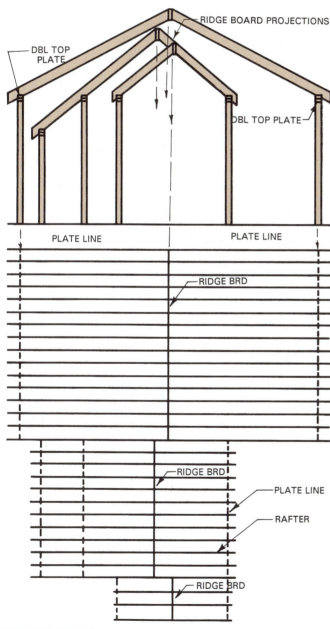

FIGURE 30.32 ■ Projection of roof framing elevation from framing plan.

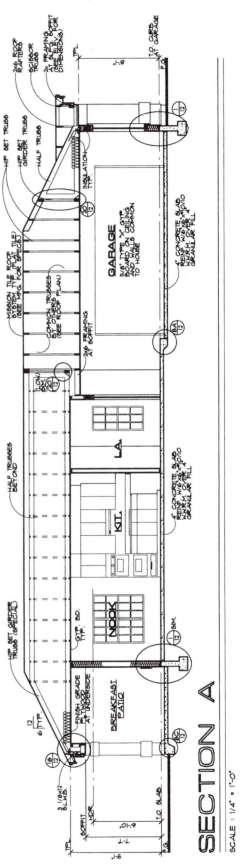

FIGURE 30.33 ■ Roof framing shown on a simplified full elevation section.

Dimensions

Dimensions on roof framing plans usually include the size and spacing of framing members and the major distances (spans) between framing components. See Figure

30.34. Regular spacing of structural members, such as roof rafters, floor joists, and wall studs, is not dimensioned if these fall on modular increments. Notes may be used on framing drawings to show the size and spacing of members. Figure 30.35 shows examples of three ceil-

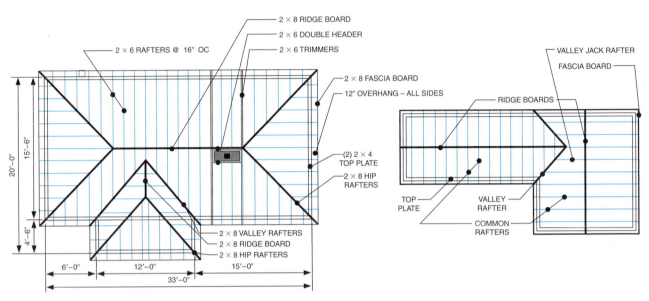

FIGURE 30.34 ■ Dimensioned roof framing plan.

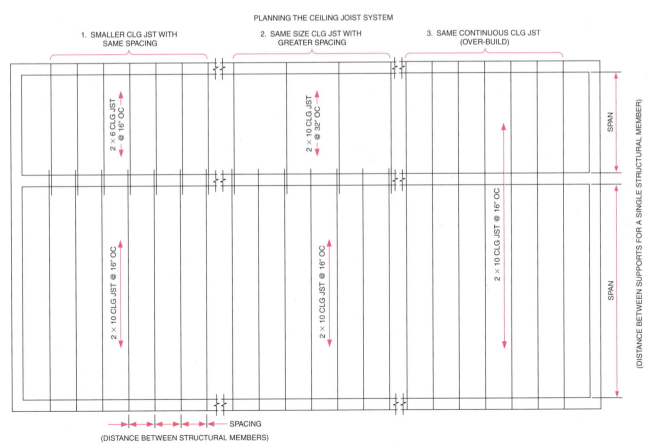

FIGURE 30.35 ■ Notation methods used to define roof and ceiling framing.

SIZE RAFTER	SPACING RAFTER	MAXIMUM SPAN			
		GROUP 1	GROUP 2	GROUP 3	GROUP 4
2 × 4	12"	10'–0"	9'–0"	7'–0"	4'–0"
	16"	9'–0"	7'–6"	6'–0"	3'–6"
	24"	7'–6"	6'–6"	5'–0"	3'–0"
	32"	6'–6"	5'–6"	4'–6"	2'–6"
2 × 6	12"	17'–6"	15'–0"	12'–6"	9'–0"
	16"	15'–6"	13'–0"	11'–0"	8'–0"
	24"	12'–6"	11'–0"	9'–0"	6'–6"
	32"	11'–0"	9'–6"	8'–0"	5'–6"
2 × 8	12"	23'–0"	20'–0"	17'–0"	13'–0"
	16"	20'–0"	18'–0"	15'–0"	11'–6"
	24"	17'–0"	15'–0"	12'–6"	9'–6"
	32"	14'–6"	13'–0"	11'–0"	8'–6"
2 × 10	12"	28'–6"	26'–6"	22'–0"	17'–6"
	16"	25'–6"	23'–6"	19'–6"	15'–6"
	24"	21'–0"	19'–6"	16'–0"	12'–6"
	32"	18'–6"	17'–0"	14'–0"	11'–0"

FIGURE 30.36 ■ Common wood rafter spans for pitch greater than 4/12.

SIZE OF CEILING JOISTS	SPACING OF CEILING JOISTS	MAXIMUM SPAN			
		WOOD GROUP 1	WOOD GROUP 2	WOOD GROUP 3	WOOD GROUP 4
2 × 4	12"	11'–6"	11'–0"	9'–6"	5'–6"
	16"	10'–6"	10'–0"	8'–6"	5'–0"
2 × 6	12"	18'–0"	16'–6"	15'–6"	12'–6"
	16"	16'–0"	15'–0"	14'–6"	11'–0"
2 × 8	12"	24'–0"	22'–6"	21'–0"	19'–0"
	16"	21'–6"	20'–6"	19'–0"	16'–6"

FIGURE 30.37 ■ Common wood ceiling joist spans.

tion for ceiling joists is listed in Figure 30.37. All roof member sizes and spacing are based on spans and loads. Refer to Appendix B, "Mathematical Calculations."

ROOF PITCH

Pitch is the angle between the roof's surface (top plate to ridge board) and the horizontal plane. The **rise** is the vertical distance from the top plate to the roof's ridge. The **run** is the horizontal distance from the top plate to the ridge. It is expressed in units of 12. The **span** is the full horizontal distance between outside supports. It is double the run.

Figure 30.38 shows how to determine units of rise (vertical distance) per units run (horizontal distance). On

ing joist framing plans using different size and spacing notations. On detail drawings, overall dimensions are not given. Instead, the key distances between structural levels and horizontal distances are shown.

If material sizes are not given on the framing drawing, they should be included in the specifications. Figure 30.36 shows typical light construction roof rafter sizes, spans, and spacing for a pitch of 4/12 or more. Similar informa-

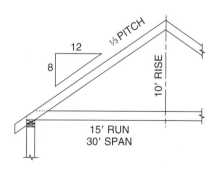

RISE = 10'
RUN = 15'
PITCH = 10 ÷ 30 = 1/3
RISE/FT RUN = $\frac{10 \times 12}{15}$ = 8"

8/24 = 1/3 PITCH

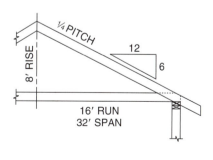

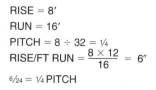

RISE = 8'
RUN = 16'
PITCH = 8 ÷ 32 = 1/4
RISE/FT RUN = $\frac{8 \times 12}{16}$ = 6"

6/24 = 1/4 PITCH

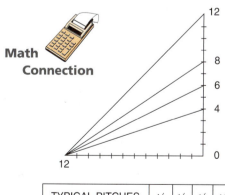

TYPICAL PITCHES	1/2	1/3	1/4	1/6
RISE/FT RUN	12	8	6	4

FIGURE 30.38 ■ Methods of determining roof pitch.

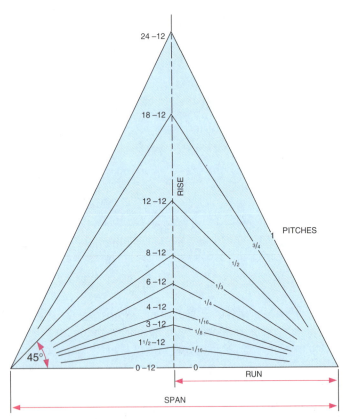

FIGURE 30.39 ■ Commonly used roof pitches.

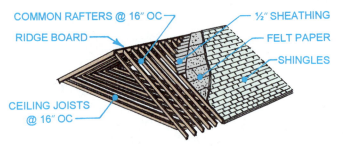

FIGURE 30.40 ■ Conventionally framed wood roof.

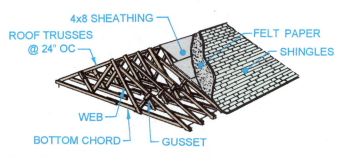

FIGURE 30.41 ■ Wood-trussed roof.

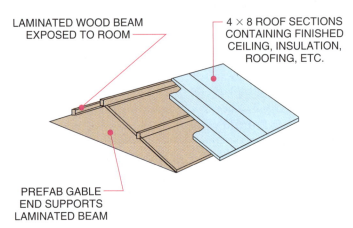

FIGURE 30.42 ■ Longitudinal post-and-beam roof construction.

drawings, this figure is shown in a **slope diagram** (triangle) near the line of the roof along with the run unit number.

Pitch is the ratio of the *actual* rise to the *actual* span. It is also the ratio of the *units* of rise to *units* of span (double the units of run). Refer again to Figure 30.38. In the drawing on the left, the pitch is 10/30 (ratio of actual rise to actual span) which reduces to 1/3. It is also 8/24 (unit ratio of rise to span) which also reduces to 1/3.

Roof pitches vary greatly. For example, a roof with a slope of 12/12 is steep. The pitch would be 12/24 or 1/2. A roof with a slope of 8/12 is moderately sloped. The pitch would be 8/24 or 1/3. A roof with a slope of 2/12 is nearly flat. The pitch would be 2/24 or 1/12. See Figure 30.39.

ROOF FRAMING METHODS FOR WOOD

Wood framing can be divided into conventional and heavy timber methods, regardless of roof shape. Both methods are used extensively in all types of wood-framed roofs.

The *conventional* method of roof framing consists of roof rafters or trusses spaced at small intervals, such as 16″ on center. See Figure 30.40. These roof rafters are

perpendicular to the ridge board and align with the exterior studs placed on the same centers.

An adaptation of this conventional method of constructing roofs is to substitute roof trusses for conventional rafters and ceiling joists. Trusses create a much more rigid roof, but they make it impossible to use space between ceiling joists and rafters for an attic or crawl-space storage. See Figure 30.41.

Heavy-timber construction, another method of roof framing, consists of posts that support beams. Longitudinal beam sizes vary with the span and spacing of beams. The beams are installed parallel to the ridge board, or ridge beam. This is called longitudinal roof beam construction. See Figure 30.42. Beams may also be placed perpendicular (transverse) to the ridge beam. Planks are then placed across the beams. The planks can serve as a ceiling, as well as a base for roofing that will shed water. When planks are selected for appearance in an open-beam ceiling, the only ceiling treatment needed may be a protective finish on the planks and beams.

ROOF FRAMING TYPES

The most common roof framing types are *gable, hip, shed,* and *flat.* Other types, such as mansard, gambrel, and A-frame, are variations of the four common types and are framed in a similar manner.

Gable Roof Framing

Gable roof framing consists of rafters that form two inclined planes extending from the outer walls to a ridge. Figure 30.43 shows gable roof framing on the right end and hip roof framing on the left end. When trusses or truss joists are substituted for common rafters, a ridge board is not used.

A-frame roof framing is similar to conventional gable framing, except the roof rafters rest on a foundation rather than on top plates.

Gable pitches range from 2/12 to 12/12. The pitch of a gable roof does not show on a roof framing plan. A framing elevation is necessary to show pitch and gable framing. See Figure 30.44. A *gable end* is the side of a building that rises to meet the ridge. See Figure 30.45. In some cases, especially on low-pitch roofs, the entire gable-end wall from the floor to the ridge can be panelized with varying lengths of studs. However, it is more common to prepare a rectangular wall panel and erect separate cripple studs an gable ends.

To design an overhang on a gable end, *lookouts* are used. Lookouts are short rafters placed perpendicular to the first or second common rafters. *Winged gables,* as shown in Figure 30.46, use lookouts of varying lengths to form a triangular overhang. Large winged gable ends may require additional columns and beam support. See Figure 30.47.

Gambrel Roof Framing

Gambrel (barn) roofs are a variation of gable roofs but use double-pitched rafters. See Figure 30.48. The pitch of the lower part is always steeper than the pitch of the top part. Purlins may be used to stabilize pitch intersections. Prefabricated truss joists may also be used for gambrel roofs.

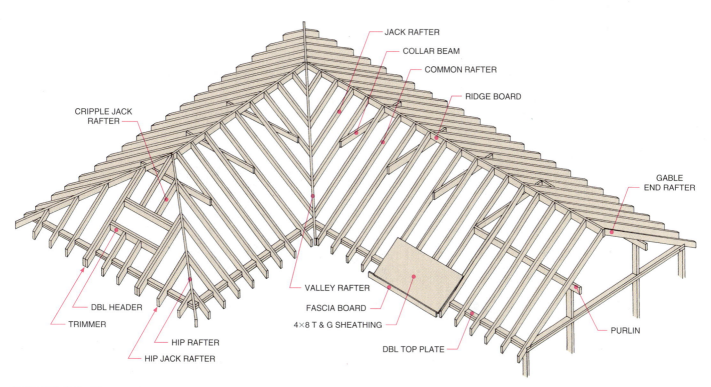

FIGURE 30.43 ■ Gable roof framing on the right and hip roof framing on the left.

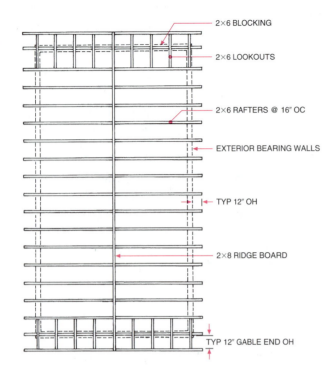

2×6 BLOCKING

2×6 LOOKOUTS

2×6 RAFTERS @ 16" OC

EXTERIOR BEARING WALLS

TYP 12" OH

2×8 RIDGE BOARD

TYP 12" GABLE END OH

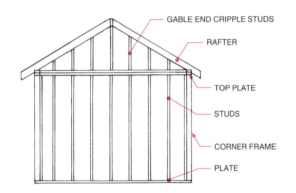

GABLE END CRIPPLE STUDS

RAFTER

TOP PLATE

STUDS

CORNER FRAME

PLATE

FIGURE 30.44 ■ Relationship of a gable framing plan and a gable framing elevation.

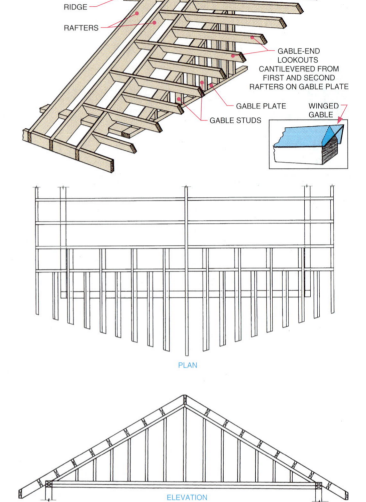

RIDGE

RAFTERS

GABLE-END LOOKOUTS CANTILEVERED FROM FIRST AND SECOND RAFTERS ON GABLE PLATE

GABLE PLATE

GABLE STUDS

WINGED GABLE

PLAN

ELEVATION

FIGURE 30.46 ■ Winged-gable lookout construction.

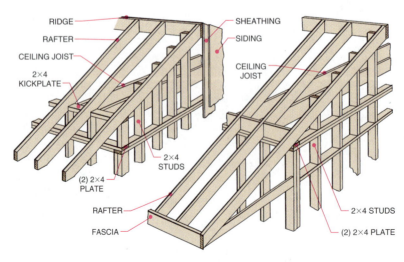

RIDGE

RAFTER

CEILING JOIST

2×4 KICKPLATE

SHEATHING

SIDING

CEILING JOIST

2×4 STUDS

(2) 2×4 PLATE

RAFTER

FASCIA

2×4 STUDS

(2) 2×4 PLATE

FIGURE 30.45 ■ Gable-end overhang construction.

FIGURE 30.47 ■ Winged gable supported by beams and columns. (*Two Creek Ranch, Fayetteville, TX—Lindi Shrovik, Coldwell Banker*)

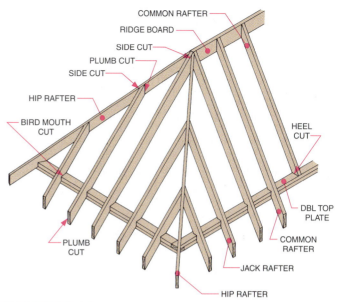

FIGURE 30.49 ■ Hip roof framing members.

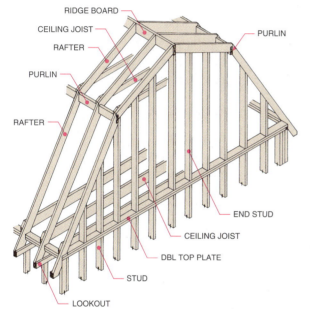

FIGURE 30.48 ■ Gambrel roof framing members.

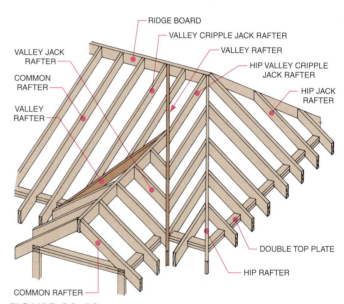

FIGURE 30.50 ■ Comparison of valley rafters and hip rafters.

Hip Roof Framing

Hip roof framing is similar to gable roof framing except that the roof slopes in four directions instead of intersecting gable-end walls. See Figure 30.49. Where two adjacent slopes meet, a hip is formed on the external angle. A *hip rafter* extends from the ridge board, over the top plate, and to the edge of the overhang. A hip rafter supports the ends of the shorter hip-jack rafters.

The internal angle formed by the intersection of two slopes of the roof is known as the *valley*. A *valley rafter* is used to form this angle. See Figure 30.50.

Hip rafters and valley rafters are normally 2″ (50 mm) deeper or 1″ (25 mm) wider than the common rafters, for spans up to 12′ (3.7 m). For spans over 12′, the hip and valley rafters should be double the width of common rafters.

Jack rafters are rafters that extend from the wall plate to the hip or valley rafter. They are always shorter than common rafters, which extend from the top plate to the ridge.

Dutch Hip Roof Framing

A Dutch hip roof is framed in the same way as a gable roof in the center and as a partial hip roof on the ends. See Figure 30.51. One of the many variations of this basic design

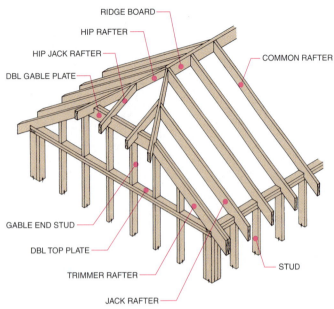

FIGURE 30.51 ■ Dutch hip roof framing.

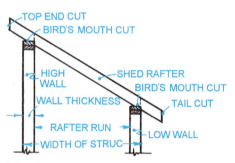

FIGURE 30.52 ■ Elements of shed roof framing.

is a mansard roof, a double-pitched hip roof. Refer back to the many roof styles shown in Chapter 15.

Shed Roof Framing

A shed roof is a roof that slants in only one direction. A gable roof is actually two shed roofs, sloping in opposite directions. Shed roof rafter design is the same as rafter design for gable roofs except that the run of the rafter is the same as the span. Some shed roofs are nearly flat. Shed roof slopes range from 2/12 to 12/12. Figure 30.52 shows the basic elements of shed roof framing.

Flat Roof Framing

In conventional flat roof framing, rafters are similar to floor joists. Rafters span from wall to wall or from exterior wall to bearing partitions, columns, or beams. In heavy-timber construction, beams may be used to span these distances. Roof decking is laid directly on the beams. The members of

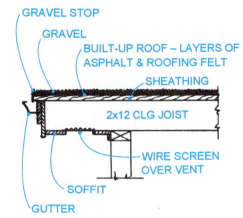

FIGURE 30.53 ■ Lightweight flat roof framing elevation.

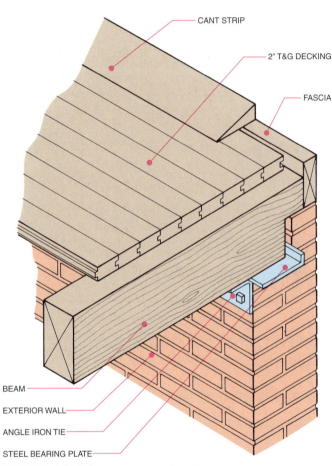

FIGURE 30.54 ■ Heavy-timber flat roof construction.

a conventional wood-frame flat roof are shown in the elevation section in Figure 30.53. Figure 30.54 shows one method of intersecting heavy-timber roof framing with a solid masonry wall. Flat roofs must be designed for maximum snow loads since the snow will not slide off. Instead, it must melt and drain away.

Flat roofs are usually not absolutely flat. Most flat roofs have a slight slope 1/8″ per foot to 1/2″ per foot to allow for

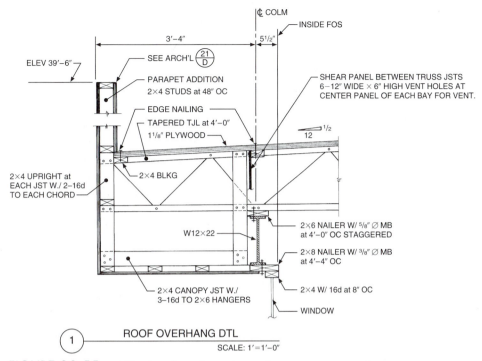

ROOF OVERHANG DTL
SCALE: 1'=1'-0"

1

FIGURE 30.55 ■ Wood and steel cornice and parapet construction.

drainage. Some roofs are flat to allow a specific level of water to remain for insulation. Flat roof drainage must be provided through internal or external downspouts. A *cant strip* should be located on flat roof perimeters to stop water from flowing over the sides rather than through downspouts. Cant strips also help waterproof joints at the wall.

A flat roof may stop at a wall intersection or continue to form an overhang. Overhangs may include lookouts on the cantilevered ends, perpendicular to the rafter direction. A flat roof may also intersect a *parapet* (short wall). Figure 30.55 shows a cornice section with a wood-framed parapet wall and intersecting steel joists. Parapets are used to hide the roofing surface vents and any roof-mounted mechanical equipment. Parapets are also used to simulate mansard roofs, although true mansard roofs are double-pitched hips.

STEEL AND CONCRETE ROOF FRAMING METHODS

Structural steel framing methods are especially appropriate for flat roofs. Steel roof framing systems are very similar to wood floor framing systems. Girders, beams, joists, and decking are used in the same manner. Only the covering and cornice intersecting details are different.

Steel joists, in light construction, rest directly on concrete or on masonry bearing walls. See Figure 30.56. In heavy construction, steel joists rest on steel girders or columns. Similar to other flat roofs, steel flat roof construction either meets an outside wall to form a right angle, intersects an outside wall to form a parapet, or extends to

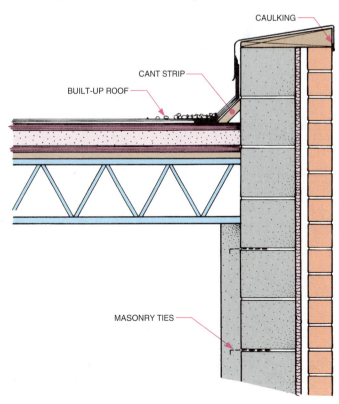

FIGURE 30.56 ■ Steel roof and masonry wall construction.

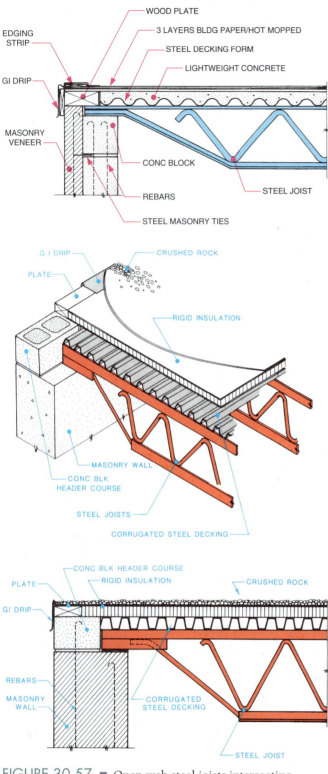

FIGURE 30.57 ■ Open web steel joists intersecting masonry exterior walls.

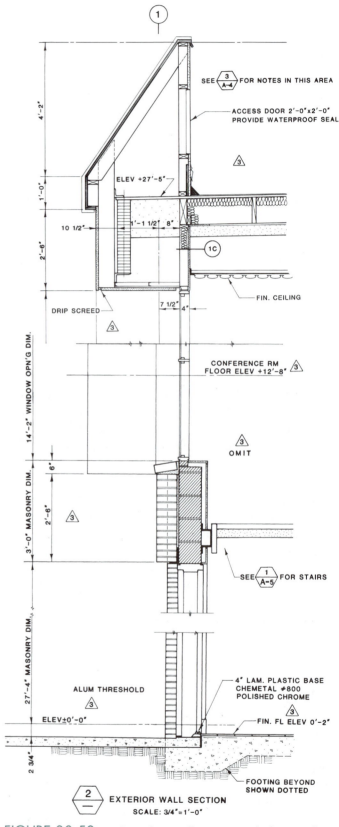

FIGURE 30.58 ■ Complete wall section including roof intersections and foundation.

form an overhang. See Figure 30.57. When complete wall sections from the foundation to the roof are prepared, they are indexed to the basic plan and elevation drawings as shown in Figure 30.58.

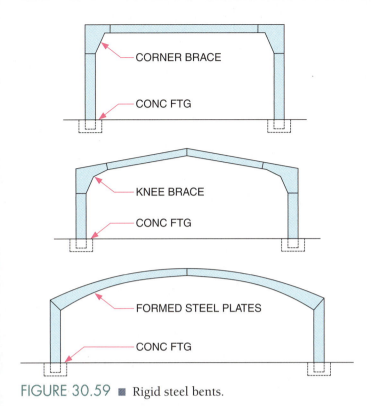

FIGURE 30.59 ■ Rigid steel bents.

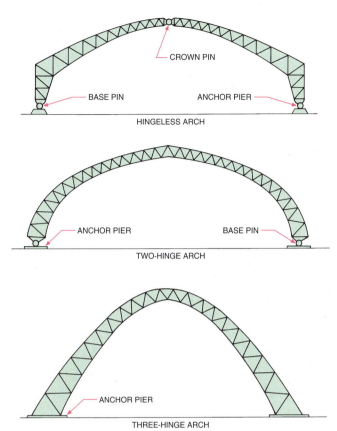

FIGURE 30.60 ■ Types of steel arches.

Steel joists can span up to 40 feet depending on load factors. For larger spans, steel bents, arches, or space frames are used. Rigid steel bent frames are either straight single-span, shaped, or multiple-span frames as shown in Figure 30.59.

Arches are bent trusses. See Figure 30.60. Arch details are shown on structural drawings the same way truss details are shown. Space frames are three-dimensional trusses formed by connecting series of triangular polyhedrons. See Figure 30.61. Space frames, because of their light weight and ability to resist bending, can span extremely large distances. Because their load-bearing capacity is limited, they are used primarily for roof and not floor systems.

Precast concrete joists are used where heavy loads and short spans exist. Filler blocks and a poured slab create a monolithic roof or floor.

Steel roof framing plans are prepared similar to single-line wood framing plans. The lines, symbols, and notations are the same as those used on steel floor framing plans.

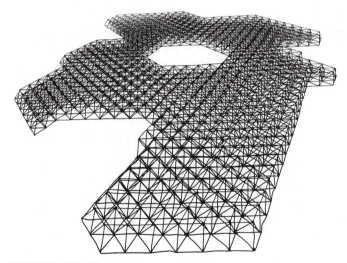

FIGURE 30.61 ■ Space frame construction.

ROOF-COVERING MATERIALS

Roof coverings protect buildings from rain, snow, wind, heat, and cold. Materials used to cover pitched roofs include wood, fiberglass, fiber, cement, asphalt, and composition shingles. On heavier roofs, ceramic tile or slate may also be used but these require stronger framing systems to compensate for the increased weight. Roll roofing or other sheet material, such as galvanized iron, aluminum, copper, and tin, may also be used for flat or low-pitched roofs. Built-up roofing of felt and gravel is

TYPE	TYPICAL FORM DESCRIPTION	TYPICAL WIDTHS	AVERAGE SPANS	REMARKS
TONGUE & GROOVE (T & G) WOOD-LAMINATE	2", 3" & 4"	6"	8' 14' 20'	MAXIMUM 20' LENGTHS.
T & G PRECAST GYPSUM	2"	15"	7'	MAXIMUM 10' LENGTHS. AVAILABLE WITH METAL EDGES.
CORRUGATED STEEL	3" / 8"	24"	15'	MAXIMUM 10' LENGTHS. ECONOMICAL FOR MEDIUM SPANS.
PRECAST CONCRETE	WWM 1" 3½" 2"	24"	8'	MAY BE CAST TO ANY LENGTH AND THICKNESS.
TONGUE & GROOVE PRECAST WOOD FIBER & CEMENT	2", 3½" & 3"	30"	3' 4' 5'	MATERIAL IS NAILABLE.

FIGURE 30.62 ■ Common roof covering characteristics.

used extensively on flat or low-pitched roofs. Figure 30.62 lists the characteristics of common roof covering materials.

The weight of roofing materials is important in computing dead loads. A heavier roofing surface makes the roof more permanent than does a lighter surface. Generally, heavier roofing materials last longer than lighter materials. Roof covering materials are classified by their weight per 100 square feet (100 sq. ft. equals 1 *square*). Thus 30-lb. roofing felt weighs 30 lbs. per 100 square feet.

Sheathing

Roof sheathing may consist of lumber boards, gypsum, fiberboard, or plywood sheets nailed directly to roof rafters. Sheathing adds rigidity to the roof and provides a surface for the attachment of waterproofing materials. In humid areas, sheathing boards or panels are sometimes spaced slightly apart to provide ventilation and to prevent shingle rot. The thickness of roof (and wall) sheathing depends on rafter (or stud) spacing and the amount of insulation required. Tongue-and-groove plywood provides the greatest amount of rigidity.

The joints of sheathing panels are always staggered, and only exterior grade material is used. Most codes in high-wind areas require sheathing to be applied with nails or screws not staples. Hurricane clips that lock the panels together may also be required. In large housing developments, a *roof sheathing plan* is often prepared to plan the best possible arrangement with the minimum amount of waste. See Figure 30.63.

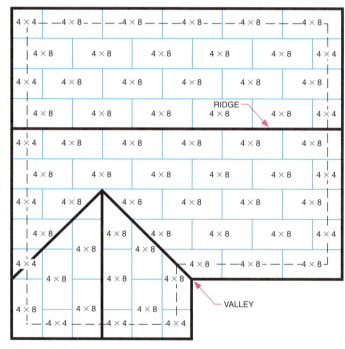

FIGURE 30.63 ■ Roof sheathing plan.

Roll Roofing

Roll (continuous membrane) roofing may be used as an underlayment for shingles or as a finished roofing material for slopes less than 3/12. Roll roofing material used as an underlayment includes asphalt and saturated felt. The underlayment serves as a barrier against moisture and wind. As a finished roofing material, copper, aluminum, or gal-

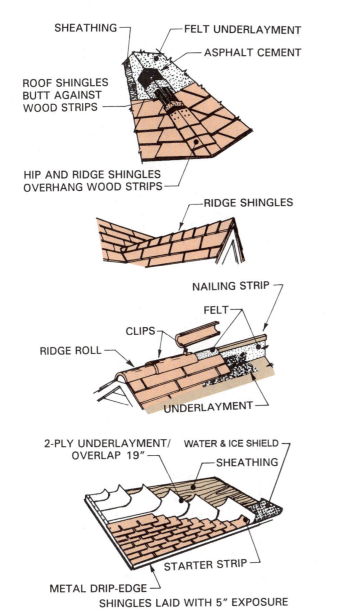

FIGURE 30.64 ■ Methods of shingle application.

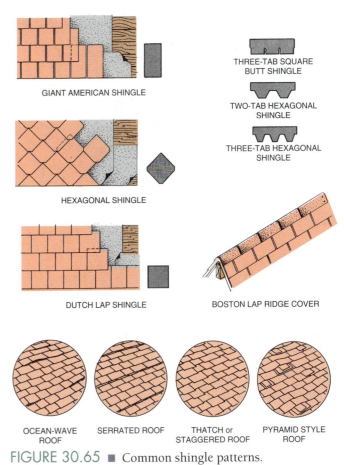

FIGURE 30.65 ■ Common shingle patterns.

Roof Shingles, Shakes, and Tiles

Roof shingles and shakes are made from asphalt, cement, fiberglass, cedar, or bonded wood fibers. Shingles are laid over building felt that covers the sheathing. See Figure 30.64. Shingles are available in a variety of patterns. See Figure 30.65. Most shingles are not recommended for pitches less than 4/12 because of the danger of wind lift.

Shingles are classified by weight per 100 square feet. Shingles range in weight from 180 lbs. to 390 lbs. per square (i.e., 100 sq. ft.) for residential roofs. The average residential asphalt shingle is 245 lbs. per square. Shingles are also classified by special features such as their resistance to mildew, wind, hail, and water and by their life expectancy, usually 20 or 25 years.

Tiles are manufactured from clay-ceramic, cement, and polystyrene. Copper, aluminum, and galvanized steel panels are also available in patterns that simulate shingles and tiles.

Built-up Roofs

Built-up roof coverings are used on flat or extremely low-pitched roofs. Built-up roofing cannot be used on high-pitched roofs because the gravel will wear off during rain and high winds. Because rain may be driven into gravel crevices and snow may not quickly melt from these roofs, complete waterproofing is essential.

Built-up roofs may have three, four, or five layers of roofing felt, sealed with hot-mopped tar or asphalt, between coatings. The final layer of tar or asphalt is then covered with roofing gravel or a top sheet of roll roofing. See Figure 30.66.

vanized steel is used in rolls or sheets. Seams are sealed to provide a watertight surface.

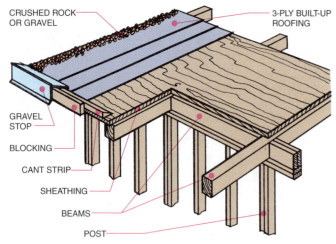

FIGURE 30.66 ■ Built-up roof covering.

ROOF APPENDAGES

Gutters, flashing, vents, fascia covers, skylights, and downspouts are additions to many roofs. Appendages perform several functions, such as to control water flow, ventilate building fumes, admit light, and cover rough lumber.

Gutters and Downspouts

Gutters are troughs designed to carry water to downspouts, where it can be emptied into a sewer system or away from the building. Materials used most commonly for gutters are sheet metal, cedar, redwood, and plastic. Gutters may be built into the roof structure, as shown in Figure 30.67. Sheet metal or wood gutters may also be attached or hung from the fascia board. All gutters should be pitched at least 1:20 to provide for drainage to downspouts. See Figure 30.68. Gutters must be kept below the

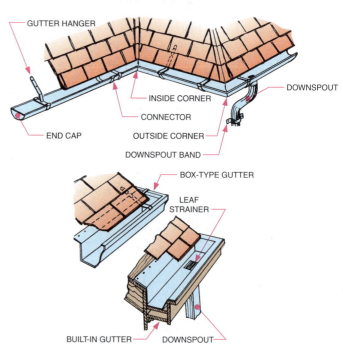

FIGURE 30.68 ■ Hanging gutter and downspout assembly.

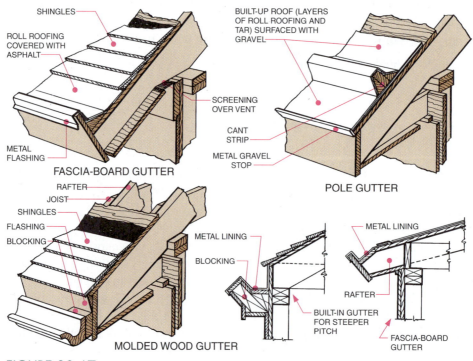

FIGURE 30.67 ■ Built-in gutters.

roof eave line to prevent snow and ice from accumulating. In selecting gutters and downspouts, care must be taken to ensure that their size is adequate for the local rainfall.

A *roof drainage plan* should show the runoff direction of water for all roof segments. The downslope is shown by arrow direction. See Figure 30.69. Gutter positions can then be planned accordingly. Some pitched roofs with large overhangs may not need gutters, if drainage is adequate and runoff doesn't fall on outdoor traffic or living areas.

Flashing

Joints where roof covering materials intersect a ridge, hip, valley, chimney, wall, vent, skylight, or parapet must be flashed. **Flashing** is additional covering used under a joint to provide complete waterproofing. Roll-roofing, galvanized sheet steel, copper sheeting, or bituthene (polyethylene film or rubberized asphalt) may be used as flashing material. On sloped roofs step flashing is used, as shown in Figure 30.70. Unless seams are sealed watertight, a second layer of counter flashing may be needed. For sheet-metal flashing, watertight sheet-metal joints should be made. Chimney flashing is frequently bonded into the mortar joint and caulked under shingles to provide a waterproof joint.

Roof Ventilation

Areas where excessive heat and moisture are trapped must be ventilated. Attics and crawl spaces can be

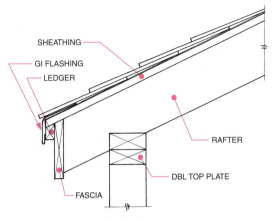

EAVE DRIP FLASHING

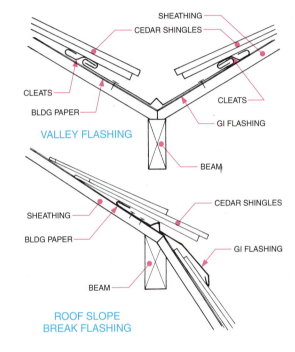

VALLEY FLASHING

ROOF SLOPE
BREAK FLASHING

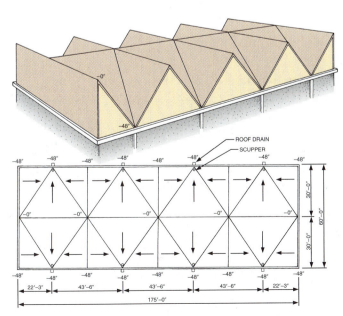

FIGURE 30.69 ■ Roof drainage plan.

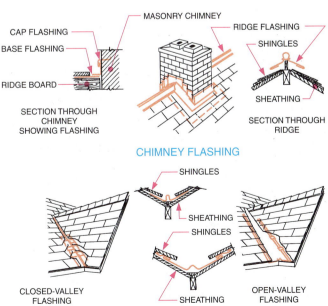

CHIMNEY FLASHING

FIGURE 30.70 ■ Flashing applications.

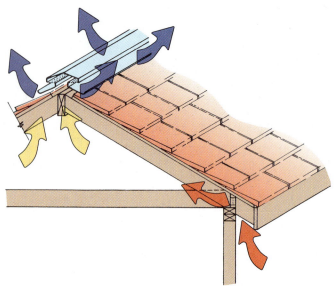

FIGURE 30.71 ■ Continuous ridge vents exhaust trapped air.

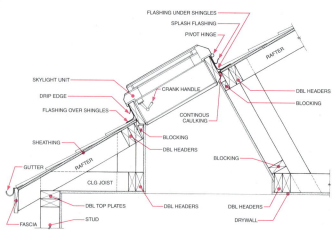

FIGURE 30.72 ■ Skylight roof framing.

ventilated by installing fans, roof vents, and/or a cupola connecting the inside area with the outside. Vent areas should be 1/150th of the enclosed area or 1/300th if the area has a vapor barrier. The space between rafters also needs to be ventilated. This is best done by creating an air flow between ventilated soffits and a continuous ridge vent. See Figure 30.71.

Skylights

Roof framing for skylights, as shown in Figure 30.72, is the same as for chimneys or any mechanical equipment requiring a break in the rafter and roof covering pattern. All openings must be flashed.

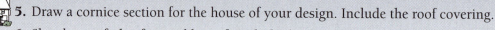

CHAPTER 30

Roof Framing Drawings Exercises

1. What are the two types of roof framing plans? What is the function of each?

2. Prepare a roof framing plan for a house of your own design. Label the members.

3. What is the roof pitch of a roof with a rise of 8′ and a run of 24′?

4. Name the main parts of a roof truss. List the advantages and disadvantages of using trusses for roofs.

 5. Draw a cornice section for the house of your design. Include the roof covering.

6. Sketch a roof plan for a gable roof. Include the dimensions of the run.

7. Specify the type of roofing to be used in the house of your design. Explain why you chose a particular type of material: wood, steel, and/or concrete.

8. How many square feet of shingles are in four squares?

9. Prepare a roof framing plan for a house of your own design, a cornice section, and/or a roof detail in this chapter.

PART 8

8

Electrical and Mechanical Design Drawings

31 CHAPTER

Electrical Design and Drawing

OBJECTIVES

In this chapter, you will learn to:

- plan and draw electrical circuits for a house on a floor plan.
- plan and draw lighting for each room in a house.
- calculate electrical measurements for each circuit.
- draw electrical symbols.
- design and draw an electronic building control system.

TERMS

ampere	ground fault circuit interrupter (GFCI)	National Electrical Code (NEC)
circuit	hard-wired	ohms
circuit breaker	incandescent	outlet
conductors	insulators	receptacle
distribution panel	kilowatt-hour	service entrance
fixture	lightning rod	switch
fluorescent	lux	volt
footcandle		watt

INTRODUCTION

Buildings cannot function as they should without electricity. The design and drawing of electrical systems requires a knowledge of electrical power distribution, wiring circuits, lighting methods, and electrical symbols and conventions.

ELECTRICAL PRINCIPLES

Understanding electrical principles is vital to designing safe and efficient architectural electrical systems.

Power Distribution

Electric power is generated from several sources of energy: wind, water, nuclear, fossil fuel, solar (photovoltaic), and geothermal. Photovoltaic cells convert solar energy directly into an electric current. All other energy sources are harnessed to produce a rotary mechanical motion that drives electric generators. The generators convert movement into electricity. Transformers are used to "step up" (increase) the electrical power to very high voltages (hundreds of thousands of volts) for transmission by wires over long distances. Wherever the transmission lines enter an industrial or residential community for local power distribution, large transformers are used to "step down" the voltage to a few thousand volts. Smaller transformers set on poles or in underground vaults are used for final distribution to small groups of houses or individual factories. Voltages of 120 and 240 are delivered to residences and small buildings. Because a voltage drop occurs in a circuit during delivery, 110 or 220 volts (V) are actually delivered to electrical outlets.

Electrical Measurements

Math Connection

Electrical properties can be measured with instruments. The terms used to describe units of electricity—*volt, am-*

pere, and *watt*—are used in both metric and customary systems.

A **volt** is the unit of electrical *pressure* or potential. This pressure makes electricity flow through a wire. For a particular electrical load, the higher the voltage, the greater the amount of electricity that will flow.

The term for flow of electricity is *current.* An **ampere,** or amp, is the unit used to measure the magnitude of an electric current. An ampere is defined as the specific quantity of electrons passing a point in one second. The amount of current, in amperes, that will flow through a circuit must be known in order to determine proper wire sizes and the current rating of circuit breakers and fuses.

The amount of *power* required to light lamps, heat water, turn motors, and do all types of work is measured in **watts.** Wattage depends on both potential and current. Current (in amperes) multiplied by potential (in volts) equals power (in watts).

$$\text{amperes} \times \text{volts} = \text{watts}$$

The actual energy used (the watts utilized) for work performed is the basis for figuring the cost of electricity. The unit used to measure the consumption of electrical energy is the **kilowatt-hour.** A kilowatt is 1,000 watts. An hour, of course, is a unit of time. A 1,000-watt hand iron operating for one hour consumes one kilowatt-hour (1 kWh). The device used to measure the kilowatt-hours consumed is the watt-hour meter.

Electricity flowing through a material always meets with some resistance. Materials such as wood, glass, and plastic have a high resistance. They are good **insulators.** Copper, aluminum, and silver have low resistance and are therefore good **conductors** of electricity. Most electrical wiring consists of copper or aluminum surrounded by plastic insulation. The plastic keeps the electricity from flowing where it isn't wanted.

The amount of electrical resistance is measured in **ohms.** The electron flow (or current in amperes) through a circuit is equal to the voltage (number of volts) divided by the resistance (ohms). This can be expressed in the formula:

$$I = \frac{E}{R} \text{ or } E = IR \text{ or } R = \frac{E}{I}$$

I = current (amperes)
E = electromotive force (volts)
R = resistance (ohms)

Example:
If the current is 10 amperes and the electromotive force is 120 volts, what is the resistance?

$$R = \frac{E}{I}$$
$$R = \frac{120}{10}$$
$$R = 12 \text{ ohms}$$

Electrical Service Entrance

Math Connection

Power is supplied to a building through a **service entrance.** Three heavy wires, together called the *drop,* extend from a utility pole or an underground source to the structure. These wires are twisted into a cable. At the building, overhead wires are fastened to the structure and spliced to service entrance wires that enter a conduit through a service head, as shown in Figure 31.1.

In planning overhead service drop paths, minimum height requirements for connector lines must be carefully followed, as shown in Figure 31.2. Designers must locate the service heads on buildings to ensure compliance with all local codes. If the minimum height distances cannot be maintained, underground rigid conduit, electrical metallic tubing, or busways (channels, ducts) must be used.

If the service is supplied underground, three wires are placed in a rigid conduit. An underground service conduit is brought to the meter socket. An underground service entrance includes a watt-hour meter, main breaker, and lightning protection. Automatic brownout equipment is also

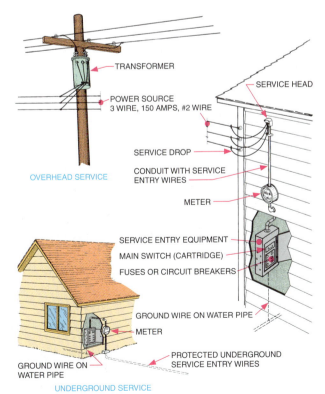

FIGURE 31.1 ■ Electrical distribution to buildings.

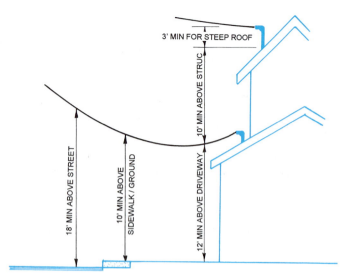

FIGURE 31.2 ■ Minimum overhead service line clearances.

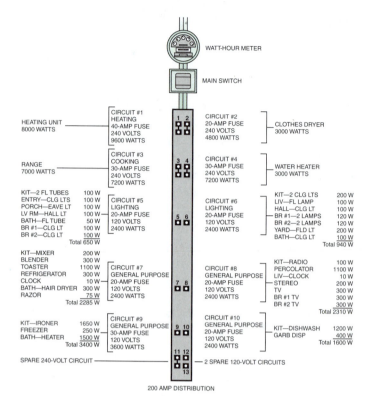

FIGURE 31.3 ■ Distribution panel circuits.

required by many codes for new construction. All electrical systems must be grounded through the service entrance. The location and path of underground service conduits is shown on plot or site plans.

Electrical Service Distribution

Electrical current is delivered throughout a building through a **distribution panel**, or service panel. See Figure 31.3. The size of a distribution panel (in amperes) is determined by the total load requirements (watts) of the entire building. Watts can be converted to amperes by dividing the total (and future) watts needed by the amount of voltage delivered to the distribution box:

$$\text{Formula: } \frac{\text{watts}}{\text{volts}} = \text{amperes}$$

W = symbol for watts
V = symbol for volts
A = symbol for amperes

$$\text{Example: } \frac{35,000\text{W}}{240\text{V}} = 145\text{A}$$

Most residences require a distribution panel with a capacity of 100 to 200 amps. The **National Electrical Code (NEC)** minimum for new residential construction is 60 amps. To compute the total load requirements, the watts needed for each circuit must first be determined.

Branch Circuits

From the distribution panel, electricity is routed to the rest of the building through *branch circuits*. A **circuit** is a

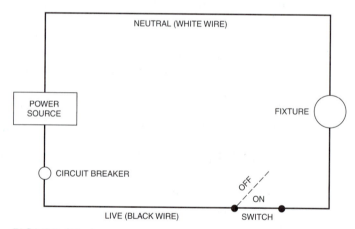

FIGURE 31.4 ■ Simple electrical circuit.

circular path that electricity follows from the power supply source to a light, appliance, or other electrical device and back again to the power supply source. See Figure 31.4. If the electrical load for an entire building were placed on one circuit, overloading would leave the entire building without power. Thus branch circuits are used. Each circuit delivers electricity to a limited number of outlets or devices.

Each circuit is protected with a **circuit breaker**. A circuit breaker is a device that opens (disconnects) a circuit when the current exceeds a certain amount. Without a circuit breaker, excessive electrical loads could cause the wiring to overheat and start a fire. When a breaker opens, or "trips," the power to the branch circuit is disconnected. Similarly, if the sum of the current drawn by the branch circuits exceeds the rating of the main circuit breaker, the main breaker will trip. This protects the service-entrance wires and equipment from overheating and damage. Older homes often have fuses instead of circuit breakers. They serve the same purpose, but overloaded fuses must be replaced. Circuit breakers that trip can be reset.

Branch circuits are divided into three types by the National Electrical Code: lighting circuits, small-appliance circuits, and individual circuits (dedicated circuits).

Lighting Circuits

Lighting circuits are connected to lighting outlets for the entire building. Different lights in each room are usually on different circuits so that if one circuit breaker trips, the room will not be in total darkness.

In all dwellings other than hotels, the NEC requires a minimum general lighting load of 3 watts per square foot of floor space. However, the amount of wattage demanded at one time (demand factor) is calculated at 100 percent only for the first 3,000 watts; 35 percent is used for the second 17,000 watts; and 25 percent is used for commercial demands over 120,000 watts. Thus, the general lighting load planned for a 1,500 sq. ft. house would be 3,525 watts, not the full 4,500 watts. It is calculated as follows:

$$1,500 \text{ sq. ft.} \times 3 \text{ W} = 4,500 \text{ W}$$

First 3,000 W × 100% = 3,000 W
Next <u>1,500 W</u> ×   35% =   <u>525 W</u>
Total 4,500 W   3,525 W

If each branch circuit supplies 2,400 watts (120V × 20A = 2,400W), a 1,500-sq.-ft. house should have two 2,400-watt general lighting circuits. See Figure 31.5. Lighting circuits are also used for small devices such as clocks and radios. However, because all lights and other items on the circuit are probably not going to be used at the same time, it is not necessary to provide a service capable of supplying the full load.

Small-Appliance Circuits

These circuits provide power to outlets wherever small appliances are likely to be connected. Small appliances include items such as toasters, electric skillets, irons, electric shavers, portable tools, and computers. Appliance circuits are not designed to also support lighting needs. See Figure 31.6. The NEC requires a minimum of two small-appliance circuits in a residence. Each circuit is usually computed as a 3,600-watt load (30A × 120V = 3,600W).

Individual Circuits

Individual dedicated circuits are designed to serve a single large electrical appliance or device, such as electric ranges, automatic heating units, built-in electric heaters, and workshop outlets. Large motor-driven appliances, such as washers, garbage disposals, and dishwashers, also use individual circuits. These circuits are designed to provide sufficient power for starting loads. When a motor starts, it needs an extra surge of power to bring it to full speed. This is called a *starting load*.

Separate circuits (20 amps) are required in a laundry area to provide power for the washing machine and the dryer. Because of the danger of water leakage, a **ground-fault circuit interrupter (GFCI)** receptacle is recommended.

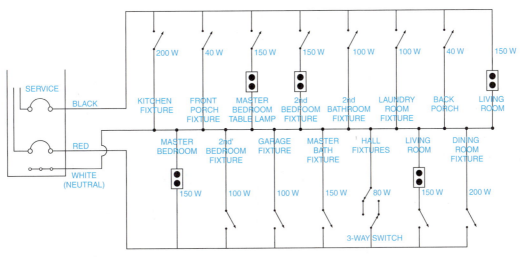

FIGURE 31.5 ■ Typical breakdown of lighting loads.

	TYPICAL CONNECTED WATTS	VOLTS	WIRES	CIRCUIT BREAKER OR FUSE	OUTLETS ON CIRCUIT	OUTLET TYPE	NOTES
KITCHEN							
Range	12,500	120/240	3 #6 + GND	50A	1	14-50R	
Oven (built-in)	4,500	120/240	3 #10 + GND	30A	1	14-30R	#1
Range top	6,000	120/240	3 #10 + GND	30A	1	14-30R	#1
Dishwasher	1,500	120	2 #12 + GND	20A	1	5-15R	#2
Waste disposer	800	120	2 #12 + GND	20A	1	5-15R	#2
Trash compactor	1,200	120	2 #12 + GND	20A	1	5-15R	#2
Microwave oven	1,450	120	2 #12 + GND	20A	1 or more	5-15R	
Broiler	1,500	120	2 #12 + GND	20A	1 or more	5-15R	#3
Fryer	1,300	120	2 #12 + GND	20A	1 or more	5-15R	#3
Coffeemaker	1,000	120	2 #12 + GND	20A	1 or more	5-15R	#3
Refrigerator/freezer 16–25 cubic feet	800	120	2 #12 + GND	20A	1 or more	5-15R	#4
Freezer chest or upright 14–25 cubic feet	600	120	2 #12 + GND	20A	1 or more	5-15R	#4
Roaster-broiler	1,500	120	2 #12 + GND	20A	1 or more	5-15R	#3
Waffle iron	1,000	120	2 #12 + GND	20A	1 or more	5-15R	#3
FIXED UTILITIES							
Fixed lighting	1,200	120	2 #12	20A	1 or more		#10
Window air conditioner 14,000 Btu	1,400	120	2 #12 + GND	20A	1	5-15R	#11
25,000 Btu	3,600	240	2 #12 + GND	20A	1	6-20R	#11
29,000 Btu	4,300	240	2 #10 + GND	30A	1	6-30R	#11
Central air conditioner 23,000 Btu	2,200	240					#6
57,000 Btu	5,800	240					#6
Heat pump	14,000	240					#6
Sump pump	300	120	2 #12	20A	1 or more	5-15R	#1
Heating plant oil or gas	600	120	2 #12	20A	—	—	#6
Fixed bathroom heater	1,500	120	2 #12	20A	—	—	#6
Attic fan	300	120	2 #12	20A	1	5-15R	
Dehumidifier	350	120	2 #12	20A	1 or more	5-15R	#1

(Continued)

FIGURE 31.6 ■ Load requirements of common electrical appliances.

	TYPICAL CONNECTED WATTS	VOLTS	WIRES	CIRCUIT BREAKER OR FUSE	OUTLETS ON CIRCUIT	OUTLET TYPE	NOTES
LAUNDRY							
Washing machine	1,200	120	2 #12 + GND	20A	1 or more	5-15R	#5
Dryer all-electric	5,200	120/240	3 #10 + GND	30A	1	14-30R	#1
Dryer gas/electric	500	120	2 #12 + GND	20A	1 or more	5-15R	#5
Ironer	1,650	120	2 #12 + GND	20A	1 or more	5-15R	
Hand iron	1,000	120	2 #12 + GND	20A	1 or more	5-15R	
Water heater	3,000–6,000					DIRECT	#6
LIVING AREAS							
Workshop	1,500	120	2 #12 + GND	20A	1 or more	5-15R	#7
Portable heater	1,300	120	2 #12 + GND	20A	1	5-15R	#3
Television	300	120	2 #12 + GND	20A	1 or more	5-15R	#8
Portable lighting	1,200	120	2 #12 + GND	20A	1 or more	5-15R	#9
Band saw	300	120	2 #12 + GND	20A	1 or more	5-15R	#6
Table saw	1,000	120/240	2 #12 + GND	20A	1	5-15R	#6

NOTES

#1   May be direct-connected.

#2   May be direct-connected on a single circuit; otherwise, grounded receptacles required.

#3   Heavy-duty appliances regularly used at one location should have a separated circuit. Only one such unit should be attached to a single circuit at a time.

#4   Separate circuit serving only refrigerator and freezer is recommended.

#5   Grounding-type receptacle required. Separate circuit is recommended.

#6   Consult manufacturer for recommended connections.

#7   Separate circuit recommended.

#8   Should not be connected to appliance circuits.

#9   Provide one circuit for each 500 sq. ft (46 m^2). Divided receptacle may be switched.

#10 Provide at least one circuit for each 1,200 watts of fixed lighting.

#11 Consider 20-amp, 3-wire circuits to all window-type air conditioners. Outlets may then be adapted to individual 120- or 240-volt units. This scheme will work for all but the very largest units.

FIGURE 31.6 ■ *Concluded.*

Ground-Fault Circuit Interrupter (GFCI)

A GFCI receptacle must be located wherever there is a possibility for people to ground themselves and be shocked by the electrical current flowing through their body to the ground. The purpose of a GFCI receptacle is to cut off the current at the outlet. When the GFCI receptacle senses any change of current, it immediately trips a switch to interrupt the current. It operates faster and is safer than the circuit breaker switch or fuse at the power entry panel. A GFCI switch will trip in 1/40 second when an extremely small current variation (ground fault) of 0.005 amps is reached.

In new and remodeling construction, GFCI receptacles must be located with each convenience outlet near water sources and/or pipes in the bathroom, kitchen, garage, laundry, and outdoors. Any receptacle located within 10′ or within 15′ of the inside of a permanently installed swimming pool must also be wired through a GFCI. GFCIs are also required if outlets are placed in unfinished crawl spaces below grade level.

FIGURE 31.7 ■ Sizes of copper wire conductors.

Electrical Conductors

Wires used to conduct electricity are classified by the type of wire material, the insulation material, and the wire size. Wire size is classified by number in reverse order of size, as shown in Figure 31.7. Figure 31.8 shows the ampere rating of different wire sizes from 0000 to 14. The size of the wire used in a circuit depends on the current to be carried by the circuit. Although the meter voltage is 120V and 240V, wiring resistance reduces the voltage at the receptacles to approximately 110V and 220V. Wire sizes 6 through 2/0 are used for 240-volt (240V) service entrance and circuits. The exact size depends on the capacity of the service panel. Sizes 10 through 14 are used for 120V and 240V lighting and small-appliance circuits. Sizes 16 and 18 are used for low-voltage items such as thermostats and doorbells.

A low-voltage switching system may be used to turn on or off any **fixture**, appliance, or light. Because of the low voltage, extremely small wires are used to attach the switch to the fixture. A step-down transformer at the switch reduces the voltage. A step-up transformer at the fixture raises the voltage level back.

Wire size is critical. If a wire is too small for the current applied, excessive resistance (overload) can result. This may cause the insulation to overheat and break down, causing a potential fire hazard. When selecting or preparing to use appliances, it is important to check the UL (Underwriters' Laboratories) ratings to learn the proper wiring requirements.

Aluminum wire is lighter and less expensive than copper, but many codes apply stricter rules to the use of aluminum for residential work. Insulation is available in flexible metal armored or nonmetal sheathed form. For underground or exterior exposed wiring, wires must be encased in rigid or flexible metal or PVC (plastic) conduits.

Calculating Total System Requirements

Math Connection

The installation of the proper size of service entrance equipment and branch circuits depends on the square footage of the residence, number of appliances, lighting, and future expansion allowances. To find the total amp

BRANCH CIRCUITS

SIZE	DIAMETER (INCHES)	CURRENT RATING (AMPERES)	
		COPPER	ALUMINUM
14	0.064	15	—
12	0.081	20	15
10	0.102	30	25
8	0.129	40	30
6	0.162	55	40
4	0.204	70	55
3	0.229	80	65
2	0.258	95	75
1	0.289	110	85
0	0.325	125	100
00 (2/0)	0.365	145	115
000 (3/0)	0.410	165	130
0000 (4/0)	0.460	195	155

SERVICE ENTRANCE

3-WIRE SERVICE SIZE (EACH WIRE)	SERVICE RATING CURRENT (AMPERES)	
	COPPER	ALUMINUM
4	100	—
3	110	—
2	125	100
1	150	110
0	175	125
00 (2/0)	200	150
000 (3/0)	—	175
0000 (4/0)	—	200

FIGURE 31.8 ■ Relationship of amperes to wire size.

service needed for an entire building, first determine the total number of watts needed for each circuit. Add these to find the total watts needed for the building. For example, to calculate the size of the service entrance for a 2,000-sq.-ft. residence, list the amount of wattage to be used as follows:

■ Lighting circuits (typical)
2,000 sq. ft. uses 3 watts per sq. ft. = 4,050 watts

■ Convenience outlets
2 circuits in service area (120V × 20A) = 4,800 watts
2 circuits in sleeping area (120V × 20A) = 4,800 watts
2 circuits in living area (120V × 20A) = 4,800 watts

- Dedicated circuits

central AC	=	4,000 watts
microwave oven	=	1,200 watts
electric range	=	10,000 watts
electric dryer	=	5,000 watts
washing machine	=	1,000 watts
dishwasher	=	1,000 watts
forced air unit	=	1,000 watts
electric water heater	=	2,000 watts
	total	43,650 watts

To find the required service panel amps needed, divide the total watts by the available voltage (240V):

$$43,650 \text{ watts} \div 240 = 182 \text{ amps}$$

Service panels are available with capacities of 30, 40, 50, 60, 70, 100, 125, 150, 175, and 200 amps. The next highest panel above the required amps should be chosen to allow for future expansion. In this case, the next highest is 200-amp service.

LIGHTING DESIGN

Functional lighting design involves the interaction among eyesight, objects, and light sources. Good lighting design provides sufficient but not excessive light. Glare from unshielded bulbs or improperly placed lighting should be avoided. Excessive contrast between light and shadows within the same room should also be avoided, especially in work areas.

For centuries, candles and oil lamps were the major source of artificial light. Although candles continue to function for special effects, the major sources of light today are incandescent and fluorescent lamps. **Incandescent** lamps have a filament (a very thin wire) that gives off light when heated. **Fluorescent** lamps have an inner coating that gives off visible light when exposed to ultraviolet light. The ultraviolet light is released by a gas inside the fluorescent tube when an electronic circuit is passed through the tube. Incandescent lamps concentrate the light source, while fluorescent lamps provide linear patterns of light. Fluorescent lamps give a uniform glareless light that is ideal for large working areas. Fluorescent lamps give more light per watt, last seven times longer, and generate less heat than incandescent lamps.

Light Measurements

Human eyes adapt to varying intensities of light. However, they must be given enough time to adjust slowly to

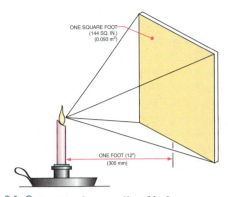

FIGURE 31.9 ■ One footcandle of light.

SUNLIGHT	
Beaches, open fields	10,000 FC (107,640 LX)
Tree shade	1,000 FC (10,764 LX)
Open park	500 FC (5,382 LX)
Inside 3' from window	200 FC (2,153 LX)
Inside center of room	10 FC (108 LX)

ACCEPTED ARTIFICIAL LIGHT LEVELS	
Casual visual tasks, conversation, watching TV, listening to music	10–20 FC (108–215 LX)
Easy reading, sewing, knitting, house cleaning	20–30 FC (215–323 LX)
Reading newspapers, kitchen & laundry work, keyboarding	30–50 FC (323–538 LX)
Prolonged reading, machine sewing, hobbies, homework	50–70 FC (538–753 LX)
Prolonged detailed tasks such as fine sewing, reading fine print, drafting	70–200 FC (753–2,153 LX)

FIGURE 31.10 ■ Comparison of sunlight and artificial light levels.

different light levels. Sudden extreme changes of light may cause discomfort.

Light intensity is measured in units called footcandles. A **footcandle** (candela) is equal to the amount of light a candle casts on an object one foot away. See Figure 31.9. Ten footcandles (10 fc) equals the amount of light that 10 candles throw on a surface one foot away. In the metric system, the standard unit of illumination is the **lux** (lx). One lux is equal to 0.093 fc. To convert footcandles to lux, multiply by 10.764. Figure 31.10 shows the accepted artificial light levels for common tasks compared to natural sunlight levels.

Types of Lighting

The three basic types of lighting are general lighting, specific lighting, and decorative lighting. Examples of all three types

FIGURE 31.11 ■ Combined effect of lighting types and color. *(Marc Michaels)*

FIGURE 31.12 ■ Well diffused indoor general lighting combined with natural lighting. *(Carl's Furniture Showrooms Inc.—Teri Kennedy, ASID, and Linda Dragin, ASID, Designers; Lee Gordon, Photographer)*

of lighting can be found throughout Part 3. Figure 31.11 shows an example of an excellent combination of all three types of lighting combined with an effective use of color. Note how this design eliminates glare and shadows.

General Lighting

General lighting provides overall illumination and radiates a comfortable level of brightness for an entire room. General lighting replaces sunlight and is provided primarily with chandeliers, ceiling or wall-mounted fixtures, and track lights. To avoid contrast and glare, general lighting should be diffused through the use of fixtures that totally hide the light source or that spread light through panels. Close spacing of hanging fixtures also creates diffuse lighting. Another solution is to use adjustable fixtures so that the light can be directed away from eye contact.

Where possible, daylight should be included as a part of the general lighting plan during daylight hours. If adequate window light is not available, the use of skylights should be considered. The general and decorative lighting in Figure 31.12 blends with natural lighting to create a well-diffused lighting environment.

The intensity of general lighting should be between 5 and 10 fc (54 to 108 lx). A higher level of general lighting should be used in the service area and bathrooms. Many general lighting fixtures can also be used for decorative lighting by a connection to dimmer switches.

Specific Lighting

Light directed to a specific area or located to support a particular task is known as specific, local, or task lighting. Specific lighting helps in performing such tasks as reading, sewing, shaving, computer work, and home theater viewing. It also adds to the general lighting level. Track lighting and portable lamps provide sources of specific indoor lighting.

Decorative Lighting

Bright lights are stimulating, while low levels of light are quieting. Decorative lighting is used to create atmosphere and interest. Indoor decorative lights are often directed on plants, bookshelves, pictures, wall textures, fireplaces, or any architectural feature worthy of emphasis. Some decorative lighting can be used as general lighting through the use of dimmer switches.

Outdoor decorative lighting can be most dramatic. Exterior structural and landscape features can be accented by well-placed lights. Outdoor lighting is used to light and accent wall textures, trees, shrubs, architectural features, pools, fountains, and sculptures. Outdoor lighting is especially needed to provide a safe view of stairs, walks, and driveways. The exterior lighting design in Figure 31.13 provides both decorative and specific lighting for outdoor activities.

Remember to conceal light sources and don't overlight. Use waterproof devices and an automatic timing device to turn lights on and off.

Light Dispersement

Light from any artificial source can be distributed (dispersed or directed) in five different ways: direct, indirect,

FIGURE 31.13 ■ Dramatic example of outdoor decorative and specific lighting. (*Frank Serpe, Owner; Franco D. Demetrio, Architect; Susan Miller, Photographer*)

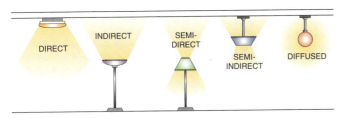

FIGURE 31.14 ■ Methods of light dispersement.

semidirect, semi-indirect, and diffused. See Figure 31.14. *Direct* light shines directly on an object from a light source. *Indirect* light is reflected from surfaces. *Semidirect* light shines mainly down as direct light, but a small portion of it is directed upward as indirect light. *Semi-indirect* light is mostly reflected, but some light shines directly. *Diffused* light is spread evenly in all directions with the light source (bulb) not visible.

Reflection

All objects absorb and reflect light. Some white surfaces reflect 94 percent of the light that strikes them. Some black surfaces reflect only 2 percent. The remainder of the light is absorbed. All surfaces in a room act as a secondary source of light when light is reflected. Excessive reflection causes glare. Glare can be eliminated from this secondary source by using matte (dull) finish surfaces and by avoiding exposed light bulbs. Eliminating excessive glare is essential in designing adequate lighting.

Structural Light Fixtures

Light fixtures are either portable plug-in lamps or structural fixtures. Structural fixtures are wired and built into a building's **hard-wired** system. These must therefore be shown on electrical plans and specifications. Structural fixtures may be located on ceilings, on interior and exterior walls, and on the grounds around the building.

Different light patterns are produced, depending on the type of light fixture. Figure 31.15 illustrates the types of structural light fixtures:

- Soffit lighting is used to direct a light source downward to "wash" over wall surfaces as general and decorative lighting. Soffit lighting can also be designed to provide light for horizontal surfaces, such as kitchen and bath countertops, wall desks, music centers, and computer centers.

- Cove lighting directs light (usually fluorescent) onto ceiling surfaces and indirectly reflects light into the center of a room. Soffits should hide the fixtures from view from any position in the room.

- Valance lighting directs light upward to the ceiling and down over the wall or window treatment. Valance faceboards can be flat, scalloped, notched, perforated, papered, upholstered, painted, or trimmed with molding.

- Cornice lighting directs all light downward. It is similar to soffit lighting, except cornice lights are totally exposed at the bottom.

Wall Fixtures

Wall fixtures are used as a source of general lighting, as well as decorative lighting when attached to a dimmer switch.

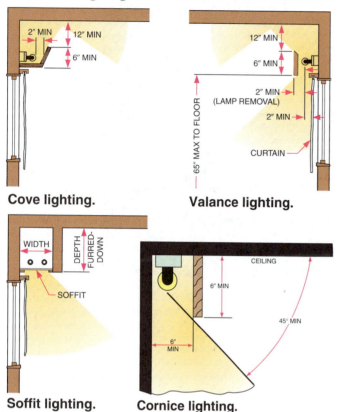

Cove lighting.

Valance lighting.

Soffit lighting.

Cornice lighting.

FIGURE 31.15 ■ Structural light fixture details.

Wall spotlights or fluorescent fixtures may also be used as task lighting. Wall spotlights for accents, diffusing fixtures for general lighting, and sconces are used extensively on walls. See Figure 31.16. Vanity lights and concealed fluorescent tube lights are also used on walls as task lighting.

Ceiling Fixtures

A wide variety of lighting fixtures are designed for ceiling installation. Many optional designs are possible within each type. See Figure 31.17. Likewise, track-lighting units are available in a variety of shapes, materials, and colors. Because track light units can be moved and rotated, the track should be placed to take full advantage of these features. The path of irregular curved tracks (wave

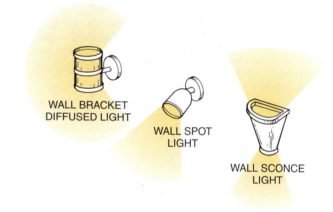

WALL BRACKET DIFFUSED LIGHT

WALL SPOT LIGHT

WALL SCONCE LIGHT

FIGURE 31.16 ■ Spot and sconce wall fixtures.

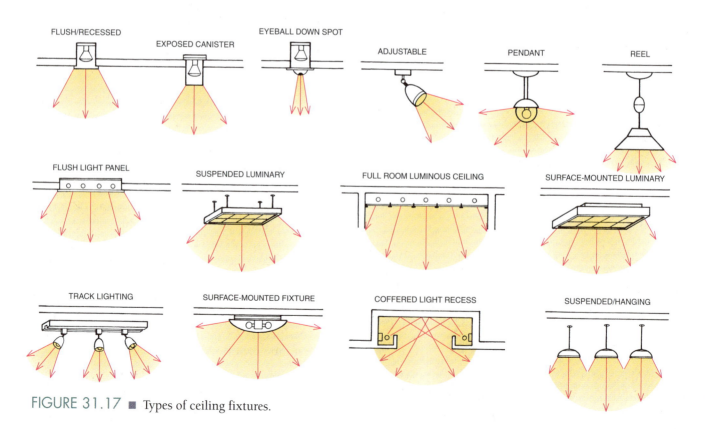

FIGURE 31.17 ■ Types of ceiling fixtures.

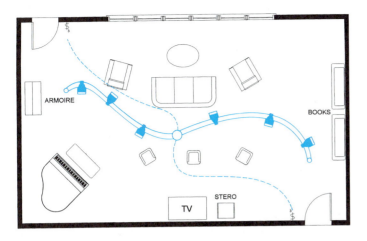

LIVING AREA
870 sq ft

FIGURE 31.18 ■ Ceiling wire rail fixture.

rails) can be designed to provide lighting for any area in a room. The path outline of these tracks is drawn on a reflected ceiling plan as shown in Figure 31.18.

When entire ceilings are to be illuminated, fluorescent fixtures are ceiling mounted. Translucent or open mesh panels are suspended below the fixtures. The position of the fixtures should be shown on a reflected ceiling plan and detailed as shown in Figure 31.19. To provide adequate lighting, downlights are often used for specific task lighting over reading, dining, and work areas. To provide the correct location for these fixtures, junction boxes in the ceiling must be dimensioned. Task lighting is ineffectual if in the wrong position.

Exterior Lighting Fixtures

Waterproof spotlights, floodlights, and wall bracket lights are used on exterior walls for both general lighting and decorative lighting. See Figure 31.20. Exterior wall

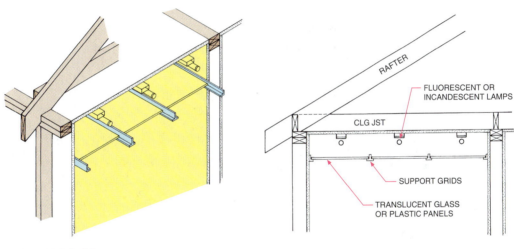

FIGURE 31.19 ■ Luminous ceiling design.

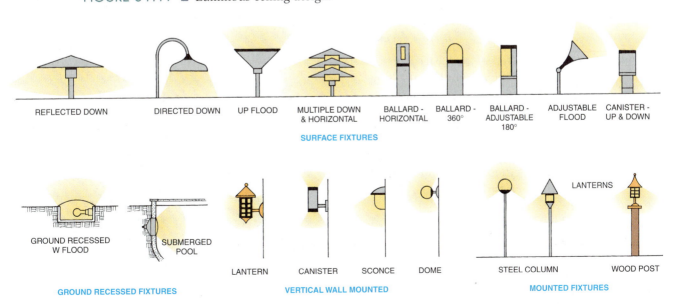

FIGURE 31.20 ■ Types of exterior lighting fixtures.

lights are often connected to motion detectors for security purposes. Lighting fixtures are used for landscaping, driveways, and walkways. These fixtures are designed to direct light at any angle to illuminate design features. Some fixtures, such as post lamps (lanterns), are designed to emit light in all directions. Other post (ballard) lights are designed with shields that can be adjusted to direct light 360 degrees or to any smaller segment. Swimming pool lights can also be used effectively for landscape lighting since the entire pool becomes a large light source when illuminated.

DEVELOPING AND DRAWING ELECTRICAL PLANS

Wiring methods are regulated by building codes, and wiring is approved and installed by licensed electricians. However, wiring plans are prepared by designers. For large structures, a consulting electrical contractor may prepare the final detailed electrical plans. Electrical plans include data on the type and location of all fixtures, devices, switches, and outlets.

Fixture and Device Selection

Before placing fixture locations on a floor plan, the number and type of fixtures needed for each room should be determined and listed. See Figure 31.21. In addition to lighting fixtures, all electrical or electronic devices should also be listed. This list becomes the basis for developing an electrical fixture and device schedule. (Schedules are discussed in Chapter 35.)

Switches

The number, type, and location of switches depends on the fixtures and devices. **Switches** control the flow of electricity to outlets and to individual devices.

Types of Switches

Small-appliance circuits and individual circuits are usually "hot," meaning that electricity is available in the outlet at all times. Lighting circuits, however, may be either hot or controlled by switches. See Figure 31.22. *Single-pole switches* control fixtures, devices, or outlets. To control lights from two different switches, a *three-way* switching circuit (three wires and two switches) is used. A three-way switching circuit is often installed for the top and bottom of stairways and at the end of long halls. Fig-

FIXTURE	LR	DR	FOYER	KIT	BFK	LAU	UT	GAR	OFFICE	BR1	MBR	GATE	PORCH	PATIO	POOL	HALL	UP FOYER	M. BATH	G. BATH	MAKEUP	EXT HOUSE
Chandelier		✓																			
Pendant				✓																	
Track	✓																				
Sconce	✓								✓												
Recess				✓													✓	✓	✓	✓	
Makeup																		✓	✓	✓	
Flood																					✓
Carriage												✓									
Valance	✓	✓			✓																
Cove				✓																	
Soffit	✓			✓																	
Exhaust Fan																		✓	✓		
Ceiling Fan	✓			✓					✓	✓											
Door Chimes			✓																		
Door Button											✓	✓									
Smoke Alarm					✓		✓	✓								✓					
Intercom			✓						✓	✓	✓										
Telephone			✓						✓		✓										
Answer Machine									✓												
Fax									✓												
Computer									✓												
Surge Protector	✓								✓	✓											
TV-Cable	✓								✓												
VCR									✓												
Security Sensor	✓	✓	✓	✓	✓	✓	✓	✓	✓	✓	✓	✓					✓				✓
Thermostat	✓					✓				✓						✓					
Low-Voltage Circuit													✓	✓							✓
Outdoor 110V Circuit															✓						✓
Pool Motor															✓						
Pool Light															✓						
Water Heater							✓												✓		
Washer							✓														
Dryer							✓														
Refrigerator				✓																	
Range				✓																	
Oven				✓																	
Dishwasher				✓																	
Disposal				✓																	
Outside Wall																					✓

FIGURE 31.21 ■ Electrical fixture and device listing.

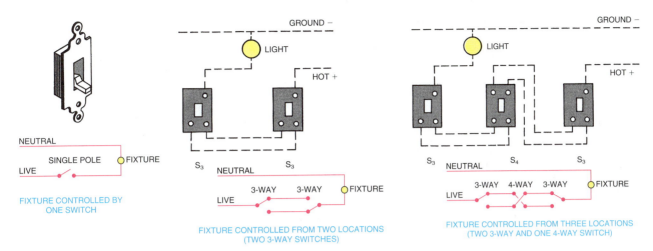

FIGURE 31.22 ■ Types of switching controls.

TYPE	DESCRIPTION
Toggle Switches	Used to control circuits; available in single-pole, three-way, or four-way types.
Mercury Switches	Silent, shockproof; available in single-pole, three-way, and four-way types.
Automatic Cycle Controls	Installed on appliances to control their functions on a time cycle.
Photoelectric Cells	Control switching by blocking a beam of light.
Automatic Controls	Adjust heating and cooling systems to desired temperatures.
Clock Thermostats	Adjust heating and cooling by adjusting both temperature and time.
Aquastats	Keep water heated or cooled to selected temperatures.
Dimmer Switches	Control the intensity of light; can be controlled by touch, slider, or rotary controls.
Time Switches	Cause lights or devices to switch on and off at specified time intervals.
Safety Alarm	Systems and switches that activate bells or lights when a circuit, usually on a door or window, is broken.
Master Switches	Control circuits throughout an entire building or area from one location.
Low-voltage Switching	Systems that provide economical long runs for low-voltage lighting.
Computer Systems	Control all mechanical-electrical devices from a master control unit.
Delayed Action Switches	Allow current to flow for a limited time (usually one minute) after the switch has been turned off.

FIGURE 31.23 ■ Switch mechanisms.

ure 31.23 lists the major types of switch mechanisms and the circuits or devices they control.

A four-way switch and two three-way switches are used to control fixtures from three different locations. Additional four-way switches may be added to the circuit in Figure 31.24, allowing the lights to be controlled from any switch. Low-voltage (24V) switching systems permit long runs that allow master control switching from remote locations.

Switch Locations

Switch symbols are located on floor plans. Connections to the outlet, fixture, or device each controls are shown with a dotted line. See Figure 31.24. Use the following guidelines in planning switch locations:

1. Include a switch for all structural fixtures and devices that need to be turned on or off.

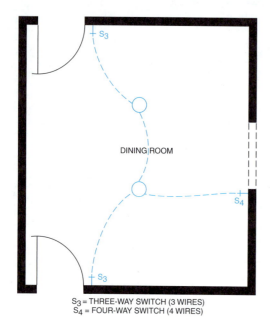

S₃ = THREE-WAY SWITCH (3 WIRES)
S₄ = FOUR-WAY SWITCH (4 WIRES)

FIGURE 31.24 ■ Three- and four-way switches used to control fixtures from three locations.

2. Indicate the height of all switches, which is usually 4' above floor level (3–6" for wheelchair access).

3. Locate switches on the latch side of doors, no closer than 21/2" from the casing. See Figure 31.25.

4. Exceptions to any standard should be dimensioned on the plan or elevation drawing.

5. Select the type of switch, switch mechanism, switch plate cover, and type of finish for each switch.

6. Plan a switch to control at least one light in each room.

7. Use three-way switches to control lights at the ends of stairwells, halls, and garages.

8. Locate garage door-closer switches at the house entry and within easy reach inside the garage door.

9. Control bedroom lights with a three-way switch at the entry and at the bed.

10. Use timer switches for garage general lighting, bathroom exhaust fans, and heatlights.

11. Use three-way switches for all large rooms that have two exits. Use additional four-way switches for rooms with more than two exits.

12. Use timer switches on closet and storage areas.

13. Specify timer switches for pool motors.

14. Locate safety alarm switches for a security system in the master control unit and in the master bedroom.

15. Switches for outdoor security lighting (motion detector lights) should be installed on all levels.

16. Locate lighted switches in all rooms to ensure that a person need not enter or leave a room in the dark.

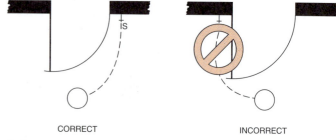

CORRECT   INCORRECT

FIGURE 31.25 ■ Light switches must be located on the latch side of doors.

Electrical Outlets and Receptacles

The terms **outlet** and **receptacle** are often used interchangeably. The NEC defines an outlet as a point in a circuit where other devices can be connected. A receptacle is a device (at an outlet box) to which any plug-in extension line, appliance, or device can be connected.

Types of Outlets and Receptacles

Different types of electrical receptacles and outlets serve different functions.

- Convenience receptacles are used for small appliances and lamps. These are available in single, double, or multiple units. See Figure 31.26.

- Lighting outlets are for the connection of lampholders, surface-mounted fixtures, flush or recessed fixtures, and all other types of lighting fixtures.

- Special-purpose (dedicated circuit) receptacles are the connection point of a circuit for only one electrical device.

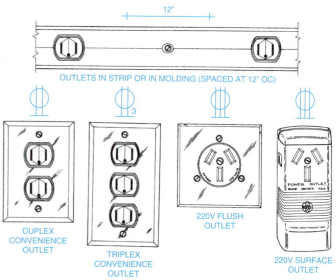

12"

OUTLETS IN STRIP OR IN MOLDING (SPACED AT 12" OC)

DUPLEX CONVENIENCE OUTLET

TRIPLEX CONVENIENCE OUTLET

220V FLUSH OUTLET

220V SURFACE OUTLET

FIGURE 31.26 ■ Types of receptacles.

Special-purpose outlets and convenience outlets are connected to hot circuits, while lighting outlets are controlled with a switching device. Remember, GFCIs are installed in all outlets and switches located near water sources.

Outlet Locations

The positioning of outlets must be consistent with local codes. In addition to code requirements, the following guidelines should be used for locating outlets:

1. Outlets (except in the kitchen) should average one every 6′ (1.8 m) of wall space.
2. Kitchen appliance outlets (with GFCIs) should average one every 4′ of wall space, be located over countertops, and include at least one countertop outlet between major appliances.
3. Hall outlets should be placed every 15′.
4. An outlet should be placed no further than 6′ from each room corner, unless a door or built-in feature occupies this space.
5. GFCI outlets should be placed as described earlier in this chapter.
6. One switch-controlled (split) outlet should be provided in each room.
7. Consider furniture placement and positioning of portable lamps when placing lighting outlets. Room-centered furniture may need floor outlets.
8. An outlet should be placed on any wall between doors regardless of space.
9. The height of all outlets should be noted on the electrical plan. Exceptions to standard dimensions should be noted at each outlet or referenced on an interior wall elevation. Normal code height for wall outlets is 12″ to 18″ from the floor. Countertop switch heights are normally 40″ to 44″ above the floor line.
10. All individual outlets should be labeled with the appliance or device served.
11. At least one GFCI outlet should be placed above each bathroom countertop or vanity table. A minimum of two GFCI outlets should be in each bathroom.
12. Provide an outlet for each fixture, device, or appliance in the plan.
13. An outdoor weatherproof outlet (GFCI) should be provided on each side of a house. Position a waterproof outlet for a patio, pool, and a grill. Position outside outlets for decorative lighting and for low-voltage circuits, such as for entry doors, garage, and security lights.

NAME	ABBREV	SYMBOL	ELEVATION	PICTORIAL
SWITCH SINGLE-POLE	S	OR $S\ \$$		
SWITCH DOUBLE-POLE	S_2	S_2		
SWITCH THREE-WAY	S_3	S_3		
SWITCH FOUR-WAY	S_4	S_4		
SWITCH WEATHERPROOF	S_{WP}	S_{WP}		
SWITCH AUTOMATIC DOOR	S_D	S_D		
SWITCH PILOT LIGHT	S_P	S_P		
SWITCH LOW-VOLTAGE SYSTEM	S	$\underline{S}$		
SWITCH LOW-VOLTAGE MASTER	MS	**MS**		
TWO SWITCHES	SS	**S S**		
THREE SWITCHES	SSS	**S S S**		
TELEVISION AERIAL OUTLET	TV AER	**TV**		
DUPLEX OUTLET	DUP OUT			
SINGLE OUTLET	S OUT	1		

FIGURE 31.27 ■ Commonly used electrical symbols.
(Continued)

14. The location of special-purposes outlets, such as cable and telephone jacks, is shown on floor plans using one of the symbols from Figure 31.27. See Chapter 13 for standards for people with disabilities.

NAME	ABBREV	SYMBOL	ELEVATION	PICTORIAL
TRIPLE OUTLET	TR OUT			
WEATHERPROOF OUTLET	WP OUT			
SPLIT WIRE OUTLET	SPT WR OUT			
FLOOR OUTLET	FL OUT			
OUTLET WITH SWITCH	OUT/S			
HEAVY-DUTY OUTLET 220 VOLTAGE	HVY DTY OUT			SURFACE FLUSH
SPECIAL-PURPOSE OUTLET 110 VOLTAGE	SP PUR OUT			
RANGE OUTLET	R OUT			
REFRIGERATOR OUTLET	REF OUT			
WATER HEATER OUTLET	WH OUT			
GARBAGE-DISPOSAL OUTLET	GD OUT			
DISHWASHER OUTLET	DW OUT			
WASHER OUTLET	W OUT			
DRYER OUTLET	D OUT			

NAME	ABBREV	SYMBOL	ELEVATION	PICTORIAL
MOTOR OUTLET	M OUT			
STRIP OUTLET	STP OUT			
GROUNDED OUTLET	GRD OUT			
LIGHTING OUTLET–CEILING	LT OUT CLG			
LIGHTING OUTLET– RECESSED	LT OUT REC			
LIGHTING OUTLET–WALL	LT OUT WALL			
LIGHTING OUTLET–CEILING PULL SWITCH	PS			
FLOOD LIGHT	FL			
SPOT LIGHT	SL			
LIGHTING OUTLET–VAPOR PROOF	VP			
WALL BRACKET LIGHT WITH SWITCH	WL BRK LT/S			
LIGHTING FLUORESCENT	LT FLUOR			
EXIT LIGHT	EXT LT			
ILLUMINATED HOUSE NUMBER	ILL HSE NO			

FIGURE 31.27 ■ Commonly used electrical symbols.

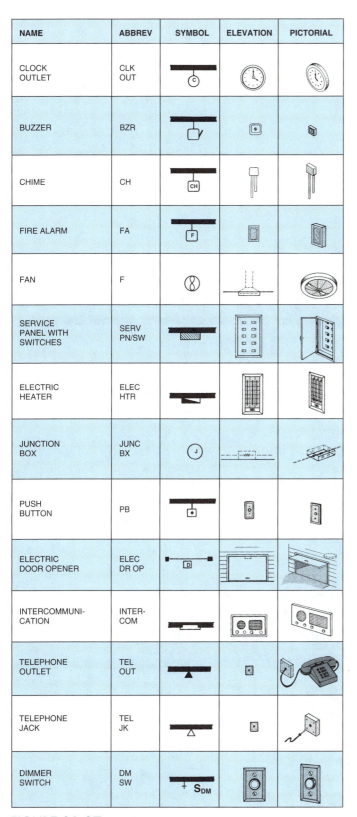

NAME	ABBREV	SYMBOL	ELEVATION	PICTORIAL
CLOCK OUTLET	CLK OUT			
BUZZER	BZR			
CHIME	CH			
FIRE ALARM	FA			
FAN	F			
SERVICE PANEL WITH SWITCHES	SERV PN/SW			
ELECTRIC HEATER	ELEC HTR			
JUNCTION BOX	JUNC BX			
PUSH BUTTON	PB			
ELECTRIC DOOR OPENER	ELEC DR OP			
INTERCOMMUNI-CATION	INTER-COM			
TELEPHONE OUTLET	TEL OUT			
TELEPHONE JACK	TEL JK			
DIMMER SWITCH	DM SW			

FIGURE 31.27 ■ *Concluded.*

ELECTRICAL WORKING DRAWINGS

Complete electrical plans ensure that electrical equipment and wiring are installed exactly as planned. If electrical plans are incomplete and sketchy, the installation depends on the judgment of the electricians. Designers should not rely on electricians to design the electrical system, only to install it. Conversely, designers do not plan the position of every wire, only the position and relationship of all fixtures, devices, switches, and controls. This is done with the use of electrical symbols. Hundreds of electrical symbols are used on floor plans to describe what and where electrical elements will be installed. About 60 symbols apply to residential or light construction. Some of the more commonly used symbols are shown in Figure 31.27.

Completely true wiring diagrams are not used on electrical floor plans. The abbreviated architectural method, which shows only the position of fixtures, switches, and connecting lines, is drawn as shown in Figure 31.28.

Preparing Electrical Plans

The following sequence is recommended for drawing electrical floor plans:

1. Prepare the base floor plan by one of the following methods:
   a. Print a reproducible floor plan without dimensions or notes. Print a CAD layer if available.
   b. Trace a floor plan including only walls and major features.

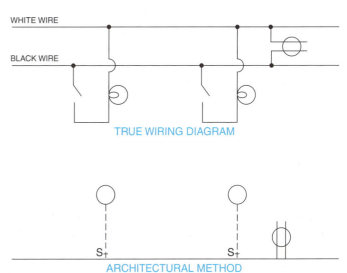

FIGURE 31.28 ■ True wiring diagram compared to the architectural method.

c. Print a reproducible floor plan with very light lines so electrical symbols will have high contrast.

d. Add electrical symbols and features on a print in red. This is the least desirable method, but is often practiced by electrical contractors in the field.

These methods are also used in preparing plumbing, HVAC, and interior design plans.

2. Select fixtures using Figure 31.21 as a guide.

3. Locate and draw the service entrance and distribution panel.

4. Locate and draw the position of each fixture and device on the plan using the symbols shown in Figure 31.27. Ceiling fixtures may be shown on a reflected ceiling plan.

5. Select the switch type (see Figure 31.23) needed to control each fixture or device.

6. Locate and draw the switch symbol for each switch on the plan.

7. Draw a curved solid or dashed line connecting each switch with the fixture the switch controls. The connecting lines may be curved to avoid confusion with straight object lines.

8. Locate the position and draw the symbol for each outlet on the plan.

9. Where switches or lines are connected to a different level or another drawing, draw the line outside the plan and label the line destination.

10. Show the location of devices and control centers that require special wiring such as TV or computer network cables. Connect these with curved lines.

11. Add notes to clarify any wiring needs not obvious on the drawing.

Figures 31.29A through 31.29J show the common application of electrical symbols to various residential room plans. Figure 31.30 illustrates the use of electrical symbols on a complete electrical floor plan. The first floor plan (Figure 31.30A) shows switch locations to fixtures and devices located outside the buiding and also to the second level. The second-level plan (Figure 31.30B) shows where these first-level connections are made. The complete floor plan used as a base for this electrical plan is found in Figure 14.24A. The electrical plan in Figures 31.30A and 31.30B was drawn on a floor plan printed before most dimensions, notes, or reference symbols were added. These plans are part of the set of plans included in Chapter 34.

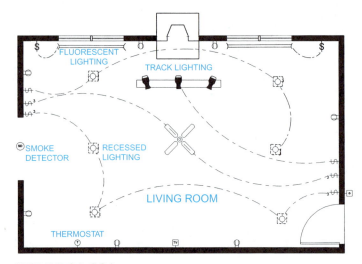

FIGURE 31.29A ■ Living room wiring plan.

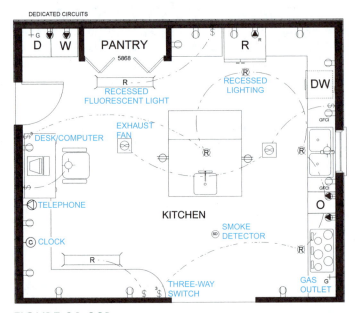

FIGURE 31.29B ■ Kitchen wiring plan.

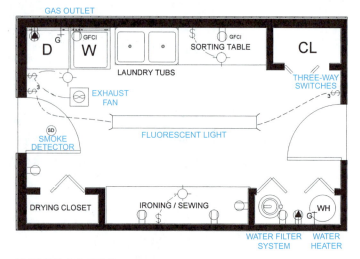

FIGURE 31.29C ■ Utility room wiring plan.

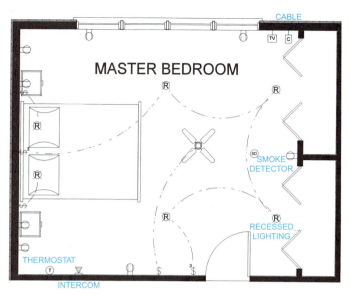

FIGURE 31.29D ■ Bedroom wiring plan.

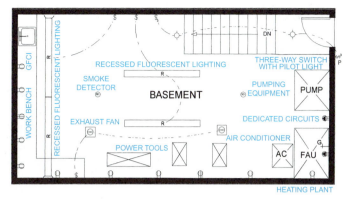

FIGURE 31.29G ■ Basement wiring plan.

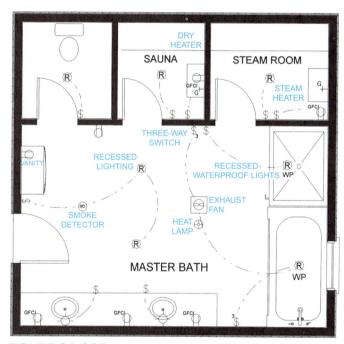

FIGURE 31.29E ■ Bathroom wiring plan.

FIGURE 31.29H ■ Hall and stairs wiring plan.

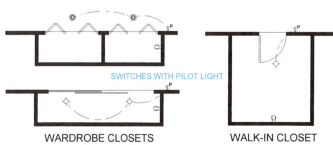

FIGURE 31.29F ■ Closet wiring plan.

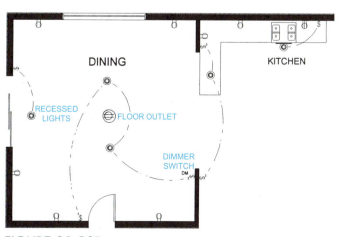

FIGURE 31.29I ■ Dining room wiring plan.

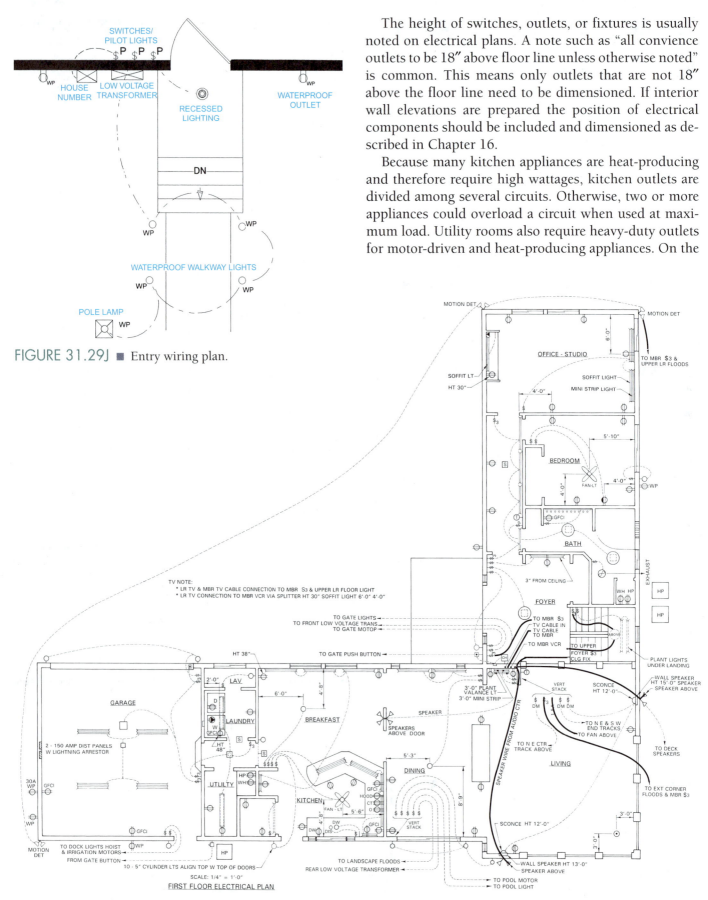

FIGURE 31.29J ■ Entry wiring plan.

The height of switches, outlets, or fixtures is usually noted on electrical plans. A note such as "all convience outlets to be 18″ above floor line unless otherwise noted" is common. This means only outlets that are not 18″ above the floor line need to be dimensioned. If interior wall elevations are prepared the position of electrical components should be included and dimensioned as described in Chapter 16.

Because many kitchen appliances are heat-producing and therefore require high wattages, kitchen outlets are divided among several circuits. Otherwise, two or more appliances could overload a circuit when used at maximum load. Utility rooms also require heavy-duty outlets for motor-driven and heat-producing appliances. On the

FIGURE 31.30A ■ First-level electrical plan showing wiring to the site and the second level.

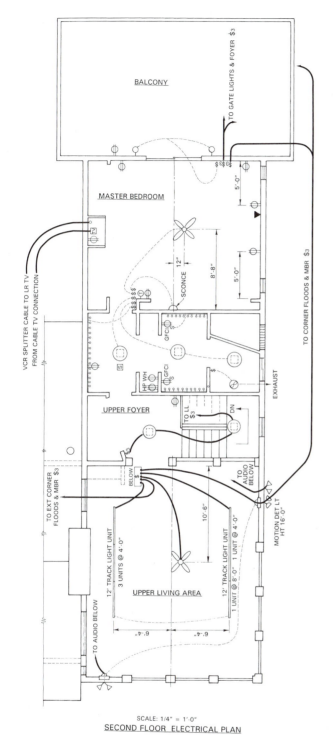

SCALE: 1/4" = 1'-0"
SECOND FLOOR ELECTRICAL PLAN

FIGURE 31.30B ■ Second-level electrical plan showing wiring from first level.

USING CAD

CAD Electrical Symbols Library

Electrical plans are prepared over a layered floor plan. On simple plans, electrical symbols are often included as a separate layer on the floor plan. For larger or more complex structures, a separate plan is prepared. In either case electrical features or plans should be prepared on a separate level. Fixtures, switches, and devices are moved from the electrical symbol library to their position on a drawing, and then connected with a curved line using the *Spline* task.

ELECTRONIC SYSTEMS

Many building systems can now be controlled electronically with computers. Such electronic systems include a centralized computer and control units (touchscreen or keyboard) near entry doors and in the master bedroom. Lights, alarms, safety devices, telephones, and heating and air conditioning units are some examples of building systems that can be controlled electronically. High-technology features that should be considered in the architectural design process are those that:

1. Monitor smoke, gas, sound, and movement and sound an alarm when established levels are exceeded.

2. Monitor and adjust heating, cooling, and humidity levels for each room.

3. Provide videophone intercom communication between entrances and selected rooms, including synthesized voice response to visitors.

4. Open and close, lock and unlock doors, windows, vents, gates, and vents from a central control center.

5. Turn on or off appliances, audio systems, or VCRs from a central control center directly or on timed sequences.

6. Monitor and/or time the opening and closing of solar energy devices, including window shades and drapes.

7. Control lighting circuits from a central location directly or on timed sequences.

other hand, bedrooms, bathrooms, and closets require comparatively low wattage levels. Stairs and halls present special problems in electrical planning. Three-way and four-way switches must be carefully located to provide control at many locations and thus eliminate unnecessary backtracking.

8. Program combinations of controls to activate systems on a timed basis for night, morning, midday, evening, work week, weekend, or vacation modes of operation.

9. Are capable of activation via outside, inside, and remote phone commands. Some systems use electronically synthesized voice commands for communication.

10. Monitor intruder action by body heat (infrared sensors), motion, light interruption, or noise levels to automatically alarm occupants and/or security command center.

11. Record images for immediate use and future reference with closed-circuit video cameras interfaced with a computer.

Automation systems can be combined, for instance, to automatically dial police and fire departments when sensors are activated. Security companies usually design total security systems as specified by the designer.

Electronic System Drawings

After outlets and switches are located on floor plans, electronic systems are added with special symbols. The control units and the devices for any combination of these systems may be drawn directly on the electrical plans. Separate system plans may also be prepared.

Security Systems

Designing security systems involves specifying the locations of controls and of sonic (sound), motion, and heat sensor devices. Electronic security systems are controlled and monitored through a central control panel. Control panels may also be used to automatically signal police, fire, and medical services. The components of an electronic security system linked to a control panel or panels include the following:

1. *Door and window sensors* trigger alarms when contact points are separated, breaking a circuit.

2. *Sonic detectors* activate an alarm when the noise of breaking glass, voices, or forced entry is sensed.

3. *Motion detectors* sense movement when placed in the path of an intruder. These detectors must be placed to cover all locations where intruders would move as shown in Figure 31.31.

4. *Fire detectors* are designed to react to smoke and/or heat and then activate an alarm and/or sprinkler system.

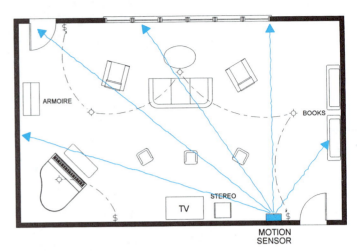

FIGURE 31.31 ■ Motion detector positioning.

5. *Carbon monoxide detectors* are set to sound an alarm when CO levels reach danger levels.

6. *Water sensors* sense water leakages and sound an alarm or notify a public utility company of a problem.

7. *Gas sensors* detect leaking gas from range or furnace burners and can sound an alarm or shut off a gas line.

8. *Line cut monitors* alert occupants that an electrical drop line or telephone line has been cut outside the building.

9. *Low-temperature sensors* are used to inform a remote location of the danger of a pipe freezing in an unoccupied building.

10. *Visual monitors* link a camera on the premises to a remote video monitoring station.

Any or all of these devices can be connected to a variety of alarms including high-decibel (more than 100 dB) indoor and outdoor sirens, voice sirens, and strobe lights. They may also provide notification of police, fire, service, or medical personnel. Security systems can be hard wired or wireless. If hard wired, security devices should be added to the electrical plan. If wireless, only the position of devices needs to be shown, as in Figure 31.32.

Surge and Lightning Protection

A power surge can damage or destroy televisions, VCRs, stereos, and computers. Surge protection devices should be provided at the distribution panel for the entire system.

FIGURE 31.32 ■ Security system device symbols.

Lightning rods (air terminals) can reduce lightning damage by providing a grounding path of least resistance to discharge the electricity. Rods are placed on roof tops and are connected to cables at diagonal corners of a building. The cables extend into the ground, two feet from the foundation.

Communication Systems

Positions and connections of doorbells, chimes, music, and any communication units need to be included on the floor plan. The bell- or chime-activating button is connected to the unit with a line drawn the same ways as a line connecting a switch to a fixture. Lines also show the connection between intercom master and remote units. Connections of audio devices with remote speakers, or television cables with TV sets and VCRs, are similarly indicated with lines.

Integrated Systems

Security, communications, and device-control units may be combined into one integrated system. Automated systems can activate particular parts of the integrated system, such as climate control or lighting, on a timed basis. Similarly, remote controls can be used to control any part or all of the system. These remote controls are low-voltage (24-volt) switching systems. In addition to controlling security system components, an integrated electronic control system (smart house) should be designed and prewired to serve the following functions:

1. Video distribution system
2. Audio distribution system
3. Multiple-line telephone system
4. Internet sharing device (router)

5. Data networking for computers, printers, scanners, etc.

6. Remote control of
   — lighting (outdoor and indoor)
   — appliances
   — thermostats
   — security arming and disconnecting
   — mode selection (occupied or unoccupied)
   — drape operation
   — car phone linkage.

Electrical Solar Systems

Unlike the active solar systems covered in Chapter 32, solar electrical systems transform solar energy into electrical currents. Solar energy is absorbed through photovoltaic panels and then applied to the electrical system of a building where it is used in a conventional manner. Photovoltaic panels create DC currents, which are connected to an inverter that changes the DC currents into usable AC currents.

CHAPTER

31 *Electrical Design and Drawing Exercises*

1. What size distribution panel is required for a total of 40,000 watts with a service supply of 240 volts?

2. How many amperes flow through a 120-volt circuit with 14 ohms resistance?

3. What are the three types of branch circuits?

4. Name five locations where GFCI receptacles are required.

5. What is the unit used to measure light intensity?

6. Name the three types of lighting functions.

7. Name five methods of light distribution.

8. Plan the lighting needs and fixtures of the house you are designing. Specify fixtures by room.

9. Determine the location and size of the distribution box for the house you are designing.

10. Locate the position of all lights, small appliances, and individual circuits on the floor plan you are designing.

11. Calculate the number of lighting circuits needed for a three-bedroom house.

12. Describe three types of switching mechanisms that you would select. Tell why.

13. Describe three types of outlets and receptacles that you would select. Why? What guidelines would you follow?

14. What is the normal spacing and height of outlets?

15. Draw an automated electronic system and a lighting system with lines between the switch symbol(s) and the fixture symbol(s) represented.

16. Draw 10 commonly used electrical symbols for a kitchen, bedroom, and bathroom.

17. Draw the complete electrical plan for the house you are designing. Show all circuits and label the capacity of each. Identify the circuits protected by a GFCI device.

CHAPTER 32

Comfort Control Systems (HVAC)

OBJECTIVES

In this chapter, you will learn to:

- ■ plan and draw a mechanical heating and cooling system on a floor plan.

- ■ use appropriate symbols to draw devices, ductwork, or piping for heating and cooling systems.

- ■ calculate heat loss to design HVAC systems needed for specific situations.

- ■ plan and draw a passive and active solar heating and cooling system.

TERMS

active solar system
Btu
conduction
convection
forced-air system
greenhouse effect
heat pump

HVAC
hydronic system
passive solar system
photovoltaic film
plenum
radiation

R-value
thermal conductivity
thermal mass
thermostat
trombe wall
U-value

INTRODUCTION

Comfort control requires more than just providing warmth or coolness. True comfort includes the correct temperature, correct humidity, and a constant supply of clean, fresh, and odorless air in motion. This is accomplished through the use of a heating system, a cooling system, air filters, and humidifiers. Climate control plans show the systems and devices used to maintain temperature, moisture, and the exchange and purification of the air supply. Effective solar design may also increase the efficiency of all climate control systems.

Air temperature, movement, pollutants, humidity, and odors can all be controlled through the use of mechanical systems. Heating, ventilating, and air conditioning systems (**HVAC**) include a wide variety of devices and delivery systems, which are shown on HVAC plans.

The types of systems designed to bring comfort control to a building are forced-air, hydronics, radiant, steam, active solar, and passive solar systems.

HVAC PLANS AND CONVENTIONS

Before HVAC plans can be drawn, heat loss must be calculated and the type and size of the HVAC system must be determined. The location of devices, pipes, and ducts must also be determined prior to drawing an HVAC plan.

It is necessary to know the standard HVAC symbols and abbreviations. Symbols are used to show the location and type of equipment. See Figure 32.1. Arrows are used to show the movement of hot and cold air and water.

The location of horizontal ducts is shown on an HVAC duct plan by outlining the position of the ducts. Because vertical ducts pass through the plane of projection, diagonal lines are used to indicate the position of vertical ducts. HVAC plans also show the position of all control devices, outlets, pipes, and heating and cooling units. See Figure 32.2.

NAME	ABBREV	SYMBOL	NAME	ABBREV	SYMBOL
DUCT SIZE & FLOW SELECTION	DCT/FD	← 10″ × 15″	HEAT REGISTER	R	R
DUCT SIZE CHANGE	DCT/SC		THERMOSTAT	T	T
DUCT LOWERING	DCT/LW	D D	RADIATOR	RAD	RAD
DUCT RISING	DCT/RS	R R	CONVECTOR	CONV	CONV
DUCT RETURN	DCT/RT		ROOM AIR CONDITIONER	RAC	RAC
DUCT SUPPLY	DCT SUP	S	HEATING PLANT FURNACE	HT PLT FUR	FURN
CEILING-DUCT OUTLET	CLG DCT OUT	○	FUEL-OIL TANK	FOT	OIL
WARM-AIR SUPPLY	WA SUP	↓ WA	HUMIDISTAT	H	H
SECOND-FLOOR SUPPLY	2nd FL SUP		HEAT PUMP	HP	HP
COLD AIR RETURN	CA RET	↓ CA	THERMOMETER	T	T
SECOND-FLOOR RETURN	2 FL RET		PUMP	P	
GAS OUTLET	G OUT	↑ G	GAGE	GA	
HEAT OUTLET	HT OUT		FORCED CONVECTION	FRC CONV	

FIGURE 32.1 ■ Climate control symbols.

USING CAD

CAD HVAC Symbol Library

Because most HVAC symbols are located on or near ductwork, all duct lines should be drawn first. This is done using the *Line* task with *Offset* command because parallel duct lines are usually too far apart to use the *Polyline* command. Once the ducts are drawn, the symbols can be moved from the HVAC library and placed on the drawing using the *Insert* command.

PRINCIPLES OF HEAT TRANSFER

As preparation for drawing HVAC plans, this chapter describes the principles of heat transfer and design considerations for an efficient HVAC system. Heat inside a building is created not only by solar heat through roofs, windows, and walls but also by heat-producing equipment and by human activity. Whether heat is inside or outside a building, it always travels from a warm area to a cool area.

FIGURE 32.2 ■ Typical HVAC floor plan.

Methods of Heat Transfer

Science Connection

Heat travels in three ways: by radiation, convection, and conduction. See Figure 32.3. In **radiation**, heat travels as waves through space in the same manner that light travels. In **convection**, heat travels through liquids or gases. For example, a warm surface heats the air around it. The warmed air rises, and cool air moves in to take its place, causing a convection current. In **conduction**, heat moves through a solid material. The denser the material, the better it will conduct heat.

Heat Measurement

Math Connection

The standard unit of measurement for heat is the British thermal unit (Btu). A **Btu** is the amount of heat needed to raise the temperature of 1 pound of water 1 degree Fahrenheit, at a constant pressure of one atmosphere (air pressure at sea level). The metric unit is *joules* (J). To convert Btu's to joules, multiply the Btu value by 1055.

A Btu is a measure of heat *generated*. The measure of heat *flow* is thermal conductivity. **Thermal conductivity** is the amount of heat that flows from one face of a material, through the material, to the opposite face. Thermal conductivity is defined mathematically as the amount of heat transferred through a 1 square foot area, 1 inch thick, for each 1 degree Fahrenheit temperature difference between the faces of the material. See Figure 32.4.

Materials that transfer heat readily are known as *conductors*. Some materials resist the transfer of heat. Materials with high resistance are known as *insulators*. The effectiveness of a material in resisting heat transfer is indicated by its R-value.

R-Values and U-Values

The **R-value** is a uniform method of rating the resistance of heat flow through building materials. The higher the R number, the greater the resistance to heat flow. All building materials have been tested and assigned a thermal resistance number, or R-value. For example, the R-value of 2.5″ thick fiberglass insulation is R-7. The R-value of 6.5″ thick fiberglass insulation is R-22.

When building materials are combined in layers, the sum of their R-values is the total R-value for the component. See Figure 32.5.

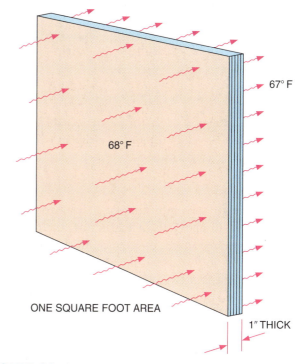

FIGURE 32.4 ■ Measurement of thermal conductivity.

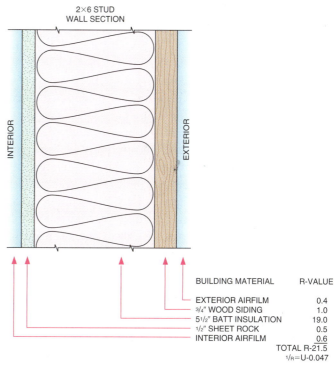

BUILDING MATERIAL	R-VALUE
EXTERIOR AIRFILM	0.4
3/4″ WOOD SIDING	1.0
5 1/2″ BATT INSULATION	19.0
1/2″ SHEET ROCK	0.5
INTERIOR AIRFILM	0.6

TOTAL R-21.5
1/R=U-0.047

FIGURE 32.5 ■ Calculating R- and U-values.

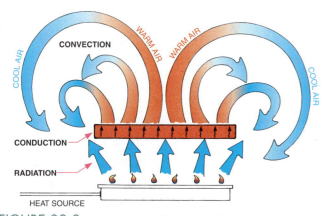

FIGURE 32.3 ■ Methods of heat transfer.

OUTDOOR TEMP	INDOOR SURFACE TEMPERATURE				
	COOL 60°F	FAIR 64°F	MEDIUM 66°F	WARM 68°F	MIN FOR FLOOR
+30°F	R-2.3	R-3.4	R-5.1	R-10.0	R-1.7
+20°F	R-2.8	R-4.2	R-6.4	R-12.5	R-2.2
+10°F	R-3.4	R-5.1	R-7.8	R-14.5	R-2.6
0°F	R-3.9	R-6.0	R-9.2	R-17.0	R-3.0
−10°F	R-4.4	R-6.8	R-10.1	R-20.0	R-3.4
−20°F	R-5.1	R-7.8	R-11.3	R-23.0	R-3.9
−30°F	R-5.7	R-8.4	R-12.8	R-25.0	R-4.4
−40°F	R-6.4	R-10.2	R-14.5	R-28.0	R-4.8

FIGURE 32.6 ■ R-values required to maintain indoor temperatures.

To more accurately indicate the combined thermal conductivity of all materials in a structure, including air spaces, the U-value is used. **U-value** is the amount of heat conducted in 1 hour, through a 1-square-foot area, for each degree Fahrenheit difference in temperature between inside and outside air. The U-value is the reciprocal (1/R) of the R-value. High R-values and low U-values indicate greater efficiency.

Different climates and seasons require different R-value (or heat resistance) levels to maintain desired indoor temperatures. R-values must be chosen for the average low temperature of a geographic area. See Figure 32.6.

The R-value of building materials varies greatly. For example, the R-value of a 1″ face brick is only .11, while the R-value of 1″ pine is 1.25. See Figure 32.7. As the thickness of any material is increased, so is its R-value.

Windows and doors account for most heat loss in cold climates. Windows alone can allow 25 percent of the heat within a house to transfer to the outside. Heat flows through windows, in both directions, through radiation, convection, conduction, and infiltration (air leakage). Double window panes (double glazing) with argon or krypton gas between the window panes slow heat transfer. See Figure 32.8. Low-emissivity (low-E) coatings reflect heat energy (invisible solar radiation) yet transmit visible light. These measures increase R-values and decrease U-values. See Figure 32.9. R- and U-values also vary greatly among exterior door types. The door surface and core material greatly affect R- and U-values. See Figure 32.10.

In a poorly constructed building, cracks around doors, windows, and fireplaces can allow all internal heat to escape in less than an hour. Effective orientation and design can help prevent much of this heat loss. The use of

TYPE OF BUILDING MATERIAL	THICKNESS	CONDUCTANCE R-VALUE (HIGH VALUE IS MORE EFFICIENT)	RESISTANCE U-FACTOR (LOW VALUE IS MORE EFFICIENT)
Roof decking insulation	2″	5.56	0.18
Mineral wool fibrous insul	1″	3.12	0.32
Loose fill insulation	1″	3.00	0.33
Acoustical tile	1″	2.86	0.35
Carpet and pad, fibrous	1″	2.08	0.48
Wood fiber sheathing	25/32″	2.06	0.49
Wood door	1 3/4″	1.96	0.51
Fiber board sheathing	1/2″	1.45	0.69
Softwoods (pine, fir, etc.)	1″	1.25	0.80
Wood subfloor	25/32″	0.98	1.02
Hardwoods	1″	0.91	1.10
Wood shingles, 16″ 71 2″ exp	standard	0.87	1.15
Hardboard, wood fiber	1″	0.72	1.39
Plywood	1/2″	0.65	1.54
Asphalt shingles	standard	0.44	2.27
Sheet rock/ plasterboard	1/2″	0.44	2.27
Built-up roofing	3/8″	0.33	3.03
Concrete/stone	4″	0.32	3.13
Gypsum plaster (light weight)	1/2″	0.32	3.13
Common brick	1″	0.20	5.00
Stucco	1″	0.20	5.00
Cement plaster	1″	0.20	5.00
Face brick	1″	0.11	9.10
Felt building paper (15 lb)	standard	0.06	16.67
Steel	1″	0.0032	312.50
Aluminum	1″	0.0007	1428.57

FIGURE 32.7 ■ R- and U-values for typical building materials.

insulation is also a significant deterrent to heat loss or heat gain.

Insulation

Insulation is any material that is used to slow the transfer of heat. When effectively placed, insulation not only

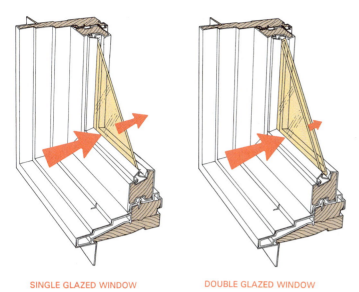

SINGLE GLAZED WINDOW   DOUBLE GLAZED WINDOW

FIGURE 32.8 ■ Effect of double glazing on heat transfer.

MATERIAL	U-FACTOR		R-VALUE	
	COLD CLIMATE (WINTER)	WARM CLIMATE (SUMMER)	COLD CLIMATE (WINTER)	WARM CLIMATE (SUMMER)
SINGLE GLASS	1.13	1.06	0.88	0.94
INSULATED GLASS 1/4″ Air space 1/2″ Air space	0.65 0.58	0.61 0.56	1.54 1.72	1.64 1.79
STORM WINDOWS 1″–4″ Air space	0.56	0.54	1.79	1.85
LOW EMITTANCE 1/2″ Air space $\varepsilon = .20$ $\varepsilon = .60$	0.32 0.43	0.38 0.51	3.13 2.33	2.63 1.96
GLASS BLOCK 6″ × 6″ × 4″ 12″ × 12″ × 4″	0.60 0.52	0.57 0.50	1.67 1.92	1.76 2.00

FIGURE 32.9 ■ R- and U-values for window types.

TYPE OF DOOR	U-FACTOR	R-VALUE
Hollow core wood	1.00	R-1.0
Hollow core wood and storm door	0.67	R-1.5
Solid core wood	0.43	R-2.3
Solid core wood and storm door	0.28	R-3.5
Metal with urethane core	0.07	R-13.5

FIGURE 32.10 ■ R- and U-values for exterior doors.

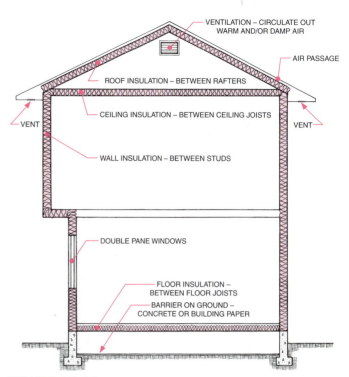

FIGURE 32.11 ■ Insulation locations.

retards the transfer of heat but also stops moisture, sound, fire, and insect penetration. To be effective, insulation must be placed everywhere that heat loss (or gain) will occur.

Without insulation an HVAC system must work harder to overcome the loss of warm air or cool air through walls, floors, and ceilings. The use of the proper insulation in walls and floors can reduce 25 percent of the heat transfer. Because most roofs cannot be sheltered from the sun, 40 percent of most heat transfer is through the roof. Insulation, with an area for ventilation above the insulation, is most effective.

To prevent excessive heat transfer, a layer of insulation should be placed between the foundation and the earth below. This insulation should be outside the structure, thus placing the building in an insulation envelope. Such an arrangement not only conserves heat but prevents rapid changes in inside temperature in all seasons. Figure 32.11 shows a house totally enveloped in insulation.

Insulation is most critical in cornice areas and exposed foundation walls and sills. Figure 32.12 shows methods of insulating cornice areas, and Figure 32.13 shows areas of greatest perimeter heat loss on a T-foundation.

Insulation is produced in a wide variety of vegetable, mineral, plastic, paper, and metal materials and in several different forms. These are illustrated in Figure 32.14A and described in Figure 32.14B. Insulation R-values vary

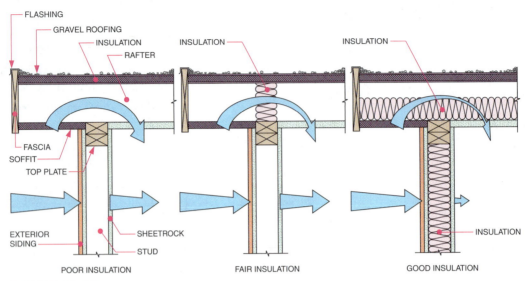

FIGURE 32.12 ■ Insulation at wall and ceiling joist junction.

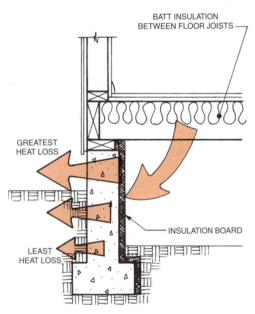

FIGURE 32.13 ■ Heat loss on a T-foundation.

greatly depending on the type of insulating material and thickness. See Figure 32.15.

Surface air film is a thin covering of air that clings to surfaces. The amount of air clinging varies with the amount of air movement and the type of surface. Figure 32.16, shows R- and U-values for 3/4″ air spaces with reflective and nonreflective surfaces. Vapor barrier film is plastic fabric sheets used to totally wrap the exterior and prevent air vapors and water from penetrating exterior walls.

Heat Loss Calculations

Math Connection

Heat loss or heat gain is the amount of heat that passes through an exterior surface of a building. Regardless of

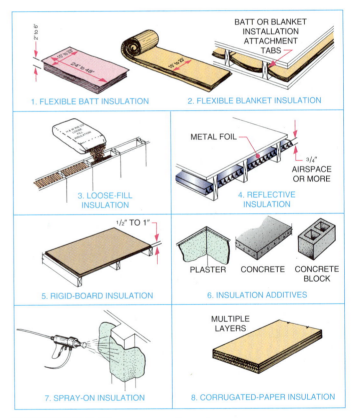

FIGURE 32.14A ■ Types of insulation.

the material, some heat gain or heat loss always occurs. When this happens, the temperature transfer is always from hot to cold. Factors influencing the amount of heat transfer in a building include the difference between indoor and outdoor temperatures, the type and thickness of building materials, the amount and type of insulation, and the amount of air leakage (infiltration) into or out of a structure.

Flexible batt	Paper-covered insulating materials that are attached between structural members. The batts are 2" to 6" (50 mm to 150 mm) thick.
Flexible blanket	Paper-covered insulating materials that are attached between structural members. The blankets are long sheets, 1" to 3" (25 mm to 75 mm) thick.
Loose fill	Materials poured or blown into walls or attic floors.
Reflective	Reflecting metal foil attached between construction members. Reflective material is often mounted on other types of insulation. It is excellent for reflecting heat, for retarding fire and decay, and for keeping out insects.
Rigid board	Thin insulating sheathing cover that is manufactured in varying sizes.
Additives	Lightweight aggregates that are mixed with construction materials to increase their insulating properties.
Spray-on	Insulating materials mixed with an adhesive.
Corrugated paper	Multiple layers of corrugated paper that are easy to cut and install.

FIGURE 32.14B ■ Insulation descriptions.

INSULATING MATERIALS	MATERIAL'S THICKNESS	R-VALUE
Glass fiber	2" Batt	R-7
	4" Batt	R-11
	6" Batt	R-19
	6" Blown	R-13
	8 1/2" Blown	R-19
	12" Blown	R-26
	18" Blown	R-38
	2-6" Batts	R-38
Rock wool	4" Blown	R-11
	6 1/2" Blown	R-19
	13" Blown	R-38
Cellulose fiber	4" Blown	R-11
	6 1/2" Blown	R-19
	13" Blown	R-38
Polystyrene	1" Board	R-5
	1 1/2" Board	R-7.5
Polyurethane foam	1 1/2" Board	R-9.3
	4" Injected	R-25
Expanded MICA	Loose	R-2.5

FIGURE 32.15 ■ R-values of insulating materials.

3/4" AIR SPACES	R-VALUE	U-FACTOR
Heat flow up		
Nonreflective	R-0.87	1.15
Reflective, one surface	R-2.23	0.45
Heat flow down		
Nonreflective	R-1.02	0.98
Reflective, one surface	R-3.55	0.28
Heat flow horizontal		
Nonreflective (also same for 4" thickness)	R-1.01	0.99
Reflective, one surface	R-3.48	0.29

AIR SURFACE FILMS (INSIDE STILL AIR)	R-VALUE	U-FACTOR
Heat flow up (through horizontal surface)		
Nonreflective	R-0.61	1.64
Reflective	R-1.32	0.76
Heat flow down (through horizontal surface)		
Nonreflective	R-0.92	1.09
Reflective	R-4.55	0.22
Heat flow horizontal (through vertical surface)		
Nonreflective	R-0.68	1.47

FIGURE 32.16 ■ R- and U-values for air spaces.

To compute heat loss in Btu's, multiply the total interior surface areas (floors, walls, ceilings) by the U-value and by the temperature difference (HL = A × U × T). See Figure 32.17. Infiltration is calculated by multiplying the interior volume by 0.018 to find the average amount of lost Btu's per hour. Special software is available to aid in computing heat loss.

HEATING SYSTEMS

Conventional heating methods use a variety of devices and distribution (delivery) systems. Devices that produce heat include warm-air, hot-water, steam, electrical, solar, fireplace, and heat pump units. Heat pumps are also used for cooling.

Heat is usually distributed throughout buildings by ducts, pipes, or wires. Ducts are used to move both heated and cooled air. Ducts are either round, square, or rectangular. They are made of sheet metal, wood, or

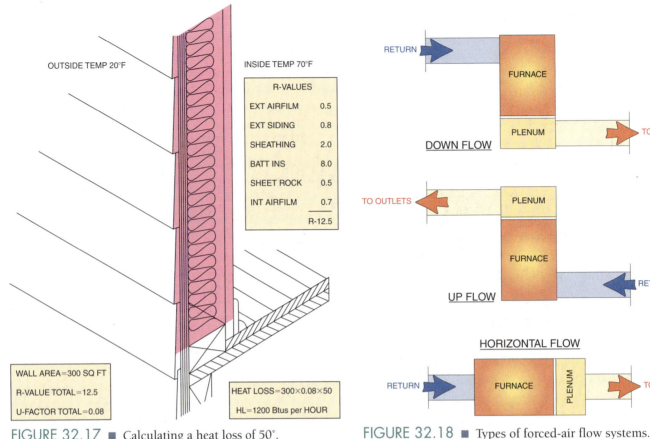

FIGURE 32.17 ■ Calculating a heat loss of 50°.

FIGURE 32.18 ■ Types of forced-air flow systems.

flexible foil-covered fiberglass. Pipes are used to carry hot water or steam to radiators or base units within each room. Electric resistance wires are either embedded in ceilings or floors for radiant heating or connected to floor or wall convection units.

Forced Warm-Air Units

In warm-air systems, air is heated in a furnace. Air ducts distribute the heated air to outlets throughout the building. Air filters and humidity control devices can be combined with warm-air units. Cooling systems can use the same ducts as the heating system if the ducts are rustproof.

Warm-air units operate either by gravity or by forced air. *Gravity systems* rely on allowing warm air to rise naturally without the use of a fan. Therefore, furnaces in a gravity system must be located on a level lower than the area to be heated. Gravity systems are rarely used today.

In **forced-air systems**, air is blown through the ducts by a fan located in the heating or cooling device. See Figure 32.18. In a downflow furnace, cool air enters the top and warmed air exits the bottom. Conversely, cool

air enters the bottom of an upflow furnace. Cool and warm air move at the same level through a horizontal flow furnace.

Forced-air ducts are either distribution ducts or return ducts. The return ducts bring cool air back to the furnace to be warmed. Most open-plan small buildings or buildings with many zones do not require return ducts. Instead, return air moves directly to the return side of a furnace.

Forced-air distribution ducts are connected to a plenum chamber, an enclosed space located between the furnace and distribution ducts. The plenum is larger than any duct and slows the flow of air through the ducts. For heating, ducts lead to floor outlets usually on outside walls. At least one outlet is needed for every 15 feet of exterior wall space. For cooling, ceiling locations are preferred.

Duct systems vary in patterns as they connect a furnace to outlets throughout a building. Figure 32.19 shows an *individual duct system*, in which separate ducts directly link the furnace, or AC unit, with each outlet. These systems provide well-balanced heat but require more duct space. Figure 32.20 illustrates an *individual*

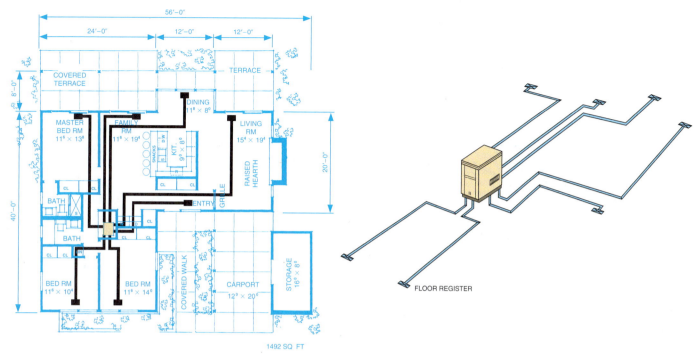

FIGURE 32.19 ■ Individual duct system.

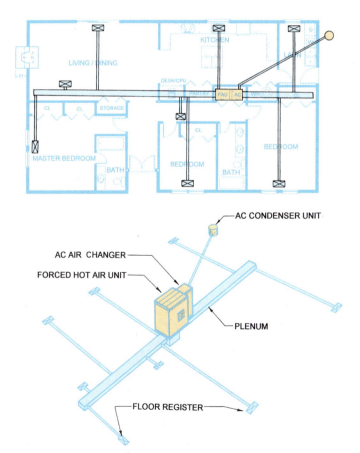

FIGURE 32.20 ■ Individual plenum system.

plenum system. In some systems of this type the plenum size is the same through its length. In some systems the plenum size is reduced as fewer outlets are served.

In a *perimeter loop system,* as shown in Figure 32.21, the perimeter duct is connected to a furnace with a feeder duct on each side. In a variation of the perimeter loop, the ducts of a *perimeter radial system* (Figure 32.22) radiate directly to each outlet like the spokes of a wheel.

Where heating ducts also serve an air-conditioning system, the locations of the inside air handlers, exhaust hoses, and outside compressor units need to be added. All warm-air systems include room outlets (registers) on either the floor, ceiling, or wall as shown in Figure 32.23. Figure 32.24 shows methods of connecting ducts to floor, wall, or ceiling outlets.

Wood Plenum System

In light-frame construction, a wood plenum system may be specified. A **plenum** is an enclosed space in which the air pressure is greater than it is outside. Plenum systems constructed of wood are based on a simple concept. Instead of using heating and cooling ducts, the entire underfloor space (crawl space) is used as a sealed plenum chamber to distribute warm or cool air to floor registers in the rooms above. See Figure 32.25. The plenum consists of wood floor construction with sealed and insulated foundation walls. A forced-air mechanical heating and/or cooling unit maintains slight air pressure in the

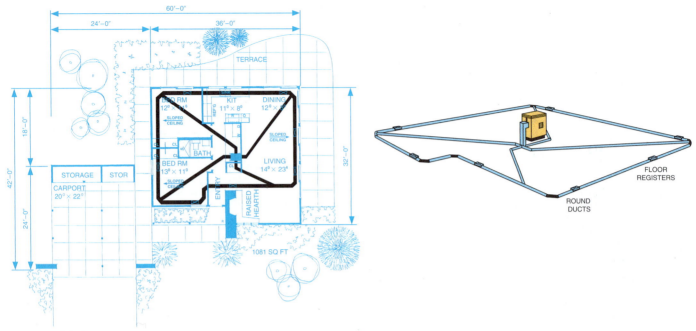

FIGURE 32.21 ■ Perimeter loop system.

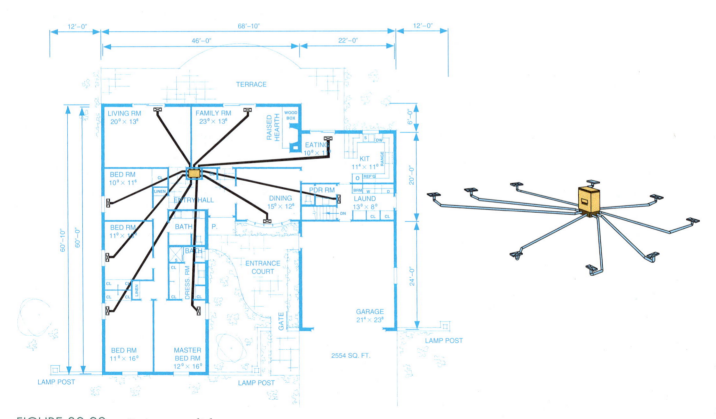

FIGURE 32.22 ■ Perimeter radial system.

plenum. This ensures a uniform distribution of conditioned air throughout the building with few or no added supply ducts. This type of system was used by the early Romans.

Forced-air system drawings include the location of each furnace, outlet, and duct. Buildings requiring more than one furnace are drawn in zones with a separate furnace and ductwork system for each. Abbreviated forced-

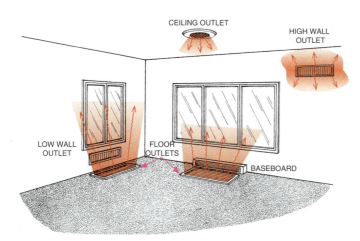

FIGURE 32.23 ■ Outlet types and locations.

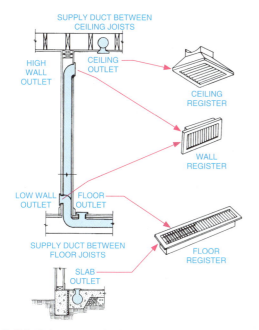

FIGURE 32.24 ■ Outlet connections.

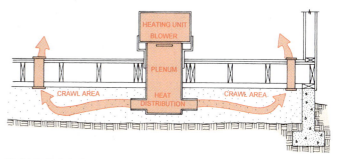

FIGURE 32.25 ■ Wood plenum system.

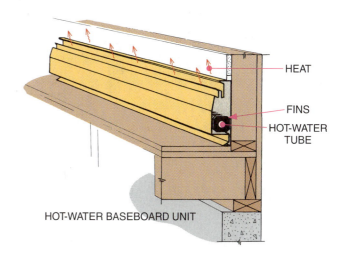

FIGURE 32.26 ■ Hydronic baseboard outlet.

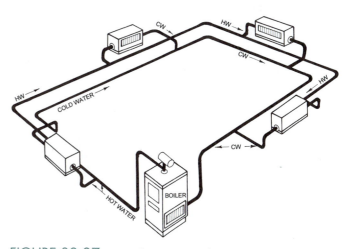

FIGURE 32.27 ■ Hydronic series-loop system.

air system drawings show only the position of outlets. This means the builder must determine the type, size, and location of all devices, controls, and ductwork.

Hydronic (Hot-Water) Units

Hot-water heating systems use an oil or gas boiler to heat water and a water pump to send the water to radiators, finned tubes, or convectors. **Hydronic systems** provide even heat and are quiet and clean. These systems do not provide air filtration or circulation and are not compatible with cooling systems that require air ducts.

The most common and most effective hydronic outlet type is the baseboard outlet shown in Figure 32.26. Some hydronic system units, such as domestic oil burners, also provide hot water for sinks and showers, thus eliminating the need for a separate water heater unit.

There are several types of hot water systems: the series-loop system, the one-pipe system, the two-pipe system, and the radiant system.

The series-loop system as shown in Figure 32.27, is a continual loop of pipes containing hot water. Hot water flows continually from the boiler through outlet units and back again to the boiler for reheating. The heat in a series-loop system cannot be controlled except at the source of the loop.

In one-pipe systems (Figure 32.28), heated water is circulated through pipes that are connected to radiators or convectors by means of bypass pipes. This allows each radiator to be individually controlled by valves. Water flows from one side of each radiator to the main line and returns to the boiler for reheating.

In a *two-pipe system,* as shown in Figure 32.29, there are two parallel pipes: one for the supply of hot water from the boiler to each radiator, and the other for the re-

turn of cooled water from each radiator to the boiler. The heated water is directed from the boiler to each radiator but returns from each radiator through the second pipe to the boiler for reheating.

A *radiant system* distributes hot water through a series of continual pipes in floors and sometimes ceilings. Ceiling systems are not often used because of the weight of the filled pipes. A radiant floor system consists of pipes laid on a concrete base then covered with a finished concrete slab. The hot pipes conduct heat to the surface where convection currents take over. Figure 32.30 shows a radiant hot-water system and a related floor plan.

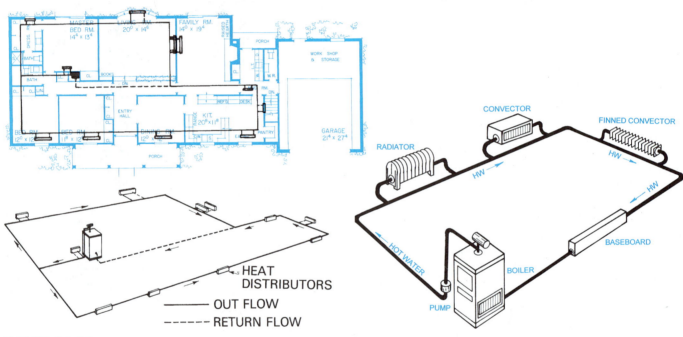

FIGURE 32.28 ■ Hydronic one-pipe system.

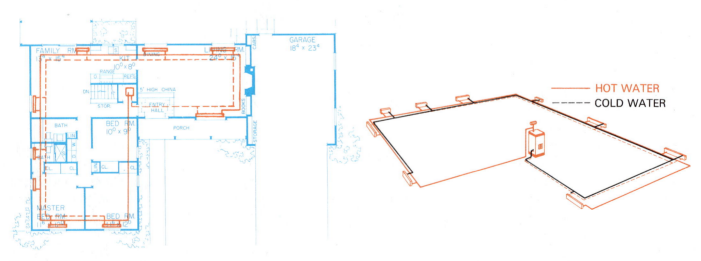

FIGURE 32.29 ■ Hydronic two-pipe system.

Steam Units

A steam-heating unit operates from a boiler that makes steam. The steam is transported by pipes to radiators or convectors and baseboards that give off the heat. The steam condenses to water, which returns to the boiler to be reheated to steam.

Although steam-heating systems function on water vapor rather than hot water, drawings for steam systems are identical with those prepared for hot-water systems. Steam systems are easy to install and maintain, but they are not suitable for use with most convector radiators. They are most popular for large apartments, commercial buildings, and industrial complexes where separate steam generation facilities are provided. Steam heat is delivered through either perimeter or radial systems.

Electric Heat

Electric heat is produced when electricity passes through resistance wires. This heat is usually radiated, although it could be fan-blown (convection). Resistance wires can be placed in panel heaters installed in the wall or ceiling, placed in baseboards, or set in plaster to heat the walls, ceilings, or floors. See Figure 32.31. Electric heaters use very little space. Because there's no flame, they require no air for combustion. Electric heat is very clean. It requires no storage or fuel and no ductwork. Complete ventilation and humidity control must accompany electric heat, however, because it provides no air circulation and tends to be very dry.

Separate plans are seldom drawn for electric heat, but notations are made on floor plans, or reflective ceiling plans, concerning the location of either resistance wires or electric panels. On electrical plans, the location of facilities for the power supply and the thermostat is shown.

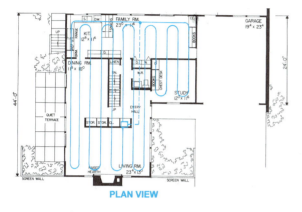

PLAN VIEW

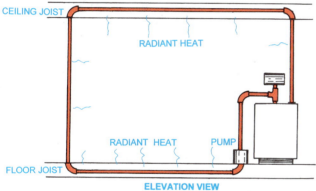

ELEVATION VIEW

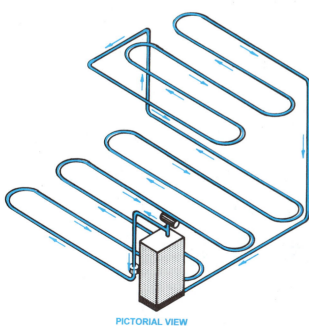

PICTORIAL VIEW

FIGURE 32.30 ■ Hydronic radiant system.

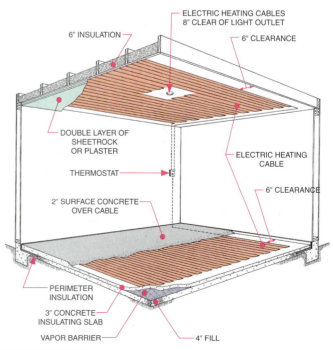

FIGURE 32.31 ■ Electric radiant heating system.

COOLING SYSTEMS

Buildings are cooled by removing heat. Heat can be transferred in one direction only, from warmer objects to cooler objects. Therefore, to cool a building, warm air is carried away from rooms to an air-conditioning unit, where a filter removes dust and other impurities. A cooling coil containing refrigerant absorbs heat from the air passing around it. Then the same blower that pulled the warm air from the rooms pushes cooled air back to the rooms.

Cooling systems may be separate systems or may share blowers, ducts, and outlets with heating systems. See Figure 32.32A. Like heating units, cooling units are positioned as upflow, downflow, or horizontal flow. Drawings of combined systems use the same duct patterns and outlets as for heating plans.

The size of air-conditioning equipment is rated in Btu's. A 2,000-sq.-ft. building can be comfortably cooled with a central air conditioning unit of 24,000 to 36,000 Btu's. Larger homes may require 60,000 or more Btu's, depending on the components specified.

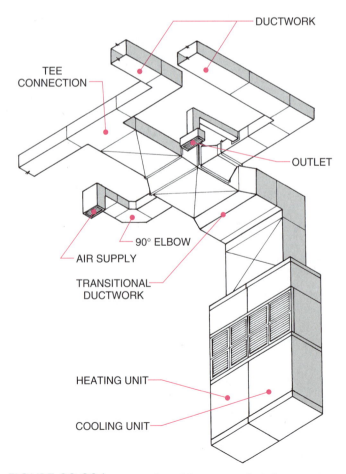

FIGURE 32.32A ■ Combined heating and cooling units.

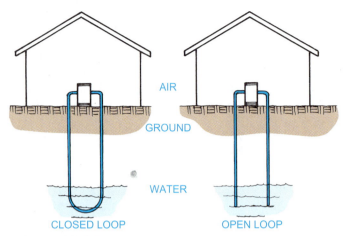

FIGURE 32.32B ■ Geothermal heating and cooling.

HEAT PUMPS

Heat pumps work like a reversible refrigeration system. They can both cool and heat buildings. In warm weather a heat pump removes heat from indoor air through a closed refrigerant cycle. In cold weather, heat is drawn from the outside air, water, or ground and transferred inside.

Air-source heat pumps are effective in warm or mild climates but do not heat efficiently when temperatures are below 30°F. A supplementary electric heat source is necessary below that temperature. Ground-source (geothermal) or water-source heat pumps extract heat from the earth and are therefore more effective for heating in cooler climates. In a geothermal system, heat exchangers extract heat from the subterranean ground or water. The heat is then delivered from the geothermal source to the heating system of a building. Geothermal systems are either open-loop or closed-loop systems as shown in Figure 32.32B.

Heat pump systems consist of an inside air handler, outside heat pump, ducts, and outlets. When a building is designed, space must be allowed for the heat pump and air handler. Ducts and outlets are drawn the same as warm air ducts and outlets. Heat pumps are classified by Btu capacity or by tons. Typical units range from 1 ton to 5 tons.

VENTILATION

Comfort control involves more than just temperature control. All buildings must be well ventilated with clean air that contains acceptable levels of humidity. Ventilation is necessary to keep fresh air circulating. Effective

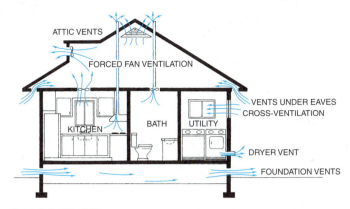

FIGURE 32.33 ■ Effective ventilation locations.

ventilation also controls moisture and keeps air relatively dry. All air in a building must be circulated gently and constantly, 24 hours a day, or physical discomfort may result. Air movement is provided by blowers in heating and cooling units. Ceiling fans and exhaust fans circulate air in attics, crawl spaces, and bathrooms.

Although structural tightness is a desirable method of controlling heat loss, some amount of fresh air must be allowed to enter the structure. See Figure 32.33. Attics and upper-level room ventilation help circulate out hot air. Crawl space ventilation helps remove excessive moist air. Structural methods that create good ventilation patterns are shown in the structural framing chapters.

AIR FILTRATION

Simply moving air does not create a supply of fresh air. To create clean air, airborne pollutants such as dust mites, chemical fumes, pollen, bacteria, mold spores, and mildew must be constantly removed from the air inside a building. Some airborne particles are just a nuisance while others—such as benzene, ammonia, radon, chloroform, formaldehyde, and carbon monoxide gases—are very dangerous. Many of these gases come from building materials, tobacco smoke, paint, carpet, and upholstery.

Air filters are placed in heating or cooling devices on the air return side of a furnace or air exchanger. Several levels of filters are available to trap pollutants. Mechanical filters filled with fiberglass or charcoal may only remove 15 percent of all pollutants. Electrostatic filters with ionizing wires can trap over 99 percent. Combinations of these filter types are most effective. To further purify the air, devices called ozonators may be used to introduce low levels of ozone (electronically charged oxygen) into the air supply.

TEMPERATURE CONTROL

Thermostatic controls keep buildings at a constant temperature by turning climate control systems on or off when a set temperature is reached. Thermostatic controls may be used with any heating or cooling system. **Thermostats** are sensing devices and should be located on interior walls away from sources of heat or cold such as fireplaces or windows. Larger buildings may need two or more separate heating or cooling zones that work on separate thermostats. One advantage of electrical heating is that each room may be thermostatically controlled. The location of thermostats is shown on electrical floor plans with a thermostat symbol.

HUMIDITY CONTROL

Humidity is moisture in the air. The proper amount of moisture in the air is important for good climate control. Adequate ventilation, especially in attics and crawl spaces, and the use of vapor barriers in construction will prevent excessive moisture.

Forced-air systems reduce humidity levels. Cooling units remove humidity from the air through condensation. If the air is too dry, a humidifier may be necessary to add more humidity to the air. Hot-water systems generally add humidity to the air.

To add more moisture to the air, a *humidifier* can be added to the plenum of a forced-air system. To remove excessive moisture from the air, a *dehumidifier* passes damp air over cold coils. The moisture-laden air then deposits excess moisture on the coils by condensation. *Humidistats* are sensing devices that monitor humidity levels and signal for more or less humidity. Planning buildings in which the correct amount of humidity can be controlled is essential for creating healthy building environments.

PASSIVE SOLAR SYSTEMS

Solar heating and/or cooling involves using the sun's energy to the fullest extent possible. **Passive solar systems** are integrated with the basic design of a structure. Passive systems operate without the use of special mechanical or electronic devices to heat or cool a structure.

Passive solar heating or cooling of a building includes four steps: collecting, storing, distributing, and controlling. These steps occur in active solar systems as well. However, the equipment, materials, and devices used differ greatly.

The earth's annual revolution around the sun and daily rotation on its axis determine how much solar energy is available at any time in any location on earth. Both the daily and seasonal paths of the sun over a building site are the first consideration in passive solar planning.

The amount of solar radiation reaching a site also depends on the atmosphere. When the sun is directly overhead, its rays travel the shortest distance through the atmosphere. As the sun moves closer to the horizon, the amount of atmosphere through which the rays travel increases. See Figure 32.34. This greater amount of atmosphere decreases the amount of solar radiation reaching the site. This means winter, early morning, and evening rays travel greater distances than summer and midday rays.

The slope of a site also affects the amount of usable solar radiation. This is because the sun's rays when striking a surface perpendicularly are more concentrated than when they intersect the surface at an angle. Therefore solar radiation striking a south-facing slope, as shown in Figure 32.35, will be more concentrated than rays striking a nearly flat terrain.

In addition to the atmospheric relationship of the sun to the earth's surface, two other principles of solar physics are used in passive solar planning. One is the **greenhouse effect.** The other is the natural law of rising warm air.

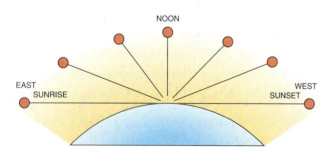

FIGURE 32.34 ■ Sun ray distances at different times of the day.

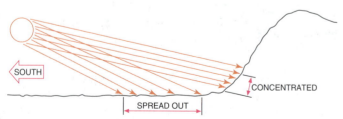

FIGURE 32.35 ■ South-facing slopes receive concentrated solar energy.

Greenhouse Effect

Science Connection

A car parked in direct sunlight with the windows closed illustrates the greenhouse effect. As sunlight enters through the windows, heat is absorbed by the interior surfaces of the car and is trapped inside the car as stored heat. This heat cannot leave the car as easily as sunlight can enter. The interior temperature of the car may reach 200°F (93°C) or more. This is why it is against the law in many states to leave children and/or pets in a parked car in the summer. They could die from the intense heat.

The greenhouse effect can be dangerous in a car, but it can be useful in a building. Heat from the sun that enters a building through windows can be stored to be used later when the sun's heat is not available.

A **thermal mass** is any material that will absorb heat from the sun and later radiate the heat back into the air. The seats in a car act as a thermal mass. Walls, floors, and masonry can also function as a thermal mass in a building designed for maximum solar effectiveness. In passive solar systems, the thermal mass is both the storage and the distribution system. Storing and dissipating the trapped heat, to either lower or raise the temperature of a building as needed, is one of the most important features of passive solar design. See Figure 32.36. A greenhouse should be attached to the south side of a structure to gain heat in winter and yet repel most of the solar heat during the summer.

Rising Warm Air

Figure 32.37 shows how heat is transferred in a passive solar system. Heat from the sun (radiation) enters the attic or upper level by conduction through the roof. The heat is radiated into the attic and generates convection currents, which spread the heat throughout the confined space. This heat is conducted through the ceiling to the lower levels, where it again radiates into the rooms, causes convection currents, and natural circulation begins.

Heated air will always rise until trapped. Therefore, recirculating heated air from high places to cooler lower areas helps heat living space. Likewise, expelling high-level warm air that would otherwise move downward helps reduce living-level temperatures. Roof-level vents may be used. Also, the level and placement of windows can provide natural convection and ventilation to both circulate and exhaust the warm air. See Figure 32.38.

Ceiling heights affect room temperature. Low ceilings tend to trap warm air in the living space. High ceilings allow it to rise.

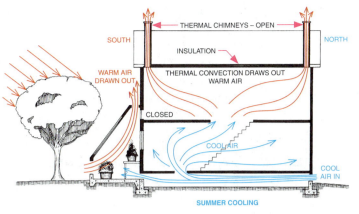
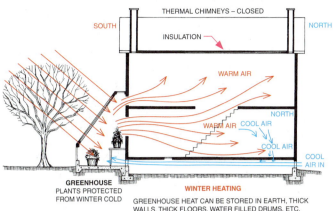

FIGURE 32.36 ■ Greenhouse effect on temperatures.

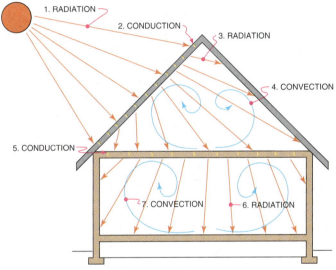

FIGURE 32.37 ■ Heat transfer in a passive solar system.

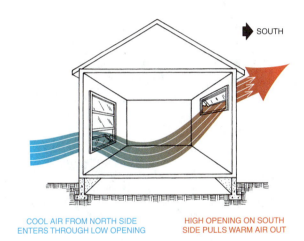

FIGURE 32.38 ■ Exhausting warm air by convection.

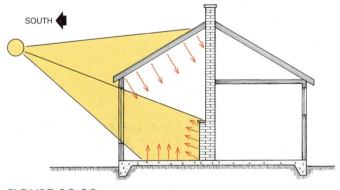

FIGURE 32.39 ■ Direct-gain heating.

Passive Solar Methods

Passive solar methods depend on environmental elements without additional assistance from electromechanical devices. Both the *direct-gain* method and the *indirect-gain* method are designed to take full advantage of the sun to provide heat when and where it is needed and to block the sun's heat when and where it is not wanted.

Direct-Gain Method

In the direct-gain method, the inside of a building is heated by the sun's rays directly as they pass through large glass areas and/or structural materials. See Figure 32.39. To maximize the amount of winter heat directly entering a building, large south-facing glass areas are used. Once the winter sun's rays enter a building through

windows, they are absorbed. The heat is stored in thermal-mass objects to be used later. These objects include floors, walls, and furniture. At night, or when clouds block the sun's rays, the stored heat in the thermal masses is slowly released, keeping the inside temperature higher. Large thermal masses such as masonry floors, walls, and fireplaces store more heat than wood and other fibrous materials do. Therefore, they hold heat longer for later use.

Water has the highest capacity for retaining heat, followed by steel, then aluminum. Masonry and rock are next. Wood and similar porous materials are the poorest retainers of heat.

Some direct-gain walls are designed with reflective insulating units that provide insulation and retain inside heat in cold weather, yet reflect most of the sun's heat in hot seasons.

Indirect-Gain Method

The indirect-gain method uses a thermal mass placed between the sun and the inside of a building. This thermal mass is heated directly by the sun. When heat is needed, the thermal mass is exposed to the inside of the building and heats the inside air. A **trombe wall** is a type of indirect-gain system. See Figure 32.40. In this system the temperature is controlled by directing varying amounts of the rising warm air to the inside or to the exterior of

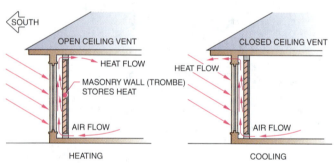

FIGURE 32.40 ■ Trombe wall indirect-gain method.

the building, depending on the comfort level needed. Movable roof panels can also be used to control, store, and distribute solar heat. See Figure 32.41.

Not all passively planned buildings use all of the passive solar features. Designers should include as many features as the design situation allows.

ACTIVE SOLAR SYSTEMS

Planning for **active solar systems** requires knowledge of both mechanical systems and thermal principles. Active solar systems use mechanical devices to drive the components needed for solar heating or cooling. This includes devices and facilities for collection, storage, distribution, and control of heat. Active solar systems operate more effectively when combined with passive solar features.

Active systems can be designed for comfort control for an entire structure but this usually requires some support from conventional HVAC systems depending on the climate. The most frequent use of active systems are for heating hot water for consumption or for heating pool and spa water. Active systems are most effective when combined with passive solar design features.

Collection

Active solar systems use south-facing solar collectors, which contain circulating water, oil, or air. Heating collectors should be set at an angle perpendicular to the rays of the sun for the maximum number of hours each day.

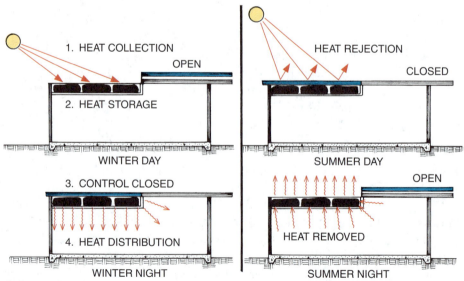

FIGURE 32.41 ■ Movable roof panels in an indirect-gain system.

Unless rotating collectors are used, this position is ideal for only a short time each day. Therefore, fixed collectors are usually positioned to face the midday sun (between 10 a.m. and 2 p.m.), since air temperatures are usually higher during these hours. In most North American areas, this means the collectors will face south-southwest.

In addition to the orientation of the collectors, the vertical tilt angle should be the same as the local latitude for maximum year-round effect. A tilt angle of 15° greater than the local latitude is best for winter heating because of the lower path of the sun.

Storage

Because heat is needed when the sun is not producing heat, storage of the absorbed heat (usually in rocks or water) is necessary. Stored heat is limited to the capacity of the storage unit. The larger the storage unit, the longer the solar system will operate without the use of auxiliary heating devices. Maximum insulation is needed in storage containers to minimize the loss of stored heat.

Distribution and Control

Thermostatic controls activate solar distribution systems. Signals cause hot air to be blown or hot water to be pumped from storage containers to parts of a building where heat is needed.

Types of Active Solar Systems

Several types of systems convert sunlight into usable energy. These systems may rely on a solar furnace, photo-voltaic (solar cell) technology, or liquid-based or air-based collectors.

Solar Furnace

A solar furnace is a collection of mirrors that focuses the sun's heat on a concentrated area. Temperatures as high as 3500°F (1927°C) can be attained, and the energy can heat or cool large clusters of buildings. Solar furnace collectors must move with the sun and are not currently practical for single building use.

Photovoltaics

A **photovoltaic film**, known as a solar cell, converts sunlight directly into electricity. Five square miles of cells can produce the same amount of electricity as a nuclear power plant. Solar cells are not yet economically feasible for individual building use but may be soon.

Liquid-Based Systems

Liquid-based systems are most popular for light construction. They use liquid to trap and distribute the heat from the sun. In designing or choosing an active solar-heating system, the type of fluid used to transport the heat must be determined. Water, antifreeze solutions, or oils may be used. See Figure 32.42. If the building is in a cold climate and antifreeze is not used, a draindown system must be installed that empties the pipes when the system is not needed.

The liquid is heated in collectors. The collectors may be attached to a roof or installed near the structure. Collectors consist of *absorbers* placed over a layer of insulation to help prevent heat loss. Liquid system absorbers may be plastic

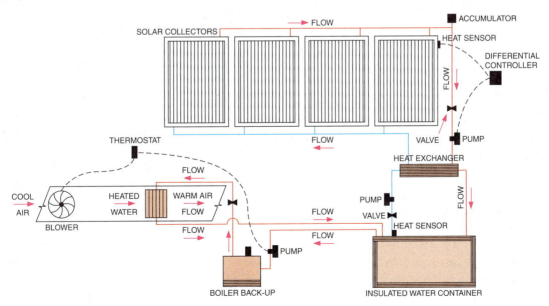

FIGURE 32.42 ■ Liquid-based house solar system.

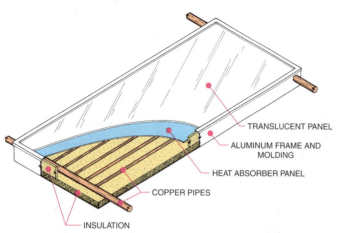

FIGURE 32.43 ■ Solar collector panel design. (*Solar Industries, Inc.*)

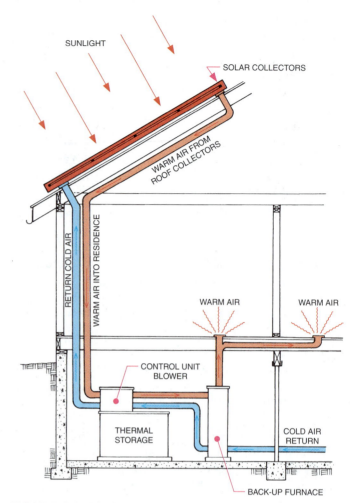

FIGURE 32.44 ■ Solar system using air to carry heat.

or metal sheets (plates) over which a liquid flows. Absorber plates are constructed of steel, copper, aluminum, or plastic because of their heat conductivity. Other absorbers consist of a network of pipes containing a circulating liquid. See Figure 32.43. The absorber pipes (or tubes) are arranged in panels, usually 4′ × 10′ or 4′ × 12′. Even on a very cold day, with bright or filtered sunlight, absorbers can be heated to 200°F (93°C).

Absorbers are designed to retain a maximum amount of heat. The amount of heat retained is called *absorptance*, whereas *emittance* is the amount of heat reflected. An efficient absorber has high absorptance and low emittance.

The heated liquid from the collectors is stored in insulated tanks. In large multiple-building complexes, heated liquid is stored seasonally in underground units. When needed, the heated liquid is pumped to liquid-to-air heat exchangers, water heaters, or swimming pools.

Air-Based Systems

Active solar systems using air as the heating element are more simple to construct and operate than liquid systems. See Figure 32.44. They are effective for space heating. However, their inability to provide hot water is a serious drawback. Air system collectors use sheet-metal (copper, aluminum, or steel) plates to heat the air trapped between a cover plate and the absorber. This heated air is then blown to the storage facility where heat is stored in a thermal mass (rocks, water, masonry). From the storage area, the heated air can be blown directly through a duct system to appropriate rooms, as in a conventional warm-air system. Because air systems have low heat capacity, large ducts (up to 6″) must be used.

With air systems, heat can also be stored in rocks or gravel. The heat can be distributed through convection

or radiant panels. For each square foot of collector for heat storage, 80 to 400 lbs. (36 to 180 kg) of rock are required. Rock storage requires 2 1/2 times as much volume as water to store the same amount of heat over the same temperature rise.

Auxiliary Heating Systems

Except in very mild climates, most solar heating storage facilities cannot keep constant pace with peak demand. For this reason, an auxiliary heating system is usually recommended, especially for hot-water production. An active solar heating unit is often integrated with an auxiliary heating unit. When public utilities are not available, a self-sustaining energy system may be used. See Figure 32.45.

Solar Cooling

Solar cooling is possible through the same absorption-cooling method that is used in gas refrigerators. However, at present, the equipment for an active solar heating and cooling system is very expensive. See Figure 32.46.

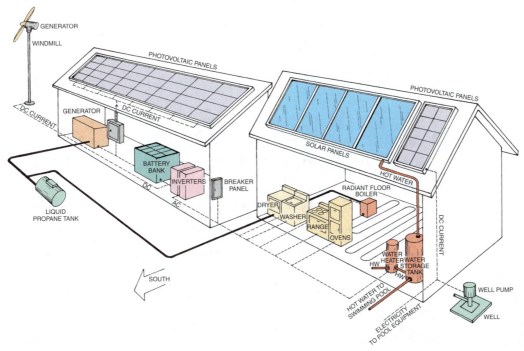

FIGURE 32.45 ■ Self-sustaining energy system.

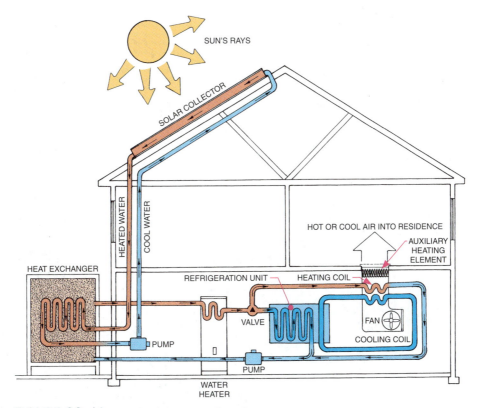

FIGURE 32.46 ■ Solar heating and cooling system.

Collectors can be used minimally to help cool buildings by exposing them to cooler night air and closing them to daytime exposure. The cooler night air cools the liquid or air that returns to the storage area. The cooled liquid is then released the next day to augment passive or conventional cooling systems.

CHAPTER

Comfort Control Systems (HVAC) Exercises

1. What factors affect heat measurement for a building's HVAC system?

2. Plan and draw an HVAC system for the plan shown in Figure 14.24A.

3. Name the four steps in the solar heating process.

4. Sketch a plan or elevation showing use of passive solar principles.

5. Evaluate the active solar systems in this chapter in terms of advantages and disadvantages.

6. Sketch a floor plan showing a mechanical heating system with a heating unit in the most appropriate location. Locate the ducts and outlets for this system.

 7. Draw a plan of a climate control system appropriate for the house of your design.

8. Sketch a floor plan and indicate the best location for solar collectors and storage.

 9. List the passive and/or active solar features you would include in a residence of your own design.

Plumbing Drawings

OBJECTIVES

In this chapter, you will learn to:

- draw plumbing fixtures on a floor plan.
- draw the water supply lines and waste discharge system on a floor plan.
- draw the water supply lines and waste discharge system on an elevation.

TERMS

branches
building main
cleanout
drainage field
fixture trap

gray-water waste
percolation
piping
riser diagram
septic tank

soil line
stacks
valve
vent stack
waste discharge

INTRODUCTION

The history of plumbing extends back to 500 B.C. when the ancient Greeks and Egyptians built tile-lined bathtubs. However, the Romans are credited with creating the first plumbing systems through their development of aqueducts. The term *plumbing* is derived from the Latin *plumbum*. Today plumbing refers to the supply, distribution, control, and discharge of all liquid and gases. Plumbing drawings are needed to design and locate the type and position of all plumbing fixtures, devices, and pipes within a plumbing system.

PLUMBING CONVENTIONS AND SYMBOLS

Like most architectural drawings, plumbing drawings must be prepared to a very small scale. Therefore, schematic symbols are used as a substitute for drawing plumbing lines, joints, and intersections. See Figure 33.1. These schematic symbols show the type and location of fixtures, joints, valves, and other plumbing devices that control the flow of liquids. Figure 33.2 shows the plan and elevation symbols for pipe intersections, Figure 33.3 shows pipe joints, Figure 33.4 shows sanitary facilities,

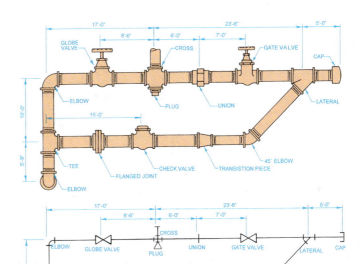

FIGURE 33.1 ■ Orthographic drawing compared to a schematic plumbing drawing.

and Figure 33.5 shows the types of piping lines used on plumbing drawings. Symbols used to represent bath and kitchen plumbing fixtures are part of a complete floor plan and are shown in Chapter 14.

NAME	ABBREV	SYMBOL	ELEVATION	PICTORIAL
DOUBLE BRANCH ELBOW	DBL BR EL			
STRAIGHT CROSS	ST X			
SAFETY VALVE	SFTY V			
GATE VALVE	GT V			
HAND VALVE	HND V			
UP ELBOW PIPE OUTLET	UP EL			
DOWN ELBOW PIPE OUTLET	DN EL			
UP TEE OUTLET	UP T			

NAME	ABBREV	SYMBOL	ELEVATION	PICTORIAL
90 DEGREE TEE	T			
45 DEGREE LATERAL	LAT			
CLEAN OUT	CO			
REDUCER	RED.			
DOWN TEE OUTLET	DN T			
COUPLING	CPLG			
90 DEGREE ELBOW	EL			
45 DEGREE ELBOW	EL			

FIGURE 33.2 ■ Pipe intersection symbols.

USING CAD

CAD Plumbing Library

Plumbing drawings contain series of lines representing pipes. Special line types for gas, water, oil, etc., can be selected using the *Line* task. The *Break* command can also be used to interrupt a solid line at regular intervals by inputting the gap size, spacing, and number. Once lines are drawn, symbols for valves, unions, elbows, tees, and joints are added. This is done by first breaking the line using the *Trim* task. Then symbols from the plumbing symbol library are inserted in the gap with the cursor.

PLUMBING SYSTEMS

Plumbing drawings are prepared to describe types of plumbing systems, such as water supply, waste discharge, and hydronic heating systems. Plumbing drawings are also used to show gas appliance lines, built-in vacuum systems, and pest-control **piping**.

Water Supply System

A water supply system consists of a network of pipes that carry fresh water to appliances, sinks, water closets (toilets), tubs, filters, showers, and water heaters. See Figure 33.6.

Water systems are designed for consumption such as drinking and washing or for circulation such as heating

NAME	ABBREV	SYMBOL	ELEVATION	PICTORIAL
FLANGED FITTING	FL FT			
SCREWED FITTING	SC FT			
BELL & SPIGOT FITTING	BL/SP FT			
WELDED FITTING	WLD FT			
SOLDERED FITTING	SLD FT			
EXPANSION JOINT	EXP JT			

FIGURE 33.3 ■ Pipe joint symbols.

NAME	ABBREV	SYMBOL	NAME	ABBRV	SYMBOL
COLD-WATER LINE	CW		AIR-PRESSURE RETURN LINE	APR	
HOT-WATER LINE	HW		ICE-WATER LINE	IW	IW
GAS LINE	G	G — G	DRAIN LINE	D	D — D
VENT	V		FUEL-OIL RETURN LINE	FOF	FOF
SOIL STACK PLAN VIEW	SS		FUEL-OIL FLOW LINE	FOR	FOR
SOIL LINE ABOVE GRADE	SL		REFRIGERANT LINE	R	+ + + + +
SOIL LINE BELOW GRADE	SL		STEAM LINE MEDIUM PRESSURE	ST	
CAST-IRON SEWER	S-CI	S-CI	STEAM RETURN LINE—MEDIUM PRESSURE	ST	
CLAY-TILE SEWER	S-CT	S-CT	PNEUMATIC TABE	PT	
LEACH LINE	LEA		INDUSTRIAL SEWAGE	IS	
SPRINKLER LINE	SPR	S — S	CHEMICAL WASTE LINE	CW	
VACUUM LINE	VAC	V — V	FIRE LINE	F	F — F
COMPRESSED AIR LINE	COMA	A — A	ACID WASTE LINE	AC WST	ACID
AIR-PRESSURE LINE FLOW	APF		HUMIDIFICATION LINE	HUM	H

FIGURE 33.5 ■ Piping line symbols.

NAME	ABBREV	SYMBOL	ELEVATION	PICTORIAL
METER	M	M		
FLOOR DRAIN	FD	FD		
CESS POOL	CP			
DRY WELL	DW			
SEPTIC TANK	SEP TNK			
SEPTIC-TANK DISTRIBUTION BOX	SEP TANK DIS BX			
SUMP PIT	SP			

FIGURE 33.4 ■ Sanitary facilities.

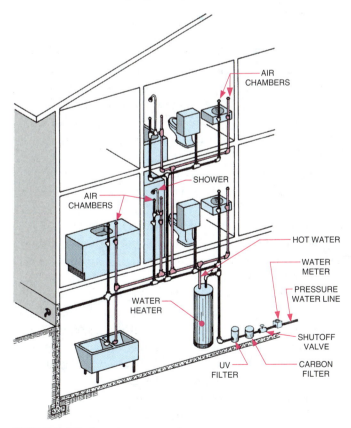

FIGURE 33.6 ■ Water supply system.

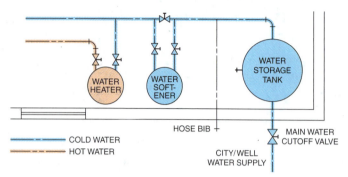

FIGURE 33.7 ■ Filter system branching.

FIGURE 33.8 ■ Filtering, softening, and purifying equipment.

FIGURE 33.9 ■ Riser pipes extending above a slab.

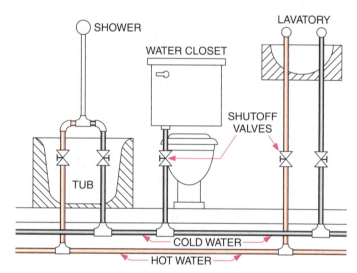

FIGURE 33.10 ■ Shutoff valve locations.

and humidity control. The storage of water for fire control and swimming pool use must also be considered.

Water is brought from a well or public water supply main through a **building main.** If public water is used, this main contains a utility company **valve,** a meter, and a building main valve. If water filters are used, the water may be directed through carbon filters, a reverse osmosis system, ultraviolet (UV) filters, and/or softeners. Figure 33.7 shows a typical pipe branching arrangement for a water filtering system. Figure 33.8 shows a water filtration, softening, and purifying system combined with the electrical delivery system for a residence. A separate line may be used to connect a special drinking water supply through a purification system. Hose bibs or sprinkler system lines not requiring purification may be connected before filtering.

After filtering, water is branched into a cold water main and a hot water main, which passes through a water heater or heaters. Because this water is under pressure, pipes can be located in any direction. Lines that pass vertically through slabs must be precisely located to align with partition and fixture locations. See Figure 33.9.

Hot water branch lines are normally located 6″ from cold water branch lines and parallel with them. Where a

fixture includes both hot and cold water outlets, the hot water line and valve are located to the left of the cold water line and valve. Every fixture must include a shutoff valve. See Figure 33.10.

The continuous hot water circulating system shown in Figure 33.11A delivers hot water because hot water remains in the pipes continuously. In the Metlund system shown in Figure 33.11B, cold water is stored in the hot water system until the hot water valve is turned on and the water pump is activated. This pumps the stored cold water back into the water heater where the thermostat controls the water temperature.

Pipes used for light construction plumbing are available in copper, plastic, brass, iron, and steel. In addition to pipe material, pipe size is important. The size of all residential water supply lines ranges from 1/2" to 3/4" as shown in Figure 33.12. All fixtures must have a free-flowing supply of water at all times. Lines that are too small cause a whistling sound as water flows through at high speeds. *Air-cushion chambers* stop hammering noises.

The location of pipes must be carefully planned. Too many changes of pipe direction cause friction and reduce water pressure. Designers must also consider energy conservation. Placing insulation around hot water lines conserves hot water and reduces the total cost of fuel for heating the water. A note on the plumbing diagram should specify the type and thickness of the insulation material.

Waste Discharge System

Although water is supplied to a building under pressure, waste water is discharged through a disposal gravity system. See Figure 33.13. **Waste discharge** lines are not under pressure. They are empty except when waste is flushed through them. Therefore, all pipes in a discharge system must slope in a downward direction, usually 1/4"

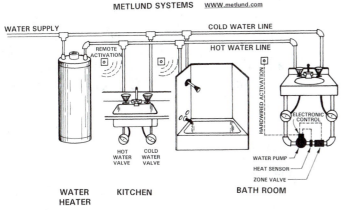

FIGURE 33.11A ■ A continuous hot water circulating system.

FIGURE 33.11B ■ Metlund hot water system.

FIXTURE	COLD WATER	HOT WATER	SOIL, WASTE	VENT
Sinks (Lav)	1/2"	1/2"	1 1/2"–2"	1 1/4"–1 1/2"
Lavatory	1/2"–3/8"	1/2"–3/8"	1 1/4"–2"	1 1/4"–1 1/2"
Water closet	1/2"–3/8"	—	3" – 4"	2"
Tub (Bath)	1/2"	1/2"	1 1/2"–2"	1 1/4"–1 1/2"
Shower	1/2"	1/2"	2"	1 1/4"
Water heater	3/4"	3/4"	—	4"
Washer	1/2"	1/2"	2"	1 1/2"
Lau Sink	1/2"	1/2"	1 1/2"	1 1/4"

FIGURE 33.12 ■ Minimum pipe sizes for common fixtures.

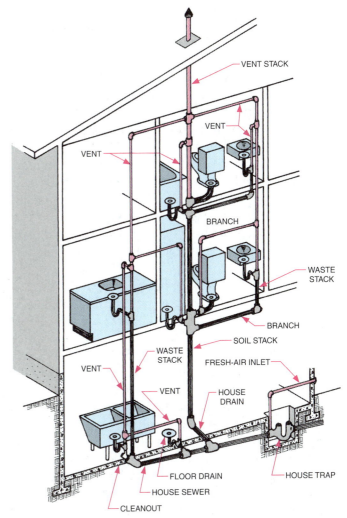

FIGURE 33.13 ■ Waste discharge system.

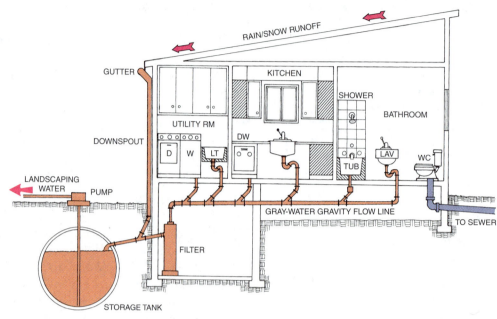

FIGURE 33.14 ■ Gray-water collection system.

per foot. Because of this gravity flow, waste (soil) lines are larger than water supply lines that are under pressure.

There are two types of **soil lines**: branches and stacks. **Branches** are nearly horizontal lines that carry waste from each fixture to the **stacks**. Stacks are vertical lines. Soil stacks (3″ to 4″ in diameter) carry waste to the house sewer. The portion of the soil stack above the highest branch intersection is known as a **vent stack**. Some vent stacks are separate and parallel to soil stacks. Vent stacks are dry pipes that extend through the roof (a minimum of 6″) and provide ventilation for the discharge system. Vent stacks permit sewer gases to escape to the outside and equalize the air pressure in the system. Stacks that intersect house sewer lines under a slab must be dimensioned accurately to ensure alignment of the stack pipe with the partition location.

The flow of all waste water begins at a fixture. Each fixture contains a **fixture trap** to prevent the backflow of sewer gas from the branch lines. Fixture traps are exposed for easy maintenance, except for water-closet traps, which are built into the fixture. A total system house trap is provided in the house sewer line outside the perimeter of the building. **Cleanouts** must be provided for the main house sewer drain. Some municipalities allow sewer waste to be separated from **gray-water waste** as shown in Figure 33.14. Filtered gray-water (nonsewage waste water) is usually acceptable for landscape irrigation.

When municipal sewage-disposal facilities are available, connection of the house sewer to the public system is shown on the plot plan. When public systems are not available, a septic system is used. Either a separate drawing is provided or the location of the system is included on the plot plan. This location drawing is usually required by the building code and the local board of health before a building permit is issued.

Septic Tanks

In a septic system, building wastes flow from the house sewer into a **septic tank** that is buried a prescribed distance from the building. A septic tank is a tank in which solid waste is processed by bacteria. Lighter liquids flow out of the septic tank into drainage fields through porous pipes. These pipes spread over an area to allow wide distribution of liquids. The solid wastes, which settle to the bottom of the septic tank, are converted to liquids by bacterial action.

The size and type of septic tank and drainage field are specified by code. Many factors such as the number of occupants, baths, bedrooms, and kitchen disposals, as well as soil type and topography, are considered to determine the size and type specified. Septic tank size in gallons must be at least 50 percent larger than the daily sewage output. Because detergents interfere with bacterial action, some codes allow smaller septic tanks if laundry waste is diverted to a separate discharge system. Septic tanks must be completely watertight to prevent contamination of surrounding soils. See Figure 33.15. Flood drains or rain water runoff should not be connected to the septic system.

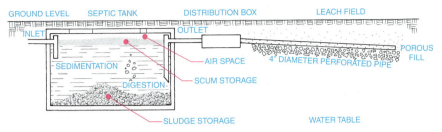

FIGURE 33.15 ■ Septic system components.

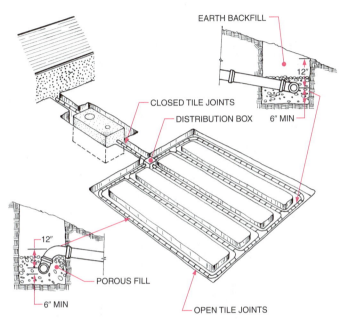

FIGURE 33.16 ■ Trench-type distribution field.

Drainage Fields

Liquid waste (effluent) flows from the top of the septic tank to a drainage field. **Drainage fields**, also called leaching or absorption fields, consist of plastic (PVC) perforated pipe or agricultural open-joint pipe laid in coarse gravel. These seepage pipes are either laid in trenches or in continuous beds 12″ to 16″ below grade level. See Figure 33.16. Fields are arranged in a variety of shapes and patterns depending on the site contour and restrictions, such as building or tree locations. See Figure 33.17.

Local codes specify the minimum distance allowed between the end of a drainage field and bodies of water, wells, roadways, right-of-ways, buildings, and property lines. See Figure 33.18. Fields cannot be located under paved areas or uphill from a well or water supply. Codes also require that fields be placed no closer than 5′ from a water table level. Regardless of location, soil under the field must be porous enough to absorb the effluent. The process of liquid absorption is known as **percolation**. A percolation test consists of digging a hole in the drainage field area and filling the hole with water. Then the rate

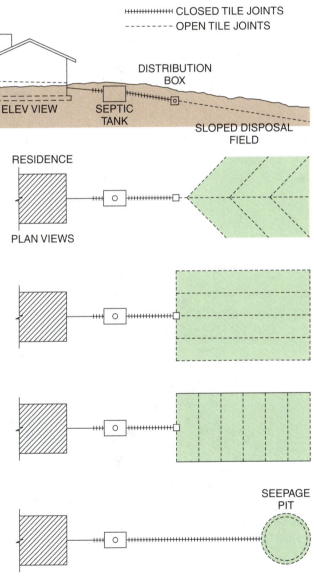

FIGURE 33.17 ■ Common distribution field shapes.

at which the water is absorbed into the soil is measured. A percolation rate of 40 to 60 minutes per inch is considered acceptable, depending on local code requirements. Figure 33.19 shows the method of determining distribution field size for a two-bedroom residence with good percolation.

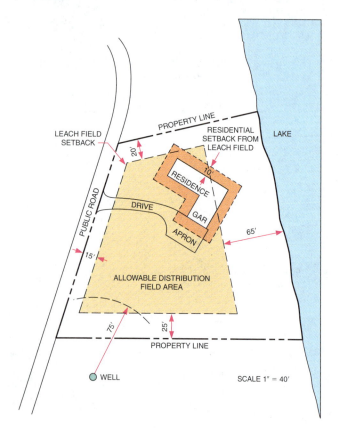

FIGURE 33.18 ■ Typical minimum code requirements for distribution field location.

PERCOLATION RATE (MIN. PER INCH)	RATING	REQUIRED SQ. FT. (PER BEDROOM)
Over 45 minutes	Will not pass code	Unacceptable
31 to 45 minutes	Poor	600 square feet
16 to 30 minutes	Acceptable	400 square feet
15 or less minutes	Very good	300 square feet

FIGURE 33.19 ■ Percolation test standards.

PLUMBING DRAWINGS

Drawings used to describe plumbing systems include plumbing plans, elevations, and in some cases pictorial drawings. Figure 33.20 shows an example of these three types of plumbing drawings.

Plumbing Floor Plans

Plumbing lines and symbols are usually added to a reproducible print of a complete floor plan. If this is not done,

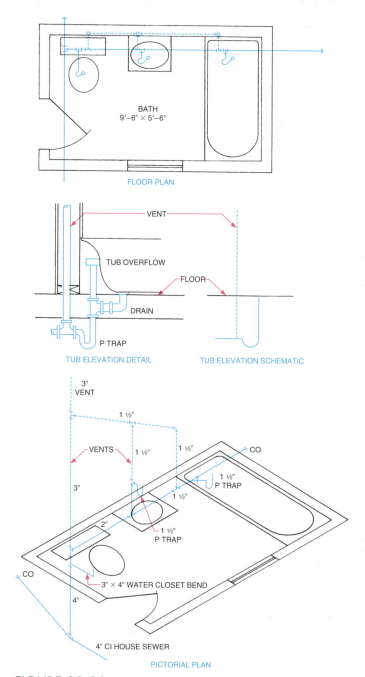

FIGURE 33.20 ■ Types of plumbing drawings.

a separate abbreviated floor plan is prepared. See Figure 33.21. Many floor plan details and dimensions can be eliminated, except for partition positions that require the placement of stacks before a slab or T-foundation is poured. Some plumbing plans are combined with HVAC and/or electrical floor plans.

The location of the following items are drawn on plumbing floor plans:

■ Fixtures

■ Vent pipes

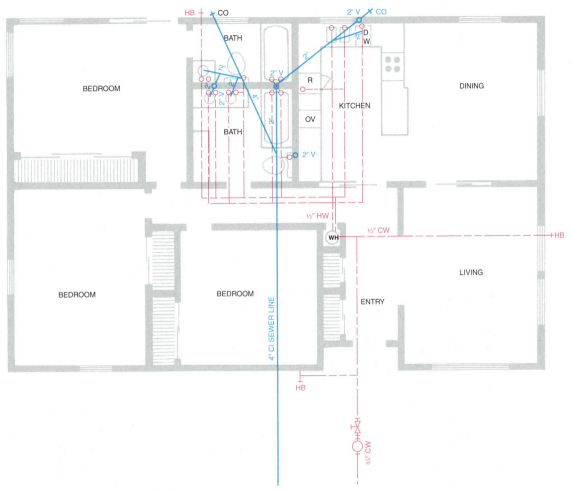

FIGURE 33.21 ■ Plumbing system floor plan.

- Floor drains
- Hose bibs
- Cold water supply lines
- Hot water supply lines
- Shutoff valves
- Water heating equipment
- Soil lines
- Cleanouts
- Gas lines
- Oil lines
- Water purification devices.

Some plumbing lines and facilities are usually located on a plot plan or site plan. These include septic system, including lines, septic tank, and field; irrigation system lines and devices; swimming pool or spa lines, heaters, filters, and pump; and public water and discharge lines and connections.

Water Supply Lines

Plumbing lines are drawn to represent connections in the water supply system. Begin with the house main and draw cold water branch lines to each fixture. Next, draw the hot water lines from the house main to water heaters and then to each fixture requiring hot water. Draw the hot water lines parallel to the cold water lines where possible. Add symbols for shutoff valves on each fixture line.

Waste Water Lines

Draw heavy waste water branch lines connecting fixtures with soil stacks. Add a trap symbol at each fixture. The positions of stack and vertical pipes are shown with a circle inside walls and partitions. The main house drain and sewer lines are next drawn connected to the stack symbols and cleanouts, and the house trap is added. The house drain is then connected to either a municipal sewer line or to a septic tank. The outline of septic tanks and fields is normally drawn on the plot plan or site plan. Since soil lines are larger than supply lines, they should be drawn heavier to show the difference.

The waste line diameter from the farthest fixture in a building to a public sewer or septic tank is 4″. The waste line behind the farthest fixture has a diameter of 3″ because it does not carry waste products. Horizontal waste lines slope at 1/4″ per foot to ensure an adequate flow.

Gas Lines

If gas appliances or hydronic heating systems are used, lines with cutoff valves are drawn from the main gas line or gas tanks to each device.

Plan Details

Because most plumbing fixtures are concentrated in the bathroom, laundry, and kitchen, detailed schematic plans are sometimes prepared for those rooms. See Figure 33.22. This enables areas where piping is dense to be drawn at a larger scale for easier reading. Where special framing adjustments must be made for piping, framing details should be prepared. See Figure 33.23.

The maximum outside pipe diameter that will fit inside a standard 4″ stud wall or an 8″ concrete block is 2″. For a 2″ × 6″ stud wall, the maximum diameter is 3″; for a 12″ concrete block its 6″. Also note that pipe notches cut into any structural member must be replaced with blocking to avoid structurally weakening the member, as shown in Figure 33.24.

Notes are added to plumbing plans to label the type and size of pipes. Dimensions are added to critical positions, such as locations of underground valves and cleanouts.

Plumbing Elevations

Plumbing plans show only the horizontal positioning of pipes and fixtures. Therefore, the amount of rise above floor level and the flow of fresh water and wastes between levels are difficult to read. Elevation drawings show vertical distances and angles that cannot be shown in floor plans.

Plumbing elevations are sometimes called **riser diagrams** because they show the amount of vertical rise of each pipe. Figure 33.25 includes a plumbing elevation drawing with the matching plumbing plan. Note that the supply system is shown with red lines and the waste discharge system in blue.

Because much of a waste discharge system contains vertical stacks, elevations are very effective for describing these systems. Figure 33.26 shows a riser diagram on a full

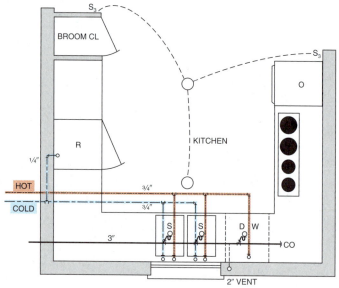

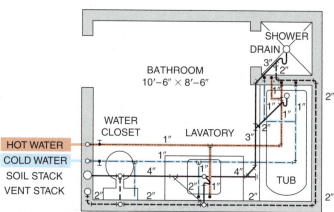

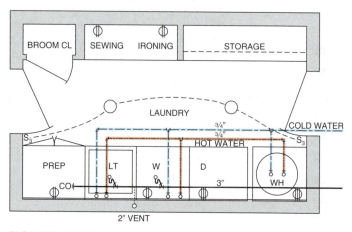

FIGURE 33.22 ■ Room plumbing plan details.

building elevation. In this illustration of a waste discharge system, the waste lines are shown in green and the vent pipes in blue. A full building riser diagram of a water supply system is shown in Figure 33.27. In this drawing hot water lines are shown in red and cold water lines in blue.

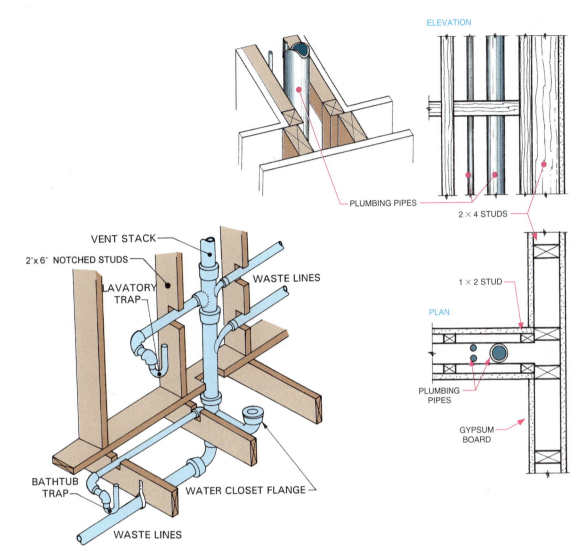

ELEVATION

PLUMBING PIPES

2 × 4 STUDS

1 × 2 STUD

PLAN

PLUMBING PIPES

GYPSUM BOARD

VENT STACK

2"x 6" NOTCHED STUDS

LAVATORY TRAP

WASTE LINES

BATHTUB TRAP

WATER CLOSET FLANGE

WASTE LINES

FIGURE 33.23 ■ Framing allowances for piping.

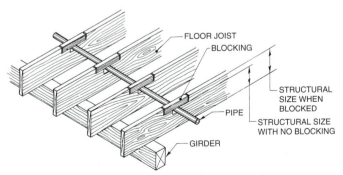

FLOOR JOIST

BLOCKING

STRUCTURAL SIZE WHEN BLOCKED

PIPE

STRUCTURAL SIZE WITH NO BLOCKING

GIRDER

FIGURE 33.24 ■ Blocking of pipe notching.

Because plumbing fixtures are located in different parts and on different planes of a building, detail riser diagrams are often required. A detailed drawing of this type is shown in Figure 33.28. Fixture valves are usually located on vertical pipes, so valve positions are also shown on elevations.

Pool and Spa Plumbing

To design and draw the plumbing facilities for a swimming pool or spa, the following components must be located to provide for maximum operating efficiency and should be hidden from view when possible.

- Filter
- Circulating pump
- Heater
- Purifier
- Strainer

(Continued page 630)

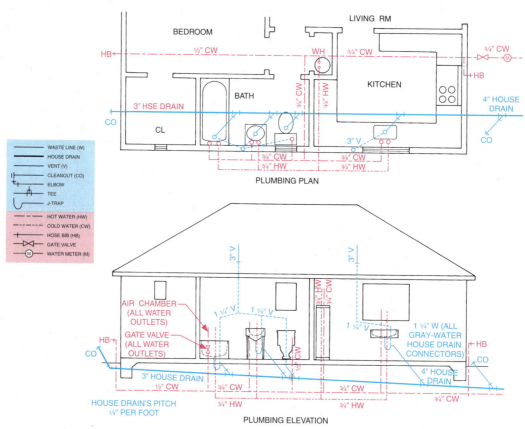

FIGURE 33.25 ■ Related plumbing plan and elevation.

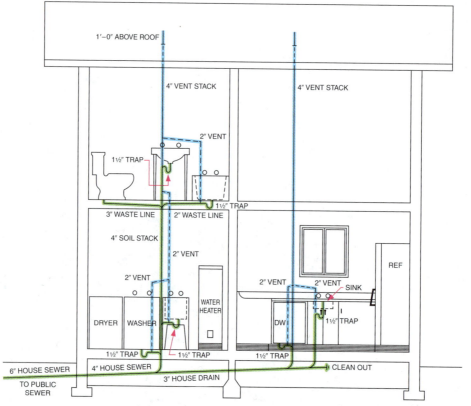

FIGURE 33.26 ■ Waste discharge system elevation.

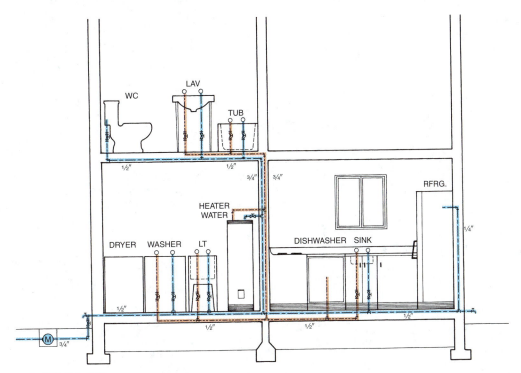

FIGURE 33.27 ■ Water supply system elevation.

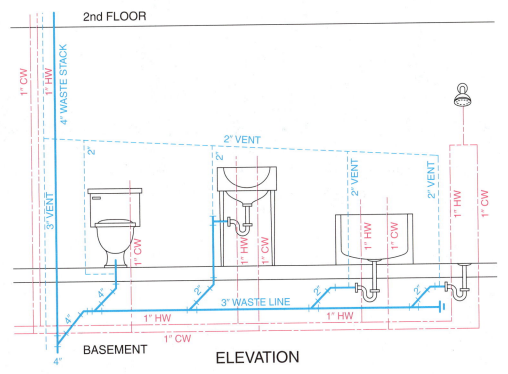

FIGURE 33.28 ■ Bath piping elevation.

- Drain lines and pump
- Circulating lines
- Control valves

- Flushing lines

Figure 33.29 shows how these components are interrelated in a typical swimming pool plumbing system.

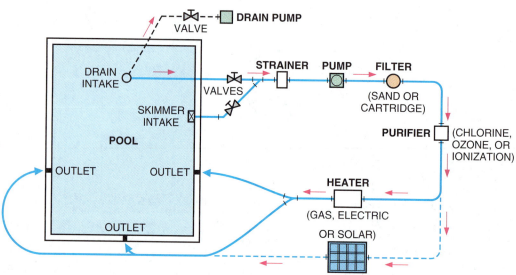

FIGURE 33.29 ■ Pool plumbing system components.

CHAPTER 33

Plumbing Drawings Exercises

1. Briefly describe a water supply system.

2. Name the types of lines in a waste discharge system.

3. Make a sketch of a plumbing plan in this chapter. Draw the position of all plumbing fixtures. Sketch water supply lines on an overlay. Sketch a waste discharge system on another overlay.

4. Add plumbing fixture symbols to your library.

 5. Prepare a plumbing plan for a house of your own design.

6. Draw an elevation of a waste discharge system.

PART 9

Drawing Management and Support Services

CHAPTER

Drawing Management

OBJECTIVES

In this chapter, you will learn:

- to organize and check a complete set of architectural drawings.

- how drawings in a set are related.

- to identify identical locations on all drawings in a set.

- to select the drawings needed to complete a set of architectural drawings.

- methods of drawing and recording changes on drawings according to change orders.

TERMS

change orders
checker
combination plans

cross-referencing symbols
drawing codes

drawing sequence
layering

INTRODUCTION

Architectural drawings must be complete and accurate. Each drawing must also be easily retrievable and consistent with all other drawings in a set. To accomplish this, drawings are organized and arranged according to established standards. The standards for coordinating, cross-referencing, coding, checking, and making changes are presented in this chapter.

DRAWING SEQUENCE

The **drawing sequence**, that is, the order in which drawings are prepared in a set, is important because the preparation of some drawings depends on the completion of others. The site plan should be developed first because the location and orientation of buildings depends on the design of the site. Next, the first-level floor plan is prepared because other level plans, specialized discipline plans, and elevations are projected from or overlaid on this plan. Once these are complete, framing plans and construction detail drawings can be drawn. Bearing partitions, plumbing walls, stairwells, chimney openings, and other compo-

nents must usually align vertically with the first-floor plan. Therefore, this plan must be carefully checked for accuracy before other drawings are completed.

DRAWING COORDINATION

Large construction projects may require thousands of drawings and documents, and even a set of residential plans may include dozens of drawings. A system of codes and symbols is necessary to organize and relate each drawing to other drawings and documents in a set.

Drawing Codes

The American Institute of Architects (AIA) recommends the use of an alphanumerical **drawing code** system to identify and classify drawings in a set. An alphabetical prefix is used to denote a specific discipline as introduced in Chapter 4 and as shown in Figure 34.1. Each discipline is divided into groups. For example the architectural discipline (A) is further divided into 10 specific groups, A0 through A9, as shown in Figure 34.2. Drawings are also identified by the work phase of each project

A	Architectural
C	Civil
D	Interior design (color schemes, furniture, furnishings)
E	Electrical
F	Fire protection (sprinkler, standpipes, CO_2, and so forth)
G	Graphics
K	Dietary (food service)
L	Site
L	Landscape
M	Mechanical (heating, ventilating, air-conditioning)
P	Plumbing
S	Structural
T	Transportation/conveying systems

FIGURE 34.1 ■ Letters used to identify drawing disciplines.

in order to separate preliminary drawings from final working drawings. These include:

SK	Sketches
PR	Programing
MP	Master plan
SC	Schematics
DD	Design development.

Cross-Referencing Symbols

To further relate drawings to other drawings and/or details in a set, **cross-referencing symbols** are used. Although different versions of these symbols are used, the symbols shown in Figure 34.3 are the most common.

SYSTEM CODE:

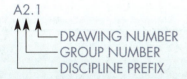

A2.1
— DRAWING NUMBER
— GROUP NUMBER
— DISCIPLINE PREFIX

Architectural Drawings

A0.1,2,3—General (Index, Symbols, Abbrev. notes, references)
A1.1,2,3—Demolition, Site Plan, Temporary Work
A2.1,2,3—Plans, Room Material Schedule, Door Schedule, Key Drawings
A3.1,2,3—Sections, Exterior Elevations
A4.1,2,3—Detailed Floor Plans
A5.1,2,3—Interior Elevations
A6.1,2,3—Reflected Ceiling Plans
A7.1,2,3—Vertical Circulation, Stairs (Elevators, Escalators)
A8.1,2,3—Exterior Details
A9.1,2,3—Interior Details

Structural Drawings

S0.1,2,3—General Notes
S1.1,2,3—Site Work
S2.1,2,3—Framing Plans
S3.1,2 —Elevations
S4.1,2 —Schedules
S5.1,2 —Concrete
S6.1,2 —Masonry
S7.1,2 —Structural Steel
S8.1,2 —Timber
S9.1,2 —Special Design

Mechanical Drawings

M0.1,2 —General Notes
M1.1,2 —Site/Roof Plans
M2.1,2 —Floor Plans
M3.1,2 —Riser Diagrams
M4.1,2 —Piping Flow Diagram
M5.1,2 —Control Diagrams
M6.1,2 —Details

Plumbing Drawings

P0.1,2 —General Notes
P1.1,2 —Site Plan
P2.1,2 —Floor Plans
P3.1,2 —Riser Diagram
P4.1,2 —Piping Flow Diagram
P5.1,2 —Details

Electrical Drawings

E0.1,2 —General Notes
E1.1,2 —Site Plan
E2.1,2 —Floor Plans, Lighting
E3.1,2 —Floor Plans, Power
E4.1,2 —Electrical Rooms
E5.1,2 —Riser Diagrams
E6.1,2 —Fixture/Panel Schedules
E7.1,2 —Details

FIGURE 34.2 ■ Code system used to identify drawing groups.

APPLIANCE	ROOM	QUANTITY	MANUFACTURER	MODEL	SIZE	TYPE	COLOR	MATERIAL	ELECTRICAL REQUIREMENTS
Dishwasher	Kitchen	1	Maytag	DWU 7400	24"	4 cycle	White on white (w/w)	Steel porcelain	110 V
Garbage disposal	Kitchen	1	GE	GFC 1000	¾ HP	Automatic	—	Stainless steel	110 V
Refrigerator	Kitchen	1	GE	TBH 25 PAS	25 CF	Top mount	w/w	Steel porcelain	110 V
Cooktop	Kitchen	1	Amana	AKE 30	30"	Euro	White	Cast iron heat units	220 V
Oven	Kitchen	1	GE	JPP11 WP	30"	Built-in	w/w	Steel porcelain	220 V
Range hood	Kitchen	1	Kenmore	52 391	30"	Ducted	w/w	Steel procelain	110 V
Washer	Laundry	1	Amada	LW 4303	7 CF	Top load	White	Steel porcelain	110 V
Dryer	Laundry	1	Amada	LE 4407	Max cap	Front load	White	Steel porcelain	220 V

FIGURE 35.9 ■ Appliance schedule.

ROOMS	FLOOR: Asphalt tile	Ceramic tile	Cork tile	Vinyl tile	Wood strip—Oak	Woods sqs.—Oak	Plywood panel	Carpeting	Slate	CEILING: Diazo	Plaster	Wood panel	Acoustical tile	Exposed beam	WALL: Plaster	Wood panel	Wall paper	WAINSCOT: Wood	Ceramic tile	Paper	BASE: Asphalt tile	Stone veneer	Sheet vinyl	Wood	Rubber	Tile—Ceramic	Asphalt	REMARKS
Entry									✓	✓		✓				✓								✓				Diazo step covering
Hall		✓									✓				✓				✓					✓				
Bedroom 1					✓								✓		✓			✓								✓		Mahogany wainscot
Bedroom 2					✓								✓		✓		✓	✓								✓		Mahogany wainscot
Bedroom 3							✓	✓					✓			✓									✓			See owner for grade carpet
Bath 1		✓									✓				✓				✓							✓		Water-seal-tile edges
Bath 2	✓										✓				✓				✓							✓		Water-seal-tile edges
Kitchen			✓										✓		✓						✓		✓					
Dining			✓									✓	✓	✓	✓						✓		✓					
Living							✓	✓					✓											✓	✓			See owner for grade carpet

FIGURE 35.10 ■ General floor and wall covering schedule.

COVERING	ROOMS	MANUFACTURER	MATERIAL	UNIT SIZE	COLOR	STYLE	BORDER	GROUT	AREA (SQ. FT.)
Carpet	Liv./Din./Bfk.	Aladdin	Nylon 100%	—	White 717	Virtuous	—	—	80
Carpet	Brs./Halls/Stairs	Aladdin	Nylon 100%	—	Beige 618	Virtuous	—	—	84
Floor tile	Foyer	Fl Tile	Shellstone	16″ × 16″	Coral	Honed	—	White	154
Floor tile	Bath 2	Fl Tile	Ceramic	12″ × 12″	Coral	1212	—	Peach 610	70
Floor tile	M. bath	Watson	Ceramic	12″ × 12″	Cameo	8402	—	White	70
Floor tile	Lau.–Util.	Natura	Vinyl	12″ × 12″	Sand	8425	—	Sand	130
Floor tile	Kitchen	Saloni	Ceramic	12″ × 12″	Beige	03016	—	White	110
Wall tile	Bath 2	Watson	Ceramic	4″ × 4″	Peach	1428	LD 474	White	40
Wall tile	M. bath	Watson	Ceramic	8″ × 10″	Uropa	180RD	½″ × 8″ gold	White	42
Wall tile	Kitchen	Watson	Ceramic	6″ × 8″	3202	190RD	½″ × 8″ gold	White	34
Int. wall paint	All	S & W Co.	Latex	—	1108	Flat	—	—	All
Door & wood paint	All	S & W Co.	Oil	—	1641	Semigloss	—	—	All
Ceilings	All	S & W Co.	Latex	—	1004	Flat	—	—	All
Ext. Stucco	All	S & W Co.	Latex	—	SW 2060	Semigloss	—	—	All
Decks	All	Thompson's	Water Seal®	—	Clear	2-coat	—	—	All

FIGURE 35.11 ■ Floor and wall covering schedule with material specifications.

complete and covering materials can be finalized. See Figure 35.11. Different materials or colors are often used on the same floor, wall, or ceiling. In these cases, a color-coded floor plan or wall elevation is prepared to indicate the exact position and pattern of each material. Figure 35.12 shows a floor plan with different floor covering materials shown in colors that can be keyed to a floor covering schedule. Figure 35.13 shows an interior wall elevation with different materials labeled and also shown in color.

Often fabric, texture, color, or other features cannot be accurately described in words on a schedule. In these cases, swatches of material or paint color chips are often attached to a sample collage. These are either keyed to the floor plan by number or related directly with leaders. This identification method eliminates much misunderstanding in the final execution of the design and selection of materials.

Paint and Finishing Schedules

Many types and grades of finishes are used on the interior and exterior of a building. A finishing schedule condenses all this information into one chart. A preliminary interior finish schedule shows the color classification for each wall in each room for design purposes. See Figure 35.14. In a final and more complete schedule, each manufacturer's exact color code, numbers of coats, and surface preparation are added to the schedule.

Furniture and Accessory Schedules

Furniture and accessories are usually not included as part of a basic architectural design. When a totally integrated plan involves all aspects of the design, furniture schedules are developed and the information is keyed to a floor plan that includes furniture outlines.

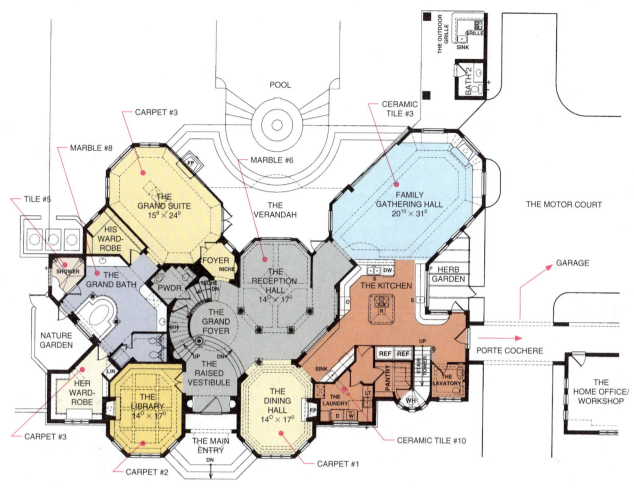

FIGURE 35.12 ■ Color coding of floor coverings. (*Home Design Services, Inc.*)

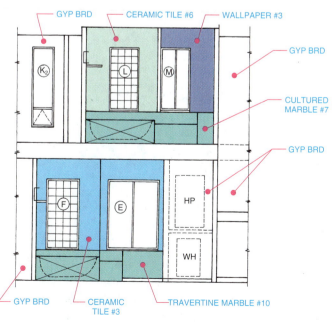

FIGURE 35.13 ■ Color coding of wall coverings.

Furniture and accessory schedules include all pertinent information relating to each piece of furniture. Numbers corresponding to the schedule numbers are placed on or near the appropriate item on the floor plan. Figure 35.15A shows an abbreviated floor plan with numbered furniture outlines keyed to the furniture schedule shown in Figure 35.15B. This system allows the designer and the contractor to find both the locations and specifications of each item. Photographs, catalog illustrations, or collage boards are sometimes keyed or attached to furniture schedules. See Figure 35.16.

Construction Component Schedules

Separate schedules can be developed for a wide variety of construction components. For example, schedules may be prepared for cabinetry, moldings and trims, beams, lintels, footings, and any other category that includes a number of different, but related items.

ROOMS	FLOOR			CEILING					WALL					BASE					TRIM					REMARKS
	Floor varnish	Unfinished	Waxed	Enamel gloss	Enamel semigloss	Enamel flat	Flat latex	Stain	Enamel gloss	Enamel semigloss	Enamel flat	Flat latex	Stain	Enamel gloss	Enamel semigloss	Enamel flat	Flat latex	Stain	Enamel gloss	Enamel semigloss	Enamel flat	Flat latex	Stain	
Entry			✓				Off Wht					Off Wht			Off Wht					Off Wht				Oil stain
Hall			✓					Lt Brn		Tan					Drk Brn					Drk Brn				Oil stain
Bedroom 1	✓						Off Wht					Off Wht				Grey				Grey				One coat primer & sealer—painted surface
Bedroom 2	✓						Off Wht				Lt Yel					Yel						Yel		One coat primer & sealer—painted surface
Bath				Wht					Wht					Lt Blue					Lt Blue					Water-resistant finishes
Closets	✓						Brn					Brn					Brn					Brn		
Kitchen			✓		Wht				Yel					Yel					Yel					
Dining			✓					Tan		Yel					Yel				Yel					Oil stain
Living		✓						Tan					Lt Brn					Lt Brn					Lt Brn	Oil stain

FIGURE 35.14 ■ Paint and finishing schedule.

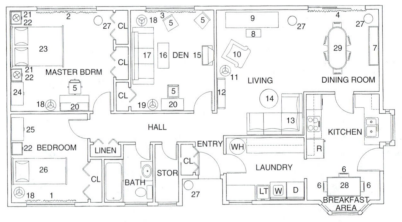

FIGURE 35.15A ■ Floor plan with furniture keyed to the schedule in Figure 35.15B.

MATERIAL LISTS

Data for building components are provided in detail on schedules. Building *materials* used for on-site construction, such as lumber, concrete, masonry, and steel, are shown on **material lists**. These lists include descriptions, quantities, and costs of materials. The creation of a material list from working drawings is known as a **takeoff**. Figure 35.17A shows a portion of a 48-page material list for the house plan featured in Chapter 34. Planting schedules (Figure 35.17B) are often called the *plant material lists* and show the botanical name, common name, quantity, and size plus the abbreviations used on the drawing. These are sometimes keyed to a landscape plan with numerals.

SYMBOL	ITEM	ROOM	LENGTH	WIDTH	HEIGHT	MATERIAL	COLOR	QUANTITY	MANUFACTURER	CAT. #	COST	REMARKS
1	Drapes	Bedroom	11'	—	7'	Cotton blend	Brown	1 set	Sears	CD101	$75	Lined
2	Drapes	M. bedroom	12'	—	7'	Cotton blend	Yellow	1 set	Sears	CD107	$85	Lined
3	Drapes	Den	7'	—	7'	Cotton blend	Yellow	1 set	Sears	CD106	$65	Lined
4	Drapes	Living/Dining	24'	—	7'	Acrylic	Brown pat.	1 set	Sears	CD203	$150	Lined
5	Chair	Mbr./Den	18"	18"	18"	Plastic	Brown	4	ID Furn. Co.	X117	$45 ea.	
6	Chair	Din./Kit.	18"	18"	18"	Oak	Natural	9	ID Furn. Co.	L217	$65 ea.	Oil finish
7	China cab.	Dining	6'	18"	5'-6"	Oak	Natural	1	Danish Furn. Co.	13712	$650	Oil finish
8	Piano bench	Living	33"	10"	18"	Mahogany	Brn. stain	1	Music Co. Inc.	23L19	$50	Piano finish
9	Piano	Living	5'-6"	2'-0"	5'-6"	Mahogany	Brn. stain	1	Music Co. Inc.	P17731	$1750	Piano finish
10	Up. wing ch.	Living	33"	30"	20"	Leather	Natural	1	Danish Furn. Co.	18979	$575	
11	Fl. lamp	Living	14" dia.	—	4'-6"	Metal/Cloth	Tan	1	Danish Furn. Co.	37111	$85	
12	Stereo	Living	30"	11"	5'-0"	Teak	Brown	1	Danish Furn. Co.	60701	$450	
13	Sofa/sec.	Living	14'	30"	18"	Velveteen	Red	3 pcs.	Danish Furn. Co.	42107	$1200	
14	Coffee tbl.	Living	30" dia.	—	15"	Teak	Natural	1	Danish Furn. Co.	77310	$110	Natural oil finish
15	Television	Den	21"	18"	30"	21" color	Brown	1	Sony	XL19	$675	
16	Coffee tbl.	Den	48"	15"	15"	Oak	Natural	1	Danish Furn. Co.	78325	$80	Natural oil finish
17	Sofa	Den	6'-6"	30"	18"	Cotton blend	Tan	1	Danish Furn. Co.	59781	$800	
18	Fl. lamp	Mbr./Br./Den	15" dia.	—	5'-0"	Wood/Cloth	Brown	3	Danish Furn. Co.	66362	$75 ea.	
19	Fl. lamp	Den	12" dia.	—	4'-6"	Wood/Plastic	Yellow	1	Danish Furn. Co.	65731	$50	
20	Desk	Mbr./Den	39"	18"	29"	Oak	Natural	2	Danish Furn. Co.	47772	$225 ea.	Natural oil finish
21	Nightstand	Mbr./Br.	18"	15"	24"	Oak	Natural	3	Danish Furn. Co.	64991	$45 ea.	Natural oil finish
22	Tbl. lamp	Mbr.	9" dia.	—	30"	Wood/Plastic	Brown	2	Danish Furn. Co.	65820	$35 ea.	
23	Full bed	Mbr.	6'-9"	46"	20"	Standard	—	1	Acme Bed Co.	AC12	$235	Box spring/Mat./Frame
24	Dresser	Mbr.	39"	20"	48"	Oak	Natural	1	Danish Furn. Co.	37452	$125	Dbl. dresser
25	Dresser	Bedroom 1	48"	20"	52"	Oak	Natural	1	Danish Furn. Co.	37471	$200	Triple dresser
26	Twin bed	Bedroom 2	6'-9"	42"	20"	Standard	—	1	Acme Bed Co.	AC08	$190	Box spring/Mat./Frame
27	Planter	Mbr./Liv./Den/Porch	10" dia.	—	12"	Terra cotta	Brown	4	Flowers Inc.	23FP	$10 ea.	
28	Table	Kitchen	36"	22"	30"	Teak	Natural	1	Danish Furn. Co.	17832	$110	Natural oil finish
29	Table	Dining	5'-0"	3'-3"	30"	Teak	Natural	1	Danish Furn. Co.	17876	$235	Natural oil finish

FIGURE 35.15B ■ Furniture and accessory schedule keyed from the floor plan in Figure 35.15A.

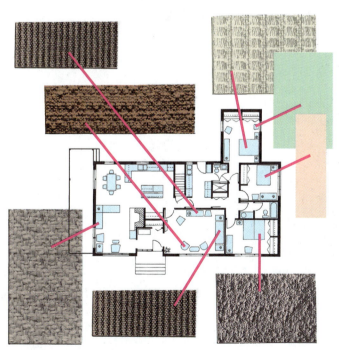

FIGURE 35.16 ■ Collage board related to floor plan furniture.

USING CAD

Spreadsheet Schedules

The *Bill of Materials* command is used to prepare a spreadsheet output in either Microsoft Excel or a generic comma-delimited format. This enables items to be entered and their cost, square footage, volume, model numbers, or quantity to be tracked by the *Attributes* command feature within the inserted blocks on a drawing. Once in a spreadsheet format, the values can be mathematically tabulated.

SPECIFICATIONS

Specifications (specs) are detailed written instructions describing the requirements for construction. Information not included in a drawing, schedule, material list, or legal document is included in the specifications.

MATERIAL LOCATION	SIZE		MATERIAL DESCRIPTION		UNIT	QUANTITY	MATERIAL COST	INSTALLATION COST	TOTAL COST	GRADE OPTIONS
GARAGE										
Bottom plate	2	4	Standard & better, pressure treated Douglas fir	R/L	Lin. feet	66	33.33	20.85	54.18	
Studs & cripples	2	4	#2 & btr., D. fir	10	Lin. ft.	160	58.62	40.44	99.06	
Studs & cripples	2	4	#2 & btr., D. fir	12	Lin. ft.	132	48.36	33.35	81.71	
Studs & cripples	2	4	#2 & btr., D. fir	14	Lin. ft.	126	46.16	31.84	78.00	
Fire blocking	2	4	#2 & btr., D. fir	R/L	Lin. ft.	24	8.79	7.75	16.54	
Trimmers	2	4	Stud grade D. fir (81 1/8")	8	Lin. ft.	16	5.86	5.05	10.91	
Top plate	2	4	#2 & btr., D. fir	R/L	Lin. ft.	138	50.56	43.59	94.15	
Ext. wall sheathing	4	8	1/2" APA RS 32/16 CDX ply	—	Sq. ft.	320	142.58	74.12	216.70	
Bottom plate	2	6	Std. & btr. PT D. fir	R/L	Lin. ft.	24	18.30	8.09	26.39	
Studs & cripples	2	6	#2 & btr., D. fir	10	Lin. ft.	110	56.64	31.65	88.29	
Studs & cripples	2	4	#2 & btr., D. fir	12	Lin. ft.	60	21.98	15.16	37.14	
Fire blocking	2	4	#2 & btr., D. fir	R/L	Lin. ft.	12	4.40	3.88	8.28	
Top plate	2	6	#2 & btr., D. fir	R/L	Lin. ft.	52	26.77	17.52	44.29	
Ext. wall sheathing	4	8	1/2" APA RS 32/16 CDX ply	—	Sq. ft.	160	71.29	37.07	108.36	
Bottom plate/Int. wall	2	4	Std. & btr. PT D. fir	R/L	Sq. ft.	14	7.07	4.43	11.50	

Credit: R.S. Means Co. and Home Planners, Inc.

FIGURE 35.17A ■ Partial material list for the plan shown in Chapter 34. (*R. S. Means Co.*)

KEY	BOTANICAL NAME	COMMON NAME	QTY.	SIZE
As	Alnus serrulata	Smooth Alder	1	12–18" #1 cont.
Aa	Aronia arbutifolia	Red Chokeberry	4	12–18" b/b
Bh	Baccharus halimifolia	Groundsel Tree	4	2 qt.
Co	Cephalanthus occidentalis	Buttonbush	14	12–18" b/b
Cl	Clethra alnifolia	Sweet Pepper Bush	3	18–24" b/b
Cr	Cornus amomum	Silky Dogwood	6	18–24" b/b
Cor	Cornus sericca 'Ruby'	R.R. Osier Dogwood	1	18–24" b/b
Hib	Hibiscus moschentos	Marsh Hibiscus	3	1 qt. pot
If	Iva frutenscens	High-Tide Bush	39	18–24" stab.
Jun	Juneus rocmerianus	Black Needle Rush	35	1 qt. pot
Kos	Kosteletzkya virginica	Seashore Mallow	5	1 qt. pot
Le	Leersia oryzoides	Rice Cutgrass	3	1–3/4" peat pot
Lc	Limonium Carolinianum	Sea Lavender	65	1 qt. pots

FIGURE 35.17B ■ Plant material list.

Purpose and Use

Specifications are written to ensure that a building will be constructed as designed and to prevent misunderstandings between the client, the architect, and the builder. Specifications are also used by lending institutions to evaluate the quality of construction. Along with drawings and schedules, specifications are used by contractors to make construction estimates and bids. Material lists are sometimes included in specifications.

It is important that the specifications agree with the information shown in the set of plans, because the specs and the drawings become a legal part of the construction contract. If a discrepancy does exist, then the information on the specs usually takes precedence over a drawing. The specs generally contain more product and material details than do drawings.

Specifications are intended to simplify and clarify, not to make the construction process longer and more difficult. Accurate specifications are critical to the contractor. If the contractor's materials or methods are inferior to those specified, the project will not be approved. If the contractor uses materials and methods that exceed the limits described in the specifications, material expense and labor costs may increase.

Specific product decisions are often not finalized at contract time. This is common for categories such as appliances, electrical fixtures, floor and wall coverings, plumbing fixtures, hardware, and fittings. If this is the case, a cost allowance is applied to each category. After the items are purchased, the contract amount is adjusted up or down.

Guidelines for Writing Specifications

Specifications entries are written to describe minimum acceptable standards or to describe specific end results. When writing specifications, these guidelines should be followed:

1. Use accepted terms and abbreviations consistently.
2. Be consistent in the format used.
3. Explain any unique or obscure terms.
4. Specify standard or manufactured components wherever possible.
5. Indicate "or equal" alternatives that do not sacrifice quality.
6. Use each manufacturer's latest catalog code numbers and titles.
7. Write brief simple sentences.
8. Use the accepted standards in each field for specifying materials.
9. Be very specific in writing criteria for work quality standards.
10. Number all pages and indicate end page.
11. Include a table of contents with headings and page numbers.
12. Capitalize proper names, rooms, legal documents, and material grades.
13. To define an obligation, use the term "shall," not "will."
14. Avoid specifying nonstandard or obscure sizes or materials.
15. Avoid repetition or overlapping.
16. Ensure specifications are reasonable, practical, and possible.
17. Be sure specs include only areas that are part of the contract.

Organization

The types of information contained in the specifications include the scope of the work, product descriptions, and methods of execution. The scope of the work describes the amount of work to be completed in each category of construction. Product descriptions, including schedules and material lists, detail the characteristics of every component and material that will become part of the finished structure. The methods of execution describe the approved methods of construction to be used. Specifications may also include restrictions on the use of some equipment or devices, delivery access routes, or processes that may potentially interfere with public safety or convenience during the construction process.

Of the many types of specifications forms, the most popular for small residences is the FHA-VA form. This form does not include scope-of-the-work or methods of execution. It is primarily a material list that includes several abbreviated schedules. It is, therefore, not comprehensive or expandable enough to use for larger residences or commercial buildings.

To ensure consistency among specifications, the American Institute of Architects (AIA) and the Construction Specification Institute (CSI) have approved and recommended the use of the **CSI format** for construction specifications. This flexible and expandable format organizes information under 16 major divisions with subdivisions included under each, as shown in Figure 35.18.

DIVISION	SECTION	DETAIL
1. General requirements		
2. Site work		
3. Concrete	06100 Rough carpentry	06131 Timber trusses
4. Masonry	06130 Heavy timber	06132 Mill-framed structures
5. Metals	06150 Trestles	06133 Pole construction
6. Wood and plastics	06170 Prefab. structural wood	
7. Thermal protection	06200 Finish carpentry	
8. Doors and windows	06300 Wood treatment	
9. Finishes	06400 Architectural woodworks	
10. Specialities	06500 Prefab. structural plastics	
11. Equipment	06600 Plastics fabrications	
12. Furnishings		
13. Special construction		
14. Conveying systems		
15. Mechanical		
16. Electrical		

FIGURE 35.18 ■ CSI Specification divisions with a sample section and detail.

USING CAD

Computer-Generated Specifications

Specifications are based on the content of the drawings and standards of construction. Master specifications programs are available from the Construction Specifications Institute in the CSI format. Specific data can then be inserted and a total set of specifications printed.

A five-digit numbering system is used to identify each specification heading. The first two digits represent the major division. The last three digits represent sections within each division. Sections are further divided into detail sections using a progression of the last three digits. In using this system only the headings that apply to each project are listed. If an entire division is not used, it is still included, but is listed with a note: "no listing in this division."

A complete set of residential specifications may cover 20 or more pages. Very brief specifications must be offset by including a great deal of detail in drawings and schedules. Specifications for large commercial or industrial designs may require hundreds of pages.

Reference Sources

Most specifications are developed using computer libraries of descriptions, material data, and cost information. This database must be updated with each project. Information to be included in sets of specifications and schedules can be found in manufacturers' literature, in

USING CAD

Accessing Resource Information on CAD

Manufacturers' websites contain valuable design information that can shorten and improve the design and drawing process. This information takes the form of drawings and data. DWG file drawings, when downloaded, can be altered to fit specific design needs. PDG or JPEG files cannot be modified and usually must be used exactly as received. However, JPEG files can be imported as stand-alone images such as details or pictures and used as guides similar to underlays. Using AUTOCAD a drafter can then "draw over" the JPEG to create a new drawing. This is typically done when an architect has hand-drawn a design which is scanned and imported into AUTOCAD as a JPEG. This version is then "drawn over" to create a computer generated drawing. The JPEG image can then be deleted.

Manufacturers websites are also good sources for detailed product specification options such as size, color, configuration, and materials.

trade association publications, or through sources such as **Sweets Catalog Files**. "Sweets files" are a compilation of information on a wide variety of architectural products from most major manufacturers. These files are organized under categories used by designers in all areas of construction. Some examples of areas are general building, engineering, interiors, home building, industrial building, electrical, kitchen, and bath design.

A CD-ROM product known as *Sweet-Source*® is an electronic supplement to Sweets files. This product compresses 250,000 pages of product information onto a single compact disc. The disc contains high-resolution color images and CAD interfacing capabilities.

Schedules and Specifications Exercises

1. What kinds of information are included on a schedule? How is a schedule referenced on a drawing?

2. Prepare a door and window schedule for the plan shown in Figure 34.7.

3. Name four uses for specifications.

4. Explain the differences between material lists, schedules, and specifications.

5. What are the three categories used in the CSI format? How would details be added to a list?

6. What division and section are represented by the CSI code 06133?

7. Name three reference sources for information that is included on specifications and schedules.

 8. Complete schedules for the home of your design. Include as much manufacturing information as you can obtain.

9. Make a specifications list for the home of your own design.

36 CHAPTER

Building Costs and Financial Planning

OBJECTIVES

In this chapter you will learn to:

- estimate building costs by the square-foot method.
- estimate building costs by cubic volume.
- make up a home budget.
- calculate monthly payments

TERMS

closing costs
down payment
escrow

foreclose
interest
mortgage

principal
soft costs

INTRODUCTION

The total cost of building and financing a new structure must be estimated before construction contracts are signed. Budgets are the framework within which designers must plan most buildings. A designer must work to create architectural plans that will provide the best facilities for the finances available.

BUILDING COSTS

Many factors affect building costs. The costs of real estate, labor, materials, and financing influence the total cost of any building.

The location of the site is extremely important. An identical house built on an identical lot can vary thousands of dollars in cost, depending on whether it is located in a city, in a suburb, or in the country. Labor costs also vary greatly from one part of the country to another. Material costs vary depending on the quality, unit cost, and quantity of materials used. These costs also depend on whether the materials are available locally or must be shipped in.

Rough estimates are usually developed using either the square-foot or cubic-foot method. When these meth-

ods are used, the cost of unique materials or components are added. Adjustments in cost may be required, such as inflationary costs and labor and material cost differences because of the geographical location.

Cost Estimating Methods

Math Connection

Different types of cost estimates are developed during various design phases. Very rough estimates are needed during the conceptual design phase. Very accurate and precise estimates are required on completion of working drawings. Accurate estimates are also needed during the bidding process.

Three basic methods of estimating building costs are the square-foot method, the cubic-foot method, and the labor-and-material-cost method. The square-foot and cubic-foot methods are quick rule-of-thumb methods. These two methods are not as accurate as itemizing the costs of labor and materials. The cost of the property, landscaping, drive, walkways, special features such as pools, and fees must be added to any of these estimates to arrive at a total cost.

Square-Foot Method

This method simply involves multiplying the square footage of a structure (sq. ft. = W × L) by a cost per square foot figure. The averages of square-foot costs

SQUARE-FOOT METHOD

Building Dimensions: W = 30'   L = 40'   H = 12'
Construction costs: $100 per square foot
Square footage: 30' × 40' = 1,200 square feet
Cost: 1,200 × $100 = $120,000

CUBIC-FOOT METHOD

Building Dimensions: W = 30'   L = 40'   H = 12'
Construction cost: $8.33 per cubic foot
Cubic volume: floor area (square feet) × height
Cubic volume: 1,200 × 12 = 14,400 cubic feet
Total cost: cubic volume × cost per cubic foot
Total costs: 14,400 × $8.33 = $119,952 = $120,000
(round off for estimate)

FIGURE 36.1 ■ Square-foot and cubic-foot methods of estimating building costs.

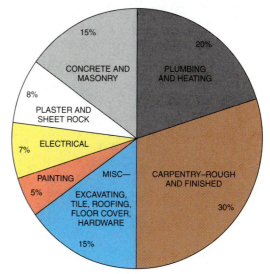

FIGURE 36.2 ■ Breakdown of construction labor and material costs.

according to building type are published for each geographical area.

If the cost of a building type is $100 per square foot, then the cost of a 3,000-sq.-ft. building would be $300,000 ($100 × 3,000). Figure 36.1 shows the calculation of building costs for a 30' × 40' cabin with 12' ceilings using both the square-foot and cubic-foot method.

If the building shape is not a square or rectangle, divide the space into rectangles, compute the square feet in each, and then add them to arrive at the total square feet.

Cubic-Foot Method

The cubic-foot method adds a height dimension factor to the square-foot method (cu. ft. = W × L × H). The height of ceilings is included in the computations. In buildings with average ceiling heights, cost estimates made by the two methods will vary little. See Figure 36.1. Much variation will occur where ceiling heights are high or rooms are double level. The estimate prepared using the cubic-foot method will be much more precise.

The cubic-foot method is especially helpful if different ceiling heights are used within a structure. The cubic footage of each area with the same ceiling height must first be calculated separately. These figures are then added to arrive at the total cubic feet. This method is particularly useful when estimating costs of multilevel structures.

Labor-and-Materials Method

The labor-and-materials method adds the total cost of all materials to the estimated cost of labor to arrive at a total

cost estimate. Approximately 40 percent of the cost of a residence is for materials. Labor costs account for another 40 percent. The remainder is used for real estate purchase, site development, and fees. These building costs are broken down into construction divisions. See Figure 36.2.

Labor costs are determined by multiplying the estimated number of hours needed to build a structure by the hourly rate. Hourly rates by trade are published for each geographical region. The number of hours needed by each trade for different building types and sizes are also available. Labor costs also vary greatly from one part of the country to another and from urban to rural areas. The amount of customized construction, such as nonmodular or unique component construction, affects the labor cost of a building most dramatically. Not only does customizing take additional time, but it requires the services of more highly paid technicians.

Building material costs vary greatly among regions. In some areas, brick is a relatively inexpensive building material. In other parts of the country, a brick home may be one of the most expensive. Climate also has an effect on the cost of building. In moderate climates, many costs are eliminated because large heating plants and frost-deep foundations are not needed. In other climates, maximum installation of heating and/or cooling equipment is needed.

One of the greatest variables affecting the cost of building materials is the use of standard sizes versus custom construction of components. For example, if built-in components are constructed on the site, the cost could easily double the price of preconstructed factory units. Designing the residence in modular units also enables the builder to use standard sizes of framing materials with a minimum of waste. A direct relationship exists

COST BY CATEGORY			PER SQ. FT. OF LIVING AREA		
			MAT.	LABOR	TOTAL
1	Site Work	Excavation for the slab and footings.		.49	.49
2	Foundation	Main house – 8″ and 12″ wide reinforced concrete stem wall on 16″ × 10″ reinforced concrete perimeter footings. Trench footings – 8″ wide reinforced concrete. Slabs – 4″ thick steel trowel finished reinforced concrete over compacted gravel.	4.78	5.58	10.36
3	Framing	Exterior walls – 2 × 6 studs, 16″ on center with ½″ plywood sheathing. Floor – 2 × 10 joists, 16″ on center with ¾″ plywood sheathing. Garage – 2 × 4 studs, 16″ on center with ½″ plywood sheathing. Roof – pre-engineered trusses and site cut rafters with ⅝″ plywood sheathing.	9.27	10.21	19.48
4	Ext. Walls	Stucco veneer siding over 15# felt building paper with R-19 and R-11 insulation. Wood awning, double-hung and fixed windows, and sliding glass patio doors.	6.72	2.46	9.18
5	Roofing	Field tile shingles over 30# felt roofing paper. Aluminum gutters, downspouts, drip edge, and flashings.	3.29	3.05	6.34
6	Interiors	Walls and ceilings – ½″ and ⅝″ taped and finished gypsum wallboard, primed and painted with one coat latex. Finger-jointed interior trim with one coat paint. Flooring – 68% carpet, 16% vinyl, and 16% ceramic tile.	9.94	10.36	20.30
7	Specialties	Plastic laminated particle board case kitchen cabinets and bathroom vanities with plastic laminate countertops. Washer, dryer, cooktop with hood, dishwasher, and refrigerator. One masonry fireplace.	3.43	.70	4.13
8	Mechanical	Oil-fired forced hot-air heat with central air conditioning. One full bath, one ½ bath, and a master suite with whirlpool, shower, and an outdoor spa. Stainless steel kitchen sink with disposal.	4.43	2.47	6.90
9	Electrical	200-amp service, branch circuit wiring with romex cable. Exterior and interior lighting fixtures, receptacles, and switches.	.88	1.53	2.41
10	Overhead Architect's Fees and Profit	Contractor's overhead and profit.	10.26	8.85	19.11
		Total Cost/sq.ft.	53.00	45.70	**98.70**

FIGURE 36.3 ■ Breakdown of labor and material costs by category. (*R.S. Means Co.*)

between the amount of on-site construction and the cost of factory-built modular components.

Specifying unique or exotic materials, such as rare stone or paneling, adds considerably to the cost. Material lists (takeoffs) are used to estimate the cost of material. The totals for each category are combined with the labor costs to arrive at the cost per square foot. See Figure 36.3. The cost of some materials and components includes labor costs. Therefore, care must be taken to ensure labor costs are not duplicated in the final sum.

Computerized Estimating

Estimating costs with computers is possible using estimating software that stores prices and descriptions of thousands of building materials and components. These figures can be combined with wage rates for each trade within each zip code area. Other factors such as building size, shape, and type can then be figured in to compute an estimate for the entire project. On CAD-generated drawings, precise takeoff cost estimates can easily be completed by scanning directly from the drawing into the computer database.

USING **CAD**

Computerized Estimating

Cost estimating software stores prices and descriptions of thousands of building materials and components. These figures can be combined with wage rates for each trade within every zip code. Other factors such as building size, shape, and type can then be used to compute an estimate for an entire project. If component symbols are coded when entered into a drawing, the final set of drawings can be scanned and entered into a computer database to automatically produce a precise takeoff estimate.

Minimizing Costs

The following construction methods and material utilization can greatly reduce the ultimate cost of a home:

1. Square or rectangular homes are less expensive to build than irregular-shaped homes.
2. Building on a flat lot is less expensive than building on a sloping or hillside lot.
3. Using locally manufactured or produced materials cuts costs greatly.
4. Using standard stock materials and standard sizes of components takes advantage of mass-production cost reductions.
5. Using materials that can be installed quickly cuts labor costs. Prefabricating large sections or panels saves time on the site.
6. Using prefinished materials saves labor costs.
7. Using prehung doors cuts considerable time from on-site construction.
8. Designing the home with a minimum amount of hall space increases the usable square footage and provides more living space for the cost.
9. Using prefabricated fireboxes for fireplaces cuts installation costs.
10. Investigating existing building codes before beginning construction eliminates unnecessary changes as construction proceeds.
11. Refraining from changing the design or any aspect of the plans after construction begins will keep costs from increasing.
12. Minimizing special jobs or custom-built items keeps costs from increasing.

13. Designing the house for short plumbing lines saves on labor and materials.
14. Proper insulation will save heating and cooling costs.
15. Using passive solar features such as correct orientation reduces future utility costs.
16. Including as many standard and/or modular components as possible will reduce expensive on-site labor time.

FINANCIAL PLANNING

Few people can accumulate enough money to pay for a home in one installment. Therefore, most home buyers pay a percentage of the cost of a home at the time of purchase. This initial or first payment is referred to as a **down payment** and is typically 5 percent to 20 percent of the cost of a home. The balance of the cost is usually acquired through a loan. After analyzing the construction costs for a home, the availability and costs of financing must be established. This includes information concerning mortgages, interest, taxes, insurance, and fees. Then a builder or owner can determine the financial feasibility of a building project.

Mortgage

A **mortgage** is an agreement (contract) for a loan. The purpose of a *mortgage loan* is to finance the purchase of a parcel of real estate, including any structures located on the property. Home mortgages usually require the loan amount, or **principal**, to be repaid over a period ranging from 15 to 30 years. A mortgage can be obtained from different sources: a mortgage company, a bank, a savings and loan association, or the seller of the property. Mortgage agreements are complex contracts that require the assistance of an attorney. Most mortgages allow the buyer (*mortgagor*) to pledge the property to the institution or individual providing the loan (*mortgagee* or *mortgage holder*). This means that if the loan payments are not made as agreed, the mortgage holder has the right to **foreclose**, or take possession of the property.

Interest

In addition to paying back the exact amount of the loan, the buyer must also pay additional amounts over the life of the loan for the use of the borrowed money. These charges are known as **interest**. The amount of interest paid is a percentage of the loan outstanding at a given point in time. Interest rates (percentages) depend on many factors, such as the availability of money and risk factors associated with the mortgagee and/or property. Interest rates can be established to remain unchanged over the life of the mortgage. These are referred to as *fixed rates*.

Interest rates can change, based on various fluctuations in lending institution rates. These rates, which change with time, are referred to as *adjustable*, *variable*, or *floating* rates. During the past several decades, the typical fixed interest rates for home mortgages have ranged from 5 percent to 15 percent. Interest rates can be a key factor in the ability to purchase a home.

The monthly payments and total amounts paid over the life of a loan differ depending on interest rates. Figure 36.4 shows how much a buyer would pay for a $200,000 home, assuming a 10 percent down payment with a $180,000 mortgage, based on various interest rates. As this table demonstrates, interest rates can have a significant impact on both the monthly mortgage payments and the total amounts paid over the life of the mortgage. Looking at the examples for a 30-year mortgage, monthly payments jump from $1,079 at 6 percent interest rate to $2133 at 14 percent. This represents a 98 percent increase. The higher payments can easily disqualify many potential buyers who might be able to afford monthly payments at lower interest rates. In the same example, total payments over the life of the loan climb from $388,509 at 6 percent to $767,797 at 14 percent. A higher down payment will reduce monthly payments and total payments over the life of the loan. For typical down payments of 20 percent or less, interest rates have the greatest influence on the ability to afford a home.

Taxes

Owners of property must also pay property taxes imposed by the various local governments. Property tax rates are usually expressed at a rate per $100 of assessed value of the property (*mil rate*). Rates vary greatly from location to location and depend on numerous factors. Property taxes can add a significant annual cost to property ownership. Residential property taxes are often several thousand dollars or more a year. For this reason, many mortgage holders collect property taxes on a regular monthly basis and hold these funds in special accounts called **escrow** accounts. This ensures that tax payments are made on a timely basis.

Insurance

The purchase of a home is a large investment and should be insured for the protection of the home buyer and the

LOAN DURATION	MONTHLY PAYMENT	TOTAL PAYMENTS
Interest rate of 6%		
15 years	1,519	273,410
30 years	1,079	388,509
Interest rate of 8%		
15 years	1,720	309,631
30 years	1,321	475,479
Interest rate of 10%		
15 years	1,934	348,172
30 years	1,580	568,666
Interest rate of 12%		
15 years	2,160	388,854
30 years	1,852	666,541
Interest rate of 14%		
15 years	2,397	431,484
30 years	2,133	767,797

FIGURE 36.4 ■ Mortgage costs at different interest rates.

mortgagee. Most mortgage agreements require that homeowner's insurance be maintained. Like property taxes, many mortgage holders collect insurance on a monthly basis, hold it in escrow accounts, and make payments directly to the insurer.

Insurance rates vary greatly depending on the cost of a home, location, type of construction, proximity to potential natural disasters, and the availability of firefighting equipment. Homes should be insured against fire, public liability, property damage, vandalism, natural destruction, and accidents to trespassers and workers.

Institutional lenders may also require insurance on their behalf, in the event a borrower is unable to pay a loan under the terms of the mortgage. This insurance is called *private mortgage insurance* (PMI). PMI is usually required when a home buyer makes less than a 20 percent down payment. Rates vary depending on the amount of the down payment, location of the property, and credit standing of the borrower.

Soft Costs and Closing Costs

The purchase of a home requires the buyer to incur many costs other than the actual cost of the property. **Soft costs** is a term used to describe the costs incurred prior to the actual construction of a home or other structure. Soft costs may include surveyor's fees, soil test fees, engineering studies, legal fees, and architectural fees. Architects usually work on an 8 percent commission for their design work and may charge fees as high as 10 percent of the project cost if they also supervise construction.

The term **closing costs** refers to costs paid by the purchaser before taking possession of a property. Closing costs don't include the actual cost of the property or buildings. Closing costs typically include legal fees, transfer taxes, title insurance, settlement of taxes and insurance with the seller, and financing fees.

Acquiring a mortgage may require the buyer to pay a fee up front to the mortgagee referred to as *points*. Points are a percentage of the mortgage amount, usually between 1 percent and 2 percent.

Budgets and Financing Qualifications

Math Connection

The ability to purchase a residence is dependent on the ability to acquire the down payment and also pay the soft costs and closing costs. The ability to make the required ongoing payments once a home is purchased is also vital.

Consider the purchase of a $200,000 home by a potential buyer who will make a 10 percent down payment as shown in Figure 36.5. The down payment in this case is 10 percent of $200,000, or $20,000. Assume there will be no

soft costs. The closing costs are estimated at 2 percent of the purchase price ($4,000). Points may be 1 percent to 2 percent of the mortage amount ($1,800 to $3,600). Finally, the lending institution will usually require the buyer to have money saved, which equals 2 or 3 months mortgage payments, taxes, and interest. Assuming a 10 percent interest rate on the mortgage, the buyer would need to have roughly $4,000 to $6,000 of savings after the home was purchased. Approximately $33,600 would be needed by the buyer to complete the purchase.

After gathering the necessary initial funds, potential home buyers must also determine their ability to meet the ongoing payments associated with owning the property. The monthly expenses associated with the home purchase include principal and interest of the monthly mortgage payment, plus property taxes and insurance (homeowner's insurance and, if applicable, PMI). Together, these items are referred to as "PITI" for principal, interest, taxes, and insurance. Most institutional lenders require that PITI not exceed 28 percent of the household income. They will also require that PITI plus other monthly loan payments not exceed 36 percent of the household income. Figure 36.6 shows the breakdown of monthly PITI expenses based on a $180,000 mortgage over 30 years at an interest rate of 10 percent.

Down Payment	
10% or $200,000 purchase price	$20,000
Closing Costs	
2% of purchase price	4,000
Points	
2% of mortgage amount (mortgage amount = $180,000)	3,600
Required Savings after Purchase	
Estimate of 3 months mortgage payments, plus taxes and insurance	6,000
Total Accumulated Funds Needed	$33,600

FIGURE 36.5 ■ Funds needed to purchase a $200,000 house.

Principal and interest (1)	$1,580
Taxes (2)	250
Insurance (3)	50
Total monthly PITI	$1,880
Minimum required monthly income (4)	$6,714

1) Monthly principal and interest on a $180,000 mortgage over 30 years at an interest rate of 10%.
2) Monthly tax escrow assuming a mil rate of $1.50 and an assessed value of $200,000.
3) Monthly homeowners insurance escrow assuming an annual premium of $600.
4) Monthly PITI cannot exceed 28% of monthly income.

FIGURE 36.6 ■ Monthly PITI requirements.

CHAPTER
36

Building Costs and Financial Planning Exercises

1. At $125 a square foot, how much will a 30′ × 40′ home cost?

2. At $15 per cubic foot, how much will a one-level, flat-roof 30′ × 40′ × 12′ home cost?

 3. Find the cost per square foot or cubic foot of a building in your area. Then compute the cost of the house you designed, based on this cost.

4. Refer to the numbers shown in Figure 36.6 and assume the potential buyer earns the minimum monthly income stated. How much could this potential buyer have in other monthly loan payments and still qualify for a mortgage?

5. What will be your total monthly PITI on a $300,000 home if you make a 10 percent down payment and pay interest of 10 percent on a 30-year mortgage for the balance? Your property taxes are $2,000 and your homeowner's insurance is $450.

6. What are property tax rates in your area? How did you find this information?

7. What is the single most significant factor affecting the ability to afford a home? Why?

8. What would your monthly mortgage payment be if you buy a 30′ × 40′ house for $150 per square foot? Your interest is 7 percent for 25 years. You make a down payment of 9 percent of the total price. Your closing costs total $2,000.

9. What is the cost of a home you could afford in your area if you were earning $40,000 a year? $75,000 a year? $100,000 a year?

CHAPTER 37

Codes and Legal Documents

OBJECTIVES

In this chapter you will learn to:

- consider building codes in architectural design.

- determine legal documents needed for building construction.

TERMS

agreements	bonds	contracts
bid forms	certificate of occupancy	deed
bids	codes	lien

INTRODUCTION

Different types of building codes are used to control the design and construction of buildings. Governmental laws are imposed on the design and construction of a building through the use of codes. Legal documents are created to protect the architect, builder, client, and the general public.

BUILDING CODES

Governmental laws that regulate building construction have existed for thousands of years. Building **codes** are collections of laws (codified) that ensure that minimum building standards are met. These laws are enacted to safeguard life, health, property, and the public welfare. The enforcement of codes has reduced the loss of life and property from earthquakes, storms, fires, and floods.

Many aspects of building construction are regulated and controlled by building codes. See Figure 37.1. Building codes include information related to building permits, fees, inspection, zoning, drawings, and legal documents required for approval. Building codes are presented through printed materials, charts, detailed drawings, and specification lists. These laws must be conformed to before beginning to design any structure.

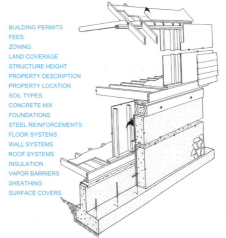

BUILDING PERMITS	WINDOW AREAS
FEES	WINDOW GLAZING
ZONING	DOOR TYPES
LAND COVERAGE	ROOM AREAS
STRUCTURE HEIGHT	FIREPLACES
PROPERTY DESCRIPTION	ELECTRICAL EQUIPMENT
PROPERTY LOCATION	SANITATION EQUIPMENT
SOIL TYPES	GAS LINES
CONCRETE MIX	PLUMBING LINES (WASTE/WATER)
FOUNDATIONS	HAVAC
STEEL REINFORCEMENTS	VENTILATION (FOUNDATIONS/ATTICS)
FLOOR SYSTEMS	ENGINEERING CALCULATIONS
WALL SYSTEMS	JOISTS
ROOF SYSTEMS	GIRDERS
INSULATION	BEAMS
VAPOR BARRIERS	HEADERS
SHEATHING	STUDS
SURFACE COVERS	

FIGURE 37.1 ■ Common construction components controlled by codes.

Types of Codes

Several types of codes are used to control building construction. These include zoning regulations, structural codes, site-related codes, restrictive codes, model codes, and health and safety codes.

Zoning Ordinances

These laws define and restrict the occupancy and use of buildings. Zoning laws may also prescribe the type, style, and location of structures on a site. (Refer back to Chapter 18 for a more detailed discussion of zoning.)

Structural Codes

Structural codes deal with the loading capacity of materials and the structural integrity of the construction. These codes involve detailed regulations related to excavations, foundations, floors, roofs, stairs, and bearing-wall construction. There are two general types of structural codes: performance criteria codes and specification criteria codes.

Performance-oriented codes do not limit or specify the use of most construction materials or methods. They establish only safety and performance requirements for the finished building.

Specification-type codes include very specific requirements for the use and location of materials and methods of construction. Options are included in many codes for some materials, but any substitutions must be equal to or better than what is specified by the code. Most building codes are specification-type codes. They are easier to enforce, and they provide a measure for compliance and evaluation.

The maximum permissible loads for each type of structure are listed in structural building codes. The sizes of structural members to support various loads are also included.

When material and size regulations are compiled for building codes, they are computed on the basis of maximum allowable loads. Engineers determine the correct size of construction members for supporting maximum loads. A safety factor is then added to the size of each material to eliminate the possibility of building failure.

Structural sizes required by building codes not only provide for the support of all weight in a vertical direction, but also allow for all possible horizontal loads, such as winds and earthquakes. Codes include precalculated sizes and types of materials used for studs, sheathing, roofing, foundations, and footings. Codes also include spans and spacing required for normal loads on rafters, joist beams, girders, and lintels.

For unique building sizes or designs, there are no predetermined code entries. In these cases, materials, sizes, and spans must be calculated by a licensed architect or engineer. See Appendix B, "Mathematical Calculations."

Site-related Codes

Many code items relate to the building site. These include specifications on soil percolation (drainage), soil support capabilities, test boring, and water runoff. Others cover environmental topics such as endangered species habitats, wetlands protection, and zoning density.

The location and size of buildings, setbacks, driveways, and road right-of-ways are also covered in these codes. When these items are not covered, a licensed landscape architect or civil engineer must approve the design.

Restrictive Codes

Some code items specify materials, processes, sizes, and locations of building materials within the structure that are prohibited. These restrictions are imposed because of potential structural or environmental safety problems.

Model Codes

To provide local authorities with a consistent and current source of code information, several organizations prepare model codes. These codes are not intended to be adopted intact. They are designed to be used as a base or guide from which local codes can be developed. The most widely used national codes are the National Building Code and the National Electrical Code. There are also specialized codes for plumbing, HVAC, and related trades. None of these codes are legally binding until passed into law by a municipality.

Health and Safety Codes

In addition to the safety prevention implications of good structural, mechanical, and electrical design, codes also contain sections dealing specifically with personal and public safety. These cover such areas as electrical hazards, swimming pool enclosures, scaffolds, elevators, number and sizes of exits, air and water pollution, health and disease prevention, and fire prevention and control. Fire codes define treatment of materials, building material size, sprinkler systems, escape routes, site security, and the functioning of alarm systems.

Building codes are extremely strict in stipulating the location, traffic patterns, and structural quality of buildings designed for public occupancy. Rigid code controls are placed on facilities such as factories and garages that may contain flammable substances or that may emit pollutants into the atmosphere. Structures in this risk category must also adhere to Environmental Protection Agency (EPA) standards in addition to local codes.

Code Compliance

Codes are enforced through a series of legal controls. *Building permits* are issued after working drawings and specifications are approved by a municipal building department. Then a *notice of commencement* must be completed when construction begins.

USING CAD

Downloading Code and Contract Information

Many municipalities store their codes, including building, zoning, planning, and land use regulations, on a website. Downloading access to these codes is restricted to licensed professionals who work in these areas. Master contract documents relating to the bidding of projects can also be downloaded by approved licensed professionals.

During the building process inspections are made of the setbacks, the foundation, electrical system, HVAC, plumbing, framing work, and insulation materials. After a final inspection, including a review of all change orders, a **certificate of occupancy** (CO) is issued. This certificate allows the building to be occupied and used.

Most sets of plans must contain a licensed architect's or engineer's seal. On larger projects, the seal of a licensed landscape architect may be required on all site design drawings. All plot plans or surveys registered with the local municipality must also contain the seal of a licensed surveyor. Licensed subcontractors are specified and are required by code on most construction jobs. Licensing requirements are usually specified in construction contracts.

Downloading Codes

Specific portions of most municipal codes can be downloaded online by architects, builders, or owner-clients. Detail drawings showing minimum structural requirements can often be downloaded and incorporated into a CAD-prepared set of plans.

LEGAL DOCUMENTS

A number of legal documents are used to protect the property and rights of architects, builders, and clients. Unlike codes, which apply to all projects within a given area, these legal documents are created individually for each building project. Documents include contracts, bonds, deeds, liens, and bids. Working drawings and specifications also become legal documents when tied to a contract.

Contracts

Legal agreements between two or more parties are known as **contracts** or **agreements**. Agreements are made between architects, builders, and owners. Architects may contract directly with an owner and/or with a builder. When an architect contracts only with an owner, the role of the architect and the builder must then be separately defined before the building process begins. Many architects supervise or monitor construction for a percentage of the construction cost.

Architects further contract with specialized consultants such as landscape architects, structural and mechanical engineers, or interior designers. Builders sign separate agreements with subcontractors. All of these documents must be consistent with the contents of the architect, owner, or builder contract.

Contracts include many conditions besides costs, such as schedules, bonds, and the responsibilities of each party. Restrictions on each party, criteria for handling changes, warranties, acts of God (unforeseeable events such as tornadoes), and cancellation conditions are also described in detail. In addition, contracts include very specific references to working drawings and specifications by citing the title, sheet numbers, and latest revision dates of each.

Bonds

A **bond** is a binding agreement that ensures that obligations are met. Two types of bonds used in construction are performance bonds and labor and materials bonds.

A *performance bond* is offered by the contractor and guarantees that the performance of responsibilities as builder will be in accordance with the conditions of the contract.

A *labor and materials bond* posted by the contractor guarantees that invoices for materials, supplies, and services of subcontractors will be paid by the prime contractor, according to the terms of the contract.

Deeds

A **deed** is a legal certificate of property ownership. Building code requirements are sometimes repeated in the contents of the property deed. These restrictions often describe the minimum setbacks, utility easements, building areas, and building types that are permitted. When building codes are updated, deeds are not necessarily changed. For this reason, many deeds may contain requirements that are no longer in existence. Designers must therefore check deeds for any obsolete restrictions and have the deed updated before proceeding with the design process.

Liens

A legal document used to take or hold the property of a debtor is known as a **lien.** Any architect, builder, subcontractor, consultant, or material supplier can petition the court to place a lien on real property for nonpayment of fees or material costs. A certificate of occupancy or bill of sale cannot be issued while a lien is in force.

Construction Bids

Bids are legal proposals to construct a project as defined in a contract, in specifications, and in a set of drawings. Contractors receive invitations to bid by mail, through newspaper advertisements, or through private resources such as the McGraw-Hill "Dodge Reports."

Construction bid forms announce the availability of documents and tell when these documents can be examined. They also provide for the resolution of questions, approval for submission of materials, specific dates for bid submission, and the form for preparing bids. The bid form includes specific instructions to the bidders, the price of the bid, allowable substitutions, restrictions, and the involvement of subcontractors.

A **bid form** is a letter sent from the owner or architect to bidders. The letter covers the following points: verification of receipt of all drawings and documents, specific length of time the bid will be held open, price quotation for the bid, and a listing of substitute materials or components if any item varies from specified requirements. When bidders sign a bid form, they are also agreeing to abide by all conditions of the bid, including the price, time, quality of work, and materials as specified in the contract documents and drawings.

CHAPTER

37 Codes and Legal Documents Exercises

1. What are building codes and what is their purpose?

2. Describe the two types of structural codes.

3. Determine any code restrictions in building the residence of your design in your community. Make and record needed changes.

4. What is the difference between codes and agreements or contracts?

5. Explain the purpose of bid forms.

APPENDIXES

The following appendixes contain information that does not align specifically with any single chapter. The scope of the following appendix material applies, in different degrees, to many subjects covered in this text. This includes information on architectural careers, mathematical calculations, architectural abbreviations, synonyms, and a glossary.

APPENDIX

Careers in Architecture and Related Fields

Career opportunities are available in a wide variety of fields related to architecture. These include careers in architectural design, engineering, and construction. Regardless of the specific field there is great reward in seeing your creations and efforts take form in structures that become part of the physical environment. A brief introduction to the major careers in architectural design, engineering, and construction is presented here.

Additional information concerning educational requirements, licensing, and professional organizations is included on the CD-ROM.

CAREERS IN ARCHITECTURAL DESIGN

Architectural design is a very broad field with many areas of specialization. Specialists contribute in different ways to the creation and completion of an architectural design. In a small firm, many tasks may overlap and be performed by one person—usually the architect. In larger firms a specialist can concentrate on one aspect of architectural design.

Professionals must first work under the direction of a licensed professional for several years and pass a state board examination before licensing. Once licensed, the professional can design and certify the safety of each design by affixing a license stamp on each approved drawing.

The following information applies to the major careers involved in the architectural design process.

Architects

Architects create original designs and are responsible for preparing or supervising the preparation of all construction drawings and documents. An architect is part artist, part engineer, and part executive. Architect Stanley Womack after 40 years of experience concludes that "a successful architect must be able to design highly functional, aesthetically pleasing structures and also possess skills in problem solving, planning, and management." A talent for creative design and skills in math and science are vital.

Architects must be state licensed to design buildings and sites that meet established standards of structural, environmental, and health safety. Licensing involves completion of a 5-year program in an accredited college of architecture, passing a rigorous licensing examination (AIA), and working under the direction of a licensed architect for several years.

Landscape Architects

Landscape architects design all aspects of a building site, from individual lots to massive buildings developments. Landscape architecture is a very diverse profession, involving the analyzing, designing, managing, and preserving of land. Landscape architects are involved with environmental hazards, civil engineering, building locations, street layout, earthwork, wildlife management, drainage, and horticulture. They must therefore be talented designers with skills in math and environmental science. Site plans, plot plans, plats, landscape plans, and details are prepared by landscape architects as covered in Chapters 13 and 18.

To become a landscape architect it is necessary to complete a bachelor's program at an accredited college of landscape architecture (ASLA), pass a state licensing examination, and work under a licensed landscape architect for several years.

City Planners

City planners plan for the overall growth and development of cities and communities. This includes planning for transportation, utilities, schools, housing, and shopping centers. City planners complete general master plans. Specific plans are then developed by architects, landscape architects, and engineers. City planners are usually architects or landscape architects who specialize in urban planning and development.

Architectural Designers

Architectural designers design buildings under the directions of an architect but are not licensed. They usually work on residential projects but may design small areas of a large project. Architectural designers must be versed in all aspects of basic architectural design including the use of CAD programs. Architectural designers are trained in colleges or technical institutes offering curricula in architectural and/or construction technology.

Architectural Drafters

Architectural drafters prepare all types of architectural drawings and documents. They draw with instruments and/or with a CAD system. Drafters work from sketches, partial or preliminary drawings, notes, and reference materials supplied by a supervising architect or architectural designer. Drafters are trained in vocational schools, technical schools, or colleges.

Architectural Detailers

Architectural detailers are drafters who specialize in preparing detail drawings such as sectional and framing drawings. They must be familiar with building codes, materials specifications, and construction methods. In small architectural offices the architectural designer, drafter, and detailer may be the same person, often the architect.

CAD Specialists

CAD specialists convert design sketches or manual drawings into CAD-generated drawings. Although architects, engineers, or drafters may prepare drawings directly on a CAD system, a CAD specialist inputs these drawings into a networked system and may further detail their work. A CAD specialist establishes and manages the information system of a firm and provides consistent standards for all drafters. More detailed drawings including 3D models and virtual reality programs are frequently developed by a CAD specialist. CAD specialists come from the ranks of architects or drafters who choose to specialize in CAD work.

Interior Designers

Interior designers deal with all aspects of a building's interior including wall construction, built-in components, and lighting systems. They also function as *interior decorators* in designing the surface treatments for walls, floors, and ceilings. This includes the design and selection of colors, fabrics, trim and moldings, window treatments, and furniture. Some specialize in designing baths and kitchens, which involves the selection and placement of fixtures, appliances, and cabinets. Interior designers receive their training as part of a college art program or in a professional school of interior design, vocational or technical school, or community college.

Lighting Designers

Lighting designers specialize in planning the illumination of building interiors, exteriors, and total sites. They must be familiar with lighting fixtures and devices and also understand electrical systems. On small projects, architects and/or interior designers design the lighting system; lighting designers are most often employed on large projects. Lighting designers are often interior designers or architectural designers who choose to specialize in lighting.

Architectural Illustrators

Architectural illustrators prepare renderings used for presentations. They need a knowledge of architectural drawings, commercial art techniques, 3D computer modeling, and often photography and photo retouching. Rendering is done manually, with a CAD system or a combination of both. The task of an illustrator is to present an architectural design as realistically as possible. Illustrators are trained as artists who specialize in architectural rendering. They hold 2- or 4-year art degrees or graduate from art institutes.

Architectural Model Makers

Architectural model makers construct scale models of buildings and sites to show the actual appearance of the finished project in three dimensions. Model makers must work from architectural drawings and be familiar with many model materials including wood, plastics, paper, and sheet metal. Model makers are drafters, finish carpenters, or cabinetmakers who specialize in model making.

Architectural Photographers

Architectural photographers specialize in producing photographs of structures or sites that reveal the best features of a design. Architectural photographers usually photograph finished buildings but models are also photographed and the photographs retouched to add realism to the environment. Photographers work closely with illustrators and work for publishers, newspapers, magazines, and real estate agencies. Many of the photographs in this text were prepared by professional photographers.

Architectural Educators

Architectural educators teach a variety of subjects relating to architecture in secondary school technology programs, vocational and technical schools, and community colleges. They also teach in colleges of architecture and engineering. A bachelor's degree and a state teaching certificate are required for teaching technology education. Vocational school teachers must hold a teaching certificate and a license in the subject taught. College professors usually hold a master's degree or doctorate.

RELATED ENGINEERING CAREERS

A wide array of engineering and engineering technology careers are directly related to architectural and construction practices. The number of engineering disciplines involved in an architectural project increases in proportion to the size and complexity of a project. Engineering careers that most directly apply to architecture and construction are introduced here. Education and licensing requirements are included with professional organizations in the accompanying CD-ROM.

Civil Engineers

Civil engineers design and calculate the structural requirements for large-scale projects such as roads, utilities, bridges, airfields, tunnels, harbors, sewage plants, and high-rise structures. Civil engineers also do site engineering and surveying. They must hold a degree in engineering and be licensed as a professional engineer to "stamp" public project drawings.

Structural Engineers

Structural engineers are civil engineers who specialize in designing the structural framework of a building. They calculate and specify the types and sizes of building materials required to make structures work. Structural engineers must possess a high level of compentency in math and science. Education and licensing requirements are the same as those for civil engineers.

Mechanical Engineers

Mechanical engineers design the mechanical components of a building. This includes the use of power and machines such as elevators, conveyer systems, and heating and cooling systems. To design components for pub-lic buildings, a degree in engineering and a state license are required.

Heating, Ventilating, and Air-Conditioning (HVAC) Engineers

HVAC engineers are mechanical engineers who specialize in designing HVAC systems. This includes oil or gas furnaces, cooling and refrigeration equipment, and air-movement systems. Education and licensing requirements are the same as those for mechanical engineers.

Surveyors

Surveyors prepare drawings that describe the size, position, and topography of a specific land area. These drawings (surveys) show the legal subdivision (plats) of properties. Surveyors also write legal descriptions of properties. This means surveyors must be proficient in geometry and site engineering and be dedicated to a high degree of accuracy in preparing documents. Surveyors are licensed civil engineers who specialize in surveying.

Electrical Engineers

Electrical engineers design and draw the electrical components of a building and site. This involves preparing electrical floor plans, details, and elevations that conform to electrical building codes. Details include lighting, power distribution, and special needs such as sound systems and computer networking. A licensed electrical engineer must hold a degree in electrical engineering and pass a state licensing board examination.

Acoustical Engineers

Acoustical engineers design shapes, materials, and devices used to control sound in a structure. This is critical in the design of concert halls and auditoriums. Noise suppression in shopping malls, medical facilities, and factories is also a vital part of this discipline. Acoustical engineers need a degree in electrical engineering with emphasis on physics, math, and architecture.

Structural Drafters

Structural drafters create the structural drawings of a building under the direction of an architect or civil engineer. Structural drafters usually prepare drawings of large structural steel buildings. A knowledge of steel structural members, fastening devices, and building methods is essential. Structural drafters are architectural drafters who specialize in steel-cage construction.

Estimators

Estimators compute the cost of a construction project by determining the amount and cost of all materials and labor based on a set of drawings and specifications. Estimators prepare *takeoffs,* which are compiled from the estimating information "taken off" the drawings and specifications. Estimators must have good math skills and be dedicated to accuracy in their work. Estimators should have an associate degree in construction technology and several years of experience in construction.

Specification Writers

Specification writers analyze construction drawings and documents and prepare written descriptions (specifications) of the materials, construction methods, finishes, and performances required in the building of a structure. Standards outlined by the Construction Specifications Institute (CSI) must be followed. A knowledge of construction materials, and processes, good analytical skills, and experience in construction are necessary. A degree in construction technology is recommended.

Military Engineers

Military engineers perform the same functions as their counterparts in the civilian construction world when not in a combat zone. Military schools teach the basics in all areas of construction in addition to combat engineering operations.

RELATED CONSTRUCTION CAREERS

Personnel on a construction site range from supervisors to highly skilled technicians to unskilled laborers. All but the unskilled work with architectural drawings in the conduct of their jobs. The careers that are served by vocational, technical, or civil technology programs are described here.

Contractors

Contractors are of two types: *general contractors* and *subcontractors.* General contractors are responsible for all phases of a construction project. General contractors must ensure that all schedules are met within the budget and according to the architectural plans and specifications. They do this by supervising subcontractors who are responsible for just one phase of construction such as carpentry, electrical, plumbing, site work, masonry, or HVAC. General contractors need years of experience in construction and come from the ranks of subcontractors, engineers, or architects. A general contractor must hold either a light construction or a commercial building license. Subcontractors need construction experience plus expertise and often licensing in their area of specialization. All must be familiar with all types of architectural plans and documents.

Carpenters

Carpenters are divided into two groups: *rough carpenters* and *finish carpenters.* Rough carpenters build the wood framing of buildings, which includes flooring, walls, and roofs. Finish carpenters complete the trim, molding, paneling, door and window installations, and on-site cabinetry. Skill in measuring and using hand and power tools is essential as is accurately reading architectural plans. Completion of a vocational, construction technology, or apprenticeship program is recommended.

Cabinetmakers, unlike carpenters, work mainly off site. Cabinets and other built-in wood components are built off site and installed during the final construction phase. This is because the precision of stationary woodworking power equipment is needed to ensure accuracy and save time in the construction and finishing of high-quality cabinetry. Training in all aspects of fine woodworking and the ability to read architectural detail drawings are essential.

Masons

Masons are divided into three categories: *stone masons, cement masons,* and *bricklayers.* Stone masons build stone walls, piers, patios, and walkways. They also build solid or stone veneer siding. Cement masons pour, shape, and smooth concrete surfaces to form slab foundations, walls, patios, walkways, and driveways. Bricklayers build chimneys, fireplaces, patio surfaces, and solid brick and brick veneer walls. Masons must be familiar with the workability and characteristics of each material. Completion of vo-tech and apprenticeship programs is recommended.

Plumbers and Pipefitters

Plumbers and pipefitters install piping systems. They must understand hydraulics and the types of piping systems needed for water, gas, steam, air, and waste disposal systems. Completion of vo-tech and apprenticeship programs is recommended.

Electricians

Electricians install the electrical system in a building including distribution panels and all wiring. They must read electrical plans to determine the type and location of all electrical fixtures and outlets. They also need to install specialized systems such as communication, security, entertainment, and computer networks. Electricians must be licensed.

HVAC Technicians

HVAC technicians install refrigeration, air flow, and heating equipment and systems. This includes creating and installing ductwork, furnaces, heat pumps, air exchangers, motors, filters, and purification systems. Vo-tech and apprenticeship programs are recommended.

Schedulers

Schedulers study the completion date goals and the set of specifications for a project. From this information they prepare timetables for the delivery of building materials and equipment to the building site. Schedulers plan for the delivery of materials prior to the date needed. If materials are delivered too soon, the site can become cluttered and work progress can be slowed. Also early delivery of materials increases the early outlay of money. If delivered too late, schedules and expensive labor time will be wasted. Schedulers must have experience in construction. Many come from the ranks of subcontractors or drafters.

Expediters

Expediters are troubleshooters who identify potential scheduling problems and communicate with material suppliers to ensure that delivery schedules are maintained. On small projects the job of an expediter is performed by the scheduler.

Marketing Specialists

Marketing specialists sell real estate in different forms. Some sell parcels in housing developments using models, plats, abbreviated floor plans, and pictorial drawings to communicate with customers. Some sell manufactured buildings to be shipped to any location. Real estate agents resell existing properties. All need a working knowledge of architectural styles, floor plan alternatives, and basic construction types.

Building Inspectors

Building inspectors check construction progress at regular intervals to determine adherence to codes and the approved set of plans and specifications. They perform a final inspection, which must be passed before the inspector approves issuing a certificate of occupancy (CO) as described in Chapter 37. Inspectors are usually builders or engineers with experience as general contractors, drafters, or specification writers.

APPENDIX B

Mathematical Calculations

Throughout the text mathematical formulas are presented where needed. Appendix B contains a summary of these for easy reference as well as other formulas not presented in the text. These are divided into arithmetic, structural, and geometric calculations.

ARITHMETIC CALCULATIONS

The majority of errors in architectural drawings are mathematical errors. Yet most calculations performed in the process of preparing architectural working drawings involve the arithmetic functions of adding, subtracting, multiplying, and dividing. Basic arithmetic follows that can be applied to the preparation of architectural drawings.

Conversions

Part of the construction industry uses customary foot, inch, and fractional dimensions. Other parts use decimal or metric dimensions. It is, therefore, important to be able to convert dimensions from one system to another.

To convert inch fractions to decimals, divide the numerator of the fraction by the denominator. For example, $7/8'' = 7 \div 8 = .875$.

To convert decimals to fractions, use the decimal number as the numerator and place a number 1 in the denominator, followed by as many zeros as there are places in the nominator. For example, $.3 = 3/10$, $.45 = 45/100$, and $.675 = 675/1000$. See Figure B.1.

To convert inches to millimeters (mm), multiply inches by 25.4 ($1'' = 25.4$ mm). For example, $1/2'' = .5''$; therefore, $1/2'' = 12.7$ mm ($.5 \times 25.4$). Likewise $6'$-$6'' = 6.5'$ ($78''$); therefore, $6'$-$6'' = 1981.2$ mm. Conversions commonly applied in architecture are listed in Figure B.2.

Adding Dimensions

Most rows of dimensions include both feet and inches and may include fractional inches. In adding rows of mixed numbers such as these, add the feet and inches separately, convert the inch total to feet and inches and then re-add the foot total. For example, in the plan

1/64	0.015625		33/64	0.515625
1/32	0.03125		17/32	0.53125
3/64	0.046875		35/64	0.546875
1/16	0.0625		9/16	0.5625
5/64	0.078125		37/64	0.578125
3/32	0.09375		19/32	0.59375
7/64	0.109375		39/64	0.609375
1/8	0.1250		5/8	0.6250
9/64	0.140625		41/64	0.640625
5/32	0.15625		21/32	0.65625
11/64	0.171875		43/64	0.671875
3/16	0.1875		11/16	0.6875
13/64	0.203125		45/64	0.703125
7/32	0.21875		23/32	0.71875
15/64	0.234375		47/64	0.734375
1/4	0.2500		3/4	0.7500
17/64	0.265625		49/64	0.765625
9/32	0.28125		25/32	0.78125
19/64	0.296875		51/64	0.796875
5/16	0.3125		13/16	0.8125
21/64	0.328125		53/64	0.828125
11/32	0.34375		27/32	0.84375
23/64	0.359375		55/64	0.859375
3/8	0.3750		7/8	0.8750
25/64	0.390625		57/64	0.890625
13/32	0.40625		29/32	0.90625
27/64	0.421875		59/64	0.921875
7/16	0.4375		15/16	0.9375
29/64	0.453125		61/64	0.953125
15/32	0.46875		31/32	0.96875
31/64	0.484375		63/64	0.984375
1/2	0.5000		1	1.0000

FIGURE B.1 ■ Decimal equivalents.

shown in Figure B.3, three dimensions combine to equal 13'-3": these are 5'-3", 5'-3", and 2'-9". To add these:

$$5'\text{-}3''$$
$$5'\text{-}3''$$
$$\underline{2'\text{-}9''}$$

Step 1: 12'-15" (Total)
Step 2: 15" = 1'-3"
Step 3: 12' + 1'-3" = 13'-3" (Total)

If fractions are involved, find the lowest common denominator in adding the fractions column. Then add the feet, inches, and fractions separately, convert the

MULTIPLY	BY	TO OBTAIN
Angles:		
Degrees	60	Minutes
Degrees	3600	Seconds
Area:		
Square feet	2.296×10^5	Acres
Square feet	144	Square inches
Square feet	3.587×10^8	Square miles
Square inches	6.452	Square centimeters
Square inches	6.944×10^3	Square feet
Square inches	645.2	Square millimeters
Square meters	2.471×10^4	Acres
Square meters	10.76	Square feet
Square miles	27.88×10^6	Square feet
Energy:		
British thermal units	2.928×10^4	Kilowatt-hours
Kilowatts	56.92	BTU/minute
Kilowatt-hours	3415	BTU
Kilowatt-hours	2.655×10^6	Foot-pounds
Watts	0.05692	BTU/minute
Watt-hours	3.415	BTU
Length:		
Centimeters	0.3937	Inches
Feet	30.48	Centimeters
Feet	0.3048	Meters
Inches	2.540	Centimeters
Kilometers	3281	Feet
Kilometers	0.6214	Miles
Meters	3.281	Feet
Meters	39.37	Inches
Miles	5280	Feet
Miles	1.609	Kilometers
Millimeters	0.03937	Inches
Volume:		
Board feet	144×1	Cubic inches
Cubic centimeters	3.531×10^5	Cubic feet
Cubic centimeters	6.102×10^2	Cubic inches
Cubic feet	1728	Cubic inches
Cubic feet	0.02832	Cubic meters
Cubic inches	5.787×10^4	Cubic feet
Cubic inches	1.639×10^5	Cubic meters
Cubic meters	35.31	Cubic feet
Cubic meters	61.023	Cubic inches
Weight/Mass:		
Pounds	16	Ounces
Tons (long)	2240	Pounds
Tons (metric)	2205	Pounds
Tons (short)	2000	Pounds

FIGURE B.2 ■ Conversion table.

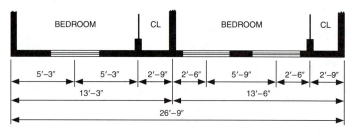

FIGURE B.3 ■ Adding dimensions.

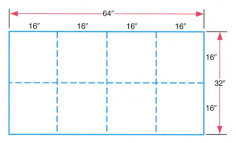

FIGURE B.4 ■ Modular fitting.

results to inches and feet and inches, and re-add. For example:

Step 1: $1'\text{-}7\frac{7}{8}'' = 1'\text{-}7\frac{14}{16}''$
$2'\text{-}8\frac{1}{4}'' = 2'\text{-}8\frac{4}{16}''$
$6'\text{-}10\frac{9}{16}'' = \underline{6'\text{-}10\frac{9}{16}''}$
Step 2: $9'\text{-}25\frac{27}{16}''$ (Total)
Step 3: $\frac{27}{16}'' = 27 \div 16 = 1\frac{11}{16}''$
Step 4: $25'' = 25 \div 12 = 2'\text{-}1''$
Step 5: $9' + 1\frac{11}{16}'' + 2'\text{-}1'' = 12'\text{-}2\frac{11}{16}''$ (Total)

Modular Measurements

In designing a structure, it is often necessary to establish modular units. To determine the number of modular units needed for a given area, divide the planned overall size by the modular unit size. For example, if a window opening dimension is planned to be approximately $36'' \times 66''$ and the modular unit is 16″, proceed as follows:

$$36'' \text{ (height)} \div 16'' = 2.25 \text{ units}$$

Use two 16″ units to cover 32″ height.

$$66'' \text{ (width)} \div 16'' = 4.125 \text{ units}$$

Use four 16″ units to cover 64″, as shown in Figure B.4. A $32'' \times 64''$ modular window is the nearest modular size window that will fit into the opening. The rough opening should be closer to $34'' \times 62''$ for the modular unit.

Construction Material Calculations

Construction materials are packaged or sold in a wide variety of set quantities and standard sizes. It is, therefore, important to convert the total amount of material needed into the standard marketing measures. Doing this often requires converting small volume units into larger units. See Figure B.5.

Board-Foot Measure

Lumber is purchased in bulk by the board foot (BF). One board foot is $1'' \times 12'' \times 12''$. To determine the number

MATERIAL	MEASUREMENT	PACKAGED OR SOLD BY
Cement	Bag (1 cubic foot)	Bag
Concrete	Cubic foot	Cubic yard
Sand	Ton/pounds	Ton/cubic yards
Blacktop	Square yards (after installation)	Cubic yard
Gravel	Size of stone ¼" to 3"	Ton/cubic yard
Concrete block	Standard height 7⅝" Standard length 15⅝" Widths 2" to 14"	Pallet, 100-block Piece
Mortar	Bag (1 cubic foot)	Bag (1 cubic foot)
Reinforcing rods	Width of bar ¼" to 1⅝"	Pounds
Welded wire mesh	Size of wire 1⁄16" to ¼" Size of grid	Roll 5 × 100 feet Sheet 5 × 20 feet
Asphalt, static	Pounds	Bucket or barrel
Plywood	Thickness of sheet ¼" to 1"	Sheet 4' × 8'
Paneling	Thickness of sheet ¼" to ⅜"	Sheet 4' × 8'
Lumber:	Board foot/piece	
2 × 4		Board foot/piece 6' to 20' lengths
2 × 6	Board foot/piece	6' to 20' lengths
2 × 8	Board foot/piece	6' to 20' lengths
2 × 10	Board foot/piece	6' to 20' lengths
2 × 12	Board foot/piece	6' to 20' lengths
Particleboard panels	Thickness of sheet 3⁄16" to 1"	Sheet 4' × 8'
Hardboard panels	Thickness of sheet ⅛" to ⅜"	Sheet 4' × 8'
Roof shingles	Square (100 square feet)	⅓ square foot per bundle or 33.3 square feet per bundle
Tar paper	Square feet	Rolls 3' × 100'
Rain gutters	Depth 4" to 5"	Lineal foot
Aluminum siding	Square foot	Square (100 square feet = 1 square)
Cedar siding	Square foot	Square foot
Drywall wallboard	Square foot	Sheet 4' × 8'
Sheathing panels	Square foot	Sheet 4' × 8'
Nails	Pennyweight	Box or keg 1 pound to 25 pounds
Pipe, copper	¼" to 2½" diameter	Lineal feet, 20' length
Pipe, plastic	¼" to 4" diameter	Lineal feet
Pipe, iron	¼" to 4" diameter	Lineal feet
Pipe, galvanized	¼" to 4" domestic ¼" to 36" industrial	Lineal feet Lineal feet
Pipe, cast iron	4" diameter	Lineal feet
Wire	Wire diameter/gauge	Roll 50' to 100'
Conduit	½" to 4" diameter	Lineal feet, 10' lengths

FIGURE B.5 ■ Standard marketing measures for construction materials. (*Leo Kwolek*)

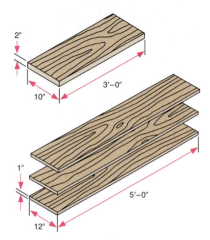

FIGURE B.6 ■ Board-foot measure.

of board feet in a given piece of lumber, multiply the thickness (in inches) × width (in inches) × length (in feet) ÷ 12:

$$BF = \frac{T'' \times W'' \times L'}{12}$$

The board feet in the top piece of lumber shown in Figure B.6 is calculated as follows:

$$\frac{2'' \times 10'' \times 3'}{12} = 5 \text{ BF}$$

When dealing with multiple pieces of the same size, either compute the BF for one piece and multiply it by the number of pieces or treat the entire package as one piece as follows:

$$\frac{1'' \times 12'' \times 5'}{12} = 5 \text{ BF each} \times 3 \text{ pieces} = 15 \text{ BF}$$

or

$$\frac{3 \times 1'' \times 12'' \times 5'}{12} = 15 \text{ BF}$$

STRUCTURAL CALCULATIONS

A well-designed building *should* be aesthetically pleasing but it *must* be structurally sound. A functional working knowledge of engineering mechanics and mathematics is a necessity in architectural design and drafting.

The field of engineering mechanics is divided into two main parts: *statics* and *dynamics*. Statics deals with objects at rest. Dynamics deals with objects in motion or potential motion. The principles of statics and dynamics are combined with information on the strength of con-

struction materials in the design of structures that will withstand the forces of nature and human use.

An effective structural design is neither underdesigned or overdesigned. When structures are underdesigned, systems fail and buildings sag or collapse. However, if a building is overdesigned, materials are wasted and costs increase greatly. Thus the primary task of the structural designer is to design all components to meet and/or exceed the safety factor without excessively increasing the cost of the building.

Designers and drafters must determine the most appropriate materials, sizes, spacing and construction methods for an architectural design. Determining building loads is the first step in this process.

Building Loads

The weight of all movable items, such as the occupants, wind, snow, and furniture, make up the *live load*. The weight of all the materials used in the construction of a building, including all permanent structures and fixtures, make up the *dead load* of a building. The total weight of the live load plus the weight of the dead load is called the *building load*. (Building loads were introduced in Chapter 22.) Loads are measured in pounds per square foot (lb/ft^2 or PSF). See Figure B.7.

Live Loads Live loads act on floors, walls, ceilings, and roofs. Floor live loads include persons and furniture. Live loads acting on walls are wind loads. Materials differ greatly in their ability to withstand loads. See Figure B.8. Lateral loads from earthquakes must also be considered. Live loads acting on roofs include wind and snow loads. Roof loads are comparatively light, but vary according to

CONSTRUCTION AREA	LIVE LOAD PSF	DEAD LOAD PSF
Roof	30	20
Ceiling joists attic/heavy storage	30	20
Ceiling joists attic no floor	0	10
Ceiling joists attic habitable rooms	30	10
Floors of rooms	40	20
Floors of bedroom	30	20
Partitions	0	20

FIGURE B.7 ■ Typical loads for a two-level frame construction building.

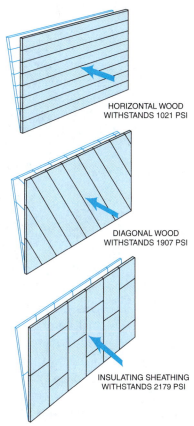

FIGURE B.8 ■ Ability of sheathing to resist horizontal (wind) loads in pressure (pounds per square inch).

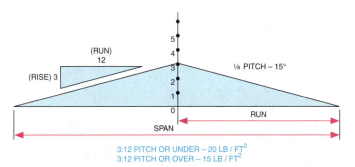

3:12 PITCH OR UNDER – 20 LB / FT^2
3:12 PITCH OR OVER – 15 LB / FT^2

FIGURE B.9 ■ Typical roof design requirements set in building codes.

the pitch of the roof. Flat roofs offer more resistance to dead loads than do pitched roofs. To meet building codes, low-pitched roofs (those below 3/12 pitch) must often be designed to support wind and snow loads of 20 lb/ft^2. See Figure B.9. The wind load increases as the pitch increases. The snow load increases as pitch decreases. High-pitched roofs (those over 3/12 pitch) need be designed to support live loads of 15 lb/ft^2.

Dead Loads Structural calculations are used to determine the size and type of materials to be used in founda-

tions to support all dead loads. The size, spacing, and type of materials used in walls that support the roof load must also be determined. Dead loads become greater from the top of the structure to the footing. For example, the load on an attic may be only 25 lb/ft^2, while the load on the first floor of the same building may be 45 lb/ft^2. The typical floor load for an average residential room usually varies from 30 to 40 lb/ft^2. Loads vary depending on the materials used in construction. See Figure B.10.

Tributary Areas

The *tributary area* of a structural member is the area of weight transmitted to a vertical support. The total load of the building is transmitted into the ground from footings and piers through a series of tributary areas. Because footings are continuous, load distribution is calculated as one lineal foot, as shown in Figure B. 11.

The tributary area over each pier extends one-half the distance to the next structural support members. For example, pier B in Figure B.12 is 9′-0″ from the nearest pier on both sides. The tributary area for pier B, therefore, extends one-half the distance to piers A and C (4′-6″) and one-half the distance to each foundation wall (5′-0″). The tributary area is, therefore, 4′-6″ + 4′-6″ by 5′-0″ × 5′-0″, or 9′-0″ × 10′-0″, or 90 ft^2. (See p. 688 for calculating the tributary area of girders and beams.)

Soil Conditions

The safety of a structure also depends on the type of soil and the load it can support. It is important that the weight of the structure not exceed the safe load capacity of the soil. See Figure B.13. Footing sizes are critical because they spread the weight of the structure on the soil. Footing area (FA) is calculated as follows (Figure B.14):

Pier footing calculation:
Structure load = 5,000 lb/ft^2
Soil-bearing capacity = 6,000 lb/ft^2 (given)

$$\text{Footing area} = \frac{\text{structure load}}{\text{soil} - \text{bearing capacity}}$$

$$FA = \frac{5000}{6000} = 0.83$$

(Convert sq. ft. to sq. in.)
$$FA = 0.83 \text{ ft}^2 \times 144^* = 119.52 \text{ in}^2$$

(Find sq. root: $\sqrt{119.52} = 10.9$)
$$FA = 11'' \times 11'' = 121 \text{ in}^2$$

(closest square to 119.52 in^2)

* 144 sq. in. = 1 sq. ft.

MATERIALS	WEIGHT	MATERIALS	WEIGHT
Roofs		**Woods** (12% moisture content, lb/ft³)	
built-up roofing, 3-ply and gravel	6 PSF	cedar	22 PCF
rafters, 2 × 4 at 16″ oc	2 PSF	douglas fir	34 PCF
rafters, 2 × 6 at 16″ oc	2.5 PSF	maple	42 PCF
rafters, 2 × 8 at 16″ oc	3.5 PSF	oak	47 PCF
sheathing, ½″ fiberboard	0.75 PSF	pine	27 PCF
sheathing, ½″ gypsum	2 PSF	poplar	28 PCF
sheathing, 1″ wood	3 PSF	redwood	28 PCF
shingles, asbestos	4 PSF	**Glass**	
shingles, asphalt	2 PSF	double strength, 1–8″	1.5 PSF
shingles, wood	2.5 PSF	insulating plate, ⅛″ with air space	3.25 PSF
skylights, glass 7 frame	11 PSF	glass block, 4″	20 PSF
tile, cement	15 PSF	plate glass, ¼″	3.25 PSF
tile, mission	13 PSF	plastic, ¼″ acrylic	1.5 PSF
Ceilings		**Masonry**	
acoustical tile, ½″	0.8 PSF	brickwork, 4″	35 PSF
plaster on wood lath	8 PSF	concrete wall, 6″	75 PSF
suspended metal lath and cement plaster	15 PSF	concrete wall, 8″	100 PSF
suspended metal lath and gypsum plaster	10 PSF	poured concrete	150 lbs./cu. ft.
Walls		concrete block, lightweight 4″	22 PSF
brick wall, 4″	35 PSF	concrete block, lightweight 6″	31 PSF
brick wall, 8″	74 PSF	concrete block, stone 6″	50 PSF
brick (4″) on 6″ concrete block	80 PSF	concrete block, stone 8″	58 PSF
brick (4″) on wood frame with sheathing & plaster	45 PSF	facing tile, 2″	16 PSF
building board, ½″	0.8 PSF	facing tile, 4″	30 PSF
concrete block wall, 6″	40 PSF	facing tile, 6″	41 PSF
concrete block wall, 8″	55 PSF	marble, 1″	13 PSF
concrete wall, 8″	100 PSF	slate, 1″	14 PSF
concrete wall, 10″	125 PSF	stone, 1″	12 PSF
gypsum block, 2″	9.5 PSF	tile, structural clay, 4″ hollow	23 PSF
gypsum block, 4″	12.5 PSF	tile, structural clay, 6″ hollow	33 PSF
gypsum wallboard	2.5 PSF	tile, structural clay, 8″ hollow	42 PSF
plaster, ½″	4.5 PSF	**Metals** (lb/ft³)	
plywood, ½″	1.5 PSF	aluminum, cast	165 PCF
tile, facing, 2″	15 PSF	brass, yellow	528 PCF
tile, glazed ⅜″	3 PSF	bronze, commercial	552 PCF
wood siding, 1″	3 PSF	copper, cast or rolled	556 PCF
wood stud wall	5 PSF	iron, cast	450 PCF
wood stud wall, plastered one side	12 PSF	iron, wrought	485 PCF
wood stud wall, plastered two sides	20 PSF	lead	710 PCF
Floors		steel, rolled	490 PCF
cement finish 1″	12 PSF	tin, cast or hammered	459 PCF
clay tile on 1″ mortar base	23 PSF	steel beam S7 × 15.3	15.3 lbs./lineal foot
concrete slab, 4″	48 PSF	**Soil, Sand, & Gravel** (lb/ft³)	
hardwood flooring, 25/32″	4 PSF	clay, damp	110 PCF
floor joist, 2 × 8, 16″ oc/subflooring	6 PSF	clay, dry	63 PCF
floor joist, 2 × 10, 16″ oc/subflooring	6.5 PSF	clay and gravel, dry	100 PCF
floor joist, 2 × 12, 16″ oc/subflooring	7.0 PSF	earth, loose, dry	76 PCF
marble on 1″ mortar base	28 PSF	earth, packed, dry	95 PCF
plywood subflooring, ½″	1.5 PSF	earth, moist, loose	78 PCF
quarry tile, ½″	6 PSF	earth, moist, packed	96 PCF
terrazzo, 2″	25 PSF	earth, mud, packed	115 PCF
vinyl asbestos tile, ⅛″	1.3 PSF	sand/gravel, dry, loose	110 PCF
4 × 6 girder/post	4 PSF	sand/gravel, dry, packed	120 PCF
Insulation		sand/gravel, wet	120 PCF
bats, blankets, 3″	0.5 PSF		
boards, vegetable fiber	1.7 PSF		
cork board, 1″	0.6 PSF		
foam board, 1″	0.1 PSF		

FIGURE B.10 ■ Building material loads.

Perimeter foundation wall footing calculation:

Structure load $\quad = 4{,}500\ \text{lb/ft}^2$

Soil-bearing capacity $\quad = 6{,}000\ \text{lb/ft}^2$ (given)

$$\text{Footing area} = \frac{\text{structure load}}{\text{soil-bearing capacity}}$$

$$\text{FA} = \frac{4{,}500}{6{,}000}$$

$$\text{FA} = 0.75\ \text{ft}^2$$

$$\text{FA} = 0.75\ \text{ft}^2 \times 144 = 108\ \text{in}^2$$
(Width is number of in² per lineal foot)

$$\text{FA} = \frac{108\ \text{in}^2}{12}$$

$$\text{FA} = 9''\ \text{wide} \times 1\ \text{lineal foot}$$

Use next standard size—12″.

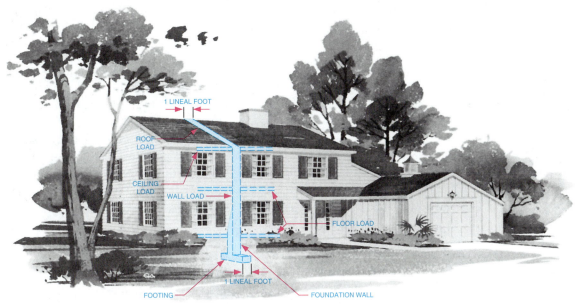

FIGURE B.11 ■ Loads are calculated on one lineal foot of footing.

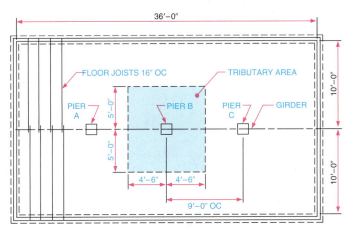

FIGURE B.12 ■ Foundation tributary area.

Roof Members

The size of all structural members used for roof framing depends on the combined loads bearing on the member and the spacing and span of each member. If the load is increased, either the spacing or span must be decreased or the size of the member increased to compensate for the increased load. Conversely, if the size of a member is decreased, the members must be spaced more closely or the span must be decreased. If the length of the span is increased, the size of the members must be increased or the spacing made closer. To compute the most appropriate size of roof rafter for a given load, spacing, and span, follow these steps:

STEP 1 To determine the total load per square foot of the roof space, add the live load and the dead

SOIL TYPE	SAFE LOAD PSF
Soft clay; sandy loam	1,000
Firm clay; sand and clay mix; fine sand, loose	2,000
Hard dry clay; fine sand, compact; sand and gravel mixtures; coarse sand, loose	3,000
Coarse sand, compact; stiff clay; gravel, loose	4,000
Gravel; sand and gravel mixtures, compact	6,000
Soft rock	8,000
Exceptionally compacted gravels and sands	10,000
Hardpan or hard shale; sillstones; sandstones	15,000
Medium hard rock	25,000
Hard, sound rock	40,000
Bedrocks, such as granite, gneiss, traprock	100,000

FIGURE B.13 ■ Safe soil loads.

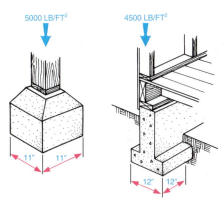

FIGURE B.14 ■ Footing sizes required by loads.

load. See Figure B.15. Figure B.16 shows the components of a roof system that are used in the calculation of roof loads as follows:

Component	Dead load (lb/ft^2)
2 × 6 roof rafter	3
3 layers felt, tar, and gravel (or shingles)	5
1″ insulation	1
1/2″ plywood sheathing	2
5/8″ gypsum board ceiling	2
2 × 6 ceiling joist	3
Dead load	16
Live load	20
Total load	36 lb/ft^2

In roof design, the live load is the safety factor allowing for weather conditions and movable objects the structure must support.

STEP 2 Loads are expressed in lineal feet. See Figure B.17. To determine the load per lineal foot on each rafter, multiply the load per square foot by the spacing of the rafters. If rafters are spaced at 12″ intervals, then the load per square foot and the load per lineal foot will be the same. However, if the rafters are spaced at 16″ intervals, then each lineal foot of rafter must support 1 1/3 of the load per square foot. See Figure B.18.

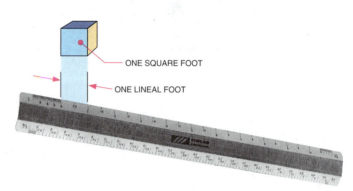

ONE SQUARE FOOT
ONE LINEAL FOOT

FIGURE B.17 ■ Comparison of a square foot and a lineal foot.

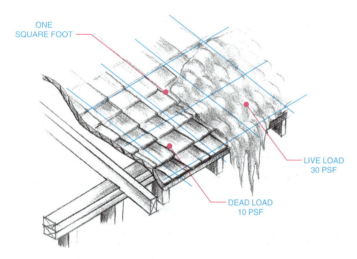

ONE SQUARE FOOT

LIVE LOAD 30 PSF

DEAD LOAD 10 PSF

DEAD LOAD PLUS LIVE LOAD EQUALS THE TOTAL LOAD
10 PSF + 30 PSF = 40 PSF

FIGURE B.15 ■ Determining the total roof load per square foot.

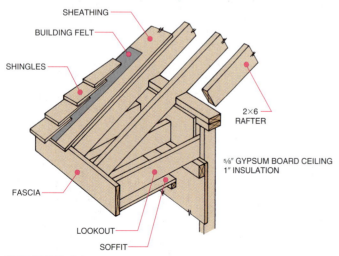

SHEATHING
BUILDING FELT
SHINGLES
2×6 RAFTER
5/8″ GYPSUM BOARD CEILING
1″ INSULATION
FASCIA
LOOKOUT
SOFFIT

FIGURE B.16 ■ Roof material loads.

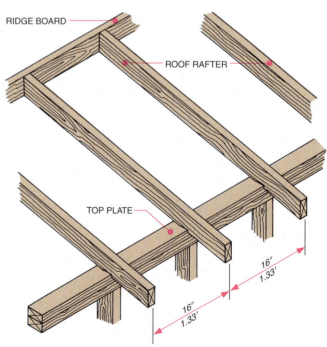

RIDGE BOARD
ROOF RAFTER
TOP PLATE
16″ 1.33′
16″ 1.33′
16″ 1.33′

TOTAL LOAD × RAFTER SPACING (FT) = LOAD/LINEAL FT
40 LB/FT^2 × 1.33′ = 53 LB/LINEAL FT

FIGURE B.18 ■ Determining the load per lineal foot on each rafter.

STEP 3 To find the total load each rafter must support, multiply the load per lineal foot by the length of the span in feet. See Figure B.19.

STEP 4 To compute the bending moment in inch-pounds, multiply total load × rafter span × 12. Divide this figure by 8 (a constant). The *bending moment* (BM) is the force needed to bend or break the rafter. When the length of the span in pounds is multiplied by the length of the span in feet, the result is expressed in foot-pounds. The span must be multiplied by 12 to convert the bending moment into inch-pounds. See Figure B.20. The bending moment (in inch-pounds) of a structural member (in pounds) equals the load (in pounds) times the length (in feet) times 12 divided by 8:

$$\text{Bending Moment} = \frac{L \times D \times 12}{8}$$

The bending moment of a beam is calculated in the same way. See Figure B.21.

$$\text{Bending Moment} = \frac{2{,}000 \text{ lb} \times 10' \times 12}{8}$$

$$BM = \frac{240{,}000}{8}$$

$$BM = 30{,}000 \text{ in-lb}$$

STEP 5 Set up the equation to determine the resisting moment. The *resisting moment* is the strength or rafter resistance the rafter must possess to

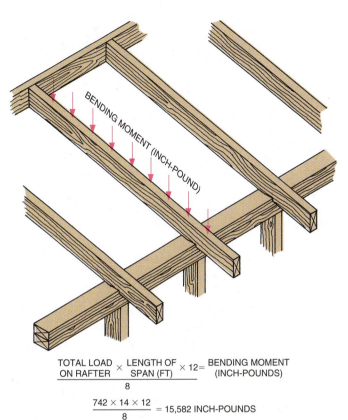

$$\frac{\text{TOTAL LOAD}}{\text{ON RAFTER}} \times \frac{\text{LENGTH OF}}{\text{SPAN (FT)}} \times 12 = \frac{\text{BENDING MOMENT}}{\text{(INCH-POUNDS)}}$$

$$\frac{742 \times 14 \times 12}{8} = 15{,}582 \text{ INCH-POUNDS}$$

FIGURE B.20 ■ Computing the bending moment.

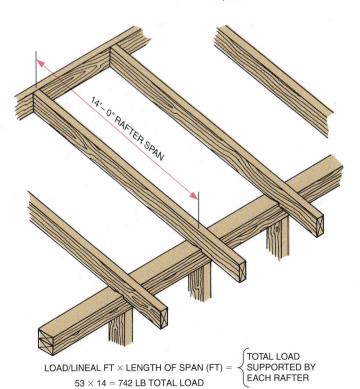

LOAD/LINEAL FT × LENGTH OF SPAN (FT) = TOTAL LOAD SUPPORTED BY EACH RAFTER

53 × 14 = 742 LB TOTAL LOAD

FIGURE B.19 ■ Determining the total load each rafter must support.

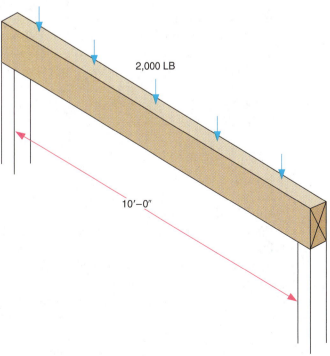

2,000 LB

10'–0"

FIGURE B.21 ■ Bending forces on a horizontal member.

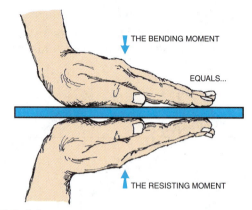

FIGURE B.22 ■ The resisting moment must equal or be greater than the bending moment.

withstand the force of the bending moment of the rafter. See Figure B.22. If the resisting moment is less than the bending moment, the member will bend or break. The resisting moment is determined by multiplying the fiber stress by the rafter width by the rafter depth squared. This figure is divided by 6 (a constant). Rafter widths should be expressed in the exact dimensions of the finished lumber. For example, if a rafter is a standard size $2'' \times 4''$, its actual width is $1\ 1/2''$ and its depth is $3\ 1/2''$. See Figure B.23.

The *fiber stress* is the tendency of the fibers of the wood to bend and become stressed as the member is loaded. Fiber stresses range from 1,750 lb/in² for southern dense pine select to 600 lb/in² for red spruce. Dense Douglas fir and southern pine have average fiber stresses of 1,200 lb/in². The rafter depth is squared, since the strength of the member increases by squares. For example, a rafter 12″ deep is not 3 times as strong as a rafter 4″ deep. It is 9 times as strong.

STEP 6 Because the bending moment is compared with the resisting moment, the formulas can be combined, to find rafter depth, as shown in Figure B.24. The formula can then be followed for any of the variables, preferably for the depth of the rafter, since varying the depth will alter the resisting moment more than any other single factor. An example of figuring the bending moment and the resisting moment together for a $2'' \times 6''$ beam is shown in Figure B.25. Computations are summarized as follows:

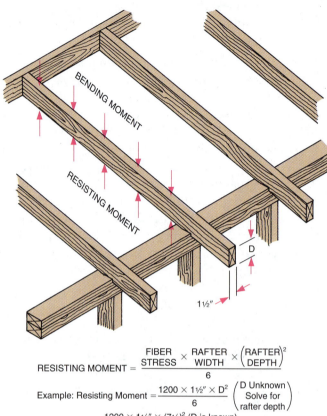

$$\text{RESISTING MOMENT} = \frac{\text{FIBER STRESS} \times \text{RAFTER WIDTH} \times \left(\text{RAFTER DEPTH}\right)^2}{6}$$

$$\text{Example: Resisting Moment} = \frac{1200 \times 1\frac{1}{2}'' \times D^2}{6} \quad \left(\begin{array}{l}\text{D Unknown} \\ \text{Solve for} \\ \text{rafter depth}\end{array}\right)$$

$$RM = \frac{1200 \times 1\frac{1}{2}'' \times (7\frac{1}{2})^2}{6} \quad \text{(D is known)}$$

$$RM = \frac{1200 \times 1.5 \times 56.25}{6}$$

$$RM = 16,875$$

FIGURE B.23 ■ Determining the resisting moment.

Bending moment	=	Resisting moment
$\dfrac{3,000 \times 10 \times 12}{8}$	=	$\dfrac{1,500 \times 1.5 \times 5.5^2}{6}$
$\dfrac{360,000}{8}$	=	$\dfrac{68,062.5}{6}$
45,000	=	11,343.75

The bending moment is greater than the resisting moment. Increase the beam size to 4×8 (standard size).

45,000	=	$\dfrac{1,500 \times 3.5 \times 7.5}{6}$
45,000	=	49,218.75

The beam is now in equilibrium. The resisting moment is greater than the bending moment. The member will not bend or break.

Care should be taken in establishing all sizes to ensure that the sizes of materials conform to manufacturers' standards and building-code allowances. Figure B.26

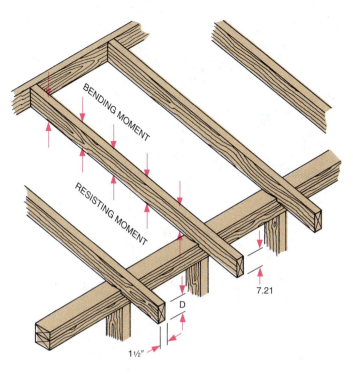

$$\text{BENDING MOMENT} = \text{RESISTING MOMENT}$$

$$\frac{742 \times 14 \times 12}{8} = \frac{1200 \times 1\frac{1}{2}'' \times D^2}{6}$$

$$15,582 = 300\,D^2$$

$$\frac{15,582 = 300\,D^2}{300}$$

$$51.94 = D^2$$

$$\sqrt{51.94} = \sqrt{D^2}$$

$$7.21 = D$$

FIGURE B.24 ■ To compare the bending moment with the resisting moment, use the formulas together.

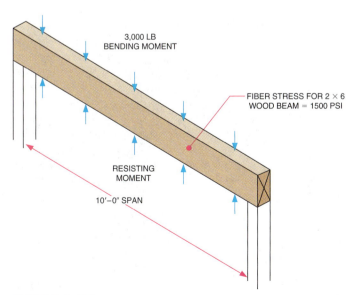

FIGURE B.25 ■ Factors used in bending and resisting moment calculations.

PITCH: 5/12		LOAD: 40 PSF

FIBER STRESS, 1200 POUNDS, FOR DOUGLAS FIR AND SOUTHERN YELLOW PINE

LUMBER SIZE	SPACING, IN INCHES	MAXIMUM SPAN
2 × 4	24	6'–6"
	20	7'–3"
	16	8'–1"
	12	9'–4"
2 × 6	24	10'–4"
	20	11'–4"
	16	12'–6"
	12	14'–2"
2 × 8	24	13'–8"
	20	15'–2"
	16	16'–6"
	12	18'–4"

FIGURE B.26 ■ Rafter spans.

shows a typical rafter space-span chart based on common lumber sizes and spacing. Figure B.27 shows common truss specifications based on normal loading for residential work.

Ceiling Joists

Information needed when selecting standard ceiling joists is provided in Figure B.28. (See Chapter 24 for information about wood groups.) If there is to be an attic above the ceiling for storage or living area, then the ceiling joists must be treated as floor joists using a floor joist table to calculate the joists for a heavier load. For a flat or very low-pitched roof, where the roof rafters are also the ceiling joists, then the table in Figure B.29 must be used.

Floors

The major structural parts of the floor system are the floor joists. See Figure B.30. The other parts of the floor system are the subfloor, finish floor, and supporting structural members. When the live load is determined, the size and spacing of joists can be established by referring to Figure B.31. This table is based on no. 1 southern white pine with a fiber stress of 1,200 lb/in^2 and a modulus of elasticity (ratio of stress and strain) of 1,600,000 lb/in^2. For other materials, such as spruce or Douglas fir lumber, with a different fiber stress and different modulus of elasticity, a different table must be used.

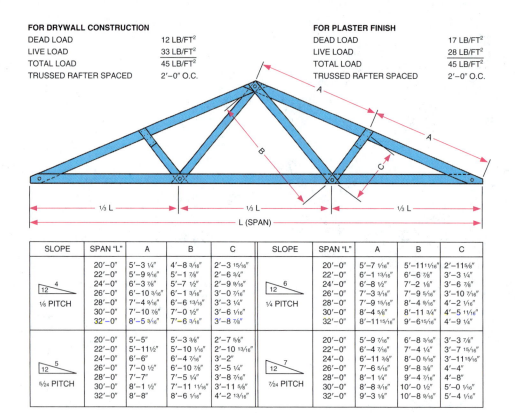

FOR DRYWALL CONSTRUCTION

DEAD LOAD	12 LB/FT^2
LIVE LOAD	33 LB/FT^2
TOTAL LOAD	45 LB/FT^2
TRUSSED RAFTER SPACED	2'–0" O.C.

FOR PLASTER FINISH

DEAD LOAD	17 LB/FT^2
LIVE LOAD	28 LB/FT^2
TOTAL LOAD	45 LB/FT^2
TRUSSED RAFTER SPACED	2'–0" O.C.

SLOPE	SPAN "L"	A	B	C	SLOPE	SPAN "L"	A	B	C
4/12 1/6 PITCH	20'–0"	5'–3 1/4"	4'–8 3/16"	2'–3 15/16"	**6/12** 1/4 PITCH	20'–0"	5'–7 1/16"	5'–11 11/16"	2'–11 5/8"
	22'–0"	5'–9 9/16"	5'–1 7/8"	2'–6 3/4"		22'–0"	6'–1 13/16"	6'–6 7/8"	3'–3 1/4"
	24'–0"	6'–3 7/8"	5'–7 1/2"	2'–9 9/16"		24'–0"	6'–8 1/2"	7'–2 1/8"	3'–6 7/8"
	26'–0"	6'–10 3/16"	6'–1 3/16"	3'–0 7/16"		26'–0"	7'–3 3/16"	7'–9 5/16"	3'–10 7/16"
	28'–0"	7'–4 9/16"	6'–6 13/16"	3'–3 1/4"		28'–0"	7'–9 15/16"	8'–4 9/16"	4'–2 1/16"
	30'–0"	7'–10 7/8"	7'–0 1/2"	3'–6 1/16"		30'–0"	8'–4 5/8"	8'–11 3/4"	4'–5 11/16"
	32'–0"	8'–5 3/16"	7'–6 3/16"	3'–8 7/8"		32'–0"	8'–11 15/16"	9'–6 15/16"	4'–9 1/4"
5/12 5/24 PITCH	20'–0"	5'–5"	5'–3 3/8"	2'–7 5/8"	**7/12** 7/24 PITCH	20'–0"	5'–9 7/16"	6'–8 3/16"	3'–3 7/8"
	22'–0"	5'–11 1/2"	5'–10 1/16"	2'–10 13/16"		22'–0"	6'–4 7/16"	7'–4 1/4"	3'–7 15/16"
	24'–0"	6'–6"	6'–4 7/16"	3'–2"		24'–0	6'–11 3/8"	8'–0 5/16"	3'–11 15/16"
	26'–0"	7'–0 1/2"	6'–10 7/8"	3'–5 1/4"		26'–0"	7'–6 5/16"	9'–8 3/8"	4'–4"
	28'–0"	7'–7"	7'–5 1/4"	3'–8 7/16"		28'–0"	8'–1 1/4"	9'–0 7/16"	4'–8"
	30'–0"	8'–1 1/2"	7'–11 11/16"	3'–11 5/8"		30'–0"	8'–8 3/16"	10'–0 1/2"	5'–0 1/16"
	32'–0"	8'–8"	8'–6 1/16"	4'–2 13/16"		32'–0"	9'–3 1/8"	10'–8 9/16"	5'–4 1/16"

RISE	MAXIMUM SPAN (APPROX)	
	2 × 4 SPAN	2 × 6 SPAN
3	25'–0"	30'–0"
4	27'–0"	32'–0"
5	30'–0"	35'–0"
6	32'–0"	38'–0"

Specifications for a Fink truss.

FIGURE B.27 ■ Truss specifications.

MAXIMUM ALLOWABLE SPAN (FEET AND INCHES)

SIZE	SPACING (OC)	GROUP I	GROUP II	GROUP III	GROUP IV
2 × 4	12"	11'–6"	11'–0"	9'–6"	5'–6"
	16"	10'–6"	10'–0"	8'–6"	5'–0"
2 × 6	12"	18'–0"	16'–6"	15'–6"	12'–6"
	16"	16'–0"	15'–0"	14'–6"	11'–0"
2 × 8	12"	24'–0"	22'–6"	21'–0"	19'–0"
	16"	21'–6"	20'–6"	19'–0"	16'–6"

FIGURE B.28 ■ Ceiling joists.

An example of the use of Figure B.31 is as follows: If the live loads are approximately 40 lb/ft^2 and 16" spaces are desired between joists, a 2 × 8 is good for a span of only 12'-1". For a span larger than 12'-1", a joist with larger cross section is necessary.

MAXIMUM SPAN

SIZE	SPACING (OC)	GROUP I	GROUP II	GROUP III	GROUP IV
2 × 6	24"	9'–4"	7'–10"	7'–2"	6'–6"
	16"	11'–4"	9'–8"	8'–8"	8'–0"
2 × 8	24"	13'–0"	11'–2"	10'–0"	9'–4"
	16"	15'–10"	13'–8"	12'–4"	11'–6"
2 × 10	24"	17'–4"	14'–10"	13'–8"	12'–4"
	16"	19'–2"	18'–2"	16'–8"	15'–0"

FIGURE B.29 ■ Low-slope roof joists.

Headers and Lintels

The header (structural wood member) and the lintel (structural steel member) are horizontal structural supports. They support the openings of windows and doors.

Engineering tables for their selection are shown in Figure B.32. This table is for celling and roof support only.

Girders and Beams

All the weight of the floor system, including live loads and dead loads, is transmitted to bearing partitions. These loads are then transmitted either to the foundation wall, to intermediate supports, or to horizontal supports known as joists, girders, or beams.

To determine the exact spacing, size, and type of girder to support the structure, follow these steps:

STEP 1 Determine the total load acting on the entire floor system in pounds per square foot. Divide the total live and dead loads by the number of square feet of floor space. For example, if the combined load for the floor system shown in Figure B.33 is 48,000 lb, then there are 50 lb/ft^2 of load acting on the floor (48,000 lb divided by 960 ft^2 equals 50 lb/ft^2).

STEP 2 Lay out the proposed position of all columns and beams. It will also help to sketch the position of the joists to be sure that the joist spans are correct.

STEP 3 Determine the number of square feet supported by the girder (girder load area). The girder load area is determined by multiplying the length of a girder, from column to column, by the girder load width. The *girder load width* is the distance extending on both sides of the center line of the girder, halfway to the nearest support, as shown in Figure B.33. The remaining distance from a girder load area to the outside wall is supported by the outside wall.

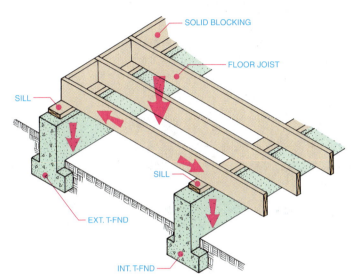

FIGURE B.30 ■ Floor joist loads are transmitted to intermediate supports.

LIVE LOAD—POUNDS PER SQUARE FOOT	SPACING	2 × 6	2 × 8	2 × 10	2 × 12	2 × 14
10	12″	12′–9″	16′–9″	21′–1″	24′–0″	—
	16″	11′–8″	15′–4″	19′–4″	23′–4″	24′–0″
	24″	10′–3″	14′–6″	17′–3″	20′–7″	24′–0″
20	12″	11′–6″	15′–3″	19′–2″	23′–0″	24′–0″
	16″	10′–5″	13′–11″	17′–6″	21′–1″	24′–0″
	24″	9′–2″	12′–3″	15′–6″	18′–7″	21′–9″
30	12″	10′–8″	14′–0″	17′–9″	21′–4″	24′–9″
	16″	9′–9″	12′–11″	16′–3″	19′–6″	22′–9″
	24″	8′–6″	11′–4″	14′–4″	17′–3″	20′–2″
40	12″	10′–0″	13′–3″	16′–8″	20′–1″	23′–5″
	16″	9′–1″	12′–1″	15′–3″	18′–5″	21′–5″
	24″	7′–10″	10′–4″	13′–1″	15′–9″	18′–5″
50	12″	9′–6″	12′–7″	15′–10″	19′–1″	22′–4″
	16″	8′–7″	11′–6″	14′–7″	17′–6″	20′–5″
	24″	7′–3″	9′–6″	12′–1″	14′–7″	17′–0″
60	12″	9′–0″	12′–0″	15′–2″	18′–3″	21′–4″
	16″	8′–1″	10′–10″	13′–8″	16′–6″	19′–3″
	24″	6′–8″	8′–11″	11′–3″	13′–7″	15′–11″
70	12″	8′–7″	11′–6″	14′–6″	17′–6″	20′–6″
	16″	7′–8″	10′–2″	12′–10″	15′–6″	18′–3″
	24″	6′–5″	8′–5″	10′–7″	12′–9″	15′–0″

FIGURE B.31 ■ Maximum floor joist spans.

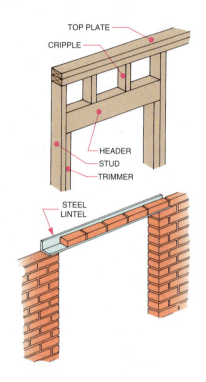

SAFE SPANS FOR WOOD HEADERS

SIZE	SPAN
4 × 4	3'–6"
4 × 6	4'–6"
4 × 8	6'–0"
4 × 10	7'–6"
4 × 12	9'–0"

SAFE SPANS FOR STEEL LINTELS 4" MASONRY WALL

SIZE	SPAN
L-3½" × 3½" × ¼"	3'–0"
L-3½" × 3½" × 5/16"	4'–0"
L-4" × 3½" × 5/16"	6'–0"
L-5" × 3½" × 5/16"	8'–0"
L-6" × 3½" × 3/8"	10'–0"

FIGURE B.32 ■ Lintel and header spans.

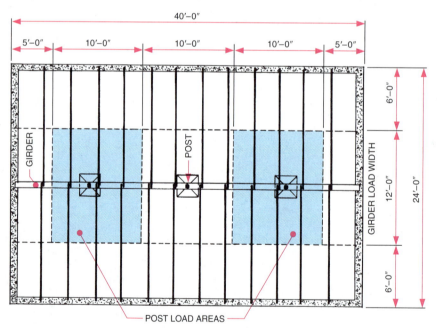

FIGURE B.33 ■ Tributary area of a girder.

STEP 4 To find the load supported by the girder load area, multiply the girder load area by the load per square foot. For example, the girder load in Figure B.33 is 6,000 lb (120 ft² × 50 lb/ft²).

STEP 5 Select the most suitable material to carry the load at the span desired. Built-up wood girders will span a greater length than a solid wood girder. However, I-beams will span a greater length without intervening support.

To compute the minimum cross-section area of a girder or beam, divide the load (in pounds) by the material's coefficient of elasticity:

$$CS = \frac{I}{E}$$

For example, if the combined live and dead loads on a wood member is 50,000 lb and the coefficient of elasticity is 1,600 the cross-section area is:

$$CS = \frac{50,000}{1,600}$$

$$= 31.25 \text{ in}^2$$

The cross section of a 4 × 10 member surfaced to 3.5″ × 9.5″ is 33.25 in². A 4 × 10 beam is therefore the smallest member possible above the minimum 31.25 in².

STEP 6   Select the exact size and classification of the beam or girder. Use Figure B.34 to select the most appropriate wood girder. For example, to support 6,000 lb over a 10′ span, either an 8 × 8 built-up girder or a 6 × 10 solid girder would suffice. The girder should be strong enough to support the load, but any size larger is a waste of materials. The only alternative to increasing the size of the girder is to decrease the size of span.

Calculating the Tributary Area   Calculating the tributary area that is supported by a girder or beam is shown in Figure B.35. The factors used to calculate a safe load that is evenly distributed on a beam are shown in Figure B.36.

The allowable fiber stress of the beam must be known to complete these calculations. for example, if a 6 × 12 beam (actual size 5 1/2″ × 11 1/4″ spans 18′-0″ and the beam has a fiber stress rated at 1,800 lb/ft², the safe evenly distributed weight is calculated as follows:

$$W = \frac{f \times b \times d^2}{9 \times L}$$

where W = weight evenly distributed, lb
f = allowable fiber stress, lb/in²
b = width of beam, in
d = depth of beam, in
L = span, ft

$$W = \frac{1,800 \times 5.5 \times 11.5^2}{9 \times 18} = 7,734 \text{ lb}$$

Calculating Deflection   The stiffness of a beam, or any structural member, is the ability of the member to resist bending. The bending force that changes a straight mem-

GIRDER SIZE	6 FT	7 FT	8 FT	9 FT	10 FT
6 × 8 built-up	8,306	7,118	6,220	5,539	4,583
6 × 8 solid	7,359	6,306	5,511	4,908	4,062
6 × 10 built-up	11,357	10,804	9,980	8,887	7,997
6 × 10 solid	10,068	9,576	8,844	7,878	7,086
8 × 8 built-up	11,326	9,706	8,482	7,553	6,250
8 × 8 solid	9,812	8,408	7,348	6,544	5,416
8 × 10 built-up	15,487	14,732	13,608	12,116	10,902
8 × 10 solid	13,424	12,768	11,792	10,504	9,448

SAFE LOAD IN POUNDS FOR SPANS FROM 6 TO 10 FEET

FIGURE B.34 ■ Wood girder safe loads.

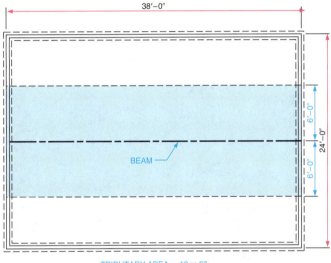

TRIBUTARY AREA = 12 × 38
TA = 456 SQUARE FEET

FIGURE B.35 ■ Calculating the tributary area of a beam.

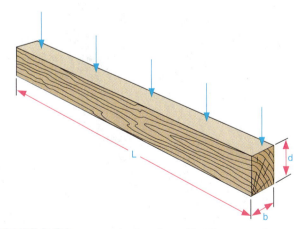

FIGURE B.36 ■ Evenly distributed load on a girder or beam.

ber to a curved member is *deflection*. The calculation of the deflection of a structural member uses Hooke's law. Data for a free support beam are shown in Figure B.37.

FIGURE B.37 ■ Beam deflection.

For example, the deflection of a 2 × 6 (1 1/2" × 5 1/2") joist which has a 20′ span and supports a central load of 4000 lb is calculated as follows:

$$D = \frac{PL^3}{48EI}$$

where
D = deflection
P = force in center of span, lb
L = length of beam span, ft
E = modulus of elasticity (given in table as 1,000,000)
I = moment of inertia (for rectangular members)
I = bd/12; b = width, d = height

$$D = \frac{PL^3}{48EI}$$

D = the amount of deflection
P = the concentrated load, lb
L = the span, ft

E = the modulus of elasticity (1,000,000) and the moment of inertia
I = the cross section of the beam

$$D = \frac{4,000 \times 20^3}{48 \times 1,000,000 \times 1.5 \times 5.5}$$

$$D = \frac{32,000,000}{396,000,000}$$

$$D = 0.08''$$

Steel Beams The method of determining the size of steel beams is the same as for determining the size of wood beams. As wood beams vary in width for a given depth, steel beams vary in weight, depth, and thickness of webs and flanges. Classifications vary accordingly. Figure B.38 shows the relationship of the span, load, depth, and weight of standard steel I-beams and channels. A steel beam may be selected by referring to the desirable span and load and then choosing the most appropriate size (depth and weight) for the I-beam. For example, a 4" × 8", 18.4 pound I-beam will support 8.5 kips at a span of 20 ft. A *kip* is equal to 1000 lb.

Columns and Posts

When girders or beams do not completely span the distance between foundation walls, then wood posts, steel-pipe columns, masonry columns, or steel columns must be used for intervening support to reduce the span. To

SIZE OF BEAM	2⅝" × 4"	2¾" × 4"	3" × 5"	3¼" × 5"	3⅜" × 6"	3⅝" × 6"	3⅝" × 7"	3⅞" × 7"	4" × 8"	4⅛" × 8"	4⅝" × 10"	5" × 10"
WEIGHT PER FOOT IN POUNDS	7.7	9.5	10.0	14.75	12.5	17.25	15.3	20.0	18.4	23.0	25.4	35.0
Span in feet												
4	9.0	10.1	14.5	18.0	21.8	26.0	31.0	36.0	42.7	48.2	73.3	87.5
5	7.2	8.0	11.6	14.4	17.4	20.8	24.8	28.7	34.1	38.5	58.6	70.0
6	6.0	6.7	9.7	12.0	14.5	17.3	20.7	24.0	28.5	32.1	48.8	58.3
7	5.1	5.7	8.3	10.3	12.5	14.9	17.7	20.5	24.4	27.5	41.9	50.0
8	4.5	5.0	7.3	9.0	10.9	13.0	15.5	18.0	21.3	24.1	36.6	43.7
9	4.0	4.5	6.5	8.0	9.7	11.6	13.8	16.0	19.0	21.4	32.6	38.9
10	3.6	4.0	5.8	7.2	8.7	10.4	12.4	14.4	17.1	19.3	29.3	35.0
11	—	—	5.3	6.5	7.9	9.5	11.3	13.1	15.5	17.5	26.6	31.8
12	—	—	—	—	7.3	8.7	10.3	12.0	14.2	16.1	24.4	29.2
13	—	—	—	—	6.7	8.0	9.5	11.1	13.1	14.8	22.5	26.9
14	—	—	—	—	6.2	7.4	8.9	10.3	12.2	13.8	20.9	25.0
15	—	—	—	—	—	—	8.3	9.6	11.4	12.8	19.5	23.3
16	—	—	—	—	—	—	7.7	9.0	10.7	12.0	18.3	21.9
17	—	—	—	—	—	—	—	—	10.0	11.3	17.2	20.6
18	—	—	—	—	—	—	—	—	9.5	10.7	16.3	19.4
19	—	—	—	—	—	—	—	—	9.0	10.1	15.4	18.4
20	—	—	—	—	—	—	—	—	8.5	9.6	14.7	17.5

FIGURE B.38 ■ Safe loads for steel I-beams in kips (1,000 lb).

determine the most appropriate size and classification of posts or columns to support girders or beams, follow these steps:

STEP 1 Determine the total load in pounds per square foot for the entire floor area. Calculate this amount by multiplying the total load by the number of square feet of floor space.

STEP 2 Determine the spacing of posts necessary to support the ends of each girder. Great distances between posts should be avoided because excessive weight would concentrate on one footing. Long spans also require extremely large girders. For example, it is possible to span a distance of 30′, but to do so a 15″ I-beam would be needed. The extreme weight and cost of this beam would be prohibitive. On the other hand, if only a 6′ span is used, the close spacing might greatly restrict the flexibility of the internal design. As a rule, use the shortest span that will not interfere with the design function of the area.

STEP 3 Find the number of square feet supported by each post. A post will carry the load on a girder to the midpoint of the span on both sides. The post also carries half the load to the nearest support wall on either side of the post tributary area.

STEP 4 Find the load supported by the post support area. Multiply the number of square feet by the load per square foot.

STEP 5 Determine the height of a post. The height of the post is related to the length of the column. The 4 × 4 post shown in Figure B.39 may be more than adequate to support a given weight if the height of the post is 6′. However, the same 4 × 4 post may be totally inadequate to support the same weight when the length is increased to 20′.

STEP 6 Determine the thickness and width of the post needed to support the load at the given height by referring to Figure B.40 for wood posts, Figure B.41 for I-columns, or Figure B.42 for the diameter of steel-pipe columns. To check the compressive stress on posts, apply the following formula:

$$F = \frac{P}{A}$$

where F = compressive stress
P = compressive force
A = cross-section area

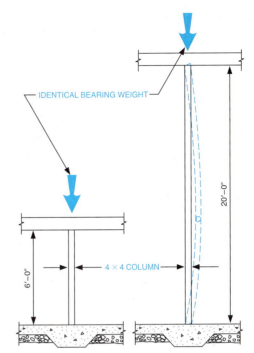

FIGURE B.39 ■ Relationship of height to stability.

STANDARD SIZE, INCHES ACTUAL SIZE, INCHES AREA IN SQUARE INCHES	3 × 4 2½ × 3½ 9.51	4 × 4 3½ × 3½ 13.14
Height of column:		
4 feet	8 720	12 920
5 feet	7 430	12 400
6 feet	5 630	11 600
6 feet 6 inches	4 750	10 880
7 feet	4 130	10 040
7 feet 6 inches	—	9 300
8 feet	—	8 350
9 feet	—	6 500
10 feet	—	—
11 feet	—	—
12 feet	—	—

FIGURE B.40 ■ Maximum loads for wood posts.

DEPTH IN INCHES WEIGHT PER POUND PER FOOT	10 25.4	9 21.8
Effective length:		
3 feet	110.7	94.8
4 feet	110.7	94.8
5 feet	109.5	91.2
6 feet	101.7	83.9
7 feet	93.8	76.7
8 feet	86.0	69.7
9 feet	78.7	63.2
10 feet	71.8	57.2
11 feet	65.5	51.8
12 feet	59.7	47.0
Area in square inches	7.38	6.32

FIGURE B.41 ■ Safe loads for I-columns.

NOMINAL SIZE, INCHES	6	5	4½	4	3½	3	2½	2	1½
EXTERNAL DIAMETER, INCHES	6.625	5.563	5.000	4.500	4.000	3.500	2.875	2.375	1.900
THICKNESS, INCHES	.280	.258	.247	.237	.226	.216	.203	.154	.145
Effective length:									
5 feet	72.5	55.9	48.0	41.2	34.8	29.0	21.6	12.2	7.5
6 feet	72.5	55.9	48.0	41.2	34.8	28.6	19.4	10.6	6.0
7 feet	72.5	55.9	48.0	41.2	34.1	26.3	17.3	9.0	5.0
8 feet	72.5	55.9	48.0	40.1	31.7	24.0	15.1	7.4	4.2
9 feet	72.5	55.9	46.4	37.6	29.3	21.7	12.9	6.6	3.5
10 feet	72.5	54.2	43.8	35.1	26.9	19.4	11.4	5.8	2.7
11 feet	72.5	51.5	41.2	32.6	24.5	17.1	10.3	5.0	—
12 feet	70.2	48.7	38.5	30.0	22.1	15.2	9.2	4.1	—
Area in square inches	5.58	4.30	3.69	3.17	2.68	2.23	1.70	1.08	0.80
Weight per pound per foot	18.9	14.6	12.5	10.7	9.11	7.58	5.79	3.65	2.72

FIGURE B.42 ■ Safe loads for steel-pipe columns.

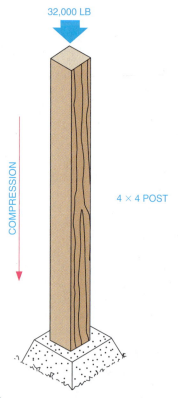

FIGURE B.43 ■ Compressive-force stress.

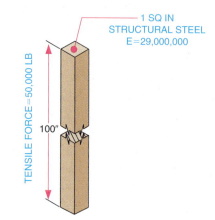

FIGURE B.44 ■ Tensile-force deformation.

The amount of deformation on a member is calculated as follows:

$$\delta = \frac{PL}{AE}$$

where
δ = deformation, in.
P = force, lb
L = length, in.
A = cross-section area
E = modulus of elasticity, lb/in² (for structural steel, E = 29,000,000 lb/in²; for wood, E = 1,000,000 to 2,000,000 lb/in²)

For example, a tensile force P of 50,000 lb is acting on the structural column shown in Figure B.44. The deformation is, therefore,

$$\delta = \frac{50{,}000 \text{ lb} \times 100''}{16'' \times 29{,}000{,}000} = \frac{5{,}000{,}000}{494{,}000{,}000}$$

δ = .0108″

For example, in Figure B.43 the compressive force on the 4 × 4 post (16 in²) is 32,000 lb; therefore,

$$F = \frac{32{,}000}{16}$$

F = 2000 lb/in² or 2 kips (compressive stress)

Calculating Deformation *Deformation* is the bending of structural members of a building caused by load stress.

HEIGHT DESCRIPTION	WOOD FRAME HOUSE		MASONRY HOUSE	
	MINIMUM FOUNDATION WALL THICKNESS	FOOTING PROJECTION EACH SIDE OF FOUNDATION WALL	MINIMUM FOUNDATION WALL THICKNESS	FOOTING PROJECTION EACH SIDE OF FOUNDATION WALL
One story–no basement	6″	2″	6″	3″
One story–with basement	6″	3″	8″	4″
Two story–no basement	6″	3″	6″	4″
Two story–with basement	8″	4″	8″	5″

FIGURE B.45 ■ Footing and foundation wall sizes.

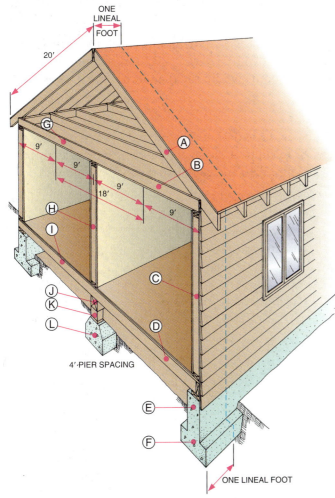

FIGURE B.46 ■ Structural members used in calculating dead loads.

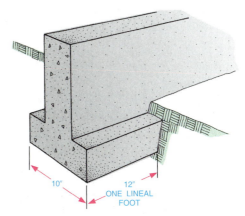

FIGURE B.47 ■ Loads calculated on one lineal foot of footing.

mended in Figure B.45. Piers support the interior floor system. In residential construction, piers are usually 12″ high and 12″ to 18″ square at the base. To compute the bearing area of the pier or the T-foundation's footing, divide the total load on one square foot by the soil-bearing capacity per square foot.

Footing Size Calculations To calculate the minimum bearing size of a lineal foot of footing refer to A through F in Figure B.46 and Figure B.47. First find the total load on the footing as follows:

A. Roof live load ....30 lb/ft^2
   Roof dead load
     Rafters 2 × 8, 16″ OC ....3.5 lb/ft^2
     Wood sheathing 1″ ....3 lb/ft^2
     Wood shingles ....3 lb/ft^2
     Batt insulation 3″...................................0.5 lb/ft^2
                          40 lb/ft^2

Rafter length = 20′ (1 side) × 40 lb/ft^2 = 800 lb

Foundation Footings and Piers

The basic T-foundation has an exterior concrete footing and a wall forming an inverted T. Generally, builders follow the sizes for foundation walls and footings as recom-

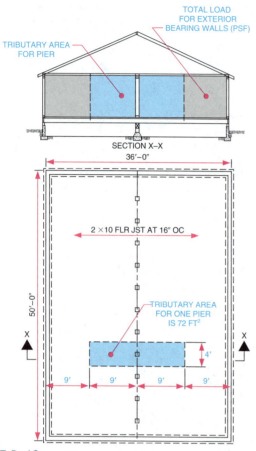

TOTAL LOAD FOR EXTERIOR BEARING WALLS (PSF)

TRIBUTARY AREA FOR PIER

SECTION X–X

36'-0"

2 ×10 FLR JST AT 16" OC

TRIBUTARY AREA FOR ONE PIER IS 72 FT²

50'-0"

4'

9'   9'   9'   9'

X   X

FIGURE B.48 ■ Tributary areas for exterior walls and piers.

B. Attic floor/ceiling live load ....20 lb/ft²
  Attic floor/ceiling dead load
    Joists 2 × 10, 16" OC ....6.5 lb/ft²
    Lath and plaster (one side) ....10 lb/ft²
    Batt insulation 3" ....0.5 lb/ft²
                                                          37 lb/ft²
          Tributary length 9' × 37 = 333 lb
(See Figure B.48.)

C. Exterior wall live load.................................0
  Exterior wall dead load
    2 × 4 wood stud/plaster
    one side ....12 lb/ft²
    Exterior wood siding 1" ....3 lb/ft²
    Batt insulation 3" ....0.5 lb/ft²
                                                        15.5 lb/ft²
          Wall height 8' × 15.5 = 124 lb

D. Floor live load..40 lb/ft²
  Floor dead load 2 × 10 floor joists
    16" OC/subfloor....................................6.5 lb/ft²
    Hardwood floor 25/32" .... 4 lb/ft²
                                                        50.5 lb/ft²
          Tributary length 9' × 50.5 = 454.5 lb
(See Figure B.48.)

E. Foundation wall live load......................................0
  Foundation wall dead load
  6" concrete..75 lb/ft²
          Approximately 3' height × 75 = 225 lb

F. Footing live load...0
  Footing dead load 6" concrete....................75 lb/ft²
          Approximately 1'-0" wide × 1 lineal foot ×
          75 = 75 lb
Total load per lineal foot of footing is 2,011.5 lb

To calculate the minimum footing size apply the following formula:

  Total load on footing:.................................2,011.5 lb
  Soil-bearing capacity (given): ....2,500 lb/ft²
          = .80 ft² (115.2 in²)

A lineal foot is 12", therefore the width of the footing should be 10" (10" × 12" = 120 in²) minimum. A 12" standard footing width can therefore be used. (See Figure B.47.)

Pier Size Calculation To calculate the size of a pier needed to support the structure refer to items G through L in Figure B.46 and calculate the load as follows:

G. Attic floor/ceiling live load ....20.0 lb/ft²
  Attic floor/ceiling dead load
    Floor joists 2 × 10, 16" OC ....6.5 lb/ft²
    Lath and plaster, one side....................10.0 lb/ft²
    3" batt insulation................................0.5 lb/ft²
                                                          37.0 lb/ft²
  (Again, see Figure B.48.)
          Tributary area (72 ft²); 72 × 37 = 2,664 lb

H. Wall live load...0
  Wall dead load
    2 × 4 wood stud/plaster
    2 sides..20 lb/ft²
          Wall height 8'; 8 × 20 = 160 lb

I. Floor live load ....40.0 lb/ft²
  Floor dead load
    2 × 10 floor joist, 16"
    OC/subfloor...6.5 lb/ft²
    Hardwood floor 25/32".......................4.0 lb/ft²
                                                          50.5 lb/ft²
          Tributary area 4' × 18' = 72 ft²; 72 × 50.5
          = 3,636 lb

J. Girder 4 × 6, dead load..............................4 lb/ft²
          4' length × 4 = 16 lb

K. Post 4 × 6, dead load ....4 lb/ft²
          Approximately 1.5' × 4 = 6 lb

L. Concrete pier..150 lb/ft^2

Approx. 24″ × 24″ × 12″ = 4 cubic feet

4 ft^3 × 150 = 600 lb

Total load per pier = 7,082 lb
(Total load on one pier)

Now use the total load and apply the following formula to compute the minimum size of pier:

Total load on pier:
$$\frac{7,082}{2,500} = 2.8 \text{ ft}^2$$

Soil-bearing capacity (given):
$$2.8 \text{ ft}^2 \times 144 = 403.2 \text{ in}^2$$

21″ × 21″ (441 in^2) is the next largest standard square size.

Stress and Dynamics Formulas

Some structural designs include construction methods, materials, and sizes that are not standard. These cannot be determined by using precalculated charts. The following formulas relate to these aspects of structural design.

Direct Stress Calculations To calculate the unit of direct stress on a structural member, divide the load by the cross-section area of the member:

$$F = \frac{P}{A}$$

where   F = unit of stress
P = load/force
A = cross-section area

For example, in Figure B.49, a 40,000 lb I-beam is supported by a 2″ × 2″ (4 in^2) support rod. The stress on the rod is

$$F = \frac{40,000}{4} = 10,000 \text{ lb/in}^2 \text{ or 10 kips}$$

Shearing Stress A unit of shearing stress equals the shearing force (in pounds) divided by the cross-section area (in square inches) of the structural member:

$$S = \frac{P}{A}$$

where S = unit of shearing stress
P = shearing force
A = cross-section area of stress object

For example, in Figure B.50, a shearing force of 10,000 lb is acting on a 1″ pin. Therefore,

$$S = \frac{10,000}{\pi r^2} = \frac{10,000}{3.14 \times .5^2} = \frac{10,000}{.785}$$
$$= 12,739 \text{ lb/in}^2 \text{ or 12.7 kips}$$

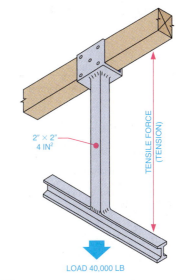

2″ × 2″
4 IN^2

TENSILE FORCE
(TENSION)

LOAD 40,000 LB

FIGURE B.49 ■ Direct stress calculation.

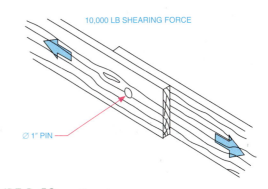

10,000 LB SHEARING FORCE

Ø 1″ PIN

FIGURE B.50 ■ Shearing stress.

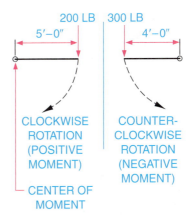

200 LB   300 LB

5′–0″   4′–0″

CLOCKWISE
ROTATION
(POSITIVE
MOMENT)

COUNTER-
CLOCKWISE
ROTATION
(NEGATIVE
MOMENT)

CENTER OF
MOMENT

FIGURE B.51 ■ Moment of force.

Moment of Force To calculate the moment of force (in foot-pounds) on a structural member, multiply the force (in pounds) by the length of the moment arm (in feet):

Moment of force = force × arm length

For example, the left side of Figure B.51 shows a force of 200 lb acting on the end of a 5′ arm. The moment of force is, therefore, 1,000 lb (200 lb × 5′). The moment of force on the 4′ arm on the right is 1,200 ft-lb.

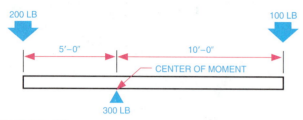

FIGURE B.52 ■ Forces in equilibrium.

Forces in Equilibrium The positive and negative moment values are always equal when there is no movement—that is, when the system is in equilibrium. The beam in Figure B.52 is in balance (equilibrium). The moment of force on the right of the center of moment is $200 \times 5 = 1,000$ ft-lb. The moment of force to the left of center is 100 lb $\times 10 = 1,000$ ft-lb.

GEOMETRIC CALCULATIONS

Architectural drawings contain a wide variety of geometric shapes. Some are two-dimensional flat surfaces and others are three-dimensional volume-containing areas. In estimating amounts of construction materials, labor, and costs, the ability to compute distances, areas, and volumes is important. The majority of geometric calculations in architectural work falls into these categories:

Perimeter of a Polygon

Formula: Perimeter = sum of all sides
Examples: See Figure B.53.

> Top drawing: $10 + 20 + 10 + 20 = 60'$

> Bottom drawing: 25'-0"
> 60'-6"
> 40'-0"
> 20'-0"
> 15'-0"
> 40'-6"
> 200'-12" = 201'

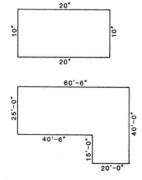

FIGURE B.53 ■ Perimeter of a polygon.

Circumference of a Circle

Formula: Circumference = $\pi \times$ diameter
$$C = \pi D$$
Example: See Figure B.54. $\pi = 22/7$ or 3.14.

> Left drawing: $C = 3.14 \times 7 = 22'$
> Right drawing: Inside-diameter
> $C = 3.14 \times 14' = 43.96'$
> Outside-diameter
> $C = 3.14 \times 22' = 69.08'$

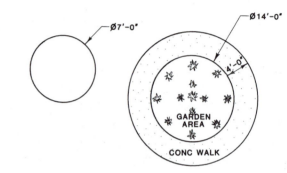

FIGURE B.54 ■ Circumference of a circle.

Area of a Square or Rectangle

Formula: Area = side $\times$ side
$$A = SS$$
Example: See Figure B.55.

Top drawing:	$50' \times 25'$ =	$1,250$ ft²
Center drawing:	175×60 =	$10,500$ ft²
Bottom drawing:	60×60 =	$3,600$
	20×20 =	400
		$4,000$ ft²

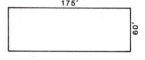

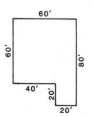

FIGURE B.55 ■ Area of a square or rectangle.

Area of a Triangle

Formula: Area = 1/2 base × altitude

$$A = 1/2 \, BA$$

Example: See Figure B.56.

Top drawing: $\dfrac{20'' \times 10''}{2} = 100 \text{ in}^2$

Center drawing: $\dfrac{15' \times 30'}{2} = 225 \text{ ft}^2$

Bottom drawing: $\dfrac{100 \times 70}{2} = 3{,}500 \text{ ft}^2$

$\dfrac{100 \times 40}{2} = 2{,}000 \text{ ft}^2$

$3{,}500 + 2{,}000 = 5{,}500 \text{ ft}^2$

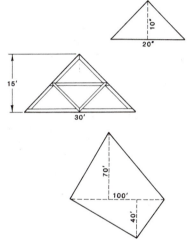

FIGURE B.56 ■ Area of a triangle.

Surface Area of a Cylinder

Formula: Surface area = π × diameter × height

$$A = \pi DH$$

Example: See Figure B.57.

Top: $3.14 \times 10'' \times 20'' = 628 \text{ in}^2$

Center: $3.14 \times 2'' \times 100'' = 628 \text{ in}^2$

Bottom: Well is 10'-0" deep and 5'-0" wide
$3.14 \times 10' \times 5' = 157 \text{ ft}^2$

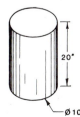 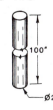

FIGURE B.57 ■ Surface area of a cylinder.

Area of a Circle

Formula: Area = π × radius squared

$$A = \pi R^2$$

Example: See Figure B.58.

Top drawing: 3.14×7^2

$3.14 \times 49 = 153.86 \text{ in}^2$

Center drawing: $10 \div 2 = 5$

3.14×5^2

$3.14 \times 25 = 78.5 \text{ in}^2$

Bottom drawing: $20 \div 2 = 10$

$3.14 \times 10^2 = 314$

$314 \div 2 = 157 \text{ in}^2$

FIGURE B.58
■ Area of a circle.

Volume of a Cube

Formula: Volume = length × height × width

$$V = LHW$$

Example: See Figure B.59.

Top: $V = 10'' \times 5'' \times 12''$
$= 600 \text{ in}^3$

Center: $V = 6 \times 50 \times 60$
$= 18{,}000 \text{ ft}^3$
$= 18{,}000 \div 27 = 666 \text{ yd}^3$

Bottom: $40' \times 24'' \times 6''$
$40' \times 2' \times 0.5' = 40 \text{ ft}^3$
$40' \times 6'' \times 12$
$40' \times 0.5' \times 1' = \underline{20} \text{ ft}^3$
60 ft^3

$60 \div 27 = 2.7 \text{ yd}^2$

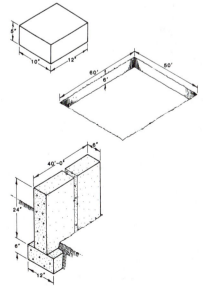 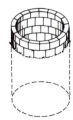

FIGURE B.59 ■ Volume of a cube.

Volume of a Cylinder

Formula: Volume = π × radius squared × height

$$V = \pi R^2 H$$

Example: See Figure B.60.

$$\begin{aligned} \text{Top: } V &= 3.14 \times 5^2 \times 7 \\ &= 3.14 \times 25 \times 7 \\ &= 550 \text{ in}^3 \end{aligned}$$

$$\begin{aligned} \text{Center: } V &= 3.14 \times 10^2 \times 14 \\ &= 3.14 \times 100 \times 14 \\ &= 4396 \text{ ft}^3 \end{aligned}$$

$$\begin{aligned} \text{Bottom: } V &= 3.14 \times 3^2 \times 7 \\ &= 3.14 \times 9 \times 7 \\ &= 197.8 \text{ yd}^3 \end{aligned}$$

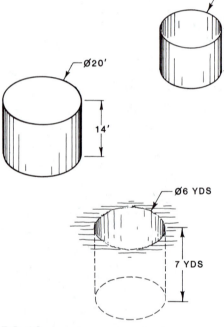

FIGURE B.60 ■ Volume of a cylinder.

Volume of a Square Pyramid

Formula: Volume = 1/3 × width of base × depth of base × height

$$V = 1/3WDH$$

Example: See Figure B.61.

Top: $V = 0.33 \times 25'' \times 30'' \times 20''$
  $= 4,950 \text{ in}^3$

Bottom: $V = 0.33 \times 10' \times 12' \times 10'$
  $= 396 \text{ ft}^3$

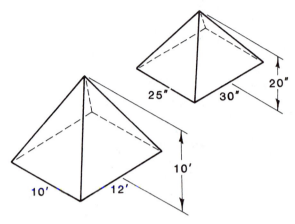

FIGURE B.61 ■ Volume of a square pyramid.

Volume of a Cone

Formula: Volume = 1/3 × π × radius squared × height

$$V = 1/3\pi R^2 H$$

Example: See Figure B.62.

$$\begin{aligned} \text{Top: } V &= .33 \times 3.14 \times 10^2 \times 30 \\ &= .33 \times 3.14 \times 100 \times 30 = 3,108.6 \text{ in}^3 \end{aligned}$$

$$\begin{aligned} \text{Bottom: } V &= .33 \times 3.14 \times 20^2 \times 20 \text{ ft}^2 \\ &= .33 \times 3.14 \times 400 \times 20 \text{ ft}^2 \\ &= 8,289.6 \text{ ft}^3 \end{aligned}$$

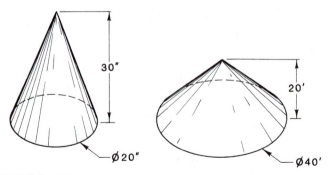

FIGURE B.62 ■ Volume of a cone.

Volume of a Sphere

Formula: Volume = 1/6 × π × diameter cubed

$$V = 1/6\pi D^3$$

Example: See Figure B.63. 1/6 = .166.

$$V = .166 \times 3.14 \times 3^3$$
$$= .166 \times 3.14 \times 27$$
$$= 14.07 \text{ in}^3$$

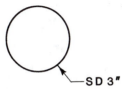

FIGURE B.63 ■ Volume of a sphere.

Right Triangle Law (Pythagorean Theorem)

Formula: Square of the hypotenuse
= the sum of the square of the two sides

$$C^2 = A^2 + B^2$$

Example: See Figure B.64.

$$C^2 = A^2 + B^2$$
$$C = \sqrt{A^2 + B^2}$$
$$C = \sqrt{32^2 + 24^2}$$
$$C = \sqrt{1024 + 576}$$
$$C = \sqrt{1600}$$
$$C = 40''$$

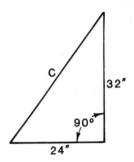

FIGURE B.64 ■ Hypotenuse of a right triangle.

Architectural Abbreviations and Professional Organizations

ARCHITECTURAL ABBREVIATIONS

Hundreds of words may be lettered on an architectural drawing. Abbreviations are often used to conserve space. By using standard abbreviations, drawings can be interpreted consistently.

Capital letters are used for abbreviations on drawings. Periods are used only when the abbreviation may be confused with a whole word. The same abbreviation is used for both singular and plural terms, and two or more terms may use the same abbreviation.

Access panel . . . . **AP**	Bedroom . . . . **BR**	Centerline . . . . **CL**
Acoustic . . . . **ACST**	Bench mark . . . . **BM**	Centimeter . . . . **CM**
Actual . . . . **ACT.**	Bending moment . . . . . **BM**	Ceramic . . . . **CER**
Addition . . . . **ADD.**	Between . . . . **BET**	Certificate of
Adhesive . . . . **ADH**	Bill of material . . . . **B/M**	occupancy . . . . **CO**
Aggregate . . . . **AGGR**	Block . . . . **BLK**	Chamfer . . . . **CHAM**
Air condition . . . . **AC**	Blocking . . . . **BLKG**	Channel . . . . **CHAN**
Alternating current . . . . **AC**	Blower . . . . **BLO**	Check . . . . **CHK**
Aluminum . . . . **AL**	Blueprint . . . . **BP**	Chimney . . . . **CHIM**
Amount . . . . **AMT**	Board . . . . **BD**	Chord . . . . **CHD**
Ampere . . . . **AMP; AP**	Board feet . . . . **BF**	Circle . . . . **CIR**
Anchor bolt . . . . **AB**	Boiler . . . . **BLR**	Circuit . . . . **CIR; CKT**
Apartment . . . . **APT.**	Both sides . . . . **BS**	Circuit breaker . . . . **CIR BKR**
Approved . . . . **APPD**	Bottom . . . . **BOT**	Circumference . . . . **CIRC**
Approximate . . . . **APPROX**	Brick . . . . **BRK**	Cleanout . . . . **CO**
Architectural . . . . **ARCH**	British thermal units . . . **BTU**	Clear . . . . **CLR**
Area . . . . **A**	Bronze . . . . **BRZ**	Closet . . . . **CL**
Asphalt . . . . **ASPH**	Broom closet . . . . **BC**	Coated . . . . **CTD**
At . . . . **@**	Building . . . . **BLDG**	Cold water . . . . **CW**
Automatic . . . . **AUTO**	Building line . . . . **BL**	Column . . . . **COL**
Auxiliary . . . . **AVE**	Cabinet . . . . **CAB.**	Combination . . . . **COMB.**
Average . . . . **AVG**	Carbon monoxide . . . . . **CO**	Common . . . . **COM**
Azimuth . . . . **AZ**	Casing . . . . **CSG**	Composition . . . . **COMP**
Balcony . . . . **BALC**	Cast concrete . . . . **C CONC**	Computer-aided
Base . . . . **B**	Cast iron . . . . **CI**	drafting . . . . **CAD**
Basement . . . . **BSMT**	Catalog . . . . **CAT.**	Concrete . . . . **CONC**
Bathroom . . . . **B**	Caulking . . . . **CLKG**	Construction . . . . **CONST**
Bathtub . . . . **BT**	Ceiling . . . . **CLG**	Continue . . . . **CONT**
Batten . . . . **BATT**	Cement . . . . **CEM**	Contractor . . . . **CONTR**
Beam . . . . **BM**	Center . . . . **CTR**	Corrugate . . . . **CORR**
Bearing . . . . **BRG**	Center to center . . . . . . **C to C**	Counter . . . . **CTR**

Course	C	Existing	EXIST.	Hardware	HDW
Cross section	X-SECT; CS	Expansion joint	EXP JT	Head	HD
		Extension	EXT	Header	HDR
Cubic foot	CU FT	Exterior	EXT	Heater	HTR
Cubic inch	CU IN.	Fabricate	FAB	Heating/ventilating/ air conditioning	HVAC
Cubic yard	CU YD	Face of studs	FOS		
Damper	DMPR	Face to face	F to F	Height	HT
Dampproofing	DP	Fahrenheit	F	Horizontal	HOR
Datum	DAT	Feet	(′) FT	Hose bib	HB
Dead load	DL	Feet board measure	FBM	Hot water	HW
Decibel	DB	Figure	FIG	Hour	HR
Decking	DK	Finish	FIN.	House	HSE
Deflection	D	Finished grade	FG	Hundred	C
Degree	(°) DEG	Fire hydrant	FH	Hypotenuse	H
Department	DEPT	Fireproof	FPRF	I-beam	I
Design	DSGN	Fixture	FIX.	Inch	(″) IN.
Design development	DD	Flashing	FL	Inch-pounds	IN-LB
Detail	DET	Floor	FL	Incinerator	INCIN
Diagonal	DIAG	Floor drain	FD	Include	INCL
Diagram	DIAG	Flooring	FLG	Information	INFO
Diameter	D; Ø	Fluorescent	FLUOR	Inside diameter	ID
Dimension	DIM	Foot	(′) FT	Insulate	INS
Dining room	DR	Foot-pounds	FT-LB	Intercommunication	INTER-COM
Direct current	DC	Footcandle	FC		
Dishwasher	DW	Footing	FTG	Interior	INT
Disk operating system	DOS	Footing area	FA	Iron	I
Disposal	DISP	Forced-air unit	FAU	Jamb	JMB
Ditto	DO.	Foundation	FDN	Joint	JT
Division	DIV	Frame	FR	Joist	JST
Door	DR	Fresh water wetlands	FWW	Joule	J
Double	DBL	Full size	FS	Junction	JCT
Double-hung	DH	Furnace	FURN	Kilowatt	kW
Dowel	DWL	Furnished by others	FBO	Kilowatt hour	kWh
Down	DN	Furred (ing)	FUR	Kip (1000 lb)	K
Downspout	DS	Furred ceiling	FC	Kitchen	KIT
Drafting	DFT	Gage	GA	Knockout	KO
Drain	DR	Gallon	GAL	Laminate	LAM
Drawing	DWG	Galvanize	GALV	Latitude	LAT
Dryer	D	Galvanized iron	GI	Laundry	LAU
Drywall	DW	Garage	GAR	Lavatory	LAV
Duplicate	DUP	Gas	G	Leader	LDR
Each	EA	Girder	G	Left	L
East	E	Glass	GL	Left hand	LH
Electric	ELEC	Glue laminated	GLU-LAM	Length	L; LG
Elevation	EL			Length overall	LOA
Enamel	ENAM	Grade	GR	Level	LEV
Engineer	ENGR	Grade line	GL	Light	LT
Entrance	ENT	Ground	GND	Linear	LIN
Equal	EQ	Ground-fault circuit interrupter	GFCI	Linen closet	L CL
Equipment	EQUIP.			Lintel	LTL
Estimate	EST	Gypsum	GYP	Live load	LL
Excavate	EXC	Hall	H	Living room	LR

Long	LG	Piece	PC	Service	SERV
Louver	LV	Plaster	PL	Sewer	SEW.
Lumber	LBR	Plate	PL	Sheathing	SHTHG
Lumen	LW	Plumbing	PLMG	Sheet	SH
Main	MN	Plywood	PLY	Shower	SH
Manual	MAN.	Pound	LB	Side	S
Manufacturing	MFG	Pounds per square		Siding	SDG
Master plan	MP	foot	LB/FT^2;	Similar	SIM
Material	MATL		PSF	Sink	S
Maximum	MAX	Pounds per square		Sketch	SK
Membrane	MEMB	inch	LB/IN^2;	Soil pipe	SP
Metal	MET.		PSI	South	S
Meter	M	Precast	PRCST	Specification	SPEC
Mile	MI	Prefabricated	PREFAB	Square	SQ
Millimeter	MM	Preferred	PFD	Square foot	FT^2; SQ
Minimum	MIN	Preliminary	PRELIM		FT
Minute	MIN	Pressure treated	PT	Square inch	IN^2; SQ
Miscellaneous	MISC	Programming	PR		IN
Mixture	MIX.	Property	PROP	Stairs	ST
Model	MOD	Property line	PL	Standard	STD
Modular	MOD	Quality	QUAL	Steel	STL
Moisture resistant	MR	Quantity	QTY	Stock	STK
Molding	MLDG	Radiator	RAD	Storage	STG
Motor	MOT	Radius	R	Street	ST
Mullion	MULL	Random-access		Supply	SUP
National Electrical		memory	RAM	Surface	SUR
Code	NEC	Range	R	Symbol	SYM
Natural	NAT	Read-only memory	ROM	Symmetrical	SYM
Nominal	NOM	Receptacle	RECP	System	SYS
North	N	Reference	REF	Tangent	TAN.
Not applicable	NA	Refrigerator	REF	Tar and gravel	T & G
Not to scale	NTS	Register	REG	Tarpaulin	TARP
Number	NO.	Reinforce	REINF	Tee	T
Obscure	OB	Reproduce	REPRO	Telephone	TEL
On center	OC	Required	REQD	Television	TV
Opening	OPNG	Resistance moment	RM	Temperature	TEMP
Opposite	OPP	Return	RET	Terra-cotta	TC
Ounce	OZ	Right hand	RH	Terrazzo	TER
Out to out	O/O	Riser	R	Thermostat	THERMO
Outside diameter	OD	Roof	RF	Thick	THK
Overall	OA	Roofing	RFG	Thousand	M
Overhead	OVHD	Room	RM	Through	THRU
Panel	PNL	Rough opening	RO	Toilet	T
Parallel	PAR.	Round	RD	Tongue and groove	T & G
Part	PT	Safety	SAF	Total	TOT.
Particle board	PBD	Sanitary	SAN	Tread	TR
Partition	PTN	Scale	SC	Tubing	TUB.
Penny (nails)	d	Schedule	SCH	Typical	TYP
Per	/	Schematics	SC	Unfinished	UNFIN
Permanent	PERM	Second	(″) SEC	Urinal	UR
Perpendicular	PERP	Section	SECT	Valve	V
Pi (3.1416)	π	Select	SEL	Vaporproof	VAP PRF

Vent pipe	VP	Water closet	WC	Width	W
Vent stack	VS	Water heater	WH	Window	WDW
Ventilate	VENT.	Watt	W	With	W/
Vertical	VERT	Weather stripping	WS	Without	W/O
Volt	V	Weatherproof	WP	Wood	WD
Volume	V; VOL	Weep hole	WH	Wrought iron	WI
Washing machine	WM	Weight	WT	Yard	YD
Water	W	West	W		

PROFESSIONAL ORGANIZATIONS RELATED TO ARCHITECTURE

American Congress on Surveying and Mapping	ACSM
American Institute of Architects	AIA
American Institute of Building Designers	AIBD
American Institute of Steel Construction	AISC
American Institute of Timber Construction	AITC
American Plywood Association	APA
American Society for Testing and Materials	ASTM
American Society of Civil Engineers	ASCE
American Society of Heating, Refrigeration, and Air Conditioning Engineers	ASHRAE
American Society of Interior Design	ASID
American Society of Landscape Architects	ASLA
American Society of Mechanical Engineers	ASME
Associated Landscape Contractors of America	ALCA
Association of Women in Architecture	AWA
Construction Specification Institute	CSI
Department of Environmental Conservation	DEC
Department of Environmental Protection	DEP
Environmental Protection Agency	EPA
Federal Housing Authority	FHA
Institute of Electrical and Electronics Engineers	IEEE
International Brotherhood of Electrical Workers	IBEW
International Masonry Institute	IMI
International Standards Organization	ISO
National Association of Plumbing, Heating, and Cooling Contractors	NAPHCC
National Association of Trade and Technical Schools	NATTS
National Homebuilders Association	NHA
National Lumber Manufacturers Association	NLMA
National Society of Professional Engineers	NSPE
Occupational Safety and Health Administration	OSHA
Society of Women Engineers	SWE
United Brotherhood of Carpenters and Joiners	UBCJA
U.S. Department of Housing and Urban Development	HUD
U.S. Geological Survey	USGS

Architectural Synonyms

Architectural terms are standard. Nevertheless, architects, drafters, and builders often use different terms for the same object. Geographic location can influence a person's word choice. For instance, what is referred to as a *faucet* in one area of the country is called a *tap* in another area. What one person calls an *attic,* another calls a *garret,* and still another a *loft.* Each entry below is followed by a word or words that may be used in some areas to refer to the entry.

Anchor bolt   securing bolt, sill bolt

Apartment   tenement, multiple dwelling, condominium

Arcade   corridor, passage

Attic   garret, loft, half story

Awning   overhang, canopy, blind

Back plaster   parget, parging

Baffle   screen

Baked clay   terra-cotta

Baseboard   mopboard, finish board, skirting, scrubboard

Basement   cellar

Base mold   shoe mold, base cap

Batten   cleat

Bead   thin molding

Bearing partition   support partition, bearing wall

Bearing plate   sill, load plate

Bearing soil   compact soil

Bibs   faucets, taps, bibcock

Bird's-mouth   seat of a rafter

Blanket insulation   sheet insulation

Bridging   bracing, joining, cross supports, strutting

Buck   door frame

Building area   setback, building lines

Building code   building regulations, building ordinances

Building lines   setback, building area

Building paper   felt, tar paper, sheathing paper, construction paper, roll roofing, underlayment

BX   conduit, tubing, metal casing, armored cable

Caps   coping, capital

Carriage   stringer

Casement window   hinged window

Casing   window frame, shell

Catch basin   cistern, dry well, reservoir

Caulking compound   grout, cogging

Cavity wall   hollow wall

Ceiling clearance   headroom

Cesspool   sewage basin, seepage pit

Circuit box   fuse box, power panel, distribution panel

Cleat   batten

Colonnade   portico

Column   post, pillar, cylinder, pile

Common wall   party wall

Coping   caps

Door frame   buck

Doorsill   saddle, threshold, cricket

Double hung   double sashed

Drainage hole   weep hole

Drywall   gypsum board, sheetrock, wallboard, building board, rocklath, plasterboard, composition board, insulating board

Duct   pipeline, vent, raceway, plenum

Dwelling   home, house, residence

Easement   right of way

Egress   exit, outlet

Escalator   motor stairs

Exit   egress, outlet

Eyebrow   dormer

Facade   exterior facing

Fillers   shims, extender

Filler   stud trimmer

Finish   work trim, millwork

Firebrick   adobe brick, fireclay brick

Flow line   drain line

Flue cap   chimney pot

Flush plate   switch plate

Footer   anchorage, footing

Foundation sill   mudsill

Gallery   ledge, platform, corridor, arcade

Gazebo   pavilion, belvedere

Glazing bar   muntin, pane frames, sash bars

Grade   ground level, ground line, grade line, material classification, slope

Ground line   grade, grade line, ground level

Hatchway   opening, trapdoor, scuttle

Header   lintel

Hoist   lift, elevator, dumbwaiter

Hung ceiling   drop ceiling, clipped ceiling, suspended ceiling

Jalousies   louvers

Lacing   lattice bars

Lavatory   sink, basin

Lintel   header

Live load   moving load

Load   weight, force

Load plate   bearing plate, sill

Lobby   vestibule, stoop, porch, portal, entry

Lot   plot, property, site

Louvers   jalousies

Mantel   shelf

Millwork   trim, finish work

Modern   contemporary

Module   standard unit

Moving load   live load

Muntin   glazing bar, pane frame, sash bar

Particleboard   composition board, fiberboard, chipboard

Partition   wall

Pilaster wall   column

Plank and beam   post and beam, post and lintel

Profile   elevation section

Plate cut   bird mouth, seat cut, seat of a rafter

Platform framing   western framing

Plenum   pipeline, vent, raceway, duct, chamber

Pressure   stress, force

Rough floor   subfloor, base floor

Rough lumber   undressed lumber

Sill   bearing plate, load plate

Sill bolt   securing bolt, anchor bolt

Slope   slant, grade, incline, pitch

Soffit   underside

Spar   lumber, beam, wood, timber, common rafter

Spiral stairs   screw stairs, winding stairs

Standard unit   module

Step   tread

Strutting   bridging

Terra-cotta   baked clay

Threshold   saddle, doorsill, cricket

Tower   turret

Tread   step

Undressed lumber   rough lumber, unsurfaced timber

Veranda   passageway, balcony, porch

Water closet   toilet, W.C.

Water table   water level

Weight   load

Western framing   platform framing

Window panes   window lights

E APPENDIX

Glossary

A

Acoustical ceilings Ceilings designed to help keep noise from spreading from one room to others.

Acrylics Water-based permanent paints.

Active solar design Design method using mechanical or electrical devices to make use of the sun's energy in heating and cooling.

Active solar systems Cooling and heating systems that use mechanical devices to collect the sun's energy.

Aesthetic value Value placed on an object because of its form, beauty, or uniqueness.

Aggregate A combination of sand and crushed rocks, slate, slag, or shale added to concrete to improve its strength or workability.

Aging in place The design concept based on maximizing the length of time people can live safely and independently in their own residence.

Air-dried lumber Lumber that is left in the open to dry rather than being dried by a kiln.

Alternating current An electric current that changes value and direction periodically.

Ampere Unit used to measure the rate of flow of an electric current; the specific quantity of electrons passing a point in one second.

Apron Part of the driveway leading to a garage.

Arch Structure in the shape of an inverted U around an opening.

Architect Person who plans and designs buildings and oversees their construction.

Architect's scale Tool used in preparing scale drawings and in checking existing architectural plans and details.

Atrium Open court within a building.

Attic The space between the roof and the ceiling of a building.

Auxiliary elevation In drawing, an additional view used to clarify the true size of the base elevation.

Azimuth The measurement of an angle measured clockwise from the north.

B

Balcony Porch having no access from the outside that is suspended from an upper level of a structure.

Balloon framing Framing in which the first-floor joists rest directly on a sill plate, and the second floor joists bear on ribbon (ledger) strips set into continuous studs; also called *eastern framing*.

Base map Map that shows all fixed factors related to the site that must be accommodated in the site plan.

Baseboard The finish board covering the interior wall where the wall and floor meet.

Basic layout model Model of a structure built to the same scale as the floor plans and elevations.

Bay window A window projecting out from the wall to form a recess in the room.

Bays Spaces between rows of structural-steel framing members in two directions.

Beam A horizontal structural member that carries a load.

Bearing-wall structures Structures with solid walls that support the weight of the walls, floors, and roof.

Bearing walls Solid walls that provide support for each other and for the roof of a structure.

Berm A continuous mound of earth.

Bids Legal proposals from contractors to construct a project according to conditions defined in a contract, in specifications, and in a set of drawings.

Bird's-eye cut A notch cut in a rafter to fit onto a top plate.

Blocking Short pieces of lumber nailed between joists to add support to subfloor joints.

Board and batten siding Originally, vertical siding consisting of a vertical board placed over studs, with the joints covered with vertical batten boards; today, plywood or strandboard sheets with battens over the 48" joints.

Board foot Unit of measure for lumber that equals $1'' \times 12'' \times 12''$.

Bond Legally binding agreement that ensures that obligations are met; bonds used in construction are performance, labor, and materials bonds.

Bracing Framing members attached at an angle to make walls rigid.

Branches In a sewerage system, horizontal lines that carry waste from fixtures to the stacks.

Break line In drawing, a line used to reduce the size of a vertical section.

Breakout sectional drawings Drawings that show the internal construction of a component.

BTU Standard unit of measurement for heat; the amount of heat needed to raise the temperature of one pound of water one degree Fahrenheit at sea level.

Building envelope Space around a building.

Building load Total weight or mass of all live and dead loads on a building.

Building main Line through which water is brought from a well or public water supply to a building.

Building permit Document issued by the local government that allows construction of a structure or dwelling.

Buttress Protruding structure added at the base of an arch or wall to give support.

C

Cabinet coding system Shortcut method of dimensioning cabinets using code numbers for standard modular units.

Cantilever Protruding horizontal structural member supported in the center or at one end.

Carbon monoxide Colorless, odorless, poisonous gas produced by combustion.

Carport Garage minus one or more of the exterior walls.

Cartesian coordinate system Means by which CAD computer software locates points on a two-dimensional surface using an imaginary grid and two axes at right angles to each other.

CD-ROM Abbreviation for "compact disc, read-only memory" computer drives and discs.

Central bath Bathroom for general use that is accessible from all the bedrooms in a house.

Central heating Single source of heat that is distributed by pipes or ducts.

Central processing unit (CPU) Basic hardware for a computer system.

Certificate of occupancy Legal document that allows a building to be occupied and used.

Change orders Requests for alteration or correction in a building under construction

Channel Structural steel shape resembling the letter U that is used for roof purlins, lintels, truss chords, and to frame-in floor and roof openings.

Checker In large architectural firms, the person who carefully checks completed drawings.

Chord Principal member of a roof or bridge truss; a straight line that connects adjacent points on a curve.

Circuit Circular path that electricity follows from the power source to a light, appliance, or other device and back again.

Circuit breaker Device that opens (disconnects) an electrical circuit when current exceeds a certain amount.

Civil engineer's scale Tool used in preparing plans to scale show the size and features of the land surrounding a building in decimal parts of a foot.

Closed plan Building plan in which rooms are completely separated by partitions and doors.

Closing costs Legal, tax, insurance, and lender's costs paid by the owner before taking possession of a property.

Codes Laws that ensure that minimum building standards are met which safeguard life, health, property, and the public welfare.

Coding system System identifying every drawing and detail for a project; method of keeping similar drawings together and organized.

Column Vertical supporting member.

Column schedule Schematic elevation drawings and specifications showing the entire height of a building and the elevation of each floor, base plate, and column splice, as well as the type, depth, weight, and length of columns.

Combination plans Plans in a set of building plans that have been combined.

Commands Instructions given to a CAD system by means of a keyboard or mouse that tell the software what function to perform.

Common wall A single wall that serves two dwelling units.

Compaction The act of compressing the volume of soils or aggregates.

Compartment plan Plan that uses partitions, such as sliding doors, glass dividers, louvers, or even plants, to divide a bathroom into the water closet area, the lavatory area, and the bathing area.

Compass Tool used to draw circles, arcs, radii.

Compression force Type of force that exerts a crushing pressure on a structure.

Computer-aided design and drafting (CADD) See *Computer-aided drafting.*

Computer-aided drafting (CAD) Computer drafting software that automates many tasks.

Conceptual design Best design response to the information from the site analysis and on the user analysis chart; two types of sketches are created: idealized and site-related.

Condensation The formation of frost or drops of water on inside walls when warm vapor inside a room meets a cold wall or window.

Condominium A structure of two or more units in which the interior space is owned in common by owners of individual units.

Conduction Process by which heat moves through a solid material. The denser the material, the better it will conduct heat.

Conductors Materials that have a low resistance to electricity flowing through them, such as copper, aluminum, and silver.

Construction document Document that contains facts, figures, and legal and financial information related to the building process.

Contour interval Vertical distance represented by the space between contour lines.

Contour-interval model Model that represents the shape and slope of a site.

Contour lines On maps or drawings showing terrain, lines connect points at the same elevation.

Contractor Manager of a construction project.

Contracts Legally binding agreements between two or more parties.

Convection Process by which heat travels through liquids or gases.

Cornice Area of the roof that intersects with the outside walls and extends to the end of the roof overhang.

Corridor kitchen Kitchen having only two walls so that appliances and work areas are directly across from one another.

Crosshatching Lines drawn closely together at an angle to show a section cut or shading.

CSI format Format approved by the American Institute of Architects (AIA) and the Construction Specification Institute (CSI) to ensure consistency among specifications; information is organized under 16 major divisions with subdivisions under each.

Cul-de-sac End of a street with a circular turnaround.

Cupola Small domed structure built on top of a roof to provide ventilation.

Curtain wall Exterior wall used to cover a structure that provides no structural support.

Cutting plane In section drawings, an imaginary plane that passes through the drawing of a building; used to show interior construction.

Cutting-plane line Long, heavy line broken by two dashes and used to indicate the cutting plane in a drawing.

Damper In a fireplace, a door that regulates the amount of draft.

Datum line Reference line on a drawing that remains constant; sea level is commonly used.

Dead load Weight of building materials and permanently installed components on a structure.

Deck Open, elevated platform attached to a building that makes outdoor areas more usable.

Decking Surface of a floor system that usually consists of a subfloor and a finished floor.

Decor General style of decoration in a room or building.

Deed Legal certificate of property ownership.

Deflection Stress that results from both compression and tension forces acting on a structural member at the same time; also called *bending*.

Den Room that functions as a reading room, writing room, hobby room, or professional office.

Density The number of people, families, or houses distributed in a specific area.

Depreciation Loss of value.

Design study model Model used to check the form of a structure, verify the basic layout, clarify construction methods, finalize orientation, or show interior design options.

Detached garage Garage that is not connected to the house but is a separate building.

Detail drawings Drawings that reveal precise information about a portion of a structure shown in its entirety.

Detail section Section drawing that shows specific parts of a building in greater detail.

Detailed model Model of a building that shows features in detail.

Dimension lines In drawings, unbroken lines with arrowheads on either end used to show the distance between two points.

Direct current An electrical circuit that flows in one constant direction.

Dividers (1) Planters, half walls, louvered walls, and even furniture used to channel hall traffic without the use of solid walls. (2) Instrument used to divide a drawn object into an equal number of parts.

Dome Hemispherical roof form.

Dormer A structure projecting from a sloping roof.

Double header Two or more timbers joined for strength.

Down payment Percentage of the cost of a home paid by the buyer at the time of purchase; typically 5% to 20% of the total cost.

Drafting brush Brush used to remove eraser and graphite particles and to keep them from being redistributed on a drawing.

Drafting machine Mechanical tool that can serve as an architect's scale, triangle, protractor, T square, or parallel slide all in one.

Drainage field Plastic (PVC) perforated pipes or agricultural open-joint pipes laid in coarse gravel, which disperse liquid waste from a septic tank into the soil; also called *leaching field* or *absorption field*.

Drawing Web Format (DWF) Autocad files used to display drawings on the Internet.

Dropleaf workbench Workbench with drop leaves that can be extended for increased work space.

Drywall construction Method of interior wall finishing using prefabricated sheets of material, such as fiberboard, gypsum wallboard, stranded lumber sheets, sheet-rock, and plywood.

Early American style Term for all styles of architecture that developed in the various regions of the American colonies.

Earth-sheltered homes Homes designed to be partially covered with earth.

Easements Right-of-way across private land, such as for utility lines.

Eave Part of a roof that projects over the outer wall.

Eave line Design term for the horizontal line created by the eaves of a structure; can be used to create emphasis on the horizontal in a design.

Eclectic The mixing of architectural styles and/or periods in one design.

Ecological planning Planning done during the design stage of a structure to protect or improve the environment.

Elements of design Elements including line, form, space, color, light (value), texture, and materials, that are used to create designs.

Elevations Exterior drawings that show the front, back, or sides of a structure.

English style Architectural style developed in England that features such things as high-pitched roofs, massive chimneys, half-timber siding, small windows, and exterior stone walls; Elizabethan and Tudor are examples.

Entourage People or objects drawn as part of a building's surroundings in order to enhance its size or realism.

Environmental Protection Agency (EPA) Federal regulatory agency concerned with the quality of the environment.

Erasing shield Thin piece of metal or plastic having small, different-shaped openings through which lines on a drawing can be erased without disturbing the rest of the drawing.

Ergonomics Science that deals with designing and arranging things for ease of use by people.

Escrow Money collected and held by mortgage holder to ensure that tax payments on property are made on a timely basis.

Exterior elevation drawings Orthographic representations of the exterior of a structure to show the design, materials, dimensions, and final appearance of doors, windows, outer surfaces, and roof.

Facade Exterior face of a building.

Family kitchen Open kitchen that provides a meeting place for the entire family in addition to the usual kitchen services.

Family room Room in a home designed for family-centered activities.

Fascia Vertical board nailed on the ends of the rafters at the eave line.

Federal Housing Administration (FHA) A government agency that insures loans made by regular lending institutions.

Fenestration Arrangement of windows or openings in a wall.

Feng shui An ancient method of design based on concepts of energy flow, balance, and harmony with nature.

Finished dimensions Actual distances between finished features in a structure, such as from the finished floor to finished ceiling levels.

Firebox Portion of the fireplace that contains and supports the fire; also called *fire chamber*.

Fixture trap Device linked to a plumbing fixture that prevents the backflow of sewer gas from the branch lines.

Fixtures Permanent items connected to plumbing, such as lavatory (sink), water closet (toilet fixture), and bathtub or shower.

Flashing Additional covering used over a roof joint to provide complete waterproofing, such as roll-roofing, shingles, or sheet metal.

Flexible curve Drawing tool used to draw irregular curves that have no true radius or series of radii and cannot be drawn with a compass.

Floor plans Plans showing locations, sizes, materials, and components contained in the interior design of a building.

Flue Opening in a chimney through which the smoke passes.

Fluorescent Type of lighting created when gas inside a fluorescent tube comes in contact with a special coating.

Flying buttress Protruding structure that helps support the sides of a wall without adding additional weight.

Footcandle Unit of measurement for light intensity; the amount of light a candle casts on an object one foot away.

Footings Bases of foundations and foundation walls that are of two types: continuous and individual; also called *footers*.

Foreclose To take possession of the property if loan payments are not made as agreed.

Foreshortened A line or area of a drawing that is not drawn to the scale of the object represented.

Foundation Lowest structural component of a building upon which all other members rest.

Foundation sills Wood or steel members that are fastened to the top of foundation walls as a base for attaching floor systems.

Foyer Inside waiting area in a house; also called an *entrance hall*.

Framing Skeleton of a building constructed one level on top of another.

Framing dimensions In drawings, numbers used to show the actual distances between framing members.

Framing elevation drawings Drawings that show vertical views of the framing, especially of interior partitions.

French curve Tool used to make curved lines that are not part of an arc, such as an irregular curve.

French style Type of architecture originating in France. French provincial houses feature steeply pitched hip roofs, long projecting windows, corner quoins, curved lintels, and towers. French chateau houses are symmetrical and have mansard roofs.

Full section Sectional drawing prepared for the entire structure; also called an *architectural section*.

Functional The quality of being useful, of serving a purpose other than that of beauty or aesthetic value.

G

Gable The triangular end of an exterior wall just above the eaves.

Gable roof Pitched roof that takes its shape from the triangular ends of exterior walls.

General-purpose drawings Architectural drawings used for sales promotion or preliminary planning purposes.

Geodesic dome A dome created by the connection of multiple triangles that combine to form hexagons and pentagons.

Glazing Placing of glass in windows or doors.

Gothic arch A pointed arch very popular in cathedrals.

Graphic accelerator Circuit board added to a computer to increase the amount of RAM and processing power dedicated to displaying graphic images.

Greenhouse effect Buildup of heat created when sunlight falls on trapped air; used in passive solar planning.

Ground line On an elevation drawing, a horizontal line that represents ground level.

Gutter A trough for carrying off water.

H

Half-bath Bathroom that contains only a lavatory and water closet.

Half-timber construction Frame construction of heavy timbers in which the spaces are filled in with masonry.

Hand tools Basic tools necessary for any type of hobby or home maintenance work, such as a claw hammer and screwdrivers.

Hardware (1) Mechanical devices used in a computer system. (2) Metal attachments to doors, windows, and cabinetry.

Hardwoods Woods that come from deciduous trees, such as oak, walnut, birch, cherry, and mahogany.

Header (1) Horizontal supporting member above openings that serves as a lintel. (2) One or more pieces of lumber supporting ends of joists.

Hip roof Roof with four sloping sides.

Horizon line On perspective drawings, a line where earth and sky meet; the observer's eye level.

Horizontal wall sections Sections of exterior and interior walls drawn to clarify how the walls are constructed.

Humidistat A device for measuring and controlling humidity levels.

HVAC Mechanical systems that control heating, ventilation, and air conditioning.

I

I-beam Steel beam with an I-shaped cross section.

Idealized drawings Sketches that designate ideal spatial relationships of the user elements as determined by the user analysis.

I-joists Joists that are structurally similar to steel I-beams (S-beams); joist weight is reduced through the use of thin webs made of stranded wood.

Incandescent Lighting produced by a filament of very thin wire that glows when heated.

Insulation Material used to stop the transfer of heat or sound in a structure.

Insulators Materials that have good resistance to the flow of electricity, such as wood, glass, and plastic.

Integral garage Garage that is connected to the house.

Interest Charges that a borrower must pay for the use of the borrowed money; usually a percentage of the loan.

Interior design model Model used to show individual room designs.

Interior elevation drawings Drawings that show the vertical design of interior walls.

Island kitchen Kitchen having a freestanding structure usually located in the center of the room, that offers additional work space, a rangetop, or a sink.

Isometric drawings Pictorial drawings the receding lines of which are projected at an angle of 30° from the horizon so that they are parallel; primarily used for smaller details and objects.

Italian style Architectural style originating in Italy and featuring columns and arches that are generally part of an entrance and windows or balconies that open onto a loggia.

J

Jalousie window Window consisting of a number of long, thin, hinged panels.

Jamb Sides of a doorway or window opening.

Joist Horizontal structural member that supports a floor or ceiling.

K

Keystone Wedge-shaped stone that locks the other stones in an arch in place. Each stone is supported by the keystone in the center.

L

Lally column Hollow steel column filled with concrete and used to support weight transferred from horizontal beams and girders.

Laminated timber Timber made from thin layers of wood glued together; used to support heavy weights or over long spans.

Lanai Covered exterior passageway; Hawaiian word for "porch."

Landings Points at which stairs change direction.

Landscape plans Drawings that show the types and locations of vegetation around a building; they may also show contour changes and the building position.

Lap siding Horizontal siding applied so that each piece overlaps part of the piece below it.

Large-span construction Type of construction that uses large trusses or arches to span long distances, such as in aircraft hangars, sports stadiums, and convention centers.

Lateral load Lateral (horizontal) force on a building.

Lavatory Bathroom washbasin; sink.

Layering The superimposition of drawings on a CAD system.

Libraries The areas of a CAD system in which frequently used symbols or details are stored.

Lien Legal document used to take or hold the property of a debtor in default of payment.

Line conventions Series of lines having special meanings that are used in architectural drawings; sometimes called the *alphabet of lines*.

Lintel Horizontal piece of wood, stone, or steel across the top of door and window openings that bears the weight of the wall above the opening.

Live load Force on a building that includes the weight of all impermanent, movable objects, such as people and furniture, snow, and the force of wind.

Living area Area in a house where the family entertains, relaxes, dines, and participates in recreational activities.

Load-bearing wall See *Bearing walls.*

Loggia Open passage covered by a roof.

Longitudinal section Section drawing the cutting plane of which runs along the length or major axis of a building.

Lookout joists Joists that are used to support a cantilevered second-floor extension.

L-shape Structural steel member rolled in the shape of the letter L with legs of equal thickness but varying length; also called an *angle*.

L-shaped kitchen Kitchen design that features continuous counters, appliances, and equipment on two adjoining, perpendicular walls; two work centers are usually located on one wall and the third center is on the other wall.

Lux In the metric system, the standard unit of illumination; one lux (lx) is equal to 0.093 footcandle.

M

Major module Large (48″) base unit in a construction system in which parts are standardized.

Mantel A shelf over a fireplace.

Marquee Covering over an entrance that is not supported by posts or columns and that connects the building with the street.

Masonry Stone, brick, tiles, or concrete.

Masonry bond Pattern of masonry units in courses (rows); different patterns can make the same material appear completely different.

Master bath Bathroom that is accessible only from the master bedroom.

Master bedroom Bedroom for adults that may also have an adjacent bath and a separate dressing room; often the largest bedroom in a house.

Material list List of materials, such as lumber, used for on-site construction; includes descriptions, quantities, and costs.

Mediterranean style Italian and Spanish architectural styles; also called *Southern European* style.

Metric scale Scale used with metric drawings; units are in millimeters (mm).

Mid-Atlantic style Architectural style common in the Mid-Atlantic states and resulting from the availability of brick, a seasonal climate, the influence of Thomas Jefferson and early Greek and Roman architecture; also known as *classical revival.*

Minor module Small (24″) base unit in a construction system in which parts are standardized.

Models Three-dimensional replicas of a structure.

Modem Telecommunications device that allows computer operators to send and receive information over standard telephone lines; acronym for *modulator/demodulator.*

Modular components Standardized building parts or sections constructed away from the building site; typical components include preassembled wall sections, windows, and molded bathrooms.

Module Base unit in a construction system in which parts are standardized.

Moisture barrier Material that retards the passage of vapor or moisture into walls and prevents condensation.

Mortar Mixture of cement, sand, and water, used to bond bricks and stone.

Mortgage Contract for a loan for the purchase of real estate property.

Mortgagee The party who lends money for property purchase.

Mortgagor The party who borrows money to build or purchase property.

M-shape Lightweight steel beam.

Mullion A vertical member that separates a series of windows or panels.

Multiple-level floor plans Floor plans for structures having more than one floor.

Muntin Thin strip separating panes (lights) in a window.

New England Colonial style Architectural style developed by the colonists who settled the New England coastal areas.

Non-load-bearing wall Interior wall that does not support a load.

O

Object lines Visible lines on a drawing, such as those describing the main exterior walls and interior partitions.

Oblique drawings Drawings in which all receding lines are commonly drawn at a very low angle of 10° to 15°.

Offset Recessed portion of a wall.

On-center (OC) Measurement from the center of one member to the center of another.

One-point perspective Drawing in which the front view is drawn to its true scale and all receding sides are projected to a single vanishing point on the horizon line.

One-wall kitchen Kitchen plan in which work centers are located in a row along one wall.

One-way slab system Foundation in which the rebars are all parallel, as are system girders.

Open plan Floor plan design in which partitions do not completely separate the rooms from one another.

Organic design Architectural concept in which all materials, forms, and surroundings are coordinated and in harmony with nature.

Orientation A building's position in relation to sun, wind, view, and noise.

Orthographic projection Drawing related views of an object so as to show it from all sides.

Overall dimensions Total length and width of a building.

Overhang Portion of a roof that projects beyond the outer walls.

Parallel slide Drawing tool used as a guide for drawing horizontal lines and a base for aligning triangles; also called a *parallel rule.*

Partition Wall that subdivides a living area into rooms.

Passive solar systems Heating and cooling systems that use the power of the sun without the aid of mechanical devices.

Pastels Light-colored, water-based drawing medium similar to chalk.

Patio Open area adjacent or directly accessible to the house; courtyard.

Peninsula kitchen Kitchen design in which a work center projects into the room like a piece of land into a body of water.

Peninsula workbench Workbench that projects into the room so that it has three working surfaces with storage compartments on each of its sides.

Perimeter The outside surfaces of a geometric space.

Permanent wood foundations Foundations constructed similarly to wood frame walls except that the plywood and lumber components are pressure treated with wood preservatives.

Perspective drawings Drawing in which receding lines appear to meet at a point on the horizon; the object appears as it would to the eye.

Phasing Prolonging the landscaping of a site over several years because of a lack of time or money; parts of the plan are completed at different times.

Phi The golden ratio of design 1:1.618.

Picture plane Imaginary plane between the station point and the object in a perspective view drawing.

Pier-and-column foundations Foundations having individual footings (piers) upon which posts and columns are placed.

Pitch Angle between the roof's surface (top plate to ridge board) and the horizontal plane; ratio of the rise to the span.

Plank-and-beam construction Floor construction system in which heavy wood planks are placed over widely spaced beams.

Plans Drawings that show views of an object from the top down, giving a "bird's-eye" view.

Planting schedules In a landscape plan, guides for the purchase and placement of each size and species of plant material.

Plat Map or chart of an area showing boundaries of lots and other parcels of property.

Platform framing Framing of a multiple-level building in which the second floor rests directly on first-floor exterior walls; also called *western framing.*

Plenum Enclosed space inside of which the air pressure is greater than it is outside; used in a type of heating and cooling system.

Plot plans Plans used to show the size and shape of a building site and the location and size of all buildings, walks, drives, pools, streams, patios, and courts on it.

Plywood Wood product manufactured from thin sheets of wood laminated together with an adhesive, under high pressure; the grain of each ply is laid perpendicular to the grain of each adjacent layer, which reduces the tendency to warp, check, split, splinter, and shrink.

Porch Covered platform leading into the entrance of a building and which may be enclosed by glass, screens, or posts and railings.

Post-and-beam construction Type of construction that uses heavy timbers for walls as well as for floors and roofs.

Post-and-lintel construction Type of construction in which a horizontal beam, called a *lintel,* is placed across two vertical posts, such as for a door or window.

Post-tensioning Method used to prestress, or compress, concrete so that both the upper and lower sides of a member remain in compression during loading; tendons are placed inside tubes embedded in the concrete and then stretched with hydraulic jacks while the ends are anchored.

Power tools Electrically powered tools, such as electric drills, belt sanders, and drill presses.

Prebuilt home House that is totally built in a factory.

Prebuilt module Construction module in the form of a complete room or independent functional area and having built-in components, such as cabinets, major appliances, and fixtures.

Prefabricated home Home in which the major components, such as the walls, decks, and partitions, are assembled at a factory; installation of electrical wiring, plumbing, and heating systems is completed on site, as is finishing work.

Presentation drawing Drawings having realistic features added for the purpose of showing the building to clients.

Presentation model Building model used for sales purposes.

Prestressing Compressing concrete so that a member remains in compression during loading.

Pretensioning Method of prestressing concrete in which tendons are stretched between anchors and the concrete is poured around them.

Principal Amount of a loan, minus any interest.

Principles of design Guidelines for how to combine the elements of design.

Prints (formerly called *blueprints*) Reproductions of architectural drawings used by contractors and workers to guide the building process.

Profile A vertical section through a site.

Profile drawings Elevation drawings of a site showing a section cut through the terrain.

Protractor Instrument used to measure angles.

Quoin Support or decorative masonry applied to building corners.

Radiant heat Method of heating surfaces to create area heat through radiation.

Radiation Process by which heat travels as waves through space.

Radon Colorless, radioactive gas produced by the natural decay of radium; radon enters a structure from the soil.

Rafter Structural member used to frame a roof.

RAM A computer's temporary memory, determining the amount of software data the CPU can process at one time; abbreviation for "random access memory."

Ranch style Architectural style adapted to the needs of settlers as they moved west, featuring a single-level, rambling plan.

Recreation room Room designed specifically for active play, exercise, and recreation; also called a *game room* or *playroom.*

Reflected ceiling plans Drawings of complex ceiling designs and multiple-lighting fixtures or levels using the floor plan as a base; the ceiling appears as though reflected in a mirror.

Reinforcing bars Steel bars added to concrete slabs, beams, and columns to help them resist tension forces; also called *rebars.*

Removed section Section drawing done at a larger scale to clarify small details.

Rendering Process of adding lights and shadows to a drawing so that it looks more realistic; usually done to 1-, 2-, or 3-point perspective drawings to show how the finished building is expected to look.

Reversed plans Plans that are a mirror image of the originals; created to provide more plan choices.

Ridge board Top member in the roof assembly; also called a *ridge beam.*

Ridge line Line formed by the roof ridge on an elevation; accenting the ridge line places emphasis on the horizontal.

Rise Vertical height of a roof from the top of the wall plate to the roof's ridge.

Riser Vertical part of a stair step.

Rolled steel Steel that has been shaped by passing it between a series of rollers.

ROM Memory in a computer containing the fixed data that the computer uses while it is operating; abbreviation for "read-only memory."

Room template Template made by cutting around the drawn outline of a room.

Rough opening Opening in the framing that is slightly larger than the door or window that will fit into it.

Run Horizontal distance covered by a roof.

R-value Rating of resistance of heat flow through building materials, such as insulation; the higher the number, the greater the resistance to heat flow.

Schedules (1) Detailed lists that contain needed information, such as size and type of windows or doors. (2) Charts that provide information for estimating purposes or to conserve drafting time and drawing space.

Sectional drawings Drawings that reveal interior construction by showing a "slice" of a planned structure; also called *sections.*

Septic tank Tank in which sewage is processed by bacteria and the liquids are allowed to seep into the ground.

Serving walls Walls that have openings with countertops for passing items back and forth between a kitchen and dining room.

Setbacks Minimum distances structures must be located from property lines as stated in zoning laws.

Shear force Force that tends to make one part of an object slide on or past another part; excessive shear may cause fractures.

Sheathing Structural covering of boards or wallboards, placed over exterior studding or rafters of a structure.

Shingles Thin pieces of wood or other materials that overlap each other in covering a roof. The number and kind needed depend on the steepness of the roof slope.

Siding Covering for an exterior wall.

Single-line drawing Simple, scaled drawing that includes both floor plan and site features.

Site analysis Study of a site that helps the designer take advantage of its positive features and minimize its negative features; site analysis also helps ensure appropriate land use.

Site-related drawing Drawing that matches the idealized drawing of a structure to its site and introduces size requirements.

Situation statement Formal statement of the agreement between a client and a designer regarding the purpose, theme, scope, budget, and schedules of the project.

Skeleton-frame construction Construction in which wall coverings are attached to an open frame in which small structural members share the loads.

Skeleton-frame structures Structures built using skeleton-frame construction.

Skylight Glazed opening in a roof.

Slab foundations Foundations made of reinforced concrete, which may be either in one piece or in several pieces.

Slope Ratio of the rise of a roof to the run.

Slope diagram Diagram drawn on the elevation after the roof pitch is established to aid the carpenter in determining the angle of the rafters.

Soffit Underside of an overhang, such as the edge of a roof.

Software Instructions by means of which a computer performs tasks, such as word processing.

Softwoods Woods from coniferous (needle-bearing) trees such as Douglas fir, pine, or cedar.

Solid form model Model of a structure created before final dimensions are applied to the design and used to check overall proportions.

Solid model Type of three-dimensional drawing done on a computer; created when properties such as mass must be considered.

Span Full distance between outer walls supporting the roof.

Specifications Detailed written instructions describing materials and requirements for construction.

S-shape Steel rolled into the shape of a capital letter I; formerly called an *I-beam*.

Stacks In a sewerage system, vertical lines 3″ to 4″ in diameter that carry waste.

Stairs Series of steps that provide access from one level of a structure to another.

Stamps Images in rubber that can be coated with ink and transferred to drawings; used for architectural features that are often repeated, such as landscape features, people, and cars.

Station point Location of the observer in a perspective drawing.

Steel cage construction Type of construction in which steel members are used in a manner similar to skeleton-frame wood members; also called *steel skeleton-frame construction*.

Stoop Projection from a building that provides shelter and access to a landing at the entrance.

Stress Any force acting on a part or member of a structure.

Structural model Model of a building in which only the structural members are shown in order to check unique structural methods or to study framing options.

Stud Upright beam in the framework of a building; usually called a "2 × 4" (1 1/2″ × 3 1/2″) and spaced at 16 inches from center to center.

Stud layout Horizontal framing section showing the position of each wall member, both exterior and interior.

Studio Room in which an engineer, architect, drafter, or artist works.

Study Room used for reading, writing, hobbies, or office duties.

Subdimensions Architectural dimensions that are subdivisions of the overall dimensions.

Surface model Three-dimensional computer drawing that consists of solid plane surfaces.

Survey Drawing showing the exact size, shape, and levels of a property; when prepared by a licensed surveyor, a survey is a legal document that establishes property rights.

Swale A continuous depression in a flat site.

Swim-out In a swimming pool, an elevated platform below the water level that allows the swimmer to get out without using a ladder.

Symbol library Computer software that contains ready-to-use symbols for doors, windows, fixtures, and other items that can be inserted easily into any computer drawing.

Symbols Standardized elements on a drawing used to identify fixtures, doors, windows, stairs, partitions, and other common items.

Takeoff Creation of a material list from working drawings.

Technical pens Ink pens used for drafting.

Tee Structural steel shape usually made by cutting through the web of an S, W, or M-shape and used for truss chords and to support concrete reinforcement rods.

Templates Cut-outs in plastic, paper, cardboard, or metal that represent various symbols, furniture, and fixtures, which can then be traced on a drawing by following the outline with a pencil or pen.

Tension force Force that pulls on objects causing them to stretch.

T-foundations Combinations of footings and a poured concrete or concrete block wall that form an inverted T; necessary in structures with basements or when the underside of the first floor must be accessible.

Thermal conductivity Ability of a material to allow heat to flow through it.

Thermal mass Any material that will absorb heat from the sun and later radiate that heat back into the air.

Three-point perspective Drawing in which horizontal lines recede to vanishing points in the distance and vertical lines also recede slightly so that the tops of extremely tall buildings appear smaller.

Timber Piece of lumber with a cross section larger than 4″ × 6″.

Title block Written information on an architectural drawing that identifies it.

Topography Drawing showing land elevation details; surface configuration of an area.

Torsion force Force that twists an object, causing it to fracture or become misshapen.

Traffic areas Areas of a building or room through which people pass to get from one place to another.

Transformer A device that converts electricity from one voltage to another.

Transverse section Section drawing, the cutting plane of which lies across the shorter, or minor axis, of the building.

Tread Horizontal part of a step; the part upon which one walks.

Triangle Drafting tool used with either a T-square or other horizontal guide to draw vertical and diagonal or inclined lines.

Trombe wall Device used in passive solar heating, consisting of an outer glass surface, a masonry wall, and vents that channel the heated air through the building.

Truss Prefabricated, triangular-shaped unit used to support roof loads over long spans.

Truss joists Joists having long, usually parallel, top and bottom chords connected by shorter pieces called webs that give the ability to span long distances.

T-square Drafting tool used primarily as a guide for drawing horizontal lines and as a base for the triangle used to draw vertical and inclined lines.

Two-point perspective Drawing in which receding lines are projected to two vanishing points on opposite ends of the horizon line.

Two-way slab system Foundation in which the rebars, girders, and beams are placed perpendicular to one another; when ribs extend in both directions, the system is known as a "waffle slab."

Underlays Drawings or parts of drawings that are placed under an original drawing and traced onto it.

Underwriters Laboratory (UL) A nonprofit agency that tests products for safety.

User analysis Study of a structure in which each goal is further refined in terms of space elements, usage, size, and the relationships between areas.

U-shaped kitchen Kitchen design in which the sink is located at the bottom of a U-shape, and the range and the refrigerator are on opposite sides.

Utility room Room that may include facilities for washing, drying, ironing, sewing, storage, and heating and air-conditioning equipment; also called *service room, all-purpose room,* and *laundry room.*

U-value Proportional heat transfer difference between inside and outside air temperatures; reciprocal (1/R) of the R-value.

Vanishing point On a perspective drawing, the point in the distance at which horizontal lines appear to meet and disappear.

Vault Passageway or room formed by a series of arches.

Vellum Medium on which architectural drawings are prepared.

Vent stack In plumbing, the portion of the soil stack above the highest branch intersection that permits sewer gases to escape to the outside and equalize the air pressure in a waste system.

Ventilated shelving Shelving made by welding steel rods together at intervals and then coating them with vinyl.

Ventilation Circulation of air through a structure.

Veranda Large porch extending around several sides of the home.

Vertical wall sections Section drawings that show exposed construction members in a wall.

Victorian Highly decorative style of the nineteenth century named for Queen Victoria.

Volt Unit of electrical pressure or potential; the higher the voltage, the greater the flow of electricity.

Walk-in closet Recessed storage facility large enough to walk around in.

Wall closet A shallow closet in the wall for cupboards, shelves, and drawers, providing access to all stored items without using an excessive amount of floor area.

Wardrobe closet Storage facility for clothing that may be either built-in or offset.

Wash drawing Rendering in which only black and gray tones are used.

Water closet Toilet fixture.

Watt Unit of measurement for electrical power; current (in amperes) multiplied by potential (in volts).

Wireframe drawing Three-dimensional computer drawing of an object that looks as though the object were shaped out of wire mesh.

Work triangle Area connecting the three main work centers of a kitchen.

Working drawings Drawings that contain all information needed to build a structure.

W-shape Type of I-beam, similar to an S-shape but with wider flanges and comparatively thinner webs, which gives a greater capacity to resist bending.

Zoning ordinance Law or regulation defining the type of structure that can be built in a certain area in order to provide safety and convenience for the public and to preserve or improve the environment.

INDEX